Springer Series in Statistics

Springer
London
Berlin
Heidelberg
New York
Hong Kong
Milan
Paris
Tokyo

Springer Series in Statistics

Andersen/Borgan/Gill/Keiding: Statistical Models Based on Counting Processes.
Atkinson/Riani: Robust Diagnostic Regression Analysis.
Berger: Statistical Decision Theory and Bayesian Analysis, 2nd edition.
Borg/Groenen: Modern Multidimensional Scaling: Theory and Applications
Brockwell/Davis: Time Series: Theory and Methods, 2nd edition.
Chan/Tong: Chaos: A Statistical Perspective.
Chen/Shao/Ibrahim: Monte Carlo Methods in Bayesian Computation.
Coles: An Introduction to Statistical Modeling of Extreme Values.
David/Edwards: Annotated Readings in the History of Statistics.
Devroye/Lugosi: Combinatorial Methods in Density Estimation.
Efromovich: Nonparametric Curve Estimation: Methods, Theory, and Applications.
Eggermont/LaRiccia: Maximum Penalized Likelihood Estimation, Volume I: Density Estimation.
Fahrmeir/Tutz: Multivariate Statistical Modelling Based on Generalized Linear Models, 2nd edition.
Fan/Yao: Nonlinear Time Series: Nonparametric and Parametric Methods.
Farebrother: Fitting Linear Relationships: A History of the Calculus of Observations 1750-1900.
Federer: Statistical Design and Analysis for Intercropping Experiments, Volume I: Two Crops.
Federer: Statistical Design and Analysis for Intercropping Experiments, Volume II: Three or More Crops.
Ghosh/Ramamoorthi: Bayesian Nonparametrics.
Glaz/Naus/Wallenstein: Scan Statistics.
Good: Permutation Tests: A Practical Guide to Resampling Methods for Testing Hypotheses, 2nd edition.
Gouriéroux: ARCH Models and Financial Applications.
Gu: Smoothing Spline ANOVA Models.
Györfi/Kohler/Krzyżak/ Walk: A Distribution-Free Theory of Nonparametric Regression.
Haberman: Advanced Statistics, Volume I: Description of Populations.
Hall: The Bootstrap and Edgeworth Expansion.
Härdle: Smoothing Techniques: With Implementation in S.
Harrell: Regression Modeling Strategies: With Applications to Linear Models, Logistic Regression, and Survival Analysis
Hart: Nonparametric Smoothing and Lack-of-Fit Tests.
Hastie/Tibshirani/Friedman: The Elements of Statistical Learning: Data Mining, Inference, and Prediction
Hedayat/Sloane/Stufken: Orthogonal Arrays: Theory and Applications.
Heyde: Quasi-Likelihood and its Application: A General Approach to Optimal Parameter Estimation.
Huet/Bouvier/Gruet/Jolivet: Statistical Tools for Nonlinear Regression: A Practical Guide with S-PLUS and R Examples, 2nd edition.
Ibrahim/Chen/Sinha: Bayesian Survival Analysis.

(continued after index)

Yury A. Kutoyants

Statistical Inference for Ergodic Diffusion Processes

Springer

Dr Yury A. Kutoyants
Laboratoire de Statistique et Processus
Université du Maine
Avenue Olivier Messiaen
72085 Le Mans
Cedex 9
France
kutoyants@univ-lemans.fr

British Library Cataloguing in Publication Data
Kutoyants, Yu.
Statistical inference for ergodic diffusion processes.
(Springer series in statistics)
1. Diffusion processes 2. Parameter estimation 3. Mathematical statistics
I. Title
519.2'3
ISBN 978-1-84996-906-2

Library of Congress Cataloging-in-Publication Data
Kutoyants, Yu. A.
Statistical inference for ergodic diffusion processes / Yury A. Kutoyants.
p. cm. -- (Springer series in statistics)
Includes bibliographical references and index.

1. Diffusion processes. 2. Ergodic theory. 3. Parameter estimation. I. Title. II. Series.
QA274.75.K92 2004
519.2'33—dc21 2003054411

Springer-Verlag London Berlin Heidelberg
a member of BertelsmannSpringer Science+Business Media GmbH
http://www.springer.co.uk

Printed in the United States of America

To my family

Preface

This work presents an introduction to the *large samples theory* of statistical inference for the model of homogeneous ergodic diffusion process

$$\mathrm{d}X_t = S(X_t)\,\mathrm{d}t + \sigma\left(X_t\right)\,\mathrm{d}W_t, \quad X_0, \quad 0 \leq t \leq T,$$

i.e., we consider several problems of parameter estimation ($S\left(x\right) = S\left(\vartheta, x\right)$, $\vartheta \in \Theta \subset \mathscr{R}^d$) and nonparametric estimation and some problems of hypotheses testing by the observations $X^T = \left\{X_t, 0 \leq t \leq T\right\}$ in the asymptotics $T \to \infty$. All statistical problems are concerned with the trend coefficient $S\left(\cdot\right)$ only. The diffusion coefficient $\sigma\left(\cdot\right)^2$ is supposed to be known and positive. Remember that the measures corresponding to the different diffusion coefficients are singular and continuous-time statistical inference for this model is often trivial.

Diffusion processes are widely used in applied problems (biomedical sciences [12], [111], [98], economics [20], [96], genetic analysis [153], mechanics [9],[130], physics [194] and especially in financial mathematics [119]).

The statements of the problems are typical for the *large samples theory* of classical mathematical statistics, i.e., statistics of independent identically distributed (i.i.d.) observations $X^n = \left\{X_1, \ldots, X_n\right\}$ in the asymptotics $n \to \infty$ (for the i.i.d. case see, for example, Rao [207], Ibragimov and Khasminskii [109], Borovkov [33], Strasser [222], van der Vaart [228] or any other book on the subject). It is well known that the maximum likelihood, Bayesian and minimum distance estimators in the regular case (the density function is sufficiently smooth with respect to the unknown parameter) are consistent and asymptotically normal and in a nonregular case (say, in the case of a discontinuous density function) the first two estimators have nonGaussian limit distributions and the rate of convergence is better than in the regular case [109]. We consider the similar (regular and nonregular) estimation problems for the diffusion process X^T and describe the properties of the corresponding estimators. In nonparametric estimation problems (for an i.i.d. model) the empirical distribution function is $\sqrt{n}$-consistent and asymptotically efficient and the kernel-type density estimators are as well consistent and asymptotically efficient with the rate of convergence depending of the *smoothness* of the model. We present the similar results for the problems of invariant distribution function and trend coefficient estimation. Special attention is paid to the notion of asymptotical efficiency. The longer title of this work which corresponds better to its contents could be "*Asymptotically efficient statistical inference for one-dimensional ergodic diffusion processes*".

Probably it will be useful to clarify the choice of continuous-time model for statistical inference (to answer one of the *most frequently asked questions*). First, many real stochastic systems "*live in continuous time*", hence a good mathematical model has to describe them in a continuous time setup and the diffusion processes are often the appropriate candidates (see above). On the basis of a continuous-time model we obtain the asymptotically optimal statistical procedures (estimators, tests) and then we have the problem of their realization. If the real data are discrete-time observations $X^n = \{X_{t_1}, \ldots, X_{t_n}\}$ of the trajectory $X^T = \{X_t, 0 \le t \le T\}$, then we have to approximate the optimal estimators or tests and there are two types of errors: the first is due to the statistical nature of the (continuous time) observations and the second is due to the approximation of the integrals by sums, etc. If the discrete observations are taken sufficiently dense, then the second error is negligible with respect to the first one and the statistical results obtained for the continuous time model are valid for discrete time observations too. Our study corresponds well to this situation. Moreover, even if the error due to discretization is not negligible, nevertheless it can be important to compare, say, the limit errors of estimators constructed by discrete observations with the limit errors of the estimators for the continuous time model. This will allow us to understand better the losses due to discretization. The case of the domination of the second error is actually studied by many authors, but as the difficulties in these two problems (continuous and discrete) are quite different, we do not discuss the second one here. Having ten years' experience with industrial contracts, I just note that, to my mind, both statements are equally close and far from the practical applications and mainly reflect the interests of the researchers in the field. If the choice of the mathematical model is restricted by the possibility of the further direct computer realization, then I doubt if the use of Gaussian random variables is permitted, because for computers all the variables are discrete. Note that the rapid increase in the computation speed enables us to realize *almost continuous-time* statistical procedures.

Remember that the trajectory of a diffusion process has *infinite length* and that one of the most important questions concerning the application of this model in a particular problem is: how close is the observed model to the diffusion model and what is the error due to this approximation? This question became of especial interest in the problems of diffusion coefficient estimation by discrete time observations.

Of course, the ergodic diffusion processes are not the only class of continuous time models and there are many others too. We mention here *signal in Gaussian noise* (see, e.g., [109], [201]), *diffusion processes with small diffusion coefficient* [139], *stochastic partial differential equations* (see, e.g., [103], [104]), *stochastic differential equations with fractional Brownian motion* (see, e.g., [121], [122]), *stochastic differential equations with stable processes* [216], *birth and death on a flow* [100], *stochastic resonance* [106], a wide class of *point processes* [4], [57], [145] *etc.*

Statistical inference for continuous time stochastic processes attracts more and more attention of researchers and at the same time this branch of *mathematical statistics* is not entirely accepted by the statistical community (statistical journals) and sometimes is judged as *too mathematical* and *not statistical enough*. This was one of the reasons for D. Bosq (University of Paris 6) and me to start in 1998 a new journal *"Statistical Inference for Stochastic Processes"*, where this class of problems is well represented.

This work is a continuation of the study of large samples theory for continuous time stochastic processes started in [136], [139], [145]. As before, we especially chose the *simplest mathematical model* and this allows us to go further in the statistical problems. Thus in the case of a multidimensional ergodic diffusion process the problems of nonparametric estimation became essentially more complicated. Here we have one-dimensional ergodic diffusion with an explicit form of the invariant density, positive diffusion coefficient (no reflecting bounds) and we even suppose that the process is stationary. Therefore, like the i.i.d. case, the distribution of X_t is always the same and, moreover, the empirical distribution function $\hat{F}_T(x), x \in \mathscr{R}$ (together with initial X_0 and final X_T values) is sufficient statistics for almost all problems considered in this work. At the same time the model is nonGaussian and nonlinear, the statistical procedures are not trivial and this allows us to construct a meaningful statistical theory. I hope that this work will attract the attention of statisticians and probabilists to this model and this will lead to further developments. Note that the problems of parametric and nonparametric estimation and some hypotheses testing problems for ergodic diffusion processes are already partially presented in several monographs devoted to statistical inference for stochastic processes. We can mention here Arato [6], Basawa and Prakasa Rao [14], Basawa and Scott [15], Bosq [36], Bosq and Nguen [35], Küchler and Sørenson [127], Kutoyants [136], Lin'kov [174], Prakasa Rao [205] and Taniguchi and Kakizawa [224]. Nevertheless, it seems that this model merits a special wider study.

This work is based on the lectures given during several years to the postgraduate students at Paris 6 University (DEA de Statistique) and at the University of Padova. We present in the introduction (Section 1.3) a relatively elementary introduction to the statistical inference for ergodic diffusion processes (asymptotic theory), which can be sufficient for the first reading.

A large part of our results are obtained in collaboration with my students S. Dachian [50], A. Dalalyan [54], [55], [56], S. Iakus [107], I. Negri [149] and my colleagues D. Dehay [59], H. Dietz [63], [64], E. Fournie [77], R. Höpfner [99],

U. Küchler [126], V. Spokoiny [150], L. Vostrikova [151] and N. Yoshida [152]. I am deeply indebted to all of them and surely without their contributions and stimulating discussions this work would have been quite different. I am particularly grateful to A. Dalalyan for his contribution in Chapter 4 and to R. Höpfner for encouragement and friendly support during all the time of preparation of this work. The different parts of the work were discussed with D. Bosq, G. Golubev, M. Kleptsyna, R. Liptser and A. Veretennikov and I am grateful for their useful advice. I would like to thank D. Dehay for a careful reading of the first version of the manuscript and many propositions, which allowed me to improve the exposition and to S. Dachian whose propositions made the typing of Latex files a simple pleasure.

Le Mans,
August 5, 2002

Yury A. Kutoyants

Contents

Introduction

In this work we are interested in the statistical estimation and hypotheses testing problems in the case of observations $X^T = \{X_t, 0 \le t \le T\}$ of a continuous-time diffusion process

$$\mathrm{d}X_t = S(X_t)\,\mathrm{d}t + \sigma(X_t)\,\mathrm{d}W_t, \quad X_0, \quad 0 \le t \le T, \tag{0.1}$$

where $\{W_t,\ t \ge 0\}$ is a standard Wiener process, $S(\cdot)$ is an (unknown to the observer) trend coefficient, $\sigma(\cdot)^2$ is a (known) diffusion coefficient and X_0 is an initial value of X_t, which is supposed to be not dependent on the Wiener process. All statistical problems concern the trend coefficient only.

We consider two types of problems: *parametric*, when the trend coefficient is known up to the value of some finite-dimensional parameter $\vartheta \in \Theta \subset \mathscr{R}^d$, i.e., $S(x) = S(\vartheta, x)$, $x \in \mathscr{R}$ with known function $S(\cdot,\cdot)$ and *nonparametric*, when $S(x)$, $x \in \mathscr{R}$ is an unknown function, i.e., it belongs to a class of functions which cannot be parameterized by a finite-dimensional parameter.

The stochastic process (0.1) has ergodic properties if the functions $S(\cdot)$ and $\sigma(\cdot)$ satisfy the following two conditions:

$$V(S,x) = \int_0^x \exp\left\{-2\int_0^y \frac{S(v)}{\sigma(v)^2}\,\mathrm{d}v\right\}\mathrm{d}y \longrightarrow \pm\infty, \quad \text{as} \quad x \to \pm\infty, \tag{0.2}$$

and

$$G(S) = \int_{-\infty}^{\infty} \sigma(x)^{-2} \exp\left\{2\int_0^x \frac{S(v)}{\sigma(v)^2}\,\mathrm{d}v\right\}\mathrm{d}x < \infty. \tag{0.3}$$

By the first condition (0.2) the diffusion process (0.1) is *recurrent*, i.e., the time to return to any bounded set $\mathbb{A}$ is finite with probability 1 and by the second condition this time has finite mathematical expectation. If both conditions (0.2) and (0.3) are fulfilled then this process is *recurrent positive*, i.e., it has ergodic properties. This means that for any measurable function $h(\cdot)$, such that $\mathbf{E}\,|h(\xi)| < \infty$ the following limit:

$$\frac{1}{T}\int_0^T h\left(X_t\right)\,\mathrm{d}t \longrightarrow \int_{-\infty}^{\infty} h(x)\, f_S\left(x\right)\,\mathrm{d}x \equiv \mathbf{E}\,h\left(\xi\right) \tag{0.4}$$

holds with probability 1. Here the function

$$f_S\left(x\right) = G(S)^{-1}\sigma\left(x\right)^{-2}\exp\left\{2\int_0^x \frac{S\left(v\right)}{\sigma\left(v\right)^2}\,\mathrm{d}v\right\} \tag{0.5}$$

is the invariant density of the process and ξ is the random variable with *stationary density function* $f_S\left(\cdot\right)$. We write $f_S\left(\cdot\right)$ to emphasize that it depends on the function $S\left(\cdot\right)$. Note as well that if the initial value X_0 has the density function $f_S\left(\cdot\right)$, then the process $\{X_t,\ t \geq 0\}$ is stationary in the strict sense.

In almost all of this work (except Sections 3.3 and 3.5) we suppose that the conditions (0.2) and (0.3) are fulfilled and so the observed stochastic process has ergodic properties. This situation probably is the closest to the traditional scheme of i.i.d. observations because the observed values X_t have the same density function $f_S\left(\cdot\right)$ for all $t \geq 0$.

The first chapter introduces the necessary notations and notions. In particular, we remind the reader of the basic properties of the stochastic Itô integral with respect to the diffusion process, likelihood ratio formula, local time of diffusion process and some elementary notions of statistical estimation theory (definitions and the first properties of estimators and tests and lower bounds on the mean square errors of all estimators). This chapter cannot be considered as an introduction to the Itô calculus or mathematical statistics and it is supposed that the reader is already familiar with diffusion processes. A knowledge of the basic theory of classical mathematical statistics is not required directly but strongly recommended because it allows for a better understanding of the statements of the problems and the results obtained. The last section contains a (relatively) elementary theory of statistical inference for ergodic diffusion processes. There we give the definitions of estimators and tests and describe their first properties for the linear models. This chapter can be recommended for the first reading or as a basis for an initial course in the field.

In Chapter 2, we consider the problems of parameter estimation for the diffusion process

$$dX_t = S(\vartheta, X_t)\,dt + \sigma\left(X_t\right)\,\mathrm{d}W_t,\quad X_0,\quad 0 \leq t \leq T, \tag{0.6}$$

where $S\left(\cdot,\cdot\right)$ is some known function and we have to estimate the finite-dimensional parameter $\vartheta \in \Theta \subset \mathscr{R}^d$. In this chapter we describe the asymptotic behavior of the MLE, Bayesian estimators, minimum distance estimators, trajectory fitting estimators, estimators of the method of moments and one-step MLE.

To study the properties of the MLE $\hat{\vartheta}_T$ and the Bayesian estimators (BE) $\tilde{\vartheta}_T$ we suppose a bit more than is given by (0.2) and (0.3). In particular, we assume that $\sigma\left(x\right)^{-2} \leq C\left(1 + |x|^p\right)$ with some $p \geq 0$ and

$$\overline{\lim_{|x|\to\infty}} \sup_{\vartheta\in\Theta} \operatorname{sgn}(x) \frac{S(\vartheta,x)}{\sigma(x)^2} < 0. \tag{0.7}$$

It is easy to see that (0.7) is sufficient for (0.2) and (0.3) and can be considered as restrictive in some problems. In particular, we do not consider the problems with $\sigma(x) = 0$ for some values of x. Note that (0.7) provides the existence of all polynomial moments of the invariant density. Nevertheless we prefer to use it in the present work because it helps us to estimate the probabilities of the large deviation for the tails of the likelihood ratio, especially in the nonregular problems.

It is shown that under regularity conditions the MLE $\hat{\vartheta}_T$ and the BE are consistent, asymptotically normal and asymptotically efficient. Particularly, we have for the MLE

$$\mathcal{L}_\vartheta\left\{\sqrt{T}\left(\hat{\vartheta}_T - \vartheta\right)\right\} \Longrightarrow \mathcal{N}\left(\mathbf{0}, \mathbf{I}(\vartheta)^{-1}\right)$$

where $\mathbf{I}(\vartheta)$ is the matrix of information

$$\mathbf{I}(\vartheta) = \mathbf{E}_\vartheta\left(\frac{\dot{S}(\vartheta,\xi)\ \dot{S}(\vartheta,\xi)^{\mathbb{T}}}{\sigma(\xi)^2}\right).$$

Here and in the following $\Rightarrow$ means the convergence in distribution (weak convergence), thus $\mathcal{L}_\vartheta\{\eta_T\} \Rightarrow \mathcal{N}(a,b)$ means the asymptotic normality of η_T whose distribution depends on a parameter ϑ, the upper index $\mathbb{T}$ means the transposition and the dot the derivation w.r.t. parameter ϑ.

The asymptotical efficiency is understood in the sense of minimax Hajek-Le Cam's lower bound:

$$\underline{\lim}_{\delta\to 0}\ \underline{\lim}_{T\to\infty}\ \inf_{\bar{\vartheta}_T}\ \sup_{|\vartheta-\vartheta_0|\leq\delta} \mathbf{E}_\vartheta\, \ell(T^{1/2}\left(\bar{\vartheta}_T - \vartheta\right)) \geq \mathbf{E}\,\ell\left(\mathbf{I}(\vartheta_0)^{-1/2}\,\zeta\right),$$

where ζ is a zero mean Gaussian vector with unit covariance matrix and $\ell(\cdot)$ is a loss function.

We study two minimum distance estimators (MDE). The first one is based on the *empirical distribution function* (EDF)

$$\hat{F}_T(x) = \frac{1}{T}\int_0^T \chi_{\{X_t<x\}}\, \mathrm{d}t \tag{0.8}$$

and is given by the formula

$$\vartheta_T^* = \arg\inf_{\vartheta\in\Theta} \int_{\mathscr{R}} \left(\hat{F}_T(x) - F(\vartheta,x)\right)^2 \mathrm{d}x.$$

The second is based on the *empirical density function* (local time estimator of the density)

$$f_T^\circ(x) = \frac{1}{\sigma(x)^2 T}\int_0^T \operatorname{sgn}(x - X_t)\,\mathrm{d}X_t + \frac{|X_T - x| - |X_0 - x|}{\sigma(x)^2 T} \tag{0.9}$$

and is given by the similar relation

$$\vartheta_T^{**} = \arg\inf_{\vartheta\in\Theta}\int_{\mathscr{R}}\left(f_T^\circ(x) - f(\vartheta, x)\right)^2\,\mathrm{d}x.$$

Here $F(\vartheta, x)$ and $f(\vartheta, x)$ are the invariant distribution function and density function (0.5) of the ergodic process (0.6) and $\chi_{\{\mathbb{A}\}}$ is the indicator function of the event $\mathbb{A}$.

We show that under regularity conditions these estimators are consistent, asymptotically normal and we have the convergence of moments. Moreover, that second estimator is asymptotically efficient as is usual in the statement of the problem of locally contaminated observations.

The next studied is the *trajectory fitting estimator* (TFE)

$$\vartheta_T^\star = \arg\inf_{\theta\in\Theta}\int_0^T\left(X_t - \hat{X}_t(\theta)\right)^2\,\mathrm{d}t,$$

where $\left\{\hat{X}_t(\theta), 0 \le t \le T\right\}$, $\theta \in \Theta$, is the family of stochastic processes defined by the equation

$$\hat{X}_t(\theta) = X_0 + \int_0^t S(\theta, X_s)\,\mathrm{d}s, \quad 0 \le t \le T, \quad \theta \in \Theta,$$

i.e., having the observations $X_t, 0 \le t \le T$, we construct this family of stochastic functions. If the condition

$$\mathbf{E}_\vartheta\, S(\vartheta_1, \xi) \neq 0$$

for $\vartheta \neq \vartheta_1, \vartheta_1 \in \Theta$ holds, then this estimator can be consistent and asymptotically normal too. Otherwise, we modify the definition of this estimator to provide consistency.

Then we study the *estimator of the method of moments* (EMM), which is defined as

$$\bar{\vartheta}_T = \arg\inf_{\vartheta\in\Theta}\left|\boldsymbol{m}(\vartheta) - \frac{1}{T}\int_0^T \boldsymbol{q}(X_t)\,\mathrm{d}t\right|$$

where the vector $\boldsymbol{m}(\vartheta) = \mathbf{E}_\vartheta\, \boldsymbol{q}(\xi)$. It is shown that under regularity conditions this estimator is consistent, asymptotically normal and the moments of the estimator converge too.

The last three estimators (MDE, TFE, EMM) are consistent and asymptotically normal but generally they are not asymptotically efficient. Hence it is reasonable to improve them to being asymptotically efficient using the *one-step MLE* defined as

$$\vartheta_T^\circ = \theta_T + \frac{\Delta\left(\theta_T, X^T\right)}{\sqrt{T}\mathrm{I}\left(\theta_T\right)},$$

where θ_T is one of the consistent estimators and

$$\Delta_T\left(\vartheta, X^T\right) = \frac{1}{\sqrt{T}} \int_0^T \frac{\dot{S}\left(\vartheta, X_t\right)}{\sigma\left(X_t\right)^2} \left[\mathrm{d}X_t - S\left(\vartheta, X_t\right)\,\mathrm{d}t\right].$$

It is shown that it is consistent, asymptotically normal

$$\mathcal{L}_\vartheta\left\{\sqrt{T}\left(\vartheta_T^\circ - \vartheta\right)\right\} \Longrightarrow \mathcal{N}\left(0, \mathrm{I}\left(\vartheta\right)^{-1}\right)$$

and asymptotically efficient. Of course we have to modify $\Delta_T\left(\vartheta, X^T\right)$ replacing the stochastic integral by an ordinary one because the estimator θ_T depends of the whole trajectory X^T and the Itô integral of the function $\dot{S}\left(\theta_T, X_t\right)$ is not well defined.

Then the properties of these estimators are described in several *non standard situations*, when some regularity conditions are not fulfilled. These situations have already been studied for diffusion processes with small diffusion coefficients [139] (Chapter 2) and for inhomogeneous Poisson processes [145] (Chapter 4) but it seems that these statements allow for a better understanding of the role of regularity conditions and their presence in the study of ergodic diffusion processes can be useful too. We consider the following models of observations.

- *No true model*, when the observed diffusion process does not belong to the prescribed parametric family, i.e., the statistician supposes that the observed process satisfies (0.6) but this process has another stochastic differential

$$dX_t = S_*(\vartheta_0, X_t)\,\mathrm{d}t + \sigma_*\left(X_t\right)\,\mathrm{d}W_t, \quad X_0, \quad 0 \leq t \leq T,$$

  where the function $S_*(\vartheta, \cdot) \notin \{S(\vartheta, \cdot),\ \vartheta \in \Theta\}$ and generally, $\sigma_*\left(\cdot\right) \neq \sigma\left(\cdot\right)$. Then under regularity conditions the MLE $\hat{\vartheta}_T$ and BE $\tilde{\vartheta}_T$ converge to the value

$$\hat{\vartheta}_0 = \arg\inf_{\vartheta\in\Theta} \mathbf{E}_{\vartheta_0}\left(\frac{S(\vartheta, \xi) - S_*(\vartheta_0, \xi)}{\sigma\left(\xi\right)}\right)^2,$$

  which minimizes the Kullback–Leibler distance between the parametric family and the true measure, and the quantity $\sqrt{T}\left(\hat{\vartheta}_T - \hat{\vartheta}_0\right)$ is asymptotically normal. In general $\hat{\vartheta}_0 \neq \vartheta_0$ but we show that in some situations of contaminated models it is possible to have consistent, asymptotically normal and even asymptotically efficient estimators. Note that if $S(\vartheta_0, \cdot) = S_*(\vartheta_0, \cdot)$, then the MLE is consistent even if we use in the likelihood ratio the wrong value of the diffusion coefficient.

- *Too many true models*, when the condition of identifiability is not fulfilled and we have k different values of ϑ which provide the same trend coefficient, i.e., the observed process is from Equation (0.6) with the true value $\vartheta = \vartheta_1$, and there are $k-1$ other values $\vartheta_2, \ldots, \vartheta_k$ such that $S(\vartheta_1, \cdot) = S(\vartheta_i, \cdot)$, $i = 2, \ldots, k$. Let $\boldsymbol{\zeta} = (\zeta_1, \ldots, \zeta_k)$ be a Gaussian vector and define the discrete random variable

$$\hat{\vartheta} = \sum_{i=1}^{k} \vartheta_i\, \chi_{\{\mathbb{H}_i\}},$$

where the sets

$$\mathbb{H}_i = \left\{\omega :\ |\zeta_i| > \max_{j \neq i} |\zeta_j| \right\}.$$

Then the MLE converges in distribution to the random variable $\hat{\vartheta}$ and the Bayesian estimator converges to another (continuous) random variable

$$\tilde{\vartheta} = \sum_{i=1}^{k} \vartheta_i\, q_i, \qquad q_i = \frac{p(\vartheta_i)\ \mathrm{I}(\vartheta_i)^{-1/2}\, e^{\zeta_i^2/2}}{\sum_{l=1}^{k} p(\vartheta_l)\ \mathrm{I}(\vartheta_l)^{-1/2}\, e^{\zeta_l^2/2}},$$

where $p(\cdot)$ is the prior density and $\mathrm{I}(\cdot)$ is the Fisher information. Both estimators are asymptotically mixing normal.

- *Null Fisher information*, when the Fisher information $\mathrm{I}(\vartheta_0) = 0$ at one point (true value). Moreover we suppose that the first $k-1$ derivatives $S^{(i)}(\vartheta_0, x) \equiv 0$ (w.r.t. ϑ) and the quantity

$$\mathrm{I}_k(\vartheta_0) = \frac{1}{(k!)^2}\, \mathbf{E}_{\vartheta_0} \left(\frac{S^{(k)}(\vartheta_0, \xi)}{\sigma(\xi)}\right)^2 > 0.$$

Then the MLE has the properties:

$$\mathcal{L}_{\vartheta_0}\left\{T^{1/2k}\left(\hat{\vartheta}_T - \vartheta_0\right)\right\} \Longrightarrow \mathcal{L}\left(\zeta_k^{1/k}\right)$$

if k is odd and

$$\mathcal{L}_{\vartheta_0}\left\{T^{1/2k}\left(\hat{\vartheta}_T - \vartheta_0\right)\right\} \Longrightarrow \mathcal{L}\left((\zeta_k)_+^{1/k}\right)$$

if k is even, where $\mathcal{L}(\zeta_k) \sim \mathcal{N}\left(0, \mathrm{I}_k(\vartheta_0)^{-1}\right)$ and $(\zeta_k)_+ = \max(\zeta_k, 0)$.

- *Optimal observation window*. Suppose that we can observe the values of the process X_t in some set (*window*) $\mathbb{A}$ of fixed Lebesgue measure $\lambda > 0$ only, and we have to choose the window $\mathbb{A}$ in the class of such windows $\mathcal{A}_\lambda$ and an estimator $\bar{\vartheta}_T$ to minimize the limit mean square risk of the estimator, i.e., we have two problems: the first one is the construction of the lower bound on the risks of all choices of windows $\bar{\mathbb{A}}_T \in \mathcal{A}_\lambda$ and estimators $\bar{\vartheta}_T$

$$\lim_{\delta\to 0}\lim_{T\to\infty}\inf_{\bar{\mathbb{A}}_T,\bar\vartheta_T}\sup_{|\vartheta-\vartheta_0|\leq\delta} T\,\mathbf{E}_\vartheta\left(\bar\vartheta_T-\vartheta\right)^2\geq \mathrm{I}_*\left(\vartheta_0\right)^{-1}$$

and the second is to propose a window and an estimator which attain this bound. Here

$$\mathrm{I}_*\left(\vartheta_0\right)=\mathbf{E}_{\vartheta_0}\,\chi_{\{\mathbb{A}_*\}}\left(\frac{\dot S\left(\vartheta_0,\xi\right)}{\sigma\left(\xi\right)}\right)^2$$

is the corresponding Fisher information and $\mathbb{A}_*$ is the optimal window.

- *Asymptotic expansions.* If we suppose that the trend coefficient is more smooth w.r.t. a parameter, then it is possible to describe the *following after asymptotically normal terms*, i.e., to obtain an expansion of the MLE, BE or MDE by the powers of $T^{-1/2}$ up to the term T^{-1} or higher. For example, the MLE admits the representation

$$\sqrt{T}\left(\hat\vartheta_T-\vartheta\right)=\frac{\eta_1}{\mathrm{I}\left(\vartheta\right)}+\frac{\eta_1\left(\eta_2-\eta_3\right)\mathrm{I}\left(\vartheta\right)-\eta_1^2\rho_{1,2}\left(\vartheta\right)-K\,\mathrm{I}\left(\vartheta\right)^2}{\mathrm{I}\left(\vartheta\right)^3\sqrt{T}}+\frac{o\left(1\right)}{\sqrt{T}},$$

where the random variables $\eta_i, i=1,2,3$ are jointly asymptotically normal. Having such an expansion it is possible to derive the expansions of the distribution function and of the moments of this estimators. We do not give detailed proofs here but just discuss the possibility of such expansions.

- *Recursive estimation.* It is interesting to have a representation of the MLE as a solution of the stochastic differential equation. This can be done with the help of the Itô–Ventzel formula and this provides us with the following stochastic differential:

$$\mathrm{d}\hat\vartheta_t=\left[\frac{\dot S(\hat\vartheta_t,X_t)\ddot S(\hat\vartheta_t,X_t)}{H_t(\hat\vartheta_t)^2}-\frac{Q_t(\hat\vartheta_t)\,\dot S(\hat\vartheta_t,X_t)^2}{2\,H_t(\hat\vartheta_t)^3}\right]dt$$
$$-\frac{\dot S(\hat\vartheta_t,X_t)}{H_t(\hat\vartheta_t)}\left[\mathrm{d}X_t-S(\hat\vartheta_t,X_t)\,\mathrm{d}t\right],$$

where $H\left(\cdot\right)$ and $Q\left(\cdot\right)$ are the second and third derivatives of the log-likelihood ratio. We discuss the possibility of such a representation and one useful modification of this formula.

In Chapter 3 we consider several particular parameter estimation problems. The first one concerns the partially observed linear model

$$\begin{aligned}\mathrm{d}Y_t&=-\,a\left(\vartheta\right)\,Y_t\,\mathrm{d}t+b\left(\vartheta\right)\,\mathrm{d}V_t,\quad Y_0,\\ \mathrm{d}X_t&=\quad c\left(\vartheta\right)\,Y_t\,\mathrm{d}t+\,\sigma\,\mathrm{d}W_t,\qquad X_0,\end{aligned}$$

where $\{V_t,W_t,\ t\geq 0\}$ are two independent Wiener processes. We observe the process $X^T=\{X_t,\ 0\leq t\leq T\}$ only and have to estimate ϑ. To calculate the MLE and BE we use the Kalman–Bucy filter and show that under certain regularity conditions these estimators are consistent, asymptotically normal and asymptotically efficient for the polynomial loss functions.

Then we consider three problems of parameter estimation of *increasing singularity*. The first one corresponds to a continuous trend coefficient with a cusp at one given point (no derivative w.r.t. a parameter at this point), then we have the problem of delay estimation, when the trend coefficient is continuous but not differentiable at all points, and the last is the parameter estimation of the discontinuous trend coefficient.

The first problem concerns the diffusion process

$$\mathrm{d}X_t = a\,\left|X_t - \vartheta\right|^{\kappa}\,\mathrm{d}t + h\left(X_t - \vartheta\right)\mathrm{d}t + \sigma\left(X_t\right)\mathrm{d}W_t,\quad 0 \le t \le T,$$

where $\kappa \in (0, 1/2)$. The Fisher information $\mathrm{I}(\vartheta) = \infty$ and the MLE and BE have different limit distributions with the rate $T^{1/2H}$, $H = \kappa + 1/2$ (the Hurst parameter). In particular,

$$T^{1/2H}\left(\hat{\vartheta}_T - \vartheta\right) \Longrightarrow \frac{\hat{u}}{\gamma_\theta},\qquad \hat{u} = \arg\sup_u Z\left(u\right),$$

where

$$Z\left(u\right) = \exp\left\{W^H\left(u\right) - \frac{\left|u\right|^{2H}}{2}\right\},\quad u \in \mathscr{R}$$

and

$$T^{1/2H}\left(\tilde{\vartheta}_T - \vartheta\right) \Longrightarrow \frac{\tilde{u}}{\gamma_\theta},\qquad \tilde{u} = \frac{\int_{\mathscr{R}} u\,Z\left(u\right)\,\mathrm{d}u}{\int_{\mathscr{R}} Z\left(v\right)\,\mathrm{d}v}.$$

Here $W^H\left(\cdot\right)$ is the fractional Brownian motion and γ_ϑ is some constant. In this problem as in the next two parameter estimation problems, the asymptotically efficient are Bayesian estimators only.

The next problem is the parameter $\boldsymbol{\vartheta} = (\gamma, \tau)$ estimation by the observations

$$\mathrm{d}X_t = -\gamma\,X_{t-\tau}\,\mathrm{d}t + \sigma\,\mathrm{d}W_t,\quad 0 \le t \le T.$$

The process $\{X_t, t \ge 0\}$ is no longer a diffusion one because the trend coefficient depends on the previous value of X_t. This process can be called a diffusion type process [175] and if $\boldsymbol{\vartheta} \in (\alpha_1, \beta_1) \times (\alpha_2, \beta_2)$, where $\alpha_1 > 0, \alpha_2 > 0$ and $\beta_2 < \pi/2\beta_1$, then this process has ergodic properties. Remember that this process is as smooth w.r.t. time t as the Wiener process, hence the trend coefficient $-\gamma\,X_{t-\tau}$ is not differentiable w.r.t. an unknown parameter τ (delay) at any point $t \in [0, T]$.

The normalized likelihood ratio converges to the stochastic process

$$Z\left(v, u\right) = \exp\left\{v\Delta - \frac{v^2}{2}r\left(\boldsymbol{\vartheta}\right)^2 + \gamma\,W\left(u\right) - \frac{u^2}{2}\,\gamma^2\right\},\quad u \in \mathscr{R},$$

where $\Delta \sim \mathcal{N}$ and $W\left(\cdot\right)$ is a two-sided Wiener process. We show that the MLE $\hat{\boldsymbol{\vartheta}}_T$ and the BE $\tilde{\boldsymbol{\vartheta}}_T$ are consistent, have two different limit distributions and only the BE are asymptotically efficient. Note that we have two different rates of convergence of estimators of the components of the vector $\hat{\boldsymbol{\vartheta}}_T$. In particular, $\sqrt{T}\left(\hat{\gamma}_T - \gamma\right)$ is asymptotically normal and $T\left(\hat{\tau}_T - \tau\right)$ converges to some random variable and these quantities are asymptotically independent.

Then we study the properties of these estimators in the case when the trend coefficient $S(\boldsymbol{\vartheta}, x)$ is a discontinuous function of the parameter $\boldsymbol{\vartheta}$. For example, if the observed process (0.1) has the form

$$\mathrm{d}X_t = -\operatorname{sgn}\left(X_t - \vartheta\right)\,\mathrm{d}t + \mathrm{d}W_t, \quad X_0, \quad 0 \le t \le T, \tag{0.10}$$

then the function $S(\vartheta, x) = -\operatorname{sgn}\left(x - \vartheta\right)$ has a jump equal to 2 at the point $\vartheta = x$, so the trend *switches* every time when the process X_t takes a value ϑ. We consider the general case of observations (0.6) with the trend coefficient $S\left(\vartheta, x\right)$ discontinuous along two curves $x_*^{(i)}\left(\vartheta\right), \vartheta \in \left[\alpha, \beta\right], i = 1, 2$, i.e., $S\left(\vartheta, x_*^{(i)}\left(\vartheta\right)+\right) - S\left(\vartheta, x_*^{(i)}\left(\vartheta\right)-\right) \neq 0$. The functions $x_*^{(i)}\left(\cdot\right), i = 1, 2$, are supposed to be smooth w.r.t. θ. It is shown that the likelihood ratio converges to the stochastic process

$$Z\left(u\right) = \exp\left\{\Gamma_\vartheta\, W\left(u\right) - \frac{u^2}{2}\Gamma_\vartheta^2\right\}, \quad u \in \mathscr{R},$$

where Γ_ϑ is some constant and $W\left(\cdot\right)$ is a two-sided Wiener process. The MLE $\hat{\vartheta}_T$ and the BE $\tilde{\vartheta}_T$ for such models are consistent, the quantities $T(\hat{\vartheta}_T - \vartheta)$ and $T(\tilde{\vartheta}_T - \vartheta)$ have different limit distributions (similar to those in the delay estimation problem) and the BE is asymptotically efficient for polynomial loss functions.

Note that in these change-point type problems, the MLE and BE are consistent even for some misspecified models. In particular, if the observed process is

$$\mathrm{d}X_t = -\operatorname{sgn}\left(X_t - \vartheta\right)\,\mathrm{d}t + h\left(X_t\right)\,\mathrm{d}t + \mathrm{d}W_t, \quad X_0, \quad 0 \le t \le T, \tag{0.11}$$

where $h\left(\cdot\right)$ is an unknown function satisfying the only condition $\sup_x \left|h\left(x\right)\right| < 1$, then the MLE and BE constructed on the basis of the model with $h\left(x\right) \equiv 0$ are consistent and have the same rate of convergence T. Another particularity of these models is the possibility to restrict observations up to the window $\mathbb{A} = \left[\alpha, \beta\right]$ and to have the same properties of MLE and BE as if we observe on the whole space, i.e., for the models (0.10) and (0.11) we can use the likelihood ratio

$$L\left(\vartheta, X^T\right) = \exp\left\{-\int_0^T \operatorname{sgn}\left(X_t - \vartheta\right)\,\chi_{\{X_t \in \mathbb{A}\}}\,\mathrm{d}X_t - \frac{1}{2}\int_0^T \chi_{\{X_t \in \mathbb{A}\}}\,\mathrm{d}t\right\}$$

and the MLE and BE will have the same asymptotical properties as if $\mathbb{A} = \mathscr{R}$.

Further we study the properties of the MLE in three nonergodic diffusion processes of the type

$$\mathrm{d}X_t = \vartheta\, S\left(X_t\right)\,\mathrm{d}t + \sigma\,\mathrm{d}W_t, \quad X_0, \quad 0 \le t \le T. \tag{0.12}$$

The first one is *null recurrent* diffusion with the stochastic differential

$$\mathrm{d}X_t = -\vartheta\,\frac{X_t}{1+X_t^2}\,\mathrm{d}t + \sigma\,\mathrm{d}W_t,\quad X_0,\quad 0\le t\le T,$$

where $\vartheta \in \left(-\sigma^2/2, \sigma^2/2\right)$. The MLE

$$\hat\vartheta_T = -\frac{\int_0^T h(X_t)\,\mathrm{d}X_t}{\int_0^T h(X_t)^2\,\mathrm{d}t},\qquad h(x) = \frac{x}{1+x^2}$$

is consistent and the quantity $T^{\gamma/2}(\hat\vartheta_T - \vartheta)$ is asymptotically mixing normal. Here $\gamma = 1/2 + \vartheta/\sigma^2$, i.e., the rate of convergence depends on the value parameter.

The second model is

$$\mathrm{d}X_t = \vartheta\,|X_t|^\kappa\,\mathrm{d}t + \sigma\,\mathrm{d}W_t,\quad X_0,\quad 0\le t\le T,$$

where $\kappa\in(0,1)$. Then the quantity $T^{\gamma/2}\left(\hat\vartheta_T - \vartheta\right)$ is asymptotically normal with $\gamma = (1+\kappa)\,/\,(1-\kappa)$.

The last model is close to the Ornstein–Uhlenbeck process in the nonergodic case. We suppose that

$$\mathrm{d}X_t = \vartheta\,S\left(X_t\right)\mathrm{d}t + \mathrm{d}W_t,\quad X_0,\quad 0\le t\le T,$$

where the function $S(x) = a\,x\;(1+o(1))$ as $|x|\to\infty$ and $a>0$. Then we see that

$$e^{a\vartheta T}\left(\hat\vartheta_T - \vartheta\right) \Longrightarrow \frac{\zeta\sqrt{2\vartheta}}{X_0+Y+Z}$$

as $T\to\infty$. Here

$$Y = \int_0^\infty \mathrm{e}^{-a\vartheta t}\,\mathrm{d}W_t,\qquad Z = \vartheta\int_0^\infty \mathrm{e}^{-a\vartheta t}\,r\left(X_t\right)\,\mathrm{d}t$$

and the random variable $\zeta\sim\mathcal{N}(0,1)$ is independent of X_0, Y and Z.

Chapter 4 is devoted to nonparametric estimation problems. The model of observation is an ergodic diffusion process

$$\mathrm{d}X_t = S(X_t)\,\mathrm{d}t + \sigma\left(X_t\right)\mathrm{d}W_t,\quad X_0,\quad 0\le t\le T, \tag{0.13}$$

where $S(\cdot)$ is a function unknown to the observer, the diffusion coefficient is positive and known. We denote by $F_S(\cdot)$ and $f_S(\cdot)$ the distribution function and the density function (0.3) of the invariant law. Both are unknown because they depend on the function $S(\cdot)$. We suppose that the functions $S(\cdot)$ are such that the conditions (0.2) and (0.3) are fulfilled, Equation (0.13) has a unique weak solution and all measures $\mathbf{P}_S^{(T)}$ are equivalent. We denote this class of functions as $\mathscr{S}$.

The first problem is the asymptotically efficient estimation of the distribution $F_S(x)$ at one point x. We have a lower minimax bound on the risks of all estimators

$$\lim_{\delta\to 0}\lim_{T\to\infty}\inf_{\bar F_T}\sup_{S(\cdot)\in V_\delta}\mathbf{E}_S\ell(T^{1/2}\left(\bar F_T(x)-F_S(x)\right))\geq \mathbf{E}\,\ell\left(\zeta\,\mathrm{I}_F(S_*,x)^{-1/2}\right)$$

where V_δ is a δ-vicinity of a fixed function $S_*(\cdot)$

$$V_\delta\equiv\left\{S(\cdot)\;:\;\sup_{y\in\mathscr{R}}|S(y)-S_*(y)|\leq\delta,\quad S(\cdot)\in\mathscr{S}\right\},$$

$\ell(\cdot)$ is a loss function, $\zeta\sim\mathcal{N}(0,1)$ and

$$\mathrm{I}_F(S,x)=\left(4\;\mathbf{E}_S\left(\frac{F_S(\xi\wedge x)\left[1-F_S(\xi\vee x)\right]}{\sigma(\xi)f_S(\xi)}\right)^2\right)^{-1}$$

is the Fisher information in this problem. Then we show that the EDF (0.8) is asymptotically efficient in the sense of this bound.

The next problem is the asymptotically efficient estimation of the invariant density. The lower bound is similar to that given above and is

$$\lim_{\delta\to 0}\lim_{T\to\infty}\inf_{\bar f_T}\sup_{S(\cdot)\in V_\delta}\mathbf{E}_S\,\ell\left(T^{1/2}(\bar f_T(x)-f_S(x))\right)\geq \mathbf{E}\,\ell\left(\zeta\,\mathrm{I}_f(S_*,x)^{-1/2}\right),$$

where

$$\mathrm{I}_f(S,x)=\left\{4\,f_S(x)^2\;\mathbf{E}_S\left(\frac{\chi_{\{\xi>x\}}-F_S(\xi)}{\sigma(\xi)f_S(\xi)}\right)^2\right\}^{-1}$$

is the corresponding Fisher information.

The first consistent, asymptotically normal and asymptotically efficient estimator is the *empirical density* or *local time estimator* (LTE) (0.9). Moreover we show that the stochastic process

$$\eta_T(x)=\sqrt{T}\left(f_T^\circ(x)-f_S(x)\right),\qquad x\in\mathscr{R}$$

converges in distribution in the space $\mathscr{C}_0(\mathscr{R})$ of continuous on $\mathscr{R}$ functions, decreasing to zero at infinity to a Gaussian process. The second asymptotically efficient estimator is an estimator of the form

$$f_T^*(x)=\frac{1}{T}\int_0^T R_x(X_t)\,\mathrm{d}X_t+\frac{1}{T}\int_0^T N_x(X_t)\,\mathrm{d}t,$$

where the functions

$$R_x(y)=\frac{2\,\chi_{\{y\leq x\}}\,h(y)}{\sigma(x)^2\,h(x)},\qquad N_x(y)=\frac{\chi_{\{y\leq x\}}\,h'(y)\,\sigma(y)^2}{\sigma(x)^2\,h(x)},$$

and $h(\cdot)$ is an arbitrary function from the class $C'(\mathscr{R})$ such that $h(x) \neq 0$. Then under mild regularity conditions we show that this estimator is unbiased, consistent, asymptotically normal and asymptotically efficient for the polynomial loss functions. Hence we construct this estimator with the help of the function $h(\cdot)$ and its limit variance does not depend on $h(\cdot)$. We have as many asymptotically efficient estimators as we have functions $h(\cdot)$.

The third estimator is the traditional *kernel-type estimator*

$$\hat{f}_T(x) = \frac{1}{\sqrt{T}} \int_0^T K\left(\sqrt{T}\,(X_t - x)\right) \,\mathrm{d}t,$$

where $K(\cdot)$ is the usual kernel. If we suppose that the function $f_S(\cdot)$ is smooth enough, then this estimator is consistent, asymptotically normal and asymptotically efficient, i.e., its limit variance is equal to $\mathrm{I}_f(S_*, x)^{-1}$ and so it does not depend on $K(\cdot)$.

Then we consider the problem of semiparametric estimation, i.e., we estimate the one-dimensional parameter

$$\vartheta_S = \mathbf{E}_S\left(R(\xi)\; S(\xi) + N(\xi)\right),$$

where $R(\cdot)$ and $N(\cdot)$ are known functions and ξ is a *stationary random variable*, i.e., its density function is $f_S(\cdot)$. The observed process is always the same (0.13) and the function $S(\cdot)$ is supposed to be unknown. In this problem we propose a lower minimax bound on the risks of all estimators (with the corresponding Fisher information) and then show that the *empirical estimator*

$$\hat{\vartheta}_T = \frac{1}{T} \int_0^T R(X_t)\; \mathrm{d}X_t + \frac{1}{T} \int_0^T N(X_t)\; \mathrm{d}t$$

under some regularity conditions is unbiased, consistent, asymptotically normal and asymptotically efficient. Note that if we put

- $R(y) \equiv 0$ and $N(y) = \chi_{\{y<x\}}$, then $\vartheta_S = F_S(x)$ and we obtain the problem of the distribution function estimation with an asymptotically efficient empirical distribution function.
- $R(y) \equiv \sigma(x)^{-2} \operatorname{sgn}(x-y)$ and $N(y) = 0$, then $\vartheta_S = f_S(x)$ and we obtain the problem of the density estimation with an asymptotically efficient unbiased estimator

$$\bar{f}_T(x) = \frac{1}{T\sigma(x)^2} \int_0^T \operatorname{sgn}(x - X_t)\; \mathrm{d}X_t.$$

- $R(y) \equiv 0$ and $N(y) = y^k$, then $\vartheta_S = \mathbf{E}_S\,\xi^k$ and we obtain the problem of the moment estimation with an asymptotically efficient empirical moment estimator

$$\bar{\vartheta}_T = \frac{1}{T} \int_0^T X_t^k \;\mathrm{d}t.$$

From the results obtained here, it follows that the local time estimator, the class of unbiased estimators and the kernel type estimators are asymptotically efficient in the problem of density estimation for the integral type risk too. That means that (under mild regularity conditions) we have for the risk

$$\mathcal{R}\left(\bar{f}_T, f_S\right) = \mathbf{E}_S \int_{\mathcal{R}} \left[\bar{f}_T(x) - f_S(x)\right]^2 \mathrm{d}x$$

the following lower bound:

$$\varliminf_{\delta\to 0}\varliminf_{T\to\infty}\inf_{\bar{f}_T}\sup_{S(\cdot)\in V_\delta} T\,\mathcal{R}\left(\bar{f}_T, f_S\right) \geq \mathcal{R}_f\left(S_*\right) = \int_{\mathcal{R}} \mathrm{I}_f(S_*, x)^{-1}\mathrm{d}x \qquad (0.14)$$

and then we show that the mentioned estimators are asymptotically efficient in the sense of this bound. At the end of this chapter we study the problem of second order asymptotically efficient estimation of the density function, but before we consider the problems of density derivative and trend coefficient estimation.

To study the problem of trend coefficient $S(\cdot)$ estimation we note that the obvious relation

$$S(x) = \frac{\left(f_S(x)\,\sigma(x)^2\right)'\sigma(x)^2}{2 f_S(x)}$$

suggests we study first the asymptotically efficient estimation of the derivative of the function $f_S(x)\,\sigma(x)^2$. For simplicity of exposition we put $\sigma(x) \equiv 1$ which provides the equality

$$S(x) = \frac{f_S'(x)}{2 f_S(x)}.$$

Therefore to have a good estimator of $S(\cdot)$ we need a good estimator of the function $f_S'(\cdot)$. Hence the problem considered next is the density derivative estimation. Note that the next three problems (derivative, trend coefficient estimation and second order efficient estimation) are studied using the so-called *Pinsker's approach* [201], developed later by Golubev and Levit [91].

Let us suppose that the trend coefficient is a k-times differentiable function and introduce the integral type risk of an estimator $\bar{\vartheta}_T(\cdot)$ of the derivative $f_S'(\cdot)$ as

$$\mathcal{R}\left(\bar{\vartheta}_T, f_S'\right) = \int_{\mathcal{R}} \mathbf{E}_S\left(\bar{\vartheta}_T(x) - f_S'(x)\right)^2 \mathrm{d}x$$

and the set

$$\Sigma(k, R) = \left\{S(\cdot) : \int_{\mathcal{R}} f_S^{(k+1)}(x)^2\,\mathrm{d}x \leq 4\,R\right\}.$$

Then we show that

$$\varliminf_{D\to\infty}\varliminf_{T\to\infty}\inf_{\bar\vartheta_T}\sup_{S(\cdot)\in\Sigma_D} T^{\frac{2k}{2k+1}}\mathcal{R}\left(\bar\vartheta_T, f'_S\right)\geq 4\,\Pi\left(k,R\right)$$

where $\Sigma_D=\Sigma\left(k,R\right)\cap\Sigma\left(D\right)$, $\Sigma\left(D\right)$ is some set increasing with D, and

$$\Pi\left(k,R\right)=\left(2k+1\right)\left(\frac{k}{\pi\ \left(k+1\right)\ \left(2k+1\right)}\right)^{\frac{2k}{2k+1}}\ R^{\frac{1}{2k+1}}$$

is the so-called *Pinsker's constant*. We call an estimator $\hat\vartheta_T\left(\cdot\right)$ asymptotically efficient if

$$\varliminf_{D\to\infty}\varliminf_{T\to\infty}\sup_{S(\cdot)\in\Sigma_D} T^{\frac{2k}{2k+1}}\mathcal{R}\left(\hat\vartheta_T, f'_S\right)=4\,\Pi\left(k,R\right).$$

The asymptotically efficient estimator in this problem is the kernel-type estimator

$$\hat\vartheta_T\left(x\right)=\frac{2\nu_T}{T}\int_0^T K^*\left(\nu_T\left(X_t-x\right)\right)\,\chi_{\left\{|X_t|\leq\sqrt{T}\right\}}\mathrm{d}X_t,$$

where the kernel

$$K^*\left(y\right)=\frac{1}{\pi}\int_0^1\left(1-u^k\right)\ \cos\left(uy\right)\ \mathrm{d}u \tag{0.15}$$

and

$$\nu_T=\left(\frac{\pi\ R\ \left(k+1\right)\left(2k+1\right)}{k}\right)^{\frac{1}{2k+1}}\ T^{\frac{1}{2k+1}}. \tag{0.16}$$

We propose as well the second lower minimax bound (local) on the $\mathscr{L}_2$ risk of any estimator and then construct the kernel-type estimator $\vartheta_T^*\left(\cdot\right)$ asymptotically efficient in the sense of this bound. This lower bound is then used in the problem of trend coefficient estimation.

The next problem is the asymptotically efficient estimation of the trend coefficient $S\left(\cdot\right)$. We take the following $\mathscr{L}_2$ risk:

$$\mathcal{R}\left(\bar S_T,S\right)=\int_{\mathscr{R}}\mathbf{E}_S\left(\bar S_T\left(x\right)-S(x)\right)^2\ f_S\left(x\right)^2\ \mathrm{d}x$$

and show that if the function $S\left(\cdot\right)$ is k-times differentiable, then the following lower bound:

$$\varliminf_{\delta\to0}\varliminf_{T\to\infty}\inf_{\bar S_T}\sup_{S(\cdot)\in\Sigma_\delta} T^{\frac{2k}{2k+1}}\mathcal{R}\left(\bar S_T,S\right)\geq\Pi\left(k,R\right)$$

holds. Here $\Sigma_\delta=\Sigma\left(k,R\right)\cap\Sigma\left(\delta\right)$ is some neighborhood of a fixed function $S\left(\cdot\right)$. Then we show that the estimator

$$\hat S_T(x)=\frac{\theta_T^*(x)}{2f_T^\circ(x)+\varepsilon_T\mathrm{e}^{-l_T|x|}}$$

is asymptotically efficient, i.e.,

$$\lim_{\delta\to 0}\lim_{T\to\infty}\sup_{S(\cdot)\in\Sigma_\delta} T^{\frac{2k}{2k+1}}\mathcal{R}\left(\hat S_T, S\right) = \Pi\left(k, R\right).$$

Here $f_T^\circ(x)$ is the local-time estimator of the density, $\vartheta_T^*(\cdot)$ is the asymptotically efficient estimator of the density derivative (without truncation),

$$\theta_T^*(x) = \frac{\nu_T}{T}\int_0^T K^*\left(\nu_T\left(X_t - x\right)\right)\,dt$$

where $K^*(\cdot)$ and ν_T are defined in (0.15) and (0.16) respectively and $\varepsilon_T \to 0$, $l_T \to 0$.

Using the same Pinsker's approach we consider the problem of asymptotically second order efficient estimation of the invariant density in the following statement. Having so many asymptotically efficient estimators in the sense of the bound (0.14) it is natural to study the risk

$$\mathscr{R}\left(\bar f_T, f_S\right) = \mathbf{E}_S\int_{\mathscr{R}}\left[\bar f_T(x) - f_S(x)\right]^2 dx - \frac{\mathcal{R}_f(S)}{T}.$$

We suppose that the trend coefficient $S(\cdot)$ is $(k-1)$-times differentiable and belongs to a specially defined set Σ_*. The lower bound in this problem is

$$\lim_{T\to\infty}\inf_{\bar f_T}\sup_{S(\cdot)\in\Sigma_*} T^{\frac{2k}{2k-1}}\mathscr{R}\left(\bar f_T, f_S\right) \geq -\hat\Pi\left(k, R\right),$$

where

$$\hat\Pi\left(k, R\right) = 2\left(2k-1\right)\left(\frac{4k}{\pi\left(k-1\right)\left(2k-1\right)}\right)^{\frac{2k}{2k-1}} R^{\frac{1}{2k-1}}.$$

Then we propose an asymptotically efficient estimator, i.e., the estimator which attains this lower bound.

The last chapter is devoted to several problems of hypotheses testing. The basic hypothesis is always simple. In the problem of testing two simple hypotheses we describe the asymptotics of the power function using the large deviation principle. Then we study the score function test, which is locally asymptotically uniformly most powerful, and the likelihood ratio test in the case of one-sided local parametric alternatives for smooth and nonregular ergodic diffusion processes (with cusp, delay and discontinuity in the trend). The nonparametric hypotheses testing is considered in the situation quite similar to the parametric one, i.e., we propose a locally asymptotically uniformly most powerful test for a class of local alternatives defined by special parameterization.

The properties of the MLE and BE estimators are studied mainly with the help of the two general theorems by I.A. Ibragimov and R.Z. Khasminskii [109]. The asymptotical efficiency of estimators is defined in the terms of lower minimax bounds. The lower bounds in regular (smooth) problems of parameter estimation follow the well-known Hajek-Le Cam minimax bound [95], [163] and in non regular (non smooth) estimation problems we use the minimax lower bound based on the Bayesian estimators (see I.A. Ibragimov and R.Z. Khasminskii [109]). The $\sqrt{T}$-nonparametric lower bounds are constructed following B. Levit's [170] approach and the $T^{\frac{k}{2k+1}}$-nonparametric bounds follow the Pinsker's [201] approach.

1

Diffusion Processes and Statistical Problems

We introduce the stochastic integral and the stochastic differential equation and present their properties. Then we give some useful formulae for the local time process and for the likelihood ratio. Finally we present an elementary theory of asymptotic inference for ergodic diffusion processes. In particular, we introduce the first definitions in the estimation and hypotheses testing problems as well as some inequalities for the risk of estimators. For several simple models of diffusion processes we describe the asymptotic behavior of the maximum likelihood, minimum distance and trajectory fitting estimators in parameter estimation problems. Then we study the asymptotic behavior of some nonparametric estimators of the invariant distribution function, density and trend coefficient. We conclude this chapter with two hypotheses testing problems.

1.1 Stochastic Differential Equation

This section cannot be considered as an introduction to stochastic calculus. We assume that the reader is already familiar with Itô calculus (stochastic integrals with respect to the Wiener process, stochastic differential equations etc.). The purpose of this is just to introduce the notations used in the present work and to remind the reader of some definitions as well as some properties of stochastic integrals.

1.1.1 Stochastic Integral

We are given a probability space $\{\Omega, \mathfrak{F}, \mathbf{P}\}$, where $\Omega = \{\omega\}$ is a space of elementary events, $\mathfrak{F}$ is a σ-algebra of subsets of Ω and $\mathbf{P}$ is some probability measure defined on the sets from $\mathfrak{F}$, i.e., $\mathbf{P}$ is a measurable mapping $\mathbf{P} : \mathfrak{F} \rightarrow [0,1]$, which is countably additive and $\mathbf{P}\{\Omega\} = 1$. We suppose that this probability space is *complete*, i.e., it contains all subsets of the sets of measure zero. Random variables $\xi(\omega)$ are defined as measurable mappings $\xi : \Omega \rightarrow \mathscr{R}$, i.e., for any set $\mathbb{B} \in \mathfrak{B}(\mathscr{R})$, $\mathfrak{B}(\mathscr{R})$ is the σ-algebra of Borel subsets of the real line $\mathscr{R}$, the inclusion $\{\omega : \ \xi(\omega) \in \mathbb{B}\} \in \mathfrak{F}$ holds. We say that the stochastic process (collection of random variables) $\{h(t,\omega), 0 \leq t \leq T\}$ is measurable if for any Borel set $\mathbb{B} \in \mathfrak{B}(\mathscr{R})$ we have

$$\{(\omega, t) : \ h(t,\omega) \in \mathbb{B}\} \in \ \mathfrak{F} \times \mathfrak{B}[0,T],$$

where $\mathfrak{B}[0,T]$ is the Borel σ-algebra of the interval $[0,T]$. Let $\{\mathfrak{F}_t, \ 0 \leq t \leq T\}$ be an increasing family of σ-algebras (filtration), i.e., for any $0 \leq s < t \leq T$ the inclusions $\mathfrak{F}_s \subset \mathfrak{F}_t \subset \mathfrak{F}$ hold. We say that the measurable stochastic process $\{h(t,\omega), \ 0 \leq t \leq T\}$ is $\mathfrak{F}_t$-adapted if the random variables $h(t,\omega)$ are $\mathfrak{F}_t$-measurable for every $t \in [0,T]$, i.e., for any Borel $\mathbb{B}$ we have $\{\omega : \ h(t,\omega) \in \mathbb{B}\} \in \mathfrak{F}_t$. An $\mathfrak{F}_t$-adapted stochastic process is *progressively measurable* if for any $t \in [0,T]$ and Borel $\mathbb{B}$ we have

$$\{(\omega, s) : \ s < t, \ h(s,\omega) \in \mathbb{B}\} \in \mathfrak{F}_t \otimes \mathfrak{B}[0,t].$$

Let $\mathscr{M}_T$ be the class of progressively measurable random functions $h(\cdot)$ such that

$$\mathbf{P}\left\{\int_0^T h(t,\omega)^2 \, \mathrm{d}t < \infty\right\} = 1.$$

We say that $h(\cdot) \in \mathscr{M}_T^2$ if $h(\cdot) \in \mathscr{M}_T$ and

$$\mathbf{E}\int_0^T h(t,\omega)^2 \, \mathrm{d}t < \infty.$$

Here and in the following $\mathbf{E}$ means a mathematical expectation. Let $W = \{W_t, \mathfrak{F}_t \ 0 \leq t \leq T\}$ be a standard Wiener process. Remember that the (standard) Wiener process is a continuous (with probability 1) Gaussian process with independent increments and with the following first two moments: $\mathbf{E}\,W_t = 0$, $\mathbf{E}\,W_t\,W_s = t \wedge s$, $t, s \in [0,T]$ $(t \wedge s = \min(s,t))$.

The stochastic Itô integral

$$\mathcal{I}_T(h) = \int_0^T h(t,\omega) \, \mathrm{d}W_t$$

is defined for the functions $h(\cdot) \in \mathscr{M}_T$ as follows. Let $\{h_n(\cdot,\omega), n = 1,2,\ldots\}$ be a sequence of elementary functions, i.e., $h_n(t,\omega) = h_{n,l}(\omega)$ for $t \in [t_l^{(n)}, t_{l+1}^{(n)})$, where the random variables $h_{n,l}(\omega)$ are $\mathfrak{F}_{t_l^{(n)}}$ measurable and $\left\{t_l^{(n)}, l = 0,1,\ldots,L_n\right\}$ is some subdivision of the interval $[0,T]$. Then the stochastic integral of $h_n(\cdot,\omega)$ is defined as

$$\mathcal{I}_T(h_n) = \sum_{l=0}^{L_n - 1} h_n(\omega) \left[W_{t_{l+1}^{(n)}} - W_{t_l^{(n)}}\right].$$

For the functions $h(\cdot,\omega) \in \mathscr{M}_T$ we can take such a sequence of elementary functions $\{h_n(\cdot,\omega), n = 1,2,\ldots\}$ that

$$\mathbf{P} - \lim_{n\to\infty} \int_0^T [h(t,\omega) - h_n(t,\omega)]^2 \, \mathrm{d}t = 0$$

and the stochastic integral is defined as the limit

$$\mathcal{I}_T(h) = \mathbf{P} - \lim_{n\to\infty} \int_0^T h_n(t,\omega)\, \mathrm{d}W_t.$$

It has the following properties.

Lemma 1.1. • *If $h(\cdot) \in \mathscr{M}_T^2$, then*

$$\mathbf{E}\, \mathcal{I}_T(h) = 0, \quad \mathbf{E}\{\mathcal{I}_T(h)\,|\mathfrak{F}_t\} = \mathcal{I}_t(h).$$

- *For any two functions $h(\cdot), g(\cdot) \in \mathscr{M}_T^2$*

$$\mathbf{E}\, \mathcal{I}_T(h)\, \mathcal{I}_T(g) = \mathbf{E} \int_0^T h(t,\omega)\, g(t,\omega)\, \mathrm{d}t.$$

In particular,

$$\mathbf{E}\, \mathcal{I}_T(h)^2 = \mathbf{E} \int_0^T h(t,\omega)^2\, \mathrm{d}t.$$

- *If $h(\cdot) \in \mathscr{M}_T$, then for any $\delta > 0$ and $\gamma > 0$*

$$\mathbf{P}\left\{\sup_{0\le t\le T} \left|\int_0^t h(s,\omega)\, \mathrm{d}W_s\right| > \delta\right\} \le \frac{\gamma}{\delta^2} + \mathbf{P}\left\{\int_0^T h(t,\omega)^2 \mathrm{d}t > \gamma\right\}. \quad (1.1)$$

- *Let $h(\cdot) \in \mathscr{M}_T^2$ and for some $m \ge 1$*

$$\mathbf{E} \int_0^T |h(t,\omega)|^{2m}\, \mathrm{d}t < \infty. \quad (1.2)$$

Then

$$\mathbf{E}\,|\mathcal{I}_T(h)|^{2m} \le [m(2m-1)]^m\, T^{m-1}\, \mathbf{E} \int_0^T |h(t,\omega)|^{2m}\, \mathrm{d}t. \quad (1.3)$$

Moreover there exists a constant $C_m \ge 0$, such that (the Burkholder–Davis–Gundy inequality)

$$\mathbf{E}\left|\sup_{0\le t\le T} \mathcal{I}_t(h)\right|^{2m} \le C_m\, \mathbf{E}\left(\int_0^T h(t,\omega)^2\, \mathrm{d}t\right)^m. \quad (1.4)$$

In particular,

$$\mathbf{E}\left|\sup_{0\le t\le T} \mathcal{I}_t(h)\right|^2 \le 4\, \mathbf{E} \int_0^T h(t,\omega)^2\, \mathrm{d}t. \quad (1.5)$$

- *If* $h(\cdot) \in \mathscr{M}_T$, *then*

$$\mathbf{E}\exp\left\{\mathcal{I}_T(h) - \frac{1}{2}\int_0^T h(t,\omega)^2\,\mathrm{d}t\right\} \le 1.$$

The proof of these properties can be found in any book on stochastic calculus (see, e.g., [69], [119], [175], [208]).

Let $g(t,\omega)$ be $\mathfrak{F}_t$-adapted for almost all $t \in [0,T]$,

$$\mathbf{P}\left\{\int_0^T |g(t,\omega)|\,\mathrm{d}t < \infty\right\} = 1,$$

and $h(\cdot) \in \mathscr{M}_T$. Then the stochastic process

$$X_t = X_0 + \int_0^t g(s,\omega)\,\mathrm{d}s + \int_0^t h(s,\omega)\,\mathrm{d}W_s, \qquad 0 \le t \le T, \tag{1.6}$$

is called the *Itô process*. Here X_0 is a $\mathfrak{F}_0$-measurable random variable. In the shortened form it is usually written as

$$\mathrm{d}X_t = g(t,\omega)\,\mathrm{d}t + h(t,\omega)\,\mathrm{d}W_t, \quad X_0, \quad 0 \le t \le T. \tag{1.7}$$

This last equality is called the *stochastic differential* of the Itô process. It can be shown that the trajectory $\{X_t,\ 0 \le t \le T\}$ is continuous with probability 1 and $\mathfrak{F}_t$-adapted.

The class of Itô processes is closed with respect to smooth transformations in the following sense. Let $\{X_t, 0 \le t \le T\}$ be an Itô process with stochastic differential (1.7) and $G(x,t)$ be a differentiable function with the following continuous derivatives: $G'_t(x,t), G'_x(x,t),\ G''_{xx}(x,t)$ (with obvious notation). Then the stochastic process $Y_t = G(X_t,t), 0 \le t \le T$ is the Itô process too with the stochastic differential

$$\begin{aligned}\mathrm{d}Y_t = &\left[G'_t(X_t,t) + G'_x(X_t,t)\,g(t,\omega) + \frac{1}{2}G''_{xx}(X_t,t)\,h^2(t,\omega)\right]\,\mathrm{d}t\\ &+ G'_x(X_t,t)\,h(t,\omega)\,\mathrm{d}W_t, \quad Y_0 = G(X_0,0), \quad 0 \le t \le T.\end{aligned} \tag{1.8}$$

This equality is called the *Itô formula* and it can be written as

$$\mathrm{d}Y_t = \left[G'_t(X_t,t) + \frac{1}{2}G''_{xx}(X_t,t)\,h^2(t,\omega)\right]\,\mathrm{d}t + G'_x(X_t,t)\,\mathrm{d}X_t, \tag{1.9}$$

with the same initial value.

A nonnegative random variable $\tau = \tau(\omega)$ is called the *stopping time* if for any $t \in [0,T]$ we have $\{\omega : \tau(\omega) < t\} \in \mathfrak{F}_t$. Let $h(\cdot) \in \mathscr{M}_T$ and τ is some stopping time, then we put

$$\int_0^{\tau\wedge T} h(t,\omega)\ \mathrm{d}W_t = \int_0^T \chi_{\{t\le\tau\}}\, h(t,\omega)\ \mathrm{d}W_t.$$

Using the Itô formula for the stochastic integral

$$\mathcal{I}_{\tau_H}(h) = \int_0^{\tau_H} h(t,\omega)\, \mathrm{d}W_t$$

with special stopping time τ_H (1.10) we give an elementary proof of the well-known property of this integral to be a Gaussian random variable.

Lemma 1.2. *Let $h(\cdot) \in \mathscr{M}_T$ and for some $H > 0$ with probability 1*

$$\int_0^T h(t,\omega)^2\, \mathrm{d}t \geq H,$$

then the stopping time

$$\tau_H = \inf\left\{t\ :\ \int_0^t h(s,\omega)^2\, \mathrm{d}s \geq H\right\} \tag{1.10}$$

is well defined and

$$\mathcal{L}\left\{\mathcal{I}_{\tau_H}(h)\right\} = \mathcal{N}(0,H), \tag{1.11}$$

i.e., $\mathcal{I}_{\tau_H}(h)$ is a Gaussian random variable with mean zero and variance H.

Proof. We have to show that for any $\lambda \in \mathscr{R}$ the characteristic function

$$\phi(\lambda) - \mathbf{E}\ \mathrm{e}^{\mathrm{i}\lambda\mathcal{I}_\tau(h)} = \mathrm{e}^{-\lambda^2 H/2}, \tag{1.12}$$

where $\tau = \tau_H$. Let us introduce the random function

$$U(t) = \exp\left\{\mathrm{i}\lambda \int_0^t \chi_{\{s\leq\tau\}}\ h(s,\omega)\ \mathrm{d}W_s + \frac{\lambda^2}{2}\int_0^t \chi_{\{s\leq\tau\}}\ h(s,\omega)^2\ \mathrm{d}s\right\}.$$

This function by the Itô formula admits the stochastic differential

$$\mathrm{d}U(t) = \mathrm{i}\lambda\, U(t)\ \chi_{\{t\leq\tau\}}\ h(t,\omega)\ \mathrm{d}W_t, \quad U(0) = 1,$$

or in integral form

$$U(\tau) = 1 + \mathrm{i}\lambda \int_0^\tau U(t)\ h(t,\omega)\ \mathrm{d}W_t.$$

Let us check the condition (1.2) for the last stochastic integral. We have

$$\begin{aligned}
\mathbf{E} & \int_0^\tau |U(t)|^2\ h(t,\omega)^2\ \mathrm{d}t \\
&= \mathbf{E} \int_0^\tau \exp\left\{\lambda^2 \int_0^t \chi_{\{s\leq\tau\}}\ h(s,\omega)^2\,\mathrm{d}s\right\}\ h(t,\omega)^2\ \mathrm{d}t \\
&= \frac{1}{\lambda^2}\,\mathbf{E}\left(\mathrm{e}^{\lambda^2 \int_0^\tau h(s,\omega)^2 \mathrm{d}s} - 1\right) < \frac{1}{\lambda^2}\,\mathrm{e}^{\lambda^2 H} < \infty.
\end{aligned}$$

Hence

$$\begin{aligned}\mathbf{E}\,U(\tau) &= \mathbf{E}\exp\left\{\mathrm{i}\lambda\int_0^{\tau} h(t,\omega)\,\mathrm{d}W_t + \frac{\lambda^2}{2}\int_0^{\tau} h(t,\omega)^2\,\mathrm{d}t\right\}\\ &= \phi(\lambda)\ \mathrm{e}^{\lambda^2 H/2} = 1,\end{aligned}$$

and this equality proves (1.12).

We will need the stochastic integral with respect to the Itô process (1.7)

$$\mathcal{I}_T^*(f) = \int_0^T f(t,\omega)\,\mathrm{d}X_t.$$

It is defined for the functions $f(\cdot)$ such that

$$f(\cdot)\,h(\cdot) \in \mathcal{M}_T, \qquad |f(\cdot)\,g(\cdot)|^{1/2} \in \mathcal{M}_T \tag{1.13}$$

in the same way as the Itô integral w.r.t. the Wiener process. Its properties follow from the representation

$$\mathcal{I}_T^*(f) = \int_0^T f(t,\omega)\ g(t,\omega)\,\mathrm{d}t + \int_0^T f(t,\omega)\ h(t,\omega)\,\mathrm{d}W_t$$

and the properties of the stochastic Itô integral w.r.t. the Wiener process (for details see [175], p. 117).

The construction of many estimators and statistical tests in this work involves such integrals with respect to observations $X^T = \{X_t, 0 \le t \le T\}$. Hence we need a pathwise definition of it that we give below following Karandrikar [118].

Lemma 1.3. *Let X^T be an Itô process (1.7) and $f(\cdot,\omega)$ be continuous from the right and having limits from the left (cadlag) random function, satisfying (1.13). For $n \ge 1$, let $\{\tau_i^n : i \ge 0\}$ be defined by $\tau_0^n = 0$ and for $i \ge 0$,*

$$\tau_{i+1}^n = \inf\left\{t \ge \tau_i^n : \ |f(t,\omega) - f(\tau_i^n,\omega)| \ge 2^{-n}\right\}.$$

Let $f^n(\cdot)$ be defined as follows. For $\tau_k^n < t \le \tau_{k+1}^n$, $k \ge 0$ we put $f^n(t,\omega) = f(\tau_k^n,\omega)$. Then for $\tau_k^n < t \le \tau_{k+1}^n$ we have

$$\mathcal{I}_t^*(f^n) = f(0,\omega)\ X_0 + \sum_{i=0}^{k-1} f(\tau_i^n,\omega)\left(X_{\tau_{i+1}^n} - X_{\tau_i^n}\right) + f(\tau_k^n,\omega)\ \left(X_t - X_{\tau_k^n}\right)$$

and

$$\sup_{0\le t\le T} |\mathcal{I}_t^*(f) - \mathcal{I}_t^*(f^n)| \longrightarrow 0 \qquad \text{a.s.} \tag{1.14}$$

as $n \to \infty$.

Proof. Note that

$$\sup_{0\leq t\leq T} \left|f\left(t,\omega\right)-f^{n}\left(t,\omega\right)\right|\leq 2^{-n}.$$

Define the stopping times τ_H, $H>0$, increasing to ∞ as

$$\tau_H=\inf\left\{\tau:\ \int_0^{\tau} h\left(t,\omega\right)^2 \mathrm{d}t\geq H\right\}$$

and put

$$Q_n=\sup_{0\leq t\leq \tau_H}\left|\int_0^t f\left(t,\omega\right)h\left(t,\omega\right)\ \mathrm{d}W_t-\int_0^t f^n\left(t,\omega\right)h\left(t,\omega\right)\ \mathrm{d}W_t\right|.$$

Then using (1.5) one has $\mathbf{E}\,Q_n^2\leq 2^{-2n+2}\,H$ and $\mathbf{E}\,Q_n\leq 2^{-n+1}\sqrt{H}$. Hence it follows that

$$\mathbf{E}\sum_{n=0}^{\infty}Q_n\leq\sum_{n=0}^{\infty}\mathbf{E}\,Q_n\leq 2\sqrt{H}\sum_{n=0}^{\infty}2^{-n}<\infty.$$

As a consequence, one has

$$\sum_{n=0}^{\infty}Q_n<\infty\qquad \text{a.s.}$$

which gives the convergence

$$\sup_{0\leq t\leq \tau_H}\left|\int_0^t f\left(t,\omega\right)h\left(t,\omega\right)\ \mathrm{d}W_t-\int_0^t f^n\left(t,\omega\right)h\left(t,\omega\right)\ \mathrm{d}W_t\right|\rightarrow 0\quad \text{a.s.}$$

Since $H\rightarrow\infty$, we get $\tau_H\rightarrow T$ and

$$\sup_{0\leq t\leq T}\left|\int_0^t f\left(t,\omega\right)h\left(t,\omega\right)\ \mathrm{d}W_t-\int_0^t f^n\left(t,\omega\right)h\left(t,\omega\right)\ \mathrm{d}W_t\right|\rightarrow 0\quad \text{a.s.}$$

The uniform convergence of $f^n\left(\cdot\right)$ to $f\left(\cdot\right)$ yields

$$\sup_{0\leq t\leq T}\left|\int_0^t f\left(t,\omega\right)g\left(t,\omega\right)\ \mathrm{d}t-\int_0^t f^n\left(t,\omega\right)g\left(t,\omega\right)\ \mathrm{d}t\right|\rightarrow 0\quad \text{a.s.}$$

The last two convergences yield the result.

1.1.2 Diffusion Process

In the present work we are interested in a particular case of the Itô processes — the so-called *homogeneous diffusion processes* defined as solutions of the *stochastic differential equation* (SDE)

$$\mathrm{d}X_t = S(X_t)\,\mathrm{d}t + \sigma(X_t)\,\mathrm{d}W_t, \quad X_0, \quad 0 \le t \le T, \tag{1.15}$$

i.e., it is the Itô process (1.9) with $g(t,\omega) = S(X_t)$ and $h(t,\omega) = \sigma(X_t)$. The functions $S(x), \sigma(x)^2$, $x \in \mathscr{R}$ are non random and are called the *trend coefficient* and *diffusion coefficient* respectively. As X^T is an Itô process we suppose, of course, that the condition

$$\mathbf{P}\left\{\int_0^T \left[|S(X_t)| + \sigma(X_t)^2\right]\,\mathrm{d}t < \infty\right\} = 1$$

holds.

Remember that (1.15) is a short version of the integral representation

$$X_t = X_0 + \int_0^t S(X_u)\,\mathrm{d}u + \int_0^t \sigma(X_u)\,\mathrm{d}W_u, \quad 0 \le t \le T. \tag{1.16}$$

This equality can be considered as an integral equation with respect to the random function $X^T = \{X_t, 0 \le t \le T\}$ and the question of the existence of the solution of this equation naturally arises. There are two types of solutions: *strong* and *weak*.

Strong Solution

Let us introduce a family of sigma-algebras

$$\mathfrak{F}_t^{X_0,W} = \sigma\{X_0, W_s, 0 \le s \le t\}, \quad 0 \le t \le T$$

generated by the initial value X_0 and by the given Wiener process up to time t.

Definition 1.4. *We say that Equation (1.15) has a strong solution X^T on the given probability space $(\Omega, \mathfrak{F}, \mathbf{P})$ with respect to the fixed Wiener process $W^T = \{W_t, 0 \le t \le T\}$ and initial condition X_0 if the random function X^T satisfies the equality (1.15), has continuous sample paths and*

$$X_t \text{ is } \mathfrak{F}_t^{X_0,W} \text{ measurable for all } t \in [0,T]. \tag{1.17}$$

We say that the SDE (1.15) has a unique strong solution if for any two solutions $\left\{X_t^{(1)}, 0 \le t \le T\right\}$ and $\left\{X_t^{(2)}, 0 \le t \le T\right\}$ the equality

$$\mathbf{P}\left\{\sup_{0 \le t \le T}\left|X_t^{(1)} - X_t^{(2)}\right| > 0\right\} = 0$$

holds.

By this definition the solution X_t depends on the initial value X_0 and the trajectory of the Wiener process up to time t.

Remark 1.5. ([119], p. 285) The crucial requirement of this definition is captured in the condition (1.17); it corresponds to our intuitive understanding of X^T as the "output" of a dynamical system described by the pair of coefficients $(S(\cdot), \sigma(\cdot))$, whose "input" is W^T and which is also fed by the initial datum X_0. The *principle of causality* for dynamical systems requires that the output X_t at time t depend only on X_0 and the values of the input $\{W_s, 0 \leq s \leq t\}$. This principle finds its mathematical expression in (1.17).

There are several conditions of existence and uniqueness of the strong solution.

$\mathcal{GL}$. (Globally Lipschitz condition) *There exists a constant L such that*

$$|S(x) - S(y)| + |\sigma(x) - \sigma(y)| \leq L\,|x - y| \tag{1.18}$$

for all $x, y \in \mathscr{R}$.

Note that by this condition the functions $S(\cdot)$ and $\sigma(\cdot)$ satisfy the linear growth condition

$$|S(x)| + |\sigma(x)| \leq |S(0)| + |\sigma(0)| + L\,|x| \leq \tilde{L}\,(1 + |x|)$$

too.

Theorem 1.6. *Let the condition $\mathcal{GL}$ be fulfilled and $\mathbf{P}\{|X_0| < \infty\} = 1$. Then Equation (1.15) has a unique (strong) solution $\{X_t, 0 \leq t \leq T\}$, continuous with probability 1. If moreover $\mathbf{E}\,X_0^{2m} < \infty$, then*

$$\mathbf{E}\,X_t^{2m} \leq \left(1 + \mathbf{E}\,X_0^{2m}\right) e^{c_m t} - 1,$$

where c_m is some positive constant.

The proof can be found, for example, in [175], Theorem 4.6.

The condition $\mathcal{GL}$ in some problems can be too restrictive and can be weakened in the following way.

$\mathcal{LL}$. (Locally Lipschitz condition) *For any $N < \infty$ and $|x|, |y| \leq N$ there exists a constant $L_N > 0$ such that*

$$|S(x) - S(y)| + |\sigma(x) - \sigma(y)| \leq L_N\,|x - y| \tag{1.19}$$

and

$$2xS(x) + \sigma(x)^2 \leq B\,(1 + x^2). \tag{1.20}$$

The condition $\mathcal{LL}$ is, of course, less restrictive and is fulfilled, for example, for the diffusion process

$$\mathrm{d}X_t = -\vartheta X_t^3\,\mathrm{d}t + \sigma\,\mathrm{d}W_t.$$

Note that for ergodic diffusion processes usually we have $xS(x) < 0$ for the large values of x, so this is just a condition on the growth for the diffusion coefficient.

Proposition 1.7. *Let the condition $\mathcal{LL}$ be fulfilled and* $\mathbf{P}\{|X_0| < \infty\} = 1$. *Then Equation (1.15) has a unique strong solution* $\{X_t, 0 \le t \le T\}$, *continuous with probability 1.*

The proof can be found in [69].

Remember as well that this is not the *minimal regularity condition* for the existence and uniqueness of the strong solution (see, e.g., Karatzas and Shreve [119], Section 5.5).

The goal of this work is to study statistical problems concerning the observed continuous trajectories X^T of diffusion processes. Say, in parametrical inference this process is a solution of the SDE

$$\mathrm{d}X_t = S\left(\vartheta, X_t\right)\,\mathrm{d}t + \sigma\left(X_t\right)\,\mathrm{d}W_t, \quad X_0, \quad 0 \le t \le T, \tag{1.21}$$

where $\boldsymbol{\vartheta} \in \boldsymbol{\Theta} \subset \mathscr{R}^{\mathrm{d}}$ is an unknown parameter. Hence the space of observations is the space of continuous on $[0,T]$ functions $\mathscr{C}_T = \mathscr{C}[0,T]$ with uniform metrics and Borel σ-algebra $\mathfrak{B}_T$. Therefore, the *statistical experiment* (see Le Cam [164], Ibragimov and Khasminskii [109]) is given by the triplet

$$\left(\mathscr{C}_T, \mathfrak{B}_T, \left\{\mathbf{P}_{\boldsymbol{\vartheta}}^{(T)}, \ \boldsymbol{\vartheta} \in \boldsymbol{\Theta}\right\}\right).$$

Here $\mathbf{P}_{\boldsymbol{\vartheta}}^{(T)}$ is the measure induced in the space $\mathscr{C}_T$ by the process (1.21). This statistical experiment is the starting point for statistical inference. Therefore all we need is the family of measures $\left\{\mathbf{P}_{\boldsymbol{\vartheta}}^{(T)}, \ \vartheta \in \boldsymbol{\Theta}\right\}$. The conditions of the existence and uniqueness of a measure corresponding to a stochastic process having the stochastic differential (1.21) are weaker and the notion of the weak solution of Equation (1.21) (given below) fits statistical problems better.

Weak Solution

The existence of the strong solution says that if we have the probability space, Wiener process and the initial random variable, then for given $S(\cdot)$ and $\sigma(\cdot)$ the process (1.15) with (1.17) can be constructed. It is possible to consider the "inverse" problem, which is described in the following definition.

Definition 1.8. *Suppose that we are given the functions $S(\cdot)$ and $\sigma(\cdot)$ and the distribution function $F(\cdot)$. We say that there exists a weak solution of Equation (1.15) if there exists a probability space $(\Omega, \mathfrak{F}, \mathbf{P})$, a nondecreasing family of sub-sigma algebras $\{\mathfrak{F}_t, 0 \le t \le T\}$, a continuous random process $\{X_t, \mathfrak{F}_t, 0 \le t \le T\}$, and a Wiener process $\{W_t, \mathfrak{F}_t, 0 \le t \le T\}$, such that*

$$\mathbf{P}\left\{\int_0^T \left[\left|S\left(X_t\right)\right| + \sigma\left(X_t\right)^2\right]\,\mathrm{d}t < \infty\right\} = 1$$

and for all $t \in [0,T]$

$$X_t = X_0 + \int_0^t S(X_s)\ \mathrm{d}s + \int_0^t \sigma(X_s)\ \mathrm{d}W_s$$

with $\mathbf{P}\{X_0 < x\} = F(x)$.

Existence of the weak solution does not imply the existence of a strong solution but the existence of the strong solution implies the existence of the weak solution [69], [119].

The conditions of the existence of a weak solution are not too restrictive. For our purposes the following one is sufficient.

$\mathcal{ES}$. *The function* $S(\cdot)$ *is locally bounded, the function* $\sigma(\cdot)^2$ *is continuous and positive and for some* $A > 0$ *the condition*

$$x\ S(x) + \sigma(x)^2 \leq A\ \left(1 + x^2\right) \tag{1.22}$$

holds.

Theorem 1.9. *Suppose that condition* $\mathcal{ES}$ *is fulfilled, then the SDE (1.15) has a unique weak solution.*

Proof. The proof can be found in [69], p. 210.

In almost all examples of ergodic diffusion processes we have $x\ S(x) < 0$ for the large values of x, so the condition (1.22) is fulfilled if the function $\sigma(\cdot)$ satisfies the linear growth condition. It is possible to give even weaker conditions (see, e.g., [119], Section 5.5), but for the statistical problems considered in the present work we find the condition $\mathcal{ES}$ quite reasonable.

Note that we do not require that the function $S(\cdot)$ satisfies the conditions like (1.19) and this is an important point in the problems of parameter estimation for *cusp* and *change-point* models of Sections 3.2 and 3.5.

In all problems considered in the present work we suppose that the condition $\mathcal{ES}$ is always fulfilled. If the trend coefficient belongs to some family of functions, this means that we consider only such families for which $\mathcal{ES}$ is fulfilled for all its elements.

1.1.3 Local Time

Let us consider a homogeneous diffusion process

$$\mathrm{d}X_t = S(X_t)\ \mathrm{d}t + \sigma(X_t)\ \mathrm{d}W_t, \quad X_0, \quad 0 \leq t \leq T \tag{1.23}$$

where $S(\cdot)$ and $\sigma(\cdot)$ satisfy the condition $\mathcal{ES}$.

The *local time* of this diffusion process denoted by $\Lambda_T(x)$ is defined as the following limit (with probability 1):

$$\Lambda_T(x) = \lim_{\varepsilon\downarrow 0} \frac{\text{meas}\{t:\ |X_t - x| \le \varepsilon,\ 0\le t\le T\}}{4\,\varepsilon}, \quad T\ge 0,\ x\in\mathscr{R}$$

where $meas\{A\}$ is the Lebesgue measure of the set A. By the *Tanaka–Meyer formula* (see [119], p. 220) it admits the representation:

$$|X_T - x| = |X_0 - x| + \int_0^T \text{sgn}\,(X_t - x)\ \mathrm{d}X_t + 2\,\Lambda_T(x), \tag{1.24}$$

i.e.,

$$\Lambda_T(x) = \frac{|X_T - x| - |X_0 - x|}{2} - \frac{1}{2}\int_0^T \text{sgn}\,(X_t - x)\ S\,(X_t)\ \mathrm{d}t$$
$$-\frac{1}{2}\int_0^T \text{sgn}\,(X_t - x)\ \sigma\,(X_t)\ \mathrm{d}W_t. \tag{1.25}$$

The random function $\{\Lambda_T(x),\ x\in\mathscr{R}\}$ is nonnegative and continuous with probability 1 and, as follows from its definition, $\Lambda_T(x) = 0$ for all $x < X_*\,(T)$ and $x > X^*\,(T)$, where $X_*\,(T) = \min_{0\le t\le T} X_t$ and $X^*\,(T) = \max_{0\le t\le T} X_t$.

Let $h\,(\cdot)$ be a measurable function. Then with probability 1

$$\int_0^T h\,(X_t)\,\sigma\,(X_t)^2\ \mathrm{d}t = 2\int_{-\infty}^{\infty} h\,(x)\,\Lambda_T\,(x)\ \mathrm{d}x. \tag{1.26}$$

We will use this equality in a different form. Let us denote

$$f_T^\circ\,(x) = \frac{2\Lambda_T\,(x)}{T\sigma\,(x)^2} \tag{1.27}$$

and remember that the function $\sigma\,(x)^2$ is supposed to be positive. The statistic $f_T^\circ\,(x)$ we call the *local time estimator* of the invariant density (see Section 1.3). Then (1.26) can be written as

$$\frac{1}{T}\int_0^T h\,(X_t)\ \mathrm{d}t = \int_{-\infty}^{\infty} h\,(x)\,f_T^\circ\,(x)\ \mathrm{d}x. \tag{1.28}$$

These and many other properties of the local time can be found in Karatzas and Shreve [119] or Revuz and Yor [208].

Definition 1.10. *Define the class* $\mathscr{P}$ *of functions as*

$$\mathscr{P} = \{h\,(\cdot):\ |h\,(x)| \le C\ (1 + |x|^p)\} \tag{1.29}$$

i.e., $\mathscr{P}$ *is the class of functions having polynomial majorants. We say that a vector function* $\boldsymbol{h}\,(\boldsymbol{\vartheta},\cdot)$ *depending on parameter* $\boldsymbol{\vartheta}$*, belongs to* $\mathscr{P}$ *if its norm* $|\boldsymbol{h}\,(\boldsymbol{\vartheta},\cdot)|$ *satisfies (1.29), where the constants* C *and* p *can be chosen not to be dependent on* $\boldsymbol{\vartheta}$*.*

Let us introduce the following two conditions

$\mathcal{A}_0$. *The function* $\sigma(\cdot)^{-1} \in \mathscr{P}$ *and*

$$\overline{\lim_{|x|\to\infty}} \quad \operatorname{sgn}(x)\ \frac{S(x)}{\sigma(x)^2} < 0.$$

The second condition is
$\mathcal{RP}$. *The functions* $S(\cdot)$ *and* $\sigma(\cdot)$ *are such that*

$$V(x) = \int_0^x \exp\left\{-2\int_0^y \frac{S(v)}{\sigma(v)^2}\mathrm{d}v\right\}\mathrm{d}y \longrightarrow \pm\infty, \qquad as \quad x\to\pm\infty \quad (1.30)$$

and

$$G = \int_{-\infty}^{\infty} \sigma(y)^{-2}\exp\left\{2\int_0^y \frac{S(v)}{\sigma(v)^2}\mathrm{d}v\right\}\mathrm{d}y < \infty. \tag{1.31}$$

It is easy to see that condition $\mathcal{A}_0$ implies $\mathcal{RP}$. If the condition $\mathcal{RP}$ is fulfilled then according to Theorem 1.16 (Section 1.2) the process $\{X_t,\ t\geq 0\}$ has ergodic properties with the density of invariant law

$$f(x) = \frac{1}{G\,\sigma(x)^2}\ \exp\left\{2\int_0^x \frac{S(v)}{\sigma(v)^2}\,\mathrm{d}v\right\}, \qquad x\in\mathscr{R}.$$

Here we are interested in a moment's inequality. We suppose that the initial random variable X_0 has $f(\cdot)$ as a density function, so the process (1.23) is stationary. Below ξ is a random variable with the same density $f(\cdot)$.

Proposition 1.11. *Let the conditions* $\mathcal{RP}$ *be fulfilled. Then the random function*

$$\eta_T(x) = \sqrt{T}\left(\frac{2\Lambda_T(x)}{T\,\sigma(x)^2} - f(x)\right), \qquad x\in\mathscr{R}$$

admits the representation

$$\begin{aligned}\eta_T(x) &= \frac{2\,f(x)}{\sqrt{T}}\int_{X_0}^{X_T}\frac{\chi_{\{v>x\}} - F(v)}{\sigma(v)^2 f(v)}\,\mathrm{d}v\\ &\quad -\frac{2\,f(x)}{\sqrt{T}}\int_0^T \frac{\chi_{\{X_t>x\}} - F(X_t)}{\sigma(X_t)\,f(X_t)}\,\mathrm{d}W_t.\end{aligned} \tag{1.32}$$

Further, if $\mathbf{E}\,\sigma(\xi)^2<\infty$ *and*

$$\int_0^x \frac{S(v)}{\sigma(v)^2}\,\mathrm{d}v \to -\infty \quad as \quad x\to\pm\infty \tag{1.33}$$

then

$$\mathbf{E}\,\Lambda_T(x) = \frac{T}{2}\, f\,(x)\,\sigma\,(x)^2\,. \tag{1.34}$$

Moreover, if the condition $\mathcal{A}_0$ is fulfilled, then for any $p > 1$ there exist constants $C > 0$ and $\gamma > 0$ such that

$$\mathbf{E}\left|\eta_T\,(x)\right|^{2p} \leq C\,\mathrm{e}^{-\gamma|x|} \tag{1.35}$$

Proof. Let us introduce the function

$$H_n\,(y,x) = 2\,f\,(x)\int_0^y \frac{\phi_n\,(v,x) - F\,(v)}{\sigma\,(v)^2\,f\,(v)}\,\mathrm{d}v,$$

where

$$\phi_n\,(y,x) = \frac{n}{\sqrt{2\pi}}\int_{-\infty}^y \mathrm{e}^{-(v-x)^2n^2/2}\,\mathrm{d}v \longrightarrow \chi_{\{y>x\}}, \qquad n\to\infty$$

is a continuous approximation of the indicator function and

$$\phi_n'\,(y,x) = \frac{\partial\phi_n\,(y,x)}{\partial y} \to \delta\,(y-x)\,, \qquad n\to\infty.$$

Here $\delta\,(y-x)$ is a Delta-function.

By the Itô formula

$$\begin{aligned}
&H_n\,(X_T,x) - H_n\,(X_0,x)\\
&\quad = f\,(x)\int_0^T \frac{\phi_n'\,(X_t,x) - f\,(X_t)}{f\,(X_t)}\,\mathrm{d}t\\
&\qquad\qquad +2f\,(x)\int_0^T \frac{\phi_n\,(X_t,x) - F\,(X_t)}{\sigma\,(X_t)\,f\,(X_t)}\,\mathrm{d}W_t\\
&\quad = 2f\,(x)\int_{-\infty}^{\infty}\frac{\phi_n'\,(y,x)\,\Lambda_T(y)}{\sigma\,(y)^2\,f\,(y)}\,\mathrm{d}y - T\,f\,(x)\\
&\qquad\qquad +2f\,(x)\int_0^T \frac{\phi_n\,(X_t,x) - F\,(X_t)}{\sigma\,(X_t)\,f\,(X_t)}\,\mathrm{d}W_t.
\end{aligned}$$

Therefore if we put

$$H\,(y,x) = 2\,f\,(x)\int_0^y \frac{\chi_{\{v>x\}} - F\,(v)}{\sigma\,(v)^2\,f\,(v)}\,\mathrm{d}v, \tag{1.36}$$

then, as $n\to\infty$, we obtain the representation

$$\begin{aligned}
&H\,(X_T,x) - H\,(X_0,x)\\
&\quad = \frac{2\Lambda_T(x)}{\sigma\,(x)^2} - Tf(x) + 2f\,(x)\int_0^T \frac{\chi_{\{X_t>x\}} - F\,(X_t)}{\sigma\,(X_t)\;f\,(X_t)}\,\mathrm{d}W_t,
\end{aligned}$$

which is equivalent to (1.32). Remember that the functions $\Lambda_T(\cdot)$, $\sigma(\cdot)$ and $f(\cdot)$ are continuous. Therefore with probability 1

$$f(x)\int_{-\infty}^{\infty}\frac{\phi_n'(y,x)\,\Lambda_T(y)}{\sigma(y)^2 f(y)}\,dy \longrightarrow \frac{\Lambda_T(x)}{\sigma(x)^2}.$$

To verify (1.36) we note first that according to the representation (1.25)

$$\mathbf{E}\,\Lambda_T(x)=\frac{T}{2}\;\mathbf{E}\Big(\operatorname{sgn}(x-\xi)\;S(\xi)\Big).$$

Then we use the equality

$$\begin{aligned}\mathbf{E}\,S(\xi)&=\int\frac{S(y)}{G\,\sigma(y)^2}\;\exp\left\{2\int_0^y\frac{S(v)}{\sigma(v)^2}\,dv\right\}\,dy\\&=\frac{1}{2G}\;\exp\left\{2\int_0^y\frac{S(v)}{\sigma(v)^2}\,dv\right\}\Bigg|_{-\infty}^{\infty}=0\end{aligned}$$

and integrating by parts we obtain

$$\begin{aligned}\mathbf{E}\Big(\operatorname{sgn}(x-\xi)\;S(\xi)\Big)&=\int_{-\infty}^{x}S(y)\;f(y)\;dy-\int_x^{\infty}S(y)\;f(y)\;dy\\&=-2\int_x^{\infty}S(y)\;f(y)\;dy=\sigma(x)^2 f(x)\,.\end{aligned}$$

Therefore $\mathbf{E}\,\eta_T(x)=0$.

To obtain the last estimate we write

$$\mathbf{E}\left|\eta_T(x)\right|^{2p}\le C_1T^{-p}\;\;\mathbf{E}\left|H(\xi,x)\right|^{2p}+C_2\;f(x)^{2p}\;\mathbf{E}\left|\frac{\chi_{\{\xi>x\}}-F(\xi)}{\sigma(\xi)\,f(\xi)}\right|^{2p},$$

where $C_i=C_i(p)\,,i=1,2$ are two positive constants (we used the estimate (1.3)). Therefore we have to estimate these two mathematical expectations for the large values of x.

Note that by the condition $\mathcal{A}_0$ there exist constants $A>0$ and $\gamma>0$, such that for all $|v|>A$

$$\operatorname{sgn}(v)\;\frac{S(v)}{\sigma(v)^2}<-\gamma.$$

Hence, for a fixed x and $v>A$ and $v>x$

$$\begin{aligned}\frac{\chi_{\{v>x\}}-F(v)}{\sigma(v)^2\;f(v)}&=\frac{1}{\sigma(v)^2}\int_v^{\infty}\frac{f(u)}{f(v)}\,du\\&=\int_v^{\infty}\sigma(u)^{-2}\exp\left\{2\int_v^{u}\frac{S(r)}{\sigma(r)^2}\,dr\right\}du\\&\le C\int_v^{\infty}(1+u^q)\exp\{-2\gamma(u-v)\}\,du\le C\;(1+v^q)\,.\end{aligned}$$

We have (for $x > A$)

$$
\begin{aligned}
\mathbf{E}\left|H(\xi,x)\right|^{2p} &= 2^{2p} f(x)^{2p} \int_{-\infty}^{\infty} \left| \int_0^y \frac{\chi_{\{v>x\}} - F(v)}{\sigma(v)^2\, f(v)} \mathrm{d}v \right|^{2p} f(y)\, \mathrm{d}y \\
&= 2^{2p} f(x)^{2p} \int_{-\infty}^{0} \left| \int_0^y \frac{F(v)}{\sigma(v)^2\, f(v)} \mathrm{d}v \right|^{2p} f(y)\, \mathrm{d}y \\
&\quad + 2^{2p} f(x)^{2p} \int_{0}^{x} \left| \int_0^y \frac{F(v)}{\sigma(v)^2\, f(v)} \mathrm{d}v \right|^{2p} f(y)\, \mathrm{d}y \\
&\quad + 2^{2p} f(x)^{2p} \int_{x}^{\infty} \left| \int_0^y \frac{\chi_{\{v>x\}} - F(v)}{\sigma(v)^2\, f(v)} \mathrm{d}v \right|^{2p} f(y)\, \mathrm{d}y \\
&\equiv 2^{2p}\Big(I_1 + I_2(x) + I_3(x)\Big)
\end{aligned}
$$

in obvious notation. As the first integral does not depend on x we have to estimate the $I_2(x)$ and $I_3(x)$ only.

Observe that

$$
\begin{aligned}
I_2(x) &< f(x)^{2p} \int_0^x \left| \int_0^y \frac{\mathrm{d}v}{\sigma(v)^2\, f(v)} \right|^{2p} f(y)\, \mathrm{d}y \\
&= f(x)^{2p} \int_0^A \left| \int_0^y \frac{\mathrm{d}v}{\sigma(v)^2\, f(v)} \right|^{2p} f(y)\, \mathrm{d}y \\
&\quad + f(x)^{2p} \int_A^x \left| \int_0^y \frac{\mathrm{d}v}{\sigma(v)^2\, f(v)} \right|^{2p} f(y)\, \mathrm{d}y \\
&\le C\, f(x)^{2p} + C\, f(x)^{2p} \int_A^x \left| \int_A^y \frac{f(y)}{\sigma(v)^2\, f(v)} \mathrm{d}v \right|^{2p} f(y)^{-2p+1}\, \mathrm{d}y \\
&\le C f(x)^{2p} \left(C + \int_A^x \left(1+y^r\right) \left| \int_A^y \mathrm{e}^{-2\gamma(y-v)} \mathrm{d}v \right|^{2p} f(y)^{-2p+1}\, \mathrm{d}y \right) \\
&\le C\, f(x)^{2p} + C\, f(x) \int_A^x \left(1+y^r\right) \left(\frac{f(x)}{f(y)} \right)^{2p-1} \mathrm{d}y \\
&\le C\, f(x)^{2p} + C\, \left(1+x^q\right) f(x) \int_A^x \left(1+y^r\right)\, \mathrm{e}^{-2\gamma(2p-1)(x-y)} \mathrm{d}y \\
&\le C\, \left(1+x^q\right) f(x) \le C\, \left(1+x^m\right)\, \mathrm{e}^{-2\gamma x}.
\end{aligned}
$$

Furthermore

$$
I_3(x) \le f(x)^{2p} \int_x^{\infty} \left| \int_0^A \frac{F(v)}{\sigma(v)^2\, f(v)} \mathrm{d}v \right.
$$

$$
\begin{aligned}
& - \int_A^y \frac{\chi_{\{v>x\}} - F(v)}{\sigma(v)^2\, f(v)} \mathrm{d}v \Bigg|^{2p} f(y)\,\mathrm{d}y \\
&\leq C\, f(x)^{2p} + C\, f(x)^{2p} \int_x^\infty \left| \int_A^y \frac{\chi_{\{v>x\}} - F(v)}{\sigma(v)^2\, f(v)} \mathrm{d}v \right|^{2p} f(y)\,\mathrm{d}y \\
&= C f(x)^{2p} + C f(x)^{2p} \int_x^\infty \Bigg| \int_A^x \frac{\mathrm{d}v}{\sigma(v)^2 f(v)} \\
&\qquad - \int_x^y \frac{1 - F(v)}{\sigma(v)^2 f(v)} \mathrm{d}v \Bigg|^{2p} f(y)\,\mathrm{d}y \\
&\leq C\, f(x)^{2p} + C f(x)^{2p} \int_x^\infty \left| \int_A^x \frac{\mathrm{d}v}{\sigma(v)^2 f(v)} \right|^{2p} f(y)\,\mathrm{d}y \\
&= C f(x)^{2p} + C \int_x^\infty \left| \int_A^x \frac{f(x)}{\sigma(v)^2 f(v)} \mathrm{d}v \right|^{2p} f(y)\,\mathrm{d}y \\
&\leq C f(x)^{2p} + C\left(1 + x^q\right) \int_x^\infty \left| \int_A^x \left(1 + v^q\right) \mathrm{e}^{-2\gamma(x-v)} \mathrm{d}v \right|^{2p} f(y)\,\mathrm{d}y \\
&\leq C\, \left(1 + x^m\right)\, \mathrm{e}^{-2\gamma x}.
\end{aligned}
$$

For $x < -A$ we have a similar estimate. The same arguments allow to establish the estimate

$$
f(x)^{2p}\, \mathbf{E} \left| \frac{\chi_{\{\xi>x\}} - F(\xi)}{\sigma(\xi)\, f(\xi)} \right|^{2p} \leq C\, \left(1 + |x|^m\right)\, \mathrm{e}^{-2\gamma|x|}.
$$

Therefore we obtain (1.35) with some constant $C > 0$.

1.1.4 Likelihood Ratio

As is usual in mathematical statistics, the statistical inference for diffusion processes is heavily based on the likelihood ratio formula, which we give below.

Let us consider three stochastic differential equations

$$
\begin{aligned}
\mathrm{d}X_t &= S_1(X_t)\,\mathrm{d}t + \sigma(X_t)\,\mathrm{d}W_t, \quad X_0^{(1)}, \quad 0 \leq t \leq T, \\
\mathrm{d}X_t &= S_2(X_t)\,\mathrm{d}t + \sigma(X_t)\,\mathrm{d}W_t, \quad X_0^{(2)}, \quad 0 \leq t \leq T, \\
\mathrm{d}X_t &= \qquad\qquad\quad \sigma(X_t)\,\mathrm{d}W_t, \quad X_0, \quad 0 \leq t \leq T,
\end{aligned}
$$

and denote by $\mathbf{P}_1^{(T)}$, $\mathbf{P}_2^{(T)}$ and $\mathbf{P}^{(T)}$ the probability measures induced in $(\mathscr{C}_T, \mathfrak{B}_T)$ by the solutions of these equations respectively. Introduce as well the function

$$
\delta(x) = \frac{S_2(x) - S_1(x)}{\sigma(x)}
$$

and the condition.

$\mathcal{EM}$. *The functions $S_1(\cdot), S_2(\cdot), \sigma(\cdot)$ satisfy condition $\mathcal{ES}$ and the densities $f_1(\cdot), f_2(\cdot), f(\cdot)$ (with respect to the Lebesgue measure) of the corresponding initial values have the same support (if the initial value is nonrandom then we suppose that it takes the same value for all processes).*

According to the following theorem this condition provides the equivalence of these three measures.

Theorem 1.12. *Suppose that condition $\mathcal{EM}$ is fulfilled, then the measures $\mathbf{P}_1^{(T)}$, $\mathbf{P}_2^{(T)}$ and $\mathbf{P}^{(T)}$ are equivalent and the corresponding Radon–Nikodym derivatives are*

$$\frac{\mathrm{d}\mathbf{P}_1^{(T)}}{\mathrm{d}\mathbf{P}^{(T)}}\left(X^T\right)=\frac{f_1(X_0)}{f(X_0)}\exp\left\{\int_0^T\frac{S_1(X_t)}{\sigma(X_t)^2}\,\mathrm{d}X_t-\frac{1}{2}\int_0^T\frac{S_1(X_t)^2}{\sigma(X_t)^2}\,\mathrm{d}t\right\}$$

and

$$\frac{\mathrm{d}\mathbf{P}_2^{(T)}}{\mathrm{d}\mathbf{P}_1^{(T)}}\left(X^T\right)=\frac{f_2(X_0)}{f_1(X_0)}\exp\left\{\int_0^T\frac{S_2(X_t)-S_1(X_t)}{\sigma(X_t)^2}\,\mathrm{d}X_t\right.$$
$$\left.-\frac{1}{2}\int_0^T\frac{S_2(X_t)^2-S_1(X_t)^2}{\sigma(X_t)^2}\,\mathrm{d}t\right\}. \tag{1.37}$$

Proof. The proof can be found in [112], [175]. We just remark that by condition $\mathcal{EM}$ we have

$$\mathbf{P}_i^{(T)}\left\{\int_0^T\delta(X_t)^2\,\mathrm{d}t<\infty\right\}=1,$$
$$\mathbf{P}^{(T)}\left\{\int_0^T\left(\frac{S_i(X_t)}{\sigma(X_t)}\right)^2\mathrm{d}t<\infty\right\}=1,$$
$$\mathbf{P}_i^{(T)}\left\{\int_0^T\left(\frac{S_i(X_t)}{\sigma(X_t)}\right)^2\mathrm{d}t<\infty\right\}=1$$

for $i=1,2$.

We denote by $\mathbf{E}_1$, $\mathbf{E}_2$ and $\mathbf{E}$ the corresponding mathematical expectations. Put

$$Z_1=\frac{\mathrm{d}\mathbf{P}_1^{(T)}}{\mathrm{d}\mathbf{P}^{(T)}}\left(X^T\right),\qquad Z_2=\frac{\mathrm{d}\mathbf{P}_2^{(T)}}{\mathrm{d}\mathbf{P}^{(T)}}\left(X^T\right),\qquad V_T=\left(\frac{Z_2}{Z_1}\left(X^T\right)\right)^{1/2m}.$$

We will often use the following lemma.

Lemma 1.13. *Let $m \geq 1$ be some integer and*

$$\mathbf{E}_2 \int_0^T |\delta(X_t)|^{4m} \, \mathrm{d}t < \infty. \tag{1.38}$$

Then there exist constants $C_1(m), C_2(m)$ and $C_3(m)$ such that

$$\mathbf{E}\left|Z_1^{1/2m} - Z_2^{1/2m}\right|^{2m} \leq C_1(m) \int \left|f_1(x)^{1/2m} - f_2(x)^{1/2m}\right|^{2m} \mathrm{d}x$$
$$+ C_2(m)\, \mathbf{E}_1 \left(\int_0^T V_t\, \delta(X_t)^2 \, \mathrm{d}t\right)^{2m} + C_3(m)\, \mathbf{E}_1 \left(\int_0^T V_t^2\, \delta(X_t)^2 \, \mathrm{d}t\right)^{m}. \tag{1.39}$$

Proof. We have the equalities

$$\mathbf{E}\left|Z_1^{1/2m} - Z_2^{1/2m}\right|^{2m} = \mathbf{E}\ Z_1 \left|1 - V_T\right|^{2m} = \mathbf{E}_1 \left|1 - V_T\right|^{2m}.$$

The stochastic process $V^T = \{V_t, 0 \leq t \leq T\}$ by the Itô formula admits (with $\mathbf{P}_1^{(T)}$ probability 1) the differential

$$\mathrm{d}V_t = -\frac{2m-1}{8m^2} V_t\, \delta(X_t)^2 \, \mathrm{d}t + \frac{1}{2m} V_t\, \delta(X_t)\, \mathrm{d}W_t, \quad V_0,$$

where

$$V_0 = \left(\frac{f_2(X_0)}{f_1(X_0)}\right)^{1/2m}.$$

Hence

$$\mathbf{E}_1 \left|1 - V_T\right|^{2m} \leq 3^{2m-1}\, \mathbf{E}_1 \left|1 - V_0\right|^{2m}$$
$$+ 3^{2m-1} \left(\frac{2m-1}{8m^2}\right)^{2m} \mathbf{E}_1 \left(\int_0^T V_t\, \delta(X_t)^2 \, \mathrm{d}t\right)^{2m}$$
$$+ 3^{2m-1} \left(\frac{1}{2m}\right)^{2m} \mathbf{E}_1 \left(\int_0^T V_t\, \delta(X_t)\, \mathrm{d}W_t\right)^{2m},$$

where we used the inequality $(a+b+c)^{2m} \leq 3^{2m-1}(a^{2m} + b^{2m} + c^{2m})$. To the last stochastic integral we apply the estimate (1.4) and obtain the desired inequality (1.39).

Note that all mathematical expectations here are finite. For example

$$\mathbf{E}_1 \left|1 - V_0\right|^{2m} = \mathbf{E}\left|\left(\frac{f_1(X_0)}{f(X_0)}\right)^{1/2m} - \left(\frac{f_2(X_0)}{f(X_0)}\right)^{1/2m}\right|^{2m}$$
$$= \int \left|f_1(x)^{1/2m} - f_2(x)^{1/2m}\right|^{2m} \mathrm{d}x \leq \int \left(f_1(x) + f_2(x)\right) \mathrm{d}x = 2.$$

Remark 1.14. If the underlying processes are stationary, then

$$\mathbf{E}\left|Z_1^{1/2m} - Z_2^{1/2m}\right|^{2m} \leq C_1(m) \int \left|f_1(x)^{1/2m} - f_2(x)^{1/2m}\right|^{2m} \mathrm{d}x$$
$$+ C_2(m)\, T^{2m}\, \mathbf{E}_2\, \delta(\xi)^{4m} + \bar{C}_3(m)\, T^m\, \mathbf{E}_2\, \delta(\xi)^{2m}. \qquad (1.40)$$

Indeed

$$\mathbf{E}_1 \left(\int_0^T V_t\, \delta(X_t)^2\, \mathrm{d}t\right)^{2m} \leq T^{2m-1}\, \mathbf{E}_1 \int_0^T V_t^{2m}\, \delta(X_t)^{4m}\, \mathrm{d}t$$
$$= T^{2m-1} \int_0^T \mathbf{E}_1\, V_t^{2m}\, \delta(X_t)^{4m}\, \mathrm{d}t = T^{2m-1} \int_0^T \mathbf{E}_2\, \delta(X_t)^{4m}\, \mathrm{d}t$$
$$= T^{2m}\, \mathbf{E}_2\, \delta(\xi)^{4m} < \infty.$$

Here ξ is a random variable with a corresponding stationary distribution function. For the last integral in (1.39) we have the similar estimate

$$\mathbf{E}_1 \left(\int_0^T V_t^2\, \delta(X_t)^2\, \mathrm{d}t\right)^{m} \leq T^{m-1}\, \mathbf{E}_1 \int_0^T V_t^{2m}\, \delta(X_t)^{2m}\, \mathrm{d}t$$
$$= T^{m-1} \int_0^T \mathbf{E}_1\, V_t^{2m}\, \delta(X_t)^{2m}\, \mathrm{d}t = T^{m-1} \int_0^T \mathbf{E}_2\, \delta(X_t)^{2m}\, \mathrm{d}t$$
$$= T^{m}\, \mathbf{E}_2\, \delta(\xi)^{2m} < \infty.$$

In the case $m = 1$ the condition (1.38) can be replaced by

$$\mathbf{E}_i\, \delta(\xi)^2 < \infty, \qquad i = 1, 2, \qquad (1.41)$$

because

$$\mathbf{E}\left|Z_1^{1/2} - Z_2^{1/2}\right|^2 = 2\left[1 - \mathbf{E}\sqrt{Z_1 Z_2}\right] = 2\left[1 - \mathbf{E}_1\, V_T\right]$$
$$= 2 - 2\int \sqrt{f_1(x)\, f_2(x)}\, \mathrm{d}x + \frac{1}{4}\mathbf{E}_1 \int_0^T V_t\, \delta(X_t)^2\, \mathrm{d}t$$
$$\leq \int \left[f_1(x)^{1/2} - f_2(x)^{1/2}\right]^2 \mathrm{d}x + \frac{T}{8}\mathbf{E}_1\, \delta(\xi)^2 + \frac{T}{8}\mathbf{E}_2\, \delta(\xi)^2, \qquad (1.42)$$

where we used the estimate $2\,\mathbf{E}_1\, V_t \delta(X_t)^2 \leq \mathbf{E}_2\, \delta(\xi)^2 + \mathbf{E}_1\, \delta(\xi)^2$.

We give below two forms of the likelihood ratio formula (1.35) for the parametric family of the diffusion processes which will be used later. Suppose that the observed diffusion process

$$\mathrm{d}X_t = S(\vartheta, X_t)\, \mathrm{d}t + \sigma(X_t)\, \mathrm{d}W_t, \quad X_0, \quad 0 \leq t \leq T, \qquad (1.43)$$

belongs to a family of processes parameterized by some parameter $\vartheta \in \Theta$ and the functions $S(\vartheta, \cdot), \vartheta \in \Theta$ and $\sigma(\cdot)$ satisfy the condition $\mathcal{EM}$. Moreover, we suppose that the following condition is fulfilled too.

$\mathcal{RP}(\Theta)$. *The functions* $S(\vartheta,\cdot)$ *and* $\sigma(\cdot)$ *are such that for all* $\vartheta \in \Theta$

$$V(\vartheta, x) = \int_0^x \exp\left\{-2\int_0^y \frac{S(\vartheta, v)}{\sigma(v)^2}\,dv\right\} dy \to \pm\infty \quad \text{as} \quad x \to \pm\infty \quad (1.44)$$

and

$$G(\vartheta) = \int_{-\infty}^{\infty} \sigma(y)^{-2} \exp\left\{2\int_0^y \frac{S(\vartheta, v)}{\sigma(v)^2}\,dv\right\} dy < \infty. \quad (1.45)$$

Then there exists an invariant distribution (see Theorem 1.16 below) with the density function

$$f(\vartheta, x) = G(\vartheta)^{-1}\ \sigma(x)^{-2} \exp\left\{2\int_0^x \frac{S(\vartheta, v)}{\sigma(v)^2}\,dv\right\}, \qquad x \in \mathscr{R}, \quad (1.46)$$

and if we suppose that the initial value X_0 has the density $f_\vartheta(\cdot) = f(\vartheta, \cdot)$, then the process $\{X_t, t \geq 0\}$ is stationary. The measures $\left\{\mathbf{P}_\vartheta^{(T)}, \vartheta \in \Theta\right\}$ are equivalent. The likelihood ratio

$$L(\vartheta, \vartheta_1; X^T) = \frac{d\mathbf{P}_\vartheta^{(T)}}{d\mathbf{P}_{\vartheta_1}^{(T)}}(X^T)$$

is given by the formula

$$L(\vartheta, \vartheta_1; X^T) = \frac{G(\vartheta_1)}{G(\vartheta)} \exp\left\{2\int_0^{X_0} \frac{S(\vartheta, v) - S(\vartheta_1, v)}{\sigma(v)^2}\,dv\right\}$$
$$\exp\left\{\int_0^T \frac{S(\vartheta, X_t) - S(\vartheta_1, X_t)}{\sigma(X_t)^2}\,dX_t - \frac{1}{2}\int_0^T \frac{S(\vartheta, X_t)^2 - S(\vartheta_1, X_t)^2}{\sigma(X_t)^2}\,dt\right\}. \quad (1.47)$$

If ϑ_1 is the true value, then with $\mathbf{P}_{\vartheta_1}^{(T)}$ probability 1

$$L(\vartheta, \vartheta_1; X^T) = \frac{G(\vartheta_1)}{G(\vartheta)} \exp\left\{2\int_0^{X_0} \frac{S(\vartheta, v) - S(\vartheta_1, v)}{\sigma(v)^2}\,dv\right\}$$
$$\exp\left\{\int_0^T \frac{S(\vartheta, X_t) - S(\vartheta_1, X_t)}{\sigma(X_t)}\,dW_t\right.$$
$$\left. -\frac{1}{2}\int_0^T \left(\frac{S(\vartheta, X_t) - S(\vartheta_1, X_t)}{\sigma(X_t)}\right)^2 dt\right\}.$$

Sometimes it is preferable to have a likelihood ratio formula without the Itô integral. Suppose that the functions $S(\vartheta,x), S(\vartheta_1,x)$ and $\sigma(x)$ are continuously differentiable in x, then by the Itô formula we obtain the equality

$$\int_0^T \frac{S(\vartheta,X_t)-S(\vartheta_1,X_t)}{\sigma(X_t)^2}\mathrm{d}X_t = \int_{X_0}^{X_T} \frac{S(\vartheta,v)-S(\vartheta_1,v)}{\sigma(v)^2}\,\mathrm{d}v - \frac{1}{2}\int_0^T \left(\frac{S(\vartheta,X_t)-S(\vartheta_1,X_t)}{\sigma(X_t)^2}\right)' \sigma(X_t)^2\,\mathrm{d}t,$$

where the prime means differentiating by x in the following sense: $h(X_t)' = h(x)'\big|_{x=X_t}$.

Now the log-likelihood ratio can be written as

$$\ln L(\vartheta,\vartheta_1;X^T) = \ln\frac{G(\vartheta_1)}{G(\vartheta)} + \int_0^{X_0} \frac{S(\vartheta,v)-S(\vartheta_1,v)}{\sigma(v)^2}\,\mathrm{d}v + \int_0^{X_T} \frac{S(\vartheta,v)-S(\vartheta_1,v)}{\sigma(v)^2}\,\mathrm{d}v - \frac{1}{2}\int_0^T \left[\frac{S(\vartheta,X_t)^2-S(\vartheta_1,X_t)^2}{\sigma(X_t)^2} + \left(\frac{S(\vartheta,X_t)-S(\vartheta_1,X_t)}{\sigma(X_t)^2}\right)' \sigma(X_t)^2\right]\mathrm{d}t. \tag{1.48}$$

This representation will be useful in those problems when one has to substitute a parameter estimate, say, $\bar\vartheta_T$ in the likelihood ratio to form the statistic

$$L\left(\bar\vartheta_T,\vartheta_1,X^T\right).$$

Such a statistic can be used, for example, in the adaptive estimation or in hypotheses testing. As the estimator $\bar\vartheta_T$ usually depends on all the trajectory $\{X_t, 0\le t\le T\}$, then the stochastic integral needs a special definition and is not easy to work with.

Another interesting point is the representation of the likelihood ratio (1.48) with the help of the *empirical density* (1.27)

$$L(\vartheta,\vartheta_1;X^T) = \frac{G(\vartheta_1)}{G(\vartheta)}\exp\left\{\int_0^{X_0} \frac{S(\vartheta,v)-S(\vartheta_1,v)}{\sigma(v)^2}\,\mathrm{d}v + \int_0^{X_T} \frac{S(\vartheta,v)-S(\vartheta_1,v)}{\sigma(v)^2}\,\mathrm{d}v\right\} \exp\left\{-\frac{T}{2}\int_{\mathscr{R}}\left[\frac{S(\vartheta,x)^2-S(\vartheta_1,x)^2}{\sigma(x)^2} + \left(\frac{S(\vartheta,x)-S(\vartheta_1,x)}{\sigma(x)^2}\right)'\sigma(x)^2\right] f_T^\circ(x)\,\mathrm{d}x\right\}. \tag{1.49}$$

This representation shows that $(X_0, X_T, f_T^\circ(x), x\in\mathscr{R})$ is a sufficient statistic in the problems of parameter estimation, when the trend and diffusion coefficients are continuously differentiable w.r.t. x.

Remember that we have as well the equality

$$\frac{1}{T}\int_0^T h\left(X_t\right)\,\mathrm{d}t=\int_{\mathscr{R}} h\left(x\right)\,\mathrm{d}\hat{F}_T\left(x\right),$$

where $\hat{F}_T\left(x\right)$ is an empirical distribution function. Hence $(X_0, X_T, \hat{F}_T(x), x\in\mathscr{R})$ is a sufficient statistic too. If the trend coefficient has discontinuities, then the representation (1.49) has to be modified. In particular, suppose that the function $S(\vartheta,x)$ is continuously differentiable on x for $x\neq x_i,\ i=1,\ldots,k$ and has jumps at the points $x_i,\ i=1,\ldots,k$, i.e., $S(\vartheta,x_i+)-S(\vartheta,x_i-)=r_i(\vartheta)\neq 0, i=1,\ldots,k$. Then the stochastic integral admits the representation

$$\begin{aligned}\frac{1}{T}\int_0^T\frac{S(\vartheta,X_t)}{\sigma(X_t)^2}\,\mathrm{d}X_t&=\frac{1}{T}\int_{X_0}^{X_T}\frac{S(\vartheta,v)}{\sigma(v)^2}\,\mathrm{d}v\\&+\int_{\mathscr{R}}\frac{S(\vartheta,x)\,\sigma'(x)-S'(\vartheta,x)\,\sigma(x)}{\sigma(x)}\,f_T^{\circ}(x)\,\mathrm{d}x-\frac{1}{2}\sum_{i=1}^{k}r_i(\vartheta)\,f_T^{\circ}(x_i).\end{aligned}$$

Hence, both the above–mentioned statistics are sufficient in the problem of parameter estimation. In nonparametric estimation problems (Chapter 4) these statistics are sufficient by the same reason.

1.2 Limit Theorems

We are given a probability space $(\Omega,\mathfrak{F},\mathbf{P})$ with filtration $\{\mathfrak{F}_t, t\geq 0\}$ and a homogeneous diffusion process

$$\mathrm{d}X_t=S\left(X_t\right)\,\mathrm{d}t+\sigma\left(X_t\right)\,\mathrm{d}W_t,\quad X_0=x_0,\ t\geq 0, \tag{1.50}$$

where the coefficients $S\left(\cdot\right)$ and $\sigma\left(\cdot\right)$ satisfy the condition $\mathcal{ES}$, so this equation has a unique weak solution $X^T=\{X_t,\mathfrak{F}_t, 0\leq t\leq T\}$ on any interval $[0,T]$.

The statistical inference for such models is essentially based on two limit theorems: the *law of large numbers* (LLN) for ordinary integrals and the *central limit theorem* (CLT) for stochastic and ordinary integrals. Note that the CLT for the ordinary integral is a consequence of the CLT for stochastic integrals (see below).

1.2.1 Law of Large Numbers

Let $\tau_a=\inf\left\{t\geq 0:\ X_t=a\right\}$, $\tau_{ab}=\inf\left\{t\geq\tau_a:\ X_t=b\right\}$. We say that the stochastic process $X=\left\{X_t,\geq 0\right\}$ is *recurrent* if $\mathbf{P}\left\{\tau_{ab}<\infty\right\}=1$ for all $a,b\in\mathscr{R}$. The recurrent process X is called *recurrent positive* if $\mathbf{E}\,\tau_{ab}<\infty$ for all $a,b\in\mathscr{R}$ and is called *null recurrent* if $\mathbf{E}\,\tau_{ab}=\infty$ for all $a,b\in\mathscr{R}$.

Proposition 1.15. *The process X is recurrent if and only if*

$$V(x) = \int_0^x \exp\left\{-2\int_0^y \frac{S(u)}{\sigma(u)^2}\,\mathrm{d}u\right\}\,\mathrm{d}y, \longrightarrow \pm\infty \quad \text{as} \quad x \to \pm\infty. \tag{1.51}$$

The recurrent process X is positive if and only if

$$G = \int_{-\infty}^{\infty} \sigma(y)^{-2}\,\exp\left\{2\int_0^y \frac{S(z)}{\sigma(z)^2}\,\mathrm{d}z\right\}\,\mathrm{d}y < \infty. \tag{1.52}$$

The process X is recurrent null if it is recurrent and

$$G = \int_{-\infty}^{\infty} \sigma(y)^{-2}\,\exp\left\{2\int_0^y \frac{S(z)}{\sigma(z)^2}\,\mathrm{d}z\right\}\,\mathrm{d}y = \infty. \tag{1.53}$$

For the proof see, e.g., Durett [69], p. 221.

As before, we call $\mathcal{RP}$ the conditions (1.51) and (1.52). Note that the condition (1.52) does not imply (1.51). Indeed, let us consider the diffusion process

$$\mathrm{d}X_t = X_t\,\left(1+X_t^2\right)\,\mathrm{d}t + \left(1+X_t^2\right)\,\mathrm{d}W_t.$$

Then

$$G = \int_{-\infty}^{\infty} \left(1+x^2\right)^{-2} \exp\left\{\ln\left(1+x^2\right)\right\}\,\mathrm{d}x = 2V\left(\infty\right) = \pi < \infty.$$

Of course, if $\sigma\left(x\right) \equiv 1$, then (1.52) implies (1.51) and in this case it is easy to see that the following condition:

$$\overline{\lim_{|x|\to\infty}}\; x\,S\left(x\right) < -1/2 \tag{1.54}$$

is sufficient for (1.51) and (1.52).

We say that the process X has ergodic properties if there exists an (invariant) distribution $F\left(\cdot\right)$, such that for any measurable function $h\left(\cdot\right)$ with $\mathbf{E}\left|h\left(\xi\right)\right| < \infty$ (here ξ has a distribution $F\left(\cdot\right)$) we have the convergence

$$\frac{1}{T}\int_0^T h\left(X_t\right)\,\mathrm{d}t \longrightarrow \mathbf{E}\,h\left(\xi\right) \tag{1.55}$$

with probability 1.

Theorem 1.16. (Law of large numbers) *Let the conditions $\mathcal{RP}$ be fulfilled. Then the stochastic process (1.50) has ergodic properties with the invariant density given by*

$$f\left(x\right) = \frac{1}{G\,\sigma(x)^2}\,\exp\left\{2\int_0^x \frac{S(y)}{\sigma(y)^2}\,\mathrm{d}y\right\}. \tag{1.56}$$

The proof of this theorem can be found, for example, in Skorohod [219], Gikhman and Skorohod [85] or Durett [69].

Let X^T be the solution of the equation

$$\mathrm{d}X_t = S(\vartheta, X_t)\ \mathrm{d}t + \sigma(X_t)\ \mathrm{d}W_t, \quad X_0 = x_0,\ t \geq 0,$$

where the parameter $\vartheta \in \Theta$. Denote by $\mathbf{P}_\vartheta^{(T)}$ and $\mathbf{E}_\vartheta$ the corresponding distribution and mathematical expectation. Remind that we suppose that the functions $S(\vartheta, \cdot), \vartheta \in \Theta$ and $\sigma(\cdot)$ are locally integrable. We need the uniform in ϑ version of this LLN.

$\mathcal{A}_0(\Theta)$. *The function* $\sigma(\cdot)^{-1} \in \mathscr{P}$ *and*

$$\overline{\lim_{|x|\to\infty}} \sup_{\vartheta\in\Theta} \operatorname{sgn}(x) \frac{S(\vartheta, x)}{\sigma(x)^2} < 0. \tag{1.57}$$

Note that by this condition we have

$$\inf_{\vartheta\in\Theta} V(\vartheta, x) \longrightarrow +\infty \quad \text{as} \quad x \to +\infty,$$

$$\sup_{\vartheta\in\Theta} V(\vartheta, x) \longrightarrow -\infty \quad \text{as} \quad x \to -\infty,$$

and

$$\sup_{\vartheta\in\Theta} G(\vartheta) < \infty$$

in obvious notation.

Lemma 1.17. *Let* $h(\vartheta, \cdot) \in \mathscr{P}$, $\mathbf{E}_\vartheta\, h(\vartheta, \xi) = 0$ *and the condition* $\mathcal{A}_0(\Theta)$ *be fulfilled. Then for any* $p > 0$ *there exists a constant* $C_p > 0$ *such that*

$$\sup_{\vartheta\in\Theta} \mathbf{E}_\vartheta \left| \frac{1}{\sqrt{T}} \int_0^T h(\vartheta, X_t)\ \mathrm{d}t \right|^p \leq C_p. \tag{1.58}$$

Proof. Introduce the function

$$H(\vartheta, x) = \int_0^x \frac{2}{\sigma(y)^2\ f(\vartheta, y)} \int_{-\infty}^y h(\vartheta, z)\ f(\vartheta, z)\ \mathrm{d}z\, \mathrm{d}y.$$

The Itô formula gives us the equality

$$\mathrm{d}H(\vartheta, X_t) = h(\vartheta, X_t)\ \mathrm{d}t + H'(\vartheta, X_t)\, \sigma(X_t)\ \mathrm{d}W_t, \qquad H(\vartheta, X_0).$$

Hence

$$\int_0^T h(\vartheta, X_t)\,\mathrm{d}t = H(\vartheta, X_T) - H(\vartheta, X_0)$$
$$- \int_0^T \frac{2}{\sigma(X_t)\, f(\vartheta, X_t)} \int_{-\infty}^{X_t} h(\vartheta, v)\, f(\vartheta, v)\,\mathrm{d}v\,\mathrm{d}W_t.$$

Note that by condition $\mathcal{A}_0(\Theta)$there exist constants $A > 0$ and $\gamma > 0$ such that for all $\vartheta \in \Theta$ and all $|x| > A$ we have the estimate

$$\operatorname{sgn}(x)\, \frac{S(\vartheta, x)}{\sigma(x)^2} \le -\gamma. \tag{1.59}$$

Hence for the function

$$\frac{1}{\sigma(y)^2\, f(\vartheta, y)} \int_{-\infty}^{y} h(\vartheta, z)\, f(\vartheta, z)\,\mathrm{d}z$$
$$= - \int_y^{\infty} \frac{h(\vartheta, z)}{\sigma(z)^2} \exp\left\{ 2 \int_y^z \frac{S(\vartheta, v)}{\sigma(v)^2}\,\mathrm{d}v \right\} \mathrm{d}z$$

we have the estimate $(y > A)$

$$\left| \frac{1}{\sigma(y)^2\, f(\vartheta, y)} \int_{-\infty}^{y} h(\vartheta, z)\, f(\vartheta, z)\,\mathrm{d}z \right|$$
$$\le \int_y^{\infty} \frac{|h(\vartheta, z)|}{\sigma(z)^2} \exp\{-2\,\gamma\,(z-y)\}\,\mathrm{d}z$$
$$\le C \int_y^{\infty} (1 + z^{p_1})\, \mathrm{e}^{-2\gamma(z-y)} \mathrm{d}z = C \int_0^{\infty} (1 + (y+v)^{p_1})\, \mathrm{e}^{-2\gamma v} \mathrm{d}v$$
$$\le C\,(1 + y^{p_1}).$$

We have a similar estimate for $y \le -A$. Therefore

$$\sup_{\vartheta \in \Theta} \mathbf{E}_\vartheta \left| H(\vartheta, \xi) \right|^p \le C_1 + C_2 \int_{|y| > A} (1 + |y|^{p_2})\; \mathrm{e}^{-2\gamma |y - A|}\, \mathrm{d}y \le C. \tag{1.60}$$

Further, let $p \ge 2$, then using (1.3) we obtain

$$\sup_{\vartheta \in \Theta} \mathbf{E}_\vartheta \left| \int_0^T \frac{1}{\sigma(X_t)\, f(\vartheta, X_t)} \int_{-\infty}^{X_t} h(\vartheta, v)\, f(\vartheta, v)\,\mathrm{d}v\,\mathrm{d}W_t \right|^p$$
$$\le C\, T^{p/2-1} \sup_{\vartheta \in \Theta} \mathbf{E}_\vartheta \int_0^T \left| \frac{1}{\sigma(X_t)\, f(\vartheta, X_t)} \int_{-\infty}^{X_t} h(\vartheta, v)\, f(\vartheta, v)\,\mathrm{d}v \right|^p \mathrm{d}t$$
$$= C\, T^{p/2} \sup_{\vartheta \in \Theta} \mathbf{E}_\vartheta \left| \frac{1}{\sigma(\xi)\, f(\vartheta, \xi)} \int_{-\infty}^{\xi} h(\vartheta, v)\, f(\vartheta, v)\,\mathrm{d}v \right|^p$$

because the process X is supposed to be stationary. Following the same arguments as in (1.60) it can be shown that

$$\sup_{\vartheta\in\Theta} \mathbf{E}_\vartheta \left| \int_{-\infty}^{\xi} \frac{h(\vartheta,v)\, f(\vartheta,v)}{\sigma(\xi)\, f(\vartheta,\xi)} \mathrm{d}v \right|^p < C. \tag{1.61}$$

The estimates (1.60) and (1.61) allow us to write

$$\sup_{\vartheta\in\Theta} \mathbf{E}_\vartheta \left| \frac{1}{\sqrt{T}} \int_0^T h(\vartheta, X_t)\, \mathrm{d}t \right|^p \leq C\, T^{-p/2} \sup_{\vartheta\in\Theta} \mathbf{E}_\vartheta \left| H(\vartheta,\xi) \right|^p$$
$$+ C \sup_{\vartheta\in\Theta} \mathbf{E}_\vartheta \left| \int_{-\infty}^{\xi} \frac{h(\vartheta,v)\, f(\vartheta,v)}{\sigma(\xi)\, f(\vartheta,\xi)} \mathrm{d}v \right|^p \leq C_p$$

with corresponding $C_p > 0$.

Proposition 1.18. *Let $h(\vartheta,\cdot) \in \mathscr{P}$, and condition $\mathcal{A}_0(\Theta)$ be fulfilled. Then for any $p > 0$ there exists a constant $C_p > 0$ such that*

$$\sup_{\vartheta\in\Theta} \mathbf{E}_\vartheta \left| \frac{1}{T} \int_0^T h(\vartheta, X_t)\, \mathrm{d}t - \mathbf{E}_\vartheta\, h(\vartheta,\xi) \right|^p \leq \frac{C_p}{T^{p/2}}. \tag{1.62}$$

This estimate gives a bit more than the uniform law of large numbers.

1.2.2 Central Limit Theorem

Let $h(\cdot,\omega), \in \mathcal{M}_T$. Then the stochastic integral

$$\int_0^T h(t,\omega)\, \mathrm{d}W_t$$

is well defined and we have the following

Theorem 1.19. (Central Limit Theorem) *Suppose that there exist a (non-random) function φ_T and a positive constant ϱ such that*

$$\mathbf{P} - \lim_{T\to\infty} \varphi_T^2 \int_0^T h(t,\omega)^2\, \mathrm{d}t = \varrho^2 < \infty. \tag{1.63}$$

Then

$$\mathcal{L}\left\{ \varphi_T \int_0^T h(t,\omega)\, \mathrm{d}W_t \right\} \Longrightarrow \mathcal{N}\left(0, \varrho^2\right). \tag{1.64}$$

Proof. Let us put

$$\mathcal{I}_T = \varphi_T \int_0^T h(t,\omega)\ \mathrm{d}W_t.$$

For a fixed T we redefine the function $h(\cdot)$ on the interval $(T, T+1]$ as $h(t,\omega) = \varphi_T^{-1}\ \varrho$ and introduce the stopping time

$$\tau_T = \inf\left\{t:\ \varphi_T^2 \int_0^t h(s,\omega)^2\ \mathrm{d}s = \varrho^2\right\}.$$

Then with probability 1 the random variable $\tau_T \in [0, T+1]$ and by Lemma 1.2 the stopped stochastic integral

$$\mathcal{I}_\tau = \varphi_T \int_0^{\tau_T} h(t,\omega)\ \mathrm{d}W_t$$

is a Gaussian random variable

$$\mathcal{L}\{\mathcal{I}_\tau\} = \mathcal{N}(0, \varrho^2).$$

Remember that if necessary we introduce an independent Wiener process on the interval $[T, T+1]$.

Fix some $\delta > 0$ and denote $\mathbb{A} = \{\omega : \tau_T < T\}$, $\mathbb{A}^c = \{\omega : \tau_T \in (T, T+1]\}$. Then according to (1.1) we can write for any $\varepsilon > 0$

$$\begin{aligned}
\mathbf{P}\{|\mathcal{I}_T - \mathcal{I}_\tau| > \delta\} &= \mathbf{P}\left\{\left|\varphi_T \int_0^{T+1} h(t,\omega)\,[\chi_{\{t\leq T\}} - \chi_{\{t\leq \tau_T\}}]\,\mathrm{d}W_t\right| > \delta\right\} \\
&\leq \frac{\varepsilon}{\delta^2} + \mathbf{P}\left\{\varphi_T^2 \int_0^{T+1} h(t,\omega)^2\ |\chi_{\{t\leq T\}} - \chi_{\{t\leq \tau_T\}}|\ \mathrm{d}t > \varepsilon\right\} \\
&= \frac{\varepsilon}{\delta^2} + \mathbf{P}\left\{\varphi_T^2 \int_0^{T+1} h(t,\omega)^2\ [\chi_{\{t\leq T\}} - \chi_{\{t\leq \tau_T\}}]\ \mathrm{d}t > \varepsilon, \mathbb{A}\right\} \\
&\quad + \mathbf{P}\left\{\varphi_T^2 \int_0^{T+1} h(t,\omega)^2\ [\chi_{\{t\leq \tau_T\}} - \chi_{\{t\leq T\}}]\ \mathrm{d}t > \varepsilon, \mathbb{A}^c\right\} \\
&= \frac{\varepsilon}{\delta^2} + \mathbf{P}\left\{\varphi_T^2 \int_0^{T} h(t,\omega)^2\ \mathrm{d}t - \varrho^2 > \varepsilon, \mathbb{A}\right\} \\
&\quad + \mathbf{P}\left\{\varrho^2 - \int_0^{T} h(t,\omega)^2\ \mathrm{d}t > \varepsilon, \mathbb{A}^c\right\} \\
&\leq \frac{\varepsilon}{\delta^2} + \mathbf{P}\left\{\left|\varphi_T^2 \int_0^{T} h(t,\omega)^2\ \mathrm{d}t - \varrho^2\right| > \varepsilon\right\}
\end{aligned}$$

because $\mathbf{P}\left\{|\mathcal{I}_T - \mathcal{I}_\tau| > \delta, \tau_T = T\right\} = 0$.

Hence, if we put $\varepsilon = \delta^3$ then

$$\mathbf{P}\left\{|\mathcal{I}_T - \mathcal{I}_\tau| > \delta\right\} \leq \delta + \mathbf{P}\left\{\left|\varphi_T^2 \int_0^T h(t,\omega)^2 \, \mathrm{d}t - \varrho^2\right| > \delta^3\right\} \leq 2\delta \quad (1.65)$$

as $T \to \infty$.

For the difference of the characteristic functions we have

$$\begin{aligned} \left|\psi_T(\lambda) - \mathrm{e}^{-\frac{\lambda^2}{2}\varrho^2}\right| &= \left|\mathbf{E}\left(\mathrm{e}^{\mathrm{i}\lambda\mathcal{I}_T} - \mathrm{e}^{\mathrm{i}\lambda\mathcal{I}_\tau}\right)\right| \leq \mathbf{E}\left|\mathrm{e}^{\mathrm{i}\lambda\mathcal{I}_T} - \mathrm{e}^{\mathrm{i}\lambda\mathcal{I}_\tau}\right| \\ &= \mathbf{E}\left|\mathrm{e}^{\mathrm{i}\lambda\mathcal{I}_T} - \mathrm{e}^{\mathrm{i}\lambda\mathcal{I}_\tau}\right| \chi_{\{|\mathcal{I}_T - \mathcal{I}_\tau| < \delta\}} + \mathbf{E}\left|\mathrm{e}^{\mathrm{i}\lambda\mathcal{I}_T} - \mathrm{e}^{\mathrm{i}\lambda\mathcal{I}_\tau}\right| \chi_{\{|\mathcal{I}_T - \mathcal{I}_\tau| \geq \delta\}} \\ &\leq |\lambda|\, \mathbf{E}\left|\mathcal{I}_T - \mathcal{I}_\tau\right| \chi_{\{|\mathcal{I}_T - \mathcal{I}_\tau| < \delta\}} + 2\,\mathbf{P}\{|\mathcal{I}_T - \mathcal{I}_\tau| \geq \delta\} \\ &\leq |\lambda|\,\delta + 2\,\mathbf{P}\{|\mathcal{I}_T - \mathcal{I}_\tau| \geq \delta\} \leq (4 + |\lambda|)\,\delta. \end{aligned} \quad (1.66)$$

Therefore the characteristic function of the stochastic integral converges to the characteristic function of the Gaussian law $\mathcal{N}(0, \varrho^2)$ and we obtain the asymptotic normality (1.64).

We present as well a uniform version of the CLT. Let Θ be some set of values of the variable ϑ and for every $\vartheta \in \Theta$ the random functions $h(\vartheta, \cdot, \omega) \in \mathcal{M}_T$ are given. We now introduce the stochastic integral

$$\mathcal{I}_T(\vartheta) = \varphi_T(\vartheta) \int_0^T h(\vartheta, t, \omega)\, \mathrm{d}W_t.$$

Here $\varphi_T(\vartheta)$ is some positive function. We have

Proposition 1.20. (Uniform CLT) *Suppose that there exists a positive function $\varrho(\vartheta)$ such that*

$$\sup_{\vartheta \in \Theta} \varrho(\vartheta) < \infty$$

and the convergence

$$\mathbf{P} - \lim_{T \to \infty} \varphi_T(\vartheta)^2 \int_0^T h(\vartheta, t, \omega)^2 \, \mathrm{d}t = \varrho(\vartheta)^2, \quad (1.67)$$

is uniform in ϑ. Then the convergence

$$\mathcal{L}\left\{\varphi_T(\vartheta) \int_0^T h(\vartheta, t, \omega)\, \mathrm{d}W_t\right\} \Longrightarrow \mathcal{N}\left(0, \varrho(\vartheta)^2\right) \quad (1.68)$$

is uniform in ϑ too.

Proof. Remember that the convergence in probability uniform in $\vartheta \in \Theta$ means that for any $\delta > 0$

$$\lim_{T\to\infty}\sup_{\vartheta\in\Theta}\mathbf{P}\left\{\left|\varphi_T(\vartheta)^2\int_0^T h\left(\vartheta,t,\omega\right)^2\,\mathrm{d}t-\varrho(\vartheta)^2\right|>\delta\right\}=0$$

and the uniform convergence in distribution (1.68) means that for any continuous bounded function $g(\cdot)$

$$\lim_{T\to\infty}\sup_{\vartheta\in\Theta}\left|\mathbf{E}\,g(\mathcal{I}_T(\vartheta))-\mathbf{E}\,g(\zeta)\right|=0,\qquad \zeta\sim\mathcal{N}(0,\varrho(\vartheta)^2).$$

From the estimates (1.62) and (1.63) we obtain

$$\sup_{\vartheta\in\Theta}\mathbf{P}\left\{\left|\mathcal{I}_T(\vartheta)-\mathcal{I}_\tau(\vartheta)\right|>\delta\right\}\leq\delta$$
$$+\sup_{\vartheta\in\Theta}\mathbf{P}\left\{\left|\varphi_T(\vartheta)^2\int_0^T h\left(\vartheta,t,\omega\right)^2\,\mathrm{d}t-\varrho(\vartheta)^2\right|>\delta^3\right\}\leq 2\delta,$$
$$\sup_{\vartheta\in\Theta}\left|\psi_T(\lambda,\vartheta)-\mathrm{e}^{-\frac{\lambda^2}{2}\varrho(\vartheta)^2}\right|\leq|\lambda|\,\delta$$
$$+2\sup_{\vartheta\in\Theta}\mathbf{P}\{\left|\mathcal{I}_T(\vartheta)-\mathcal{I}_\tau(\vartheta)\right|\geq\delta\}\leq(4+|\lambda|)\,\delta$$

for sufficiently large T. Hence the characteristic function of the integral converges uniformly in $\vartheta\in\Theta$ to the Gaussian characteristic function.

Note as well that the family of the random variables $\{\mathcal{I}_T(\vartheta),\vartheta\in\Theta\}$ is uniformly tight in the following sense:

$$\sup_{\vartheta\in\Theta}\mathbf{P}\left\{\left|\mathcal{I}_T(\vartheta)\right|>H\right\}\leq\frac{1}{H}+\sup_{\vartheta\in\Theta}\mathbf{P}\left\{\varphi_T(\vartheta)^2\int_0^T h\left(\vartheta,t,\omega\right)^2\,\mathrm{d}t>H\right\}\to 0$$

as $H\to\infty$.

The uniform convergence of the characteristic functions together with this condition provides the uniform asymptotic normality (1.68) (see [109], p. 365).

There are several generalizations of this CLT which can be obtained directly from Theorem 1.19. The first one is a multidimensional version of this theorem for the vector-valued integral $\boldsymbol{\mathcal{I}}_T(\boldsymbol{\vartheta})$ in the so-called "scheme of series", i.e., we have a sequence of problems indexed by T and $T\to\infty$. We are given for every $T>0$ and $\boldsymbol{\vartheta}\in\boldsymbol{\Theta}$ the functions $h_T^{(i,j)}(\boldsymbol{\vartheta},\cdot,\omega)\in\mathscr{M}_T$, $i=1,\ldots,d_1$, $j=1,\ldots,d_2$. Let us define

$$\boldsymbol{\mathcal{I}}_T(\boldsymbol{\vartheta})=\left(\mathcal{I}_T^{(1)}(\boldsymbol{\vartheta}),\ldots,\mathcal{I}_T^{(d_1)}(\boldsymbol{\vartheta})\right),\qquad \mathcal{I}_T^{(i)}(\boldsymbol{\vartheta})=\sum_{j=1}^{d_2}\int_0^T h_T^{(i,j)}(\boldsymbol{\vartheta},t,\omega)\,\mathrm{d}W_t^{(j)},$$

where $\{W_t^{(1)},\ldots,W_t^{(d_2)},0\leq t\leq T\}$ are d_2 independent Wiener processes.

Proposition 1.21. *Suppose that there exists a (nonrandom) positive definite matrix* $\boldsymbol{\varrho}^2(\boldsymbol{\vartheta})=\left(\varrho^{(i,m)}(\boldsymbol{\vartheta})\right)_{d_1\times d_1}$ *such that the convergence*

$$\mathbf{P}-\lim_{T\to\infty}\sum_{l=1}^{d_2}\int_0^T h_T^{(i,l)}(\vartheta,t,\omega)\, h_T^{(m,l)}(\vartheta,t,\omega)\, dt=\varrho^{(i,m)}(\vartheta) \tag{1.69}$$

is uniform with respect to $\vartheta\in\Theta$. *Then the convergence in distribution*

$$\mathcal{L}\left\{\mathcal{I}_T(\vartheta)\right\}\Longrightarrow\mathcal{N}\left(\mathbf{0},\varrho^2(\vartheta)\right) \tag{1.70}$$

is uniform w.r.t. ϑ *too.*

Proof. It is sufficient to note that (1.70) is equivalent to the asymptotic normality of the scalar product

$$\left(\boldsymbol{\lambda},\mathcal{I}_T(\vartheta)\right)=\int_0^T h_T(\boldsymbol{\lambda},\vartheta,t,\omega)\, dW_t\Longrightarrow\mathcal{N}\left(0,\varrho_\lambda^2(\vartheta)\right)$$

for any vector $\boldsymbol{\lambda}\in\mathscr{R}^{d_1}$. Here $\{W_t, 0\le t\le T\}$ is a Wiener process,

$$\varrho_\lambda^2(\vartheta)=\sum_{i,j=1}^{d}\lambda^{(i)}\,\lambda^{(m)}\,\varrho^{(i,m)}(\vartheta)$$

and the function $h_T(\lambda,\vartheta,t,\omega), 0\le t\le T$ is specially defined. Details can be found in [136], Theorem 3.3.6.

Another generalization concerns the case when ρ^2 in the condition (1.63) is a $\mathfrak{F}_0$-measurable random variable, then the stochastic integral is conditionally asymptotically normal

$$\mathcal{L}\left\{\varphi_T\int_0^T h(t,\omega)\, dW_t\,\middle|\,\mathfrak{F}_0\right\}\Longrightarrow\mathcal{N}\left(0,\varrho^2\right). \tag{1.71}$$

Details can be found in [136], Theorem 3.3.5.

The limit distribution of the stochastic integral in the case of a null recurrent diffusion process is described in Section 3.5.1 and below we consider the CLT in an ergodic (recurrent positive) case.

Let $\{X_t,\ t\ge 0\}$ be a solution of the stochastic differential equation

$$dX_t=S(X_t)\, dt+\sigma(X_t)\, dW_t,\quad X_0,\quad 0\le t\le T.$$

The next proposition provides the CLT for the stochastic integral

$$\mathcal{I}_T=\frac{1}{\sqrt{T}}\int_0^T g(X_t)\, dW_t \tag{1.72}$$

in the case of a recurrent positive process X. Until the end of this section we suppose that the condition $\mathcal{RP}$ is fulfilled.

Proposition 1.22. *Suppose that $g(\cdot)$ is a measurable function, such that*

$$\rho^2 = \mathbf{E}\, g(\xi)^2 < \infty. \tag{1.73}$$

Then

$$\mathcal{L}\left\{\frac{1}{\sqrt{T}}\int_0^T g(X_t)\, \mathrm{d}W_t\right\} \Longrightarrow \mathcal{N}\left(0,\rho^2\right).$$

Proof. This follows immediately from Theorem 1.19 if we put $\varphi_T = T^{-1/2}$ and remember that by the LLN (Theorem 1.16)

$$\mathbf{P} - \lim_{T\to\infty} \frac{1}{T}\int_0^T g(X_t)^2\, \mathrm{d}t = \mathbf{E}\, g(\xi)^2.$$

Another central limit theorem can be formulated for the ordinary integral

$$\mathcal{J}_T = \frac{1}{\sqrt{T}}\int_0^T h(X_t)\, \mathrm{d}t \tag{1.74}$$

as follows.

Proposition 1.23. (CLT for ordinary integral) *Let $h(\cdot)$ be a measurable function, such that $\mathbf{E}\left|h(\xi)\right| < \infty$ and $\mathbf{E}\, h(\xi) = 0$. Then if*

$$\delta^2 = 4\,\mathbf{E}\left(\int_{-\infty}^{\xi} \frac{h(v)\ f(v)}{\sigma(\xi)\, f(\xi)}\, \mathrm{d}v\right)^2 < \infty, \tag{1.75}$$

then

$$\mathcal{L}\left\{\frac{1}{\sqrt{T}}\int_0^T h(X_t)\, \mathrm{d}t\right\} \Longrightarrow \mathcal{N}\left(0,\delta^2\right). \tag{1.76}$$

Proof. This follows immediately from the representation by the Itô formula

$$\frac{1}{\sqrt{T}}\int_0^T h(X_t)\, \mathrm{d}t = \frac{H(X_T) - H(X_0)}{\sqrt{T}} - \frac{1}{\sqrt{T}}\int_0^T H'(X_t)\,\sigma(X_t)\, \mathrm{d}W_t \tag{1.77}$$

with the function

$$H(x) = \int_0^x \frac{2}{\sigma(y)^2 f(y)}\int_{-\infty}^{y} h(v)\ f(v)\ \mathrm{d}v\, \mathrm{d}y$$

and the asymptotic normality of the stochastic integral in this representation. Note that

$$\mathbf{P}\left\{\left|\frac{H(X_T)}{\sqrt{T}}\right| > \varepsilon\right\} \to 0$$

as $T \to \infty$ because the function $H(\cdot)$ is continuous and X is a stationary process.

In a similar way it is easy to formulate conditions for the joint asymptotic normality of the ordinary and stochastic integrals. Indeed, let us consider the vector of integrals $(\mathcal{I}_T, \mathcal{J}_T)$ defined by (1.72) and (1.74) respectively. We have the following

Proposition 1.24. *Let the functions $g(\cdot)$ and $h(\cdot)$ be such that the conditions (1.73) and (1.75) be fulfilled. Then the vector $(\mathcal{I}_T, \mathcal{J}_T)$ is asymptotically normal with the limit covariance matrix*

$$\begin{pmatrix} \rho^2, & \gamma \\ \gamma, & \delta^2 \end{pmatrix},$$

where

$$\gamma = 2\,\mathbf{E}\left(\frac{g(\xi)}{\sigma(\xi)\, f(\xi)} \int_{-\infty}^{\xi} h(v)\, f(v)\, \mathrm{d}v\right).$$

Proof. This follows from the above-mentioned representation (1.77) of the ordinary integral and the central limit theorem for stochastic integrals.

The CLT for the stochastic integral and the representation (1.32) for the local time yield the following

Proposition 1.25. (CLT for local time) *Let the conditions $\mathcal{RP}$ be fulfilled, $\mathbf{E}\,\sigma(\xi)^2 < \infty$ and*

$$d_f(x)^2 = 4f(x)^2\, \mathbf{E}\left(\frac{\chi_{\{\xi > x\}} - F(\xi)}{\sigma(\xi)\; f(\xi)}\right)^2 < \infty.$$

Then the local time $\Lambda_T(x)$ is asymptotically normal

$$\mathcal{L}\left\{\sqrt{T}\left(\frac{2\Lambda_T(x)}{T\,\sigma(x)^2} - f(x)\right)\right\} \Longrightarrow \mathcal{N}\left(0, d_f(x)^2\right). \qquad (1.78)$$

Proof. Indeed, the function $H(y,x)$ (see (1.36)) is continuous in y. Hence, for any $\nu > 0$

$$\mathbf{P}\left\{|H(\xi, x)| > \sqrt{T}\nu\right\} \to 0,$$

and by the central limit theorem

$$\mathcal{L}\left\{\frac{2f(x)}{\sqrt{T}} \int_0^T \frac{\chi_{\{X_t > x\}} - F(X_t)}{\sigma(X_t)\, f(X_t)}\, \mathrm{d}W_t\right\} \Longrightarrow \mathcal{N}\left(0, d_f(x)^2\right).$$

1.3 Statistical Inference

In the present work we consider two types of statistical problems: estimation and hypotheses testing. That is why we introduce below several notions and definitions of the mathematical statistics and describe the first properties of the statistical procedures. In both problems the underlying observation models can be *parametric*, when the trend coefficient of the observed ergodic diffusion process belongs to a family of functions described by a finite-dimensional parameter, or *nonparametric*, when the set of all possible trend coefficients cannot be parameterized using such a parameter.

In the parametric case the observed process $X^T = \{X_t, 0 \le t \le T\}$ is a solution of the stochastic differential equation

$$\mathrm{d}X_t = S(\vartheta, X_t)\ \mathrm{d}t + \sigma(X_t)\ \mathrm{d}W_t, \quad X_0, \quad t \ge 0, \tag{1.79}$$

where the unknown parameter $\vartheta = \left(\vartheta^{(1)}, \ldots, \vartheta^{(d)}\right)$ belongs to an open set $\Theta \in \mathscr{R}^d$. The initial value X_0 is a $\mathcal{F}_0$-measurable random variable with the density function $f(\cdot)$ which can depend on ϑ. We have to estimate ϑ or to test some statistical hypotheses concerning the value of this parameter by the observations X^T.

The second class of problems (nonparametric) concerns the situations when the observed process X^T has the stochastic differential

$$\mathrm{d}X_t = S(X_t)\ \mathrm{d}t + \sigma(X_t)\ \mathrm{d}W_t, \quad X_0, \quad t \ge 0, \tag{1.80}$$

where the trend coefficient $S(\cdot)$ is an unknown function. In the nonparametric case we consider the problems of estimation of the invariant distribution function $F(\cdot)$, invariant density $f(\cdot)$ and trend coefficient $S(\cdot)$ by the observations $X^T = \{X_t,\ 0 \le t \le T\}$ or we test some hypotheses concerning the values of these functions.

In this work we do not consider the statistical problems for diffusion processes with an unknown diffusion coefficient $\sigma(\cdot)^2$. It is known that the measures induced in $(\mathscr{C}_T, \mathfrak{B}_T)$ by diffusion processes with different diffusion coefficients are singular and so the statistical inference can be trivial. So, if the observed process is

$$\mathrm{d}X_t = S(\vartheta, X_t)\ \mathrm{d}t + \sigma(\vartheta, X_t)\ \mathrm{d}W_t, \quad X_0, \quad 0 \le t \le T, \tag{1.81}$$

then under mild identifiability conditions the parameter ϑ can be estimated without error by observations on any time interval. The problem became much more interesting if the process (1.81) is observed in discrete times, i.e., $X^T = \{X_{t_1}, \ldots, X_{t_n}\}$ and $n \to \infty$. This statement of the problem actually attracts the attention of many statisticians due to its importance in financial mathematics (see [81] and the references therein), but is not treated in the present work. Note that difficulties of continuous and discrete time statements are quite different. All problems considered here are in *continuous-time setup* and the diffusion coefficient $\sigma(\cdot)^2$ is supposed to be a known positive function.

Below we give an "elementary theory of estimation and hypotheses testing" for ergodic diffusion processes. "Elementary" means that we consider the situations (mainly linear models) when it is possible to have explicit expressions for estimators and then we describe the asymptotic properties of these estimators as $T \to \infty$ by direct calculations. Starting from Chapter 2 we consider more general nonlinear models and describe the properties of the same estimators with special attention paid to the notion of asymptotical efficiency.

1.3.1 Parameter Estimation

Suppose that the observed process X^T is a weak solution of Equation (1.79) and the functions $S(\cdot)$ and $\sigma(\cdot)$ satisfy the condition $\mathcal{RP}(\boldsymbol{\Theta})$. Therefore, the process X^T has ergodic properties with the density of the invariant law

$$f(\boldsymbol{\vartheta},\cdot) = \frac{1}{G(\boldsymbol{\vartheta})\,\sigma(x)^2} \exp\left\{2\int_0^x \frac{S(\boldsymbol{\vartheta},v)}{\sigma(v)^2}\,\mathrm{d}v\right\}.$$

We suppose that the initial value has this density function.

Our goal is to estimate the parameter $\boldsymbol{\vartheta} \in \boldsymbol{\Theta} \subset \mathscr{R}^d$ and then to describe the asymptotic properties of this estimator as $T \to \infty$. An estimator ϑ_T is any measurable mapping from $\mathscr{C}_T$ to $\boldsymbol{\Theta}$. Sometimes it is convenient to extend this definition to mapping on the whole $\mathscr{R}^d$, if, for example, we construct an unbiased estimator, when the set $\boldsymbol{\Theta}$ is bounded.

We consider the following methods of estimation: maximum likelihood, Bayesian, minimum distance, trajectory fitting and method of moments. The first two are based on the likelihood ratio formula, so we suppose that the condition $\mathcal{EM}$ is fulfilled, so the measures $\{\mathbf{P}_{\boldsymbol{\vartheta}}^{(T)}, \boldsymbol{\vartheta} \in \boldsymbol{\Theta}\}$ are equivalent and the likelihood ratio

$$L(\boldsymbol{\theta},\boldsymbol{\theta}_1;X^T) = \frac{\mathrm{d}\mathbf{P}_{\boldsymbol{\theta}}^{(T)}}{\mathrm{d}\mathbf{P}_{\boldsymbol{\theta}_1}^{(T)}}(X^T), \qquad \boldsymbol{\vartheta} \in \boldsymbol{\Theta}$$

is given by (1.47) (or by (1.48)). Here $\boldsymbol{\vartheta}_1$ is some fixed value.

Examples

We consider several relatively simple models of diffusion processes and then we discuss the properties of parameter estimators for these models.

Example 1.26. (*Ornstein–Uhlenbeck process*) Let the observed process X^T be from the linear stochastic differential equation

$$\mathrm{d}X_t = -\,(a\,X_t - b)\,\mathrm{d}t + \sigma\,\mathrm{d}W_t, \qquad X_0,\ t \geq 0 \tag{1.82}$$

where $a > 0$. Note that the condition $\mathcal{RP}(\Theta)$ is fulfilled and the invariant distribution is $\mathcal{N}\left(b/a, \sigma^2/2a\right)$ with the density function

$$f(\vartheta, x) = \sqrt{\frac{a}{\pi\sigma^2}}\ \exp\left\{-\frac{(ax-b)^2}{a\sigma^2}\right\}. \tag{1.83}$$

We have the statistical problems with $\vartheta = a$, $\vartheta = b$ and $\boldsymbol{\vartheta} = (a, b)$.

Example 1.27. Let the observed process $\{X_t,\ 0 \le t \le T\}$ be

$$\mathrm{d}X_t = -\frac{\vartheta\, X_t}{1+X_t^2}\,\mathrm{d}t + \sigma\,\mathrm{d}W_t, \quad X_0, \quad t \ge 0. \tag{1.84}$$

If $\vartheta \in (\alpha, \beta)$ and $\alpha > \sigma^2/2$, then the condition $\mathcal{RP}(\Theta)$ is fulfilled and the invariant density is

$$f(\vartheta, x) = \frac{1}{G(\vartheta)}\,\frac{1}{(1+x^2)^{\vartheta/\sigma^2}}, \qquad G(\vartheta) = \int_{-\infty}^{\infty} \frac{1}{(1+y^2)^{\vartheta/\sigma^2}}\,\mathrm{d}y. \tag{1.85}$$

Note that if $\vartheta = \sigma^2$ then we have the Cauchy distribution.

Example 1.28. Suppose that

$$\mathrm{d}X_t = \left(\boldsymbol{\vartheta}, \mathbf{h}(X_t)\right)\mathrm{d}t + \sigma(X_t)\,\mathrm{d}W_t, \quad X_0, \quad t \ge 0, \tag{1.86}$$

where $\boldsymbol{\vartheta} = \left(\vartheta^{(1)}, \ldots, \vartheta^{(d)}\right) \in \boldsymbol{\Theta} \subset \mathscr{R}^d$, the function $\sigma(x)^2$ and the vector function $\mathbf{h}(x) = (h_1(x), \ldots, h_d(x))$ are such that the condition $\mathcal{RP}(\boldsymbol{\Theta})$ is fulfilled and so the process has the invariant density

$$f(\boldsymbol{\vartheta}, x) = \frac{1}{G(\boldsymbol{\vartheta})\,\sigma(x)^2}\exp\left\{2\left(\boldsymbol{\vartheta}, \mathbf{H}(x)\right)\right\}, \quad \mathbf{H}(x) = \int_0^x \frac{\mathbf{h}(y)}{\sigma(y)^2}\mathrm{d}y \tag{1.87}$$

with the corresponding normalization function $G(\boldsymbol{\vartheta})$. Here $(\boldsymbol{\vartheta}, \mathbf{H})$ is the scalar product in $\mathscr{R}^d$.

Example 1.29. Let $\kappa \in (0,1)$ and we have the process

$$\mathrm{d}X_t = -\,\mathrm{sgn}(X_t - \vartheta)\ |X_t - \vartheta|^{\kappa}\ \mathrm{d}t + \sigma\,\mathrm{d}W_t, \quad X_0, \quad t \ge 0. \tag{1.88}$$

Then this process is ergodic for any $\vartheta \in (\alpha, \beta)$ and the invariant density is

$$f(\vartheta, x) = G^{-1}\exp\left\{-\frac{2\ |x-\vartheta|^{1+\kappa}}{(1+\kappa)\ \sigma^2}\right\}. \tag{1.89}$$

Note that such singularities are called *cusps* [203] and that the constant G does not depend on ϑ.

Example 1.30. Let the observed stochastic process be

$$\mathrm{d}X_t = -\gamma X_{t-\tau}\,\mathrm{d}t + \sigma\,\mathrm{d}W_t, \quad X_0, \quad t \geq 0, \tag{1.90}$$

where $\vartheta = (\gamma, \tau) \in \Theta = (\alpha_1, \beta_1) \times (\alpha_2, \beta_2)$, $\alpha > 0$ is an unknown parameter. Note that the process $\{X_t,\ t \geq 0\}$ is not diffusion, because the trend coefficient depends on the past and is called a *diffusion type* process ([175], p. 118). It can be shown (see, e.g., [125]) that it is a Gaussian process and under the condition $\beta_2 < \pi/2\beta_1$ it has ergodic properties with the invariant density

$$f(\vartheta, x) = \frac{1}{\sqrt{2\pi d(\vartheta)^2}} \exp\left\{-\frac{x^2}{2\,d(\vartheta)^2}\right\}, \tag{1.91}$$

where

$$d(\vartheta)^2 = \sigma^2 \int_0^\infty x_0(t)^2\,\mathrm{d}t. \tag{1.92}$$

The function $x_0(\cdot)$ (called the *fundamental solution*) is a solution of the equation (see Section 3.3 for details)

$$\frac{\mathrm{d}x_0(t)}{\mathrm{d}t} = -\gamma\, x_0(t-\tau), \quad x_0(s) = 0,\ s < 0, \quad x_0(0) = 1.$$

Of course, $x_0(t) = x_0(\vartheta, t)$. Note that the parameter τ is called *delay* and the trend coefficient is as smooth w.r.t. this parameter as the Wiener process w.r.t. time.

Example 1.31. Consider the diffusion process with a discontinuous trend coefficient

$$\mathrm{d}X_t = -\operatorname{sgn}(X_t - \vartheta)\,\mathrm{d}t + \sigma\,\mathrm{d}W_t, \quad X_0, \quad t \geq 0, \tag{1.93}$$

where $\vartheta \in (\alpha, \beta)$. This process has ergodic properties with the invariant density

$$f(\vartheta, x) = \mathrm{e}^{-2|x-\vartheta|}. \tag{1.94}$$

Below we describe the first properties of some estimators of the parameters for these models. A more detailed study of these and other parameter estimators can be found in Chapters 2 and 3.

Lower Bounds

Before we start the study of different estimators we remind two well known *lower bounds* on the mean square risks of all the estimators. The first one is the Cramér-Rao bound and the second is the van Trees bound. These bounds are derived in different situations. The first one supposes that the unknown parameter is nonrandom and the second bound is in the *Bayesian framework*, i.e., ϑ is a random variable with known prior distribution. For simplicity of exposition we suppose that the parameter ϑ is one-dimensional.

Suppose that the observed diffusion process is

$$\mathrm{d}X_t = S(\vartheta, X_t)\ \mathrm{d}t + \sigma(X_t)\ \mathrm{d}W_t, \quad X_0, \quad 0 \le t \le T$$

and we have to estimate some continuously differentiable function $\psi(\vartheta)$, $\vartheta \in \Theta \subset \mathscr{R}$. Let us denote by $\bar\psi_T = \bar\psi_T(X^T)$ an estimator of this function and suppose that its mathematical expectation can be differentiable over ϑ (below the dot means differentiation over ϑ) and

$$\frac{\partial}{\partial\vartheta}\mathbf{E}_\vartheta\,\bar\psi_T = \mathbf{E}_\vartheta\left(\Delta_T(\vartheta, X^T)\,\bar\psi_T\right) \tag{1.95}$$

where $\Delta_T(\vartheta, X^T)$ is the formal derivative of the log-likelihood ratio function (1.47) and is equal to

$$\Delta_T(\vartheta, X^T) = \frac{\dot f(\vartheta, X_0)}{f(\vartheta, X_0)} + \int_0^T \frac{\dot S(\vartheta, X_t)}{\sigma(X_t)^2}\left[\mathrm{d}X_t - S(\vartheta, X_t)\ \mathrm{d}t\right]. \tag{1.96}$$

The validity of differentiating (1.95) and (1.96) will follow if we suppose that the *statistical experiment is regular*, i.e., the likelihood ratio function is continuous w.r.t. ϑ with probability 1 and the Fisher information exists and is a continuous function (see [109], p. 65 for details).

Then we can write

$$\begin{aligned}\mathbf{E}_\vartheta\left(\Delta_T(\vartheta, X^T)\,\bar\psi_T\right) &= \mathbf{E}_\vartheta\left(\Delta_T(\vartheta, X^T)\left[\bar\psi_T - \mathbf{E}_\vartheta\,\bar\psi_T\right]\right)\\ &\le \left(\mathbf{E}_\vartheta\left[\bar\psi_T - \mathbf{E}_\vartheta\,\bar\psi_T\right]^2\right)^{1/2}\left(\mathbf{E}_\vartheta\,\Delta_T(\vartheta, X^T)^2\right)^{1/2}.\end{aligned}$$

Remember that X_0 is independent of the Wiener process, hence

$$\mathbf{E}_\vartheta\,\Delta_T(\vartheta, X^T)^2 = \mathbf{E}_\vartheta\left(\frac{\dot f(\vartheta, X_0)}{f(\vartheta, X_0)}\right)^2 + T\ \mathbf{E}_\vartheta\left(\frac{\dot S(\vartheta, \xi)}{\sigma(\xi)}\right)^2.$$

Direct calculation provides

$$\mathbf{E}_\vartheta\left(\frac{\dot f(\vartheta, X_0)}{f(\vartheta, X_0)}\right)^2 = 4\,\mathbf{E}_\vartheta\left(\int_{\mathscr{R}}\left[\chi_{\{v>\xi\}} - F(\vartheta, v)\right]\frac{\dot S(\vartheta, v)}{\sigma(v)^2}\mathrm{d}v\right)^2 \equiv \mathrm{I}_0(\vartheta).$$

Hence

$$\mathbf{E}_\vartheta\left[\bar\psi_T - \mathbf{E}_\vartheta\,\bar\psi_T\right]^2 \ge \frac{\left[\dot\psi(\vartheta) + \dot b(\vartheta)\right]^2}{\mathrm{I}_T(\vartheta)},$$

where we denote the bias $b(\vartheta) = \mathbf{E}_\vartheta\,\bar\psi_T - \psi(\vartheta)$ and the *Fisher information*

$$\mathrm{I}_T(\vartheta) = \mathrm{I}_0(\vartheta) + T\mathrm{I}(\vartheta), \qquad \mathrm{I}(\vartheta) = \mathbf{E}_\vartheta\left(\frac{\dot{S}(\vartheta,\xi)}{\sigma(\xi)}\right)^2. \tag{1.97}$$

Using the equality $\mathbf{E}_\vartheta\left[\bar{\psi}_T - \psi(\vartheta) - b(\vartheta)\right]^2 = \mathbf{E}_\vartheta\left[\bar{\psi}_T - \psi(\vartheta)\right]^2 - b(\vartheta)^2$ we obtain finally

$$\mathbf{E}_\vartheta\left[\bar{\psi}_T - \psi(\vartheta)\right]^2 \geq \frac{\left[\dot{\psi}(\vartheta) + \dot{b}(\vartheta)\right]^2}{\mathrm{I}_T(\vartheta)} + b(\vartheta)^2 \tag{1.98}$$

which is called the *Cramér–Rao inequality*. Of course, we have to suppose that $\mathrm{I}_T(\vartheta) > 0$.

If the estimator $\bar{\psi}_T$ is unbiased ($b(\vartheta) = 0$), then it becomes

$$\mathbf{E}_\vartheta\left[\bar{\psi}_T - \psi(\vartheta)\right]^2 \geq \frac{\dot{\psi}(\vartheta)^2}{\mathrm{I}_T(\vartheta)} \tag{1.99}$$

and this last inequality is sometimes used to define an *asymptotically efficient estimator* $\bar{\psi}_T$ as an estimator satisfying for any $\vartheta \in \Theta$ the relation

$$\lim_{T\to\infty} T\ \mathbf{E}_\vartheta\left[\bar{\psi}_T - \psi(\vartheta)\right]^2 = \frac{\dot{\psi}(\vartheta)^2}{\mathrm{I}(\vartheta)}. \tag{1.100}$$

Due to the well-known Hodges example (see [109], p. 91) this definition is not entirely satisfactory. Therefore we use in our work another bound (inequality) called the Hajek–Le Cam bound to define asymptotically efficient estimators (see Section 2.1). Note here that for a quadratic loss function this lower bound is: for any estimator $\bar{\psi}_T$ and any $\vartheta_0 \in \Theta$

$$\underline{\lim}_{\delta\to 0}\ \underline{\lim}_{T\to\infty}\ \sup_{|\vartheta-\vartheta_0|<\delta} T\ \mathbf{E}_\vartheta\left[\bar{\psi}_T - \psi(\vartheta)\right]^2 \geq \frac{\dot{\psi}(\vartheta_0)^2}{\mathrm{I}(\vartheta_0)} \tag{1.101}$$

with a corresponding definition of the asymptotically efficient estimators. In particular, if we estimate the parameter ϑ, then the bound is

$$\underline{\lim}_{\delta\to 0}\ \underline{\lim}_{T\to\infty}\ \sup_{|\vartheta-\vartheta_0|<\delta} T\ \mathbf{E}_\vartheta\left[\bar{\vartheta}_T - \vartheta\right]^2 \geq \mathrm{I}(\vartheta_0)^{-1}, \tag{1.102}$$

and can be considered as an asymptotic minimax version of the Cramér–Rao inequality.

To derive the van Trees bound (inequality) we suppose that the unknown parameter $\vartheta \in \Theta = (\alpha, \beta)$ is a random variable with the prior density function $p(\theta)$, $\theta \in \Theta$. Suppose as well that the function $p(\cdot)$ is absolutely continuous, $p(\alpha) = 0$, $p(\beta) = 0$ and the Fisher information of the prior law is

$$\mathrm{I}_p = \int_\alpha^\beta \frac{\dot{p}(\theta)}{p(\theta)}\, \mathrm{d}\theta < \infty.$$

Further we suppose that the regularity conditions allowing differentiating with probability 1

$$\frac{\partial}{\partial \vartheta} \ln L\left(\vartheta, \vartheta_1; X^T\right) = \Delta\left(\vartheta, X^T\right)\ L\left(\vartheta, \vartheta_1; X^T\right) \tag{1.103}$$

are fulfilled. Then we can write

$$\int_\alpha^\beta \psi(\vartheta)\ \frac{\partial}{\partial \vartheta}\left[L\left(\vartheta, \vartheta_1; X^T\right)\ p(\vartheta)\right] \mathrm{d}\vartheta = \psi(\vartheta)\, L\left(\vartheta, \vartheta_1; X^T\right)\ p(\vartheta)\Big|_\alpha^\beta$$
$$- \int_\alpha^\beta \dot\psi(\vartheta)\ L\left(\vartheta, \vartheta_1; X^T\right)\ p(\vartheta)\,\mathrm{d}\vartheta = - \int_\alpha^\beta \dot\psi(\vartheta)\ L\left(\vartheta, \vartheta_1; X^T\right)\ p(\vartheta)\,\mathrm{d}\vartheta.$$

In a similar way

$$\mathbf{E}_{\vartheta_1} \int_\alpha^\beta \left(\bar\psi_T - \psi(\vartheta)\right)\ \frac{\partial}{\partial \vartheta}\left[L\left(\vartheta, \vartheta_1; X^T\right)\ p(\vartheta)\right] \mathrm{d}\vartheta$$
$$= \mathbf{E}_{\vartheta_1} \int_\alpha^\beta \dot\psi(\vartheta)\ L\left(\vartheta, \vartheta_1; X^T\right)\ p(\vartheta)\,\mathrm{d}\vartheta = \int_\alpha^\beta \dot\psi(\vartheta)\ p(\vartheta)\,\mathrm{d}\vartheta = \mathbf{E}_P\ \dot\psi(\vartheta)$$

because $\mathbf{E}_{\vartheta_1} L\left(\vartheta, \vartheta_1; X^T\right) = 1$. We write below $\mathbb{E}$ for the expectation with respect to the joint distribution of X^T and ϑ.

The Cauchy–Schwarz inequality gives us

$$\left(\mathbf{E}_P\ \dot\psi(\vartheta)\right)^2 \leq \mathbf{E}_{\vartheta_1} \int_\alpha^\beta \left(\bar\psi_T - \psi(\vartheta)\right)^2\ L\left(\vartheta, \vartheta_1; X^T\right)\ p(\vartheta)\,\mathrm{d}\vartheta$$
$$\times\, \mathbf{E}_{\vartheta_1} \int_\alpha^\beta \left(\frac{\partial}{\partial \vartheta}\left[L\left(\vartheta, \vartheta_1; X^T\right)\ p(\vartheta)\right]\right)^2\ L\left(\vartheta, \vartheta_1; X^T\right)\ p(\vartheta)\,\mathrm{d}\vartheta.$$

For the first integral we have

$$\mathbf{E}_{\vartheta_1} \int_\alpha^\beta \left(\bar\psi_T - \psi(\vartheta)\right)^2\ L\left(\vartheta, \vartheta_1; X^T\right)\ p(\vartheta)\,\mathrm{d}\vartheta$$
$$= \int_\alpha^\beta \mathbf{E}_\vartheta \left(\bar\psi_T - \psi(\vartheta)\right)^2\ p(\vartheta)\,\mathrm{d}\vartheta = \mathbb{E}\ \left(\bar\psi_T - \psi(\vartheta)\right)^2,$$

and for the second the direct calculation provides the equality

$$\mathbf{E}_{\vartheta_1} \int_\alpha^\beta \left(\frac{\partial}{\partial \vartheta}\left[L\left(\vartheta, \vartheta_1; X^T\right)\ p(\vartheta)\right]\right)^2\ L\left(\vartheta, \vartheta_1; X^T\right)\ p(\vartheta)\,\mathrm{d}\vartheta$$
$$= \mathbf{E}_P\ \mathrm{I}_T(\vartheta) + \mathrm{I}_p.$$

Therefore

$$\mathbb{E}\ \left(\bar\psi_T - \psi(\vartheta)\right)^2 \geq \frac{\left(\mathbf{E}_P\ \dot\psi(\vartheta)\right)^2}{\mathbf{E}_P\ \mathrm{I}_T(\vartheta) + \mathrm{I}_p}. \tag{1.104}$$

This lower bound on the mean square error due to van Trees [230] is called the *van Trees inequality* [86], *global Cramér–Rao bound* [30], *integral type Cramér–Rao inequality* [33] or *Bayesian Cramér–Rao bound* [86]. If we need to estimate ϑ only, then it becomes

$$\mathbb{E}\,\left(\bar{\vartheta}_T-\vartheta\right)^2\geq\frac{1}{\mathbf{E}_p\ \mathrm{I}_T\left(\vartheta\right)+\mathrm{I}_p}. \tag{1.105}$$

The main advantage of this inequality is that the right hand side does not depend on the properties of the estimators (say, bias) and so is the same for all estimators. It is widely used in asymptotic nonparametric statistics. In particular, it gives the Hajek–Le Cam inequality (1.101) in the following elementary way. Let us introduce a random variable η with density function $p\left(v\right),\ v\in\left[-1,1\right]$ such that $p\left(-1\right)=p\left(1\right)=0$ and the Fisher information $\mathrm{I}_p<\infty$. Fix some $\delta>0$, put $\vartheta=\theta_0+\delta\eta$ and write $\mathbb{E}$ for the expectation with respect to the joint distribution of X^T and η. Then we have

$$\begin{aligned}\varliminf_{T\to\infty}\sup_{\left|\theta-\theta_0\right|<\delta}T\ \mathbf{E}_\theta\left(\bar{\psi}_T-\psi\left(\theta\right)\right)^2&\geq\varliminf_{T\to\infty}T\,\mathbb{E}\ \left(\bar{\psi}_T-\psi\left(\vartheta\right)\right)^2\\ \geq\varliminf_{T\to\infty}T\ \frac{\left(\mathbf{E}_p\ \dot{\psi}\left(\vartheta\right)\right)^2}{\mathbf{E}_p\ \mathrm{I}_T\left(\vartheta\right)+\delta^{-2}\mathrm{I}_p}&=\frac{\left(\int_{-1}^{1}\dot{\psi}\left(\theta_0+\delta u\right)\ p\left(u\right)\ du\right)^2}{\int_{-1}^{1}\mathrm{I}\left(\theta_0+\delta u\right)\ p\left(u\right)\ du}.\end{aligned} \tag{1.106}$$

Hence from the continuity of the functions $\dot{\psi}\left(\cdot\right)$ and $\mathrm{I}\left(\cdot\right)$ as $\delta\to 0$ we obtain

$$\varliminf_{\delta\to 0}\varliminf_{T\to\infty}\sup_{\left|\theta-\theta_0\right|<\delta}T\ \mathbf{E}_\theta\left(\bar{\psi}_T-\psi\left(\theta\right)\right)^2\geq\frac{\dot{\psi}\left(\theta_0\right)^2}{\mathrm{I}\left(\theta_0\right)} \tag{1.107}$$

which coincides with (1.101).

In nonparametric estimation problems we need a multidimensional version of this inequality in the situation when we estimate a function $\psi\left(\boldsymbol{\vartheta}\right),\ \boldsymbol{\vartheta}=\left(\vartheta^{(1)},\ldots,\vartheta^{(1)}\right)\in\boldsymbol{\Theta}\subset\mathscr{R}^d$. The van Trees inequality is

$$\mathbb{E}\left(\bar{\psi}_T-\psi\left(\boldsymbol{\vartheta}\right)\right)^2\geq\frac{\left(\mathbf{E}_p\ \mathrm{div}\ \psi\left(\boldsymbol{\vartheta}\right)\right)^2}{\mathbf{E}_p\ \mathrm{Tr}\,\mathbf{I}_T\left(\boldsymbol{\vartheta}\right)+\mathrm{Tr}\,\mathbf{I}_p} \tag{1.108}$$

where $\mathbf{I}_T\left(\boldsymbol{\vartheta}\right)$ and $\mathbf{I}_p$ are the information matrices

$$\begin{aligned}\mathbf{I}_T\left(\boldsymbol{\vartheta}\right)_{ij}=\mathbf{E}_{\boldsymbol{\vartheta}}&\left(\frac{1}{f\left(\boldsymbol{\vartheta},\xi\right)^2}\frac{\partial f\left(\boldsymbol{\vartheta},\xi\right)}{\partial\vartheta^{(i)}}\frac{\partial f\left(\boldsymbol{\vartheta},\xi\right)}{\partial\vartheta^{(j)}}\right)\\ &+T\ \mathbf{E}_{\boldsymbol{\vartheta}}\left(\frac{1}{\sigma\left(\xi\right)^2}\frac{\partial S\left(\boldsymbol{\vartheta},\xi\right)}{\partial\vartheta^{(i)}}\ \frac{\partial S\left(\boldsymbol{\vartheta},\xi\right)}{\partial\vartheta^{(j)}}\right)\end{aligned}$$

and

$$\mathbf{I}_{p,ij} = \mathbb{E}\left(\frac{1}{p\left(\boldsymbol{\vartheta}\right)^2}\frac{\partial p\left(\boldsymbol{\vartheta}\right)}{\partial \vartheta^{(i)}}\frac{\partial p\left(\boldsymbol{\vartheta}\right)}{\partial \vartheta^{(j)}}\right).$$

Here div $\psi\left(\boldsymbol{\vartheta}\right) = \sum_{i=1}^{d}\frac{\partial \psi(\boldsymbol{\vartheta})}{\partial \vartheta^{(i)}}$ and Tr $\mathbf{I}$ is the trace of the matrix $\mathbf{I}$. To obtain the inequality (1.108) it is sufficient to suppose that the functions $\psi\left(\boldsymbol{\vartheta}\right),\ p\left(\boldsymbol{\vartheta}\right),\ S\left(\boldsymbol{\vartheta},\cdot\right)$ are continuously differentiable over $\boldsymbol{\vartheta}$, the Fisher information matrix $\mathbf{I}_T\left(\boldsymbol{\vartheta}\right)$ is continuous on $\boldsymbol{\Theta}$, $\boldsymbol{\Theta}$ is compact with a boundary which is piecewise $\mathscr{C}^1$ and $p\left(\cdot\right)$ is positive on the interior of $\boldsymbol{\Theta}$ and zero on its boundary. Moreover we suppose that differentiating (1.103) is also valid for the partial derivatives. The proof (and more general conditions and forms) can be found in [86].

Maximum Likelihood Estimator

Below we suppose that the conditions $\mathcal{ES}$, $\mathcal{EM}$ and $\mathcal{RP}\left(\boldsymbol{\Theta}\right)$ are always fulfilled. The *maximum likelihood estimator* (MLE) $\hat{\boldsymbol{\vartheta}}_T$ is defined as a solution of the equation

$$L(\hat{\boldsymbol{\vartheta}}_T, \boldsymbol{\theta}_1; X^T) = \sup_{\boldsymbol{\theta}\in\boldsymbol{\Theta}} L(\boldsymbol{\theta}, \boldsymbol{\theta}_1; X^T). \tag{1.109}$$

If this equation has more than one solution then any one can be taken as the MLE. The function $L(\boldsymbol{\theta}, \boldsymbol{\theta}_1; X^T)$, $\boldsymbol{\theta} \in \boldsymbol{\Theta}$ in all our problems will be continuous with probability 1 and the set $\boldsymbol{\Theta}$ will be bounded (see, for example, Lemmae 2.10 and 3.19 below). Therefore the MLE $\hat{\boldsymbol{\vartheta}}_T \in \bar{\boldsymbol{\Theta}}$ with $\mathbf{P}_{\boldsymbol{\vartheta}}^{(T)}$ probability 1, where $\bar{\boldsymbol{\Theta}}$ is the closure of $\boldsymbol{\Theta}$.

Sometimes the MLE can be found in explicit form with the help of the *maximum likelihood equation* obtained from (1.109) by differentiating it with respect to $\boldsymbol{\vartheta}$:

$$\dot{\boldsymbol{L}}\left(\boldsymbol{\theta}, \boldsymbol{\theta}_1; X^T\right) = \mathbf{0}, \qquad \boldsymbol{\theta} \in \boldsymbol{\Theta}. \tag{1.110}$$

Of course this system of equations can have many more solutions than (1.109) or to not have even one.

In the ergodic (stationary) case it became (see (1.47))

$$\frac{\dot{\boldsymbol{G}}(\boldsymbol{\vartheta})}{G(\boldsymbol{\vartheta})} + 2\int_0^{X_0}\frac{\dot{\boldsymbol{S}}\left(\boldsymbol{\vartheta},v\right)}{\sigma(v)^2}\mathrm{d}v + \int_0^T \frac{\dot{\boldsymbol{S}}\left(\boldsymbol{\vartheta},X_t\right)}{\sigma\left(X_t\right)^2}\left[\mathrm{d}X_t - S\left(\boldsymbol{\vartheta},v\right)\,\mathrm{d}t\right] = \mathbf{0} \tag{1.111}$$

or without a stochastic integral (see (1.48))

$$\begin{aligned}\frac{\dot{\boldsymbol{G}}(\boldsymbol{\vartheta})}{G(\boldsymbol{\vartheta})} &+ \int_0^{X_0}\frac{\dot{\boldsymbol{S}}\left(\boldsymbol{\vartheta},v\right)}{\sigma(v)^2}\mathrm{d}v + \int_0^{X_T}\frac{\dot{\boldsymbol{S}}\left(\boldsymbol{\vartheta},v\right)}{\sigma(v)^2}\mathrm{d}v \\ &- \int_0^T\left[\frac{S\left(\boldsymbol{\vartheta},X_t\right)\dot{\boldsymbol{S}}\left(\boldsymbol{\vartheta},X_t\right)}{\sigma\left(X_t\right)^2} + \left(\frac{\dot{\boldsymbol{S}}\left(\boldsymbol{\vartheta},X_t\right)}{2\sigma\left(X_t\right)^2}\right)'\sigma\left(X_t\right)^2\right]\mathrm{d}t = \mathbf{0}.\end{aligned} \tag{1.112}$$

Remember that the prime means differentiation w.r.t. x, i.e., $h\left(X_t\right)' = h\left(x\right)'\big|_{x=X_t}$. If the initial value X_0 does not depend on $\boldsymbol{\vartheta}$, then these two equations are simplified:

$$\int_0^T \frac{\dot{\boldsymbol{S}}(\boldsymbol{\vartheta}, X_t)}{\sigma(X_t)^2}\left[\mathrm{d}X_t - S(\boldsymbol{\vartheta}, v)\ \mathrm{d}t\right] = \mathbf{0}, \tag{1.113}$$

and

$$\int_{X_0}^{X_T} \frac{\dot{\boldsymbol{S}}(\boldsymbol{\vartheta}, v)}{\sigma(v)^2}\mathrm{d}v - \int_0^T \left[\frac{S(\boldsymbol{\vartheta}, X_t)\,\dot{\boldsymbol{S}}(\boldsymbol{\vartheta}, X_t)}{\sigma(X_t)^2} + \left(\frac{\dot{\boldsymbol{S}}(\boldsymbol{\vartheta}, X_t)}{2\sigma(X_t)^2}\right)' \sigma(X_t)^2\right]\mathrm{d}t = \mathbf{0}, \tag{1.114}$$

respectively (the last equation was used by Lanska [154] and Yoshida [241] to study the minimum contrast estimators, a class of estimators including MLE)).

Note that Equations (1.109) and (1.110) are not equivalent, because (1.110) contains all local minima and maxima of the likelihood ratio function and in nonlinear cases their number can be far from 1. At the same time, even in the linear case, Equation (1.110) sometimes cannot have a solution at all with positive probability. For example, if

$$\mathrm{d}X_t = \vartheta\, h(X_t)\,\mathrm{d}t + \mathrm{d}W_t, \quad X_0 = x, \quad 0 \le t \le T$$

and $\vartheta \in (\alpha, \beta)$, where α and β are finite, then the solution of Equation (1.109) is the MLE

$$\hat{\vartheta}_T = \alpha\, \chi_{\{\eta_T \le \alpha\}} + \eta_T\, \chi_{\{\alpha < \eta_T < \beta\}} + \beta\, \chi_{\{\eta_T \ge \beta\}}, \tag{1.115}$$

where

$$\eta_T = \frac{\int_0^T h(X_t)\,\mathrm{d}X_t}{\int_0^T h(X_t)^2\,\mathrm{d}t}$$

or (without a stochastic integral)

$$\eta_T = \frac{\int_x^{X_T} h(v)\ \mathrm{d}v - \frac{1}{2}\int_0^T h(X_t)'\,\mathrm{d}t}{\int_0^T h(X_t)^2\,\mathrm{d}t}.$$

It is clear that if $\eta_T \le \alpha$ or $\eta_T \ge \beta$, then Equation (1.110) has no solution.

To describe the asymptotic behavior of the MLE we start with its consistency. For simplicity of exposition we suppose that $\vartheta \in [\alpha, \beta]$ where α and β are finite and that the initial value does not depend on ϑ. Let us define the norm

$$\|h(\cdot)\|_T^2 = \frac{1}{T}\int_0^T h(X_t)^2\ \mathrm{d}t = \int_{\mathscr{R}} h(x)^2\ \mathrm{d}\hat{F}_T(x)$$

and denote

$$\delta(\vartheta_1, \vartheta_2, x) = \frac{S(\vartheta_1, x) - S(\vartheta_2, x)}{\sigma(x)}, \qquad h(x) = \frac{S(\vartheta_1, x) + S(\vartheta_2, x) - 2S(\vartheta_0, x)}{\sigma(x)}.$$

Below ξ is the random variable with the density $f(\vartheta_0, \cdot)$.

Proposition 1.32. *Let the function*

$$g(\vartheta,\vartheta_0)=\mathbf{E}_{\vartheta_0}\,\delta(\vartheta,\vartheta_0,\xi)^2,\qquad \vartheta\in\bar\Theta=[\alpha,\beta]$$

have a unique minimum at the point $\vartheta=\vartheta_0$ *and*

$$\mathbf{E}_{\vartheta_0}\left(\|\delta(\vartheta_1,\vartheta_2,\cdot)\|_T^{2p}+\|\delta(\vartheta_1,\vartheta_2,\cdot)h(\cdot)\|_T^{2p}\right)\leq C\left|\vartheta_1-\vartheta_2\right|^q, \tag{1.116}$$

with some constants $C>0, p\geq 1$ *and* $q>1$. *Then the MLE* $\hat\vartheta_T$ *is consistent.*

Proof. Let us introduce the random function

$$z_T(\vartheta)=\frac{\ln L\left(\vartheta,\vartheta_0,X^T\right)}{T}=\frac{1}{T}\int_0^T\frac{S(\vartheta,X_t)-S(\vartheta_0,X_t)}{\sigma(X_t)}\,\mathrm{d}W_t$$
$$-\frac{1}{2T}\int_0^T\left(\frac{S(\vartheta,X_t)-S(\vartheta_0,X_t)}{\sigma(X_t)}\right)^2\mathrm{d}t$$

and note that for any $\nu>0$

$$\mathbf{P}_{\vartheta_0}^{(T)}\left\{\left|\hat\vartheta_T-\vartheta_0\right|>\nu\right\}=\mathbf{P}_{\vartheta_0}^{(T)}\left\{\sup_{|\vartheta-\vartheta_0|>\nu}z_T(\vartheta)>\sup_{|\vartheta-\vartheta_0|\leq\nu}z_T(\vartheta)\right\}.$$

Hence we have to study $\sup z_T(\vartheta)$. By the law of large numbers for any set $\vartheta_1,\ldots,\vartheta_k$ we have the convergence (with probability 1) of the vectors

$$\left(z_T(\vartheta_1),\ldots,z_T(\vartheta_k)\right)\longrightarrow\left(z(\vartheta_1),\ldots,z(\vartheta_k)\right),$$

where $z(\vartheta)=-g(\vartheta,\vartheta_0)/2$. Furthermore

$$\mathbf{E}_{\vartheta_0}\left|z_T(\vartheta_1)-z_T(\vartheta_2)\right|^{2p}\leq C\,\mathbf{E}_{\vartheta_0}\left(\frac{1}{T}\int_0^T\frac{S(\vartheta_1,X_t)-S(\vartheta_2,X_t)}{\sigma(X_t)}\,\mathrm{d}W_t\right)^{2p}$$
$$+C\,\mathbf{E}_{\vartheta_0}\left(\frac{1}{2T}\int_0^T\left[\delta(\vartheta_1,\vartheta_0,X_t)^2-\delta(\vartheta_2,\vartheta_0,X_t)^2\right]\mathrm{d}t\right)^{2p}$$
$$\leq C\;\mathbf{E}_{\vartheta_0}\|\delta(\vartheta_1,\vartheta_2,\cdot)\|_T^{2p}+C\;\mathbf{E}_{\vartheta_0}\|\delta(\vartheta_1,\vartheta_2,\cdot)h(\cdot)\|_T^{2p}\leq C\;|\vartheta_1-\vartheta_2|^q,$$

where we used the estimate (1.4) for stochastic integral and the Cauchy–Schwarz inequality for the ordinary integral.

Let us denote $\mathscr{C}(\Theta)$ the space of continuous functions on $[\alpha,\beta]$ with uniform metric and $\mathfrak{B}$ be the corresponding sigma algebra of Borel subsets. Let $\mathbf{Q}_T$ and $\mathbf{Q}$ be measures induced in $(\mathscr{C}(\Theta),\mathfrak{B})$ by the stochastic processes $y_T=\left\{y_T(\vartheta),\alpha\leq\vartheta\leq\beta\right\}$ and $y=\left\{y(\vartheta),\alpha\leq\vartheta\leq\beta\right\}$ respectively. We need the following lemma.

Lemma 1.33. *Let the finite-dimensional distributions of y_T converge to the finite-dimensional distributions of y (as $T \to \infty$) and for some $C > 0, p > 0$ and $q > 1$*

$$\mathbf{E}\left|y_T(\vartheta_1) - y_T(\vartheta_2)\right|^p \leq C\left|\vartheta_1 - \vartheta_2\right|^q$$

then

$$\mathbf{Q}_T \Longrightarrow \mathbf{Q}.$$

The proof can be found in Billingsley [23], Theorem 12.3.

As the conditions of Lemma 1.33 are fulfilled for the process $z_T(\cdot)$, we have the weak convergence of the corresponding measures and, in particular, the convergence of the distributions of the continuous in the uniform metric functionals of the process $z_T(\cdot)$. Therefore

$$\begin{aligned}\mathbf{P}_{\vartheta_0}^{(T)}&\left\{\sup_{|\vartheta-\vartheta_0|>\nu} z_T(\vartheta) > \sup_{|\vartheta-\vartheta_0|\leq\nu} z_T(\vartheta)\right\}\\ &\longrightarrow \mathbf{P}_{\vartheta_0}\left\{\sup_{|\vartheta-\vartheta_0|>\nu} z(\vartheta) > \sup_{|\vartheta-\vartheta_0|\leq\nu} z(\vartheta)\right\}\\ &= \chi_{\left\{\inf_{|\vartheta-\vartheta_0|>\nu} g(\vartheta,\vartheta_0)<\inf_{|\vartheta-\vartheta_0|\leq\nu} g(\vartheta,\vartheta_0)\right\}} = 0\end{aligned}$$

because $g(\vartheta_0, \vartheta_0) = 0$ and this minimum is unique. Here $\chi_{\{A\}}$ is the indicator of the event A.

Therefore, for any $\nu > 0$

$$\mathbf{P}_{\vartheta_0}^{(T)}\left\{\left|\hat\vartheta_T - \vartheta_0\right| > \nu\right\} \longrightarrow 0.$$

The asymptotic normality of the MLE we establish for the regular case only and just note that in other cases the MLE is not asymptotically normal and the rate of convergence is better than in the regular case (see Sections 3.2–3.4).

Below $\dot S(\vartheta,\cdot)$ and $\ddot S(\vartheta,\cdot)$ are the first two derivatives of $S(\vartheta,\cdot)$ w.r.t. $\vartheta \in \Theta = (\alpha,\beta)$ and $S'(\vartheta,\cdot)$ and $\dot S'(\vartheta,\cdot)$ are the derivatives of the functions $S(\vartheta,\cdot)$, $\dot S(\vartheta,\cdot)$ w.r.t. x.

Proposition 1.34. *Suppose that the functions $S(\vartheta,\cdot), \sigma(\cdot)^{-1} \in \mathcal{P}$ have the following continuous derivatives:*

$$S'(\vartheta,\cdot), \dot S(\vartheta,\cdot), \dot S'(\vartheta,\cdot), \ddot S(\vartheta,\cdot), \sigma'(\cdot) \in \mathcal{P},$$

the condition $\mathcal{A}_0(\Theta)$ is fulfilled and the function $g(\vartheta,\vartheta_0)$ has a unique minimum at the point $\vartheta = \vartheta_0$. Then the MLE $\hat\vartheta_T$ is consistent and asymptotically normal

$$\mathcal{L}_{\vartheta_0}\left\{\sqrt{T}\left(\hat\vartheta_T - \vartheta_0\right)\right\} \Longrightarrow \mathcal{N}\left(0, \mathrm{I}(\vartheta_0)^{-1}\right).$$

Proof. The consistency of the MLE follows from Proposition 1.32 because

$$\begin{aligned}\mathbf{E}_{\vartheta_0}\left\|\delta(\vartheta_1,\vartheta_2,\cdot)\right\|_T^2&=\mathbf{E}_{\vartheta_0}\int_{\mathscr{R}}\left(\frac{S\left(\vartheta_1,x\right)-S\left(\vartheta_2,x\right)}{\sigma\left(x\right)}\right)^2 f_T^{\circ}\left(x\right)\,\mathrm{d}x\\&=\int_{\mathscr{R}}\left(\frac{S\left(\vartheta_1,x\right)-S\left(\vartheta_2,x\right)}{\sigma\left(x\right)}\right)^2 f\left(\vartheta_0,x\right)\,\mathrm{d}x\leq C\left|\vartheta_1-\vartheta_2\right|^2\end{aligned}$$

and similarly

$$\mathbf{E}_{\vartheta_0}\left\|\delta(\vartheta_1,\vartheta_2,\cdot)h(\cdot)\right\|_T^2\leq C\left|\vartheta_1-\vartheta_2\right|^2.$$

To verify its asymptotic normality we use the log-likelihood ratio formula without a stochastic integral (1.114). We have

$$\begin{aligned}0=&\int_{X_0}^{X_T}\frac{\dot S(\hat\vartheta_T,v)}{\sigma\left(v\right)^2}\,\mathrm{d}v-\int_0^T\frac{\dot S(\hat\vartheta_T,X_t)\,S\left(\vartheta_0,X_t\right)}{\sigma\left(X_t\right)^2}\,\mathrm{d}t\\&-\left(\hat\vartheta_T-\vartheta_0\right)\int_0^T\frac{\dot S(\hat\vartheta_T,X_t)\dot S(\bar\vartheta_T,X_t)}{\sigma\left(X_t\right)^2}\,\mathrm{d}t\\&-\frac12\int_0^T\left(\frac{\dot S(\hat\vartheta_T,X_t)}{\sigma\left(X_t\right)^2}\right)'\sigma\left(X_t\right)^2\mathrm{d}t,\end{aligned}$$

where $\left|\bar\vartheta_T-\vartheta_0\right|\leq\left|\hat\vartheta_T-\vartheta_0\right|$. Hence for $\hat u_T=\sqrt{T}\left(\hat\vartheta_T-\vartheta_0\right)$ we have the representation

$$\begin{aligned}\hat u_T=&\sqrt{T}\left(\int_0^T\frac{\dot S(\hat\vartheta_T,X_t)\dot S(\bar\vartheta_T,X_t)}{\sigma\left(X_t\right)^2}\,\mathrm{d}t\right)^{-1}\left[\int_{X_0}^{X_T}\frac{\dot S(\hat\vartheta_T,v)}{\sigma\left(v\right)^2}\,\mathrm{d}v\right.\\&\left.-\int_0^T\frac{\dot S(\hat\vartheta_T,X_t)\,S\left(\vartheta_0,X_t\right)}{\sigma\left(X_t\right)^2}\,\mathrm{d}t-\frac12\int_0^T\left(\frac{\dot S(\hat\vartheta_T,X_t)}{\sigma\left(X_t\right)^2}\right)'\sigma\left(X_t\right)^2\mathrm{d}t\right]\\=&\sqrt{T}\left(\int_0^T\frac{\dot S(\vartheta_0,X_t)\dot S(\bar\vartheta_T,X_t)}{\sigma\left(X_t\right)^2}\,\mathrm{d}t\right)^{-1}\left[\int_{X_0}^{X_T}\frac{\dot S(\vartheta_0,v)}{\sigma\left(v\right)^2}\,\mathrm{d}v\right.\\&-\int_0^T\frac{\dot S\left(\vartheta_0,X_t\right)S\left(\vartheta_0,X_t\right)}{\sigma\left(X_t\right)^2}\,\mathrm{d}t\\&\left.-\frac12\int_0^T\left(\frac{\dot S(\vartheta_0,X_t)}{\sigma\left(X_t\right)^2}\right)'\sigma\left(X_t\right)^2\mathrm{d}t\right]\left(1+o\left(1\right)\right)\\=&\left(\frac1T\int_0^T\left(\frac{\dot S(\vartheta_0,X_t)}{\sigma\left(X_t\right)}\right)^2\mathrm{d}t\right)^{-1}\frac{1}{\sqrt{T}}\int_0^T\frac{\dot S(\vartheta_0,X_t)}{\sigma\left(X_t\right)}\,\mathrm{d}W_t\left(1+o\left(1\right)\right)\end{aligned}$$

where we used the consistency of the MLE. Now the asymptotic normality

$$\mathcal{L}\left\{\hat u_T\right\}\Longrightarrow\mathcal{N}\left(0,\mathrm{I}\left(\vartheta_0\right)^{-1}\right)$$

follows from the last representation, the law of large numbers and the CLT.

We illustrate the construction of the MLE on several models of diffusion processes. For simplicity of exposition we usually suppose that the initial value X_0 does not depend on the unknown parameter. Hence the observed processes cannot be stationary, but it has ergodic properties like the existence of the invariant distribution and the law of large numbers for ordinary integrals. In the examples below, for the sake of brevity we do not write the representation (1.111) every time and we will suppose that $\hat{\vartheta}_T = \eta_T$.

Example 1.35. (*Ornstein–Uhlenbeck process*) We are given the observations X^T of the process (1.82) with the invariant law $\mathcal{N}\left(b/a, \sigma^2/2a\right)$. Consider three cases: $\vartheta = a$, $\vartheta = b$ and $\boldsymbol{\vartheta} = (a, b)$ separately. In every case to construct the MLE we solve the corresponding maximum likelihood Equation (1.109).

- Let $\vartheta = a \in (\alpha_1, \alpha_2)$ $\alpha_1 > 0$ and b is known. Then the MLE of a is

$$\hat{a}_T = -\frac{\int_0^T X_t\ [\mathrm{d}X_t - b\ \mathrm{d}t]}{\int_0^T X_t^2\ \mathrm{d}t} = a - \sigma\,\frac{\int_0^T X_t\ \mathrm{d}W_t}{\int_0^T X_t^2\ \mathrm{d}t}.$$

Hence

$$\sqrt{T}\left(\hat{a}_T - a\right) = -\sigma\,\frac{T^{-1/2}\int_0^T X_t\ \mathrm{d}W_t}{T^{-1}\int_0^T X_t^2\ \mathrm{d}t}.$$

By the LLN and the CLT we have

$$\mathbf{P}_a - \lim_{T\to\infty}\frac{1}{T}\int_0^T X_t^2\ \mathrm{d}t = \frac{b^2}{a^2} + \frac{\sigma^2}{2a},$$

$$\mathcal{L}_a\left\{\frac{1}{\sqrt{T}}\int_0^T X_t\ \mathrm{d}W_t\right\} \Longrightarrow \mathcal{N}\left(0, \frac{b^2}{a^2} + \frac{\sigma^2}{2a}\right).$$

Therefore

$$\mathcal{L}_a\left\{\sqrt{T}\left(\hat{a}_T - a\right)\right\} \Longrightarrow \mathcal{N}\left(0, \frac{2\sigma^2 a^2}{2b^2 + a\sigma^2}\right).$$

- Let $a > 0$ be known and we have to estimate $\vartheta = b \in (\beta_1, \beta_2)$. Then the MLE is

$$\hat{b}_T = T^{-1}\left(X_T - X_0 + a\int_0^T X_t\ \mathrm{d}t\right) = b + \frac{\sigma\, W_T}{T}$$

and

$$\mathcal{L}_b\left\{\sqrt{T}\left(\hat{b}_T - b\right)\right\} = \mathcal{N}\left(0, \sigma^2\right)$$

If the initial value X_0 has a density function (1.83), then

$$\hat{b}_T = \frac{a\,(X_T - X_0)}{(a+2)\,T} + \frac{1}{T}\int_0^T X_t\,\mathrm{d}t.$$

This estimator will be asymptotically normal with the same parameters.

- If both parameters are unknown and $\boldsymbol{\vartheta} = (a,b) \in (\alpha_1,\alpha_2)\times(\beta_1,\beta_2)$, then

$$\hat{a}_T = \frac{T\,Z + Y_2\,(X_T - x)}{T\,Y_1 - Y_2^2},$$
$$\hat{b}_T = \frac{Y_2\,Z + Y_1\,(X_T - x)}{T\,Y_1 - Y_2^2},$$

where

$$Y_1 = \int_0^T X_t^2\,\mathrm{d}t,\quad Y_2 = \int_0^T X_t\,\mathrm{d}t,\quad Z = -\int_0^T X_t\,\mathrm{d}X_t$$

and

$$\sqrt{T}\left(\hat{\boldsymbol{\vartheta}}_T - \boldsymbol{\vartheta}\right) \Longrightarrow \mathcal{N}\left(\mathbf{0}, \mathbf{I}\left(\boldsymbol{\vartheta}\right)^{-1}\right)$$

with the information matrix

$$\mathbf{I}\left(\boldsymbol{\vartheta}\right) = \frac{1}{a^2\sigma^2}\begin{pmatrix} b^2 + \frac{a\sigma^2}{2}, & -ab \\ -ab, & a^2 \end{pmatrix}.$$

Example 1.36. Let the observed process X^T be (1.84) with $\alpha > \sigma^2/2$, invariant distribution (1.85) and the deterministic initial value $X_0 = x$. Then the MLE is

$$\hat{\vartheta}_T = -\left(\int_0^T \left(\frac{X_t}{1+X_t^2}\right)^2 \mathrm{d}t\right)^{-1} \int_0^T \frac{X_t}{1+X_t^2}\,\mathrm{d}X_t$$
$$= \vartheta - \sigma\left(\int_0^T \left(\frac{X_t}{1+X_t^2}\right)^2 \mathrm{d}t\right)^{-1} \int_0^T \frac{X_t}{1+X_t^2}\,\mathrm{d}W_t.$$

By the LLN and the CLT we have immediately the asymptotic normality

$$\mathcal{L}_\vartheta\left\{\sqrt{T}\left(\hat{\vartheta}_T - \vartheta\right)\right\} \Longrightarrow \mathcal{N}\left(0, \frac{\sigma^2}{c(\vartheta)^2}\right),$$

where

$$c(\vartheta)^2 = \mathbf{E}_\vartheta\left(\frac{\xi}{1+\xi^2}\right)^2.$$

Note that if $\vartheta = \sigma^2$, then the invariant law is Cauchy and for $\vartheta \in \left(0, \sigma^2/2\right)$ the process $\{X_t,\ t \geq 0\}$ is recurrent null. The properties of the MLE in this case are described in Section 3.5.1.

Example 1.37. In the case of observations (1.86) the MLE is

$$\hat{\boldsymbol{\vartheta}}_T = \mathbf{I}_T\left(X^T\right)^+ \int_0^T \frac{\mathbf{h}(X_t)}{\sigma(X_t)^2}\,\mathrm{d}X_t = \boldsymbol{\vartheta}\,\chi_{\left\{\det \mathbf{I}_T(X^T)\neq 0\right\}}$$
$$+\mathbf{I}_T\left(X^T\right)^+ \int_0^T \frac{\mathbf{h}(X_t)}{\sigma(X_t)}\,\mathrm{d}W_t,$$

where $\mathbf{I}_T\left(X^T\right)$ is a $d \times d$ matrix:

$$\mathbf{I}_T\left(X^T\right)_{ij} = \int_0^T \frac{h_i(X_t)\,h_j(X_t)}{\sigma(X_t)^2}\mathrm{d}t, \qquad i,j = 1,\ldots,d$$

and $\mathbf{I}^+ = \mathbf{I}^{-1}$ if $\mathbf{I}$ is nondegenerate, otherwise $\mathbf{I}^+ = \mathbf{O}$ (matrix null). Therefore if we suppose that the matrix

$$\mathrm{I}(\boldsymbol{\vartheta})_{ij} = \mathbf{E}_{\boldsymbol{\vartheta}}\left(\frac{h_i(\xi)\,h_j(\xi)}{\sigma(\xi)^2}\right), \qquad i,j = 1,\ldots,d$$

is positive definite, then

$$\mathbf{P}_{\boldsymbol{\vartheta}} - \lim_{T\to\infty} \frac{1}{T}\,\mathbf{I}_T\left(X^T\right) = \mathbf{I}(\boldsymbol{\vartheta})$$

and by the CLT for the vector stochastic integrals we obtain the asymptotic normality of the estimator

$$\mathcal{L}_{\boldsymbol{\vartheta}}\left\{\sqrt{T}\left(\hat{\boldsymbol{\vartheta}}_T - \boldsymbol{\vartheta}\right)\right\} \Longrightarrow \mathcal{N}\left(0, \mathbf{I}(\boldsymbol{\vartheta})^{-1}\right).$$

Example 1.38. (*Cusp estimation*) If the observed process satisfies Equation (1.88), then the MLE cannot be found from Equation (1.110) because the trend coefficient is not differentiable w.r.t. ϑ and is defined by Equation (1.109) only. Hence it has no explicit expression. To prove its consistency we verify the condition (1.116).

$$\mathbf{E}_{\vartheta_0} \int_{\mathscr{R}} \delta(\vartheta_1, \vartheta_2, x)^2\, f_T^\circ(x)\,\mathrm{d}x = \int_{\mathscr{R}} \delta(\vartheta_1, \vartheta_2, x)^2\, f(\vartheta_0, x)\,\mathrm{d}x$$
$$= \sigma^{-2} \int_{\mathscr{R}} \left[\mathrm{sgn}(x-\vartheta_1)\,|x-\vartheta_1|^\kappa - \mathrm{sgn}(x-\vartheta_2)\,|x-\vartheta_2|^\kappa\right]^2 f(\vartheta_0, x)\,\mathrm{d}x$$
$$\leq C\,|\vartheta_1 - \vartheta_2|^{1+2\kappa} \int_{\mathscr{R}} \left[\mathrm{sgn}(z)\,|z|^\kappa - \mathrm{sgn}(z-1)\,|z-1|^\kappa\right]^2 \mathrm{d}z$$
$$\leq C\,|\vartheta_1 - \vartheta_2|^{1+2\kappa},$$

where we changed the variable $x = \vartheta_1 + z(\vartheta_2 - \vartheta_1)$. Now the consistency follows from Proposition 1.32.

The limit distribution of $T^{\kappa+1/2}\left(\hat{\vartheta}_T - \vartheta_0\right)$ is described in Section 3.2.

Example 1.39. (*Delay estimation*) In the case of observations (1.90) the MLE $\hat{\tau}_T$ of delay τ cannot be written in explicit form too and is defined by Equation (1.109) only. The "trend coefficient" $S(\tau, X) = -\gamma X_{t-\tau}$ is not differentiable w.r.t. τ because the process $X_{t-\tau}$ is as smooth w.r.t. t as a Wiener process. To prove its consistency we check the conditions of Proposition 1.32. The process $X_t, t \geq 0$ is Gaussian and stationary, therefore

$$\mathbf{E}_{\tau_0} \left(X_{t-\tau_1} - X_{t-\tau_2}\right)^4 \leq 8\, \mathbf{E}_{\tau_0} \left(\int_{t-\tau_1}^{t-\tau_2} X_{s-\tau_0}\, \mathrm{d}s\right)^4 + \sigma^4\, \mathbf{E}_{\tau_0} \left(W_{t-\tau_1} - W_{t-\tau_2}\right)^4 \leq C\left|\tau_1 - \tau_2\right|^2 .$$

Using this estimate and the proof of Proposition 1.32 we obtain the consistency of the MLE $\hat{\tau}_T$.

We study the limit distribution of $T\left(\hat{\tau}_T - \tau_0\right)$ in Section 3.3.

If the unknown parameter is $\vartheta = \gamma$ and $\tau > 0$ is known, then the MLE is

$$\hat{\gamma}_T = -\frac{\int_0^T X_{t-\tau}\, \mathrm{d}X_t}{\int_0^T X_{t-\tau}^2\, \mathrm{d}t} = \gamma - \sigma\, \frac{\int_0^T X_{t-\tau}\, \mathrm{d}W_t}{\int_0^T X_{t-\tau}^2\, \mathrm{d}t}.$$

By the LLN

$$\frac{1}{T}\int_0^T X_{t-\tau}^2\, \mathrm{d}t \longrightarrow \mathbf{E}_\vartheta\, \xi^2 = d\left(\gamma\right)^2,$$

and hence

$$\mathcal{L}_\gamma\left\{\sqrt{T}\left(\hat{\gamma}_T - \gamma\right)\right\} \Longrightarrow \mathcal{N}\left(0, \sigma^2 d\left(\gamma\right)^{-2}\right).$$

Example 1.40. (*Discontinuity estimation*) The trend coefficient of processes (1.93) is a discontinuous function of the unknown parameter. Therefore as in the preceding two examples the MLE $\hat{\vartheta}_T$ of the point ϑ of jump has no explicit expression. To show its consistency we use once more Proposition 1.32. We have

$$\begin{aligned}
&\mathbf{E}_{\vartheta_0} \left\|\delta\left(\vartheta_1, \vartheta_2, \cdot\right)\right\|_T^4 \\
&\quad = \sigma^{-4}\, \mathbf{E}_{\vartheta_0} \left(\int_{\mathscr{R}} \left[\operatorname{sgn}\left(x - \vartheta_1\right) - \operatorname{sgn}\left(x - \vartheta_2\right)\right]^2\, f_T^\circ\left(x\right)\, \mathrm{d}x\right)^2 \\
&\quad = \sigma^{-4}\, \mathbf{E}_{\vartheta_0} \left(\int_{\mathscr{R}} \chi_{\{\vartheta_1 \leq x \leq \vartheta_2\}}\, f_T^\circ\left(x\right)\, \mathrm{d}x\right)^2 \\
&\quad \leq \frac{\left(\vartheta_2 - \vartheta_1\right)}{\sigma^4} \int_{\vartheta_1}^{\vartheta_2} \mathbf{E}_{\vartheta_0}\, f_T^\circ\left(x\right)^2\, \mathrm{d}x \leq C\left|\vartheta_2 - \vartheta_1\right|^2 .
\end{aligned}$$

The estimate

$$\mathbf{E}_{\vartheta_0} \left\|\delta\left(\vartheta_1, \vartheta_2, \cdot\right) h\left(\cdot\right)\right\|_T^4 \leq 4\, \mathbf{E}_{\vartheta_0} \left\|\delta\left(\vartheta_1, \vartheta_2, \cdot\right)\right\|_T^4 \leq C\left|\vartheta_2 - \vartheta_1\right|^2$$

follows from that obtained above.

We study the asymptotic behavior of $T\left(\hat{\vartheta}_T - \vartheta_0\right)$ in Section 3.4.

Bayesian Estimator

Let us suppose that the unknown parameter $\boldsymbol{\vartheta}$ is a random vector with known probability density $p(\boldsymbol{v})$, $\boldsymbol{v} \in \Theta$. We suppose as well that the loss function $\ell(\boldsymbol{u}) = |\boldsymbol{u}|^2$ is given too. We introduce the mean risk of the estimator $\bar{\boldsymbol{\vartheta}}_T$ as

$$\mathcal{R}(\bar{\boldsymbol{\vartheta}}_T) = \int_\Theta \mathbf{E}_{\boldsymbol{\theta}}\, \ell\left(\bar{\boldsymbol{\vartheta}}_T - \boldsymbol{\theta}\right)\, p(\boldsymbol{\theta})\, \mathrm{d}\boldsymbol{\theta} = \mathbb{E}\left|\bar{\boldsymbol{\vartheta}}_T - \boldsymbol{\vartheta}\right|^2,$$

where $\mathbb{E}$ means the expectation with respect to the joint distribution of X^T and $\boldsymbol{\vartheta}$. Define a *Bayesian estimator* $\tilde{\boldsymbol{\vartheta}}_T = \tilde{\boldsymbol{\vartheta}}_T(X^T)$ as an estimator which minimizes the mean risk, i.e.,

$$\mathcal{R}(\tilde{\boldsymbol{\vartheta}}_T) = \inf_{\bar{\boldsymbol{\vartheta}}_T} \mathbb{E}\left|\bar{\boldsymbol{\vartheta}}_T - \boldsymbol{\vartheta}\right|^2.$$

Let us denote by $\tilde{\boldsymbol{\vartheta}}_T$ the conditional expectation

$$\tilde{\boldsymbol{\vartheta}}_T = \boldsymbol{E}\left(\boldsymbol{\vartheta} \mid X^T\right) = \int_\Theta \boldsymbol{\theta}\, p(\boldsymbol{\theta} \mid X^T)\, \mathrm{d}\boldsymbol{\theta} \tag{1.117}$$

and show that $\tilde{\boldsymbol{\vartheta}}_T$ is a Bayesian estimator. Here $p(\boldsymbol{\theta} \mid X^T)$ is the density *a posteriori* of the random vector $\boldsymbol{\vartheta}$:

$$p(\boldsymbol{\theta} \mid X^T) = \frac{L(\boldsymbol{\theta}, \boldsymbol{\theta}_1, X^T)\, p(\boldsymbol{\theta})}{\int_\Theta L(\boldsymbol{u}, \boldsymbol{\theta}_1, X^T)\, p(\boldsymbol{u})\, \mathrm{d}\boldsymbol{u}},$$

where $\boldsymbol{\theta}_1$ is some fixed value and $L(\boldsymbol{\theta}, \boldsymbol{\theta}_1, X^T)$ is the likelihood ratio function.

The mean risk of an estimator can be written as

$$\begin{aligned}\mathcal{R}(\bar{\boldsymbol{\vartheta}}_T) &= \mathbb{E}\left|\bar{\boldsymbol{\vartheta}}_T - \tilde{\boldsymbol{\vartheta}}_T + \tilde{\boldsymbol{\vartheta}}_T - \boldsymbol{\vartheta}\right|^2 \\ &= \mathbb{E}\left|\bar{\boldsymbol{\vartheta}}_T - \tilde{\boldsymbol{\vartheta}}_T\right|^2 + 2\,\mathbb{E}\left(\bar{\boldsymbol{\vartheta}}_T - \tilde{\boldsymbol{\vartheta}}_T, \tilde{\boldsymbol{\vartheta}}_T - \boldsymbol{\vartheta}\right) + \mathbb{E}\left|\tilde{\boldsymbol{\vartheta}}_T - \boldsymbol{\vartheta}\right|^2.\end{aligned}$$

Using the properties of the conditional expectations we can write

$$\begin{aligned}\mathbb{E}\left(\bar{\boldsymbol{\vartheta}}_T - \tilde{\boldsymbol{\vartheta}}_T, \tilde{\boldsymbol{\vartheta}}_T - \boldsymbol{\vartheta}\right) &= \mathbb{E}\,\boldsymbol{E}\left(\left(\bar{\boldsymbol{\vartheta}}_T - \tilde{\boldsymbol{\vartheta}}_T, \tilde{\boldsymbol{\vartheta}}_T - \boldsymbol{\vartheta}\right) \middle| X^T\right) \\ &= \mathbb{E}\left(\bar{\boldsymbol{\vartheta}}_T - \tilde{\boldsymbol{\vartheta}}_T, \tilde{\boldsymbol{\vartheta}}_T - \boldsymbol{E}\left(\boldsymbol{\vartheta} \middle| X^T\right)\right) = 0\end{aligned}$$

because the estimators $\bar{\boldsymbol{\vartheta}}_T$ and $\tilde{\boldsymbol{\vartheta}}_T$ are measurable with respect to the observations X^T. Therefore

$$\inf_{\bar{\boldsymbol{\vartheta}}_T} \mathcal{R}(\bar{\boldsymbol{\vartheta}}_T) \geq \mathbb{E}\left|\tilde{\boldsymbol{\vartheta}}_T - \boldsymbol{\vartheta}\right|^2$$

and if $\bar{\boldsymbol{\vartheta}}_T = \tilde{\boldsymbol{\vartheta}}_T$ with probability 1 then the equality holds. Hence $\tilde{\boldsymbol{\vartheta}}_T$ is a Bayesian estimator.

The integral (1.117) sometimes can be calculated explicitly.

Example 1.41. For example, suppose that

$$\mathrm{d}X_t = \vartheta\, h(X_t)\,\mathrm{d}t + \mathrm{d}W_t, \quad X_0 = x, \quad 0 \le t \le T,$$

where $\vartheta \in \mathscr{R}_+$ and $p(\vartheta) = \lambda\, \mathrm{e}^{-\lambda\vartheta}$, $\vartheta \ge 0$. Then the direct calculation provides (see [6])

$$\tilde{\vartheta}_T = \frac{\Delta_T - \lambda}{\mathrm{I}_T} + \sqrt{\frac{2\pi}{\mathrm{I}_T}}\, \frac{\exp\left\{-\frac{(\Delta_T-\lambda)^2}{2\mathrm{I}_T}\right\}}{1 - \Phi\left(\frac{\lambda-\Delta_T}{\sqrt{\mathrm{I}_T}}\right)}, \tag{1.118}$$

where

$$\Delta_T = \int_0^T h(X_t)\ \mathrm{d}X_t, \qquad \mathrm{I}_T = \int_0^T h(X_t)^2\ \mathrm{d}t,$$
$$\Phi(a) = \mathbf{P}\{\zeta < a\}, \quad \zeta \sim \mathcal{N}(0,1).$$

Hence if we suppose that the condition $\mathcal{RP}(\Theta)$ is fulfilled and

$$d(\vartheta)^2 \equiv \mathbf{E}_\vartheta\, h(\xi)^2 < \infty,$$

then

$$\tilde{\vartheta}_T = \vartheta + \frac{\int_0^T h(X_t)\ \mathrm{d}W_t}{\mathrm{I}_T}\ (1 + o(1))$$

as $T \to \infty$ and

$$\mathcal{L}_\vartheta\left\{\sqrt{T}\left(\tilde{\vartheta}_T - \vartheta\right)\right\} \Longrightarrow \mathcal{N}\left(0, d(\vartheta)^{-2}\right).$$

Therefore the BEs in Examples 1.26 ($\vartheta = a$, $b = 0$) and 1.27 for this prior distribution can be calculated by (1.118) and are asymptotically normal with the same parameters as the MLEs.

If ϑ is a Gaussian random variable, $\vartheta \sim \mathcal{N}(m, \sigma^2)$, then for the same diffusion process we have

$$\tilde{\vartheta}_T = \frac{\Delta_T + m/\sigma^2}{\mathrm{I}_T + 1/\sigma^2}. \tag{1.119}$$

Note as well that we cannot use (1.119) in our examples because, for the negative values of ϑ, these processes are not ergodic.

In the ergodic regular case the asymptotic behavior of the Bayesian estimators is similar to the asymptotic behavior of MLE, i.e., they are consistent, asymptotically normal and asymptotically efficient. In non regular cases like Examples 1.29–1.31 the MLE and BE have different limit distributions and the BE are asymptotically efficient. Moreover the BE estimator has the following advantage: it can be written explicitly in Examples 1.29–1.31 as a ratio of two integrals (for quadratic loss function), when the MLE has no explicit expression.

Minimum Distance Estimator

The minimum distance estimator can be defined in many ways. In particular, if $X^{(n)} = \{X_1, \ldots, X_n\}$ are n independent observations of a random variable with a distribution function belonging to the family $\{F(\vartheta, x), \vartheta \in \Theta\}$ and

$$\hat{F}_n(x) = \frac{1}{n} \sum_{j=1}^{n} \chi_{\{X_j < x\}} \tag{1.120}$$

is the empirical distribution function, then the minimum distance estimator ϑ_n^* can be defined by the equation

$$\left\| \hat{F}_n(\cdot) - F(\vartheta_n^*, \cdot) \right\| = \inf_{\theta \in \Theta} \left\| \hat{F}_n(\cdot) - F(\theta, \cdot) \right\|, \tag{1.121}$$

where $\|\cdot\|$ is a norm in some Banach space $\mathscr{B}$ such that $\hat{F}_n(\cdot) - F(\vartheta, \cdot) \in \mathscr{B}$. By the Glivenko Cantelli theorem $\hat{F}_n(\cdot) \to F(\vartheta_0, \cdot)$ uniformly in $x \in \mathscr{R}$ with probability 1 (here ϑ_0 is the true value). Therefore (under the identifiability condition) the estimator ϑ_n^* is consistent and (under additional regularity conditions and for Hilbert space $\mathscr{B}$) it is asymptotically normal. Moreover MDE ϑ_T^* is asymptotically efficient in a certain sense as well (see Millar [183] for more general statements and references).

We study two different minimum distance estimators motivated by (1.120)–(1.121) and based on invariant distribution and density functions. The observed process is ergodic diffusion

$$\mathrm{d}X_t = S(\vartheta, X_t)\,\mathrm{d}t + \sigma(X_t)\,\mathrm{d}W_t, \quad X_0,\ 0 \le t \le T, \tag{1.122}$$

where $\vartheta \in \Theta \subset \mathscr{R}^d$ is an unknown parameter, Θ is an open bounded set and $\sigma(\cdot)^2$ is a known positive function.

Remember that the invariant distribution function is

$$F(\vartheta, x) = \frac{1}{G(\vartheta)} \int_{-\infty}^{x} \frac{1}{\sigma(y)^2} \exp\left\{ 2 \int_0^y \frac{S(\vartheta, v)}{\sigma(v)^2} \mathrm{d}v \right\} \mathrm{d}y$$

and the *empirical distribution function* is

$$\hat{F}_T(x) = \frac{1}{T} \int_0^T \chi_{\{X_t < x\}}\, \mathrm{d}t.$$

The first MDE ϑ_T^* is defined as a solution of the equation

$$\left\| \hat{F}_T(\cdot) - F(\vartheta_T^*, \cdot) \right\|_\mu = \inf_{\vartheta \in \Theta} \left\| \hat{F}_T(\cdot) - F(\vartheta, \cdot) \right\|_\mu, \tag{1.123}$$

where $\|\cdot\|_\mu$ is the $\mathscr{L}_2(\mu)$ norm and $\mu(\cdot)$ is some finite measure on $\mathscr{R}$, $(\mu(\mathscr{R}) < \infty)$. If Equation (1.123) has many solutions then we call any of them MDE .

This estimator can be written as well as

$$\vartheta_T^* = \arg\inf_{\theta\in\Theta} \int_{\mathscr{R}} \left[\hat{F}_T(x) - F(\theta, x)\right]^2 \mu(\mathrm{d}x).$$

The second MDE can be constructed with the help of an invariant density function $f(\vartheta,\cdot)$ and the *local time estimator* of this density (which can be called the *empirical density*)

$$f_T^\circ(x) = \frac{1}{\sigma(x)^2 T} \int_0^T \operatorname{sgn}(x - X_t)\ \mathrm{d}X_t + \frac{|X_T - x| - |X_0 - x|}{\sigma(x)^2 T}. \qquad (1.124)$$

In this case the MDE $\bar{\vartheta}_T^{**}$ is a solution of the problem

$$\left\| f_T^\circ(\cdot) - f(\vartheta_T^{**}, \cdot)\right\|_\mu = \inf_{\theta\in\Theta} \left\| f_T^\circ(\cdot) - f(\theta, \cdot)\right\|_\mu. \qquad (1.125)$$

Like the empirical distribution function, the estimator $f_T^\circ(x)$ is unbiased: $\mathbf{E}_\vartheta f_T^\circ(x) = f(\vartheta, x)$ and $\sqrt{T}$-asymptotically normal (see (1.34) and (1.32)). Note that as in the case of MLE Equations (1.89) and (1.90) can be replaced by the systems of *minimum distance equations*

$$\int_{\mathscr{R}} \left[\hat{F}_T(x) - F(\theta_T^*, x)\right] \dot{F}(\theta_T^*, x)\, \mu(\mathrm{d}x) = \mathbf{0} \qquad (1.126)$$

and

$$\int_{\mathscr{R}} \left[f_T^\circ(x) - f(\theta_T^{**}, x)\right] \dot{f}(\theta_T^{**}, x)\, \mu(\mathrm{d}x) = \mathbf{0}$$

respectively.

These estimators have no explicit presentation even for the simplest stochastic processes. For example, if the observed process is (1.82) with $\vartheta = (a, b)$, then the invariant distribution is $\mathcal{N}(b/a, \sigma^2/2a)$ and the MDE is

$$\vartheta_T^{**} = \arg\inf_{\vartheta\in\Theta} \int_{\mathscr{R}} \left(\sigma\sqrt{\pi}\, f_T^\circ(x) - \sqrt{a}\exp\left\{-\frac{(ax-b)^2}{a\sigma^2}\right\}\right)^2 \mathrm{d}x.$$

At the same time these estimators have several advantages. First, the construction of ϑ_T^* is not based on the stochastic integral and can be applied for a wider class of ergodic stochastic processes. Further, its calculation can be easier than MLE and it has good asymptotical properties. The problems of parameter estimation by the minimum distance method in Examples 1.29–1.31 are regular because the invariant distribution functions and densities are differentiable w.r.t. ϑ.

For example, the process (1.90) has the invariant density function (1.91) and the MDE of the parameter $\vartheta = \tau$ is

$$\tau_T^{**} = \arg \inf_{\tau \in (\alpha,\beta)} \int_{\mathscr{R}} \left(d(\tau)\; f_T^{\circ}(x) - (2\pi)^{-1/2} \exp\left\{ -\frac{x^2}{2\,d(\tau)^2} \right\} \right)^2 \mathrm{d}x.$$

The function $d(\tau)$ is monotonic. Hence we can first find

$$d_T^{**} = \arg \inf_{d \in \mathbb{D}} \int_{\mathscr{R}} \left(d\, f_T^{\circ}(x) - (2\pi)^{-1/2} \exp\left\{ -\frac{x^2}{2\,d^2} \right\} \right)^2 \mathrm{d}x$$

with the corresponding set $\mathbb{D}$ and then to solve the equation $d(\tau) = d_T^{**}$. Thus $\tau_T^{**} = d^{-1}\left(d_T^{**}\right)$. Here $d^{-1}(\cdot)$ is the function inverse to $d(\cdot)$.

In the case of the process (1.93) the invariant density is a double exponential (1.94) and the distribution function and density function are differentiable in the $\mathscr{L}_2$ sense.

Of course in both examples the rate of convergence of the MDE is worse than that of the MLE or BE.

The consistency of the MDE follows easily from the identifiability condition.

Let us denote

$$H(\vartheta, x, y) = 2 \int_0^y \frac{F(\vartheta, v \wedge x) - F(\vartheta, v)\; F(\vartheta, x)}{\sigma(v)^2\; f(\vartheta, v)}\, \mathrm{d}v \tag{1.127}$$

and

$$d_F(\vartheta, x)^2 = 4\; \mathbf{E}_\vartheta \left(\frac{F(\vartheta, \xi \wedge x) - F(\vartheta, \xi)\; F(\vartheta, x)}{\sigma(\xi)\; f(\vartheta, \xi)} \right)^2.$$

Proposition 1.42. *Suppose that for any $\vartheta_0 \in \Theta$ and $\nu > 0$*

$$g(\vartheta_0, \nu) \equiv \inf_{|\vartheta - \vartheta_0| > \nu} \left\| F(\vartheta, \cdot) - F(\vartheta_0, \cdot) \right\|_\mu > 0 \tag{1.128}$$

and

$$\mathbf{P}_{\vartheta_0}\left\{ \left\| H(\vartheta_0, \cdot, \xi) \right\|_\mu < \infty \right\} = 1, \qquad \left\| d_F(\vartheta_0, \;) \right\|_\mu < \infty, \tag{1.129}$$

then

$$\mathbf{P}_{\vartheta_0}^{(T)}\left\{ \left| \vartheta_T^* - \vartheta_0 \right| > \nu \right\} \to 0$$

as $T \to \infty$.

Proof. We can write

$$\begin{aligned}
&\mathbf{P}_{\vartheta_0}^{(T)}\left\{ \left| \vartheta_T^* - \vartheta_0 \right| > \nu \right\} \\
&\quad = \mathbf{P}_{\vartheta_0}^{(T)}\left\{ \inf_{|\vartheta - \vartheta_0| \le \nu} \left\| \hat F_T(\cdot) - F(\vartheta, \cdot) \right\|_\mu > \inf_{|\vartheta - \vartheta_0| > \nu} \left\| \hat F_T(\cdot) - F(\vartheta, \cdot) \right\|_\mu \right\} \\
&\quad \le \mathbf{P}_{\vartheta_0}^{(T)}\left\{ 2\, \left\| \eta_T(\cdot) \right\|_\mu > g(\vartheta_0, \nu)\, \sqrt{T} \right\},
\end{aligned}$$

where we have denoted

$$\eta_T(x) = \sqrt{T}\left(\hat{F}_T(x) - F(\vartheta_0, x)\right),$$

and have used the following properties of norm:

$$\left\|\hat{F}_T(\cdot) - F(\vartheta,\cdot)\right\|_\mu \leq \left\|\hat{F}_T(\cdot) - F(\vartheta_0,\cdot)\right\|_\mu + \left\|F(\vartheta,\cdot) - F(\vartheta_0,\cdot)\right\|_\mu,$$
$$\left\|\hat{F}_T(\cdot) - F(\vartheta,\cdot)\right\|_\mu \geq \left\|F(\vartheta,\cdot) - F(\vartheta_0,\cdot)\right\|_\mu - \left\|\hat{F}_T(\cdot) - F(\vartheta_0,\cdot)\right\|_\mu$$

and the identity

$$\inf_{|\vartheta-\vartheta_0|\leq\nu} \left\|F(\vartheta,\cdot) - F(\vartheta_0,\cdot)\right\|_\mu = 0.$$

For the process $Y_t = H(\vartheta_0, x, X_t)$ by the Itô formula we obtain the following representation:

$$\mathrm{d}Y_t = \left[\chi_{\{X_t<x\}} - F(\vartheta_0, x)\right]\,\mathrm{d}t + 2\frac{F(\vartheta_0, X_t\wedge x) - F(\vartheta_0, X_t)\,F(\vartheta_0,x)}{\sigma(X_t)\;f(\vartheta_0, X_t)}\,\mathrm{d}W_t,$$

which is equivalent to

$$\begin{aligned}\eta_T(x) &= \frac{H(\vartheta_0, x, X_T) - H(\vartheta_0, x, X_0)}{\sqrt{T}}\\ &\quad - \frac{2}{\sqrt{T}}\int_0^T \frac{F(\vartheta_0, X_t\wedge x) - F(\vartheta_0, X_t)\;F(\vartheta_0,x)}{\sigma(X_t)\;f(\vartheta_0, X_t)}\,\mathrm{d}W_t\\ &= \eta_T^{(1)}(x) - \eta_T^{(2)}(x) \qquad (1.130)\end{aligned}$$

with obvious notation. Note that the second derivative w.r.t. y of the function $H(\vartheta_0, x, y)$ is not continuous, but using standard arguments it can be approximated by a sequence of continuous functions and so the equality will be proved. Hence

$$\begin{aligned}&\mathbf{P}_{\vartheta_0}^{(T)}\left\{2\left\|\eta_T^{(1)}(\cdot) - \eta_T^{(2)}(\cdot)\right\|_\mu > g(\vartheta_0,\nu)\sqrt{T}\right\}\\ &\quad\leq 2\,\mathbf{P}_{\vartheta_0}^{(T)}\left\{\left\|H(\vartheta_0,\cdot,\xi)\right\|_\mu > g(\vartheta_0,\nu)\,T\right\}\\ &\qquad + \mathbf{P}_{\vartheta_0}^{(T)}\left\{2\left\|\eta_T^{(2)}(\cdot)\right\|_\mu > g(\vartheta_0,\nu)\sqrt{T}\right\}\\ &\quad\leq 2\,\mathbf{P}_{\vartheta_0}^{(T)}\left\{\left\|H(\vartheta_0,\cdot,\xi)\right\|_\mu > g(\vartheta_0,\nu)\,T\right\} + \frac{4\left\|d_F(\vartheta_0,\cdot)\right\|_\mu^2}{Tg(\vartheta_0,\nu)^2}.\end{aligned}$$

Now the consistency of ϑ_T^* follows from the conditions (1.128) and (1.129).

The proof of the consistency of the MDE ϑ_T^{**} is quite similar. Note that the identifiability condition is obviously the following: for any ϑ_0 and $\nu > 0$

$$\inf_{|\vartheta-\vartheta_0|>\nu} \left\| f(\vartheta,\cdot) - f(\vartheta_0,\cdot) \right\|_\mu > 0.$$

The conditions (1.129) are not too restrictive. In Section 2.2 we show that they are fulfilled if the condition $\mathcal{A}_0(\Theta)$ is fulfilled.

To prove the asymptotic normality of this estimator we can use the minimum distance equation (1.126) as follows. Let us denote $u_T^* = \sqrt{T}\left(\vartheta_T^* - \vartheta_0\right)$ and write (1.126) as

$$\int_{\mathscr{R}} \left[\eta_T(x) - \left(u_T^*, \dot{F}\left(\tilde{\theta}_T^*, x\right)\right)\right] \dot{F}\left(\theta_T^*, x\right) \mu(dx) = \mathbf{0}.$$

Therefore u_T^* admits the representation

$$u_T^* = \left(\int_{\mathscr{R}} \dot{F}\left(\tilde{\theta}_T^*, x\right) \dot{F}\left(\theta_T^*, x\right)^{\mathrm{T}} \mu(dx)\right)^{-1} \int_{\mathscr{R}} \eta_T(x)\, \dot{F}\left(\theta_T^*, x\right) \mu(dx).$$

As the estimator θ_T^* is consistent, we can write

$$u_T^* = \left(\int_{\mathscr{R}} \dot{F}(\vartheta_0, x)\, \dot{F}(\vartheta_0, x)^{\mathrm{T}} \mu(dx)\right)^{-1} \times \int_{\mathscr{R}} \eta_T(x)\, \dot{F}(\vartheta_0, x)\, \mu(dx)\, (1 + o(1)).$$

In the last integral we use the representation (1.130) and exchange the order of integration

$$J_T^{(1)} = \int_{\mathscr{R}} \eta_T^{(1)}(x)\, \dot{F}(\vartheta_0, x)\, \mu(dx) = \frac{2}{\sqrt{T}} \int_{X_0}^{X_T} \int_{\mathscr{R}} \frac{F(\vartheta_0, v \wedge x) - F(\vartheta_0, v)\, F(\vartheta_0, x)}{\sigma(v)^2\, f(\vartheta_0, v)} \dot{F}(\vartheta_0, x)\, \mu(dx)\, dv,$$

$$J_T^{(2)} = \int_{\mathscr{R}} \eta_T^{(2)}(x)\, \dot{F}(\vartheta_0, x)\, \mu(dx) = \frac{2}{\sqrt{T}} \int_0^T \int_{\mathscr{R}} \frac{F(\vartheta_0, X_t \wedge x) - F(\vartheta_0, X_t)\, F(\vartheta_0, x)}{\sigma(X_t)\, f(\vartheta_0, X_t)} \dot{F}(\vartheta_0, x)\, \mu(dx)\; dW_t.$$

Therefore if we denote the vector integral

$$M(\vartheta_0, X_t) = 2 \int_{\mathscr{R}} \frac{F(\vartheta_0, X_t \wedge x) - F(\vartheta_0, X_t)\, F(\vartheta_0, x)}{\sigma(X_t)\, f(\vartheta_0, X_t)} \dot{F}(\vartheta_0, x)\, \mu(dx)$$

and suppose that

$$\mathbf{E}_{\vartheta_0} \left|\boldsymbol{M}\left(\vartheta_0, \xi\right)\right|^2 < \infty$$

then $J_T^{(2)}$ is asymptotically normal

$$\mathcal{L}_{\vartheta_0}\left\{J_T^{(2)}\right\} \Longrightarrow \mathcal{N}\left(\mathbf{0}, \mathbf{E}_{\vartheta_0} \boldsymbol{H}\left(\vartheta_0, \xi\right) \boldsymbol{M}\left(\vartheta_0, \xi\right)^{\mathbb{T}}\right).$$

Hence the MDE $\boldsymbol{\vartheta}_T^*$ is asymptotically normal:

$$\mathcal{L}_{\vartheta_0}\left\{\sqrt{T}\left(\boldsymbol{\vartheta}_T^* - \boldsymbol{\vartheta}_0\right)\right\} \Longrightarrow \mathcal{N}\left(\mathbf{0}, \boldsymbol{R}\left(\boldsymbol{\vartheta}_0\right)\right)$$

with corresponding covariance matrix $\boldsymbol{R}\left(\boldsymbol{\vartheta}_0\right)$.

Trajectory Fitting Estimator

Another method of parameter estimation widely used in time-series analysis is the so-called least squares method. It can be described as follows. Suppose that the observed discrete-time process is autoregressive of order d, i.e.,

$$X_j = \sum_{i=1}^{d} \vartheta^{(i)} X_{j-i} + \varepsilon_j, \quad j = 1, \ldots, n$$

(with specially defined first $d < n$ values). Here ε_j, $j = 1, \ldots, n$ are independent identically distributed random variables with $\mathbf{E}\,\varepsilon_j = 0$ and $\mathbf{E}\,\varepsilon_j^2 < \infty$. Then the parameter $\boldsymbol{\vartheta} = \left(\vartheta^{(1)}, \ldots, \vartheta^{(d)}\right)$ can be estimated as a solution of the minimization problem

$$\sum_{j=1}^{n}\left[X_j - \sum_{i=1}^{d} \vartheta_T^{\star(i)} X_{j-i}\right]^2 = \inf_{\theta \in \Theta} \sum_{j=1}^{n}\left[X_j - \sum_{i=1}^{d} \theta^{(i)} X_{j-i}\right]^2.$$

The estimator $\boldsymbol{\vartheta}_n^\star$ (called the *least squares estimator*) is easy to calculate

$$\boldsymbol{\vartheta}_T^\star = \boldsymbol{A}_T^{-1} \boldsymbol{y}_T,$$

where the matrix $\boldsymbol{A}_T$ and the vector $\boldsymbol{y}_T$ are

$$\left(\boldsymbol{A}_T\right)_{i,l} = \sum_{j=1}^{n} X_{j-i} X_{j-l}, \qquad \left(\boldsymbol{y}_T\right)_i = \sum_{j=1}^{n} X_{j-i} X_j.$$

To have an analog of this estimator for continuous-time diffusion processes we first write the observations in the integral form

$$X_t - X_0 = \int_0^t S\left(\boldsymbol{\vartheta}, X_s\right) \mathrm{d}s + \int_0^t \sigma\left(\boldsymbol{\vartheta}, X_s\right) \mathrm{d}W_s.$$

The comparison of this equation with the autoregressive process shows that the role of the sum $\sum_{i=1}^{d} \vartheta^{(i)} X_{j-i}$ can be played by the integral

$$\int_0^t S(\vartheta, X_s)\, ds.$$

Therefore we can introduce a family of stochastic processes

$$\hat{X}_t(\theta) = X_0 + \int_0^t S(\theta, X_s)\, ds, \quad t \in [0,T], \quad \theta \in \Theta,$$

and then define the estimator $\vartheta^\star_T$ as a solution of the following problem:

$$\int_0^T \left[X_t - \hat{X}_t(\vartheta^\star_T)\right]^2 dt = \inf_{\theta\in\Theta} \int_0^T \left[X_t - \hat{X}_t(\theta)\right]^2 dt. \tag{1.131}$$

We call this estimator the *trajectory fitting estimator* (TFE) because the choice of it is based on the fitting of the observed trajectory $\{X_t, 0 \le t \le T\}$ by an *artificial* one $\left\{\hat{X}_t(\vartheta^\star_T), 0 \le t \le T\right\}$. This estimator can be found with the help of the corresponding *trajectory fitting equation*

$$\int_0^T \left[X_t - \hat{X}_t(\vartheta^\star_T)\right] \dot{\hat{\boldsymbol{X}}}_t(\vartheta^\star_T) dt = \mathbf{0},$$

where

$$\dot{\hat{\boldsymbol{X}}}_t(\theta) = \int_0^t \dot{\boldsymbol{S}}(\theta, X_s)\, ds.$$

The difference between the TFE and least squares estimator is in the nature of the conditions providing the consistency. The identifiability condition for the TFE is

$$\mathbf{E}_\vartheta\, S(\vartheta_1, \xi) \neq 0, \quad \text{for} \quad \text{all} \quad \vartheta_1 \neq \vartheta \tag{1.132}$$

and if this condition is not satisfied then the estimator in general is not consistent.

The general results concerning this estimator can be found in Section 2.3.

Example 1.43. (*Ornstein–Uhlenbeck process*) In the linear case

$$dX_t = -(a\, X_t - b)\, dt + \sigma dW_t, \quad X_0,\ 0 \le t \le T$$

the TFE of the parameter $\vartheta = (a, b)$ is not consistent because

$$\mathbf{E}_\vartheta\,(a_1\xi - b_1) = \frac{a_1\, b}{a} - b_1 = 0, \quad \text{for} \quad \vartheta_1 = (\kappa a, \kappa b)$$

and all $\kappa > 0$. But if $\vartheta = a$ and $b \neq 0$ is known, then the TFE

$$a^\star_T = -\frac{\int_0^T (X_t - X_0 - bt)\, Y_t\, dt}{\int_0^T Y_t^2 dt}, \qquad Y_t = \int_0^t X_s\, ds$$

can have good properties. In particular, we have

$$a^\star_T = a - \sigma\,\frac{\int_0^T W_t\, Y_t\, \mathrm{d}t}{\int_0^T Y_t^2 \mathrm{d}t}.$$

By the LLN

$$Z_t = \frac{1}{t}\int_0^t X_s\, \mathrm{d}s \longrightarrow \mathbf{E}_a\,\xi = \frac{b}{a}.$$

Hence using the Tœplitz lemma we obtain

$$\sqrt{T}\left(a^\star_T - a\right) = -\frac{\int_0^T t\, W_t\, Z_t\, \mathrm{d}t}{\int_0^T t^2\, Z_t^2\, \mathrm{d}t} = -\frac{3a\sigma}{b}\, T^{-5/2}\int_0^T t\, W_t\, \mathrm{d}t\, \left(1 + o(1)\right).$$

The last integral is

$$\int_0^T t\int_0^t \mathrm{d}W_s\, \mathrm{d}t = \int_0^T\int_s^T t\, \mathrm{d}t\, \mathrm{d}W_s = \int_0^T \frac{T^2 - s^2}{2}\, \mathrm{d}W_s.$$

Hence, the TFE is asymptotically normal

$$\mathcal{L}_a\left\{\sqrt{T}\left(a^\star_T - a\right)\right\} \Longrightarrow \mathcal{N}\left(0, \frac{6\, a^2\sigma^2}{5\, b^2}\right).$$

Comparison with the limit variance of the MLE

$$\frac{6}{5}\,\frac{a^2\sigma^2}{b^2} > \frac{1}{1 + a\sigma^2/2b^2}\,\frac{a^2\sigma^2}{b^2}$$

shows that the MLE is better.

In the second case with $a > 0$ known and $\vartheta = b$ we have

$$b^\star_T = \frac{3}{T^3}\int_0^T \left(X_t - X_0 + aY_t\right) t\, \mathrm{d}t = b + \frac{3\,\sigma}{T^3}\int_0^T W_t\, t\, \mathrm{d}t.$$

Hence this TFE is consistent, has Gaussian distribution and the limit variance

$$\mathcal{L}_b\left\{\sqrt{T}\left(b^\star_T - b\right)\right\} \Longrightarrow \mathcal{N}\left(0, \frac{6\,\sigma^2}{5}\right)$$

is also larger than that of the MLE equal to σ^2.

Of course, if the value $\vartheta^\star_T$ does not belong to $\boldsymbol{\Theta}$, then we have to write an expression like (1.115).

Example 1.44. If the observed process is (1.84), then the condition (1.132) is not fulfilled and we have to modify the TFE. For example, put

$$Y_t(\theta) = X_0^2 - 2\theta \int_0^t \frac{X_s^2}{1+X_s^2}\, ds + \sigma^2 t, \quad 0 \le t \le T, \quad \theta \in \Theta$$

and define the TFE by the relation

$$\vartheta_T^\star = \arg\inf_{\theta\in\Theta} \int_0^T \left(X_t^2 - Y_t(\theta)\right)^2 dt.$$

This estimator has the following explicit representation:

$$\vartheta_T^\star = -\frac{\int_0^T \left(X_t^2 - X_0^2 - \sigma^2 t\right)\, h_t\, dt}{2\int_0^T h_t^2\, dt}, \qquad h_t = \int_0^t \frac{X_s^2}{1+X_s^2}\, ds.$$

Hence

$$\vartheta_T^\star = \vartheta - \sigma \frac{\int_0^T h_t \int_0^t X_s\, dW_s\, dt}{\int_0^T h_t^2\, dt}.$$

To study this estimator we have to suppose that the unknown parameter $\vartheta \in (\alpha, \beta)$, where $\alpha > 3\sigma^2/2$, because we need the condition $\mathbf{E}_\vartheta \xi^2 < \infty$ for all $\vartheta \in \Theta$. We have the convergence

$$Z_t = \frac{1}{t}\int_0^t \frac{X_s^2}{1+X_s^2}\, ds \to \mathbf{E}_\vartheta \left(\frac{\xi^2}{1+\xi^2}\right)^2 \equiv d(\vartheta) < 1$$

and

$$T^{-3}\int_0^T h_t^2\, dt = T^{-3}\int_0^T t^2 Z_t^2\, dt \to \frac{d(\vartheta)^2}{3}.$$

Further, let us denote $Y_t = t^{-1}\int_0^t X_s^2\, ds$, then we can write

$$\begin{aligned}\int_0^T h_t \int_0^t X_s\, dW_s\, dt &= d(\vartheta)\int_0^T t \int_0^t X_s\, dW_s dt\ (1+o(1))\\ &= \frac{d(\vartheta)}{2}T^2\int_0^T X_t\, dW_t - \frac{d(\vartheta)}{2}\int_0^T t^2\, X_t\, dW_t\ (1+o(1))\\ &= \frac{d(\vartheta)}{2}T^{5/2}\left(J_T^{(1)} - J_T^{(2)}\right)\ (1+o(1))\end{aligned}$$

where

$$J_T^{(1)} = \frac{1}{\sqrt{T}}\int_0^T X_t\, dW_t, \qquad J_T^{(2)} = \frac{1}{T^{5/2}}\int_0^T t^2\, X_t\, dW_t.$$

We have to show the asymptotic normality of the vector $\boldsymbol{J}_T = \left(J_T^{(1)}, J_T^{(2)}\right)$. To apply the CLT we calculate the limits of the integrals

$$\frac{1}{T^5}\int_0^T t^4\,X_t^2\,{\rm d}t,\qquad \frac{1}{T^3}\int_0^T t^2\,X_t^2\,{\rm d}t,\qquad \frac{1}{T}\int_0^T X_t^2\,{\rm d}t.$$

The last one by the LLN converges to $\mathbf{E}_\vartheta\,\xi^2$. For the first one integrating by parts we have

$$\begin{aligned}\int_0^T t^4\,X_t^2\,{\rm d}t &= T^4\int_0^T X_t^2\,{\rm d}t-4\int_0^T t^3\int_0^t X_s^2\,{\rm d}s\,{\rm d}t\\ &= T^5\,Y_T-4\int_0^T t^4\,Y_t\,{\rm d}t=\frac{1}{5}\,T^5\,\mathbf{E}_\vartheta\,\xi^2\,\left(1+o(1)\right).\end{aligned}$$

Similarly

$$\frac{1}{T^3}\int_0^T t^2\,X_t^2\,{\rm d}t=\frac{1}{3}\,T^3\,\mathbf{E}_\vartheta\,\xi^2\,\left(1+o(1)\right).$$

Hence the vector $\boldsymbol{J}_T$ is asymptotically normal with limit covariance matrix

$$\begin{pmatrix}1/5 & 1/3\\ 1/3 & 1\end{pmatrix}\,\mathbf{E}_\vartheta\,\xi^2$$

and

$$\mathcal{L}_\vartheta\left\{J_T^{(1)}-J_T^{(2)}\right\}\Longrightarrow\mathcal{N}\left(0,\frac{8}{15}\,\mathbf{E}_\vartheta\,\xi^2\right).$$

Finally we obtain the asymptotic normality of the TFE:

$$\mathcal{L}_\vartheta\left\{\sqrt{T}\left(\vartheta_T^\star-\vartheta\right)\right\}\Longrightarrow\mathcal{N}\left(0,\frac{6}{5}\frac{\sigma^2}{d\left(\vartheta\right)^2}\,\mathbf{E}_\vartheta\,\xi^2\right).$$

Remember that the MLE of this parameter (for wider ϑ) has limit variance $\sigma^2/c\left(\vartheta\right)^2$.

Note that in Example 1.28 condition (1.132) is not fulfilled and the definition of the TFE needs to be modified.

In nonregular problems (Examples 1.29–1.31) application of the trajectory fitting method is quite difficult and it is better to use another one, say the method of moments.

Estimator of the Method of Moments

The traditional definition of the (generalized) method of moments in classical statistics (statistics of i.i.d. observations) is the following. Let $\{X_1,\ldots,X_n\}$ be i.i.d., with a distribution function $F\left(\boldsymbol{\vartheta},\cdot\right)$ of one observation X_j depending on $\boldsymbol{\vartheta}\in\boldsymbol{\Theta}\subset\mathscr{R}^d$. Introduce a vector function $\boldsymbol{q}(x)=\left(q_1(x),\ldots,q_d(x)\right)$ and denote by $\boldsymbol{m}\left(\boldsymbol{\vartheta}\right)=\mathbf{E}_{\boldsymbol{\vartheta}}\,\boldsymbol{q}(X_1)$ its mathematical expectation. Define the set of values of $\boldsymbol{m}\left(\boldsymbol{\vartheta}\right)$ as $\mathbb{M}=\left\{\boldsymbol{m}:\,\boldsymbol{m}\left(\boldsymbol{\vartheta}\right),\,\boldsymbol{\vartheta}\in\boldsymbol{\Theta}\right\}$ and suppose that the function $\boldsymbol{q}(\cdot)$ is such that the system of equations

$$m(\vartheta) = m, \qquad \vartheta \in \Theta$$

has a unique solution for any $m \in \mathbb{M}$. Then the estimator of the method of moments $\bar{\vartheta}_n$ is defined as the solution of the following system of equations:

$$m(\bar{\vartheta}_n) = \frac{1}{n}\sum_{j=1}^{n} q(X_j).$$

Of course, if the value of the right hand side is out of the set $\mathbb{M}$ then $\bar{\vartheta}_n$ is defined by the value of $m(\vartheta)$ which is the closest one to this sum. This estimator is consistent and asymptotically normal (see, for example, [33]). We describe below the behavior of the continuous-time analog of this estimator.

The observed process as before is ergodic diffusion

$$\mathrm{d}X_t = S(\vartheta, X_t)\,\mathrm{d}t + \sigma(X_t)\,\mathrm{d}W_t, \quad X_0, \quad 0 \le t \le T,$$

where $\vartheta \in \Theta \subset \mathscr{R}^d$, Θ is an open bounded set. The functions $S(\cdot,\cdot)$ and $\sigma(\cdot)$ are such that this equation has a unique weak solution and this solution has ergodic properties with the invariant distribution function $F(\vartheta,\cdot)$ for all $\vartheta \in \Theta$.

Let $q(x) = (q_1(\cdot), \ldots, q_d(\cdot))$ be a vector function and $m(\vartheta) = \mathbf{E}_\vartheta\, q(\xi)$ be its mathematical expectation. Here ξ is a "stationary random variable". The set $\mathbb{M}$ contains all values of $m(\vartheta)$ and the function $q(\cdot)$ is chosen in such a way that the equation

$$m(\vartheta) = m, \qquad \vartheta \in \Theta$$

has a unique solution for any $m \in \mathbb{M}$. The *estimator of the method of moments* (EMM) $\bar{\vartheta}_T$ is defined by the equality

$$\left| m(\bar{\vartheta}_T) - \frac{1}{T}\int_0^T q(X_t)\,\mathrm{d}t \right| = \inf_{\vartheta\in\Theta} \left| m(\vartheta) - \frac{1}{T}\int_0^T q(X_t)\,\mathrm{d}t \right|. \qquad (1.133)$$

If this equation has more than one solution then any of them can be taken as the EMM $\bar{\vartheta}_T$. In Section 2.4 we show that as in classical statistics this EMM is consistent and asymptotically normal.

Note that this estimator admits the representation

$$\bar{\vartheta}_T = \vartheta_T\, \chi_{\{\hat{m}_T \in \mathbb{M}\}} + \vartheta_T^o\, \chi_{\{\hat{m}_T \in \mathbb{M}^c\}}, \qquad (1.134)$$

where

$$\hat{m}_T = \frac{1}{T}\int_0^T q(X_t)\,\mathrm{d}t$$

and ϑ_T is the solution of the system of equations $m(\vartheta_T) = \hat{m}_T$ when $\hat{m}_T \in \mathbb{M}$. The value of ϑ_T^o corresponds to $m \in \mathbb{M}$ closest to $\hat{m}_T$, when $\hat{m}_T \in \mathbb{M}^c$.

Say, in a one-dimensional case with $\Theta = (\alpha, \beta)$ and monotonically increasing function $m(\cdot)$ we have $\mathbb{M} = (m(\alpha), m(\beta))$ and similarly to (1.115)

$$\bar{\vartheta}_T = \alpha \chi_{\{\hat{m}_T \leq m(\alpha)\}} + m^{-1}\left(\hat{m}_T\right) \chi_{\{\alpha < \hat{m}_T < \beta\}} + \beta \chi_{\{\hat{m}_T \geq m(\beta)\}},$$

where $m^{-1}(\cdot)$ is inverse to $m(\cdot)$ function.

Example 1.45. In the case of the Ornstein–Uhlenbeck process

$$\mathrm{d}X_t = -\left(a\, X_t - b\right)\, \mathrm{d}t + \sigma\, \mathrm{d}W_t, \quad X_0,\ 0 \leq t \leq T,$$

with $\vartheta = (a, b)$ and $a > 0$ we take $\boldsymbol{q}(x) = \left(x, x^2\right)$. We have $\mathbf{E}_\vartheta\, \xi = b/a$ and $\mathbf{E}_\vartheta\, \xi^2 = b^2/a^2 + \sigma^2/2a$. Hence the EMM $\bar{\boldsymbol{\vartheta}}_T = \left(\bar{a}_T, \bar{b}_T\right)$ is

$$\bar{a}_T = \frac{\sigma^2}{2\,\left(Y_2 - Y_1^2\right)}, \tag{1.135}$$

$$\bar{b}_T = \frac{Y_1\, \sigma^2}{2\,\left(Y_2 - Y_1^2\right)}, \tag{1.136}$$

where

$$Y_1 = \frac{1}{T}\int_0^T X_t\, \mathrm{d}t \longrightarrow \frac{b}{a}, \qquad Y_2 = \frac{1}{T}\int_0^T X_t^2\, \mathrm{d}t \longrightarrow \frac{b^2}{a^2} + \frac{\sigma^2}{2a}.$$

Hence this estimator is consistent. Let us put $\delta_T = T^{-1/2}$ and

$$\eta_T = \frac{1}{\sqrt{T}}\int_0^T \left[X_t^2 - \frac{b^2}{a^2} - \frac{\sigma^2}{2a}\right]\, \mathrm{d}t, \qquad \zeta_T = \frac{1}{\sqrt{T}}\int_0^T \left[X_t - \frac{b}{a}\right]\, \mathrm{d}t.$$

Then we can write

$$\begin{aligned}\bar{a}_T &= \frac{\sigma^2}{2\left(\frac{\sigma^2}{2a} + \delta_T\, \eta_T - \frac{2b}{a}\delta_T\, \zeta_T\right)}\left(1 + o(1)\right)\\ &= a\left(1 - \delta_T\left[\frac{2a}{\sigma^2}\eta_T - \frac{4b}{\sigma^2}\zeta_T\right]\right)\left(1 + o(1)\right).\end{aligned}$$

Therefore

$$\sqrt{T}\left(\bar{a}_T - a\right) = -\frac{2a}{\sigma^2}\,\frac{1}{\sqrt{T}}\int_0^T \left[aX_t^2 - 2bX_t + \frac{b^2}{a} - \frac{\sigma^2}{2}\right]\, \mathrm{d}t\left(1 + o(1)\right).$$

Note that for $h(\vartheta, x) = ax^2 - 2bx + b^2/a - \sigma^2/2$ we have $\mathbf{E}_\vartheta\, h(\vartheta, \xi) = 0$. Hence by the Itô formula

$$\begin{aligned}\int_0^T h\left(X_t\right)\, \mathrm{d}t &= H(\vartheta, X_T) - H(\vartheta, X_0)\\ &\quad -2\int_0^T \frac{1}{\sigma f\left(\vartheta, X_t\right)}\int_{-\infty}^{X_t} h(\vartheta, v) f\left(\vartheta, v\right)\, \mathrm{d}v\, \mathrm{d}W_t.\end{aligned}$$

Finally

$$\mathcal{L}_{\vartheta}\left\{\sqrt{T}\left(\bar{a}_T - a\right)\right\} \Longrightarrow \mathcal{N}\left(0, d_a\left(\vartheta\right)^2\right)$$

with

$$d_a\left(\vartheta\right)^2 = \frac{16\,a^2}{\sigma^6}\,\mathbf{E}_a\left(\int_{-\infty}^{\xi}\frac{h(\vartheta,v)f\left(\vartheta,v\right)}{f\left(\vartheta,\xi\right)}\mathrm{d}v\right)^2.$$

In a similar way we obtain the representation of EMM $\bar{b}_T$ as

$$\begin{aligned}
&\sqrt{T}\left(\bar{b}_T - b\right)\\
&\quad= -\frac{1}{\sqrt{T}}\int_0^T\left[\frac{2a}{\sigma^2}X_t^2 - \frac{4b^2 + a\sigma^2}{\sigma^2}X_t + \frac{2b^3}{a\sigma^2} - b\right]\,\mathrm{d}t\ \left(1 + o(1)\right)\\
&\quad= -\frac{1}{\sqrt{T}}\int_0^T g\left(\vartheta, X_t\right)\,\mathrm{d}t\ \left(1 + o(1)\right)\\
&\quad= -\frac{2}{\sqrt{T}}\int_0^T\frac{1}{\sigma f\left(\vartheta, X_t\right)}\int_{-\infty}^{X_t} g(\vartheta,v)f\left(\vartheta,v\right)\mathrm{d}v\ \mathrm{d}W_t\ \left(1 + o(1)\right),
\end{aligned}$$

which provides the asymptotic normality

$$\mathcal{L}_{\vartheta}\left\{\sqrt{T}\left(\bar{b}_T - b\right)\right\} \Longrightarrow \mathcal{N}\left(0, d_b\left(\vartheta\right)^2\right)$$

with

$$d_b\left(\vartheta\right)^2 = \frac{4}{\sigma^2}\,\mathbf{E}_{\vartheta}\left(\int_{-\infty}^{\xi}\frac{g(\vartheta,v)f\left(\vartheta,v\right)}{f\left(\vartheta,\xi\right)}\mathrm{d}v\right)^2.$$

Of course these representations of the estimators give us the joint asymptotic normality

$$\mathcal{L}_{\vartheta}\left\{\sqrt{T}\left(\bar{\vartheta}_T - \vartheta\right)\right\} \Longrightarrow \mathcal{N}\left(0, \mathbf{d}\left(\vartheta\right)^2\right),$$

where the matrix

$$\mathbf{d}\left(\vartheta\right)^2 = \begin{pmatrix} d_a\left(\vartheta\right)^2, & d_{ab}\left(\vartheta\right)\\ d_{ab}\left(\vartheta\right), & d_b\left(\vartheta\right)^2\end{pmatrix}$$

and

$$d_{ab}\left(\vartheta\right) = \frac{4a}{\sigma^4}\,\mathbf{E}_{\vartheta}\left(\int_{-\infty}^{\xi}\frac{h(\vartheta,v)f\left(\vartheta,v\right)}{f\left(\vartheta,\xi\right)}\mathrm{d}v\int_{-\infty}^{\xi}\frac{g(\vartheta,u)f\left(\vartheta,u\right)}{f\left(\vartheta,\xi\right)}\mathrm{d}u\right).$$

Example 1.46. In the case of the observations

$$\mathrm{d}X_t = -\frac{\vartheta\, X_t}{1+X_t^2}\,\mathrm{d}t + \sigma\,\mathrm{d}W_t, \quad X_0, \quad 0 \le t \le T,$$

and invariant density (1.85) the choice of the function $q\left(\cdot\right)$ depends on the values of ϑ. If $\vartheta \in (\alpha, \beta)$ with $\alpha > 3\sigma^2$, then $\mathbf{E}_\vartheta\, \xi^2 < \infty$ and we can take $q\left(x\right) = x^2$. Otherwise (the ergodic case with $\alpha > \sigma^2/2$), it can be $q\left(x\right) = x^2\, \chi_{\{|x|<A\}}$ with any $A > 0$. We have no explicit expression for EMM $\bar\vartheta_T$.

Example 1.47. In the case of observations (1.86) the choice of the function $\boldsymbol{q}\left(\cdot\right)$ depends on $\boldsymbol{h}\left(\cdot\right)$ and $\sigma\left(\cdot\right)$. In particular, if $d = 1$ and $h_1\left(x\right) = -x$ then we have Example 1.26 with $b = 0$.

Example 1.48. (*Cusp estimation*) Note that the diffusion process (1.88) has a mean ϑ. Hence we can take $q(x) = x$ and define the EMM

$$\bar\vartheta_T = \frac{1}{T}\int_0^T X_t\,\mathrm{d}t \longrightarrow \vartheta.$$

Its asymptotic normality follows from the representation

$$\sqrt{T}\left(\bar\vartheta_T - \vartheta\right) = -\frac{2}{\sigma\sqrt{T}}\int_0^T\int_{-\infty}^{X_t}\frac{(x-\vartheta)\; f\left(\vartheta, x\right)}{f\left(\vartheta, X_t\right)}\mathrm{d}x\;\mathrm{d}W_t + o(1), \qquad (1.137)$$

and the CLT:

$$\mathcal{L}\left\{\sqrt{T}\left(\bar\vartheta_T - \vartheta\right)\right\} \Longrightarrow \mathcal{N}\left(0, d\left(\vartheta\right)^2\right). \qquad (1.138)$$

Here $f\left(\vartheta, \cdot\right)$ is the invariant density (1.89) of the process and

$$d\left(\vartheta\right)^2 = \frac{4}{\sigma^2}\;\mathbf{E}_\vartheta\left(\frac{g\left(\xi\right) - \vartheta\; F\left(\vartheta, \xi\right)}{f\left(\vartheta, \xi\right)}\right)^2$$

with

$$g(x) = \mathbf{E}_\vartheta\left(\xi\chi_{\{\xi<x\}}\right).$$

Example 1.49. (*Delay estimation*) If the process (1.90) is observed we can take $q\left(x\right) = x^2$ and to define the EMM as

$$\bar\tau_T = \arg\inf_{\tau\in\Theta}\left|d\left(\tau\right)^2 - \frac{1}{T}\int_0^T X_t^2\;\mathrm{d}t\right|,$$

where for the function $d\left(\tau\right)^2$ see (1.92).

Example 1.50. (*Discontinuity estimation*) For the process (1.93) we have $\mathbf{E}_\vartheta\,\xi = \vartheta$. Hence the EMM can be

$$\bar\vartheta_T = \frac{1}{T}\int_0^T X_t\,\mathrm{d}t \longrightarrow \vartheta$$

and this estimator, like EMM in Example 1.29, is asymptotically normal (see (1.137) and (1.138))

$$\mathcal{L}_\vartheta\left\{\sqrt{T}\left(\bar\vartheta_T - \vartheta\right)\right\} \Longrightarrow \mathcal{N}\left(0, \frac{5}{4\sigma^2}\right).$$

Surely, its calculation is much simpler than that of the MLE, but its rate of convergence will be $\sqrt{T}$ only and the MLE has the better rate (see section 3.4).

1.3.2 Nonparametric Estimation

Suppose that the observed stochastic process is

$$\mathrm{d}X_t = S(X_t)\,\mathrm{d}t + \sigma(X_t)\,\mathrm{d}W_t, \quad X_0, \quad 0 \le t \le T, \tag{1.139}$$

where the trend coefficient $S(\cdot)$ is an *unknown function*. As before we suppose that the functions $S(\cdot)$ and $\sigma(\cdot)^2$ are such that the conditions $\mathcal{ES}$ and $\mathcal{EM}$ and the conditions $\mathcal{RP}$:

$$V(S,x) = \int_0^x \exp\left\{-2\int_0^y \frac{S(z)}{\sigma(z)^2}\,\mathrm{d}z\right\}\mathrm{d}y \longrightarrow \pm\infty \quad \text{as} \quad x \to \pm\infty, \tag{1.140}$$

$$G(S) = \int_{\mathscr{R}} \sigma(y)^{-2}\exp\left\{2\int_0^y \frac{S(z)}{\sigma(z)^2}\,\mathrm{d}z\right\}\mathrm{d}y < \infty \tag{1.141}$$

are fulfilled.

The class of such functions $S(\cdot)$ (for a fixed function $\sigma(\cdot)$) we denote as

$$\mathscr{S}_\sigma = \left\{S(\cdot)\ :\ \text{conditions } \mathcal{ES},\ \mathcal{EM},\ \mathcal{RP} \text{ are fulfilled}\right\}. \tag{1.142}$$

Remember that if $\sigma(x) \equiv 1$, then (1.54) is a simply verified sufficient condition for (1.140) and (1.141).

For all $S(\cdot) \in \mathscr{S}_\sigma$ the process (1.139) is ergodic with the invariant distribution function

$$F_S(x) = \frac{1}{G(S)}\int_{-\infty}^x \sigma(y)^{-2}\exp\left\{2\int_0^y \frac{S(v)}{\sigma(v)^2}\,\mathrm{d}v\right\}\mathrm{d}y$$

and its density function is

$$f_S(x) = \frac{1}{G(S)\,\sigma(x)^2} \exp\left\{2\int_0^x \frac{S(v)}{\sigma(v)^2}\,dv\right\}.$$

We suppose for simplicity that the initial value X_0 is a *stationary random variable*, i.e., it has a density function $f_S(\cdot)$, so the process (1.139) is stationary and consider the following three problems: estimation of the invariant distribution function $F_S(x)$, estimation of the density $f_S(x)$ and estimation of the trend coefficient $S(x)$.

Invariant Distribution Function Estimation

Remember that in the case of independent identically distributed observations $X^{(n)} = \{X_1, \ldots, X_n\}$ the empirical distribution function

$$\hat{F}_n(x) = \frac{1}{n}\sum_{j=1}^{n} \chi_{\{X_j<x\}}$$

is uniformly consistent (the Glivenko–Cantelli theorem): for any $\varepsilon > 0$

$$\mathbf{P}\left\{\lim_{n\to\infty}\sup_x \left|\hat{F}_n(x) - F(x)\right| > \varepsilon\right\} = 0,$$

asymptotically normal

$$\mathcal{L}\left\{\sqrt{n}\left(\hat{F}_n(x) - F(x)\right)\right\} \Longrightarrow \mathcal{N}\left(0, \mathrm{I}_F^{-1}\right), \qquad \mathrm{I}_F^{-1} = F(x)\ (1 - F(x)),$$

and is asymptotically efficient (see, e.g., Dvoretsky *et al.* [66], Levit [171], Millar [183]). The problem of the estimation of $F_S(x)$ is quite close to the problem of distribution function estimation in the i.i.d. case. In particular, we have

Proposition 1.51. *Let $S(\cdot) \in \mathscr{S}_\sigma$, then the empirical distribution function*

$$\hat{F}_T(x) = \frac{1}{T}\int_0^T \chi_{\{X_t<x\}}\,dt$$

is unbiased and consistent. Moreover, if in addition

$$d_F(S,x)^2 = 4\ \mathbf{E}_S\left(\frac{F_S(\xi\wedge x) - F_S(\xi)\,F_S(x)}{\sigma(\xi)\ f_S(\xi)}\right)^2 < \infty, \tag{1.143}$$

then it is asymptotically normal too

$$\mathcal{L}_S\left\{\sqrt{T}\left(\hat{F}_T(x) - F_S(x)\right)\right\} \Longrightarrow \mathcal{N}\left(0, d_F(S,x)^2\right). \tag{1.144}$$

Proof. The consistency follows from the law of large numbers. As we supposed that the initial value X_0 is a *stationary random variable*, we have $\mathbf{E}_S \chi_{\{X_t<x\}} = F_S(x)$.

Remember that (see (1.130))

$$\sqrt{T}\left(\hat{F}_T(x) - F_S(x)\right) = \frac{2}{\sqrt{T}} \int_{X_0}^{X_T} \frac{F_S(v \wedge x) - F_S(v) F_S(x)}{\sigma(v)^2 f_S(v)} \, dv$$
$$- \frac{2}{\sqrt{T}} \int_0^T \frac{F_S(X_t \wedge x) - F_S(X_t) F_S(x)}{\sigma(X_t) \, f_S(X_t)} \, dW_t.$$

Further, the function

$$H_S(x,y) = 2 \int_0^y \frac{F_S(v \wedge x) - F_S(v) F_S(x)}{\sigma(v)^2 f_S(v)} \, dv$$

is continuous. Hence for the first integral we have

$$\mathbf{P}_S^{(T)}\left\{|H_S(x, X_T)| > \delta\sqrt{T}\right\} = \mathbf{P}_S\left\{|H_x(\xi)| > \delta\sqrt{T}\right\} \longrightarrow 0$$

as $T \to \infty$ for any $\delta > 0$. The second integral by condition (1.143) and CLT is asymptotically normal and this provides (1.144).

Note that the well-known proof of the Glivenko–Cantelli theorem (see, e.g., [33], p. 5) can be applied here too.

Theorem 1.52. (Glivenko–Cantelli) *Let $S(\cdot) \in \mathscr{S}_\sigma$, then*

$$\mathbf{P}_S\left\{\lim_{T\to\infty} \sup_x \left|\hat{F}_T(x) - F_S(x)\right| = 0\right\} = 1. \qquad (1.145)$$

Proof. Let us introduce the subdivision $x_0 = -\infty, x_1, \ldots, x_{n-1}, x_n = \infty$ of the real line , such that $F_S(x_k) = \varepsilon\, k$, where ε is small and $n = \frac{1}{\varepsilon}$ is an integer. Denote by $\mathbb{A}_k$ the event $\left\{\omega : \hat{F}_T(x_k) \to F_S(x_k)\right\}$, then by LLN $\mathbf{P}_S\{\mathbb{A}_k\} = 1$. Hence, for $\omega \in \mathbb{A} = \cap \mathbb{A}_k$ there exists a $T(\omega, \varepsilon)$ such that for all $T \geq T(\omega, \varepsilon)$

$$\left|\hat{F}_T(x_k) - F_S(x_k)\right| \leq \varepsilon. \qquad (1.146)$$

As the functions $\hat{F}_T(\cdot)$ and $F_S(\cdot)$ are continuous nondecreasing we have for all $x \in [x_k, x_{k+1})$

$$\hat{F}_T(x) - F_S(x) \leq \hat{F}_T(x_{k+1}) - F_S(x_k),$$
$$\hat{F}_T(x) - F_S(x) \geq \hat{F}_T(x_k) - F_S(x_{k+1}).$$

These estimates together with (1.146) imply that

$$\sup_x \left| \hat{F}_T(x) - F_S(x) \right| \leq 2\varepsilon.$$

Thus, this inequality holds for an arbitrary $\varepsilon > 0$, for all $\omega \in \mathbb{A}$, and for all $T \geq T(\omega, \varepsilon)$. Since $\mathbf{P}_S\{\mathbb{A}\} = 1$, the theorem is proved.

Using the Itô formula we construct a class of unbiased estimators of the invariant distribution function with the same asymptotic variance as follows. Let $R_x(\cdot)$ and $N_x(\cdot)$ be such functions that

$$\mathbf{E}_S\big(R_x(\xi)\; S(\xi) + N_x(\xi)\big) = F_S(x)$$

and $\mathbf{E}_S\left(R_x(\xi)\sigma(\xi)\right)^2 < \infty$. Then

$$\tilde{F}_T(x) = \frac{1}{T}\int_0^T R_x(X_t)\,\mathrm{d}X_t + \frac{1}{T}\int_0^T N_x(X_t)\,\mathrm{d}t$$

will be an *unbiased estimator of the distribution function* , i.e., $\mathbf{E}_S\,\tilde{F}_T(x) = F_S(x)$ for all $S(\cdot) \in \mathscr{S}_\sigma$ and (by the law of large numbers) consistent:

$$\mathbf{P}_S\left\{\lim_{T\to\infty} \tilde{F}_T(x) = F_S(x)\right\} = 1.$$

The function $S(\cdot)$ is supposed to be unknown. Nevertheless such functions $R_x(\cdot)$ and $N_x(\cdot)$ can be constructed as follows. Let $h(\cdot)$ be a positive continuously differentiable function, then the integral

$$K_x(y) = \int_y^x \frac{\mathrm{d}v}{\sigma(v)^2\; h(v)}$$

is well defined. Put

$$R_x(y) = 2\;\chi_{\{y<x\}}\, K_x(y)\; h(y), \qquad N_x(y) = \chi_{\{y<x\}}\, K_x(y)\; h'(y)\; \sigma(y)^2,$$

and introduce the conditions

$$\mathbf{E}_S\left(R_x(\xi)\;\sigma(\xi)\right)^2 < \infty, \quad \mathbf{E}_S\left|N_x(\xi)\right| < \infty, \quad \lim_{y\to-\infty} R_x(y)\;\sigma(y)^2\; f_S(y) = 0. \tag{1.147}$$

Then we have

Proposition 1.53. *Let $S(\cdot) \in \mathscr{S}_\sigma$, the function $h(y) \neq 0$, $y \in \mathscr{R}$, $h(\cdot) \in \mathscr{C}'(\mathscr{R})$ and the conditions (1.143) and (1.147) be fulfilled. Then the estimator*

$$\tilde{F}_T(x) = \frac{2}{T}\int_0^T \chi_{\{X_t<x\}}\, h(X_t)\; K_x(X_t)\; \mathrm{d}X_t + \frac{1}{T}\int_0^T \chi_{\{X_t<x\}}\, h'(X_t)\; K_x(X_t)\; \sigma(X_t)^2\; \mathrm{d}t \tag{1.148}$$

is unbiased, consistent and asymptotically normal

$$\mathcal{L}_S\left\{\sqrt{T}\left(\tilde{F}_T(x) - F_S(x)\right)\right\} \Longrightarrow \mathcal{N}\left(0, d_F(S,x)^2\right),$$

where $d_F(S,x)^2$ *is defined in (1.143).*

Proof. Below we integrate by parts and use the condition (1.147).

$$\begin{aligned}\mathbf{E}_S\,\tilde{F}_T(x) &= 2\;\mathbf{E}_S\,\chi_{\{\xi<x\}}\,K_x(\xi)\;h(\xi)\;S(\xi) + \mathbf{E}_S\,\chi_{\{\xi<x\}}\,K_x(\xi)\;h'(\xi)\;\sigma(\xi)^2\\ &= K_x(x)\;h(x)\;\sigma(x)^2\;f_S(x) - \int_{-\infty}^{x}\sigma(y)^2\;\left[K_x(y)\;h(y)\right]'\;f_S(y)\,\mathrm{d}y\\ &\quad+\mathbf{E}_S\,\chi_{\{\xi<x\}}\,K_x(\xi)\;h'(\xi)\;\sigma(\xi)^2 = F_S(x).\end{aligned}$$

Hence the estimator $\tilde{F}_T(x)$ is unbiased. The consistency follows immediately from the law of large numbers because

$$\begin{aligned}\tilde{F}_T(x) &= \frac{2}{T}\int_0^T \chi_{\{X_t<x\}}\,h(X_t)\;K_x(X_t)\;\sigma(X_t)\;\mathrm{d}W_t\\ &\quad+\frac{1}{T}\int_0^T \chi_{\{X_t<x\}}\,K_x(X_t)\left[2\;h(X_t)\;S(X_t) + h'(X_t)\;\sigma(X_t)^2\right]\;\mathrm{d}t,\end{aligned}$$

where the first term tends to zero and the second converges to its mean value equal to $F_S(x)$.

To show the asymptotic normality we note that

$$\begin{aligned}&\int_{-\infty}^{y}\chi_{\{v<x\}}\,K_x(v)\left[2\;h(v)\;S(v) + h'(v)\;\sigma(v)^2\right]\;f_S(v)\,\mathrm{d}v\\ &\qquad= K_x(y\wedge x)\,h(y\wedge x)\,\sigma(y\wedge x)^2\,f_S(y\wedge x) + F_S(y\wedge x)\end{aligned}$$

and

$$\begin{aligned}&\int_0^T \frac{1}{\sigma(X_t)\,f_S(X_t)}\left[K_x(X_t\wedge x)\,h(X_t\wedge x)\,\sigma(X_t\wedge x)^2\,f_S(X_t\wedge x)\right.\\ &\qquad\left.+F_S(X_t\wedge x) - F_S(x)F_S(X_t)\right]\;\mathrm{d}W_t\\ &\quad= \int_0^T \frac{F_S(X_t\wedge x) - F_S(X_t)F(x)}{\sigma(X_t)\;f_S(X_t)}\;\mathrm{d}W_t\\ &\qquad+\int_0^T \chi_{\{X_t<x\}}\,K_x(X_t)\;h(X_t)\;\sigma(X_t)\,\mathrm{d}W_t.\end{aligned}$$

Therefore by the Itô formula we can write

$$\begin{aligned}\sqrt{T}\left(\tilde{F}_T(x) - F_S(x)\right) &= \frac{M_S(x,X_T) - M_S(x,X_0)}{\sqrt{T}}\\ &\quad-\frac{2}{\sqrt{T}}\int_0^T \frac{F_S(X_t\wedge x) - F_S(X_t)\;F_S(x)}{\sigma(X_t)\;f(X_t)}\;\mathrm{d}W_t,\end{aligned}$$

where

$$M_S(x,y) = 2\,\chi_{\{y<x\}} \int_x^y K_x(v)\,h(v)\,dv$$
$$+2\int_x^y \frac{F_S(v\wedge x) - F_S(v)\,F_S(x)}{\sigma(v)^2\,f_S(v)}\,dv.$$

The stochastic integral is asymptotically normal

$$\frac{2}{\sqrt{T}}\int_0^T \frac{F_S(X_t\wedge x) - F_S(X_t)F_S(x)}{\sigma(X_t)\,f_S(X_t)}\,dW_t \Longrightarrow \mathcal{N}\left(0, d_F(S,x)^2\right).$$

The first term tends to zero because $M_S(x,\cdot)$ is a continuous function and for any $\delta > 0$

$$\mathbf{P}_S^{(T)}\left\{|M_S(x,X_T)| > \delta\sqrt{T}\right\} = \mathbf{P}_S\left\{|M_S(x,\xi)| > \delta\sqrt{T}\right\} \longrightarrow 0$$

as $T \to \infty$.

Example 1.54. Let $\sigma(y) \equiv 1$, $h(y) = 1 + y^2$ and the conditions (1.147) and (1.143) be fulfilled. Then the estimator

$$\tilde{F}_T(x) = \frac{2}{T}\int_0^T \chi_{\{X_t<x\}}\,(\text{arctg}\,x - \text{arctg}\,X_t)\left[(1+X_t^2)\,dX_t + X_t\,dt\right]$$

according to Proposition 1.53 is unbiased and asymptotically normal with the same limit variance as EDF (under additional conditions).

Invariant Density Estimation

Consider the problem of invariant density $f_S(x)$ estimation by the observations X^T of the stationary diffusion process (1.139) in the same situation as above, i.e., the trend coefficient $S(\cdot)$ is an unknown function from the set $\mathscr{S}_\sigma$, etc. First note that the density estimation problems in the i.i.d. case and in diffusion processes case are entirely different. Remember that in the case of discrete-time i.i.d. models $X_1, \ldots, X_n$ the rate of convergence of any estimator $\bar{f}_n(x)$ of the density function $f(x)$ is usually worse than $n^{1/2}$. In particular, suppose that the density function $f(x)$ of X_j has k continuous derivatives and define the *kernel-type estimator*

$$\hat{f}_n(x) = \frac{1}{n\varphi_n}\sum_{j=1}^n K\left(\frac{X_j - x}{\varphi_n}\right) = \frac{1}{\varphi_n}\int_{-\infty}^{\infty} K\left(\frac{y-x}{\varphi_n}\right) d\hat{F}_n(y) \qquad (1.149)$$

with $\varphi_n = n^{-1/2k+1}$. The kernel $K(\cdot)$ has compact support (supp $K = [A, B]$) and satisfies the conditions:

$$\int_A^B K(u)\,du = 1, \qquad \int_A^B u^j\,K(u)\,du = 0, \quad j = 1, \ldots, k. \tag{1.150}$$

Then the estimator (1.149) is consistent and has the rate of convergence $n^{k/(2k+1)}$, i.e.,

$$\lim_{n\to\infty} \mathbf{E}\left|n^{\frac{k}{2k+1}}\left(\hat{f}_n(x) - f(x)\right)\right|^p < C.$$

Moreover this rate cannot be improved (see, e.g., [109], Theorem IV.5.1). Therefore this estimator is asymptotically efficient (in the rate of convergence).

The rate $n^{1/2}$ in this problem can be attained if the support of the characteristic function is a compact set, hence $f(\cdot)$ is infinitely differentiable (see Watson and Leadbetter [239] and Ibragimov and Khasminskii [110]). It was natural to think that in continuous-time models we will have similar properties of density estimators, but this is not the case and for ergodic diffusion processes it is possible to have many estimators of the density which converge to the true density with the *parametric rate* $\sqrt{T}$.

Let us introduce the *kernel-type estimator*:

$$\hat{f}_T(x) = \frac{1}{T\,\varphi_T}\int_0^T K\left(\frac{X_t - x}{\varphi_T}\right)\,dt = \frac{1}{\varphi_T}\int_{-\infty}^{\infty} K\left(\frac{y-x}{\varphi_T}\right)\,d\hat{F}_T(y), \tag{1.151}$$

where $\hat{F}_T(\cdot)$ is the empirical distribution function and the kernel $K(\cdot)$ is a nonnegative function with compact support $[A, B]$ satisfying the usual conditions:

$$\int_A^B K(u)\,du = 1, \qquad \int_A^B u\,K(u)\,du = 0. \tag{1.152}$$

This estimator is similar to the kernel-type estimator (1.149) (where the sum is replaced by the integral) with only one essential difference: we can take the normalizing function $\varphi_T = T^{-1/2}$.

The estimator (1.151) can be written in other equivalent forms :

$$\begin{aligned}\hat{f}_T(x) &= \frac{1}{\varphi_T}\int_{-\infty}^{\infty} K\left(\frac{y-x}{\varphi_T}\right)\,d\hat{F}_T(y) = \frac{2}{T}\int_{-\infty}^{\infty} K\left(\frac{y-x}{\varphi_T}\right)\,\frac{\Lambda_T(y)}{\sigma(y)^2}\,dy \\ &= \frac{1}{T}\int_{-\infty}^{\infty} K\left(\frac{y-x}{\varphi_T}\right)\,f_T^\circ(y)\,dy, \end{aligned}\tag{1.153}$$

where $\Lambda_T(\cdot)$ is the local time of the diffusion process (1.25), $f_T^\circ(\cdot)$ is the empirical density (1.124) and the first integral is understood as a Lebesgue–Stieltjes integral with respect to a continuous, monotonically increasing function $\hat{F}_T(\cdot)$. Recall that Nguyen and Pham [190] proposed to use the local time in the study of kernel type estimators. This equality shows the main advantage of the diffusion process model: the EDF $\hat{F}_T(\cdot)$ is absolutely continuous w.r.t. the Lebesgue measure and its derivative $f_T^\circ(\cdot)$ (with probability 1) can be used as a density estimator.

Note that if the function $\sigma(\cdot)$ is continuous at the point x (and this is our case), then this estimator is asymptotically unbiased:

$$\begin{aligned}\left|\mathbf{E}_S \hat{f}_T(x) - f_S(x)\right| &= \varphi_T^{-1}\left|\int K\left(\frac{y-x}{\varphi_T}\right) f_S(y)\,\mathrm{d}y - f_S(x)\right| \\ &\leq \int_A^B K(u)\left|f_S(x+u\varphi_T) - f_S(x)\right|\mathrm{d}u \\ &\leq \sup_{|h|\leq c\varphi_T}\left|f_S(x+h) - f_S(x)\right| \to 0 \qquad (1.154)\end{aligned}$$

as $T \to \infty$ because the function $f_S(\cdot)$ is continuous at the point x. Using Bochner's lemma it is easy to show that this result holds true for the wider class of kernels.

Lemma 1.55. (Bochner) *Let $K(\cdot)$ be a Borel function such that*

$$\int |K(y)|\,\mathrm{d}y < \infty, \qquad \sup_y |K(y)| < \infty, \qquad \lim_{y\to\pm\infty} y\,K(y) = 0.$$

Then for any function $g(\cdot) \in \mathscr{L}_1(\mathscr{R})$ we have

$$\lim_{h\to 0}\frac{1}{h}\int_{-\infty}^{\infty} g(y)\,K\left(\frac{y-x}{h}\right)\mathrm{d}y = g(x)\int_{-\infty}^{\infty} K(y)\,\mathrm{d}y.$$

Proof. This can be found in [31].

To describe the asymptotic properties of the kernel-type estimators we first remember the result by Castellana and Leadbetter [45] and then show that for ergodic diffusion processes the proof can be simplified.

Below $\{X_t,\ t \geq 0\}$ is some stationary stochastic process with invariant density $f(\cdot)$ and two-dimensional density $f(\tau,\cdot,\cdot)$, i.e.,

$$f(\tau, y, z) = \frac{\partial^2 \mathbf{P}\{X_t < y,\ X_{t+\tau} < z\}}{\partial y\,\partial z}.$$

Introduce the following condition.

$\mathcal{CL}$ (Castellana and Leadbetter). *The functions $f(y), f(\tau, y, z), \tau > 0$ are continuous at point x and*

$$|f(\tau, y, z) - f(y)f(z)| \leq \psi(\tau) \in \mathscr{L}_1(\mathscr{R}_+). \qquad (1.155)$$

Put

$$A(x) = 2\int_0^{\infty}\left[f(\tau, x, x) - f(x)^2\right]\mathrm{d}\tau.$$

The properties of the variance of the estimator (1.151) are described by the following proposition.

Proposition 1.56. ([45]) *Let the condition $\mathcal{CL}$ be fulfilled, then*

$$\lim_{T\to\infty} T\,\mathbf{E}\left(\hat f_T(x)-\mathbf{E}\,\hat f_T(x)\right)^2 = A(x). \tag{1.156}$$

Proof. We have

$$\begin{aligned}
T\mathbf{E}\left(\hat f_T(x)-\mathbf{E}\hat f_T(x)\right)^2 &= \varphi_T^2\int_0^T\int_0^T\left[\mathbf{E}\,K\left(\frac{X_t-x}{\varphi_T}\right)K\left(\frac{X_S-x}{\varphi_T}\right)\right.\\
&\quad\left.-\,\mathbf{E}\,K\left(\frac{X_t-x}{\varphi_T}\right)\mathbf{E}\,K\left(\frac{X_S-x}{\varphi_T}\right)\right]\,\mathrm{d}s\,\mathrm{d}t\\
&=2\int_A^B\int_A^B K(u)\,K(v)\int_0^T\left(1-\frac{\tau}{T}\right)\left[f(\tau,x+u\varphi_T,x+v\varphi_T)\right.\\
&\quad\left.-f(x+u\varphi_T)f(x+v\varphi_T)\right]\,\mathrm{d}\tau\,\mathrm{d}u\,\mathrm{d}v\\
&=2\int_0^\infty\left[f(\tau,x,x)-f(x)^2\right]d\tau\,\left(1+o(1)\right)=A(x)(1+o(1))
\end{aligned}$$

because the function $\Psi(\tau,y,z)=f(\tau,y,z)-f(y)f(z),\quad \tau>0$ is continuous at the point (x,x) and is majorized by the function $\psi(\tau)\in\mathscr{L}_1(0,\infty)$.

Moreover, it can be shown ([45], Theorem 5.5) that these estimators are asymptotically normal

$$\mathcal{L}\left\{T^{1/2}\left(\hat f_T(x)-\mathbf{E}\,\hat f_T(x)\right)\right\}\Longrightarrow\mathcal{N}\left(0,A(x)\right). \tag{1.157}$$

If we suppose that the derivative $f'(x)$ is a continuous function at the point x then we can calculate the limit of the mean-square error as well:

$$\begin{aligned}
&\lim_{T\to\infty}\mathbf{E}\left(\sqrt{T}(\hat f_T(x)-f(x))\right)^2\\
&=\lim_{T\to\infty}\left\{\mathbf{E}\left(\sqrt{T}(\hat f_T(x)-\mathbf{E}\,\hat f(x))\right)^2+\left(\sqrt{T}(\mathbf{E}\,\hat f_T(x)-f(x))\right)^2\right\}\\
&=2\int_0^\infty\left[f(\tau,x,x)-f(x)^2\right]d\tau\,+f'(x)^2\left(\int_A^B u\,K(u)\,du\right)^2\\
&=2\int_0^\infty\left[f(\tau,x,x)-f(x)^2\right]d\tau.
\end{aligned}$$

Therefore for the continuous-time models satisfying the condition $\mathcal{CL}$ it is possible to have a $T^{1/2}$ rate under less restrictive smoothness conditions. This effect was explained by the following way "... *the sampling collects a whole continuum of 'somewhat independent' random variables*" and "... *the irregular nature of the paths corresponds to less correlation and hence 'more information' in the measurement of X_t leading to the maximal rate of convergence of the variance to zero*" (see Castellana and Leadbetter [45], p. 190).

This result was further generalized in several directions by Leblanc [158], Bosq [36], Bosq and Davydov [37] and Sköld and Hössjer [218]. If the trajectories $X_t, t \geq 0$ are differentiable with respect to t and the class of processes contains the Gaussian processes, then the rate of convergence is less than $T^{1/2}$ [36], [178].

To study the properties of kernel-type estimators for ergodic diffusion processes and to verify (1.156) we can either check the condition (1.155) or use (1.153) and the properties of the local time.

Note that condition (1.155) corresponds to two different conditions: the first one is *the integrability of the function* $f_S(\tau, y, z)$ *in the vicinity of zero* (remember that $f_S(\tau, y, z) \to \delta(y - z)$ ($\delta(\cdot)$ is a delta function) as $\tau \to 0$), say,

$$\int_0^1 \sup_{y,z} \left| f_S(\tau, y, z) - f_S(y) f_S(z) \right| \, \mathrm{d}\tau < \infty, \tag{1.158}$$

and the second is *the good decrease to zero at infinity* of the function $\psi(\cdot)$ (*the good mixing* of the observed process)

$$\int_1^\infty \sup_{y,z} \left| f_S(\tau, y, z) - f_S(y) f_S(z) \right| \, \mathrm{d}\tau < \infty. \tag{1.159}$$

The first condition (1.158) follows from the estimates

$$f_S(\tau, y, z) \leq C\, \tau^{-1/2}, \qquad 0 < \tau \leq 1, \qquad \sup_x f_S(x) < \infty$$

and these estimates hold if, for example, $\sigma(\cdot)$ *is non-degenerate and* $\sigma(\cdot)$ *and* $S(\cdot)$ *are bounded of class* $\mathscr{C}^\alpha$*-Hölder continuous.*

The second condition (1.159) holds if

$$\inf_y \sigma(y)^2 > 0, \qquad \overline{\lim_{|x| \to \infty}} \frac{x\, S(x)}{\sigma(x)^2} < -\frac{3}{2}$$

and

$$\sigma(\cdot), \ S(\cdot) \in \mathscr{C}^m$$

for any $m > 0$, (see Veretennikov [237] for a proof and detailed discussion of these conditions).

These conditions seem to be quite restrictive. Hence we follow the second approach based on the representation (1.153). As estimators we will take the *local time estimators*, a class of unbiased estimators and the traditional *kernel-type estimators.* We show that all these estimators are $\sqrt{T}$-consistent and asymptotically normal with the same limit variance [142].

Local-time Estimator

Note that the *local time estimator* (LTE) (1.124)

$$f_T^\circ(x) = \frac{2\Lambda_T(x)}{\sigma(x)^2\, T}$$

is a nonnegative function and $f_T^\circ(x) = 0$ for $x > \max_{0\leq t\leq T} X_t$ and $x < \min_{0\leq t\leq T} X_t$. We have as well

$$\int_{\mathscr{R}} f_T^\circ(x)\ \mathrm{d}x = 1,$$

i.e., the LTE is a density function because it is the derivative of the EDF.

The asymptotic properties of LTE were already described in Proposition 1.25 and we just remember them here.

Proposition 1.57. *Let $S(\cdot) \in \mathscr{S}_\sigma$, $\mathbf{E}_S\, \sigma(\xi)^2 < \infty$ and*

$$d_f(S,x)^2 = 4\, f_S(x)^2\, \mathbf{E}_S \left(\frac{\chi_{\{\xi>x\}} - F_S(\xi)}{\sigma(\xi)\ f_S(\xi)}\right)^2 < \infty. \tag{1.160}$$

Then the LTE is an unbiased, consistent and asymptotically normal

$$\mathcal{L}_S\left\{\sqrt{T}\left(f_T^\circ(x) - f_S(x)\right)\right\} \Longrightarrow \mathcal{N}\left(0, d_f(S,x)^2\right) \tag{1.161}$$

estimator of the density.

Proof. This follows immediately from Proposition 1.25.

Unbiased Estimators

Note that the invariant density is a solution of the integral equation

$$f_S(x) = \frac{2}{\sigma(x)^2}\int_{-\infty}^{x} \frac{S(y)}{\sigma(y)^2}\, f_S(y)\ \mathrm{d}y.$$

Hence we can construct the estimator based on this representation as

$$\bar{f}_T(x) = \frac{2}{\sigma(x)^2}\int_0^T \frac{\chi_{\{X_t<x\}}}{\sigma(X_t)^2}\,\mathrm{d}X_t.$$

Obviously

$$\mathbf{E}_S\, \bar{f}_T(x) = \frac{2}{\sigma(x)^2}\int_{-\infty}^{x} \frac{S(y)}{\sigma(y)^2}\, f_S(y)\ \mathrm{d}y = f_S(x)\,.$$

Of course, this is not the only unbiased estimator.

Using the same idea as in the construction of the unbiased distribution function estimators (1.148) we now construct the class of unbiased density estimators. Hence we have to find two functions $R_x(\cdot)$ and $N_x(\cdot)$ such that

$$\mathbf{E}_S\left(R_x\left(\xi\right)\ S\left(\xi\right)+N_x\left(\xi\right)\right)=f_S\left(x\right).$$

Then the estimator

$$\tilde{f}_T\left(x\right)=\frac{1}{T}\int_0^T R_x\left(X_t\right)\,\mathrm{d}X_t+\frac{1}{T}\int_0^T N_x\left(X_t\right)\,\mathrm{d}t \tag{1.162}$$

will be the *unbiased estimator of the density.*

Let $h(\cdot)$ be a continuously differentiable function. Introduce the functions

$$R_x(y)=\frac{2\,\chi_{\{y\leq x\}}\,h(y)}{\sigma(x)^2\,h(x)},\qquad N_x(y)=\frac{\chi_{\{y\leq x\}}\,h'(y)\,\sigma(y)^2}{\sigma(x)^2\,h(x)}$$

and the conditions

$$\mathbf{E}_S\left(R_x\left(\xi\right)\sigma\left(\xi\right)\right)^2<\infty,\quad \mathbf{E}_S\left|N_x\left(\xi\right)\right|<\infty,\quad \lim_{y\rightarrow-\infty}h(y)\sigma(y)^2 f_S(y)=0. \tag{1.163}$$

Then we have

Proposition 1.58. *Let $S(\cdot)\in\mathscr{S}_\sigma$, the function $h(\cdot)\in\mathscr{C}^1(\mathscr{R})$, $h(x)\neq 0$ and the conditions (1.160) and (1.163) be fulfilled. Then the estimator*

$$\tilde{f}_T(x)=\frac{2}{T}\int_0^T\frac{\chi_{\{X_t\leq x\}}\,h(X_t)}{\sigma(x)^2\,h(x)}\,\mathrm{d}X_t+\frac{1}{T}\int_0^T\frac{\chi_{\{X_t\leq x\}}\,h'(X_t)\,\sigma(X_t)^2}{\sigma(x)^2\,h(x)}\,\mathrm{d}t \tag{1.164}$$

is unbiased, consistent and asymptotically normal:

$$\mathcal{L}_S\left\{\sqrt{T}\left(\tilde{f}_T(x)-f_S(x)\right)\right\}\Longrightarrow\mathcal{N}\left(0,d_f(S,x)^2\right). \tag{1.165}$$

Proof. The integration by parts gives us the equality

$$\mathbf{E}_S\,\tilde{f}_T(x)=f_S(x)$$

and the law of large numbers provides the consistency of the estimator.

Further, the estimator $\tilde{f}_T(x)$ admits the representation

$$\begin{aligned}\sqrt{T}\left(\tilde{f}_T(x)-f_S(x)\right)&=\frac{1}{\sqrt{T}\sigma(x)^2\,h(x)}\left\{\int_0^T 2\,\chi_{\{X_t\leq x\}}\,h(X_t)\sigma(X_t)\,\mathrm{d}W_t\right.\\&+\int_0^T\Big[2\,\chi_{\{X_t\leq x\}}\,h(X_t)S(X_t)\\&\left.\quad+\chi_{\{X_t\leq x\}}\,h'(X_t)\sigma(X_t)^2-f_S(x)\sigma(x)^2\,h(x)\Big]\,\mathrm{d}t\right\}.\end{aligned}$$

Let us denote

$$H_S(x,y) = \frac{2\,\chi_{\{y\le x\}}\, h(y)\, S(y) + \chi_{\{y\le x\}}\, h'(y)\,\sigma(y)^2}{\sigma(x)^2\, h(x)} - f_S(x).$$

Then direct calculation yields the equality

$$\int_{-\infty}^{y} H_S(x,v)\, f_S(v)\,\mathrm{d}v = \frac{h(y\wedge x)\,\sigma(y\wedge x)^2\, f_S(y\wedge x)}{\sigma(x)^2\, h(x)} - f_S(x)\, F_S(y).$$

Therefore by Itô formula we have

$$\int_0^T H_S(x,X_t)\,\mathrm{d}t = M_S(x,X_T) - M_S(x,X_0)$$
$$-2\int_0^T \left(\frac{h(X_t\wedge x)\,\sigma(X_t\wedge x)^2\, f_S(X_t\wedge x)}{\sigma(X_t)\, f_S(X_t)\,\sigma(x)^2\, h(x)} - \frac{f_S(x)\, F_S(X_t)}{\sigma(X_t)\, f_S(X_t)}\right)\mathrm{d}W_t,$$

where

$$M_S(x,z)$$
$$= \int_x^z \frac{2}{\sigma(y)^2 f_S(y)}\left(\frac{h(y\wedge x)\,\sigma(y\wedge x)^2\, f_S(y\wedge x)}{\sigma(x)^2\, h(x)} - f_S(x)\, F_S(y)\right)\mathrm{d}y$$
$$= 2f_S(x)\,\chi_{\{z\ge x\}}\int_x^z \frac{1-F_S(y)}{\sigma(y)^2 f_S(y)}\mathrm{d}y +$$
$$+\,2\,\chi_{\{z<x\}}\int_x^z \left[\frac{h(y)}{\sigma(x)^2\, h(x)} - \frac{f_S(x)}{\sigma(y)^2 f_S(y)}F_S(y)\right]\mathrm{d}y. \tag{1.166}$$

Hence the estimator $\tilde{f}_T(x)$ can be written in the following form

$$\sqrt{T}\left(\tilde{f}_T(x) - f_S(x)\right) = \frac{M_S(x,X_T) - M_S(x,X_0)}{\sqrt{T}}$$
$$+\frac{2}{\sqrt{T}}\int_0^T \frac{\chi_{\{X_t\le x\}}\, h(X_t)\,\sigma(X_t)}{\sigma(x)^2\, h(x)}\mathrm{d}W_t$$
$$-\frac{2}{\sqrt{T}}\int_0^T \left(\frac{h(X_t\wedge x)\,\sigma(X_t\wedge x)^2\, f_S(X_t\wedge x)}{\sigma(x)^2\, h(x)\sigma(X_t)\, f_S(X_t)} - \frac{f_S(x)\, F_S(X_t)}{\sigma(X_t)\, f_S(X_t)}\right)\mathrm{d}W_t$$
$$= \frac{M_S(x,X_T) - M_x(X_0)}{\sqrt{T}} - \frac{2\, f_S(x)}{\sqrt{T}}\int_0^T \frac{\chi_{\{X_t\ge x\}} - F_S(X_t)}{\sigma(X_t)\, f_S(X_t)}\mathrm{d}W_t.$$

The function $M_S(x,y)$, $y\in\mathscr{R}$ is continuous, hence

$$\frac{M_S(x,X_T) - M_S(x,X_0)}{\sqrt{T}} \longrightarrow 0$$

and we have the convergence

$$\mathcal{L}_S\left\{\frac{2\, f_S(x)}{\sqrt{T}}\int_0^T \left(\frac{\chi_{\{X_t\ge x\}} - F_S(X_t)}{\sigma(X_t)\, f_S(X_t)}\right)\mathrm{d}W_t\right\} \Longrightarrow \mathcal{N}\left(0, d_f(S,x)^2\right)$$

which provides (1.165).

Example 1.59. If we put $h(y) = 1$, then we obtain the estimator

$$\tilde{f}_T(x) = \frac{2}{T\sigma(x)^2} \int_0^T \chi_{\{X_t \le x\}} \, dX_t \tag{1.167}$$

which is (under conditions (1.160) and (1.163)) unbiased, consistent and asymptotically normal.

Note that the LTE $f_T^\circ(x)$ cannot be represented as (1.164).

Example 1.60. Let $\sigma(y) \equiv 1$ and $h(x) = x^2$, then at any point $x \neq 0$ we have the estimator of the density

$$\tilde{f}_T(x) = \frac{2}{T\,x^2} \int_0^T \chi_{\{X_t \le x\}} \, X_t^2 \, dX_t + \frac{2}{T\,x^2} \int_0^T \chi_{\{X_t \le x\}} \, X_t \, dt,$$

which is consistent and asymptotically normal:

$$\mathcal{L}_S\left\{\sqrt{T}\left(\tilde{f}_T(x) - f_S(x)\right)\right\} \Longrightarrow \mathcal{N}\left(0, d_f(S,x)^2\right).$$

Of course, we need the conditions (1.160) and (1.163).

The same can be said about the estimator

$$\tilde{f}_T(x) = \frac{1}{T\,[2+\sin x]} \left[2\int_0^T \chi_{\{X_t \le x\}} \, [2+\sin(X_t)] \, dX_t + \int_0^T \chi_{\{X_t \le x\}} \, \cos(X_t) \, dt\right].$$

Kernel-type Estimators

We study the kernel-type estimator $\hat{f}_T(x)$ (1.151) with the nonnegative kernel $K(\cdot)$ satisfying (1.152). Below we will show that $A(x) = d_f(S,x)^2$.

Proposition 1.61. *Let $S(\cdot) \in \mathscr{S}_\sigma$, $\mathbf{E}_S\,\sigma(\xi)^2 < \infty$, the function $f_S(\cdot)$ be continuously differentiable at the point x and the condition (1.160) be fulfilled. Then the kernel-type estimator $\hat{f}_T(x)$ is consistent and asymptotically normal:*

$$\mathcal{L}_S\left\{\sqrt{T}\left(\hat{f}_T(x) - f_S(x)\right)\right\} \Longrightarrow \mathcal{N}\left(0, d_f(S,x)^2\right). \tag{1.168}$$

Proof. We have the following representation for the estimator

$$
\begin{aligned}
&\sqrt{T}\left(\hat f_T(x)-f_S(x)\right)\\
&\quad=\sqrt{T}\int_A^B K(u)\,\left[f_S(x-u\varphi_T)-f_S(x)\right]\,\mathrm{d}u+\left[\frac{2\Lambda_T(x)}{T\sigma(x)^2}-f_S(x)\right]\sqrt{T}\\
&\quad+\sqrt{T}\int_A^B K(u)\left[2\frac{\Lambda_T(x+u\varphi_T)}{T\sigma(x+u\varphi_T)^2}-f_S(x+u\varphi_T)\right.\\
&\quad\left.+f_S(x)-2\frac{\Lambda_T(x)}{T\sigma(x)^2}\right]\mathrm{d}u.
\end{aligned}
$$

The first term can be estimated as follows:

$$
\begin{aligned}
\sqrt{T}\left|\int_A^B K(u)\,\left[f_S(x-u\varphi_T)-f_S(x)\right]\,\mathrm{d}u\right|&\\
\le\int_A^B K(u)\,\left|f'(x)-f'(x+\tilde u\varphi_T)\right|\,\mathrm{d}u&\\
\le C\sup_{|h|<(B-A)T^{-1/2}}\left|f'(x)-f'(x+h)\right|\longrightarrow 0&
\end{aligned}
$$

because the derivative $f'(\cdot)$ is continuous at the point x.

The second term is equal to $\sqrt{T}\left(f_T^{\circ}(x)-f_S(x)\right)$ and by Proposition 1.57 is asymptotically normal (it gives the main contribution to the limit variance).

To estimate the last term we denote

$$
\begin{aligned}
W_T(y,x)&=\sqrt{T}\left[2\frac{\Lambda_T(y)}{T\sigma(y)^2}-f_S(y)+f_S(x)-2\frac{\Lambda_T(x)}{T\sigma(x)^2}\right]\\
&=\sqrt{T}\left(f_T^{\circ}(y)-f_T^{\circ}(x)\right)
\end{aligned}
$$

and will show that

$$
\sup_{S(\cdot)\in\Theta_\beta}\int_A^B K(u)^2\,\mathbf{E}_S\left(W_T(x+u\varphi_T,x)\right)^2\mathrm{d}u\longrightarrow 0. \tag{1.169}
$$

Using the Tanaka–Meyer formula (1.24) we can write

$$
\begin{aligned}
W_T(y,x)&=\frac{1}{\sqrt{T}}\int_0^T\left[\frac{\operatorname{sgn}(y-X_t)}{\sigma(y)^2}-\frac{\operatorname{sgn}(x-X_t)}{\sigma(x)^2}\right]\mathrm{d}W_t\\
&\quad+\frac{1}{\sqrt{T}}\int_0^T\left[\frac{\operatorname{sgn}(y-X_t)}{\sigma(y)^2}\,S(X_t)-f_S(y)\right.\\
&\quad\left.-\frac{\operatorname{sgn}(x-X_t)}{\sigma(x)^2}\,S(X_t)+f_S(x)\right]\mathrm{d}t.
\end{aligned}\tag{1.170}
$$

Let us denote

$$h_x(z) = \frac{\operatorname{sgn}(x-z)}{\sigma(x)^2}\, S(z) - f_S(x),$$

$$H_{x,y}(z) = 2f_S(x)\int_0^z \frac{\chi_{\{v\in[x,y]\}}}{\sigma(v)^2 f_S(v)}\mathrm{d}v$$
$$+2\left(f_S(x)-f_S(y)\right)\int_0^z \frac{\chi_{\{v>x\}}-F_S(v)}{\sigma(v)^2 f_S(v)}\mathrm{d}v.$$

Then by the Itô formula we have

$$\frac{1}{\sqrt{T}}\int_0^T \left[h_x(X_t)-h_y(X_t)\right]\,\mathrm{d}t = \frac{H_{x,y}(X_T)-H_{x,y}(X_0)}{\sqrt{T}}$$
$$-\frac{2}{\sqrt{T}}\int_0^T \frac{1}{\sigma(X_t)f_S(X_t)}\int_{-\infty}^{X_t}\left[h_x(v)-h_y(v)\right] f_S(v)\,\mathrm{d}v\,\mathrm{d}W_t.$$

Direct calculation yields

$$\int_{-\infty}^{z}\left[h_x(v)-h_y(v)\right] f_S(v)\,\mathrm{d}v$$
$$= f_S(x)[\chi_{\{z>x\}}-F_S(z)] - f_S(y)[\chi_{\{z>y\}}-F_S(z)]$$
$$+\frac{\sigma(z)^2 f_S(z)}{2}\left(\frac{\operatorname{sgn}(x-z)}{\sigma(x)^2}-\frac{\operatorname{sgn}(y-z)}{\sigma(y)^2}\right)$$
$$= f_S(x)\,\chi_{\{z\in[x,y]\}} + \left(f_S(x)-f_S(y)\right)[\chi_{\{z>y\}}-F_S(z)]$$
$$+\frac{\sigma(z)^2 f_S(z)}{2}\left(\frac{\operatorname{sgn}(x-z)}{\sigma(x)^2}-\frac{\operatorname{sgn}(y-z)}{\sigma(y)^2}\right).$$

The last term will compensate for the last integral in (1.170). Hence

$$\mathbf{E}_S\left|\frac{1}{\sqrt{T}}\int_0^T \frac{1}{\sigma(X_t)f_S(X_t)}\int_{-\infty}^{X_t}\left[h_x(v)-h_y(v)\right] f_S(v)\,\mathrm{d}v\,\mathrm{d}W_t\right|^p$$
$$\le C\,\mathbf{E}_S\left|\frac{1}{T}\int_0^T\left(\frac{f_S(x)\chi_{\{X_t\in[x,y]\}}}{\sigma(X_t)f_S(X_t)}\right.\right.$$
$$\left.\left.+\frac{\left(f_S(x)-f_S(y)\right)[\chi_{\{X_t>y\}}-F_S(X_t)]}{\sigma(X_t)f_S(X_t)}\right)^2\mathrm{d}t\right|^{p/2}$$
$$\le C\,f_S(x)^p\int_x^y \frac{1}{\sigma(v)^p f_S(v)^{p-1}}\mathrm{d}v$$
$$+C\left|f_S(x)-f_S(y)\right|^p\,\mathbf{E}_S\left|\frac{\chi_{\{\xi>y\}}-F_S(\xi)}{\sigma(\xi)f_S(\xi)}\right|^p.$$

We have as well

$$\int_x^y \frac{1}{\sigma(v)^p f_S(v)^{p-1}} \mathrm{d}v = (y-x) \frac{1}{\sigma(x)^p f_S(x)^{p-1}} \left(1 + O(y-x)\right).$$

Furthermore

$$\begin{aligned}\mathbf{E}_S \left|H_{x,y}(\xi)\right|^p \leq 2^{2p-1} f_S(x)^p \mathbf{E}_S \left(\int_0^\xi \frac{\chi_{\{v\in[x,y]\}}}{\sigma(v)^2 f_S(v)} \mathrm{d}v\right)^p \\ + 2^{2p-1} |f_S(x) - f_S(y)|^p \mathbf{E}_S \left|\int_0^\xi \frac{\chi_{\{v>y\}} - F_S(v)}{\sigma(v)^2 f_S(v)}\right|^p\end{aligned}$$

and

$$\mathbf{E}_S \left(\int_0^\xi \frac{\chi_{\{v\in[x,y]\}}}{\sigma(v)^2 f_S(v)} \mathrm{d}v\right)^p \leq \left(\int_x^y \frac{\mathrm{d}z}{\sigma(z)^2 f_S(z)}\right)^p = (y-x)^p \frac{(1+O(y-x))}{\sigma(x)^{2p} f_S(x)^p}.$$

Now the convergence (1.169) follows directly from the inequalities obtained above if we put $y = x + u\varphi_T$.

Remark 1.62. Remember that the only unknown function of the model is the trend coefficient $S(\cdot)$. Hence, we can first estimate $S(\cdot)$ with the help of the estimator $\hat{S}_T(\cdot)$ proposed below (1.172) and then to construct the *plug-in estimator* of the density

$$f_{\hat{S}_T}(x) = \frac{1}{\hat{G}\left(\hat{S}_T\right) \sigma(x)^2} \exp\left\{2 \int_0^x \frac{\hat{S}_T(v)}{\sigma(v)^2} \mathrm{d}v\right\}.$$

Here $\hat{G}\left(\hat{S}_T\right)$ is a specially truncated integral and the estimator $\hat{S}_T(\cdot)$ has to be slightly modified to avoid the zeroes of the local time estimator of the density (see, e.g., (4.136)). It can be shown that this estimator of the density is asymptotically normal as well with the same limit variance $d_f(S,x)^2$.

Remark 1.63. Comparison of the convergence (1.157) and (1.168) gives us the equality

$$2\int_0^\infty \left[f_S(\tau,x,x) - f_S(x)^2\right] \mathrm{d}\tau = 4 f_S(x)^2 \mathbf{E}_S \left(\frac{\chi_{\{\xi>x\}} - F_S(\xi)}{\sigma(\xi) f_S(\xi)}\right)^2$$

which, probably, can be obtained in another (direct) way as well.

Remark 1.64. The limit variances (1.143) and (1.160) of the EDF and LTE can be written as follows:

$$d_F(S,x)^2 = 4\ \mathbf{E}_S\left(\frac{F_S(\xi\wedge x) - F_S(\xi)\,F_S(x)}{\sigma(\xi)\ f_S(\xi)}\right)^2$$
$$= 4\ (1-F_S(x))^2 \int_{-\infty}^{x}\left(\frac{F_S(y)}{\sigma(y)\ f_S(y)}\right)^2 f_S(y)\ \mathrm{d}y$$
$$+4\,F_S(x)^2\int_x^{\infty}\left(\frac{1-F_S(y)}{\sigma(y)\ f_S(y)}\right)^2 f_S(y)\ \mathrm{d}y$$

and

$$d_f(S,x)^2 = 4\ f_S(x)^2\ \mathbf{E}_S\left(\frac{\chi_{\{\xi>x\}} - F_S(\xi)}{\sigma(\xi)\ f_S(\xi)}\right)^2$$
$$= 4\ f_S(x)^2\int_{-\infty}^{x}\left(\frac{F_S(y)}{\sigma(y)\ f_S(y)}\right)^2 f_S(y)\ \mathrm{d}y$$
$$+4\ f_S(x)^2\int_x^{\infty}\left(\frac{1-F_S(y)}{\sigma(y)\ f_S(y)}\right)^2 f_S(y)\ \mathrm{d}y.$$

Therefore these quantities are finite or infinite simultaneously.

It is easy to verify that if $\sigma(x)\equiv 1$, then the condition

$$\overline{\lim_{|x|\to\infty}}\ x\,S(x) < -3/2 \tag{1.171}$$

is sufficient for the finiteness of these quantities. Note as well that (1.171) provides the existence of the second moment of the invariant distribution, i.e., $\mathbf{E}_S\,\xi^2<\infty$. In particular, if $S(x) = -\frac{c}{x}(1+o(1))$ as $|x|\to\infty$, then

$$\frac{1-F_S(y)}{f_S(y)} = y - \int_y^{\infty} v\,\frac{f_S'(v)}{f_S(y)}\ \mathrm{d}v = y\ (1+o(1)),$$

and

$$\int_x^{\infty}\left(\frac{1-F_S(y)}{f_S(y)}\right)^2 f_S(y)\ \mathrm{d}y = \mathbf{E}_S\left(\chi_{\{\xi>x\}}\,\xi^2\right)\ (1+o(1))$$

for the large values of x. Hence $c>3/2$.

Trend Coefficient Estimation

The model of observation is the same ergodic diffusion process

$$\mathrm{d}X_t = S(X_t)\ \mathrm{d}t + \sigma(X_t)\ \mathrm{d}W_t,\quad X_0,\quad 0\le t\le T$$

where $S(\cdot)\in\mathscr{S}_\sigma$ is an unknown function and the diffusion coefficient $\sigma(\cdot)^2$ is a known positive function. We consider now the problem of estimation of the where $S(\cdot) \in \mathscr{S}_\sigma$ is an unknown function and the diffusion coefficient $\sigma(\cdot)^2$ is a known positive function. We consider now the problem of estimation of the function $S(x)$ and study the asymptotic ($T \to \infty$) behavior of the following *trend coefficient estimator*:

$$\hat{S}_T(x) = \frac{\bar{\vartheta}_T(x)}{f_T^\circ(x)}. \tag{1.172}$$

Here

$$\bar{\vartheta}_T(x) = \frac{1}{T\varphi_T}\int_0^T K\left(\frac{X_t - x}{\varphi_T}\right) \mathrm{d}X_t$$

with the kernel $K(\cdot)$ satisfying the conditions (1.150) and $f_T^\circ(x)$ is the LTE. If $f_T^\circ(x) = 0$ we put $\hat{S}_T(x) = 0$ too.

Note that this problem is similar to the problem of density estimation in the i.i.d. case, i.e., the rate of convergence of the estimator depends on the *smoothness* of the unknown function.

Suppose that the function $S(\cdot) \in \mathscr{S}_\sigma$ and it is k times continuously differentiable. The set of such functions we denote by $\mathscr{S}_\sigma^k$ and suppose that the diffusion coefficient $\sigma(x)^2$ is k times continuously differentiable too.

Proposition 1.65. *Let $S(\cdot) \in \mathscr{S}_\sigma^k$, then the estimator $\hat{S}_T(x)$ with $\varphi_T = T^{-\frac{1}{2k+1}}$ is consistent and asymptotically normal*

$$\mathcal{L}_S\left\{T^{\frac{k}{2k+1}}\left(\hat{S}_T(x) - S(x)\right)\right\} \Longrightarrow \mathcal{N}\left(0, d_S(x)^2\right) \tag{1.173}$$

where

$$d_S(x)^2 = \frac{\sigma(x)^2}{f_S(x)}\int_A^B K(u)^2\,\mathrm{d}u.$$

Proof. Note that $\bar{\vartheta}_T(x)$ is an estimator of the product $Z(x) = S(x)\,f_S(x)$. For its mean we have

$$\begin{aligned}
\mathbf{E}_S\,\bar{\vartheta}_T(x) &= \frac{1}{\varphi_T}\,\mathbf{E}_S\left[S(\xi)\,K\left(\frac{\xi - x}{\varphi_T}\right)\right] \\
&= \int_{\mathscr{R}} K\left(\frac{y - x}{\varphi_T}\right)\,S(y)\,f_S(y)\,\frac{\mathrm{d}y}{\varphi_T} \\
&= \int_A^B K(u)\,S(x + u\varphi_T)\,f_S(x + u\varphi_T)\,\mathrm{d}u.
\end{aligned}$$

The function $Z(\cdot)$ belongs to the same set $\mathscr{S}_\sigma^k$. Hence we can expand it in the vicinity of the point x and use the conditions (1.150):

$$\int_A^B K(u)\, S(x+u\varphi_T)\, f_S(x+u\varphi_T)\, \mathrm{d}u = S(x)\, f_S(x)$$
$$+\sum_{m=1}^{k} \frac{\varphi_T^m}{m!}\, Z^{(m)}(x) \int_A^B K(u)\, u^m\, \mathrm{d}u$$
$$+\frac{\varphi_T^k}{k!} \int_A^B K(u)\, u^k \left[Z^{(k)}(x+\tilde{u}\varphi_T) - Z^{(k)}(x)\right]\, \mathrm{d}u.$$

The function $Z^{(k)}(\cdot)$ is continuous at the point x. Therefore the last integral tends to zero as $T\to\infty$ and we obtain the convergence

$$\varphi_T^{-k}\left(\mathbf{E}_S\,\bar{\vartheta}_T(x) - S(x)\, f_S(x)\right) \to 0.$$

Further,

$$\bar{\vartheta}_T(x) - \mathbf{E}_S\,\bar{\vartheta}_T(x)$$
$$= \frac{1}{T\varphi_T}\int_0^T \left[K\left(\frac{X_t-x}{\varphi_T}\right) S(X_t) - \mathbf{E}_S\left(K\left(\frac{\xi-x}{\varphi_T}\right) S(\xi)\right)\right] \mathrm{d}t$$
$$+\frac{1}{T\varphi_T}\int_0^T K\left(\frac{X_t-x}{\varphi_T}\right) \sigma(X_t)\, \mathrm{d}W_t$$
$$= \frac{H_T(X_T)-H_T(X_0)}{T\varphi_T} + \frac{1}{T\varphi_T}\int_0^T \left[K\left(\frac{X_t-x}{\varphi_T}\right)\sigma(X_t)\right.$$
$$\left. - \frac{2}{\sigma(X_t)\, f_S(X_t)}\int_{-\infty}^{X_t} h_T(v)\, f_S(v)\, \mathrm{d}v\right] \mathrm{d}W_t$$
$$= \frac{H_T(X_T)-H_T(X_0)}{T\varphi_T} + \frac{1}{T\varphi_T}\int_0^T g_T(X_t)\, \mathrm{d}W_t,$$

where

$$h_T(v) = K\left(\frac{v-x}{\varphi_T}\right) S(v) - \mathbf{E}_S\left(K\left(\frac{\xi-x}{\varphi_T}\right) S(\xi)\right)$$

and $H_T(\cdot)$ is the corresponding integral.

We have the convergence

$$\frac{1}{T\varphi_T}\int_0^T g_T(X_t)^2\, \mathrm{d}t \to \sigma(x)^2 f_S(x) \int_A^B K(u)^2\, \mathrm{d}u = d(x)^2.$$

Finally,

$$T^{\frac{k}{2k+1}}\left(\bar{\vartheta}_T(x) - S(x)\, f_S(x)\right) = T^{\frac{k}{2k+1}}\left(\bar{\vartheta}_T(x) - \mathbf{E}_S\,\bar{\vartheta}_T(x)\right)$$
$$+T^{\frac{k}{2k+1}}\left(\mathbf{E}_S\,\bar{\vartheta}_T(x) - S(x)\, f_S(x)\right)$$
$$= \frac{1}{\sqrt{T\varphi_T}}\int_0^T g_T(X_t)\, \mathrm{d}W_t + o(1) \Longrightarrow \mathcal{N}\left(0, d(x)^2\right).$$

Remark 1.67. Using similar but cumbersome calculus it is possible to show that for any collection of points $x_1, \ldots, x_l$ the vector of estimators $\left(\hat{S}_T(x_1), \ldots, \hat{S}_T(x_l)\right)$ and the vector of the kernel-type estimators of derivatives of the density function are asymptotically normal with asymptotically independent components (see [78], [233]).

1.3.3 Hypotheses Testing

We consider just two problems of testing a simple hypothesis against simple and one-sided composite alternatives. The observed process is diffusion:

$$\mathrm{d}X_t = S(\vartheta, X_t)\,\mathrm{d}t + \sigma(X_t)\,\mathrm{d}W_t, \quad X_0, \quad 0 \le t \le T,$$

where the unknown parameter ϑ can take two values $\vartheta = \vartheta_0$ and $\vartheta = \vartheta_1$ or $\vartheta = \vartheta_0$ and $\vartheta > \vartheta_0$. Of course, we suppose that the trend coefficient $S(\vartheta, \cdot) \in \mathscr{S}_\sigma$ under the hypothesis and alternatives. The diffusion coefficient $\sigma(\cdot)$ is supposed to be a positive known function and the initial random variable X_0 has the stationary density $f(\vartheta, \cdot)$ with a corresponding value of ϑ. We suppose as well that the measures $\mathbf{P}_{\vartheta_0}^{(T)}$ and $\mathbf{P}_{\vartheta_1}^{(T)}$ are equivalent.

Simple Hypothesis and Alternative

We have to check a simple hypothesis

$$\mathscr{H}_0: \quad \vartheta = \vartheta_0,$$

against a simple alternative

$$\mathscr{H}_1: \quad \vartheta = \vartheta_1.$$

Fix a number $\varepsilon \in (0,1)$ and define the class $\mathscr{K}_\varepsilon$ of *tests of level* $1-\varepsilon$ (or *tests of size* ε) as follows. Denote by $\phi_T = \phi_T\left(X^T\right)$ the *statistical decision function* of the test, i.e., $\phi_T\left(X^T\right)$ is the probability to accept the hypothesis $\mathscr{H}_1$ having the observations $X^T = \{X_t, 0 \le t \le T\}$. Then

$$\mathscr{K}_\varepsilon = \left\{\phi_T \ : \ \mathbf{E}_{\vartheta_0}\phi_T\left(X^T\right) \le \varepsilon\right\}.$$

Denote by $\beta_T(\phi_T)$ the probability of the true decision under $\mathscr{H}_1$, i.e.,

$$\beta_T(\phi_T) = \mathbf{E}_{\vartheta_1}\phi_T\left(X^T\right).$$

The value $\beta_T(\phi_T)$ is called the *power* of the test ϕ_T.

Definition 1.68. *We say that a test $\hat{\phi}_T \in K_\varepsilon$ is the most powerful in the class $\mathscr{K}_\varepsilon$ if for any other test $\phi_T \in \mathscr{K}_\varepsilon$ we have*

$$\beta_T(\hat{\phi}_T) \ge \beta_T(\phi_T).$$

Denote by $\beta_T(\phi_T)$ the probability of the true decision under $\mathscr{H}_1$, i.e.,

$$\beta_T(\phi_T) = \mathbf{E}_{\vartheta_1} \phi_T\left(X^T\right).$$

The value $\beta_T(\phi_T)$ is called the power of the test ϕ_T.

Definition 1.68. *We say that a test $\hat{\phi}_T \in \mathcal{K}_\varepsilon$ is the most powerful in the class $\mathscr{K}_\varepsilon$ if for any other test $\phi_T \in \mathscr{K}_\varepsilon$ we have*

$$\beta_T(\hat{\phi}_T) \geq \beta_T(\phi_T).$$

The likelihood ratio in this problem is

$$L\left(\vartheta_1, \vartheta_0, X^T\right) = \frac{G(\vartheta_1)}{G(\vartheta_0)} \exp\left\{ \int_0^T \frac{S(\vartheta_1, X_t) - S(\vartheta_0, X_t)}{\sigma(X_t)^2} \, dX_t \right.$$
$$\left. -\frac{1}{2}\int_0^T \frac{S(\vartheta_1, X_t)^2 - S(\vartheta_0, X_t)^2}{\sigma(X_t)^2} \, dt + 2\int_0^{X_0} \frac{S(\vartheta_1, y) - S(\vartheta_0, y)}{\sigma(y)^2} \, dy \right\}.$$

According to the *Neyman–Pearson lemma* the solution of the problem is given by the following proposition.

Proposition 1.69. *The test*

$$\hat{\phi}_T\left(X^T\right) = \chi_{\left\{L\left(\vartheta_1, \vartheta_0, X^T\right) \geq c_\varepsilon\right\}}, \tag{1.174}$$

where the constant $c_\varepsilon = c_\varepsilon(T)$ is defined by the equation

$$\mathbf{E}_{\vartheta_0} \hat{\phi}_T\left(X^T\right) = \mathbf{P}_{\vartheta_0}^{(T)}\left\{L\left(\vartheta_1, \vartheta_0, X^T\right) \geq c_\varepsilon\right\} = \varepsilon \tag{1.175}$$

is the most powerful in the class $\mathcal{K}_\varepsilon$.

Proof. (Lehman [167], p. 75) Let ϕ_T be any other test with $\mathbf{E}_{\vartheta_0} \phi_T\left(X^T\right) \leq \varepsilon$. Introduce the sets

$$\mathbb{S}_+ = \left\{x^T : L\left(\vartheta_1, \vartheta_0, x^T\right) \geq c_\varepsilon\right\}, \qquad \mathbb{S}_- = \left\{x^T : L\left(\vartheta_1, \vartheta_0, x^T\right) < c_\varepsilon\right\}.$$

If $x^T \in \mathbb{S}_+$, then $\hat{\phi}_T\left(x^T\right) > 0$ and $L\left(\vartheta_1, \vartheta_0, x^T\right) - c_\varepsilon \geq 0$. Otherwise, $\hat{\phi}_T\left(x^T\right) - \phi_T\left(x^T\right) \leq 0$ and $L\left(\vartheta_1, \vartheta_0, x^T\right) - c_\varepsilon < 0$. Therefore

$$\mathbf{E}_{\vartheta_0}\left(\hat{\phi}_T\left(X^T\right) - \phi_T\left(X^T\right)\right)\left(L\left(\vartheta_1, \vartheta_0, X^T\right) - c_\varepsilon\right) \geq 0$$

and this implies

$$\begin{aligned}\beta_T(\hat{\phi}_T) - \beta_T(\phi_T) &= \mathbf{E}_{\vartheta_0} L\left(\vartheta_1, \vartheta_0, X^T\right)\left[\hat{\phi}_T\left(X^T\right) - \phi_T\left(X^T\right)\right] \\ &\geq c_\varepsilon \, \mathbf{E}_{\vartheta_0}\left[\hat{\phi}_T\left(X^T\right) - \phi_T\left(X^T\right)\right] \geq 0\end{aligned}$$

because $\mathbf{E}_{\vartheta_0}\,\phi_T\left(X^T\right)\leq\varepsilon$.

Therefore the problem is solved but nevertheless the proposed solution is not satisfactory because it can be difficult to find the constant c_ε from Equation (1.175). To simplify the construction of the test we can either take the asymptotical approach with the class of tests of *asymptotic level* $1-\varepsilon$, i.e., define

$$\mathscr{K}_\varepsilon'=\left\{\phi_T\;:\;\varlimsup_{T\to\infty}\mathbf{E}_{\vartheta_0}\,\phi_T\left(X^T\right)\leq\varepsilon\right\},$$

or introduce the randomized tests with exact level $1-\varepsilon$.

We start with the asymptotic approach and seek such constants c_ε that the test $\hat\phi_T\in\mathscr{K}_\varepsilon'$. It is easy to see that under $\mathscr{H}_0$

$$\frac{\ln L\left(\vartheta_1,\vartheta_0,X^T\right)}{T}\longrightarrow-\frac{1}{2}\,\mathbf{E}_{\vartheta_0}\left(\frac{S\left(\vartheta_1,\xi\right)-S\left(\vartheta_0,\xi\right)}{\sigma\left(\xi\right)}\right)^2\equiv-J\left(\vartheta_0,\vartheta_1\right).$$

Remember that the *Kullback–Leibler distance* between two measures $\mathbf{P}_{\vartheta_1}^{(T)}$ and $\mathbf{P}_{\vartheta_0}^{(T)}$ is

$$\mathbf{E}_{\vartheta_0}\ln L\left(\vartheta_0,\vartheta_1,X^T\right)=T\;J\left(\vartheta_0,\vartheta_1\right)+\mathbf{E}_{\vartheta_0}\ln\frac{f\left(\vartheta_0,\xi\right)}{f\left(\vartheta_1,\xi\right)}.$$

Let us denote

$$g\left(x\right)=\int_{-\infty}^{x}\left(\left[\frac{S\left(\vartheta_1,v\right)-S\left(\vartheta_0,v\right)}{\sigma\left(v\right)}\right]^2-2\;J\left(\vartheta_0,\vartheta_1\right)\right)\frac{2f\left(\vartheta_0,v\right)}{\sigma\left(x\right)\;f\left(\vartheta_0,x\right)}\;\mathrm{d}v.$$

Let $X_0=x_0$ (nonrandom). Then (under hypothesis $\mathscr{H}_0$) by the Itô formula we can write

$$\int_0^T\left(\frac{S\left(\vartheta_1,X_t\right)-S\left(\vartheta_0,X_t\right)}{\sigma\left(X_t\right)}\right)^2\;\mathrm{d}t=2TJ\left(\vartheta_0,\vartheta_1\right)+\int_{x_0}^{X_T}\frac{g\left(x\right)}{\sigma\left(x\right)}\;\mathrm{d}x$$
$$-\int_0^T g\left(X_t\right)\;\mathrm{d}W_t.$$

Hence

$$\mathbf{P}_{\vartheta_0}^{(T)}\left\{L\left(\vartheta_1,\vartheta_0,X^T\right)>c_\varepsilon\right\}=\mathbf{P}_{\vartheta_0}^{(T)}\left\{\int_0^T\left[\frac{S\left(\vartheta_1,X_t\right)-S\left(\vartheta_0,X_t\right)}{\sigma\left(X_t\right)}\right]\mathrm{d}W_t\right.$$
$$\left.-\frac{1}{2}\int_0^T\left[\frac{S\left(\vartheta_1,X_t\right)-S\left(\vartheta_0,X_t\right)}{\sigma\left(X_t\right)}\right]^2\mathrm{d}t>\ln c_\varepsilon\right\}$$
$$=\mathbf{P}_{\vartheta_0}^{(T)}\left\{\frac{1}{\sqrt{T}}\int_0^T\left[\frac{S\left(\vartheta_1,X_t\right)-S\left(\vartheta_0,X_t\right)}{\sigma\left(X_t\right)}+\frac{1}{2}\,g\left(X_t\right)\right]\mathrm{d}W_t\right.$$
$$\left.-\frac{1}{2\sqrt{T}}\int_{x_0}^{X_T}\frac{g\left(x\right)}{\sigma\left(x\right)}\;\mathrm{d}x>\frac{\ln c_\varepsilon+T\;J\left(\vartheta_0,\vartheta_1\right)}{\sqrt{T}}\right\}.$$

The stochastic integral by the central limit theorem is asymptotically normal

$$\mathcal{L}_{\vartheta_0}\left\{\frac{1}{\sqrt{T}}\int_0^T\left[\frac{S(\vartheta_1,X_t)-S(\vartheta_0,X_t)}{\sigma(X_t)}+\frac{g(X_t)}{2}\right]\mathrm{d}W_t\right\}\Longrightarrow\mathcal{N}\left(0,d^2\right)$$

with

$$d^2=\int_{-\infty}^{\infty}\left[\frac{S(\vartheta_1,x)-S(\vartheta_0,x)}{\sigma(x)}+\frac{g(x)}{2}\right]^2 f(\vartheta_0,x)\;\mathrm{d}x.$$

Therefore the constant c_ε satisfies the condition

$$\lim_{T\to\infty}\frac{\ln c_\varepsilon+T\,J(\vartheta_0,\vartheta_1)}{\sqrt{T}}=z_\varepsilon\,d,$$

where z_ε is $1-\varepsilon$ quantile of the standard Gaussian law, i.e., $\mathbf{P}\{\zeta>z_\varepsilon\}=\varepsilon$, where $\zeta\sim\mathcal{N}(0,1)$. Finally we put

$$c_\varepsilon=\exp\left\{-T\,J(\vartheta_0,\vartheta_1)+z_\varepsilon\,d\sqrt{T}\right\}. \tag{1.176}$$

Then the test $\hat{\phi}_T\in\mathcal{K}'_\varepsilon$. The asymptotics of the power function $\beta(\hat{\phi}_T)$ is described in Section 5.1.

Using randomization and an asymptotical ($T\to\infty$) approach we construct a test of exact level $1-\varepsilon$ for all T and asymptotically equivalent in a certain sense to the optimal one.

Remember that we observe a diffusion process

$$\mathrm{d}X_t=S(\vartheta,X_t)\;\mathrm{d}t+\sigma(X_t)\;\mathrm{d}W_t,\quad X_0,\quad 0\le t\le T,$$

where ϑ can take one of the values ϑ_0 or ϑ_1 and we have to check the hypotheses $\mathscr{H}_0$ and $\mathscr{H}_1$ defined above.

We construct the test as follows. We put $\sigma(X_t)\equiv 1$ for $t\in[T,T+1]$ and

$$S(\vartheta_1,X_t)-S(\vartheta_0,X_t)\equiv J(\vartheta_1,\vartheta_0)^{1/2}\,\sqrt{2T}.$$

Note that we have observations X_t on the interval $[0,T]$ only and these quantities are defined on $[T,T+1]$. Introduce the stopping time

$$\tau_T=\inf\left\{\tau\le T+1\,:\frac{1}{T}\int_0^\tau\left(\frac{S(\vartheta_1,X_t)-S(\vartheta_0,X_t)}{\sigma(X_t)}\right)^2\mathrm{d}t\ge 2\,J(\vartheta_1,\vartheta_0)\right\}$$

and note that with probabilities $\mathbf{P}_{\vartheta_0}^{(T)}$ and $\mathbf{P}_{\vartheta_1}^{(T)}$ of 1, we have $\tau_T\le T+1$. Then we define the statistic

$$\Delta_\tau\left(\vartheta_1,\vartheta_0,X^T\right)=\frac{1}{\sqrt{T}}\int_0^{\tau_T}\frac{S(\vartheta_1,X_t)-S(\vartheta_0,X_t)}{\sigma(X_t)^2}\left[\mathrm{d}X_t-S(\vartheta_0,X_t)\;\mathrm{d}t\right] \tag{1.177}$$

where for $t \in [T, T+1]$ we introduce an independent Wiener process W_t, $T \leq t \leq T+1$ and put

$$\mathrm{d}X_t - S(\vartheta_0, X_t)\ \mathrm{d}t \equiv \mathrm{d}W_t,$$

i.e., for $\tau_{_T} > T$ we have

$$\Delta_\tau\left(\vartheta_1, \vartheta_0, X^T\right) = \frac{1}{\sqrt{T}} \int_0^T \frac{S(\vartheta_1, X_t) - S(\vartheta_0, X_t)}{\sigma(X_t)^2} \left[\mathrm{d}X_t - S(\vartheta_0, X_t)\ \mathrm{d}t\right]$$
$$+\sqrt{2}\, J(\vartheta_1, \vartheta_0)^{1/2} \left(W_{\tau_{_T}} - W_T\right).$$

Now the statistic $\Delta_\tau\left(\vartheta_1, \vartheta_0, X^T\right)$ under hypothesis $\mathscr{H}_0$ is

$$\Delta_\tau\left(\vartheta_1, \vartheta_0, X^T\right) = \frac{1}{\sqrt{T}} \int_0^{\tau_T} \frac{S(\vartheta_1, X_t) - S(\vartheta_0, X_t)}{\sigma(X_t)}\, \mathrm{d}W_t$$

and by Lemma 1.2 is a Gaussian random variable, i.e.,

$$\mathcal{L}_{\vartheta_0}\Big(\Delta_\tau\left(\vartheta_1, \vartheta_0, X^T\right)\Big) = \mathcal{N}\Big(0, 2\, J(\vartheta_1, \vartheta_0)\Big). \tag{1.178}$$

This property of the stochastic integral yields the following proposition.

Proposition 1.70. *The test*

$$\phi^*_{_T}\left(X^T\right) = \chi_{\left\{\Delta_\tau\left(\vartheta_1, \vartheta_0, X^T\right) \geq z_\varepsilon \sqrt{2J(\vartheta_1, \vartheta_0)}\right\}}$$

belongs to $\mathscr{K}_\varepsilon$ *and*

$$\beta_T\left(\phi^*_{_T}\right) = \mathbf{P}\left\{\zeta \geq z_\varepsilon - \sqrt{2T}\, J(\vartheta_1, \vartheta_0)^{1/2}\right\},$$

where $\mathcal{L}\{\zeta\} = \mathcal{N}(0,1)$.

Proof. The proof follows immediately from (1.178) and the equality (under $\mathscr{H}_1$)

$$\Delta_\tau\left(\vartheta_1, \vartheta_0, X^T\right) = \frac{1}{\sqrt{T}} \int_0^{\tau_T} \frac{S(\vartheta_1, X_t) - S(\vartheta_0, X_t)}{\sigma(X_t)}\, \mathrm{d}W_t$$
$$+ \frac{1}{\sqrt{T}} \int_0^{\tau_T} \left(\frac{S(\vartheta_1, X_t) - S(\vartheta_0, X_t)}{\sigma(X_t)}\right)^2 \mathrm{d}t$$
$$= \sqrt{2}\, J(\vartheta_1, \vartheta_0)^{1/2} \zeta + 2\sqrt{T}\, J(\vartheta_1, \vartheta_0).$$

Note as well that under $\mathscr{H}_1$ with probability 1 we have

$$\lim_{T \to \infty} \frac{\ln L\left(\vartheta_1, \vartheta_0, X^T\right)}{T} = J(\vartheta_1, \vartheta_0),$$

and the same time

$$\beta_T\left(\hat{\phi}_T\right) = 1 - \exp\left\{-T\,J\left(\vartheta_1,\vartheta_0\right) + z_\varepsilon\,\sqrt{2T}\,J\left(\vartheta_1,\vartheta_0\right)^{1/2}\left(1+o(1)\right)\right\}. \tag{1.179}$$

Simple Hypothesis and One-sided Alternative

Suppose that the observed process is

$$\mathrm{d}X_t = \vartheta\,h\left(X_t\right)\,\mathrm{d}t + \sigma\left(X_t\right)\,\mathrm{d}W_t, \qquad X_0, \quad 0 \le t \le T,$$

where $h\left(\cdot\right)$ and $\sigma\left(\cdot\right)$ are known functions and such that $S\left(\vartheta,\cdot\right) = \vartheta\,h\left(\cdot\right) \in \mathscr{S}_\sigma$ for all $\vartheta \ge \vartheta_0$. The initial value X_0 has invariant density $f\left(\vartheta,\cdot\right)$. We consider the problem of hypotheses testing

$$\begin{aligned}\mathscr{H}_0 &: \quad \vartheta = \vartheta_0,\\ \mathscr{H}_1 &: \quad \vartheta > \vartheta_0,\end{aligned}$$

i.e., the alternative is composite one-sided.

As before, $\mathscr{K}_\varepsilon$ is the class of tests of *level* $1-\varepsilon$. The power function $\beta_T\left(\vartheta,\phi_T\right) = \mathbf{E}_\vartheta\,\phi_T\left(X^T\right)$ now depends on $\vartheta > \vartheta_0$.

Definition 1.71. *We say that a test $\hat{\phi}_T \in \mathscr{K}_\varepsilon$ is uniformly most powerful in the class $\mathscr{K}_\varepsilon$ if for any other test $\phi_T \in \mathscr{K}_\varepsilon$ we have*

$$\inf_{\vartheta>\vartheta_0}\left[\beta_T\left(\vartheta,\hat{\phi}_T\right) - \beta_T\left(\vartheta,\phi_T\right)\right] \ge 0.$$

The log-likelihood ratio formula is

$$\begin{aligned}\ln L\left(\vartheta,\vartheta_0,X^T\right) &= \ln\frac{f\left(\vartheta,X_0\right)}{f\left(\vartheta_0,X_0\right)} + \\ &= \left(\vartheta-\vartheta_0\right)\int_0^T \frac{h\left(X_t\right)}{\sigma\left(X_t\right)^2}\,\mathrm{d}X_t - \frac{\vartheta^2-\vartheta_0^2}{2}\int_0^T\left(\frac{h\left(X_t\right)}{\sigma\left(X_t\right)}\right)^2\,\mathrm{d}t.\end{aligned}$$

It is clear that the construction of the uniformly most powerful test even in this linear case is in general impossible. Hence we turn to the asymptotical approach.

Definition 1.72. *We say that a test $\hat{\phi}_T \in \mathscr{K}_\varepsilon'$ is asymptotically uniformly most powerful in the class $\mathscr{K}_\varepsilon'$ if for any other test $\phi_T \in \mathscr{K}_\varepsilon$ we have*

$$\varliminf_{T\to\infty}\,\inf_{\vartheta>\vartheta_0}\left[\beta_T\left(\vartheta,\hat{\phi}_T\right) - \beta_T\left(\vartheta,\phi_T\right)\right] \ge 0.$$

Let us introduce the statistic

$$\hat{\delta}_T\left(X^T\right) = \frac{1}{\sqrt{T}}\int_0^T \frac{h\left(X_t\right)}{\sigma\left(X_t\right)^2}\,\left[\mathrm{d}X_t - \vartheta_0\,h\left(X_t\right)\,\mathrm{d}t\right]$$

and the corresponding test

Let us introduce the statistic

$$\hat{\delta}_T\left(X^T\right) = \frac{1}{\sqrt{T}} \int_0^T \frac{h\left(X_t\right)}{\sigma\left(X_t\right)^2}\, \left[\mathrm{d}X_t - \vartheta_0\, h\left(X_t\right)\, \mathrm{d}t\right]$$

and the corresponding test

$$\hat{\phi}_T\left(X^T\right) = \chi_{\left\{\hat{\delta}_T\left(X^T\right) \geq r_\varepsilon\right\}}$$

with

$$r_\varepsilon = z_\varepsilon \left(\mathbf{E}_{\vartheta_0}\left(\frac{h\left(\xi\right)}{\sigma\left(\xi\right)}\right)^2\right)^{1/2} \equiv z_\varepsilon\, \mathrm{I}\left(\vartheta_0\right)^{1/2}.$$

Then we have the following proposition.

Proposition 1.73. *The test $\hat{\phi}_T(\cdot) \in \mathscr{K}_\varepsilon'$ and is asymptotically uniformly most powerful in the class $\mathscr{K}_\varepsilon'$.*

Proof. Under hypothesis $\mathscr{H}_0$

$$\mathbf{P}_{\vartheta_0}^{(T)}\left\{\hat{\delta}_T \geq r_\varepsilon\right\} = \mathbf{P}_{\vartheta_0}^{(T)}\left\{\frac{1}{\sqrt{T}} \int_0^T \frac{h\left(X_t\right)}{\sigma\left(X_t\right)}\, \mathrm{d}W_t \geq r_\varepsilon\right\}$$

$$= \mathbf{P}_{\vartheta_0}^{(T)}\left\{\frac{1}{\sqrt{T}\,\mathrm{I}\left(\vartheta_0\right)^{1/2}} \int_0^T \frac{h\left(X_t\right)}{\sigma\left(X_t\right)}\, \mathrm{d}W_t \geq z_\varepsilon\right\} \longrightarrow \mathbf{P}\left\{\zeta \geq z_\varepsilon\right\} = \varepsilon,$$

because the stochastic integral is asymptotically normal. Hence $\hat{\phi}_T(\cdot) \in \mathscr{K}_\varepsilon'$.

Fix some $\vartheta_1 > \vartheta_0$, and compare the asymptotic behavior of the power functions of the test $\hat{\phi}_T(\cdot)$ and that of the test

$$\phi_T^\star\left(X^T\right) = \chi_{\left\{L\left(\vartheta_1, \vartheta_0, X^T\right) \geq c_\varepsilon\right\}}, \qquad \mathbf{P}_{\vartheta_0}^{(T)}\left\{L\left(\vartheta_1, \vartheta_0, X^T\right) \geq c_\varepsilon\right\} = \varepsilon,$$

with c_ε given by (1.98) which is the most powerful in the problem of testing these two simple hypotheses. It is evident that for any two fixed values $\vartheta_1 \neq \vartheta_0$ both power functions tend to 1 because under $\mathscr{H}_1$

$$L\left(\vartheta_1, \vartheta_0, X^T\right) \longrightarrow +\infty, \qquad \hat{\delta}_T\left(X^T\right) \longrightarrow \infty.$$

Therefore we consider the close alternatives only. In particular, we put $\vartheta_1 = \vartheta_0 + u/\sqrt{T}$ and denote

$$\hat{\beta}_T(u) = \hat{\beta}_T\left(\vartheta_0 + u/\sqrt{T}, X^T\right) \quad \text{and} \quad \beta_T^\star(u) = \beta_T^\star\left(\vartheta_0 + u/\sqrt{T}, X^T\right)$$

respectively.

We have

$$\hat{\beta}_T(u) = \mathbf{P}_{\vartheta_1}^{(T)} \left\{ \frac{1}{\sqrt{T}} \int_0^T \frac{h(X_t)}{\sigma(X_t)} \, \mathrm{d}W_t + \frac{u}{T} \int_0^T \left(\frac{h(X_t)}{\sigma(X_t)} \right)^2 \mathrm{d}t \geq z_\varepsilon \mathrm{I}(\vartheta_0)^{1/2} \right\}$$
$$\longrightarrow \mathbf{P}\left\{ \zeta \geq z_\varepsilon - u\, \mathrm{I}(\vartheta_0)^{1/2} \right\}.$$

It can be shown that this convergence is uniform on the compacts $u \in [0, K]$ for any $K > 0$.

Further, the condition $\phi_T^\star \in \mathscr{K}_\varepsilon'$ provides us with the value

$$c_\varepsilon = \exp\left\{ -\frac{u^2}{2} \mathrm{I}(\vartheta_0) + u\, z_\varepsilon\, \mathrm{I}(\vartheta_0)^{1/2} \right\}$$

and under hypothesis $\mathscr{H}_1$ we have

$$\beta_T^\star(u) = \mathbf{P}_{\vartheta_1}^{(T)} \left\{ \frac{u}{\sqrt{T}} \int_0^T \frac{h(X_t)}{\sigma(X_t)} \, \mathrm{d}W_t + \frac{u^2}{2T} \int_0^T \left(\frac{h(X_t)}{\sigma(X_t)} \right)^2 \mathrm{d}t \geq c_\varepsilon \right\}$$
$$\longrightarrow \mathbf{P}\left\{ \zeta \geq z_\varepsilon - u\, \mathrm{I}(\vartheta_0)^{1/2} \right\}.$$

Therefore, the test $\hat{\phi}_T(\cdot)$ is asymptotically uniformly most powerful.

2

Parameter Estimation

The asymptotic behavior of the MLE, Bayesian, minimum distance and some other estimators is studied in the asymptotics of large samples. It is shown that under regularity conditions these estimators are consistent, asymptotically normal and in a certain sense are asymptotically efficient. Then we describe what happens if some of the regularity conditions are not fulfilled (misspecified and non identifiable models, null Fisher information etc.).

2.1 Maximum Likelihood and Bayesian Estimators

We are given a probability space $\{\Omega, \mathfrak{F}, \mathbf{P}\}$ with filtration $\{\mathfrak{F}_t,\ t \geq 0\}$, a standard Wiener process $\{W_t, \mathfrak{F}_t, t \geq 0\}$, an open set $\Theta \subset \mathscr{R}^d$ and for each $\vartheta \in \Theta$ the diffusion process

$$\mathrm{d}X_t = S(\vartheta, X_t)\,\mathrm{d}t + \sigma(X_t)\,\mathrm{d}W_t, \quad X_0, \quad t \geq 0 \tag{2.1}$$

is defined. Here X_0 is a $\mathfrak{F}_0$-measurable random variable. We suppose that the functions $S(\vartheta, \cdot)$ and $\sigma(\cdot)$ for all $\vartheta \in \Theta$ satisfy the conditions $\mathcal{ES}$ of the existence of a unique weak solution of Equation (2.1) and the condition $\mathcal{RP}(\Theta)$ provides the ergodic properties of the solutions of this equation.

We consider the problem of estimation of the parameter ϑ by the observations $X^T = \{X_t,\ 0 \leq t \leq T\}$ and describe the properties of estimators as $T \to \infty$, that is, in the *asymptotics of large samples*.

Let us denote by $\mathbf{P}_\vartheta^{(T)}$ and $\mathbf{E}_\vartheta$ the probability measure induced in the measurable space $(\mathscr{C}_T, \mathfrak{B}_T)$ of continuous on $[0,T]$ functions by the process (2.1) and the expectation with respect to this measure, respectively. We suppose as well that the condition $\mathcal{EM}$ of the equivalence of the measures $\left\{\mathbf{P}_\vartheta^{(T)},\ \vartheta \in \Theta\right\}$ is fulfilled too and we denote the corresponding likelihood ratio by $L\left(\vartheta, \vartheta_1; X^T\right)$ (see (1.47)).

2.1.1 Lower Bound

To define the asymptotically ($T \to \infty$) best estimators in regular problems of parameter estimation we use the Hajek–Le Cam inequality on the risks of all estimators. This inequality is established for *locally asymptotically normal* families of measures [162].

Definition 2.1. *A family of measures $\left\{\mathbf{P}_{\vartheta}^{(T)}, \vartheta \in \Theta\right\}$ is called locally asymptotically normal (LAN) at a point $\vartheta_0 \in \Theta$ if for some nondegenerate $d \times d$ matrix $\varphi_T(\vartheta)$ and any $u \in \mathscr{R}^d$, the likelihood ratio*

$$Z_T(u) = L\left(\vartheta_0 + \varphi_T(\vartheta_0)\, u, \vartheta; X^T\right)$$

admits the representation

$$Z_T(u) = \exp\left\{\left(u, \Delta_T\left(\vartheta_0, X^T\right)\right) - \frac{1}{2}|u|^2 + r_T\left(\vartheta_0, u, X^T\right)\right\}, \qquad (2.2)$$

where

$$\mathcal{L}_{\vartheta_0}\left\{\Delta_T\left(\vartheta_0, X^T\right)\right\} \Longrightarrow \mathcal{N}(\mathbf{0}, J)) \qquad (2.3)$$

and

$$\mathbf{P}_{\vartheta_0} - \lim_{T\to\infty} r_T\left(\vartheta_0, u, X^T\right) = 0. \qquad (2.4)$$

We say that this family is LAN at Θ if it is LAN at every point $\vartheta_0 \in \Theta$ and uniformly LAN at Θ if the convergences (2.3) and (2.4) are uniform on the compacts $\mathbb{K} \subset \Theta$.

Here J is the unit $d \times d$ matrix and $\left(u, \Delta_T\right)$ and $|u|$ are the scalar product and the norm in $\mathscr{R}^d$, respectively.

Sometimes it is more convenient to check the equivalent representation

$$Z_T(u) = \exp\left\{\left(u, \Delta_T\left(\vartheta_0, X^T\right)\right) - \frac{1}{2}\left(\mathbf{I}(\vartheta_0)\, u, u\right) + r_T\left(\vartheta_0, u, X^T\right)\right\} \quad (2.5)$$

with

$$\mathcal{L}_{\vartheta_0}\left\{\Delta_T\left(\vartheta_0, X^T\right)\right\} \Longrightarrow \mathcal{N}(\mathbf{0}, \mathbf{I}(\vartheta_0))) \qquad (2.6)$$

and (2.4), where $\mathbf{I}(\vartheta_0)$ is a nondegenerate information matrix. Indeed, we can always put $\tilde\varphi_T(\vartheta_0) = \varphi_T(\vartheta_0)\,\mathbf{I}(\vartheta_0)^{-1/2}$ and obtain from (2.4), (2.5), the representation (2.2) and (2.3) (with new normalizing matrix $\tilde\varphi_T(\vartheta_0)$).

Note that for bounded (open or closed) sets Θ the LAN property cannot hold uniformly at Θ because for the points close to the border, the representation (2.2)–(2.4) is no longer valid (see, e.g., [139], Section 2.8). Sometimes the limit matrix $\mathbf{I}(\vartheta_0)$ in the representation can be random and (2.6) is replaced by

$$\mathcal{L}_{\vartheta_0}\left\{\Delta_T\left(\vartheta_0, X^T\right)\right\} \Longrightarrow \mathcal{L}_{\vartheta_0}\left\{\zeta_0\right\}, \quad \mathcal{L}_{\vartheta_0}\left\{\zeta_0 \middle| \mathbf{I}(\vartheta_0)\right\} = \mathcal{N}\left(\mathbf{0}, \mathbf{I}(\vartheta_0)\right), \ (2.7)$$

i.e., $\boldsymbol{\Delta}_T\left(\boldsymbol{\vartheta}_0, X^T\right)$ is asymptotically conditionally Gaussian. Then the family of measures is called *locally asymptotically mixing normal* (LAMN)(see [115], [163], [165] for definition and properties). We have two such nonergodic models in Sections 3.5.1 and 3.5.3.

For ergodic diffusion processes the conditions of LAN can be the following.

Proposition 2.2. *Suppose that the function $S(\boldsymbol{\vartheta},\cdot)\,\sigma(\cdot)^{-1}, \boldsymbol{\vartheta} \in \boldsymbol{\Theta}$ is differentiable over $\boldsymbol{\vartheta}$ in the $\mathscr{L}_2(f_{\boldsymbol{\vartheta}})$ sense: there exists a vector function $\dot{\boldsymbol{S}}(\boldsymbol{\vartheta},\cdot)$ such that for all $\boldsymbol{\vartheta}_0 \in \boldsymbol{\Theta}$*

$$\mathbf{E}_{\boldsymbol{\vartheta}_0}\left(\frac{S(\boldsymbol{\vartheta}_0+\boldsymbol{h},\xi)-S(\boldsymbol{\vartheta}_0,\xi)-\left(\boldsymbol{h},\dot{\boldsymbol{S}}(\boldsymbol{\vartheta}_0,\xi)\right)}{\sigma(\xi)}\right)^2 = o\left(|\boldsymbol{h}|^2\right) \qquad (2.8)$$

and the $d\times d$ matrix

$$\mathbf{I}(\boldsymbol{\vartheta}_0)=\mathbf{E}_{\boldsymbol{\vartheta}_0}\left(\frac{\dot{\boldsymbol{S}}(\boldsymbol{\vartheta}_0,\xi)\;\dot{\boldsymbol{S}}(\boldsymbol{\vartheta}_0,\xi)^{\mathrm{T}}}{\sigma(\xi)^2}\right)$$

is nondegenerate. Then the family of measures $\{\mathbf{P}_{\boldsymbol{\vartheta}}, \boldsymbol{\vartheta}\in\boldsymbol{\Theta}\}$ is LAN at $\boldsymbol{\Theta}$ with the normalizing matrix $\varphi_T(\boldsymbol{\vartheta}_0)=T^{-1/2}\mathbf{I}(\boldsymbol{\vartheta}_0)^{-1/2}$ and the vector

$$\boldsymbol{\Delta}_T\left(\boldsymbol{\vartheta}_0,X^T\right)=T^{-1/2}\mathbf{I}(\boldsymbol{\vartheta}_0)^{-1/2}\int_0^T\frac{\dot{\boldsymbol{S}}(\boldsymbol{\vartheta}_0,X_t)}{\sigma(X_t)^2}\left[\mathrm{d}X_t-S(\boldsymbol{\vartheta}_0,X_t)\;\mathrm{d}t\right].$$

Proof. This follows immediately from the representation (1.32) of the likelihood ratio and the central limit theorem for stochastic integral $\boldsymbol{\Delta}_T\left(\boldsymbol{\vartheta}_0, X^T\right)$. Remember that the random variable ξ has *stationary distribution*, i.e., it has $f(\boldsymbol{\vartheta},x)$ as the density function. For example, $\mathbf{E}_{\boldsymbol{\vartheta}_0}\Phi(\xi)$ means that the density function of ξ is $f(\boldsymbol{\vartheta}_0,x)$.

Remark 2.3. In the one-dimensional case ($d=1$) using a stopping time we can take the random variable $\Delta_T\left(\vartheta,X^T\right)$ in (2.5) and (2.6) a Gaussian for all T as follows (as was done in the proof of the CLT). Let us introduce a Wiener process $\tilde{W}_t,\ T\leq t\leq T+1$ independent of the Wiener process $W_t,\ 0\leq t\leq T$. Further, put $\sigma(X_t)=1, T<t\leq T+1$ and define the function $\dot{S}(\vartheta,X_t)=(T\ \mathrm{I}(\vartheta))^{1/2}$ for $T<t\leq T+1$. Then introduce the stopping time

$$\tau_T=\inf\left\{\tau:\ \frac{1}{T}\int_0^\tau\left(\frac{\dot{S}(\vartheta,X_t)}{\sigma(X_t)}\right)^2\,\mathrm{d}t\geq \mathrm{I}(\vartheta)\right\}$$

and the random variable

$$\tilde{\Delta}_\tau\left(\vartheta,X^T\right)=T^{-1/2}\int_0^{\tau_T}\frac{\dot{S}(\vartheta,X_t)}{\sigma(X_t)^2}\left[\mathrm{d}X_t-S(\vartheta,X_t)\;\mathrm{d}t\right], \qquad (2.9)$$

where we put $\mathrm{d}X_t - S(\vartheta, X_t)\ \mathrm{d}t = \mathrm{d}\tilde{W}_t$ for $T < t \leq T+1$. Obviously $\tau_T \leq T+1$ with probability 1 and the random variable $\tilde{\Delta}_\tau(\vartheta, X^T)$ is Gaussian $\mathcal{N}(0, \mathrm{I}(\vartheta))$ for all $T > 0$ (see Lemma 1.2).

It can be easily shown that $Z_T(u) = L\left(\vartheta_0 + T^{-1/2}u, \vartheta_0, X^T\right)$ admits the representation

$$Z_T(u) = \exp\left\{u\,\tilde{\Delta}_\tau\left(\vartheta_0, X^T\right) - \frac{u^2}{2}\mathrm{I}(\vartheta_0) + \tilde{r}_T\left(\vartheta_0, u, X^T\right)\right\}, \tag{2.10}$$

where $\tilde{r}_T\left(\vartheta_0, u, X^T\right) \to 0$.

Introduce the function $\ell(\boldsymbol{u})$, $\boldsymbol{u} \in \mathscr{R}^d$ with the following properties:

- *the function $\ell(\cdot)$ is symmetric, i.e., $\ell(\boldsymbol{u}) = \ell(-\boldsymbol{u})$, continuous at $\boldsymbol{u} = \mathbf{0}$, $\ell(\mathbf{0}) = 0$ but is not identically 0, $\ell(\boldsymbol{u}) \geq 0$,*
- *the sets $\{\boldsymbol{u} : \ell(\boldsymbol{u}) < c\}$ are convex sets for all $c > 0$,*
- *the growth of $\ell(\boldsymbol{u})$ as $|\boldsymbol{u}| \to \infty$ is less than any one of the functions $\exp\{\varepsilon|\boldsymbol{u}|^2\}$, $\varepsilon > 0$.*

The class of such functions we denote as $\mathscr{W}_{e,2}$ and the subclass of functions having polynomial majorants will be denoted as $\mathscr{W}_p$, i.e., if $\ell(\cdot) \in \mathscr{W}_p$, then $\ell(\cdot) \in \mathscr{W}_{e,2}$ and there exist constants $C > 0$ and $p > 0$ such that $\ell(\boldsymbol{u}) \leq C\ (1 + |\boldsymbol{u}|^p)$.

For LAN families we have the following general result due to Hajek [95] and Le Cam [163].

Theorem 2.4. (Hajek–Le Cam) *Let the family of measures $\{\mathbf{P}_{\boldsymbol{\vartheta}}^{(T)},\ \boldsymbol{\vartheta} \in \Theta\}$ be LAN at the point $\boldsymbol{\vartheta}_0$ with (2.5) and (2.6). Then for any loss function $\ell(\cdot) \in \mathscr{W}_{e,2}$ the inequality*

$$\varliminf_{\delta\to 0}\ \varliminf_{T\to\infty}\ \inf_{\bar{\boldsymbol{\vartheta}}_T}\ \sup_{|\boldsymbol{\vartheta}-\boldsymbol{\vartheta}_0|<\delta} \mathbf{E}_{\boldsymbol{\vartheta}}\,\ell\left(\varphi_T(\boldsymbol{\vartheta}_0)^{-1}(\bar{\boldsymbol{\vartheta}}_T - \boldsymbol{\vartheta})\right) \geq \mathbf{E}\ \ell\left(\mathbf{I}(\boldsymbol{\vartheta}_0)^{-1/2}\zeta\right) \tag{2.11}$$

holds. Here $\mathscr{L}(\boldsymbol{\zeta}) = \mathcal{N}(\mathbf{0}, \boldsymbol{J})$.

If, moreover, $\Theta \subset \mathscr{R}^1$ and the sets $\{u : \ell(u) < c\}$ are bounded for all $c > 0$ sufficiently small, then the equality

$$\lim_{\delta\to 0}\ \lim_{T\to\infty}\ \sup_{|\vartheta-\vartheta_0|<\delta} \mathbf{E}_\vartheta\,\ell\left(\varphi_T^{-1}(\bar{\vartheta}_T - \vartheta)\right) = \mathbf{E}\ \ell\left(\mathrm{I}(\vartheta_0)^{-1/2}\zeta\right)$$

is possible if and only if

$$\mathbf{P}_{\vartheta_0} - \lim_{T\to\infty}\left(\varphi_T^{-1}(\bar{\vartheta}_T - \vartheta) - \Delta_T\left(\vartheta, X^T\right)\right) = 0. \tag{2.12}$$

Proof. See [109], Theorem 12.1.

In the one-dimensional case and a quadratic loss function an elementary proof was given in Section 1.3 (see (1.102), (1.106) and (1.107)). For a more general class of loss functions but Gaussian Δ_T (see (2.10) and (2.11)) another (slightly simplified) proof can be found in [139], Theorem 2.1. The relation (2.11) is called the *Hajek–Le Cam inequality.*

Following [109] we define an asymptotically efficient estimator with the help of the inequality (2.11) as follows.

Definition 2.5. *Let the family of measures be LAN at Θ. Then we say that an estimator $\bar\vartheta_T$ is asymptotically efficient (or locally asymptotically minimax) for the loss function $\ell(\cdot)$ if*

$$\lim_{\delta\to 0}\lim_{T\to\infty}\sup_{|\vartheta-\vartheta_0|<\delta}\mathbf{E}_\vartheta\,\ell(T^{1/2}(\bar\vartheta_T-\vartheta))=\mathbf{E}\;\ell\left(\mathbf{I}\left(\vartheta_0\right)^{-1/2}\zeta\right)\tag{2.13}$$

for all $\vartheta_0\in\Theta$.

We need a bit more than the condition $\mathcal{RP}\,(\Theta)$ providing the ergodicity of the observed process. We suppose that the condition $\mathcal{A}_0\,(\Theta)$ is fulfilled and remember that by this condition the process X has ergodic properties with the invariant density

$$f\left(\vartheta,x\right)=G\left(\vartheta\right)^{-1}\,\sigma(x)^{-2}\exp\left\{2\int_0^x\frac{S(\vartheta,v)}{\sigma(v)^2}\,\mathrm{d}v\right\},\quad x\in\mathscr{R}\tag{2.14}$$

and the likelihood ratio is equal to

$$L(\vartheta,\vartheta_1;X^T)=\frac{G\left(\vartheta_1\right)}{G\left(\vartheta\right)}\exp\left\{2\int_0^{X_0}\frac{S(\vartheta,v)-S(\vartheta_1,v)}{\sigma(v)^2}\,\mathrm{d}v\right\}$$
$$\exp\left\{\int_0^T\frac{S(\vartheta,X_t)-S(\vartheta_1,X_t)}{\sigma(X_t)^2}\mathrm{d}X_t-\frac{1}{2}\int_0^T\frac{S(\vartheta,X_t)^2-S(\vartheta_1,X_t)^2}{\sigma(X_t)^2}\,\mathrm{d}t\right\}.\tag{2.15}$$

Using this expression for the likelihood ratio we study below the asymptotical behavior of the MLE and BE as $T\to\infty$. First we describe the properties of these estimators in the regular case, when the trend coefficient is a differentiable function of the unknown parameter. A little later (in Sections 2.6 and 3.2–3.4) we discuss the properties of the same estimators in nonregular cases, i.e., in the situations when the regularity conditions given below are not fulfilled.

Regularity conditions $\mathcal{A}$.

$\mathcal{A}_1$. *The function $S(\vartheta,\cdot)$ is continuously differentiable* w.r.t. ϑ, *the derivative (vector) $\dot S(\vartheta,\cdot)\in\mathscr{P}$ and is uniformly continuous in the following sense: for any compact $\mathbb{K}\subset\Theta$*

$$\lim_{\delta\to 0}\sup_{\vartheta_0\in\mathbb{K}}\sup_{|\vartheta-\vartheta_0|<\delta}\mathbf{E}_{\vartheta_0}\left|\frac{\dot{\mathbf{S}}(\vartheta,\xi)-\dot{\mathbf{S}}(\vartheta_0,\xi)}{\sigma(\xi)}\right|^2=0. \tag{2.16}$$

$\mathcal{A}_2$. *The information matrix*

$$\mathbf{I}\left(\vartheta\right)=\mathbf{E}_\vartheta\left(\frac{\dot{S}(\vartheta,\xi)\,\dot{S}(\vartheta,\xi)^{\mathbb{T}}}{\sigma\left(\xi\right)^2}\right)$$

is positive definite

$$\inf_{\vartheta\in\mathbb{K}}\inf_{e:|e|=1}\left(\mathbf{I}\left(\vartheta\right)e,e\right)>0 \tag{2.17}$$

(here $e\in\mathscr{R}^d$*) and for any* $\nu>0$

$$\inf_{\vartheta_0\in\mathbb{K}}\inf_{|\vartheta-\vartheta_0|>\nu}\mathbf{E}_{\vartheta_0}\left(\frac{S(\vartheta,\xi)-S(\vartheta_0,\xi)}{\sigma(\xi)}\right)^2>0. \tag{2.18}$$

The identifiability condition in this problem is (2.18).

2.1.2 Maximum Likelihood Estimator

The *maximum likelihood estimator* (MLE) $\hat{\vartheta}_T$ is defined as a solution of the equation

$$L(\hat{\vartheta}_T,\theta_1;X^T)=\sup_{\theta\in\Theta}L(\theta,\theta_1;X^T).$$

If this equation has more than one solution, then any one can be taken as the MLE. The function $L(\vartheta,\vartheta_1;X^T)$, $\vartheta\in\Theta$ in all our problems will be continuous with probability 1 and the set Θ is bounded (see, e.g., Lemmae 2.10, 3.9 and 3.20, 3.28 below). Therefore the MLE $\hat{\vartheta}_T\in\bar{\Theta}$ with $\mathbf{P}_\vartheta^{(T)}$ probability 1, where $\bar{\Theta}$ is the closure of Θ. Remember that with positive probability the MLE can take its values on the border of the set Θ.

To describe the asymptotic behavior of the MLE in regular and nonregular (nonsmooth) cases we use the method by Ibragimov–Khasminskii. More precisely, we use two general theorems : Theorem I.10.1 (for MLE) and Theorem I.10.2 (for BE) in [109]), which states that if the likelihood ratio satisfies certain conditions, then uniformly on compacts the MLE and BE are consistent, asymptotically normal in the regular case (or have another limit distribution in non regular cases) and the moments converge. Below we verify these conditions but first for convenience of referencing we repeat these two results (here for MLE and in the next section for BE).

Let us denote by $\left(\mathscr{X}^{(T)},\mathfrak{B}^{(T)}\right)$ a measurable space and consider a family of measures $\left\{\mathbf{P}_\vartheta^{(T)},\ \vartheta\in\Theta\right\}$ induced by X^T on this space. Here $\Theta\subset\mathscr{R}^d$ is an open set. Suppose that there exists a matrix $\varphi_T(\vartheta)$, such that its norm $|\varphi_T(\vartheta)| \to 0$ as $T \to \infty$ for all $\vartheta \in \Theta$ and the process

$$Z_{T,\vartheta}(u) = \frac{\mathrm{d}\mathbf{P}^{(T)}_{\vartheta+\varphi_T(\vartheta)u}}{\mathrm{d}\mathbf{P}^{(T)}_{\vartheta}}\left(X^T\right), \qquad u \in \mathbb{U}_{T,\vartheta} = \left\{u:\ \vartheta + \varphi_T(\vartheta)\,u \in \Theta\right\} \tag{2.19}$$

has a nondegenerate limit as $T \to \infty$.

Denote by $\mathscr{C}_0 = \mathscr{C}_0\left(\mathscr{R}^d\right)$ the space of continuous on $\mathscr{R}^d$ functions $z(u), u \in \mathscr{R}^d$ decreasing to zero at infinity, i.e., $\lim_{|u|\to\infty} z(u) = 0$ with the norm $\|z\| = \sup_u |z(u)|$ and denote by $\mathfrak{B}_0$ the corresponding σ-algebra of Borel sets.

Further, denote by $\mathscr{G}$ the set of families of functions $k_T(y)$ with the properties:

- For a fixed $T > 0$, $k_T(y)$ is a positive function on $[0,\infty)$ and $k_T(y) \uparrow \infty$ as $y \uparrow \infty$.
- For any $N > 0$,

$$\lim_{T\to\infty, y\to\infty} y^N\, \mathrm{e}^{-k_T(y)} = 0. \tag{2.20}$$

Theorem 2.6. (Ibragimov–Khasminskii) *Let the functions $Z_{T,\vartheta}(u)$, $u \in \mathbb{U}_{T,\vartheta}$ be continuous with probability 1 and possessing the following properties:*

1. *For any compact $\mathbb{K} \subset \Theta$, there correspond numbers $a(\mathbb{K}) = a$ and $B(\mathbb{K}) = B$ and functions $k_T^{\mathbb{K}}(\cdot) = k_T(\cdot) \in \mathscr{G}$, such that*
   - *there exist numbers $q > d$ and $m \geq q$ such that for $|u_1| < R$, $|u_2| < R$ and any $R > 0$*

$$\sup_{\vartheta\in\mathbb{K}} \mathbf{E}_\vartheta \left|Z_{T,\vartheta}(u_2)^{1/m} - Z_{T,\vartheta}(u_1)^{1/m}\right|^m \leq B\left(1 + R^a\right)|u_2 - u_1|^q, \tag{2.21}$$

   - *for all $u \in \mathbb{U}_{T,\vartheta}$*

$$\sup_{\vartheta\in\mathbb{K}} \mathbf{E}_\vartheta\ Z_{T,\vartheta}(u)^{1/2} \leq \mathrm{e}^{-k_T(|u|)}. \tag{2.22}$$

2. *Uniformly in $\vartheta \in \mathbb{K}$ the marginal (finite-dimensional) distributions of the random functions $Z_{T,\vartheta}(u)$ converge to marginal distributions of the random functions $Z_\vartheta(u)$ where $Z_\vartheta(\cdot) \in \mathscr{C}_0\left(\mathscr{R}^d\right)$.*
3. *The limit functions $Z_\vartheta(\cdot)$ with probability 1 attain the maximum at the unique point $\hat{u}(\vartheta) = \hat{u}$.*

Then the MLE $\hat{\vartheta}_T$ is uniformly in $\vartheta \in \mathbb{K}$ consistent, i.e., for any $\nu > 0$

$$\lim_{T\to\infty} \sup_{\vartheta\in\mathbb{K}} \mathbf{P}^{(T)}_\vartheta \left\{\left|\hat{\vartheta}_T - \vartheta\right| > \nu\right\} = 0,$$

the distributions of the random variables $\hat{\mathbf{u}}_T = \varphi_T(\vartheta)^{-1}\left(\hat{\vartheta}_T - \vartheta\right)$ *converge uniformly in* $\vartheta \in \mathbb{K}$ *to the distribution of* $\hat{\mathbf{u}}$ *and for any loss function* $\ell(\cdot) \in \mathscr{W}_p$ *we have uniformly in* $\vartheta \in \mathbb{K}$

$$\lim_{T\to\infty} \mathbf{E}_\vartheta\ \ell\left(\varphi_T(\vartheta)^{-1}\left(\hat{\vartheta}_T - \vartheta\right)\right) = \mathbf{E}_\vartheta\ \ell(\hat{u}). \tag{2.23}$$

For the detailed proof see [109], Theorem I.10.1, and we give here just some comments concerning the method.

Remark 2.7. In our problems we will check not the condition (2.22) but another one such as: *for any* $M > 1$ *there exist constants* $\kappa > 0$ *and* $C(M) > 0$ *such that*

$$\sup_{\vartheta\in\mathbb{K}} \mathbf{P}_\vartheta^{(T)}\left\{Z_{T,\vartheta}(\boldsymbol{u}) \geq \mathrm{e}^{-\frac{\kappa}{4}|\boldsymbol{u}|^2}\right\} \leq \frac{C(M)}{|\boldsymbol{u}|^M}$$

(see Lemmas 2.11, 3.12, 3.29 etc.). We just note that, say,

$$\begin{aligned}
&\mathbf{E}_\vartheta\ Z_{T,\vartheta}^{1/2}(\boldsymbol{u}) \\
&\quad= \mathbf{E}_\vartheta\ Z_{T,\vartheta}^{1/2}(\boldsymbol{u})\ \chi_{\left\{\ln Z_{T,\vartheta}(\boldsymbol{u}) < -\frac{\kappa}{4}|\boldsymbol{u}|^2\right\}} \\
&\qquad\quad + \mathbf{E}_\vartheta\ Z_{T,\vartheta}^{1/2}(\boldsymbol{u})\ \chi_{\left\{\ln Z_{T,\vartheta}(\boldsymbol{u}) \geq -\frac{\kappa}{4}|\boldsymbol{u}|^2\right\}} \\
&\quad\leq \mathrm{e}^{-\frac{\kappa}{8}|\boldsymbol{u}|^2} + \left(\mathbf{E}_\vartheta\ Z_{T,\vartheta}(\boldsymbol{u})\right)^{1/2}\left(\mathbf{P}_\vartheta^{(T)}\left\{Z_{T,\vartheta}(\boldsymbol{u}) \geq \mathrm{e}^{-\frac{\kappa}{4}|\boldsymbol{u}|^2}\right\}\right)^{1/2} \\
&\quad\leq \frac{C}{|\boldsymbol{u}|^{\frac{M}{2}}} = \mathrm{e}^{-k_T(|\boldsymbol{u}|)}.
\end{aligned}$$

The function $k_T(\cdot)$ in the last inequality depends on M, but for any $N = M/2$ there exists a function $k_T(\cdot)$ such that the condition (2.20) is fulfilled and for the proof of Theorems 2.6 and 2.12 below this estimate is sufficient.

If the set $\boldsymbol{\Theta}$ is bounded, then the process $Z_{T,\vartheta}(\boldsymbol{u})$ is defined continuously decreasing to zero on the set $\mathbb{U}_{T,\vartheta}^c$, so $Z_{T,\vartheta}(\cdot) \in \mathscr{C}_0\left(\mathscr{R}^d\right)$ with probability 1. Let us denote by $\mathfrak{Q}_T^\vartheta$ the measure induced in the measurable space $(\mathscr{C}_0, \mathfrak{B}_0)$ by the process $Z_{T,\vartheta}(\cdot)$ and let $\mathfrak{Q}^\vartheta$ be the corresponding measure in this space of the limiting process $Z_\vartheta(\cdot)$. Obviously, the random vector $\hat{\boldsymbol{u}}_T$ satisfies the equality

$$Z_{T,\vartheta}(\hat{\boldsymbol{u}}_T) = \sup_{\boldsymbol{u}\in\mathscr{R}^d} Z_{T,\vartheta}(\boldsymbol{u}).$$

Therefore $\hat{\boldsymbol{u}}_T$ is a continuous functional ($\hat{\boldsymbol{u}}_T = \Phi(Z_{T,\vartheta})$) on the space $\mathscr{C}_0$. Suppose that we already proved the weak convergence uniform on $\vartheta \in \mathbb{K}$

$$\mathfrak{Q}_T^\vartheta \Longrightarrow \mathfrak{Q}^\vartheta. \tag{2.24}$$

Then the distributions of the continuous functionals converge as well and we have for any parallelepiped $\mathbb{B} \in \mathfrak{B}\left(\mathscr{R}^d\right)$ with $\mathfrak{Q}^\vartheta(\partial\mathbb{B}) = 0$ the convergence

$$\mathbf{P}_{\vartheta}^{(T)}\left\{\varphi_T(\vartheta)^{-1}\left(\hat{\vartheta}_T-\vartheta\right)\in\mathbb{B}\right\}=\mathbf{P}_{\vartheta}^{(T)}\left\{\sup_{u\in\mathbb{B}}Z_{T,\vartheta}(u)>\sup_{u\in\mathbb{B}^c}Z_{T,\vartheta}(u)\right\}$$

$$=\mathbf{P}_{\vartheta}^{(T)}\left\{\sup_{u\in\mathbb{B}}Z_{T,\vartheta}(u)-\sup_{u\in\mathbb{B}^c}Z_{T,\vartheta}(u)>0\right\}$$

$$\longrightarrow\mathbf{P}_{\vartheta}\left\{\sup_{u\in\mathbb{B}}Z_{\vartheta}(u)-\sup_{u\in\mathbb{B}^c}Z_{\vartheta}(u)>0\right\}$$

$$=\mathbf{P}_{\vartheta}\left\{\sup_{u\in\mathbb{B}}Z_{\vartheta}(u)>\sup_{u\in\mathbb{B}^c}Z_{\vartheta}(u)\right\}=\mathbf{P}_{\vartheta}\left\{\hat{u}\in\mathbb{B}\right\}. \tag{2.25}$$

In the regular cases of the ergodic diffusion process, we can take $\varphi_T(\vartheta)=T^{-1/2}\mathbf{J}$, $\mathbf{J}$ is a unit $d\times d$ matrix and the limiting process for $Z_{T,\vartheta}(u)=L\left(\vartheta+T^{-1/2}u,\vartheta,X^T\right)$ is

$$Z_{\vartheta}(u)=\exp\left\{\left(u,\boldsymbol{\Delta}(\vartheta)\right)-\frac{1}{2}\left(\mathbf{I}(\vartheta)u,u\right)\right\},\qquad u\in\mathscr{R}^d, \tag{2.26}$$

where $\boldsymbol{\Delta}(\vartheta)\sim\mathcal{N}\left(\mathbf{0},\mathbf{I}(\vartheta)\right)$ and $\mathbf{I}(\vartheta)$ is the information matrix. Hence

$$\hat{u}=\arg\sup_{u\in\mathscr{R}^d}Z_{\vartheta}(u)=\mathbf{I}(\vartheta)^{-1}\boldsymbol{\Delta}(\vartheta)$$

and the MLE is asymptotically normal. This convergence, together with the uniform integrability of the random variables $\left|\varphi_T(\vartheta)^{-1}\left(\hat{\vartheta}_T-\vartheta\right)\right|^p$ for any $p>0$, provides the convergence of moments. The uniform integrability follows from the estimate on the tails of the likelihood ratio: for any $R>0$

$$\sup_{\vartheta\in\mathbb{K}}\mathbf{P}_{\vartheta}^{(T)}\left\{\sup_{|u|>R}Z_{T,\vartheta}(u)>\frac{1}{R^N}\right\}<\frac{C_N}{|R|^N}, \tag{2.27}$$

which have to be proved too (see [109], Remark I.5.1). Here $N>0$ can be chosen as large as we want. Indeed, for any $R>0$

$$\sup_{\vartheta\in\mathbb{K}}\mathbf{P}_{\vartheta}^{(T)}\left\{\left|\varphi_T(\vartheta)^{-1}\left(\hat{\vartheta}_T-\vartheta\right)\right|^p>R\right\}$$

$$=\sup_{\vartheta\in\mathbb{K}}\mathbf{P}_{\vartheta}^{(T)}\left\{\sup_{|u|>R^{1/p}}Z_{T,\vartheta}(u)>\sup_{|u|\leq R^{1/p}}Z_{T,\vartheta}(u)\right\}$$

$$\leq\sup_{\vartheta\in\mathbb{K}}\mathbf{P}_{\vartheta}^{(T)}\left\{\sup_{|u|>R^{1/p}}Z_{T,\vartheta}(u)>1\right\}\leq\frac{C_N}{R^{N/p}}.$$

Hence denoting by $H_T(u)$ the distribution function of the random variable

$$u_T=\left|\varphi_T(\vartheta)^{-1}\left(\hat{\vartheta}_T-\vartheta\right)\right|$$

we can write

$$\sup_{\vartheta\in\mathbb{K}} \mathbf{E}_\vartheta u_T^p = \sup_{\vartheta\in\mathbb{K}} \int_0^\infty u^p \,\mathrm{d}H_T(u) \le \sup_{\vartheta\in\mathbb{K}} \left(1 - \int_1^\infty u^p \,\mathrm{d}\left[1 - H_T(u)\right]\right)$$
$$\le 1 - \left[1 - H_T(1)\right] + p\int_1^\infty u^{p-1} \frac{C_N}{u^{N/p}}\,\mathrm{d}u \le C$$

because we can always choose N sufficiently large. This uniform integrability implies the convergence of moments.

To verify a uniform on compacts $\mathbb{K}$ convergence (2.24) one needs to check the uniform on compacts convergence of marginal distributions of $Z_{T,\vartheta}(\cdot)$ (the weak convergence of the vectors $\left(Z_{T,\vartheta}(\boldsymbol{u}_1),\ldots,Z_{T,\vartheta}(\boldsymbol{u}_k)\right)$ to $(Z_\vartheta(\boldsymbol{u}_1),\ldots,Z_\vartheta(\boldsymbol{u}_k))$ for any collection $\boldsymbol{u}_1,\ldots,\boldsymbol{u}_k$ and $k\ge 1$) and the tightness of the family of measures $\left\{\mathfrak{Q}_T^\vartheta,\ \vartheta\in\mathbb{K}, T>0\right\}$, i.e., to check that for any $\varepsilon>0$ and any $\mathbb{K}\subset\Theta$ there exists a compact $\mathbb{D}_\varepsilon\in\mathfrak{B}_0$, such that

$$\sup_{\vartheta\in\mathbb{K}} \mathfrak{Q}_T^\vartheta(\mathbb{D}_\varepsilon) \ge 1-\varepsilon \tag{2.28}$$

for all T.

The convergence of marginal distributions is just condition 2 of Theorem 2.6 and the tightness is (2.28) together with the estimate (2.27) is verified with the help of condition 1 of this theorem (for details see [109], Theorem I.5.1).

It is important to note that the similar approach works in the case of nonregular problems too and this makes the Ibragimov–Khasminskii method quite attractive in many problems of parameter estimation. Say, in Section 3.2 the limiting process for the normalized likelihood ratio $Z_{T,\vartheta}(u) = L\left(\vartheta + T^{-1/2H}u, \vartheta, X^T\right)$, $H\in(1/2,1)$ is

$$Z_\vartheta(u) = \exp\left\{\Gamma_\vartheta\, W^H(u) - \frac{\Gamma_\vartheta^2}{2}\,|u|^{2H}\right\},\qquad u\in\mathscr{R},$$

where $W^H(\cdot)$ is the *fractional Brownian motion* and Γ_ϑ is some constant. In Sections 3.3 and 3.4 the normalized likelihood ratio

$$Z_{T,\vartheta}(u) = L\left(\vartheta + T^{-1}u, \vartheta, X^T\right),\qquad u\in\mathscr{R}$$

converges to the process

$$Z_\vartheta(u) = \exp\left\{\Gamma W(u) - \frac{\Gamma^2}{2}\,|u|\right\},\qquad u\in\mathscr{R},$$

where $W(\cdot)$ is a *two-sided Wiener process* and Γ is some constant. If the weak convergence (2.24) is proved for the corresponding distributions, then we have the convergence similar to (2.25). In particular, for the MLE in Sections 3.3 and 3.4 we have

$$\begin{aligned}
\mathbf{P}_{\vartheta}^{(T)}\left\{T\left(\hat{\vartheta}_T-\vartheta\right)\in\mathbb{B}\right\} &= \mathbf{P}_{\vartheta}^{(T)}\left\{\sup_{u\in\mathbb{B}} Z_{T,\vartheta}\left(u\right) > \sup_{u\in\mathbb{B}^c} Z_{T,\vartheta}\left(u\right)\right\} \\
&= \mathbf{P}_{\vartheta}^{(T)}\left\{\sup_{u\in\mathbb{B}} Z_{T,\vartheta}\left(u\right) - \sup_{u\in\mathbb{B}^c} Z_{T,\vartheta}\left(u\right) > 0\right\} \\
&\longrightarrow \mathbf{P}_{\vartheta}\left\{\sup_{u\in\mathbb{B}} Z_{\vartheta}\left(u\right) - \sup_{u\in\mathbb{B}^c} Z_{\vartheta}\left(u\right) > 0\right\} \\
&= \mathbf{P}_{\vartheta}\left\{\sup_{u\in\mathbb{B}} Z_{\vartheta}\left(u\right) > \sup_{u\in\mathbb{B}^c} Z_{\vartheta}\left(u\right)\right\} = \mathbf{P}_{\vartheta}\left\{\hat{u}\in\mathbb{B}\right\}, \qquad (2.29)
\end{aligned}$$

where

$$\begin{aligned}
\hat{u} &= \arg\sup_{u\in\mathscr{R}} Z_{\vartheta}\left(u\right) = \arg\sup_{u\in\mathscr{R}}\left\{\Gamma\, W\left(u\right) - \frac{\Gamma^2}{2}\,\left|u\right|\right\} \\
&= \Gamma^{-2}\arg\sup_{v\in\mathscr{R}}\left\{W\left(v\right) - \frac{\left|v\right|}{2}\right\} = \frac{\hat{v}}{\Gamma^2}. \qquad (2.30)
\end{aligned}$$

Therefore we obtain the limit distribution of the MLE

$$\mathcal{L}_{\vartheta}\left\{T\left(\hat{\vartheta}_T-\vartheta\right)\right\} \Longrightarrow \mathcal{L}\left\{\hat{u}\right\} \qquad (2.31)$$

in these situations too.

Now we return to the problem of parameter estimation by the observations of the ergodic diffusion process (2.1). We have the following result.

Theorem 2.8. *Let the conditions $\mathcal{A}_0\left(\Theta\right)$, $\mathcal{A}$ be fulfilled. Then the MLE $\hat{\boldsymbol{\vartheta}}_T$ is uniformly on compacts $\mathbb{K}\subset\boldsymbol{\Theta}$ consistent, i.e., for any $\nu>0$*

$$\lim_{T\to\infty}\sup_{\boldsymbol{\vartheta}\in\mathbb{K}}\mathbf{P}_{\boldsymbol{\vartheta}}^{(T)}\left\{\left|\hat{\boldsymbol{\vartheta}}_T-\boldsymbol{\vartheta}\right|>\nu\right\}=0,$$

uniformly asymptotically normal

$$\mathcal{L}_{\boldsymbol{\vartheta}}\left\{T^{1/2}\left(\hat{\boldsymbol{\vartheta}}_T-\boldsymbol{\vartheta}\right)\right\}\Longrightarrow\mathcal{N}\left(\mathbf{0},\mathbf{I}\left(\boldsymbol{\vartheta}\right)^{-1}\right),$$

and we have the uniform on $\boldsymbol{\vartheta}\in\mathbb{K}$ convergence of moments: for any $p>0$

$$\lim_{T\to\infty}\mathbf{E}_{\boldsymbol{\vartheta}}\left|T^{1/2}\left(\hat{\boldsymbol{\vartheta}}_T-\boldsymbol{\vartheta}\right)\right|^p=\mathbf{E}\left|\mathbf{I}\left(\boldsymbol{\vartheta}\right)^{-1/2}\zeta\right|^p,\qquad \zeta\sim\mathcal{N}\left(0,\boldsymbol{J}\right), \qquad (2.32)$$

where $\boldsymbol{J}$ is a unit $d\times d$ matrix. Moreover, the MLE $\hat{\boldsymbol{\vartheta}}_T$ is asymptotically efficient for any loss function $\ell\left(\cdot\right)\in\mathscr{W}_p$.

Proof. Let us introduce the normalized likelihood ratio

$$Z_{T,\boldsymbol{\vartheta}}\left(\boldsymbol{u}\right)\equiv L\left(\boldsymbol{\vartheta}+T^{-1/2}\boldsymbol{u},\boldsymbol{\vartheta},X^T\right),\quad \boldsymbol{u}\in\mathbb{U}_T=\left\{\boldsymbol{u}:\ \boldsymbol{\vartheta}+T^{-1/2}\boldsymbol{u}\in\boldsymbol{\Theta}\right\} \qquad (2.33)$$

and check the conditions of Theorem 2.6 for the process $Z_{T,\boldsymbol{\vartheta}}\left(\cdot\right)$.

We need the following three lemmae.

Lemma 2.9. *Let the conditions $\mathcal{A}_0(\Theta), \mathcal{A}_1$ be fulfilled, then the family of measures $\left\{\mathbf{P}_{\vartheta}^{(T)},\ \vartheta\in\Theta\right\}$ is uniformly at Θ LAN, i.e., the process*

$$Z_{T,\vartheta}(\boldsymbol{u}), \qquad \boldsymbol{u}\in \boldsymbol{U}_T=\left\{\boldsymbol{u}:\boldsymbol{\vartheta}+T^{-1/2}\boldsymbol{u}\in\Theta\right\}$$

admits the representation (2.5), where the random vector

$$\boldsymbol{\Delta}_T(\boldsymbol{\vartheta})=\frac{1}{\sqrt{T}}\int_0^T\frac{\dot{\boldsymbol{S}}(\boldsymbol{\vartheta},X_t)}{\sigma(X_t)^2}\,[\mathrm{d}X_t-S(\boldsymbol{\vartheta},X_t)\,\mathrm{d}t]$$

is asymptotically normal uniformly on compacts $\mathbb{K}\subset\Theta$

$$\mathcal{L}_{\boldsymbol{\vartheta}}\left\{\boldsymbol{\Delta}_T(\boldsymbol{\vartheta})\right\}\Longrightarrow\mathcal{N}(\mathbf{0},\mathbf{I}(\boldsymbol{\vartheta})) \tag{2.34}$$

and for any $\delta>0$

$$\lim_{T\to\infty}\sup_{\boldsymbol{\vartheta}\in\mathbb{K}}\mathbf{P}_{\boldsymbol{\vartheta}}^{(T)}\left\{\left|r_T(\boldsymbol{\vartheta},\boldsymbol{u})\right|\geq\delta\right\}=0. \tag{2.35}$$

Proof. Note that by condition $\mathcal{A}_1$ the measures $\mathbf{P}_{\boldsymbol{\vartheta}}^{(T)}$, $\vartheta\in\Theta$ are equivalent and the likelihood ratio is $\left(\text{here } \boldsymbol{\vartheta}_{\boldsymbol{u}}=\boldsymbol{\vartheta}+T^{-1/2}\boldsymbol{u}\right)$

$$\begin{aligned}Z_{T,\boldsymbol{\vartheta}}(\boldsymbol{u})&=\frac{G(\boldsymbol{\vartheta})}{G(\boldsymbol{\vartheta}_{\boldsymbol{u}})}\exp\left\{2\int_0^{X_0}\frac{S(\boldsymbol{\vartheta}_{\boldsymbol{u}},v)-S(\boldsymbol{\vartheta},v)}{\sigma(v)^2}\,\mathrm{d}v\right\}\\&\times\exp\left\{\int_0^T\frac{S(\boldsymbol{\vartheta}_{\boldsymbol{u}},X_t)-S(\boldsymbol{\vartheta},X_t)}{\sigma(X_t)}\mathrm{d}W_t\right.\\&\qquad\left.-\frac{1}{2}\int_0^T\left(\frac{S(\boldsymbol{\vartheta}_{\boldsymbol{u}},X_t)-S(\boldsymbol{\vartheta},X_t)}{\sigma(X_t)}\right)^2\mathrm{d}t\right\}.\end{aligned}$$

The invariant density $f(\boldsymbol{\vartheta},X_0)$ is a continuous function of $\boldsymbol{\vartheta}$. Therefore it is sufficient to study only the last two integrals.

We write

$$\ln Z_{T,\boldsymbol{\vartheta}}(\boldsymbol{u})=\left(\boldsymbol{u},\boldsymbol{\Delta}_T(\boldsymbol{\vartheta})\right)-\frac{1}{2}\left(\mathbf{I}(\boldsymbol{\vartheta})\,\boldsymbol{u},\boldsymbol{u}\right)+r_T(\boldsymbol{\vartheta},\boldsymbol{u}),$$

where $r_T(\boldsymbol{\vartheta},\boldsymbol{u})=r_{T,1}(\boldsymbol{\vartheta},\boldsymbol{u})+r_{T,2}(\boldsymbol{\vartheta},\boldsymbol{u})+o(1)$ with

$$r_{T,1}(\boldsymbol{\vartheta},\boldsymbol{u})=\int_0^T\frac{S(\boldsymbol{\vartheta}_{\boldsymbol{u}},X_t)-S(\boldsymbol{\vartheta},X_t)-T^{-1/2}(\boldsymbol{u},\dot{\boldsymbol{S}}(\boldsymbol{\vartheta},X_t))}{\sigma(X_t)}\,\mathrm{d}W_t$$

and

$$r_{T,2}(\vartheta,u)=\frac{1}{2}\left(\mathbf{I}(\vartheta)u,u\right)-\frac{1}{2T}\int_0^T\frac{\left|\left(u,\dot S(\vartheta,X_t)\right)\right|^2}{\sigma(X_t)^2}\,dt$$
$$+\frac{1}{2T}\int_0^T\frac{\left|\left(u,\dot S(\vartheta,X_t)\right)\right|^2}{\sigma(X_t)^2}\,dt-\frac{1}{2}\int_0^T\left(\frac{S(\vartheta_u,X_t)-S(\vartheta,X_t)}{\sigma(X_t)}\right)^2dt.$$

The function $S(\vartheta,x)$ is continuously differentiable. Hence

$$S(\vartheta_u,x)-S(\vartheta,x)=T^{-1/2}\int_0^1\left(u,\dot S\left(\vartheta+T^{-1/2}us,x\right)\right)ds. \tag{2.36}$$

For the first integral we have

$$\sup_{\vartheta\in\mathbb{K}}\mathbf{E}_\vartheta\left(r_{T,1}(\vartheta,u)\right)^2$$
$$=T\sup_{\vartheta\in\mathbb{K}}\mathbf{E}_\vartheta\left(\frac{S(\vartheta_u,\xi)-S(\vartheta,\xi)-T^{-1/2}(u,\dot{\mathbf{S}}(\vartheta,\xi))}{\sigma(\xi)}\right)^2$$
$$=\sup_{\vartheta\in\mathbb{K}}\mathbf{E}_\vartheta\left(\int_0^1\frac{(u,[\dot S(\vartheta+T^{-1/2}u\,s,\xi)-\dot{\mathbf{S}}(\vartheta,\xi)])}{\sigma(\xi)}\,ds\right)^2$$
$$\leq\sup_{\vartheta\in\mathbb{K}}\int_0^1\mathbf{E}_\vartheta\left(\frac{(u,[\dot S(\vartheta+T^{-1/2}u\,s,\xi)-\dot{\mathbf{S}}(\vartheta,\xi)])}{\upsilon(\xi)}\right)^2ds\leq|u|^2\,o(1).$$

Further

$$\left|\int_0^T\frac{\left(S(\vartheta_u,X_t)-S(\vartheta,X_t)\right)^2-T^{-1}\left|\left(u,\dot S(\vartheta,X_t)\right)\right|^2}{\sigma(X_t)^2}\,dt\right|^2$$
$$\leq\int_0^T\left(\frac{S(\vartheta_u,X_t)-S(\vartheta,X_t)-T^{-1/2}(u,\dot S(\vartheta,X_t))}{\sigma(X_t)}\right)^2dt$$
$$\int_0^T\left(\frac{S(\vartheta_u,X_t)-S(\vartheta,X_t)+T^{-1/2}(u,\dot S(\vartheta,X_t))}{\sigma(X_t)}\right)^2dt.$$

The first integral on the last right hand side tends to zero and the second is bounded in probability. Therefore $r_T(\vartheta,u)\to 0$ uniformly in $\vartheta\in\mathbb{K}$. To verify (2.35) we need the uniform on $\vartheta\in\mathbb{K}$ law of large numbers: for any $\delta>0$ and $\boldsymbol{\lambda}\in\mathscr{R}^d$

$$\lim_{T\to\infty}\sup_{\vartheta\in\mathbb{K}}\mathbf{P}_\vartheta\left\{\left|\frac{1}{T}\int_0^T\frac{\left(\boldsymbol{\lambda},\dot S(\vartheta,X_t)\right)^2}{\sigma(X_t)^2}\,dt-\left(\mathbf{I}(\vartheta)\boldsymbol{\lambda},\boldsymbol{\lambda}\right)\right|>\delta\right\}=0. \tag{2.37}$$

Put

$$h(\vartheta, x) = \frac{\left(\boldsymbol{\lambda}, \dot{S}(\vartheta, x)\right)^2}{\sigma(x)^2} - \left(\mathbf{I}(\vartheta)\,\boldsymbol{\lambda}, \boldsymbol{\lambda}\right).$$

Note that $\mathbf{E}_\vartheta\, h(\vartheta, \xi) = 0$. Therefore we can apply Proposition 1.18 to verify (2.37). Convergence (2.37) provides as well the uniform asymptotic normality of the stochastic integral $\boldsymbol{\Delta}_T(\vartheta)$ (see Proposition 1.20). Hence we obtain the representation (2.5) with (2.34) and (2.35).

Note that this representation provides us with the convergence of marginal distributions (condition 2 of Theorem 2.6) as follows. For any collection $\boldsymbol{u}_1, \ldots, \boldsymbol{u}_k$ and $k \geq 1$ we have

$$\begin{aligned}
&\mathbf{P}_\vartheta^{(T)}\left\{Z_{T,\vartheta}(\boldsymbol{u}_1) < z_1, \ldots, Z_{T,\vartheta}(\boldsymbol{u}_k) < z_k\right\} \\
&= \mathbf{P}_\vartheta^{(T)}\left\{\left(\boldsymbol{u}_1, \boldsymbol{\Delta}_T(\vartheta)\right) - \frac{1}{2}\left(\mathbf{I}(\vartheta)\,\boldsymbol{u}_1, \boldsymbol{u}_1\right) + r_T(\vartheta, \boldsymbol{u}_1) < \ln z_1,\right. \\
&\qquad \left. \ldots, \left(\boldsymbol{u}_k, \boldsymbol{\Delta}_T(\vartheta)\right) - \frac{1}{2}\left(\mathbf{I}(\vartheta)\,\boldsymbol{u}_1, \boldsymbol{u}_k\right) + r_T(\vartheta, \boldsymbol{u}_k) < \ln z_k\right\} \\
&\longrightarrow \mathbf{P}_\vartheta\left\{\left(\boldsymbol{u}_1, \boldsymbol{\Delta}(\vartheta)\right) - \frac{1}{2}\left(\mathbf{I}(\vartheta)\,\boldsymbol{u}_1, \boldsymbol{u}_1\right) < \ln z_1,\right. \\
&\qquad \left. \ldots, \left(\boldsymbol{u}_k, \boldsymbol{\Delta}(\vartheta)\right) - \frac{1}{2}\left(\mathbf{I}(\vartheta)\,\boldsymbol{u}_1, \boldsymbol{u}_k\right) < \ln z_k\right\} \\
&= \mathbf{P}_\vartheta\left\{Z_\vartheta(\boldsymbol{u}_1) < z_1, \ldots, Z_\vartheta(\boldsymbol{u}_k) < z_k\right\}
\end{aligned}$$

and this convergence is uniform on compacts.

Lemma 2.10. *Let the conditions* $\mathcal{A}_0(\Theta)$, $\mathcal{A}_1$ *be fulfilled. Then for any* $R > 0$

$$\sup_{\vartheta \in \mathbb{K}} \sup_{|\boldsymbol{u}|+|\boldsymbol{v}|<R} |\boldsymbol{u} - \boldsymbol{v}|^{-2d}\, \mathbf{E}_\vartheta \left| Z_{T,\vartheta}^{\frac{1}{2d}}(\boldsymbol{u}) - Z_{T,\vartheta}^{\frac{1}{2d}}(\boldsymbol{v}) \right|^{2d} \leq C\left(1 + R^{2d}\right). \tag{2.38}$$

Proof. Let us denote

$$\delta(\boldsymbol{u}, \boldsymbol{v}, x) = \frac{S(\vartheta_{\boldsymbol{u}}, x) - S(\vartheta_{\boldsymbol{v}}, x)}{\sigma(x)}$$

and apply Lemma 1.1. Then according to (1.23) we obtain the estimate

$$\begin{aligned}
&\mathbf{E}_\vartheta \left| Z_{T,\vartheta}^{\frac{1}{2d}}(\boldsymbol{u}) - Z_{T,\vartheta}^{\frac{1}{2d}}(\boldsymbol{v}) \right|^{2d} \\
&\leq C_1(d) \int_{\mathscr{R}} \left| f(\vartheta_{\boldsymbol{u}}, x)^{1/2d} - f(\vartheta_{\boldsymbol{v}}, x)^{1/2d} \right|^{2d} \mathrm{d}x \\
&\quad + C_2(d)\, T^{2d}\, \mathbf{E}_{\vartheta_{\boldsymbol{u}}} \delta(\boldsymbol{u}, \boldsymbol{v}, \xi)^{4d} + C_3(d)\, T^d\, \mathbf{E}_{\vartheta_{\boldsymbol{u}}} \delta(\boldsymbol{u}, \boldsymbol{v}, \xi)^{2d}.
\end{aligned}$$

For the first term by conditions $\mathcal{A}_0(\Theta)$, $\mathcal{A}_1$ we have

$$\int_{\mathscr{R}} \left| f(\vartheta_{u}, x)^{1/2d} - f(\vartheta_{v}, x)^{1/2d} \right|^{2d} \mathrm{d}x \leq C\, T^{-d}\, |u - v|^{2d},$$

and then we use the equality (2.36)

$$\begin{aligned} T^{d}\, \mathbf{E}_{\vartheta_{u}} &\left(\frac{S(\vartheta_{u}, \xi) - S(\vartheta_{v}, \xi)}{\sigma(\xi)} \right)^{2d} \\ &= \mathbf{E}_{\vartheta_{u}} \left(\int_0^1 \frac{(u - v, \dot{S}(\vartheta + (u - v)s, \xi))}{\sigma(\xi)}\, \mathrm{d}s \right)^{2d} \\ &\leq |u - v|^{2d} \sup_{\vartheta, \tilde{\vartheta} \in \Theta} \mathbf{E}_{\vartheta} \left| \frac{\dot{S}(\tilde{\vartheta}, \xi)}{\sigma(\xi)} \right|^{2d} \leq C\, |u - v|^{2d}. \end{aligned}$$

In a similar way we have

$$T^{2d}\, \mathbf{E}_{\vartheta_{u}} \left(\frac{S(\vartheta_{u}, \xi) - S(\vartheta_{v}, \xi)}{\sigma(\xi)} \right)^{4d} \leq C\, |u - v|^{4d}.$$

Remember that by conditions $\mathcal{A}_0(\Theta), \mathcal{A}_1$ the function $\dot{S}(\vartheta, \cdot)\, \sigma(\cdot)^{-1} \in \mathscr{P}$ and the invariant density $f(\vartheta, x)$ has exponentially decreasing tails.

Therefore for $|u| + |v| < R$

$$\begin{aligned} \sup_{\vartheta \in \mathbb{K}} \mathbf{E}_{\vartheta} \left| Z_{T,\vartheta}^{\frac{1}{2d}}(u) - Z_{T,\vartheta}^{\frac{1}{2d}}(v) \right|^{2d} &\leq C \left(|u - v|^{4d} + |u - v|^{2d} \right) \\ &\leq C\, |u - v|^{2d} \left(1 + R^{2d}\right) \end{aligned}$$

and (2.38) is proved.

Note that by this lemma the random functions $Z_{T,\vartheta}(\cdot)$ are continuous with probability 1.

Lemma 2.11. *Let conditions $\mathcal{A}_0(\Theta), \mathcal{A}$ be fulfilled. Then for any $N > 1$ there exist constants $\kappa > 0$ and $C(N) > 0$ such that*

$$\sup_{\vartheta \in \mathbb{K}} \mathbf{P}_{\vartheta} \left\{ Z_{T,\vartheta}(u) \geq e^{-\frac{\kappa}{4}|u|^2} \right\} \leq \frac{C(N)}{|u|^N}. \tag{2.39}$$

Proof. We have

$$\mathbf{E}_{\vartheta} \left(\frac{S(\vartheta + h, \xi) - S(\vartheta, \xi)}{\sigma(\xi)} \right)^2 = \left(\mathbf{I}(\vartheta)\, h, h\right) + o\left(|h|^2\right).$$

Therefore by condition (2.16) there exists $\nu > 0$ such that for $|\boldsymbol{h}| \leq \nu$

$$\inf_{\vartheta\in\mathbb{K}} \mathbf{E}_\vartheta \left(\frac{S(\vartheta+\boldsymbol{h},\xi)-S(\vartheta,\xi)}{\sigma(\xi)}\right)^2 \geq \frac{1}{2}\inf_{\vartheta\in\mathbb{K}}\Big(\mathbf{I}(\vartheta)\,\boldsymbol{h},\boldsymbol{h}\Big) \geq \kappa_1\,|\boldsymbol{h}|^2 .$$

Let us denote

$$g(\nu) = \inf_{\vartheta\in\mathbb{K}}\inf_{|\boldsymbol{h}|>\nu} \mathbf{E}_\vartheta \left(\frac{S(\vartheta+\boldsymbol{h},\xi)-S(\vartheta,\xi)}{\sigma(\xi)}\right)^2 .$$

For $|\boldsymbol{h}| > \nu$ by condition (2.18) we can write

$$\mathbf{E}_\vartheta \left(\frac{S(\vartheta+\boldsymbol{h},\xi)-S(\vartheta,\xi)}{\sigma(\xi)}\right)^2 \geq g(\nu) \geq g(\nu)\,\frac{|\boldsymbol{h}|^2}{(\mathrm{Diam}\,\Theta)^2} \geq \kappa_2\,|\boldsymbol{h}|^2,$$

where

$$\mathrm{Diam}\,\Theta = \sup_{\vartheta,\vartheta'\in\Theta}|\vartheta-\vartheta'|.$$

Hence for all $\boldsymbol{u}\in \boldsymbol{U}_T$

$$\inf_{\vartheta\in\mathbb{K}} T\,\mathbf{E}_\vartheta \left(\frac{S(\vartheta_{\boldsymbol{u}},\xi)-S(\vartheta,\xi)}{\sigma(\xi)}\right)^2 \geq \kappa\,|\boldsymbol{u}|^2, \tag{2.40}$$

where $\kappa = \min(\kappa_1,\kappa_2)$.

Remember that

$$\begin{aligned}\frac{1}{T}\int_0^T \delta(\boldsymbol{u},0,X_t)^2\,\mathrm{d}t &= \int_{-\infty}^{\infty}\delta(\boldsymbol{u},0,x)^2\,\frac{2\Lambda_T(\vartheta,x)}{T\,\sigma(x)^2}\,\mathrm{d}x\\ &= \mathbf{E}_\vartheta\left(\frac{S(\vartheta_{\boldsymbol{u}},\xi)-S(\vartheta,\xi)}{\sigma(\xi)}\right)^2 + \frac{1}{\sqrt{T}}\int_{-\infty}^{\infty}\delta(\boldsymbol{u},0,x)^2\,\eta_T(\vartheta,x)\,\mathrm{d}x\end{aligned} \tag{2.41}$$

where $\Lambda_T(\vartheta,x)$ is the local time of the diffusion process (see (1.25)) and for a stochastic process

$$\eta_T(\vartheta,x) = \sqrt{T}\left(\frac{2\Lambda_T(\vartheta,x)}{T\,\sigma(x)^2} - f(\vartheta,x)\right), \qquad x\in\mathscr{R}$$

we have the estimate (1.35), which in our assumptions can be written as

$$\sup_{\vartheta\in\Theta}\mathbf{E}_\vartheta\left|\eta_T(\vartheta,x)\right|^M \leq C\,\mathrm{e}^{-\gamma|x|}.$$

It can be shown that under condition $\mathcal{A}_0(\Theta)$

$$f(\vartheta,x)^p\ \mathbf{E}_\vartheta\left|\int_0^\xi\left(\frac{\chi_{\{v>x\}}-F(\vartheta,v)}{\sigma(v)^2 f(\vartheta,v)}\right)\mathrm{d}v\right|^p\leq C\,f(\vartheta,x)\left(1+|x|^{\bar p}\right),\quad(2.42)$$

$$f(\vartheta,x)^p\ \mathbf{E}_\vartheta\left|\frac{\chi_{\{\xi>x\}}-F(\vartheta,\xi)}{\sigma(\xi)\,f(\vartheta,\xi)}\right|^p<C\,f(\vartheta,x)\left(1+|x|^{\tilde p}\right)\qquad(2.43)$$

with some positive $\bar p$, $\tilde p$ (see the proof of (1.35)).

Note that as $\dot{\boldsymbol{S}}(\vartheta,\cdot)\,\sigma(\cdot)^{-1}\in\mathscr{P}$ we have the estimate

$$\delta(\boldsymbol{u},0,x)^2\leq C^2\,\frac{|\boldsymbol{u}|^2}{T}\ \left(1+|x|^p\right)^2.$$

Therefore, for any $M>1$ we can write

$$\begin{aligned}
&\mathbf{E}_\vartheta\left(\int_{-\infty}^{\infty}\delta(\boldsymbol{u},0,x)^2\ \frac{\eta_T(\vartheta,x)}{\sqrt{T}}\ \mathrm{d}x\right)^M\\
&\quad\leq C\left(\frac{|\boldsymbol{u}|^2}{T^{3/2}}\right)^M\left(\int_{-\infty}^{\infty}f(\vartheta,x)^{1/M}\,\mathrm{d}x\right)^{M-1}\\
&\qquad\times\int_{-\infty}^{\infty}\left(1+|x|^p\right)^{2M}\ f(\vartheta,x)^{M-1}\,\mathbf{E}_\vartheta\left|\eta_T(\vartheta,x)\right|^M\ f(\vartheta,x)^{1/M}\,\mathrm{d}x\\
&\quad\leq C\left(\frac{|\boldsymbol{u}|}{T}\right)^M
\end{aligned}\qquad(2.44)$$

because $T^{-1/2}\,|\boldsymbol{u}|\leq\mathrm{Diam}\,\boldsymbol{\Theta}$.

Further, put

$$\delta_0=\ln\sup_{\vartheta,\vartheta'\in\Theta}\frac{G(\vartheta)}{G(\vartheta')}\geq 0.$$

Then according to (2.40) and (2.41) we have

$$\begin{aligned}
&\mathbf{P}_\vartheta^{(T)}\left\{Z_{T,\vartheta}(\boldsymbol{u})>\mathrm{e}^{-\frac{\kappa}{4}|\boldsymbol{u}|^2}\right\}\leq\mathbf{P}_\vartheta^{(T)}\left\{2\int_0^{X_0}\frac{\delta(\boldsymbol{u},0,x)}{\sigma(x)}\,\mathrm{d}x+\delta_0\right.\\
&\qquad\left.+\int_0^T\delta(\boldsymbol{u},0,X_t)\ \mathrm{d}W_t-\frac{1}{2}\int_0^T\delta(\boldsymbol{u},0,X_t)^2\ \mathrm{d}t>-\frac{\kappa}{4}|\boldsymbol{u}|^2\right\}\\
&\quad\leq\mathbf{P}_\vartheta^{(T)}\left\{2\int_0^{X_0}\frac{\delta(\boldsymbol{u},0,x)}{\sigma(x)}\,\mathrm{d}x+\delta_0+\int_0^T\delta(\boldsymbol{u},0,X_t)\ \mathrm{d}W_t\right.\\
&\qquad\left.-\frac{\sqrt{T}}{2}\int_{-\infty}^{\infty}\delta(\boldsymbol{u},0,x)^2\,\eta_T(\vartheta,x)\ \mathrm{d}x>\frac{\kappa}{4}\,|\boldsymbol{u}|^2\right\}\\
&\quad\leq\mathbf{P}_\vartheta^{(T)}\left\{2\int_0^{X_0}\frac{\delta(\boldsymbol{u},0,x)}{\sigma(x)}\,\mathrm{d}x+\delta_0>\frac{\kappa}{12}\,|\boldsymbol{u}|^2\right\}
\end{aligned}$$

$$+ \mathbf{P}_{\vartheta}^{(T)} \left\{ \int_0^T \delta\left(\boldsymbol{u},0,X_t\right)\,\mathrm{d}W_t > \frac{\kappa}{12}\left|\boldsymbol{u}\right|^2 \right\}$$

$$+ \mathbf{P}_{\vartheta}^{(T)} \left\{ \frac{\sqrt{T}}{2}\int_{-\infty}^{\infty} \delta\left(\boldsymbol{u},0,x\right)^2 \eta_T\left(\boldsymbol{\vartheta},x\right)\,\mathrm{d}x > \frac{\kappa}{12}\left|\boldsymbol{u}\right|^2 \right\}.$$

We estimate the last three probabilities with the help of the Chebyshev inequality as follows. We have

$$\mathbf{P}_{\vartheta}^{(T)} \left\{ 2\int_0^{X_0} \frac{\delta\left(\boldsymbol{u},0,x\right)}{\sigma\left(x\right)}\,\mathrm{d}x + \delta_0 > \frac{\kappa}{12}\left|\boldsymbol{u}\right|^2 \right\}$$

$$\leq \left(\frac{12}{\kappa\left|\boldsymbol{u}\right|^2}\right)^M \mathbf{E}_{\vartheta}\left| 2\int_0^{\xi} \frac{\delta\left(\boldsymbol{u},0,x\right)}{\sigma\left(x\right)}\,\mathrm{d}x + \delta_0\right|^M \leq \frac{C}{\left|\boldsymbol{u}\right|^{2M}}.$$

Further

$$\mathbf{P}_{\vartheta}^{(T)} \left\{ \int_0^T \delta\left(\boldsymbol{u},0,X_t\right)\,\mathrm{d}W_t > \frac{\kappa}{12}\left|\boldsymbol{u}\right|^2 \right\}$$

$$\leq \delta\left(\frac{12}{\kappa\left|\boldsymbol{u}\right|^2}\right)^M \mathbf{E}_{\vartheta}\left|\int_0^T \delta\left(\boldsymbol{u},0,X_t\right)\,\mathrm{d}W_t\right|^M$$

$$\leq \frac{C\,T^{M/2}}{\left|\boldsymbol{u}\right|^{2M}}\,\mathbf{E}_{\vartheta}\left|\delta\left(\boldsymbol{u},0,\xi\right)\right|^M \leq \frac{C}{\left|\boldsymbol{u}\right|^{M}}.$$

Finally

$$\mathbf{P}_{\vartheta}^{(T)} \left\{ \frac{\sqrt{T}}{2}\int_{-\infty}^{\infty} \delta\left(\boldsymbol{u},0,x\right)^2 \eta_T\left(\boldsymbol{\vartheta},x\right)\,\mathrm{d}x > \frac{\kappa}{12}\left|\boldsymbol{u}\right|^2 \right\}$$

$$\leq \left(\frac{6\sqrt{T}}{\kappa\left|\boldsymbol{u}\right|^2}\right)^M \mathbf{E}_{\vartheta}\left|\int_{-\infty}^{\infty} \delta\left(\boldsymbol{u},0,x\right)^2 \eta_T\left(\boldsymbol{\vartheta},x\right)\,\mathrm{d}x\right|^M \leq \frac{C}{\left|\boldsymbol{u}\right|^{M}},$$

where we used the estimate (2.44).

The properties of the likelihood ratio established in Lemmas 2.9–2.11 correspond well to the conditions of Theorem 2.6 and so we obtain the consistency, asymptotic normality and convergence of moments (2.32).

The asymptotic efficiency for loss functions with polynomial majorants follows from the uniform convergence of moments and from the continuity of the information matrix $\mathbf{I}\left(\vartheta\right)$. Indeed, we have

$$\lim_{T\rightarrow\infty}\ \sup_{\left|\vartheta-\vartheta_0\right|<\delta}\ \mathbf{E}_{\vartheta}\ \ell\big(T^{1/2}(\hat{\vartheta}_T-\vartheta)\big) = \sup_{\left|\vartheta-\vartheta_0\right|<\delta}\ \mathbf{E}\ \ell\big(\mathbf{I}\left(\vartheta\right)^{-1/2}\zeta\big).$$

The continuity of the matrix $\mathbf{I}\left(\vartheta\right)$ (which follows from conditions $\mathcal{A}_0\left(\Theta\right)$, $\mathcal{A}_1$) allows us to write

$$\lim_{\delta\to 0}\sup_{|\vartheta-\vartheta_0|<\delta} \mathbf{E}\,\ell(\mathbf{I}\,(\vartheta)^{-1/2}\,\zeta) = \mathbf{E}\,\ell(\mathbf{I}\,(\vartheta_0)^{-1/2}\,\zeta).$$

Hence the MLE satisfies the equality (2.13) and is asymptotically efficient (see [109], Theorem III.1.1 and Theorem III.1.3 for a more general statement).

2.1.3 Bayesian Estimators

We observe the same ergodic diffusion process

$$\mathrm{d}X_t = S\,(\vartheta, X_t)\;\mathrm{d}t + \sigma\,(X_t)\;\mathrm{d}W_t, \quad X_0, \quad 0 \le t \le T$$

but the unknown parameter ϑ is supposed to be a random vector with known density *a priori* $\{p\,(v)\,,\; v \in \Theta\}$. We suppose as well that the loss function $\ell\,(u) = |u|^p\,,\; p > 0$ is given too. Introduce the mean (Bayesian) risk of an estimator $\bar{\vartheta}_T$ as

$$\mathcal{R}(\bar{\vartheta}_T) = \int_\Theta \mathbf{E}_\theta\,\ell\left(\sqrt{T}\left(\bar{\vartheta}_T - \theta\right)\right)\,p(\theta)\,\mathrm{d}\theta,$$

and define a *Bayesian estimator* $\tilde{\vartheta}_T = \tilde{\vartheta}_T(X^T)$ as an estimator which minimizes the mean risk, i.e.,

$$\mathcal{R}(\tilde{\vartheta}_T) = \min_{\bar{\vartheta}_T} \mathcal{R}(\bar{\vartheta}_T).$$

If this equation has more than one solution then we can take any of them as the Bayesian estimator. Using the *chain rule* for measures $\mathrm{d}\,\mathbf{P}_\theta^{(T)} = L(\theta,\theta_1,X^T)\,\mathrm{d}\,\mathbf{P}_{\theta_1}^{(T)}$ (here θ_1 is some fixed value and $L(\theta,\theta_1,X^T)$ is the likelihood ratio function) and the Fubini theorem we can write

$$\begin{aligned}\mathcal{R}(\bar{\vartheta}_T) &= \int_\Theta\int_{\mathscr{C}_T} \ell\left(\sqrt{T}\left(\bar{\vartheta}_T - \theta\right)\right)\,\mathrm{d}\,\mathbf{P}_\theta^{(T)}\;p(\theta)\,\mathrm{d}\theta\\ &= \int_{\mathscr{C}_T}\int_\Theta \ell\left(\sqrt{T}\left(\bar{\vartheta}_T - \theta\right)\right)\,L(\theta,\theta_1,X^T)\,p(\theta)\,\mathrm{d}\theta\,\mathrm{d}\,\mathbf{P}_{\theta_1}^{(T)}.\end{aligned}$$

Remember that the observed trajectory $X^T \in \mathscr{C}_T$ with probability 1. Therefore, if we find an estimator $\tilde{\vartheta}_T = \tilde{\vartheta}_T(X^T)$, which minimizes the integral

$$\int_\Theta \ell\left(\sqrt{T}\left(\bar{\vartheta}_T - \theta\right)\right)\,L(\theta,\theta_1,X^T)\,p(\theta)\,\mathrm{d}\theta,$$

then it will be Bayesian. This estimator can as well be found as a solution of the *Bayesian equation*

$$\int_\Theta \dot{\ell}\left(\sqrt{T}\left(\tilde{\vartheta}_T - \theta\right)\right)\,L(\theta,\theta_1,X^T)\,p(\theta)\,\mathrm{d}\theta = \mathbf{0}.$$

Here $\dot{\boldsymbol{\ell}}(\cdot)$ is the vector of derivatives and we suppose that this derivation of the integral is admissible. Note that this system of equations can have many solutions or no solution with positive probability (remember that the set $\boldsymbol{\Theta}$ is bounded).

For a quadratic loss function ($p = 2$) this minimization gives us the conditional expectation (1.117)

$$\tilde{\boldsymbol{\vartheta}}_T = \int_{\boldsymbol{\Theta}} \boldsymbol{\theta}\, p(\boldsymbol{\theta} \mid X^T)\, \mathrm{d}\boldsymbol{\theta}$$

where $p(\boldsymbol{\theta} \mid X^T)$ is the *density a posteriori*,

$$p(\boldsymbol{\theta} \mid X^T) = \frac{L(\boldsymbol{\theta}, \boldsymbol{\theta}_1, X^T)\, p(\boldsymbol{\theta})}{\int_{\boldsymbol{\Theta}} L(\boldsymbol{y}, \boldsymbol{\theta}_1, X^T)\, p(\boldsymbol{y})\, \mathrm{d}\boldsymbol{y}}.$$

We suppose that the set $\boldsymbol{\Theta} \subset \mathscr{R}^d$ is open and bounded and that the prior density $p(\cdot)$ is a continuous positive function on $\bar{\boldsymbol{\Theta}}$.

We describe the asymptotic properties of the Bayesian estimators with the help of another general result by Ibragimov and Khasminskii ([109], Theorem I.10.2) which we formulate here.

Consider on the measurable space $\left(\mathscr{X}^{(T)}, \mathfrak{B}^{(T)}\right)$ and the family of probability measures $\left\{\mathbf{P}_{\boldsymbol{\vartheta}}^{(T)}, \boldsymbol{\vartheta} \in \boldsymbol{\Theta}\right\}$ defined on this space. Here $\boldsymbol{\Theta} \subset \mathscr{R}^d$ is an open set. Denote by $\tilde{\boldsymbol{\vartheta}}_T$ a family of Bayesian estimators with respect to the loss function $\ell\left(\varphi_T(\boldsymbol{\vartheta})^{-1}(\boldsymbol{\theta} - \boldsymbol{\vartheta})\right)$ and prior density $p(\boldsymbol{\theta}), \boldsymbol{\vartheta} \in \boldsymbol{\Theta}$. We assume that $\ell(\cdot) \in \mathscr{W}_p$ and that, moreover, for all $H > 0$ sufficiently large and γ sufficiently small,

$$\inf_{|\boldsymbol{u}|>H} \ell(\boldsymbol{u}) - \sup_{|\boldsymbol{u}|\le H^\gamma} \ell(\boldsymbol{u}) \ge 0.$$

For example, any polynomial loss function $\ell(\boldsymbol{u}) = |\boldsymbol{u}|^p$ with $p > 0$ satisfies this condition.

Denote by $\mathscr{P}_c$ the set of continuous positive functions $p(\boldsymbol{\theta})$, $\boldsymbol{\theta} \in \mathscr{R}^d$ possessing a polynomial majorant. Remember that the Bayesian approach can be applied even in situations when the function $p(\cdot)$ is not a density of some probability law, but just a nonnegative function (of course, supposing that the corresponding integrals exist). In such cases we obtain, the so-called, *generalized Bayesian estimators.*

Theorem 2.12. (Ibragimov–Khasminskii) *Let $\tilde{\boldsymbol{\vartheta}}_T$ be a family of Bayesian estimators and prior density $p(\cdot) \in \mathscr{P}_c$. Assume that the normalized likelihood ratios $Z_{T,\boldsymbol{\vartheta}}(\cdot)$ (2.19) possess the following properties:*

1. *For any compact $\mathbb{K} \subset \boldsymbol{\Theta}$, there correspond numbers $a(\mathbb{K}) = a$ and $B(\mathbb{K}) = B$ and functions $k_T^{\mathbb{K}}(\cdot) = k_T(\cdot) \in \mathscr{G}$, such that*

- *For some $q > 0$, all $|u_1| < R$, $|u_2| < R$ and any $R > 0$*

$$\sup_{\vartheta\in\mathbb{K}} \mathbf{E}_\vartheta \left|Z_{T,\vartheta}(u_2)^{1/2} - Z_{T,\vartheta}(u_1)^{1/2}\right|^2 \leq B\,(1+R^a)\,|u_2-u_1|^q, \tag{2.45}$$

- *for all $u \in \mathbb{U}_{T,\vartheta}$*

$$\sup_{\vartheta\in\mathbb{K}} \mathbf{E}_\vartheta\; Z_{T,\vartheta}(u)^{1/2} \leq \mathrm{e}^{-k_T(|u|)}. \tag{2.46}$$

2. *The marginal distributions of the random functions $Z_{T,\vartheta}(u)$ uniformly in $\vartheta \in \mathbb{K}$ converge to marginal distributions of the random functions $Z_\vartheta(u)$.*
3. *The random function*

$$\psi(v) = \int_{\mathscr{R}^d} \ell(v-u)\,\frac{Z_\vartheta(u)}{\int_{\mathscr{R}^d} Z_\vartheta(y)\,\mathrm{d}y}\,\mathrm{d}u$$

with probability 1 attains its absolute minimum value at the unique point $\tilde{u}(\vartheta) = \tilde{u}$.

Then the BE $\tilde{\vartheta}_T$ is consistent uniformly in $\vartheta \in \mathbb{K}$, i.e., for any $\nu > 0$

$$\lim_{T\to\infty}\sup_{\vartheta\in\mathbb{K}} \mathbf{P}_\vartheta^{(T)}\left\{\left|\tilde{\vartheta}_T - \vartheta\right| > \nu\right\} = 0,$$

the distributions of the random variables $\tilde{u}_T = \varphi_T(\vartheta)^{-1}\left(\tilde{\vartheta}_T - \vartheta\right)$ converge uniformly in $\vartheta \in \mathbb{K}$ to the distribution of $\tilde{u}(\vartheta) - \tilde{u}$ and for any loss function $\ell(\cdot) \in \mathscr{W}_p$ we have uniformly in $\vartheta \in \mathbb{K}$

$$\lim_{T\to\infty} \mathbf{E}_\vartheta\; \ell\left(\varphi_T(\vartheta)^{-1}\left(\tilde{\vartheta}_T - \vartheta\right)\right) = \mathbf{E}_\vartheta\; \ell(\tilde{u}). \tag{2.47}$$

(For the proof see [109], Theorem I.10.2.) As in the case of the MLE we give here some explications concerning this proof.

Below, ϑ is the true value of the parameter. We have the relation

$$\begin{aligned}
\tilde{\vartheta}_T &= \arg\inf_{\bar{\vartheta}_T\in\Theta} \int_\Theta \ell\left(\varphi_T(\vartheta)^{-1}\left(\bar{\vartheta}_T - \theta\right)\right)\, L(\theta,\theta_1,X^T)\,p(\theta)\,\mathrm{d}\theta\\
&= \arg\inf_{\bar{\vartheta}_T\in\Theta} \int_\Theta \ell\left(\varphi_T(\vartheta)^{-1}\left(\bar{\vartheta}_T - \theta\right)\right)\,\frac{L(\theta,\vartheta,X^T)\,p(\theta)}{\int_{y\in\Theta} L(y,\vartheta,X^T)\,p(y)\,\mathrm{d}y}\,\mathrm{d}\theta\\
&= \vartheta + \varphi_T(\vartheta)\arg\inf_{\bar{u}\in\mathbb{U}_{T,\vartheta}} \int_{\mathbb{U}_{T,\vartheta}} \frac{\ell(\bar{u}_T - v)\; Z_{T,\vartheta}(v)\; p(\theta_v)}{\int_{w\in\mathbb{U}_{T,\vartheta}} Z_{T,\vartheta}(w)\; p(\theta_w)\mathrm{d}w}\,\mathrm{d}v\\
&= \vartheta + \varphi_T(\vartheta)\,\tilde{u}_T,
\end{aligned}$$

where we have changed the variables $\theta = \vartheta + \varphi_T(\vartheta)\,v$ and denoted $\theta_v = \vartheta + \varphi_T(\vartheta)\,v$, $\bar{u}_T = \varphi_T(\vartheta)^{-1}\left(\bar{\vartheta}_T - \vartheta\right)$. Therefore to describe the asymptotic behavior of the random variables $\tilde{u}_T = \varphi_T(\vartheta)^{-1}\left(\tilde{\vartheta}_T - \vartheta\right)$ we need to study the random function

$$\psi_T(u) = \int_{\mathbb{U}_{T,\vartheta}} \ell(u-v)\,\frac{Z_{T,\vartheta}(v)\; p(\theta_v)}{\int_{w\in\mathbb{U}_{T,\vartheta}} Z_{T,\vartheta}(w)\; p(\theta_w)\,dw}\,dv.$$

Introduce as well the random function

$$\psi_{T,M}(u) = \int_{|v|\le M} \ell(u-v)\,\frac{Z_{T,\vartheta}(v)\; p(\theta_v)}{\int_{|w|\le M} Z_{T,\vartheta}(w)\; p(\theta_w)\,dw}\,dv.$$

First it is shown that

$$\psi_T(u) = \psi_{T,M}(u)\;(1+\gamma_T(M))$$

where the random variables $\gamma_T(M) \to 0$ as $M \to \infty$ uniformly w.r.t. $\vartheta \in \mathbb{K}$ and

$$p(\vartheta + \varphi_T(\vartheta)\,v) = p(\vartheta)\;(1+o(1))$$

because the function $p(\cdot)$ is continuous. Hence

$$\psi_T(u) = \int_{|v|\le M} \ell(u-v)\,\frac{Z_{T,\vartheta}(v)}{\int_{|w|\le M} Z_{T,\vartheta}(w)\,dw}\,dv\;(1+o(1)).$$

According to the Theorem I.5.7 [109] for any $N>0$ there exists a constant $C = C(N)$ such that for all $M>0$

$$\sup_{\vartheta\in\mathbb{K}} \mathbf{P}_\vartheta^{(T)}\left\{\left|\tilde{u}_T\right| > M\right\} \le \frac{C}{M^N}. \tag{2.48}$$

Therefore it is sufficient to study the convergence of the random functions $\psi_T(u), u \in \mathbb{M}$, where $\mathbb{M} = \{u:\ |u| \le M\} \subset \mathscr{R}^d$.

Let us denote by $(\mathscr{C}_{\mathbb{M}},\ \mathfrak{B}_M)$ the measurable space of functions continuous on $\mathbb{M}$ and by $\mathcal{P}_T^\vartheta$ and $\mathcal{P}^\vartheta$ the distributions induced on this space by the random functions $\left\{\psi_T(u),\ u\in\mathbb{M}\right\}$ and

$$\psi_M(u) = \int_{|v|\le M} \ell(u-v)\,\frac{Z_\vartheta(v)}{\int_{|w|\le M} Z_\vartheta(w)\,dw}\,dv, \qquad u\in\mathbb{M}.$$

respectively. Suppose as well that we proved the uniform in $\vartheta \in \mathbb{K}$ weak convergence

$$\mathcal{P}_T^\vartheta \Longrightarrow \mathcal{P}^\vartheta, \tag{2.49}$$

in the space $\mathscr{C}_{\mathbb{M}}$, then the distributions of the continuous on $\mathscr{C}_{\mathbb{M}}$ functionals converge and we have the following relations. Let $\mathbb{B} \subset \mathscr{R}^d$ be a Borel set and the constant M be such that $\mathbb{B} \subset \mathbb{M} = \{u:\ |u| \le M\}$.

Below we consider the random functions $\psi_T(\cdot)$ and $\psi_M(\cdot)$ on the set $\mathbb{M}$ only. We have

$$\begin{aligned}\mathbf{P}_{\vartheta}^{(T)}&\left\{\varphi_T(\vartheta)^{-1}\left(\tilde{\vartheta}_T-\vartheta\right)\in\mathbb{B}\right\}\\&=\mathbf{P}_{\vartheta}^{(T)}\left\{\inf_{\boldsymbol{u}\in\mathbb{B}}\psi_T(\boldsymbol{u})<\inf_{\boldsymbol{u}\in\mathbb{B}^c}\psi_T(\boldsymbol{u}),\left|\tilde{\boldsymbol{u}}_T\right|<M\right\}\\&\geq\mathbf{P}_{\vartheta}^{(T)}\left\{\inf_{\boldsymbol{u}\in\mathbb{B}}\psi_T(\boldsymbol{u})<\inf_{\boldsymbol{u}\in\mathbb{B}^c}\psi_T(\boldsymbol{u})\right\}-\mathbf{P}_{\vartheta}^{(T)}\left\{\left|\tilde{\boldsymbol{u}}_T\right|>M\right\}\end{aligned}$$

and

$$\mathbf{P}_{\vartheta}^{(T)}\left\{\varphi_T(\vartheta)^{-1}\left(\tilde{\vartheta}_T-\vartheta\right)\in\mathbb{B}\right\}\leq\mathbf{P}_{\vartheta}^{(T)}\left\{\inf_{\boldsymbol{u}\in\mathbb{B}}\psi_T(\boldsymbol{u})<\inf_{\boldsymbol{u}\in\mathbb{B}^c}\psi_T(\boldsymbol{u})\right\}.$$

Hence

$$\begin{aligned}\lim_{T\to\infty}\mathbf{P}_{\vartheta}^{(T)}\left\{\varphi_T(\vartheta)^{-1}\left(\tilde{\vartheta}_T-\vartheta\right)\in\mathbb{B}\right\}\geq&\mathbf{P}_{\vartheta}\left\{\inf_{\boldsymbol{u}\in\mathbb{B}}\psi_M(\boldsymbol{u})<\inf_{\boldsymbol{u}\in\mathbb{B}^c}\psi_M(\boldsymbol{u})\right\}\\&-\sup_T\mathbf{P}_{\vartheta}^{(T)}\left\{\left|\tilde{\boldsymbol{u}}_T\right|>M\right\}\end{aligned}$$

and

$$\lim_{T\to\infty}\mathbf{P}_{\vartheta}^{(T)}\left\{\varphi_T(\vartheta)^{-1}\left(\tilde{\vartheta}_T-\vartheta\right)\in\mathbb{B}\right\}\leq\mathbf{P}_{\vartheta}\left\{\inf_{\boldsymbol{u}\in\mathbb{B}}\psi_M(\boldsymbol{u})<\inf_{\boldsymbol{u}\in\mathbb{B}^c}\psi_M(\boldsymbol{u})\right\}$$

because $\Phi(\psi_T)=\inf_{\boldsymbol{u}\in\mathbb{B}}\psi_T(\boldsymbol{u})-\inf_{\boldsymbol{u}\in\mathbb{B}^c}\psi_T(\boldsymbol{u})$ is a continuous on $\mathscr{C}_{\mathbb{M}}$ functional and we have

$$\mathbf{P}_{\vartheta}^{(T)}\left\{\Phi(\psi_T)>0\right\}\longrightarrow\mathbf{P}_{\vartheta}\left\{\Phi(\psi_M)>0\right\}.$$

Note that

$$\lim_{M\to\infty}\left\{\mathbf{P}_{\vartheta}^{(T)}\left\{\left|\tilde{\boldsymbol{u}}_T\right|>M\right\}+\mathbf{P}_{\vartheta}\left\{\left|\tilde{\boldsymbol{u}}\right|>M\right\}\right\}=0.$$

Therefore

$$\mathbf{P}_{\vartheta}^{(T)}\left\{T^{1/2}\left(\tilde{\vartheta}_T-\vartheta\right)\in\mathbb{B}\right\}\to\mathbf{P}_{\vartheta}\left\{\tilde{\boldsymbol{u}}\in\mathbb{B}\right\}.$$

Moreover it can be shown that all convergences are uniform on compacts $\mathbb{K}\subset\Theta$.

To prove the weak convergence (2.49) we first check the convergence in distribution of the vectors

$$\int_{|\vartheta|\leq M}Z_{T,\vartheta}(\boldsymbol{v})\,\mathrm{d}\boldsymbol{v},\ \int_{|\vartheta|\leq M}\ell(\boldsymbol{u}_i-\boldsymbol{v})\,Z_{T,\vartheta}(\boldsymbol{v})\,\mathrm{d}\boldsymbol{v},\ i=1,\ldots,k,$$

to

$$\int_{|\vartheta|\leq M}Z_{\vartheta}(\boldsymbol{v})\,\mathrm{d}\boldsymbol{v},\ \int_{|\vartheta|\leq M}\ell(\boldsymbol{u}_i-\boldsymbol{v})\,Z_{\vartheta}(\boldsymbol{v})\,\mathrm{d}\boldsymbol{v},\ i=1,\ldots,k,$$

for any integer $k>0$ and then we verify the tightness of this family of measures.

We return now to the problem of parameter estimation by the observations of the ergodic diffusion process (2.1).

Theorem 2.13. *Let the conditions* $\mathcal{A}_0(\Theta), \mathcal{A}$ *be fulfilled. Then the BE* $\tilde{\vartheta}_T$ *is uniformly consistent on compacts* $\mathbb{K} \subset \Theta$, *i.e., for any* $\nu > 0$

$$\lim_{T\to\infty} \sup_{\vartheta\in\mathbb{K}} \mathbf{P}_\vartheta^{(T)} \left\{ \left| \tilde{\vartheta}_T - \vartheta \right| > \nu \right\} = 0,$$

uniformly asymptotically normal

$$\mathcal{L}_\vartheta \left\{ T^{1/2} \left(\tilde{\vartheta}_T - \vartheta \right) \right\} \Longrightarrow \mathcal{N}\left(\mathbf{0}, \mathbf{I}(\vartheta)^{-1} \right)$$

and the moments converge: for any $p > 0$ *uniformly on compacts* $\mathbb{K}$

$$\lim_{T\to\infty} \mathbf{E}_\vartheta \left| T^{1/2} \left(\tilde{\vartheta}_T - \vartheta \right) \right|^p = \mathbf{E} \left| \mathbf{I}(\vartheta)^{-1/2} \zeta \right|^p, \quad \zeta \sim \mathcal{N}(\mathbf{0}, \boldsymbol{J}), \qquad (2.50)$$

where $\boldsymbol{J}$ *is a unit* $d \times d$ *matrix. Moreover the BE is asymptotically efficient for loss functions* $\ell(\cdot) \in \mathscr{W}_p$.

Proof. The proof of this theorem is just a verification of the conditions of Theorem 2.12 in our case, but it was already done in the proof of Theorem 2.2.

Remember that the limiting process for the likelihood ratio $Z_{T,\vartheta}(\cdot)$ is

$$Z_\vartheta(\boldsymbol{v}) = \exp\left\{ (\boldsymbol{v}, \boldsymbol{\Delta}(\vartheta)) - \frac{1}{2} (\mathbf{I}(\vartheta)\boldsymbol{v}, \boldsymbol{v}) \right\}, \qquad \boldsymbol{\Delta}(\vartheta) \sim \mathcal{N}(\mathbf{0}, \mathbf{I}(\vartheta))$$

and the function $\psi(\cdot)$ for this process and any $\ell(\cdot) \in \mathscr{W}_p$ attains its absolute minimum at the point

$$\tilde{\boldsymbol{u}} = \mathbf{I}(\vartheta)^{-1} \boldsymbol{\Delta}(\vartheta) \qquad (2.51)$$

(see [109], p. 180). For a quadratic loss function this can be obtained by direct calculation.

2.1.4 Examples

We consider several particular cases of the stochastic differential equation

$$\mathrm{d}X_t = S(\vartheta, X_t)\,\mathrm{d}t + \sigma(X_t)\,\mathrm{d}W_t, \quad X_0, \quad 0 \le t \le T$$

in the situations when the regularity conditions of the preceding sections are fulfilled. We discuss as well estimation problems for some ergodic diffusion processes without condition $A_0(\Theta)$.

Example 2.14. (*Ornstein–Uhlenbeck process*) Let the observed process be

$$\mathrm{d}X_t = -\left(\vartheta^{(1)}X_t - \vartheta^{(2)}\right)\,\mathrm{d}t + \sigma\,\mathrm{d}W_t,\quad X_0,\quad 0\le t\le T, \tag{2.52}$$

where $\boldsymbol{\vartheta} = \left(\vartheta^{(1)},\vartheta^{(2)}\right)\in\boldsymbol{\Theta}$, and the set $\boldsymbol{\Theta} = (\alpha_1,\beta_1)\times(\alpha_2,\beta_2)$, α_i,β_i are finite, $\alpha_1>0$.

The condition $\mathcal{A}_0\left(\boldsymbol{\Theta}\right)$ of course is fulfilled and the process $\{X_t,\ t\ge 0\}$ has ergodic properties. The invariant density is Gaussian $\mathcal{N}\left(\frac{\vartheta^{(2)}}{\vartheta^{(1)}},\frac{\sigma^2}{2\vartheta^{(1)}}\right)$.

We just check the condition of identifiability (2.18) because the others are trivial. Direct calculation gives us

$$\begin{aligned}
&\mathbf{E}_{\boldsymbol{\vartheta}_0}\left|\left(\vartheta^{(1)}-\vartheta_0^{(1)}\right)\xi-\vartheta^{(2)}+\vartheta_0^{(2)}\right|^2\\
&\quad=\left(\vartheta^{(1)}-\vartheta_0^{(1)}\right)^2\frac{\sigma^2}{2\vartheta_0^{(1)}}+\left(\vartheta^{(2)}-\vartheta_0^{(2)}\right)^2\\
&\quad\ge\min\left(\frac{\sigma^2}{2\beta_1},1\right)\left[\left(\vartheta^{(1)}-\vartheta_0^{(1)}\right)^2+\left(\vartheta^{(2)}-\vartheta_0^{(2)}\right)^2\right]=\kappa\left|\boldsymbol{\vartheta}-\boldsymbol{\vartheta}_0\right|^2.
\end{aligned}$$

The information matrix is asymptotically diagonal,

$$\mathbf{I}(\boldsymbol{\vartheta})^{-1}=\begin{pmatrix}2\vartheta^{(1)} & 0\\ 0 & \sigma^2\end{pmatrix}\left(1+o(1)\right).$$

Therefore the MLE and Bayesian estimators according to Theorems 2.8 and 2.13 are consistent, asymptotically normal

$$\mathcal{L}_{\boldsymbol{\vartheta}}\left\{\sqrt{T}\left(\hat{\boldsymbol{\vartheta}}_T-\boldsymbol{\vartheta}\right)\right\}\Longrightarrow\mathcal{N}\left(\mathbf{0},\mathbf{I}(\boldsymbol{\vartheta})^{-1}\right)$$

and the moments converge too.

Example 2.15. Suppose that the observed process is

$$\mathrm{d}X_t=-\frac{\vartheta^{(1)}\,X_t^3}{1+\vartheta^{(2)}\,X_t^2}\,\mathrm{d}t+\sigma\,\mathrm{d}W_t,\quad X_0,\quad 0\le t\le T, \tag{2.53}$$

where $\boldsymbol{\vartheta}=\left(\vartheta^{(1)},\vartheta^{(2)}\right)\in\boldsymbol{\Theta}$, the set $\boldsymbol{\Theta}=(\alpha_1,\beta_1)\times(\alpha_2,\beta_2)$, α_i,β_i are finite and the values, $\alpha_1>0,\alpha_2>0$.

It is easy to see that the process is ergodic and the invariant density is

$$f\left(\boldsymbol{\vartheta},x\right)=G\left(\boldsymbol{\vartheta}\right)^{-1}\left(1+\vartheta^{(2)}\,x^2\right)^{\frac{\vartheta^{(1)}}{\vartheta^{(2)2}\,\sigma^2}}\exp\left\{-\frac{\vartheta^{(1)}\,x^2}{\vartheta^{(2)}\,\sigma^2}\right\},\qquad x\in\mathscr{R}.$$

We check the condition (2.18) as follows:

$$\mathbf{E}_{\vartheta_0}\left(\frac{\vartheta^{(1)}\,\xi^3}{1+\vartheta^{(2)}\,\xi^2}-\frac{\vartheta_0^{(1)}\,\xi^3}{1+\vartheta_0^{(2)}\,\xi^2}\right)^2$$
$$\geq \mathbf{E}_{\vartheta_0}\,\xi^6\left(\vartheta^{(1)}-\vartheta_0^{(1)}+\left(\vartheta_0^{(2)}\,\vartheta^{(1)}-\vartheta_0^{(1)}\,\vartheta^{(2)}\right)\xi^2\right)^2.$$

If

$$\mathbf{E}_{\vartheta_0}\,\xi^6\left(\vartheta^{(1)}-\vartheta_0^{(1)}+\left(\vartheta_0^{(2)}\,\vartheta^{(1)}-\vartheta_0^{(1)}\,\vartheta^{(2)}\right)\xi^2\right)^2=0$$

for some values $\boldsymbol{\vartheta}_0$ and $\boldsymbol{\vartheta}$, then we have with probability 1 the equality

$$\vartheta^{(1)}-\vartheta_0^{(1)}+\left(\vartheta_0^{(2)}\,\vartheta^{(1)}-\vartheta_0^{(1)}\,\vartheta^{(2)}\right)\xi^2=0,$$

which is possible if and only if $\boldsymbol{\vartheta}_0=\boldsymbol{\vartheta}$.

Denote

$$\psi\left(n,m\right)=\mathbf{E}_{\boldsymbol{\vartheta}}\,\frac{\xi^{2n}}{\left(1+\vartheta^{(2)}\,\xi^2\right)^m},$$

then the information matrix can be written as

$$\mathbf{I}(\boldsymbol{\vartheta})=\begin{pmatrix}\frac{\psi(3,2)}{\sigma^2} & -\frac{\vartheta^{(1)}\psi(4,3)}{\sigma^2}\\ -\frac{\vartheta^{(1)}\psi(4,3)}{\sigma^2} & \frac{\vartheta^{(1)2}\psi(5,4)}{\sigma^2}\end{pmatrix}\left(1+o(1)\right).$$

The Cauchy–Schwarz inequality shows that this matrix is nondegenerate, i.e.,

$$\psi\left(4,3\right)^2<\psi\left(3,2\right)\psi\left(5,4\right),$$

and equality is possible iff ξ is nonrandom.

Therefore the MLE and BE are asymptotically normal with the corresponding limit covariance matrix.

Note as well that the conditions $\mathcal{A}_0\left(\Theta\right)$, $\mathcal{A}$ are fulfilled in the case of a *hyperbolic diffusion process*

$$\mathrm{d}X_t=-\frac{\vartheta\,X_t}{\sqrt{1+X_t^2}}\,\mathrm{d}t+\sigma\,\mathrm{d}W_t,\quad X_0,\quad 0\leq t\leq T$$

with $\vartheta\in\Theta=\left(\alpha,\beta\right),\ \alpha>0$.

Example 2.16. Let the observed process be

$$\mathrm{d}X_t=-\vartheta^{(1)}\,X_t\left[1+\gamma\,\sin\left(\vartheta^{(2)}X_t\right)\right]\,\mathrm{d}t+\sigma\mathrm{d}W_t,\quad X_0,\quad 0\leq t\leq T,\quad (2.54)$$

where $\boldsymbol{\vartheta}\in\boldsymbol{\Theta}=\left(\alpha_1,\beta_1\right)\times\left(\alpha_2,\beta_2\right)$ and γ is known, $|\gamma|<1$. The invariant density is

$$f(\vartheta, x) = \frac{1}{G(\vartheta)} \exp\left\{ -\frac{\vartheta^{(1)} x^2}{\sigma^2} + \frac{2\gamma\vartheta^{(1)} \left[\vartheta^{(2)} x \cos\left(\vartheta^{(2)} x\right) - \sin\left(\vartheta^{(2)} x\right)\right]}{\sigma^2 \left(\vartheta^{(2)}\right)^2} \right\}.$$

If

$$\inf_{|\vartheta - \vartheta_0| > \delta} \mathbf{E}_{\vartheta_0} \left(\vartheta^{(1)} \xi \left[1 + \gamma \sin\left(\vartheta^{(2)} \xi\right)\right] - \vartheta_0^{(1)} \xi \left[1 + \gamma \sin\left(\vartheta_0^{(2)} \xi\right)\right] \right)^2 = 0,$$

then there exist $\vartheta_* \neq \vartheta_0$ such that for almost all $x \in \mathscr{R}$

$$\frac{\vartheta_*^{(1)} - \vartheta_0^{(1)}}{\gamma} = \vartheta_*^{(1)} \sin\left(\vartheta_*^{(2)} x\right) - \vartheta_0^{(1)} \sin\left(\vartheta_0^{(2)} x\right)$$

and this equality is of course impossible.

Note that if $\vartheta^{(2)} \in \mathscr{R}$ then the condition of identifiability is not fulfilled. This condition is not fulfilled as well if $\vartheta^{(2)} \in (\alpha_2, \beta_2)$ with $\beta_2 - \alpha_2 \geq 2\pi$ for the model

$$\mathrm{d}X_t = -\vartheta^{(1)} X_t \left[1 + \gamma \sin\left(\vartheta^{(2)} + X_t\right)\right] \mathrm{d}t + \sigma \mathrm{d}W_t, \quad X_0, \quad 0 \leq t \leq T,$$

because for different values $\vartheta_0^{(2)}$ and $\vartheta_*^{(2)} = \vartheta_0^{(2)} + 2\pi$ we have the equality

$$\sin\left(\theta_*^{(2)} + \xi\right) - \sin\left(\vartheta_0^{(2)} + \xi\right) = 0.$$

We put $\vartheta_*^{(1)} = \vartheta_0^{(1)}$ here.

Consider two examples with nonconstant diffusion coefficients.

Example 2.17. Let the observed process be

$$\mathrm{d}X_t = \left[-\vartheta^{(1)} X_t^3 + \vartheta^{(2)} X_t\right] \mathrm{d}t + \sqrt{1 + X_t^2}\, \mathrm{d}W_t, \quad X_0, \quad 0 \leq t \leq T, \quad (2.55)$$

where $\vartheta \in \Theta \subset \mathscr{R}_+ \times \mathscr{R}$, Θ is a bounded open set.

The condition $\mathcal{A}_0(\Theta)$ is fulfilled and the invariant density is similar to that of Example 2.15

$$f(\vartheta, x) = G(\vartheta)^{-1} (1 + x^2)^{\vartheta^{(1)} + \vartheta^{(2)}} \exp\left\{-\vartheta^{(1)} x^2\right\}, \qquad x \in \mathscr{R}.$$

In this example the MLE (for the deterministic initial value) can be written in an explicit form, the condition of identifiability can be easily checked and the information matrix is non degenerate. Therefore, the MLE and BE have all the properties mentioned in Theorems 2.8 and 2.13.

Example 2.18. For the diffusion process

$$\mathrm{d}X_t = \vartheta \cos(X_t)\ \chi_{\{|X_t|\leq \pi/2\}}\,\mathrm{d}t + \sqrt{1+X_t^2}\,\mathrm{d}W_t, \quad X_0, \quad 0 \leq t \leq T, \tag{2.56}$$

with $\vartheta \in (\alpha, \beta)$ the condition $\mathcal{A}_0(\Theta)$ is not fulfilled but nevertheless

$$G(\vartheta) = 2\int_0^{\pi/2} \frac{e^{2\vartheta\psi(x)}}{1+x^2}\mathrm{d}x + 2\int_{\pi/2}^{\infty} \frac{1}{1+x^2}\mathrm{d}x < \infty,$$

and the process $\{X_t,\ t \geq 0\}$ has ergodic properties. Here

$$\psi(x) = \int_0^x \frac{\cos y}{1+y^2}\,\mathrm{d}y.$$

The invariant density is

$$f(\vartheta, x) = \frac{e^{2\vartheta\psi(x)}}{G(\vartheta)\,(1+x^2)}\,\chi_{\{|x|\leq\pi/2\}} + \frac{1}{G(\vartheta)\,(1+x^2)}\,\chi_{\{|x|>\pi/2\}}, \qquad x \in \mathscr{R}.$$

The MLE (with X_0 not depending on ϑ) has the explicit representation

$$\begin{aligned}\hat\vartheta_T &= \left(\int_0^T \frac{\cos^2(X_t)}{1+X_t^2}\,\chi_{\{|X_t|\leq\pi/2\}}\mathrm{d}t\right)^{-1} \int_0^T \frac{\cos(X_t)}{1+X_t^2}\,\chi_{\{|X_t|\leq\pi/2\}}\mathrm{d}X_t \\ &= \vartheta + \left(\int_0^T \frac{\cos^2(X_t)}{1+X_t^2}\,\chi_{\{|X_t|\leq\pi/2\}}\mathrm{d}t\right)^{-1} \int_0^T \frac{\cos(X_t)}{\sqrt{1+X_t^2}}\,\chi_{\{|X_t|\leq\pi/2\}}\mathrm{d}W_t,\end{aligned}$$

and is consistent and asymptotically normal

$$\mathcal{L}_\vartheta\left\{\sqrt{T}\left(\hat\vartheta_T - \vartheta\right)\right\} \Longrightarrow \mathcal{N}\left(0, \mathrm{d}(\vartheta)^{-2}\right),$$

because by the law of large numbers

$$\begin{aligned}\mathbf{P}_\vartheta - \lim_{T\to\infty} \frac{1}{T}\int_0^T \frac{\cos^2(X_t)}{1+X_t^2}\,\chi_{\{|X_t|\leq\pi/2\}}\mathrm{d}t \\ = \int_{-\pi/2}^{\pi/2} \frac{\cos^2(x)}{1+x^2}\, f(\vartheta, x)\,\mathrm{d}x \equiv \mathrm{d}(\vartheta)^2\end{aligned}$$

and the stochastic integral is asymptotically normal

$$\frac{1}{\sqrt{T}}\int_0^T \frac{\cos(X_t)}{\sqrt{1+X_t^2}}\,\chi_{\{|X_t|\leq\pi/2\}}\mathrm{d}W_t \Longrightarrow \mathcal{N}\left(0, \mathrm{d}(\vartheta)^2\right).$$

Note that in this example the estimator is based on the observations X_t in the *window* $[-\pi/2, \pi/2]$ only and $\mathbf{E}_\vartheta\,|\xi| = \infty$.

2.2 Minimum Distance Estimators

The observed diffusion process is

$$\mathrm{d}X_t = S(\vartheta, X_t)\ \mathrm{d}t + \sigma(X_t)\ \mathrm{d}W_t, \quad X_0, \quad 0 \le t \le T, \tag{2.57}$$

where the trend and diffusion coefficients satisfy the conditions $\mathcal{ES}$ and $\mathcal{A}_0(\Theta)$. Hence there exists a weak solution of this equation and the process has ergodic properties. We have to estimate the unknown parameter $\vartheta \in \Theta$ by observations $X^T = \{X_t, 0 \le t \le T\}$. The set $\Theta \subset \mathscr{R}^d$ is open and bounded.

The invariant law has the density function

$$f(\vartheta, x) = \frac{1}{G(\vartheta)\ \sigma(x)^2} \exp\left\{2\int_0^x \frac{S(\vartheta, v)}{\sigma(v)^2}\ \mathrm{d}v\right\}, \quad x \in \mathscr{R},$$

and as usual we denote by $F(\vartheta, \cdot)$ the corresponding distribution function. Remember that the empirical distribution function and empirical density (the LTE of the density) we denote as

$$\hat{F}_T(x) = \frac{1}{T}\int_0^T \chi_{\{X_t < x\}}\ \mathrm{d}t,$$

and

$$f_T^\circ(x) = \frac{1}{\sigma(x)^2 T}\int_0^T \operatorname{sgn}(x - X_t)\ \mathrm{d}X_t + \frac{|X_T - x| - |X_0 - x|}{\sigma(x)^2 T}$$

respectively.

Let $\mathscr{L}_2$ be a Hilbert space of quadratically integrable on $\mathscr{R}$ functions and denote by $\|\cdot\|$ and $\langle\cdot,\cdot\rangle$ the norm and scalar product in this space. We study two minimum distance estimators. The first one ϑ_T^* is defined by the equation

$$\left\|\hat{F}_T(\cdot) - F(\vartheta_T^*, \cdot)\right\| = \inf_{\vartheta \in \Theta}\left\|\hat{F}_T(\cdot) - F(\vartheta, \cdot)\right\|, \tag{2.58}$$

and the second ϑ_T^{**} comes from the similar relation

$$\left\|f_T^\circ(\cdot) - f(\vartheta_T^{**}, \cdot)\right\| = \inf_{\theta \in \Theta}\left\|f_T^\circ(\cdot) - f(\theta, \cdot)\right\|. \tag{2.59}$$

If these equations have more than one solution, then we call any of them the minimum distance estimator.

We show (under regularity conditions) that the first estimator is uniformly consistent, asymptotically normal and the moments converge. One of the remarkable properties of minimum distance estimators is their asymptotic efficiency for *slightly contaminated models* [183]. To show this type of efficiency for our estimators we study in such a framework the asymptotic behavior of the second MDE. Of course, it is true for the first estimator too, but for another definition of the estimated parameter (see (2.79)).

2.2.1 First MDE

To describe the asymptotic properties of the MDE ϑ_T^* we need regularity conditions, which are less restrictive than the regularity conditions of the preceding section. In particular, as it is shown in the examples below the function $S(\vartheta,\cdot)$ need not be differentiable.

Regularity conditions $\mathscr{B}^*$.

$\mathscr{B}_1^*$. *The function* $F(\vartheta,\cdot)$ *is differentiable over* ϑ *for all* $x \in \mathscr{R}$ *and the derivative* $\dot{F}(\vartheta,\cdot) \in \mathscr{P}$ *is uniformly continuous in the following sense:*

$$\lim_{\nu\to 0}\sup_{\vartheta_0\in\Theta}\sup_{|\vartheta-\vartheta_0|\leq\nu}\left\|\dot{F}(\vartheta,\cdot)-\dot{F}(\vartheta_0,\cdot)\right\| = 0. \tag{2.60}$$

$\mathscr{B}_2^*$. *For any* $\nu > 0$ *the function*

$$g^*(\nu) = \inf_{\vartheta_0\in\Theta}\inf_{|\vartheta-\vartheta_0|>\nu}\|F(\vartheta,\cdot)-F(\vartheta_0,\cdot)\| > 0 \tag{2.61}$$

and the matrix

$$\mathbf{J}(\vartheta) = \left\langle \dot{F}(\vartheta,\cdot), \dot{F}(\vartheta,\cdot)^{\mathrm{T}}\right\rangle$$

is positive definite:

$$\inf_{\vartheta\in\Theta}\inf_{e\in\mathscr{R}^d,\ |e|=1}(\mathbf{J}(\vartheta)e,e) > 0. \tag{2.62}$$

Note that $\mathscr{B}_2^*$ is the global identifiability condition in this problem and (2.61) is quite close to condition (2.18). Indeed, suppose that (2.61) is not fulfilled. Then there exists a value $\vartheta_* \neq \vartheta_0$ such that $F(\vartheta_*,x) = F(\vartheta_0,x)$ for all $x \in \mathscr{R}$ (remember that $F(\vartheta,x)$ is a continuous function of x). This equality implies $f(\vartheta_*,x) = f(\vartheta_0,x)$ and finally $S(\vartheta_*,x) = S(\vartheta_0,x)$, which contradicts (2.18).

Introduce the matrix

$$\boldsymbol{R}(\vartheta) = \mathbf{J}(\vartheta)^{-1}\int_{\mathscr{R}}\int_{\mathscr{R}}\dot{F}(\vartheta,x)\,\dot{F}(\vartheta,y)^{\mathrm{T}}\,D_\vartheta(x,y)\,\mathrm{d}x\,\mathrm{d}y\,\mathbf{J}(\vartheta)^{-1},$$

where

$$D_\vartheta(x,y) = 4\,\mathbf{E}_\vartheta\left(\frac{F(\vartheta,\xi\wedge x)\left[1-F(\vartheta,\xi\vee x)\right]F(\vartheta,\xi\wedge y)\left[1-F(\vartheta,\xi\vee y)\right]}{\sigma(\xi)^2 f(\vartheta,\xi)^2}\right). \tag{2.63}$$

We have the following result.

Theorem 2.19. *Let the conditions $\mathcal{A}_0(\Theta)$ and $\mathcal{B}^*$ be fulfilled. Then the first MDE ϑ_T^* is uniformly consistent: for any $\nu > 0$*

$$\lim_{T\to\infty} \sup_{\vartheta\in\Theta} \mathbf{P}_\vartheta^{(T)} \left\{\left|\vartheta_T^* - \vartheta\right| > \nu\right\} = 0,$$

uniformly on compacts $\mathbb{K} \subset \Theta$ asymptotically normal

$$\mathcal{L}_\vartheta\left\{\sqrt{T}\left(\vartheta_T^* - \vartheta\right)\right\} \Longrightarrow \mathcal{N}\Big(0, \boldsymbol{R}(\vartheta)\Big)$$

and for any $p > 0$ the moments converge

$$\lim_{T\to\infty} \mathbf{E}_\vartheta \left|\sqrt{T}\left(\vartheta_T^* - \vartheta\right)\right|^p = \mathbf{E}\left|\zeta\right|^p$$

uniformly on $\vartheta \in \mathbb{K}$. Here $\mathcal{L}(\zeta) = \mathcal{N}\Big(0, \boldsymbol{R}(\vartheta)\Big)$.

Proof. Let us denote

$$\eta_T(x) = \sqrt{T}\left(\hat{F}_T(x) - F(\vartheta, x)\right),$$

where ϑ is the true value. Then we can write

$$\begin{aligned}
&\mathbf{P}_\vartheta^{(T)}\left\{\left|\vartheta_T^* - \vartheta\right| > \nu\right\}\\
&\quad = \mathbf{P}_\vartheta^{(T)}\left\{\inf_{|\theta-\vartheta|<\nu}\left\|\hat{F}_T(\cdot) - F(\theta,\cdot)\right\| > \inf_{|\theta-\vartheta|>\nu}\left\|\hat{F}_T(\cdot) - F(\theta,\cdot)\right\|\right\}\\
&\quad \leq \mathbf{P}_\vartheta^{(T)}\left\{2\left\|\eta_T(\cdot)\right\| > g^*(\nu)\sqrt{T}\right\} \leq \frac{4\,\mathbf{E}_\vartheta\left\|\eta_T(\cdot)\right\|^2}{T\,g^*(\nu)^2},
\end{aligned} \tag{2.64}$$

where we used the following properties of norm:

$$\begin{aligned}
\left\|\hat{F}_T(\cdot) - F(\theta,\cdot)\right\| &\leq \left\|\hat{F}_T(\cdot) - F(\vartheta,\cdot)\right\| + \left\|F(\theta,\cdot) - F(\vartheta,\cdot)\right\|,\\
\left\|\hat{F}_T(\cdot) - F(\theta,\cdot)\right\| &\geq \left\|F(\theta,\cdot) - F(\vartheta,\cdot)\right\| - \left\|\hat{F}_T(\cdot) - F(\vartheta,\cdot)\right\|
\end{aligned}$$

and the identity

$$\inf_{|\theta-\vartheta|\leq\nu}\left\|F(\theta,\cdot) - F(\vartheta,\cdot)\right\| = 0.$$

By the Itô formula

$$\begin{aligned}
\eta_T(x) &= \frac{2}{\sqrt{T}}\int_{X_0}^{X_T} \frac{F(\vartheta, v\wedge x) - F(\vartheta, v)\,F(\vartheta, x)}{\sigma(v)^2\,f(\vartheta, v)}\,\mathrm{d}v\\
&\quad - \frac{2}{\sqrt{T}}\int_0^T \frac{F(\vartheta, X_t\wedge x) - F(\vartheta, X_t)\,F(\vartheta, x)}{\sigma(X_t)\,f(\vartheta, X_t)}\,\mathrm{d}W_t.
\end{aligned} \tag{2.65}$$

We show now that under the condition $\mathcal{A}_0(\Theta)$ for any $p > 0$

$$\sup_{\vartheta\in\Theta}\int_{\mathscr{R}}\mathbf{E}_\vartheta\left|\frac{[F(\vartheta,\xi\wedge x)-F(\vartheta,\xi)\,F(\vartheta,x)]}{\sigma(\xi)\,f(\vartheta,\xi)}\right|^p\mathrm{d}x<\infty \tag{2.66}$$

$$\sup_{\vartheta\in\Theta}\int_{\mathscr{R}}\mathbf{E}_\vartheta\left|\int_0^\xi\frac{F(\vartheta,v\wedge x)-F(\vartheta,v)\,F(\vartheta,x)}{\sigma(v)^2\,f(\vartheta,v)}\,\mathrm{d}v\right|^p\mathrm{d}x<\infty,. \tag{2.67}$$

The proof of these estimates is similar to that of (2.42) and (2.43) obtained above. First remember that by the condition $\boldsymbol{A}_0(\boldsymbol{\Theta})$ there exist constants $\gamma>0$ and $A>0$ such that for all $\vartheta\in\boldsymbol{\Theta}$ and all $|x|>A$

$$f(\vartheta,x)\le C\,\mathrm{e}^{-2\gamma|x|},\qquad\left|\chi_{\{x>0\}}-F(\vartheta,x)\right|\le C\,\mathrm{e}^{-2\gamma|x|}. \tag{2.68}$$

For the first integral and $x>A$ we have

$$\begin{aligned}
d_F(\vartheta,x)^2&\equiv 4\,\mathbf{E}_\vartheta\left|\frac{F(\vartheta,\xi\wedge x)\,[1-F(\vartheta,\xi\vee x)]}{\sigma(\xi)\,f(\vartheta,\xi)}\right|^p\\
&=4\int_{\mathscr{R}}\left|\frac{F(\vartheta,y\wedge x)\,[1-F(\vartheta,y\vee x)]}{\sigma(y)\,f(\vartheta,y)}\right|^p f(\vartheta,y)\,\mathrm{d}y\\
&=4\,[1-F(\vartheta,x)]^p\int_{-\infty}^{A}\left|\frac{F(\vartheta,y)}{\sigma(y)\,f(\vartheta,y)}\right|^p f(\vartheta,y)\,\mathrm{d}y\\
&\quad+4\int_A^x\left|\int_x^\infty\frac{F(\vartheta,y)\,f(\vartheta,z)}{\sigma(y)\,f(\vartheta,y)}\mathrm{d}z\right|^p f(\vartheta,y)\,\mathrm{d}y\\
&\quad+4\,F(\vartheta,x)^p\int_x^\infty\left|\int_y^\infty\frac{f(\vartheta,z)}{\sigma(y)\,f(\vartheta,y)}\,\mathrm{d}z\right|^p f(\vartheta,y)\,\mathrm{d}y\\
&\le C\,\mathrm{e}^{-2\gamma p x}+C\int_A^x\left|\int_x^\infty\frac{\sigma(y)}{\sigma(z)^2}\,\mathrm{e}^{-2\gamma(z-y)}\,\mathrm{d}z\right|^p\mathrm{e}^{-2\gamma y}\,\mathrm{d}y\\
&\quad+C\int_x^\infty\left|\int_y^\infty\frac{\sigma(y)}{\sigma(z)^2}\,\mathrm{e}^{-2\gamma(z-y)}\,\mathrm{d}z\right|^p\mathrm{e}^{-2\gamma y}\,\mathrm{d}y\le C\,\mathrm{e}^{-\gamma x}
\end{aligned} \tag{2.69}$$

because $\sigma(\cdot)^{\pm1}$ has a polynomial majorant. For $x\le -A$ we have a similar estimate. Therefore the first estimate (2.66) is proved.

Further, for the second estimate (2.67) we can write

$$\begin{aligned}
&\mathbf{E}_\vartheta\left|\int_0^\xi\frac{F(\vartheta,v\wedge x)\,[1-F(\vartheta,v\vee x)]}{\sigma(v)^2\,f(\vartheta,v)}\,\mathrm{d}v\right|^p\\
&=\left(\int_{-\infty}^{-A}+\int_{-A}^{A}+\int_A^\infty\right)\left|\int_0^y\frac{F(\vartheta,v\wedge x)\,[1-F(\vartheta,v\vee x)]}{\sigma(v)^2\,f(\vartheta,v)}\mathrm{d}v\right|^p f(\vartheta,y)\,\mathrm{d}y.
\end{aligned}$$

Obviously

$$\begin{aligned}
\int_{-A}^{A}\left|\int_0^y\frac{F(\vartheta,v\wedge x)\,[1-F(\vartheta,v\vee x)]}{\sigma(v)^2\,f(\vartheta,v)}\,\mathrm{d}v\right|^p f(\vartheta,y)\,\mathrm{d}y&\\
\le C\left|\chi_{\{x>0\}}-F(\vartheta,x)\right|^p\le C\,\mathrm{e}^{-2\gamma p|x|}&.
\end{aligned}$$

Let $x > A$, then

$$\int_A^\infty \left|\int_0^y \frac{F(\vartheta, v\wedge x)\,[1-F(\vartheta, v\vee x)]}{\sigma(v)^2\, f(\vartheta, v)}\,\mathrm{d}v\right|^p f(\vartheta, y)\,\mathrm{d}y$$

$$\leq \int_A^\infty \left(2^{p-1}\left|\int_0^A \frac{F(\vartheta, v\wedge x)\,[1-F(\vartheta, v\vee x)]}{\sigma(v)^2\, f(\vartheta, v)}\,\mathrm{d}v\right|^p\right.$$

$$\left. +2^{p-1}\left|\int_A^y \frac{F(\vartheta, v\wedge x)\,[1-F(\vartheta, v\vee x)]}{\sigma(v)^2\, f(\vartheta, v)}\,\mathrm{d}v\right|^p\right) f(\vartheta, y)\,\mathrm{d}y$$

$$\leq C\,[1-F(\vartheta, x)]^p + 2^{p-1}\int_A^x \left|\int_A^y \int_x^\infty \frac{f(\vartheta, z)}{\sigma(v)^2 f(\vartheta, v)}\mathrm{d}z\mathrm{d}v\right|^p f(\vartheta, y)\,\mathrm{d}y$$

$$+\int_x^\infty \left|\int_A^y \int_{x\vee v}^\infty \frac{f(\vartheta, z)}{\sigma(v)^2\, f(\vartheta, v)}\,\mathrm{d}z\,\mathrm{d}v\right|^p f(\vartheta, y)\,\mathrm{d}y \leq C\,\mathrm{e}^{-\gamma x}.$$

Hence the estimates (2.66) and (2.67) are true.

Therefore applying these estimates with $p = 2$ in (2.64) we obtain the uniform consistency of the MDE.

$$\sup_{\vartheta\in\Theta} \mathbf{P}_\vartheta^{(T)}\left\{\left|\vartheta_T^* - \vartheta\right| > \nu\right\} \leq \frac{4}{T\, g^*(\nu)^2}\sup_{\vartheta\in\Theta}\int_{\mathscr{R}} \mathbf{E}_\vartheta\, \eta_T(x)^2\,\mathrm{d}x$$
$$\leq \frac{C}{T\, g^*(\nu)^2} \to 0$$

as $T\to\infty$.

Note as well that

$$D_\vartheta(x, y)^2 \leq d_F(\vartheta, x)^2\, d_F(\vartheta, y)^2 \leq C\,\mathrm{e}^{-2\gamma|x|}\,\mathrm{e}^{-2\gamma|y|}.$$

Therefore the matrix $\boldsymbol{R}(\vartheta)$ is finite too.

Let us introduce the random vector

$$\tilde{\boldsymbol{u}}_T = \mathbf{J}(\vartheta)^{-1/2}\int_{\mathscr{R}} \eta_T(x)\,\dot{\boldsymbol{F}}(\vartheta, x)\,\mathrm{d}x$$

$$= \frac{1}{\sqrt{T}}\int_0^T \boldsymbol{h}(\vartheta, X_t)\,\mathrm{d}t = \frac{\boldsymbol{H}(\vartheta, X_T) - \boldsymbol{H}(\vartheta, X_0)}{\sqrt{T}}$$

$$+\frac{1}{\sqrt{T}}\int_0^T \frac{1}{\sigma(X_t)\, f(\vartheta, X_t)}\int_{X_t}^\infty \boldsymbol{h}(\vartheta, v)\, f(\vartheta, v)\,\mathrm{d}v\,\mathrm{d}W_t,$$

where

$$\boldsymbol{h}(\vartheta, y) = 2\,\mathbf{J}(\vartheta)^{-1/2}\int_{\mathscr{R}}\left[\chi_{\{y<x\}} - F(\vartheta, x)\right]\dot{\boldsymbol{F}}(\vartheta, x)\,\mathrm{d}x$$

and

$$\boldsymbol{H}(\vartheta, y) = \int_0^y \frac{1}{\sigma(z)^2 f(\vartheta, z)}\int_z^\infty \boldsymbol{h}(\vartheta, v)\, f(\vartheta, v)\,\mathrm{d}v\,\mathrm{d}z.$$

Note that $\mathbf{E}_\vartheta\, \boldsymbol{h}(\vartheta,\xi)=\mathbf{0}$. Let us put

$$g(\vartheta,v)=2\frac{1-F(\vartheta,v)}{\sigma(v)\; f(\vartheta,v)}\,\mathbf{J}(\vartheta)^{-1/2}\int_{-\infty}^{v} F(\vartheta,x)\;\dot{\boldsymbol{F}}(\vartheta,x)\;\mathrm{d}x$$
$$+\,2\frac{F(\vartheta,v)}{\sigma(v)\; f(\vartheta,v)}\,\mathbf{J}(\vartheta)^{-1/2}\int_{v}^{\infty}\left[1-F(\vartheta,x)\right]\;\dot{\boldsymbol{F}}(\vartheta,x)\;\mathrm{d}x.$$

Then

$$\frac{1}{\sqrt{T}}\int_0^T\frac{1}{\sigma(X_t)\,f(\vartheta,X_t)}\int_{X_t}^{\infty}\boldsymbol{h}(\vartheta,v)\;f(\vartheta,v)\,\mathrm{d}v\;\mathrm{d}W_t$$
$$=\frac{1}{\sqrt{T}}\int_0^T \boldsymbol{g}(\vartheta,X_t)\;\mathrm{d}W_t$$

and

$$\mathbf{E}_\vartheta\;\boldsymbol{g}(\vartheta,\xi)\,\boldsymbol{g}(\vartheta,\xi)^{\mathsf{T}}=\boldsymbol{R}(\vartheta).$$

The estimates established above allow us to verify that the random vector $\tilde{\boldsymbol{u}}_T$ is uniformly asymptotically normal:

$$\mathcal{L}_\vartheta\left\{\mathbf{J}(\vartheta)^{-1/2}\,\tilde{\boldsymbol{u}}_T\right\}\Longrightarrow\mathcal{N}\Big(0,\boldsymbol{R}(\vartheta)\Big). \tag{2.70}$$

We show that for any $\varepsilon>0$

$$\sup_{\vartheta\in\mathbb{K}}\mathbf{P}_\vartheta^{(T)}\left\{\left|\tilde{\boldsymbol{u}}_T-\boldsymbol{u}_T^*\right|>\varepsilon\right\}\to 0,$$

as $T\to\infty$. Here $\boldsymbol{u}_T^*=\sqrt{T}\,\mathbf{J}(\vartheta)^{1/2}\left(\vartheta_T^*-\vartheta\right)$.

Denote

$$r_T(\boldsymbol{u},x)=\sqrt{T}\left[F\left(\vartheta+\frac{\mathbf{J}(\vartheta)^{-1/2}\,\boldsymbol{u}}{\sqrt{T}},x\right)-F(\vartheta,x)-\frac{1}{\sqrt{T}}\Big(\boldsymbol{u},\dot{\mathbf{F}}(\vartheta,x)\Big)\right],$$

where

$$\dot{\mathbf{F}}(\vartheta,x)=\mathbf{J}(\vartheta)^{-1/2}\,\dot{\boldsymbol{F}}(\vartheta,x)\,.$$

Note that

$$\langle\dot{\mathbf{F}}(\vartheta,\cdot),\dot{\mathbf{F}}(\vartheta,\cdot)^{\mathsf{T}}\rangle=\mathbf{J}(\vartheta)^{-1/2}\,\langle\dot{\boldsymbol{F}}(\vartheta,\cdot),\dot{\boldsymbol{F}}(\vartheta,\cdot)^{\mathsf{T}}\rangle\;\mathbf{J}(\vartheta)^{-1/2}=\mathbf{J}, \tag{2.71}$$

where $\mathbf{J}$ is a unit $d\times d$ matrix. Introduce as well the random function

$$Z_T(\boldsymbol{u},x)=\eta_T(x)-\Big(\boldsymbol{u},\dot{\mathbf{F}}(\vartheta,x)\Big)+r_T(\boldsymbol{u},x),\quad x\in\mathscr{R},\;\;(\boldsymbol{\Theta}-\vartheta)$$

where $\boldsymbol{u}\in\sqrt{T}\,\mathbf{J}(\vartheta)^{1/2}$ and define the sets

$$\begin{aligned}
\mathbb{B}_1 &= \left\{\omega : \inf_{|\boldsymbol{u}|<\lambda_T} \left\|Z_T\left(\boldsymbol{u},\cdot\right)\right\| < \inf_{|\boldsymbol{u}|\geq\lambda_T} \left\|Z_T\left(\boldsymbol{u},\cdot\right)\right\|\right\},\\
\mathbb{B}_2 &= \left\{\omega : \left|\tilde{\boldsymbol{u}}_T\right| < \lambda_T\right\},\\
\mathbb{B} &= \mathbb{B}_1 \cap \mathbb{B}_2,
\end{aligned}$$

where $\lambda_T \to \infty$ but $T^{-1/2}\lambda_T \to 0$. Obviously

$$Z_T\left(\boldsymbol{u},x\right) = \sqrt{T}\left[\hat{F}_T(x) - F\left(\boldsymbol{\vartheta} + T^{-1/2}\mathbf{J}\left(\boldsymbol{\vartheta}\right)^{-1/2}\boldsymbol{u},x\right)\right].$$

For $\omega \in \mathbb{B}_1$ we have

$$\left|\boldsymbol{u}_T^*\right| < \lambda_T$$

and for $\omega \in \mathbb{B}_2$ the vector $\tilde{\boldsymbol{u}}_T$ minimizes the quadratic form

$$\int_{\mathscr{R}} \left[\eta_T(x) - \left(\boldsymbol{u}, \dot{\mathbf{F}}\left(\boldsymbol{\vartheta},x\right)\right)\right]^2 \mathrm{d}x, \qquad |\boldsymbol{u}| < \lambda_T$$

and is the unique solution of the system of equations

$$\int_{\mathscr{R}} \left[\eta_T(x) - \left(\tilde{\boldsymbol{u}}_T, \dot{\mathbf{F}}\left(\boldsymbol{\vartheta},x\right)\right)\right] \dot{\mathbf{F}}\left(\boldsymbol{\vartheta},x\right)\mathrm{d}x = 0. \tag{2.72}$$

Using the representations

$$\begin{aligned}
Z_T\left(\boldsymbol{u}_T^*,x\right) &= \eta_T(x) - \left(\boldsymbol{u}_T^*, \dot{\mathbf{F}}\left(\boldsymbol{\vartheta},x\right)\right) + r_T\left(\boldsymbol{u}_T^*,x\right),\\
Z_T\left(\tilde{\boldsymbol{u}}_T,x\right) &= \eta_T(x) - \left(\tilde{\boldsymbol{u}}_T, \dot{\mathbf{F}}\left(\boldsymbol{\vartheta},x\right)\right) + r_T\left(\tilde{\boldsymbol{u}}_T,x\right),
\end{aligned}$$

we can write

$$\begin{aligned}
&\left(\tilde{\boldsymbol{u}}_T - \boldsymbol{u}_T^*, \dot{\mathbf{F}}\left(\boldsymbol{\vartheta},x\right)\right) + Z_T\left(\tilde{\boldsymbol{u}}_T,x\right)\\
&\qquad = Z_T\left(\boldsymbol{u}_T^*,x\right) - r_T\left(\boldsymbol{u}_T^*,x\right) + r_T\left(\tilde{\boldsymbol{u}}_T,x\right),
\end{aligned}$$

and

$$\begin{aligned}
&\left\|\left(\tilde{\boldsymbol{u}}_T - \boldsymbol{u}_T, \dot{\mathbf{F}}\left(\boldsymbol{\vartheta},x\right)\right) + Z_T\left(\tilde{\boldsymbol{u}}_T^*,x\right)\right\|^2\\
&\quad = \left|\tilde{\boldsymbol{u}}_T - \boldsymbol{u}_T^*\right|^2 + \left\|Z_T\left(\tilde{\boldsymbol{u}}_T,\cdot\right)\right\|^2 + 2\left(\tilde{\boldsymbol{u}}_T - \boldsymbol{u}_T^*, \langle\dot{\mathbf{F}}\left(\boldsymbol{\vartheta},\cdot\right), r_T\left(\tilde{\boldsymbol{u}}_T,\cdot\right)\rangle\right).
\end{aligned}$$

Therefore

$$\begin{aligned}
\left|\tilde{\boldsymbol{u}}_T - \boldsymbol{u}_T^*\right|^2 &= \left\|Z_T\left(\boldsymbol{u}_T^*,\cdot\right) - r_T\left(\boldsymbol{u}_T^*,\cdot\right) + r_T\left(\tilde{\boldsymbol{u}}_T,\cdot\right)\right\|^2 - \left\|Z_T\left(\tilde{\boldsymbol{u}}_T,\cdot\right)\right\|^2\\
&\quad -2\left(\tilde{\boldsymbol{u}}_T - \boldsymbol{u}_T^*, \langle\dot{\mathbf{F}}\left(\boldsymbol{\vartheta},\cdot\right), r_T\left(\tilde{\boldsymbol{u}}_T,\cdot\right)\rangle\right).
\end{aligned}$$

We have the elementary estimates

$$\begin{aligned}\left\|Z_T\left(u_T^*,\cdot\right)\right\| &= \left\|Z_T^o\left(u_T^*,\cdot\right)+r_T\left(u_T^*,\cdot\right)\right\| \\ &\geq \left\|Z_T^o\left(u_T^*,\cdot\right)\right\| - \left\|r_T\left(u_T^*,\cdot\right)\right\| \\ &\geq \left\|Z_T^o\left(\tilde u_T,\cdot\right)\right\| - \left\|r_T\left(u_T^*,\cdot\right)\right\|, \\ \left\|Z_T\left(u_T^*,\cdot\right)\right\| &\leq \left\|Z_T\left(\tilde u_T,\cdot\right)\right\| = \left\|Z_T^o\left(\tilde u_T,\cdot\right)+r_T\left(\tilde u_T,\cdot\right)\right\| \\ &\leq \left\|Z_T^o\left(\tilde u_T,\cdot\right)\right\| + \left\|r_T\left(u_T^*,\cdot\right)\right\|,\end{aligned}$$

where $Z_T^o\left(\boldsymbol{u},x\right)=V_T(x)-\left(\boldsymbol{u},\dot{\mathbf{F}}\left(\vartheta,x\right)\right)$. Hence

$$\left|\left\|Z_T\left(u_T^*,\cdot\right)\right\| - \left\|Z_T^o\left(\tilde u_T,\cdot\right)\right\|\right| \leq \left\|r_T\left(u_T^*,\cdot\right)\right\| + \left\|r_T\left(\tilde u_T,\cdot\right)\right\|.$$

By condition (2.60) and identity (2.71) we have

$$T\left\|F\left(\vartheta+T^{-1/2}\mathbf{J}\left(\vartheta\right)^{-1/2}\boldsymbol{u},\cdot\right)-F\left(\vartheta,\cdot\right)\right\|^2=\left|\boldsymbol{u}\right|^2\,\left(1+o(1)\right).$$

Therefore there exists a $M>0$ such that for $|\boldsymbol{u}|<M$ and all $\vartheta\in\mathbb{K}$

$$\sqrt{T}\left\|F\left(\vartheta+T^{-1/2}\mathbf{J}\left(\vartheta\right)^{-1/2}\boldsymbol{u},\cdot\right)-F\left(\vartheta,\cdot\right)\right\|\geq\frac{1}{2}\left|\boldsymbol{u}\right|,$$

which together with (2.61) gives us the estimate

$$\inf_{\vartheta\in\mathbb{K}}\sqrt{T}\left\|F\left(\vartheta+T^{-1/2}\mathbf{J}\left(\vartheta\right)^{-1/2}\boldsymbol{u},\cdot\right)-F\left(\vartheta,\cdot\right)\right\|\geq\kappa\,\left|\boldsymbol{u}\right| \tag{2.73}$$

with some positive κ.

Using the same arguments as in (2.64) and the estimate (2.73) in the form

$$\sqrt{T}\left\|F\left(\vartheta+T^{-1/2}\mathbf{J}\left(\vartheta\right)^{-1/2}\boldsymbol{u},\cdot\right)-F\left(\vartheta,\cdot\right)\right\|\geq\kappa\,\left|\boldsymbol{u}\right|$$

we have

$$\begin{aligned}&\mathbf{P}_\vartheta^{(T)}\left\{\left|u_T^*\right|>\frac{3}{\kappa}\left\|\eta_T(\cdot)\right\|\right\}\\ &\quad=\mathbf{P}_\vartheta^{(T)}\left\{\inf_{|\boldsymbol{u}|<\frac{3}{\kappa}\left\|\eta_T(\cdot)\right\|}\left\|Z_T\left(\boldsymbol{u},\cdot\right)\right\|>\inf_{|\boldsymbol{u}|\geq\frac{3}{\kappa}\left\|\eta_T(\cdot)\right\|}\left\|Z_T\left(\boldsymbol{u},\cdot\right)\right\|\right\}\\ &\quad\leq\mathbf{P}_\vartheta^{(T)}\left\{2\,\left\|\eta_T(\cdot)\right\|\geq3\,\left\|\eta_T(\cdot)\right\|\right\}=0.\end{aligned}$$

Therefore

$$\left|u_T^*\right|<\frac{3}{\kappa}\left\|\eta_T(\cdot)\right\| \tag{2.74}$$

with $\mathbf{P}_\vartheta^{(T)}$ probability 1.

Further,

$$\mathbf{E}_\vartheta \left(\chi_{\{\mathbb{B}\}} \left|\boldsymbol{u}_T^*\right| \left\|r_T\left(\tilde{\boldsymbol{u}}_T, \cdot\right)\right\|\right)^2 \leq \mathbf{E}_\vartheta \left|\boldsymbol{u}_T^*\right|^2 \sup_{|\boldsymbol{u}|<\lambda_T} \left\|r_T\left(\boldsymbol{u}, \cdot\right)\right\|^2 = o(1),$$

and

$$\mathbf{E}_\vartheta \left(\chi_{\{\mathbb{B}^c\}} \left|\boldsymbol{u}_T^*\right|\right)^2 \leq \frac{9}{\kappa^2}\, \mathbf{E}_\vartheta \left(\chi_{\left\{\left\|\eta_T(\cdot)\right\| \geq \frac{\kappa}{3}\lambda_T\right\}} \left\|\eta_T(\cdot)\right\|\right)^2 \to 0.$$

These and similar estimates allow us to show that

$$\sup_{\vartheta\in\mathbb{K}} \mathbf{E}_\vartheta \left|\boldsymbol{u}_T^* - \tilde{\boldsymbol{u}}_T\right|^2 \to 0.$$

Hence we have the asymptotic normality of $\boldsymbol{u}_T^*$ (due to (2.70)) and the convergence of the second moments as well.

Convergence of the higher moments follows from the uniform integrability of the random variables $\left|\boldsymbol{u}_T^*\right|^p$ which can be verified as follows. We can write (like (2.64))

$$\begin{aligned} \mathbf{P}_\vartheta^{(T)} \left\{\left|\boldsymbol{u}_T^*\right| > R\right\} &= \mathbf{P}_\vartheta^{(T)} \left\{\left|\sqrt{T}\left(\vartheta_T^* - \vartheta\right)\right| > R\right\} \\ &\leq \frac{2^p\ \mathbf{E}_\vartheta \left\|\eta_T\left(\cdot\right)\right\|^p}{T^{p/2}\, g^*\left(R/\sqrt{T}\right)^p} < \frac{C_p}{R^p} \end{aligned} \qquad (2.75)$$

where we used the estimates (2.66), (2.67) and (2.73).

2.2.2 Examples

As we wrote in Section 1.3 there are no examples with explicit expressions for the MDE but these estimators have some advantages. Remember that in the study of the MLE we needed the differentiability of the trend coefficient $S(\vartheta, \cdot)$ on ϑ and here we use the differentiability of the distribution function $F(\vartheta, \cdot)$, which is a less restrictive condition because *between* $S(\vartheta, \cdot)$ *and* $F(\vartheta, \cdot)$ *there are two integrations:*

$$F(\vartheta, x) = \frac{1}{G(\vartheta)} \int_{-\infty}^{x} \sigma(y)^{-2} \exp\left\{2\int_0^y \frac{S(\vartheta, v)}{\sigma(v)^2}\, dv\right\} dy.$$

Therefore the function $F(\vartheta, x)$ can be differentiable on ϑ, when $S(\vartheta, x)$ is even discontinuous.

Note that conditions $\boldsymbol{\mathcal{A}}_0(\Theta)$ and the differentiability of $S(\vartheta, x)$ with $\dot{S}(\vartheta, \cdot) \in \mathscr{P}$ yield the differentiability of $F(\vartheta, \cdot)$ as follows.

$$\begin{aligned}\dot{F}(\vartheta,x) &= \frac{\partial}{\partial\vartheta}\,\mathbf{E}_\vartheta\,\chi_{\{\xi<x\}} = \frac{\partial}{\partial\vartheta}\int_{-\infty}^{x} f(\vartheta,y)\,\mathrm{d}y\\ &= 2\int_{-\infty}^{x}\left[\int_0^y \frac{\dot{S}(\vartheta,v)}{\sigma(v)^2}\mathrm{d}v - \int_{-\infty}^{\infty}\int_0^z \frac{\dot{S}(\vartheta,v)}{\sigma(v)^2}\mathrm{d}v\, f(\vartheta,z)\,\mathrm{d}z\right] f(\vartheta,y)\,\mathrm{d}y\\ &= 2\int_{-\infty}^{x} f(\vartheta,y)\,\mathbf{E}_\vartheta\int_\xi^y \frac{\dot{S}(\vartheta,v)}{\sigma(v)^2}\mathrm{d}v\,\mathrm{d}y. \end{aligned}\tag{2.76}$$

Hence

$$\begin{aligned}\left|\dot{F}(\vartheta,x)\right| &\le C\int_{-\infty}^{x} f(\vartheta,y)\,\mathbf{E}_\vartheta\left|\int_\xi^y\left(1+|v|^p\right)\mathrm{d}v\right|\mathrm{d}y\\ &\le C\int_{-\infty}^{x} f(\vartheta,y)\left(\mathbf{E}_\vartheta\,|\xi|^{p+1}+|y|^{p+1}\right)\mathrm{d}y < C.\end{aligned}$$

Therefore the derivative $\dot{F}(\vartheta,x)$ exists and is a bounded function of x.

It can be shown that the regularity conditions of Theorem 2.23 are fulfilled in Examples 2.14, 2.15, etc.

We mention here several nonregular (for maximum likelihood and Bayesian approaches) problems in which the regularity conditions $\boldsymbol{\mathcal{B}}^*$ nevertheless are fulfilled.

Example 2.20. (*Discontinuity estimation*) Suppose that the observed process is *switching diffusion*:

$$\mathrm{d}X_t = -\operatorname{sgn}(X_t-\vartheta)\,\mathrm{d}t + \sigma\,\mathrm{d}W_t,\quad X_0,\quad 0\le t\le T,$$

where the trend coefficient $S(\vartheta,x) = -\operatorname{sgn}(x-\vartheta)$ is discontinuous function. The density function of the stationary distribution is

$$f(\vartheta,x) = \frac{1}{\sigma^2}\,\mathrm{e}^{-\frac{2}{\sigma^2}|x-\vartheta|},\quad x\in\mathscr{R}.$$

The conditions $\boldsymbol{\mathcal{A}}$ of course are not fulfilled but the stationary distribution function is differentiable and the MDE ϑ_T^* is consistent and $\sqrt{T}\left(\vartheta_T^*-\vartheta\right)$ is asymptotically normal. Note that the MLE and BE for this model have better rates of convergence and, say, $T\left(\hat{\vartheta}_T-\vartheta\right)$ has a nondegenerate limit distribution (see Section 3.4).

Example 2.21. (*Delay estimation*) The similar situation we have in the case of a delayed diffusion-type process

$$\mathrm{d}X_t = -\gamma\,X_{t-\vartheta}\,\mathrm{d}t + \sigma\,\mathrm{d}W_t,\quad X_0(s),\quad -\tau\le s\le 0,\quad 0\le t\le T.$$

We study this model in Section 3.3. The stationary distribution is a zero mean Gaussian with the variance

$$d(\vartheta)^2 = \sigma^2 \int_0^\infty x_0(s)^2 \, ds.$$

Here $x_0(\cdot) = x_0(\vartheta, \cdot)$ is the solution of the deterministic equation

$$\frac{\partial x_0(t)}{\partial t} = -\gamma\, x_0(t-\vartheta), \quad x_0(s) = 0,\ s < 0, \quad x_0(0) = 1.$$

Therefore the invariant distribution is differentiable on ϑ and the trend coefficient $S(\vartheta, X_{t-\vartheta}) = -\gamma\, X_{t-\vartheta}$ is as smooth as the Wiener process.

The conditions of Theorem 2.19 can be checked and so the MDE is consistent and $\sqrt{T}\left(\vartheta^*_T - \vartheta\right)$ is asymptotically normal. Note that as in the case of switching diffusion the MLE and BE for this model have a nondegenerate limit distribution with better normalizing factor (see Section 3.3).

Example 2.22. (*Cusp estimation*) The conditions of Theorem 2.19 can be checked for the model of observation

$$dX_t = -\gamma \operatorname{sgn}(X_t - \vartheta)\, |X_t - \vartheta|^\kappa \, dt + \sigma \, dW_t, \quad X_0, \quad 0 \le t \le T,$$

where $\gamma > 0$ and $\kappa \in (0, 1/2)$. The trend coefficient $S(\vartheta, x)$ has a *cusp* and is not differentiable over ϑ at a point $\vartheta = x$, but the invariant density

$$f(\vartheta, x) = G^{-1} \exp\left\{-\frac{2\gamma}{\sigma^2\,(\kappa+1)}\, |x-\vartheta|^{\kappa+1}\right\}, \qquad x \in \mathscr{R},$$

is sufficiently smooth for $\boldsymbol{\mathscr{B}}^*$.

2.2.3 Second MDE

As was shown in Section 2.1 the asymptotically efficient estimators for the parametrical model (2.57) are the MLE and BE. The MDE has greater limiting variance and so is not asymptotically efficient. Nevertheless there are situations when even the asymptotic variance of MDE can be better than that of the MLE and BE. We consider below the model of observations (2.57) perturbed by a *small contamination*, i.e., we suppose that the observed diffusion process has the stochastic differential

$$dX_t = [S(\vartheta, X_t) + h(X_t)]\, dt + \sigma(X_t)\, dW_t, \quad X_0, \quad 0 \le t \le T \qquad (2.77)$$

where the contamination function $h(\cdot) \in V_\delta$ is supposed to be unknown and small. Here V_δ is some small vicinity of zero. We suppose that the functions

$S(\cdot)$ and $\sigma(\cdot)$ satisfy the condition $\mathcal{A}_0(\Theta)$. Hence for sufficiently small functions $h(\cdot)$ this condition will be fulfilled too and the corresponding invariant distribution function is

$$F_h(x) = G(\vartheta, h)^{-1} \int_{-\infty}^{x} \sigma(y)^{-2} \exp\left\{2\int_0^y \frac{S(\vartheta, v) + h(v)}{\sigma(v)^2} \mathrm{d}v\right\} \mathrm{d}y$$

where $G(\vartheta, h)$ is the normalizing constant

$$G(\vartheta, h) = \int_{-\infty}^{\infty} \sigma(y)^{-2} \exp\left\{2\int_0^y \frac{S(\vartheta, v) + h(v)}{\sigma(v)^2} \mathrm{d}v\right\} \mathrm{d}y < \infty.$$

The next question is crucial for the following: what do we want to do with the estimator of the parameter ϑ? Suppose our goal is to find the best approximation of the unknown invariant distribution function $F_h(\cdot)$ by the family $\{F(\vartheta, \cdot), \Theta\}$, i.e., to construct an estimator $\bar{\vartheta}_T$ such that $F(\bar{\vartheta}_T, \cdot)$ well approximates $F_h(\cdot)$ for all $h(\cdot) \in V_\delta$. If the *best* means *best in the* $\mathcal{L}_2(\mu)$ *sense*, then it is reasonable to estimate

$$\vartheta_h = \arg\inf_{\theta\in\Theta} \left\|F_h(\cdot) - F(\theta, \cdot)\right\|_\mu, \tag{2.78}$$

which will play the role of the true value in this problem. Here $\mu(\cdot)$ is some σ-finite measure.

If we need a similar approximation for the invariant density function $f_h(,\cdot)$ then it is natural to seek the value

$$\vartheta_h = \arg\inf_{\theta\in\Theta} \left\|f_h(\cdot) - f(\theta, \cdot)\right\|_\mu. \tag{2.79}$$

Note that there are many other possibilities to define ϑ_h. For example, it can be

$$\vartheta_h = \arg\inf_{\theta\in\Theta} \sup_x \left|f_h(x) - f(\theta, x)\right|, \tag{2.80}$$

or sometimes it is interesting to find the value ϑ_h which approximates well the trend coefficient in the following sense:

$$\vartheta_h = \arg\inf_{\theta\in\Theta} \mathbf{E}_h \left|\frac{S(\vartheta, \xi) + h(\xi) - S(\theta, \xi)}{\sigma(\xi)}\right|^2 \tag{2.81}$$

which corresponds to the minimization of the Kullback–Leibler distance between the distribution $\mathbf{P}_h^{(T)}$ induced by the process (2.77) and the family $\left\{\mathbf{P}_\theta^{(T)}, \theta \in \Theta\right\}$, i.e.,

$$\vartheta_h = \arg\inf_{\theta\in\Theta} \mathbf{E}_h \ln \frac{\mathrm{d}\,\mathbf{P}_\mathrm{h}^{(\mathrm{T})}}{\mathrm{d}\,\mathbf{P}_\theta^{(T)}}.$$

Of course we have to verify that $\vartheta_h \to \vartheta$ as $h \to 0$.

The choice of ϑ_h tells us which estimator we have to choose. For ϑ_h defined by (2.78) we can show the asymptotic optimality of the estimator ϑ_T^* defined by (2.58) with $\left\|\cdot\right\| = \left\|\cdot\right\|_\mu$ (see (1.123)) and for (2.79) it will be ϑ_T^{**} defined by (2.59). It can be shown that for (2.80) we can take the following consistent estimator:

$$\vartheta_T = \arg\inf_{\theta\in\Theta}\sup_x \left|f_T^\circ\left(x\right) - f\left(\vartheta, x\right)\right|$$

and for (2.81) the MLE can be used as the consistent estimator (see Section 2.6.1 below).

All these statements are particular cases of the general problem of asymptotically efficient estimation of special (smooth) functionals in the following nonparametric estimation problem. Let $\Phi_h = \Phi\left(h\right)$ be a smooth functional of the function $h\left(\cdot\right)$ and we have to estimate Φ_h. Then in some situations it is possible to construct a lower minimax bound of the following type:

$$\underline{\lim}_{\delta\to 0}\ \underline{\lim}_{T\to\infty}\ \inf_{\bar\vartheta_T}\ \sup_{h(\cdot)\in V_\delta} \mathbf{E}_h\, \ell\left(\sqrt{T}\left(\bar\Phi_T - \Phi_h\right)\right) \geq \mathbf{E}\,\ell\left(\zeta\, \mathrm{I}_*^{-1/2}\right),$$

where V_δ defines a small nonparametric vicinity of zero, $\ell\left(\cdot\right)$ is a loss function, $\zeta \sim \mathcal{N}\left(0,1\right)$ and I_* plays the role of *Fisher information* in this nonparametric problem. Having this lower bound we define the asymptotically optimal estimators as estimators attaining this bound asymptotically and often the empirical estimators $\bar\Phi_T = \Phi\left(\hat F_T\right)$ are asymptotically efficient. This approach initiated by Levit [170] can be found in Ibragimov and Khasminskii [109], Chapter 4 or in Bickel *et al.* [24], where further developments and applications can be found as well.

We consider the problem of estimation of the value (2.79) by the observations (2.77). Therefore $f\left(\vartheta_h, \cdot\right)$ is the best (in the family $\left\{f\left(\vartheta,\cdot\right), \vartheta\in\Theta\right\}$) in the $\left\|\cdot\right\|$ approximation of the unknown invariant density function $f_h\left(\cdot\right)$. As an estimator we take the second MDE (2.59):

$$\vartheta_T^{**} = \arg\inf_{\theta\in\vartheta}\int_{-\infty}^{\infty}\left[f_T^\circ(x) - f\left(\theta, x\right)\right]^2 \mathrm{d}x$$

and suppose that $\Theta = \left(\alpha,\beta\right)$, where α and β are finite.

The function $h\left(\cdot\right)$ is such that the condition $\mathcal{ES}$ is fulfilled and we assume that $h\left(\cdot\right)$ belongs to the set

$$V_\delta = \left\{h\left(\cdot\right):\ \sup_{x\in\mathscr{R}}\frac{\left|h\left(x\right)\right|}{\sigma\left(x\right)^2} \leq \delta\right\}.$$

It is easy to see that for sufficiently small δ the condition $\mathcal{A}_0\left(\Theta\right)$ is fulfilled for the trend coefficient $S\left(\vartheta,\cdot\right) + h\left(\cdot\right)$ too.

Regularity conditions $\mathcal{B}^{**}$.

$\mathcal{B}_1^{**}$. *The invariant density function* $f(\vartheta,\cdot)$ *is differentiable over* ϑ *for all* $x \in \mathscr{R}$ *and the derivative* $\dot f(\vartheta,\cdot) \in \mathscr{P}$ *is uniformly continuous in the following sense:*

$$\lim_{\nu\to 0} \sup_{\vartheta_0\in\Theta} \sup_{|\vartheta-\vartheta_0|\le\nu} \left\| \dot f(\vartheta,\cdot) - \dot f(\vartheta_0,\cdot)\right\| = 0. \tag{2.82}$$

$\mathcal{B}_2^{**}$. *For any* $\nu > 0$ *the function*

$$g^{**}(\nu,\delta) = \inf_{\vartheta\in\Theta} \inf_{h(\cdot)\in V_\delta} \inf_{|\theta-\vartheta_h|>\nu} \left(\|f_h(\cdot) - f(\theta,\cdot)\| - \|f_h(\cdot) - f(\vartheta_h,\cdot)\|\right) > 0 \tag{2.83}$$

and the function

$$\mathrm{J}^*(\vartheta) = \left\|\dot f(\vartheta,\cdot)\right\|^2 > 0$$

is positive uniformly in ϑ.

Let us define the functions

$$D_\vartheta^*(x,y) = 4\ \mathbf{E}_\vartheta \left(\frac{[\chi_{\{\xi>x\}} - F(\vartheta,\xi)][\chi_{\{\xi>y\}} - F(\vartheta,\xi)]}{\sigma(\xi)^2 f(\vartheta,\xi)^2}\right),$$

$$D_h^*(x,y) = 4\ \mathbf{E}_h \left(\frac{[\chi_{\{\xi>x\}} - F_h(\xi)][\chi_{\{\xi>y\}} - F_h(\xi)]}{\sigma(\xi)^2 f_h(\xi)^2}\right),$$

and put

$$\mathrm{I}_*(\vartheta) = \mathrm{J}^*(\vartheta)^2 \left(\int_{\mathscr{R}}\int_{\mathscr{R}} \dot f(\vartheta,x)\ \dot f(\vartheta,y)\, D_\vartheta^*(x,y)\ f(\vartheta,x)\ f(\vartheta,y)\ \mathrm{d}x\,\mathrm{d}y\right)^{-1}.$$

This last quantity plays the role of Fisher information in this problem. We have the following lower bound in this problem.

Theorem 2.23. *Let the condition* $\mathcal{A}_0(\Theta)$ *and* $\mathcal{B}^{**}$ *be fulfilled and* $\mathrm{I}_*(\vartheta) > 0$, *then for any* $\vartheta \in \Theta$

$$\lim_{\delta\to 0} \lim_{T\to\infty} \inf_{\bar\vartheta_T} \sup_{h(\cdot)\in V_\delta} \mathbf{E}_h\, \ell\left(\sqrt{T}\left(\bar\vartheta_T - \vartheta_h\right)\right) \ge \mathbf{E}\,\ell\left(\zeta\, \mathrm{I}_*(\vartheta)^{-1/2}\right), \tag{2.84}$$

where inf *is taken over all estimators* $\bar\vartheta_T$ *of* ϑ_h *and the loss function* $\ell(\cdot) \in \mathscr{W}_p$.

Moreover the MDE ϑ_T^{**} *is uniformly consistent, i.e., for any* $\nu > 0$

$$\lim_{T\to\infty} \sup_{h(\cdot)\in V_\delta} \mathbf{P}_h^{(T)} \left\{\left|\vartheta_T^{**} - \vartheta_h\right| > \nu\right\} = 0,$$

is uniformly asymptotically normal

$$\mathcal{L}_h\left\{\sqrt{T}\left(\vartheta_T^{**} - \vartheta_h\right)\right\} \Longrightarrow \mathcal{N}(0, R_h)$$

with

$$R_h = \mathrm{J}^*\left(\vartheta_h\right)^{-2} \int_{\mathscr{R}} \int_{\mathscr{R}} \dot f\left(\vartheta_h, x\right)\ \dot f\left(\vartheta_h, y\right)\ D_h^*\left(x,y\right)\ f_h\left(x\right)\ f_h\left(y\right)\ \mathrm{d}x\,\mathrm{d}y,$$

and is asymptotically efficient for the loss functions $\ell\left(\cdot\right) \in \mathcal{W}_p$.

Proof. Let us introduce a parametric family of diffusion processes

$$\mathrm{d}X_t = \left[S\left(\vartheta, X_t\right) + \left(\tau - \tau_0\right)\ \psi\left(X_t\right)\sigma\left(X_t\right)^2\right]\mathrm{d}t + \sigma\left(X_t\right)\ \mathrm{d}W_t, \quad 0 \le t \le T, \tag{2.85}$$

where τ_0 is some fixed positive number, $\tau \in \left(\tau_0 - \delta_1, \tau_0 + \delta_1\right)$, the function $\psi\left(\cdot\right)$ is such that $\delta_1\,\psi\left(\cdot\right)\sigma\left(\cdot\right)^2 \in V_\delta$, i.e., we put $h\left(\cdot\right) = \left(\tau - \tau_0\right)\ \psi\left(\cdot\right)\sigma\left(\cdot\right)^2$ in (2.77). Moreover we suppose that $\psi\left(\cdot\right)$ has a compact support. Then we can expand the function $f_h\left(x\right)$ by the powers of $\tau - \tau_0$ and obtain

$$f_h\left(x\right) = f\left(\vartheta, x\right) + 2\ \left(\tau - \tau_0\right) f\left(\vartheta, x\right)\ \mathbf{E}_\vartheta \int_\xi^x \psi\left(v\right)\ \mathrm{d}v + o\left(\tau - \tau_0\right).$$

The corresponding expansion of the function ϑ_h gives us the representation

$$\vartheta_h = \vartheta + 2\left(\tau - \tau_0\right)\mathrm{J}\left(\vartheta\right)^{-1} \int_{\mathscr{R}} f\left(\vartheta, x\right)\dot f\left(\vartheta, x\right)\mathbf{E}_\vartheta \int_\xi^x \psi\left(v\right)\ \mathrm{d}v\ \mathrm{d}x.$$

Let us introduce the following class of functions:

$$\mathcal{K} = \left\{\psi\left(\cdot\right):\ 2\int_{\mathscr{R}} f\left(\vartheta, x\right)\dot f\left(\vartheta, x\right)\mathbf{E}_\vartheta \int_\xi^x \psi\left(v\right)\ \mathrm{d}v\ \mathrm{d}x = \int_{\mathscr{R}} \dot f\left(\vartheta, x\right)^2 \mathrm{d}x\right\}.$$

Therefore for a function $\psi\left(\cdot\right) \in \mathcal{K}$

$$\vartheta_h = \vartheta + \tau - \tau_0 + o\left(\tau - \tau_0\right)$$

and if we put $\tau_0 = \vartheta$, then $\vartheta_h = \tau + o\left(\tau - \tau_0\right)$ and the problem of estimation ϑ_h is close to the problem of estimation τ.

Let us denote

$$\mathrm{I}_\psi = \int_{\mathscr{R}} \psi\left(v\right)^2 \sigma\left(v\right)^2 f\left(\vartheta, v\right)\mathrm{d}v.$$

This is the Fisher information in the problem of estimation τ by observations (2.85). We have the obvious inequality

$$\begin{aligned}
&\inf_{\bar\vartheta_T}\ \sup_{h(\cdot)\in V_\delta}\ \mathbf{E}_h\ \ell\left(\sqrt{T}\left(\bar\vartheta_T - \vartheta_h\right)\right)\\
&\qquad\qquad \ge \inf_{\bar\tau_T}\ \sup_{\left|\tau-\tau_0\right|\le\delta_1}\ \mathbf{E}_\tau\ \ell\left(\sqrt{T}\left(\bar\tau_T - \tau + o\left(\tau - \tau_0\right)\right)\right).
\end{aligned}$$

The family of measures $\left\{\mathbf{P}_\tau^{(T)},\ \tau \in \left(\tau_0 - \delta_1, \tau_0 + \delta_1\right)\right\}$ is LAN at the point $\tau = \tau_0$ and we can write the inequality of Hajek–Le Cam (see Theorem 2.4)

$$\varliminf_{\delta\to 0}\varliminf_{T\to\infty}\inf_{\bar\tau_T}\sup_{|\tau-\tau_0|\leq\delta_1}\mathbf{E}_\tau\,\ell\left(\sqrt{T}\left(\bar\tau_T-\tau+o\left(\tau-\tau_0\right)\right)\right)\geq\mathbf{E}_\vartheta\,\ell\left(\zeta\,\mathrm{I}_\psi^{-1/2}\right),$$

where $\zeta\sim\mathcal{N}(0,1)$. Note that the term $o(\tau-\tau_0)$ plays an asymptotically negligible role in this problem. This estimate holds for all functions $\psi(\cdot)\in\mathcal{K}$. Therefore it holds as well for the function $\psi^*(\cdot)$ such that

$$\inf_{\psi(\cdot)\in\mathcal{K}}\mathrm{I}_\psi=\mathrm{I}_{\psi^*},\tag{2.86}$$

i.e., we seek the *least favorable parametric family* defined by the function $\psi^*(\cdot)$ and this will give us the lower bound in the original problem.

Exchanging the order of integration we obtain

$$\mathbf{E}_\vartheta\int_\xi^x\psi(v)\;\mathrm{d}v=\int_{\mathscr{R}}\psi(v)\;\left[F(\vartheta,v)-\chi_{\{v>x\}}\right]\mathrm{d}v.$$

Hence for any $\psi(\cdot)\in\mathcal{K}$ we can write

$$\begin{aligned}
&\int_{\mathscr{R}}\dot f(\vartheta,x)^2\,\mathrm{d}x\\
&=2\int\int\psi(v)\left[F(\vartheta,v)-\chi_{\{v>x\}}\right]\mathrm{d}v\,f(\vartheta,x)\;\dot f(\vartheta,x)\;\mathrm{d}x\\
&=2\int\psi(v)\,\sigma(v)\left(\int\frac{\left[F(\vartheta,v)-\chi_{\{v>x\}}\right]}{\sigma(v)\,f(\vartheta,v)}f(\vartheta,x)\;\dot f(\vartheta,x)\,\mathrm{d}x\right)f(\vartheta,v)\,\mathrm{d}v\\
&\leq\mathrm{I}_\psi^{1/2}\left(\int\left(2\int\frac{\left[F(\vartheta,v)-\chi_{\{v>x\}}\right]}{\sigma(v)\,f(\vartheta,v)}f(\vartheta,x)\,\dot f(\vartheta,x)\,\mathrm{d}x\right)^2f(\vartheta,v)\,\mathrm{d}v\right)^{1/2}
\end{aligned}$$

which is equivalent to

$$\mathrm{I}_\psi\geq\frac{\mathrm{J}^*(\vartheta)^2}{\int_{\mathscr{R}}\int_{\mathscr{R}}\dot f(\vartheta,x)\;\dot f(\vartheta,y)\;D_\vartheta^*(x,y)\,f(\vartheta,x)\;f(\vartheta,y)\;\mathrm{d}x\;\mathrm{d}y}=\mathrm{I}_{\psi^*}.$$

The function $\psi^*(\cdot)$ is defined by the equality in the Cauchy–Schwarz inequality, i.e.,

$$\psi^*(v)=\frac{C}{\sigma(v)^2\,f(\vartheta,v)}\int[F(\vartheta,v)-\chi_{\{v>x\}}]f(\vartheta,x)\;\dot f(\vartheta,x)\;\mathrm{d}x$$

where $C>0$ is the corresponding normalizing constant. Note that this function has no compact support, but can be approximated by a sequence of functions $\psi_n^*(\cdot)$ with compact supports and such that

$$\lim_{n\to\infty}\int\left(\psi_n^*(v)-\psi^*(v)\right)^2\sigma(v)^2\,f(\vartheta,v)\;\mathrm{d}v=0.$$

Therefore we can write (2.86) and

$$\begin{aligned}\lim_{\delta\to 0}\lim_{T\to\infty}\inf_{\bar\vartheta_T}\sup_{h(\cdot)\in V_\delta}\mathbf{E}_h\,\ell\left(\sqrt{T}\left(\bar\vartheta_T-\vartheta_h\right)\right)\\ \geq \sup_{\psi(\cdot)\in\mathcal{K}}\mathbf{E}\,\ell\left(\zeta\,\mathrm{I}\left(\psi\right)^{-1/2}\right)=\mathbf{E}\,\ell\left(\zeta\,\mathrm{I}_*\left(\vartheta\right)^{-1/2}\right)\end{aligned}$$

and the theorem is proved with $\mathrm{I}_*=\mathrm{I}\left(\psi^*\right)$.

The next step is to show that the MDE ϑ_T^{**} is consistent, asymptotically normal and asymptotically efficient, i.e.,

$$\lim_{\delta\to 0}\lim_{T\to\infty}\sup_{h(\cdot)\in V_\delta}\mathbf{E}_h\,\ell\left(\sqrt{T}\left(\vartheta_T^{**}-\vartheta_h\right)\right)=\mathbf{E}\,\ell\left(\zeta\,\mathrm{I}_*\left(\vartheta\right)^{-1/2}\right). \tag{2.87}$$

To prove the consistency we repeat the calculations (2.64) of Theorem 2.19 with $\hat F_T\left(\cdot\right)$, $F\left(\boldsymbol{\vartheta},\cdot\right)$ replaced by $f_T^\circ\left(\cdot\right)$, $f\left(\vartheta,\cdot\right)$ respectively and η_T replaced by

$$\begin{aligned}\lambda_T^h(x)=&\sqrt{T}\left(f_T^\circ(x)-f_h(x)\right)=\frac{2f_h\left(x\right)}{\sqrt{T}}\int_{X_0}^{X_T}\frac{\chi_{\{v\geq x\}}-F_h(v)}{\sigma\left(v\right)^2f_h(v)}\mathrm{d}v\\ &-\frac{2f_h\left(x\right)}{\sqrt{T}}\int_0^T\frac{\chi_{\{X_t\geq x\}}-F_h(X_t)}{\sigma\left(X_t\right)f_h(X_t)}\mathrm{d}W_t.\end{aligned} \tag{2.88}$$

The boundedness of the moments (for any $p>0$)

$$\sup_{h(\cdot)\in V_\delta}\int f_h\left(x\right)^p\mathbf{E}_h\left|\int_0^\xi\frac{\chi_{\{v\geq x\}}-F_h(v)}{\sigma\left(v\right)^2f_h(v)}\mathrm{d}v\right|^p\mathrm{d}x<\infty,$$

$$\sup_{h(\cdot)\in V_\delta}\int f_h\left(x\right)^p\mathbf{E}_h\left|\frac{\chi_{\{\xi\geq x\}}-F_h(\xi)}{\sigma\left(\xi\right)f_h(\xi)}\right|^p\mathrm{d}x<\infty$$

follows directly from condition $\boldsymbol{\mathcal{A}}_0\left(\Theta\right)$ (see Proposition 1.11). Therefore we have the estimate

$$\begin{aligned}&\sup_{h(\cdot)\in V_\delta}\mathbf{P}_h^{(T)}\left\{\left|\vartheta_T^{**}-\vartheta_h\right|>\nu\right\}\\ &=\sup_{h(\cdot)\in V_\delta}\mathbf{P}_h^{(T)}\left\{\inf_{|\vartheta-\vartheta_h|<\nu}\left\|f_T^\circ\left(\cdot\right)-f\left(\vartheta,\cdot\right)\right\|\right.\\ &\qquad\left.>\inf_{|\vartheta-\vartheta_h|\geq\nu}\left\|f_T^\circ\left(\cdot\right)-f\left(\vartheta,\cdot\right)\right\|\right\}\\ &\leq\sup_{h(\cdot)\in V_\delta}\mathbf{P}_h^{(T)}\left\{\left\|f_T^\circ\left(\cdot\right)-f_h\left(\cdot\right)\right\|+\left\|f_h\left(\cdot\right)-f\left(\vartheta_h,\cdot\right)\right\|\right.\\ &\qquad\left.>\inf_{|\vartheta-\vartheta_h|\geq\nu}\left\|f\left(\vartheta,\cdot\right)-f_h\left(\cdot\right)\right\|-\left\|f_T^\circ\left(\cdot\right)-f_h\left(\cdot\right)\right\|\right\}\end{aligned}$$

$$\leq \sup_{h(\cdot)\in V_\delta} \mathbf{P}_h^{(T)} \left\{2\left\|\lambda_T^h(\cdot)\right\| \geq \sqrt{T}\, g^{**}(\nu,\delta)\right\}$$

$$\leq \frac{2^p}{g^{**}(\nu,\delta)^p\, T^{p/2}} \sup_{h(\cdot)\in V_\delta} \mathbf{E}_h \left\|\lambda_T^h(\cdot)\right\|^p \leq \frac{C(\nu,\delta)}{T^{p/2}} \to 0. \qquad (2.89)$$

Hence the MDE is a consistent estimator of the value ϑ_h.

The proof of the asymptotic normality does not contain any additional difficulties too. Note only that using the orthogonality

$$\int_{\mathscr{R}} \left[f(\vartheta_h,x) - f_h(x)\right] \dot f(\vartheta_h,x)\, \mathrm{d}x = 0$$

we obtain for $u_T^{**} = \sqrt{T}\left(\vartheta_T^{**} - \vartheta_h\right)$ the representation

$$u_T^{**} = \mathrm{J}^*(\vartheta_h)^{-1} \int_{\mathscr{R}} \eta_T^h(x)\, \dot f(\vartheta_h,x)\, \mathrm{d}x\, (1+o(1))$$

which together with (2.88) provides the desired asymptotic normality.

To check the uniform integrability of the random variables $\left|u_T^{**}\right|^p$ we proceed as in (2.75). First we expand the function $f(\theta,\cdot)$ in the vicinity of the point ϑ_h and obtain the estimate

$$\left\|f_h(\cdot) - f(\theta,\cdot)\right\| - \left\|f_h(\cdot) - f(\vartheta_h,\cdot)\right\| \geq |\theta-\vartheta_h| \inf_{\theta\in\mathbb{K}} \mathrm{J}^*(\theta)^{1/2} + o(\theta-\vartheta_h).$$

Therefore for some $\nu>0$ and $|\theta-\vartheta_h|<\nu$ we have

$$\left\|f_h(\cdot) - f(\theta,\cdot)\right\| - \left\|f_h(\cdot) - f(\vartheta_h,\cdot)\right\| \geq \frac{1}{2}\, |\theta-\vartheta_h| \inf_{\theta\in\mathbb{K}} \mathrm{J}^*(\theta)^{1/2},$$

which together with condition $\mathcal{B}_2^*$ gives us the estimate

$$g^{**}(\nu,\delta) \geq \kappa\, \nu$$

with some $\kappa = \kappa(\delta) > 0$. Hence we have for any $N>0$ and $p>0$

$$\sup_{h(\cdot)\in V_\delta} \mathbf{P}_h^{(T)} \left\{\left|u_T^{**}\right| > N\right\} \leq \frac{C_p}{N^p}.$$

Therefore we have the convergence of moments too. The asymptotic efficiency follows from the uniform over V_δ convergence of moments and the continuity of the function R_h at the point $h(\cdot)=0$.

Example 2.24. (*Shift estimation.*) Let us consider the problem of parameter estimation in the situation when ϑ is a shift parameter of the invariant density $f(\vartheta,x) = f(x-\vartheta)$. Suppose that $\sigma(\cdot)$ is a positive continuously differentiable function. Then it is easy to see, that

$$S(\vartheta, x) = \frac{\left(\sigma(x)^2 \, f(x-\vartheta)\right)'}{2f(x-\vartheta)} = \sigma(x) \; \sigma(x)' + \sigma(x)^2 \left(\ln f(x-\vartheta)\right)'.$$

Hence, the diffusion process

$$\mathrm{d}X_t = g(X_t - \vartheta)\, \sigma(X_t)^2 \, \mathrm{d}t + \sigma(X_t) \; \sigma(X_t)' \, \mathrm{d}t + \sigma(X_t)\, \mathrm{d}W_t$$

has invariant density

$$f(x-\vartheta) = G^{-1} \; \exp\left\{2 \int_0^{x-\vartheta} g(y) \; \mathrm{d}y\right\}.$$

Therefore, under $\boldsymbol{\mathcal{A}}_0$ and corresponding regularity conditions the second MDE

$$\vartheta_T^{**} = \arg \inf_{\vartheta \in \Theta} \int_{\mathscr{R}} \left[f_T^{\circ}(x) - f(x-\vartheta)\right]^2 \mathrm{d}x$$

is consistent and asymptotically normal

$$\sqrt{T}\left(\vartheta_T^{**} - \vartheta\right) \Longrightarrow \mathcal{N}\left(0, d^2(\vartheta)\right),$$

where

$$d^2(\vartheta) = \left(4\, \mathbf{E}_0 \left[f(\xi)\, g(\xi)^2\right]\right)^{-2} \mathbf{E}_\vartheta \left(\frac{f(\xi - \vartheta)}{\sigma(\xi)}\right)^2.$$

In particular, if $\sigma(\cdot) \equiv 1$ then the limit variance does not depend on ϑ

$$d^2 = \left(4\, \mathbf{E}_0 \left[f(\xi)\, g(\xi)^2\right]\right)^{-2} \mathbf{E}_0\, f(\xi)^2.$$

Indeed, in the representation

$$\sqrt{T}\left(\vartheta_T^{**} - \vartheta\right) = \frac{\int_{\mathscr{R}} \eta_T(x)\, \dot f(x-\vartheta)\, \mathrm{d}x}{\int_{\mathscr{R}} \dot f(x-\vartheta)^2 \, \mathrm{d}x} + o(1)$$

the integral

$$\int_{\mathscr{R}} \dot f(x-\vartheta)^2 \, \mathrm{d}x = 4 \int_{\mathscr{R}} f(y-\vartheta)^2 \, g(x-\vartheta)^2 \, \mathrm{d}x = 4\, \mathbf{E}_0 \left[f(\xi)\, g(\xi)^2\right].$$

Further

$$\begin{aligned}
&\int_{\mathscr{R}} \eta_T(x)\, \dot f(x-\vartheta)\, \mathrm{d}x \\
&\quad = \frac{2}{\sqrt{T}} \int_0^T \int_{\mathscr{R}} \frac{f(x-\vartheta)\, \dot f(x-\vartheta)}{\sigma(X_t)\, f(X_t - \vartheta)} \left[F(X_t - \vartheta) - \chi_{\{X_t > x\}}\right] \mathrm{d}x \, \mathrm{d}W_t + o(1) \\
&\quad = \frac{2}{\sqrt{T}} \int_0^T \frac{f(X_t - \vartheta)}{\sigma(X_t)} \, \mathrm{d}W_t + o(1).
\end{aligned}$$

Now the mentioned asymptotic normality follows from the central limit theorem for stochastic integrals.

2.3 Trajectory Fitting Estimator

The MDE $\boldsymbol{\vartheta}_T^*$ has several good properties such as asymptotic efficiency (for contaminated models) and (relatively) simple calculation (it does not involve the calculation of the Itô integral), but even in the linear case

$$\mathrm{d}X_t = -\vartheta\, X_t\, \mathrm{d}t + \mathrm{d}W_t, \quad X_0, \quad 0 \le t \le T \quad \vartheta \in \Theta \subset \mathscr{R}_+,$$

when the MLE $\hat{\vartheta}_T$ can be written explicitly, the MDE ϑ_T^* has no explicit expression and is defined as a solution of the corresponding nonlinear equation. Moreover, the construction of the MDE ϑ_T^* requires the knowledge of the invariant distribution function $F(\vartheta, x)$, which sometimes, say in the case of multidimensional observations, can be difficult to calculate. The trajectory fitting estimator introduced in Section 1.3 has several advantages and one of them is the possibility to have explicit expressions for linear systems. Another advantage is that its calculation does not use the Itô stochastic integral. Remember that for the process

$$\mathrm{d}X_t = S\left(\vartheta, X_t\right)\, \mathrm{d}t + \sigma\left(X_t\right)\, \mathrm{d}W_t, \quad X_0, \quad 0 \le t \le T,$$

and TFE

$$\vartheta_T^\star = \arg\inf_{\vartheta\in\Theta} \int_0^T \left[X_t - \hat{X}_t\left(\vartheta\right)\right]^2\, \mathrm{d}t$$

is consistent if (1.132)

$$\mathbf{E}_\vartheta\, S\left(\vartheta_1, \xi\right) \ne 0, \quad \text{for} \quad \text{all} \quad \vartheta_1 \ne \vartheta.$$

Remember that for the ergodic diffusion process we have

$$\mathbf{E}_\vartheta\, S\left(\vartheta, \xi\right) = 0.$$

To have a consistent estimation by this method we modify the definition of the estimator $\vartheta_T^\star$ and present the result in a more general framework.

The observed process is ergodic diffusion

$$\mathrm{d}X_t = S\left(\boldsymbol{\vartheta}, X_t\right)\, \mathrm{d}t + \sigma\left(X_t\right)\, \mathrm{d}W_t, \quad X_0, \quad 0 \le t \le T, \tag{2.90}$$

where $\boldsymbol{\vartheta} \in \boldsymbol{\Theta}$, $\boldsymbol{\Theta}$ is an open bounded subset of $\mathscr{R}^d$ and the conditions $\mathcal{ES}$ and $\boldsymbol{\mathcal{A}}(\boldsymbol{\Theta})$ are fulfilled. We have to estimate the parameter $\boldsymbol{\vartheta}$ by the observations $X^T = \{X_t, 0 \le t \le T\}$.

Let us introduce a two times continuously differentiable vector function $\mathbf{G}\left(x\right) = \left(G_1\left(x\right), \ldots, G_d\left(x\right)\right), x \in \mathscr{R}$ and denote by $\mathbf{g}\left(x\right) = \left(g_1\left(x\right), \ldots, g_d\left(x\right)\right), x \in \mathscr{R}$ the vector of its first derivative. Introduce as well a family of stochastic processes

$$\left\{\hat{\mathbf{Y}}_t(\boldsymbol{\theta}) = \left(\hat{Y}_t^{(1)}\left(\boldsymbol{\theta}\right), \ldots, \hat{Y}_t^{(d)}\left(\boldsymbol{\theta}\right)\right), \quad 0 \le t \le T\right\}, \quad \boldsymbol{\theta} \in \boldsymbol{\Theta}$$

as follows:

$$\hat{\mathbf{Y}}_t(\boldsymbol{\theta}) = \mathbf{G}(X_0) + \int_0^t \left[\mathbf{g}(X_s)\ S(\boldsymbol{\theta}, X_s) + \frac{1}{2}\,\mathbf{g}'(X_s)\ \sigma(X_s)^2\right] \mathrm{d}s. \qquad (2.91)$$

Define the corresponding *vector of observed processes* as

$$\{\mathbf{Y}_t = (G_1(X_t), \ldots, G_d(X_t)), \quad 0 \le t \le T\},$$

i.e., Equation (2.91) is obtained from the Itô formula for $\mathbf{Y}_t = \mathbf{G}(X_t)$:

$$\mathbf{Y}_t = \mathbf{G}(X_0) + \int_0^t \left[\mathbf{g}(X_s)\ S(\vartheta, X_s) + \frac{1}{2}\,\mathbf{g}'(X_s)\ \sigma(X_s)^2\right] \mathrm{d}s + \int_0^t \mathbf{g}(X_s)\ \sigma(X_s)\, \mathrm{d}W_s$$

by eliminating the last integral.

The *trajectory fitting estimator* (TFE) $\vartheta_T^\star$ we define as the solution of the equation

$$\int_0^T \left|\mathbf{Y}_t - \hat{\mathbf{Y}}_t(\vartheta_T^\star)\right|^2 \mathrm{d}t = \inf_{\boldsymbol{\theta}\in\Theta} \int_0^T \left|\mathbf{Y}_t - \hat{\mathbf{Y}}_t(\boldsymbol{\theta})\right|^2 \mathrm{d}t. \qquad (2.92)$$

The choice of the function $\mathbf{G}(\cdot)$ depends on the underlying stochastic differential equation and has to provide the consistency of the estimator. In particular, we require that

$$\left|\mathbf{E}_\vartheta \left[\mathbf{g}(\xi)\ S(\boldsymbol{\theta}_1, \xi) + \frac{1}{2}\,\mathbf{g}'(\xi)\ \sigma(\xi)^2\right]\right| \ne 0$$

for all $\vartheta \ne \boldsymbol{\theta}_1$. It is easy to see that

$$\mathbf{E}_\vartheta \left[\mathbf{g}(\xi)\ S(\vartheta, \xi) + \frac{1}{2}\,\mathbf{g}'(\xi)\ \sigma(\xi)^2\right] = \mathbf{0}. \qquad (2.93)$$

Note that the idea to construct an estimator by adjusting a quite irregular trajectory $\{X_t,\ 0 \le t \le T\}$ of an ergodic diffusion process on the growing interval $[0, T]$ by a one-parameter family of artificial trajectories $\left\{\hat{X}^T(\boldsymbol{\theta}),\ \boldsymbol{\theta} \in \Theta\right\}$ seems not to be so reasonable and the consistency of this estimator is not evident too. Of course we have the divergence $\left\|X_\cdot - \hat{X}_\cdot(\vartheta_T^\star)\right\| \to \infty$.

2.3.1 Properties of TFE

Below we describe the asymptotic behavior of the TFE $\vartheta_T^\star$.

The first *regularity condition* is as in Section 2.1.

$\mathcal{A}_1$. *The function* $S(\boldsymbol{\theta},\cdot)$ *is continuously differentiable w.r.t.* $\boldsymbol{\theta}$, *the derivative (vector)* $\dot{S}(\boldsymbol{\theta},\cdot)\in\mathscr{P}$ *and is uniformly continuous in the following sense: for any compact* $\mathbb{K}\subset\boldsymbol{\Theta}$

$$\lim_{\delta\to 0}\sup_{\vartheta\in\mathbb{K}}\sup_{|\theta-\vartheta|<\delta}\mathbf{E}_{\vartheta}\left|\frac{\dot{\mathbf{S}}(\theta,\xi)-\dot{\mathbf{S}}(\vartheta,\xi)}{\sigma(\xi)}\right|^{2}=0. \tag{2.94}$$

The *identifiability conditions* we give in terms of the vector function

$$\boldsymbol{K}(\boldsymbol{\theta},\boldsymbol{\vartheta})=\mathbf{E}_{\vartheta}\Big(\mathbf{g}(\xi)\,\left[S(\boldsymbol{\theta},\xi)-S(\boldsymbol{\vartheta},\xi)\right]\Big).$$

Using (2.93) we can write

$$\boldsymbol{K}(\boldsymbol{\theta},\boldsymbol{\vartheta}_0)=\mathbf{E}_{\vartheta}\left[\mathbf{g}(\xi)\,S(\boldsymbol{\theta},\xi)+\frac{1}{2}\,\mathbf{g}'(\xi)\,\sigma(\xi)^2\right].$$

Introduce the following $d\times d$ matrices:

$$\dot{\mathbf{K}}(\vartheta,\vartheta)=\frac{\partial}{\partial\theta}\boldsymbol{K}(\theta,\vartheta)\Big|_{\theta=\vartheta},\qquad \mathbf{Q}(\vartheta)=\mathbf{E}_{\vartheta}\Big(g(\xi)\,g(\xi)^{\mathrm{T}}\,\sigma(\xi)^2\Big),$$

$$\mathbf{D}(\vartheta)=\mathbf{E}_{\vartheta}\left(g(\xi)\,\frac{\partial S(\theta,\xi)^{\mathrm{T}}}{\partial\theta}\right),$$

$$\mathbf{N}(\vartheta)=\mathbf{D}(\vartheta)\,\mathbf{D}(\vartheta)^{\mathrm{T}},\qquad \mathbf{V}(\vartheta)=3\,\mathbf{N}(\vartheta)^{-1},$$

$$\mathbf{M}(\vartheta)=\frac{6}{5}\,\mathbf{V}(\vartheta)\,\mathbf{D}(\vartheta)\,\mathbf{Q}(\vartheta)\,\mathbf{D}(\vartheta)^{\mathrm{T}}\,\mathbf{V}(\vartheta)^{\mathrm{T}}.$$

$\mathcal{C}$. *The vector functions* $\boldsymbol{G}(\cdot)$, $\boldsymbol{g}(\cdot)$,, $\boldsymbol{g}'(\cdot)\in\mathscr{P}$. *For any compact* $\mathbb{K}\subset\boldsymbol{\Theta}$ *and* $\delta>0$

$$\inf_{\vartheta\in\mathbb{K}}\inf_{|\theta-\vartheta|>\delta}|\boldsymbol{K}(\theta,\vartheta)|>0, \tag{2.95}$$

and

$$\inf_{\vartheta\in\mathbb{K}}\inf_{|\mathbf{e}|=1}\left|\left(\dot{\mathbf{K}}(\vartheta,\vartheta)\,\mathbf{e},\mathbf{e}\right)\right|>0, \tag{2.96}$$

$$\inf_{\vartheta\in\mathbb{K}}\inf_{|\mathbf{e}|=1}|(\mathbf{N}(\vartheta)\,\mathbf{e},\mathbf{e})|>0 \tag{2.97}$$

where $\mathbf{e}\in\mathscr{R}^d$.

The main result of this section is the following theorem.

Theorem 2.25. *Let the conditions* $\boldsymbol{\mathcal{A}}_0$, $\mathcal{A}_1$, $\boldsymbol{\mathcal{C}}$ *be fulfilled. Then the TFE* $\vartheta_T^\star$ *is uniformly consistent on compacts: for any* $\nu>0$

$$\lim_{T\to\infty}\sup_{\vartheta\in\mathbb{K}}\mathbf{P}_{\vartheta}^{(T)}\left\{\left|\vartheta_T^\star-\vartheta\right|>\nu\right\}=0,$$

asymptotically normal

$$\mathcal{L}_\vartheta\left\{\sqrt{T}\left(\vartheta_T^\star-\vartheta\right)\right\}\Longrightarrow\mathcal{N}\left(\mathbf{0},\mathbf{M}(\vartheta)\right),$$

and for any $p>0$ *the moments converge*

$$\lim_{T\to\infty}\mathbf{E}_\vartheta\left|\sqrt{T}\left(\vartheta_T^\star-\vartheta\right)\right|^p=\mathbf{E}\left|\zeta_*\right|^p, \tag{2.98}$$

where $\mathcal{L}\left(\zeta_*\right)=\mathcal{N}\Big(\mathbf{0},\mathbf{R}\left(\vartheta\right)\Big)$.

Proof. We begin with the consistency. Using elementary inequalities we obtain the estimate

$$\begin{aligned}
&\mathbf{P}_\vartheta^{(T)}\left\{\left|\vartheta_T^\star-\vartheta\right|>\delta\right\}\\
&\quad=\mathbf{P}_\vartheta^{(T)}\left\{\inf_{|\theta-\vartheta|<\delta}\left\|\mathbf{Y}_\cdot-\hat{\mathbf{Y}}_\cdot\left(\theta\right)\right\|>\inf_{|\theta-\vartheta|\geq\delta}\left\|\mathbf{Y}_\cdot-\hat{\mathbf{Y}}_\cdot\left(\theta\right)\right\|\right\}\\
&\quad\leq\mathbf{P}_\vartheta^{(T)}\left\{\inf_{|\theta-\vartheta|<\delta}\left(\left\|\mathbf{Y}_\cdot-\hat{\mathbf{Y}}_\cdot\left(\vartheta\right)\right\|+\left\|\hat{\mathbf{Y}}_\cdot\left(\theta\right)-\hat{\mathbf{Y}}_\cdot\left(\vartheta\right)\right\|\right)\right.\\
&\qquad\left.\geq\mu_T\inf_{|\theta-\vartheta|\geq\delta}\left|\boldsymbol{K}\left(\theta,\vartheta\right)\right|-\left\|\mathbf{Y}_\cdot-\hat{\mathbf{Y}}_\cdot\left(\vartheta\right)\right\|-\sup_{|\theta-\vartheta|\geq\delta}\left\|\mathbf{Z}_\cdot\left(\theta\right)\right\|\right\}\\
&\quad\leq\mathbf{P}_\vartheta^{(T)}\left\{2\left\|\mathbf{W}_\cdot\right\|+\sup_{\theta\in\Theta}\left\|\mathbf{Z}_\cdot\left(\theta\right)\right\|\geq\mu_T\inf_{|\theta-\vartheta|\geq\delta}\left|\boldsymbol{K}\left(\theta,\vartheta\right)\right|\right\},
\end{aligned}$$

where $\mu_T=T^{3/2}/\sqrt{3}$, $\mathbf{W}_t=\mathbf{Y}_t-\hat{\mathbf{Y}}_t\left(\vartheta\right)$; i.e.,

$$\left\|\mathbf{W}_\cdot\right\|^2=\int_0^T\sum_{i=1}^d\left(\int_0^t g_i\left(X_s\right)\sigma(X_s)\,\mathrm{d}W_s\right)^2\,\mathrm{d}t$$

and

$$\left\|\mathbf{Z}_\cdot\left(\theta\right)\right\|^2=\int_0^T\sum_{i=1}^d\left(\int_0^t\Big(g_i\left(X_v\right)\left[S\left(\theta,X_v\right)-S\left(\vartheta,X_v\right)\right]-K_i(\theta,\vartheta)\Big)\,\mathrm{d}v\right)^2\,\mathrm{d}t.$$

Let us denote

$$\kappa_1=\inf_{\vartheta\in\mathbb{K}}\inf_{|\mathbf{e}|=1}\left|\left(\dot{\mathbf{K}}\left(\vartheta,\vartheta\right)\mathbf{e},\mathbf{e}\right)\right|.$$

By conditions $\boldsymbol{\mathcal{A}}_0$ and $\mathcal{A}_1$ the vector $\boldsymbol{K}\left(\theta,\vartheta\right)$ is differentiable w.r.t. $\boldsymbol{\theta}$ and we have

$$\boldsymbol{K}\left(\theta,\vartheta\right)=\dot{\mathbf{K}}\left(\vartheta,\vartheta\right)\left(\theta-\vartheta\right)+o\left(\left|\theta-\vartheta\right|\right).$$

Hence by condition (2.96) there exists a $\delta_*>0$ such that for $\left|\theta-\vartheta\right|<\delta_*$

$$\left|\boldsymbol{K}\left(\theta,\vartheta\right)\right|\geq\frac{\kappa_1\left|\theta-\vartheta\right|}{2}.$$

For the other values of $\boldsymbol{\theta}$ we have the estimate

$$|\boldsymbol{K}(\boldsymbol{\theta},\boldsymbol{\vartheta})| \geq \kappa_2 \geq \kappa_2 \frac{|\boldsymbol{\theta}-\boldsymbol{\vartheta}|}{D(\boldsymbol{\Theta})},$$

where

$$\kappa_2 = \inf_{\boldsymbol{\vartheta}\in\mathbb{K}} \inf_{|\boldsymbol{\theta}-\boldsymbol{\vartheta}|>\delta_*} |\mathbf{K}(\boldsymbol{\theta},\boldsymbol{\vartheta})|, \qquad D(\boldsymbol{\Theta}) = \sup_{\boldsymbol{\theta},\boldsymbol{\vartheta}\in\boldsymbol{\Theta}} |\boldsymbol{\theta}-\boldsymbol{\vartheta}|.$$

Therefore

$$|\boldsymbol{K}(\boldsymbol{\theta},\boldsymbol{\vartheta})| \geq \kappa\,|\boldsymbol{\theta}-\boldsymbol{\vartheta}|$$

with

$$\kappa = \min\left(\frac{\kappa_1}{2}, \frac{\kappa_2}{D(\Theta)}\right).$$

We have to estimate the following two probabilities:

$$\mathbf{P}_{\boldsymbol{\vartheta}}^{(T)}\left\{\|\mathbf{W}_{\cdot}\| \geq \frac{1}{4}\mu_T\,\kappa\,\delta\right\}, \qquad \mathbf{P}_{\boldsymbol{\vartheta}}^{(T)}\left\{\sup_{\boldsymbol{\theta}\in\boldsymbol{\Theta}}\|\mathbf{Z}_{\cdot}(\boldsymbol{\theta})\| \geq \frac{1}{2}\mu_T\,\kappa\,\delta\right\}.$$

The first one we estimate with the help of the Chebyshev inequality ($p>2$)

$$\begin{aligned}
&\mathbf{P}_{\boldsymbol{\vartheta}}^{(T)}\left\{\|\mathbf{W}_{\cdot}\| \geq \frac{1}{4}\mu_T\kappa\,\delta\right\}\\
&\quad\leq \left(\frac{4}{\mu_T\,\kappa\,\delta}\right)^{2p}\mathbf{E}_{\boldsymbol{\vartheta}}\left(\int_0^T\sum_{i=1}^d\left(\int_0^t g_i'(X_s)\;\sigma(X_s)\,\mathrm{d}W_s\right)^2\mathrm{d}t\right)^p\\
&\quad\leq \left(\frac{4}{\mu_T\,\kappa\,\delta}\right)^{2p}(d\,T)^{p-1}\int_0^T\mathbf{E}_{\boldsymbol{\vartheta}}\sum_{i=1}^d\left(\int_0^t g_i'(X_s)\;\sigma(X_s)\,\mathrm{d}W_s\right)^{2p}\mathrm{d}t\\
&\quad\leq C_p\left(\frac{1}{\mu_T\,\kappa\,\delta}\right)^{2p}T^{p-1}\int_0^T t^p\sum_{i=1}^d\mathbf{E}_{\boldsymbol{\vartheta}}\Big(g_i'(\xi)\;\sigma(\xi)\Big)^{2p}\mathrm{d}t \leq \frac{C}{\left(\delta\sqrt{T}\right)^{2p}}.
\end{aligned}$$

To estimate the second probability we introduce the subdivision $\boldsymbol{\Theta}_k, k=1,\ldots,K$, where $\boldsymbol{\Theta}_k$ are cubes of side h, such that $\boldsymbol{\Theta}\subset\cup_{k=1}^K\boldsymbol{\Theta}_k$. Denote by $\{\boldsymbol{\theta}_1,\ldots,\boldsymbol{\theta}_K\}$, $\boldsymbol{\theta}_k\in\boldsymbol{\Theta}_k$ the set of centers of these cubes and note that the number K of cubes is less than $K_*=\left(\frac{D(\boldsymbol{\Theta})}{h}\right)^d$. We have

$$\begin{aligned}
&\mathbf{P}_{\boldsymbol{\vartheta}}^{(T)}\left\{\sup_{\boldsymbol{\theta}\in\Theta}\|\mathbf{Z}_{\cdot}(\boldsymbol{\theta})\| \geq \frac{1}{2}\mu_T\,\kappa\,\delta\right\}\\
&\quad\leq \mathbf{P}_{\boldsymbol{\vartheta}}^{(T)}\left\{\max_k\left(\|\mathbf{Z}_{\cdot}(\boldsymbol{\theta}_k)\| + \sup_{|\boldsymbol{\theta}-\boldsymbol{\theta}_k|\leq h}\|\mathbf{Z}_{\cdot}(\boldsymbol{\theta})-\mathbf{Z}_{\cdot}(\boldsymbol{\theta}_k)\|\right) \geq \frac{1}{2}\mu_T\,\kappa\,\delta\right\}\\
&\quad\leq \mathbf{P}_{\boldsymbol{\vartheta}}^{(T)}\left\{\max_k\|\mathbf{Z}_{\cdot}(\boldsymbol{\theta}_k)\| \geq \frac{1}{4}\mu_T\,\kappa\,\delta\right\}\\
&\qquad+\mathbf{P}_{\boldsymbol{\vartheta}}^{(T)}\left\{\sup_{|\boldsymbol{\theta}-\boldsymbol{\theta}'|\leq h}\|\mathbf{Z}_{\cdot}(\boldsymbol{\theta})-\mathbf{Z}_{\cdot}(\boldsymbol{\theta}')\| \geq \frac{1}{4}\mu_T\,\kappa\,\delta\right\}.
\end{aligned}\tag{2.99}$$

Using elementary inequalities we obtain

$$\mathbf{P}_{\vartheta}^{(T)}\left\{\max_k \|\mathbf{Z}_\cdot(\boldsymbol{\theta}_k)\|\geq \frac{1}{4}\,\mu_T\,\kappa\,\delta\right\}\leq \sum_{k=1}^{K}\mathbf{P}_{\vartheta}^{(T)}\left\{\|\mathbf{Z}_\cdot(\boldsymbol{\theta}_k)\|\geq \frac{1}{4}\,\mu_T\,\kappa\,\delta\right\}$$
$$\leq \left(\frac{4}{\mu_T\,\kappa\,\delta}\right)^{2p_1}\sum_{k=1}^{K}\mathbf{E}_{\vartheta}\,\|\mathbf{Z}_\cdot(\boldsymbol{\theta}_k)\|^{2p_1}.$$

Put $\boldsymbol{h}(\boldsymbol{\theta},x)=\mathbf{g}(x)\left[S(\boldsymbol{\theta},x)-S(\boldsymbol{\vartheta},x)\right]-\boldsymbol{K}(\boldsymbol{\theta},\boldsymbol{\vartheta})$ and remember that

$$\mathbf{E}_{\vartheta}\,\boldsymbol{h}(\boldsymbol{\theta},\xi)=0$$

for all $\boldsymbol{\theta}\in\boldsymbol{\Theta}$. Then introduce the functions

$$\tilde{\boldsymbol{h}}(\boldsymbol{\theta},x)=\frac{2}{\sigma(x)\,f(\boldsymbol{\vartheta},x)}\int_{-\infty}^{x}\boldsymbol{h}(\boldsymbol{\theta},y)\,f(\boldsymbol{\vartheta},y)\,\mathrm{d}y,$$
$$\boldsymbol{H}(\boldsymbol{\theta},x)=\int_0^x\frac{\tilde{\boldsymbol{h}}(\boldsymbol{\theta},y)}{\sigma(y)}\,\mathrm{d}y$$

and by the Itô formula write the equality

$$\int_0^t\boldsymbol{h}(\boldsymbol{\theta},X_v)\,\mathrm{d}v=\boldsymbol{H}(\boldsymbol{\theta},X_t)-\boldsymbol{H}(\boldsymbol{\theta},X_0)-\int_0^t\tilde{\boldsymbol{h}}(\boldsymbol{\theta},X_s)\,\mathrm{d}W_s.$$

Therefore we can estimate the mathematical expectation as follows:

$$\mathbf{E}_{\vartheta}\,\|\mathbf{Z}_\cdot(\boldsymbol{\theta}_k)\|^{2p_1}=\mathbf{E}_{\vartheta}\left(\int_0^T\left|\int_0^t\boldsymbol{h}(\boldsymbol{\theta}_k,X_v)\,\mathrm{d}v\right|^2\mathrm{d}t\right)^{p_1}$$
$$\leq C_1\,T^{p_1}\sup_{\boldsymbol{\theta}\in\boldsymbol{\Theta}}\mathbf{E}_{\vartheta}\left|\boldsymbol{H}(\boldsymbol{\theta},\xi)\right|^{2p_1}+C_2\,T^{2p_1}\sup_{\boldsymbol{\theta}\in\boldsymbol{\Theta}}\mathbf{E}_{\vartheta}\left|\tilde{\boldsymbol{h}}(\boldsymbol{\theta},\xi)\right|^{2p_1}\leq C\,T^{2p_1}$$

(see Lemma 1.17) and we have the estimate

$$\mathbf{P}_{\vartheta}^{(T)}\left\{\max_k \|\mathbf{Z}_\cdot(\boldsymbol{\theta}_k)\|\geq \frac{1}{4}\,\mu_T\,\kappa\,\delta\right\}\leq\frac{C\,\left(\delta\sqrt{T}\right)^{q}}{\left(\delta\sqrt{T}\right)^{2p_1}}\leq\frac{C}{\left(\delta\sqrt{T}\right)^{p_1}},\qquad(2.100)$$

where we set $h=(\delta T)^{-q/d}$ and $q=p_1$. The value p_1 will be chosen later.

To estimate the last probability in (2.99) we use the following result by Ibragimov and Khasminskii.

Lemma 2.26. *Let $\eta(\boldsymbol{\theta})$ be a real-valued random function defined on a closed subset $\boldsymbol{\Theta}$ of Euclidean space $\mathscr{R}^d$. We assume the random process $\eta(\boldsymbol{\theta})$ is measurable and separable. Assume that the following condition is fulfilled: there exist numbers $m\geq r>d$ and a function $H(\boldsymbol{\theta}):\mathscr{R}^d\to\mathscr{R}$ bounded on compact sets such that for all $\boldsymbol{\theta}\in\boldsymbol{\Theta}$, $\boldsymbol{\theta}+\boldsymbol{h}\in\boldsymbol{\Theta}$,*

$$\mathbf{E}\left|\eta(\boldsymbol{\theta})\right|^m \leq H(\boldsymbol{\theta}), \qquad \mathbf{E}\left|\eta(\boldsymbol{\theta}+\boldsymbol{h})-\eta(\boldsymbol{\theta})\right|^m \leq H(\boldsymbol{\theta})\,\left|\boldsymbol{h}\right|^r.$$

Then with probability 1 the realizations of $\eta(\boldsymbol{\theta})$ are continuous functions on $\boldsymbol{\Theta}$. Moreover, set

$$\omega(\nu,\eta,L) = \sup_{|\boldsymbol{\theta}-\boldsymbol{\theta}'|\leq\nu} \left|\eta(\boldsymbol{\theta})-\eta(\boldsymbol{\theta}')\right|,$$

where the upper bound is taken over $\boldsymbol{\theta}$, $\boldsymbol{\theta}' \in \boldsymbol{\Theta}$, $|\boldsymbol{\theta}| \leq L$, $|\boldsymbol{\theta}'| \leq L$. Then

$$\mathbf{E}\,\omega(\nu,\eta,L) \leq B_0 \left(\sup_{|\boldsymbol{\theta}|\leq L} H(\boldsymbol{\theta})\right)^{1/m} L^d\,\nu^{(r-d)/m},$$

where the constant B_0 depends on m, r, d.

Proof. See [109], Theorem A.19.

Let us introduce the random function

$$\eta_T(\boldsymbol{\theta}) = \frac{1}{T^2}\int_0^T \sum_{i=1}^d \left(\int_0^t \frac{2}{\sigma(X_s)\, f(\vartheta, X_s)} \int_{-\infty}^{X_s} h_i(\boldsymbol{\theta}, v)\, f(\vartheta, v)\,\mathrm{d}v\,\mathrm{d}W_s\right)^2 \mathrm{d}t$$

and note that

$$\left\|\boldsymbol{Z}_\cdot(\boldsymbol{\theta})\right\| \leq \left\|\boldsymbol{H}_\cdot(\boldsymbol{\theta}, X_\cdot)\right\| + T^{1/2}\left(\sum_{i=1}^d H_i(\boldsymbol{\theta}, X_0)^2\right)^{1/2} + T\,\eta_T(\boldsymbol{\theta})^{1/2}.$$

We check the conditions of Lemma 2.26 for the process $\eta_T(\boldsymbol{\theta})$. Fix $m > d$. We have

$$\begin{aligned}
&\mathbf{E}_\vartheta\left|\eta_T(\boldsymbol{\theta})-\eta_T(\boldsymbol{\theta}')\right|^m \\
&= T^{-2m}\,\mathbf{E}_\vartheta\left|\int_0^T\left(\sum_{i=1}^d\left(\int_0^t \tilde h_i(\boldsymbol{\theta}, X_s)\,\mathrm{d}W_s\right)^2\right.\right. \\
&\qquad\qquad \left.\left. -\left(\int_0^t \tilde h_i(\boldsymbol{\theta}', X_s)\,\mathrm{d}W_s\right)^2\right)\mathrm{d}t\right|^m \\
&\leq T^{-m-1}\int_0^T \mathbf{E}_\vartheta\left|\sum_{i=1}^d\left(\int_0^t \tilde h_i(\boldsymbol{\theta}, X_s)\,\mathrm{d}W_s\right)^2\right. \\
&\qquad\qquad \left. -\left(\int_0^t \tilde h_i(\boldsymbol{\theta}', X_s)\,\mathrm{d}W_s\right)^2\right|^m \mathrm{d}t \\
&\leq \left(\frac{d}{T}\right)^{m-1}\int_0^T\sum_{i=1}^d \mathbf{E}_\vartheta\left|\left(\int_0^t \tilde h_i(\boldsymbol{\theta}, X_s)\,\mathrm{d}W_s\right)^2\right.
\end{aligned}$$

$$-\left(\int_0^t \tilde h_i\left(\theta',X_s\right)\mathrm{d}W_s\right)^2\Bigg|^m \mathrm{d}t$$

$$\leq \left(\frac{d}{T}\right)^{m-1}\int_0^T\sum_{i=1}^d\left(\mathbf{E}_\vartheta\left(\int_0^t\left[\tilde h_i\left(\theta,X_s\right)-\tilde h_i\left(\theta',X_s\right)\right]\mathrm{d}W_s\right)^{2m}\right)^{1/2}$$

$$\left(\mathbf{E}_\vartheta\left(\int_0^t\left[\tilde h_i\left(\theta,X_s\right)+\tilde h_i\left(\theta',X_s\right)\right]\mathrm{d}W_s\right)^{2m}\right)^{1/2}\mathrm{d}t\leq C\left|\theta-\theta'\right|^m,$$

because

$$\mathbf{E}_\vartheta\left(\int_0^t \tilde h_i\left(\theta,X_s\right)\mathrm{d}W_s\right)^{2m}\leq C\,t^m\,\mathbf{E}_\vartheta\left|\tilde h_i\left(\theta,\xi\right)\right|^{2m}$$

and

$$\mathbf{E}_\vartheta\left(\int_0^t\left[\tilde h_i\left(\theta,X_s\right)-\tilde h_i\left(\theta',X_s\right)\right]\mathrm{d}W_s\right)^{2m}\leq C\,t^m\left|\theta-\theta'\right|^{2m}.$$

To verify the last two inequalities we remember that under conditions $\mathscr{A}_0$ and $\mathscr{C}_1$ we have

$$\sup_{\vartheta\in\mathbb{K}}\sup_{\theta\in\Theta}\mathbf{E}_\vartheta\left|\tilde{\boldsymbol{h}}\left(\theta,\xi\right)\right|^{2m}\leq C.$$

Further, the function $\boldsymbol{h}\left(\theta,\xi\right)$ is continuously differentiable w.r.t. θ. Hence

$$\tilde{\boldsymbol{h}}\left(\theta,\xi\right)-\tilde{\boldsymbol{h}}\left(\theta',\xi\right)=\left(\left(\theta-\theta'\right),\int_0^1\frac{2}{\sigma\left(\xi\right)f\left(\vartheta,\xi\right)}\int_0^\xi\dot{\boldsymbol{h}}\left(\theta_s,v\right)f\left(\vartheta,v\right)\mathrm{d}v\,\mathrm{d}s\right),$$

where $\theta_s=\theta+s\left(\theta-\theta'\right)$. Therefore

$$\mathbf{E}_\vartheta\left|\tilde{\boldsymbol{h}}\left(\theta,\xi\right)-\tilde{\boldsymbol{h}}\left(\theta',\xi\right)\right|^{2m}$$

$$\leq\left|\theta-\theta'\right|^{2m}\int_0^1\mathbf{E}_\vartheta\left|\frac{2}{\sigma\left(\xi\right)f\left(\vartheta,\xi\right)}\int_0^\xi\dot{\boldsymbol{h}}\left(\theta_s,v\right)f\left(\vartheta,v\right)\mathrm{d}v\right|^{2m}\mathrm{d}s.$$

The elements of the matrix $\dot{\boldsymbol{h}}\left(\theta_s,v\right)$ have polynomial majorants, therefore the last expectation is finite as well (see Lemma 1.17).

Finally we have the estimate

$$\sup_{\vartheta\in\mathbb{K}}\mathbf{E}_\vartheta\left|\eta_T(\theta)-\eta_T(\theta')\right|^m\leq C\left|\theta-\theta'\right|^m$$

and by Lemma 2.26

$$\sup_{\vartheta\in\mathbb{K}}\mathbf{P}_\vartheta\left\{\sup_{\left|\theta-\theta'\right|\leq a}\left|\eta_T(\theta)-\eta_T(\theta')\right|>\frac{1}{4}\kappa\delta\sqrt{T}\right\}\leq\frac{C}{\delta\sqrt{T}}\left(\delta\sqrt{T}\right)^{-p_1\frac{m-d}{m}}.$$

Hence if we put $p_1 > m/d$ then

$$\sup_{\vartheta\in\mathbb{K}} \mathbf{P}_\vartheta \left\{ \sup_{|\theta-\theta'|\leq a} \left\|\boldsymbol{Z}_\cdot(\boldsymbol{\theta}) - \boldsymbol{Z}_\cdot(\boldsymbol{\theta}')\right\| > \frac{1}{4}\kappa\mu_T\delta \right\} \leq \frac{C}{\left(\delta\sqrt{T}\right)^{p_1}}. \tag{2.101}$$

All these estimates allow us to write the inequality

$$\sup_{\vartheta\in\mathbb{K}} \mathbf{P}_\vartheta \left\{\left|\vartheta_T^\star - \vartheta\right| > \delta\right\} \leq \frac{C}{\left(\delta\sqrt{T}\right)^{p_1}} \tag{2.102}$$

which provides the uniform consistency of TFE. A similar inequality

$$\sup_{\vartheta\in\mathbb{K}} \mathbf{P}_\vartheta \left\{\sqrt{T}\left|\vartheta_T^\star - \vartheta\right| > N\right\} \leq \frac{C}{N^{p_1}} \tag{2.103}$$

gives us the uniform integrability of the random variables $\sqrt{T}\left|\vartheta_T^\star - \vartheta\right|$; i.e., the asymptotic normality of $\sqrt{T}\left(\vartheta_T^\star - \vartheta\right)$ and (2.103) will give us the convergence of moments.

To prove the asymptotic normality we at first localize the problem, i.e., we introduce the set

$$\mathbb{B} = \left\{\omega : \left|\vartheta_T^\star - \vartheta\right| < T^{-1/4}\right\},$$

and note that for the complement of $\mathbb{B}$ we have

$$\mathbf{P}_\vartheta \left\{\mathbb{B}^c\right\} \leq C T^{-p_1/2}. \tag{2.104}$$

For $\omega \in \mathbb{B}$ the solution $\vartheta_T^\star$ of the equation

$$\left\|\boldsymbol{Y}_\cdot - \hat{\boldsymbol{Y}}_\cdot\left(\vartheta_T^\star\right)\right\| = \inf_{\boldsymbol{\theta}\in\Theta} \left\|\boldsymbol{Y}_\cdot - \hat{\boldsymbol{Y}}_\cdot\left(\boldsymbol{\theta}\right)\right\|$$

is in the vicinity of the true value ϑ and is the same as one of the solutions of the system of equations

$$\frac{\partial}{\partial\boldsymbol{\theta}} \int_0^T \left|\boldsymbol{Y}_t - \hat{\boldsymbol{Y}}_t\left(\boldsymbol{\theta}\right)\right|^2 \mathrm{d}t = \mathbf{0}.$$

Rewrite this system in the form

$$\int_0^T \sum_{i=1}^d \left[Y_{t,i} - \hat{Y}_{t,i}\left(\vartheta_T^\star\right)\right] \frac{\partial \hat{Y}_{t,i}}{\partial\theta_j}\left(\vartheta_T^\star\right) \mathrm{d}t = 0, \quad j = 1,\ldots,d. \tag{2.105}$$

Remember that

$$\frac{\partial \hat{Y}_{t,i}}{\partial\theta_j}\left(\boldsymbol{\theta}\right) = \int_0^t g_i\left(X_s\right) \frac{\partial S\left(\boldsymbol{\theta}, X_s\right)}{\partial\theta_j} \mathrm{d}s \equiv y_{i,j}\left(\boldsymbol{\theta}, t\right)$$

and

$$Y_{t,i} - \hat{Y}_{t,i}(\vartheta) = \int_0^t g_i(X_s)\,\sigma(X_s)\,\mathrm{d}W_s \equiv \eta_i(t),$$

(we introduced the corresponding notation). We can write the system (2.105) as

$$\begin{aligned}
&\int_0^T \sum_{i=1}^d \left[\eta_i(t) - \sum_{l=1}^d \left(\vartheta^\star_{T,l} - \vartheta_l\right) y_{i,l}\left(\tilde{\boldsymbol{\vartheta}}^\star_T, t\right)\right] y_{i,j}\left(\boldsymbol{\vartheta}^\star_T, t\right)\,\mathrm{d}t \\
&\quad = \int_0^T \sum_{i=1}^d \eta_i(t)\, y_{i,j}\left(\boldsymbol{\vartheta}^\star_T, t\right)\,\mathrm{d}t \\
&\quad\quad - \sum_{l=1}^d \left(\vartheta^\star_{T,l} - \vartheta_l\right) \int_0^T \sum_{i=1}^d y_{i,l}\left(\tilde{\boldsymbol{\vartheta}}^\star_T, t\right) y_{i,j}\left(\boldsymbol{\vartheta}^\star_T, t\right)\,\mathrm{d}t = 0.
\end{aligned}$$

Let us denote as $V_{l,j}(\boldsymbol{\theta},\boldsymbol{\theta}',T), l,j = 1\ldots,d$ the matrix inverse to the matrix

$$\int_0^T \sum_{i=1}^d y_{i,l}(\boldsymbol{\theta},t)\, y_{i,j}\left(\boldsymbol{\theta}',t\right)\,\mathrm{d}t$$

and put $V_{l,j}(\boldsymbol{\theta},\boldsymbol{\theta}',T) = 0,\ l,j = 1\ldots,d$ if it is degenerate. Then we have the following representation:

$$\begin{aligned}
&\sqrt{T}\left(\vartheta^\star_{T,l} - \vartheta_l\right) \\
&\quad = \sum_{j=1}^d T^3\, V_{l,j}(\boldsymbol{\vartheta}^\star_T, \tilde{\boldsymbol{\vartheta}}^\star_T, T)\, T^{-5/2} \int_0^T \sum_{i=1}^d \eta_i(t)\, y_{i,j}\left(\boldsymbol{\vartheta}^\star_T, t\right)\,\mathrm{d}t \\
&\quad = \sum_{j=1}^d T^3\, V_{l,j}(\boldsymbol{\vartheta}, \boldsymbol{\vartheta}, T)\, T^{-5/2} \int_0^T \sum_{i=1}^d \eta_i(t)\, y_{i,j}(\boldsymbol{\vartheta}, t)\,\mathrm{d}t\ (1 + o(1)).
\end{aligned}$$

The term $o(1)$ tends to zero in probability due to consistency. By the law of large numbers we have

$$\frac{y_{i,j}(\boldsymbol{\vartheta},t)}{t} = \frac{1}{t}\int_0^t g_i(X_s)\,\frac{\partial S(\boldsymbol{\theta},X_s)}{\partial \theta_j}\mathrm{d}s \rightarrow \mathbf{E}_{\boldsymbol{\vartheta}}\left(g_i(\xi)\,\frac{\partial S(\boldsymbol{\theta},\xi)}{\partial \theta_j}\right) \equiv D_{i,j}(\boldsymbol{\vartheta}).$$

Hence by the Tœplitz lemma

$$\frac{1}{T^3}\int_0^T t^2 \sum_{i=1}^d \frac{y_{i,l}(\boldsymbol{\theta},t)}{t}\,\frac{y_{i,j}(\boldsymbol{\theta}',t)}{t}\,\mathrm{d}t \rightarrow \frac{1}{3}\sum_{i=1}^d D_{i,l}(\boldsymbol{\vartheta})\ D_{i,j}(\boldsymbol{\vartheta}) \equiv \frac{1}{3}\, S_{l,j}(\boldsymbol{\vartheta}).$$

Hence we have the convergence of matrices

$$T^3\,\mathbf{V}(\boldsymbol{\vartheta},\boldsymbol{\vartheta},T) \longrightarrow 3\,\mathbf{N}(\boldsymbol{\vartheta})^{-1} \equiv \mathbf{V}(\boldsymbol{\vartheta}).$$

Note that

$$\int_0^T \eta_i(t)\, y_{i,j}(\vartheta,t)\, \mathrm{d}t = D_{i,j}(\vartheta)\int_0^T t\,\eta_i(t)\,\mathrm{d}t\,\Big(1+o(1)\Big).$$

Therefore it is sufficient to study the integral

$$T^{-5/2}\int_0^T t\,\eta_i(t)\,\mathrm{d}t = \int_0^1 s^{3/2}\,\eta_i^{(\varepsilon)}(s)\,\mathrm{d}s,$$

where we put $t = sT$, $\varepsilon = 1/T$ and

$$\eta_i^{(\varepsilon)}(s) = \frac{1}{\sqrt{sT}}\int_0^{sT} g_i(X_t)\,\sigma(X_t)\,\mathrm{d}W_t, \qquad 0\le s\le 1.$$

The finite-dimensional distributions of the vector process

$$\boldsymbol{\eta}^{(\varepsilon)}(s) = \left\{\eta_1^{(\varepsilon)}(s),\ldots,\eta_d^{(\varepsilon)}(s)\right\},\quad 0\le s\le 1$$

converge to the finite-dimensional distributions of the Gaussian process

$$\boldsymbol{\eta}(s) = \{\eta_1(s),\ldots,\eta_d(s)\},\quad 0\le s\le 1$$

with parameters

$$\mathbf{E}_{\vartheta}\,\eta_i(s) = 0,\qquad \mathbf{E}_{\vartheta}\,\eta_i(s)\,\eta_j(s') = \frac{s\wedge s'}{\sqrt{s\,s'}}\,Q_{i,j}(\vartheta),$$

where

$$Q_{i,j}(\vartheta) = \mathbf{E}_{\vartheta}\Big(g_i(\xi)\,g_j(\xi)\,\sigma(\xi)^2\Big),\qquad i,j=1,\ldots,d.$$

Further we have the estimate

$$\mathbf{E}_{\vartheta}\left|\eta_i^{(\varepsilon)}(s')-\eta_i^{(\varepsilon)}(s)\right|^2 = 2\left(1-\frac{s\wedge s'}{\sqrt{s\,s'}}\right)Q_{i,i}(\vartheta)\le 2\,Q_{i,i}(\vartheta)\,\frac{|s-s'|}{\sqrt{s\,s'}}.$$

Therefore we can apply Theorem A22, [109] and obtain the convergence of the distributions of the integrals

$$\left\{\int_0^1 s^{3/2}\,\eta_i^{(\varepsilon)}(s)\,\mathrm{d}s,\ i=1,\ldots,d\right\}\Longrightarrow\left\{\int_0^1 s^{3/2}\,\eta_i(s)\,\mathrm{d}s,\ i=1,\ldots,d\right\}.$$

Note that

$$\gamma = \left\{\int_0^1 s^{3/2}\,\eta_1(s)\,\mathrm{d}s,\ldots,\int_0^1 s^{3/2}\,\eta_d(s)\,\mathrm{d}s\right\}$$

is a Gaussian vector with parameters

$$\mathbf{E}_{\vartheta}\,\gamma_i = 0,\qquad \mathbf{E}_{\vartheta}\,\gamma_i\,\gamma_j = \frac{2}{15}\,Q_{i,j}(\vartheta).$$

Finally we have

$$\sqrt{T}\left(\vartheta_T^\star - \vartheta\right) \Longrightarrow 3\,\mathbf{V}(\vartheta)\,\mathbf{D}(\vartheta)\,\gamma,$$

i.e., TFE is asymptotically normal with the limit covariance matrix $\mathbf{M}(\vartheta)$. It can be shown (as in the preceding section) that this convergence is uniform on the compacts $\mathbb{K} \subset \Theta$. The uniform integrability of the random variables $\sqrt{T}\left|\vartheta_T^\star - \vartheta\right|^p$ for any $p > 0$ follows from (2.103). Hence we have the convergence of moments (2.98) too.

2.3.2 Example

We already considered two examples of TFE for linear systems in Section 1.3. Note that conditions of Theorem 2.25 are fulfilled for both examples and therefore for these TFE's we have the convergence of moments as well.

Example 2.27. (*Ornstein–Uhlenbeck process*) Consider the linear process

$$\mathrm{d}X_t = -\left(\vartheta^{(1)}\,X_t - \vartheta^{(2)}\right)\,\mathrm{d}t + \sigma \mathrm{d}W_t, \quad X_), \qquad 0 \le t \le T,$$

where $\vartheta = (\vartheta_1, \vartheta_2) \in \Theta = (\alpha_1, \beta_1) \times (\alpha_2, \beta_2)$, $\alpha_1 > 0$.

To construct the TFE we put $G(x) = \left(x, x^2\right)$ and obtain the vector process

$$\hat{\mathbf{Y}}_t(\theta) = \left(\hat{Y}_t^{(1)}(\theta), \hat{Y}_t^{(2)}(0)\right), 0 \le t \le T, \quad \theta \subset \Theta$$

where

$$\hat{Y}_t^{(1)}(\theta) = X_0 + \theta^{(2)}\,t - \theta^{(1)}\int_0^t X_s\,\mathrm{d}s,$$

$$\hat{Y}_t^{(2)}(\theta) = X_0^2 + 2\,\theta^{(2)}\int_0^t X_s\,\mathrm{d}s - 2\,\theta^{(1)}\int_0^t X_s^2\,\mathrm{d}s + \sigma^2\,t.$$

The TFE is

$$\vartheta_T^\star = \mathbf{A}_T^{-1}\,b_T$$

where the matrix

$$\mathbf{A}_T = \begin{pmatrix} \left\|H^{(1)}\right\|^2 + 2\left\|H^{(2)}\right\|^2, & \langle t, H^{(1)}\rangle + 2\langle H^{(1)}, H^{(2)}\rangle \\ \langle t, H^{(1)}\rangle + 2\langle H^{(1)}, H^{(2)}\rangle, & \|t\|^2 + 2\left\|H^{(1)}\right\|^2 \end{pmatrix}.$$

Here $\langle\cdot\rangle$ and $\|\cdot\|$ are the inner product and norm in $\mathscr{L}_2(0,T)$ and

$$H^{(k)} = \int_0^T X_t^k\,\mathrm{d}t.$$

The vector

$$b_T = \begin{pmatrix} \langle X - X_0, H^{(1)} \rangle + 2\langle X^2 - X_0^2 - \sigma^2 t, H^{(2)} \rangle \\ \langle X - X_0, t \rangle + 2\langle X^2 - X_0^2 - \sigma^2 t, H^{(1)} \rangle \end{pmatrix}.$$

The direct calculation gives us the functions

$$K_1(\boldsymbol{\theta}, \boldsymbol{\vartheta}) = \vartheta^{(2)} \left(\frac{\theta^{(2)}}{\vartheta^{(2)}} - \frac{\theta^{(1)}}{\vartheta^{(1)}} \right),$$

$$K_2(\boldsymbol{\theta}, \boldsymbol{\vartheta}) = 2\left(\theta^{(2)} - \vartheta^{(2)}\right) \frac{\vartheta^{(2)}}{\vartheta^{(1)}} + 2\left(\vartheta^{(1)} - \theta^{(1)}\right) \left(\frac{\vartheta^{(2)}}{\vartheta^{(1)}}\right)^2$$

$$+ \left(\vartheta^{(1)} - \theta^{(1)}\right) \frac{\sigma^2}{\vartheta^{(1)}}.$$

If $K_1(\boldsymbol{\theta}, \boldsymbol{\vartheta}) = 0$, then $\frac{\theta^{(2)}}{\vartheta^{(2)}} = \frac{\theta^{(1)}}{\vartheta_1} = c$ and $K_2(\boldsymbol{\theta}, \boldsymbol{\vartheta}) = (1 - c)\,\sigma^2$. Therefore the identifiability condition (2.95) is fulfilled. The matrix

$$\dot{\mathbf{K}}(\boldsymbol{\vartheta}, \boldsymbol{\vartheta}) = \begin{pmatrix} -\frac{\vartheta^{(2)}}{\vartheta^{(1)}} & -2\left(\frac{\vartheta^{(2)}}{\vartheta^{(1)}}\right)^2 + \frac{\sigma^2}{\vartheta^{(1)}} \\ 1 & 2\,\frac{\vartheta^{(2)}}{\vartheta^{(1)}} \end{pmatrix}$$

Hence

$$\det\left(\dot{\mathbf{K}}(\boldsymbol{\vartheta}, \boldsymbol{\vartheta})\right) = -\frac{\sigma^2}{\vartheta^{(1)}}$$

and the condition (2.96) is fulfilled too.

2.4 Estimator of the Method of Moments

We observe a trajectory $X^T = \{X_t, 0 \le t \le T\}$ of the diffusion process

$$\mathrm{d}X_t = S(\boldsymbol{\vartheta}, X_t)\,\mathrm{d}t + \sigma(X_t)\,\mathrm{d}W_t, \quad X_0, \quad 0 \le t \le T, \tag{2.106}$$

where the unknown parameter $\boldsymbol{\vartheta} \in \boldsymbol{\Theta} \subset \mathscr{R}^d$, the set $\boldsymbol{\Theta}$ is open and bounded and $S(\cdot,\cdot)$ and $\sigma(\cdot)$ are known functions, such that the conditions $\mathcal{ES}$ and $\boldsymbol{\mathcal{A}}_0(\boldsymbol{\Theta})$ are fulfilled.

We remember here the definition from Section 1.3. Let $\boldsymbol{q}(x) = (q_1(\cdot), \ldots, q_d(\cdot))$ be a vector-function and $\boldsymbol{m}(\boldsymbol{\vartheta}) = \mathbf{E}_{\boldsymbol{\vartheta}}\, \boldsymbol{q}(\xi)$ be its mathematical expectation. Here ξ as before is a *stationary random variable*, i.e., ξ has a density function $f(\boldsymbol{\vartheta}, \cdot)$ of the invariant law. Denote by $\mathbb{M} = \{\boldsymbol{m}(\boldsymbol{\vartheta}),\ \boldsymbol{\vartheta} \in \boldsymbol{\Theta}\}$. The function $\boldsymbol{q}(\cdot)$ is chosen in such a way that the equation

$$\boldsymbol{m}(\boldsymbol{\vartheta}) = \boldsymbol{m}, \qquad \boldsymbol{\vartheta} \in \boldsymbol{\Theta}$$

has a unique solution for any $\boldsymbol{m} \in \mathbb{M}$. The *estimator of the method of moments* (EMM) $\bar{\boldsymbol{\vartheta}}_T$ is defined by the equality

$$\left|m\left(\bar{\vartheta}_T\right)-\frac{1}{T}\int_0^T q(X_t)\,\mathrm{d}t\right|=\inf_{\vartheta\in\Theta}\left|m\left(\vartheta\right)-\frac{1}{T}\int_0^T q(X_t)\,\mathrm{d}t\right|. \qquad (2.107)$$

If this equation has more than one solution then any of them can be taken as EMM $\bar{\vartheta}_T$. Remember that this estimator admits the representation

$$\bar{\vartheta}_T=\vartheta_T\,\chi_{\left\{\hat{m}_T\in\mathbb{M}\right\}}+\vartheta_T^o\,\chi_{\left\{\hat{m}_T\in\mathbb{M}^c\right\}}, \qquad (2.108)$$

where

$$\hat{m}_T=\frac{1}{T}\int_0^T q(X_t)\,\mathrm{d}t$$

and ϑ_T is a solution of the equation $m(\vartheta_T)=\hat{m}_T$ when $\hat{m}_T\in\mathbb{M}$. The value ϑ_T^o corresponds to $m\in\partial\mathbb{M}$ closest to $\hat{m}_T$, when $\hat{m}_T\in\mathbb{M}^c$.

2.4.1 Properties of EMM

Let us denote by $\dot{m}\left(\vartheta\right)$ the $d\times d$ matrix of the derivatives

$$\dot{m}\left(\vartheta\right)=\frac{\partial m\left(\vartheta\right)}{\partial\vartheta}.$$

We study the asymptotic behavior of EMM $\bar{\vartheta}_T$ under the following

Regularity condition $\mathcal{D}$.

$\mathcal{D}_1$. *The function $q\left(\cdot\right)\in\mathscr{P}$ and the vector function $m\left(\vartheta\right)$, $\vartheta\in\Theta$ is continuously differentiable over ϑ.*

$\mathcal{D}_2$. *For any $\nu>0$ and any compact $\mathbb{K}\subset\Theta$*

$$\inf_{\vartheta_0\in\mathbb{K}}\inf_{|\vartheta-\vartheta_0|>\nu}\left|m\left(\vartheta\right)-m\left(\vartheta_0\right)\right|>0.$$

The $d\times d$ matrix

$$M\left(\vartheta\right)=\dot{m}\left(\vartheta\right)\,\dot{m}\left(\vartheta\right)^{\mathrm{T}}$$

is uniformly positive definite:

$$\inf_{\vartheta_*\in\mathbb{K}}\inf_{|e|=1}\left(M\left(\vartheta\right)e,e\right)>0.$$

Introduce as well the matrix

$$\bar{R}\left(\vartheta\right)=M\left(\vartheta\right)^{-1}\,\mathbf{E}_\vartheta\,\tilde{q}(\vartheta,\xi)\,\tilde{q}(\vartheta,\xi)^{\mathrm{T}}\,M\left(\vartheta\right)^{-1},$$

where

$$\tilde{q}(\vartheta,x)=\frac{2}{\sigma\left(x\right)f\left(\vartheta,x\right)}\int_{-\infty}^{x}\left[q(y)-m\left(\vartheta\right)\right]f\left(\vartheta,y\right)\,dy.$$

The main result of this section is the following theorem.

Theorem 2.28. *Let the conditions $\mathcal{A}_0(\Theta)$, $\mathcal{D}$ be fulfilled. Then the EMM $\bar\vartheta_T$ is uniformly consistent: for any $\nu > 0$*

$$\lim_{T\to\infty}\sup_{\vartheta\in\mathbb{K}}\mathbf{P}_\vartheta^{(T)}\left\{\left|\bar\vartheta_T-\vartheta\right|>\nu\right\}=0$$

is uniformly asymptotically normal

$$\mathcal{L}_\vartheta\left\{\sqrt{T}\left(\bar\vartheta_T-\vartheta\right)\right\}\Longrightarrow\mathcal{N}\left(0,\bar{R}\left(\vartheta\right)\right), \tag{2.109}$$

and for any $p > 0$ the moments converge

$$\lim_{T\to\infty}\mathbf{E}_\vartheta\left|\sqrt{T}\left(\bar\vartheta_T-\vartheta\right)\right|^p=\mathbf{E}\left|\bar\zeta\right|^p, \tag{2.110}$$

uniformly on $\vartheta\in\mathbb{K}$. Here $\mathcal{L}\left(\bar\zeta\right)=\mathcal{N}\left(0,\bar{R}\left(\vartheta\right)\right)$.

Proof. The proof of the consistency is similar to the proofs of the consistency in Sections 2.2 and 2.3. We have for any $\nu > 0$

$$\begin{aligned}
&\mathbf{P}_{\vartheta_0}^{(T)}\left\{\left|\bar\vartheta_T-\vartheta_0\right|>\nu\right\}\\
&\quad\leq\mathbf{P}_{\vartheta_0}^{(T)}\left\{\inf_{\left|\vartheta-\vartheta_0\right|<\nu}\left|\hat{m}_T-m\left(\vartheta\right)\right|>\inf_{\left|\hat{m}_T-m(\vartheta)\right|\geq\nu}\left|m\left(\vartheta\right)-\hat{m}_T\right|\right\}\\
&\quad\leq\mathbf{P}_{\vartheta_0}^{(T)}\left\{2\,\left|\hat{m}_T-m\left(\vartheta_0\right)\right|>\kappa\,\nu\right\},
\end{aligned}$$

where we used the inequality

$$\inf_{\left|\vartheta-\vartheta_0\right|\geq\nu}\left|m\left(\vartheta\right)-m\left(\vartheta_0\right)\right|\geq\kappa\,\nu, \tag{2.111}$$

which follows from the condition $\mathcal{D}$ as follows. At the vicinity of the point ϑ_0 we have the expansion

$$m\left(\vartheta\right)-m\left(\vartheta_0\right)=\left(\dot{m}\left(\vartheta_0\right),\left(\vartheta-\vartheta_0\right)\right)+o\left(\vartheta-\vartheta_0\right).$$

Hence for sufficiently small $\nu > 0$ and $\left|\vartheta-\vartheta_0\right|<\nu$

$$\left|m\left(\vartheta\right)-m\left(\vartheta_0\right)\right|^2\geq\frac{1}{2}\inf_{\left|e\right|=1}\left(M\left(\vartheta_0\right)e,e\right)\,\left|\vartheta-\vartheta_0\right|^2=\kappa_1^2\left|\vartheta-\vartheta_0\right|^2.$$

Further, by condition $\mathcal{D}_2$

$$\inf_{\left|\vartheta-\vartheta_0\right|>\nu}\left|m\left(\vartheta\right)-m\left(\vartheta_0\right)\right|=\gamma\geq\gamma\,\frac{\left|\vartheta-\vartheta_0\right|}{D(\Theta)}=\kappa_2\left|\vartheta-\vartheta_0\right|.$$

Here $D(\Theta)=\sup_{\vartheta,\vartheta_0}\left|\vartheta-\vartheta_0\right|$. Hence we can put $\kappa=\min\left(\kappa_1,\kappa_2\right)$ in (2.111).

Remember that

$$\hat{m}_T - m(\vartheta_0) = \frac{Q(\vartheta_0, X_T) - Q(\vartheta_0, X_0)}{T} - \frac{1}{T}\int_0^T \tilde{q}(\vartheta_0, X_t)\,\mathrm{d}W_t, \quad (2.112)$$

where

$$Q(\vartheta, z) = \int_0^z \frac{\tilde{q}(\vartheta, x)}{\sigma(x)}\,\mathrm{d}x.$$

Finally, we obtain for any $p > 0$

$$\begin{aligned}\mathbf{P}_{\vartheta_0}^{(T)}\left\{\left|\bar{\vartheta}_T - \vartheta_0\right| > \nu\right\} &\le \left(\frac{2}{\kappa\,\nu}\right)^{2p} \mathbf{E}_{\vartheta_0}\left|m(\vartheta_0) - \hat{m}_T\right|^{2p} \\ &\le \frac{C_1}{(\nu\,T)^{2p}}\left(\mathbf{E}_{\vartheta_0}\left|Q(\vartheta_0, \xi)\right|^{2p} + T^p\,\mathbf{E}_{\vartheta_0}\left|\tilde{q}(\vartheta_0, \xi)\right|^{2p}\right)\end{aligned}$$

because by conditions $\mathcal{A}_0(\Theta)$ and $\mathcal{D}_1$ these expectations are bounded (see Lemma 1.17).

Hence

$$\sup_{\vartheta_0\in\mathbf{K}} \mathbf{P}_{\vartheta_0}^{(T)}\left\{\left|\bar{\vartheta}_T - \vartheta_0\right| > \nu\right\} \le \frac{C}{\left(\nu\sqrt{T}\right)^{2p}} \to 0. \quad (2.113)$$

In a similar way we can show that

$$\sup_{\vartheta_0\in\mathbf{K}} \mathbf{P}_{\vartheta_0}^{(T)}\left\{\hat{m}_T \in \mathbb{M}^c\right\} \le \frac{C}{\left(\nu\sqrt{T}\right)^{2p}}$$

too and according to the representation (2.108) it is sufficient to show the asymptotic normality of $\bar{u}_T = \sqrt{T}\left(\vartheta_T - \vartheta_0\right)$. On the set $\mathbb{B} = \left\{\omega:\ \hat{m}_T \in \mathbb{M}\right\}$ the vector $\hat{\vartheta}_T = \vartheta_T$ is a solution of the system of equations

$$\dot{m}(\vartheta_T)\left[\hat{m}_T - m(\vartheta_T)\right] = \mathbf{0},$$

which can be rewritten as

$$\dot{m}(\vartheta_T)\left[\bar{W}_T - \dot{m}\left(\tilde{\vartheta}_T\right)\bar{u}_T\right] = \mathbf{0}. \quad (2.114)$$

Here the random vector $\bar{W}_T = \hat{m}_T - m(\vartheta_0)$ admits the representation (2.112) and by the central limit theorem is asymptotically normal

$$\mathcal{L}_{\vartheta_0}\left\{\bar{W}_T\right\} \Longrightarrow \mathcal{N}\left(\mathbf{0}, \mathbf{E}_{\vartheta_0}\,\tilde{q}(\vartheta_0, \xi)\,\tilde{q}(\vartheta_0, \xi)^{\mathrm{T}}\right).$$

Equation (2.114) yields the representation

$$\bar{u}_T = \left(\dot{m}(\vartheta_T)\ \dot{m}\left(\tilde{\vartheta}_T\right)^{\mathrm{T}}\right)^{-1} \bar{W}_T = M(\vartheta_0)^{-1}\,\bar{W}_T + o(1),$$

where we used the consistency of the EMM and the continuity of the matrix $\dot{m}(\vartheta)$.

Hence we obtain the asymptotic normality (2.109)

$$\mathcal{L}_{\vartheta_0}\left\{\bar{\boldsymbol{u}}_T\right\} \Longrightarrow \mathcal{L}\left\{\bar{\zeta}\right\}.$$

By the condition $\boldsymbol{\mathcal{A}}_0\left(\Theta\right)$ all convergences are uniform w.r.t. $\vartheta \in \mathbb{K}$ for any compact $\mathbb{K} \subset \Theta$.

The estimate (2.113) gives us the uniform integrability of $\left|\bar{\boldsymbol{u}}_T\right|^p$

$$\mathbf{P}_{\vartheta_0}^{(T)}\left\{\left|\bar{\boldsymbol{u}}_T\right| > N\right\} \leq \frac{C}{N^p}$$

and this provides the convergence of moments (2.110).

2.4.2 Examples

The differentiability of the vector function $\boldsymbol{m}\left(\cdot\right)$ follows from the regularity conditions $\boldsymbol{\mathcal{A}}_0\left(\Theta\right)$, $\mathcal{A}_1$ and $\boldsymbol{q}\left(\cdot\right) \in \mathscr{P}$ because we can write

$$\begin{aligned}\frac{\partial \boldsymbol{m}\left(\vartheta\right)}{\partial \vartheta} &= \frac{\partial}{\partial \vartheta}\int_{\mathscr{R}} \boldsymbol{q}\left(x\right)\, f\left(\vartheta, x\right)\,\mathrm{d}x = \int_{\mathscr{R}} \boldsymbol{q}\left(x\right)\,\frac{\partial f\left(\vartheta, x\right)}{\partial \vartheta}^{\mathbb{T}}\,\mathrm{d}x\\ &= 2\int_{\mathscr{R}} \boldsymbol{q}\left(x\right)\, f\left(\vartheta, x\right)\,\mathbf{E}_{\vartheta}\int_{\xi}^{x}\frac{\dot{\boldsymbol{S}}\left(\vartheta, v\right)^{\mathbb{T}}}{\sigma\left(v\right)^2}\,\mathrm{d}v\,\mathrm{d}x.\end{aligned}$$

Of course, $\boldsymbol{m}\left(\cdot\right)$ can be continuously differentiable over ϑ when the function $S\left(\vartheta, \cdot\right)$ is even discontinuous.

Example 2.29. (*Ornstein–Uhlenbeck process*) Let the observed process be

$$\mathrm{d}X_t = -\left(\vartheta^{(1)}\, X_t - \vartheta^{(2)}\right)\,\mathrm{d}t + \sigma\,\mathrm{d}W_t, \quad X_0, \quad 0 \leq t \leq T,$$

where the unknown parameter $\vartheta = \left(\vartheta^{(1)}, \vartheta^{(2)}\right) \in \Theta$, the set $\Theta = \left(\alpha_1, \beta_1\right) \times \left(\alpha_2, \beta_2\right)$ and $\alpha_1 > 0$. Then we can take $\boldsymbol{q}(x) = \left(x, x^2\right)$ and the conditions of Theorem 2.28 are fulfilled. Hence the EMM $\bar{\vartheta}_T = \left(\bar{\vartheta}_T^{(1)}, \bar{\vartheta}_T^{(2)}\right)$ (see (1.135) and (1.136)) is

$$\begin{aligned}\bar{\vartheta}_T^{(1)} &= \frac{\sigma^2}{2\,\left(Y_2 - Y_1^2\right)},\\ \bar{\vartheta}_T^{(2)} &= \frac{Y_1\,\sigma^2}{2\,\left(Y_2 - Y_1^2\right)},\end{aligned}$$

where

$$Y_1 = \frac{1}{T}\int_0^T X_t\,\mathrm{d}t, \qquad Y_2 = \frac{1}{T}\int_0^T X_t^2\,\mathrm{d}t$$

is consistent and asymptotically normal and the moments $T^{p/2}\,\mathbf{E}_{\vartheta}\left|\bar{\vartheta}_T - \vartheta\right|^p$ converge too.

Example 2.30. (*Shift estimation*) Let the observed process be one of the following:

$$\begin{aligned}
\mathrm{d}X_t &= -\left(X_t-\vartheta\right)^3\,\mathrm{d}t+\sigma\,\mathrm{d}W_t, && X_0,\quad 0\le t\le T,\\
\mathrm{d}X_t &= -\operatorname{sgn}\left(X_t-\vartheta\right)\,\left|X_t-\vartheta\right|^\kappa\,\mathrm{d}t+\sigma\,\mathrm{d}W_t, && X_0,\quad 0\le t\le T,\\
\mathrm{d}X_t &= -\operatorname{sgn}\left(X_t-\vartheta\right)\,\mathrm{d}t+\sigma\,\mathrm{d}W_t, && X_0,\quad 0\le t\le T,
\end{aligned}$$

where $\kappa\in(0,1)$ and the unknown parameter $\vartheta\in\Theta=(\alpha,\beta)$. All these processes are ergodic and can be written as

$$\mathrm{d}X_t = S\left(X_t-\vartheta\right)\,\mathrm{d}t+\sigma\,\mathrm{d}W_t,\qquad X_0,\quad 0\le t\le T,\tag{2.115}$$

i.e., we have the problem of estimation of the shift parameter $\vartheta=\mathbf{E}_\vartheta\,\xi$. The invariant density is

$$f\left(\vartheta,x\right)=\frac{1}{G}\exp\left\{\frac{2}{\sigma^2}\int_0^{x-\vartheta}S\left(v\right)\,\mathrm{d}v\right\},\qquad x\in\mathscr{R}.$$

In particular, for the first process

$$f\left(\vartheta,x\right)=\frac{1}{a\,\sqrt{\sigma}}\exp\left\{-\frac{\left(x-\vartheta\right)^4}{2\sigma^2}\right\},\quad a=\int_{-\infty}^{\infty}\mathrm{e}^{-z^4/2}\mathrm{d}z=2^{-3/4}\Gamma\left(\frac{1}{4}\right).$$

The MLE for these models cannot be written explicitly, and the EMM with $q(x)=x$ is, of course, consistent

$$\bar\vartheta_T=\frac{1}{T}\int_0^T X_t\,\mathrm{d}t\longrightarrow\mathbf{E}_\vartheta\,\xi=\vartheta$$

and asymptotically normal

$$\sqrt{T}\left(\bar\vartheta_T-\vartheta\right)\Longrightarrow\mathcal{N}\left(0,d\left(\vartheta\right)^2\right)$$

with

$$d\left(\vartheta\right)^2=\frac{4}{\sigma^2}\,\mathbf{E}_\vartheta\left(\frac{\left(\xi-\vartheta\right)F(\vartheta,\xi)-\tilde F(\vartheta,\xi)}{f\left(\vartheta,\xi\right)}\right)^2.$$

Here $F(\vartheta,\cdot)$ is the invariant distribution function and

$$\tilde F(\vartheta,x)=\int_{-\infty}^{x}F(\vartheta,y)\,\mathrm{d}y.$$

Note that the conditions of Theorem 2.28 are fulfilled ($\dot m\left(\vartheta\right)=1$) and we have the convergence of moments too:

$$\lim_{T\to\infty}T^{p/2}\,\mathbf{E}_\vartheta\left|\bar\vartheta_T-\vartheta\right|^p=d\left(\vartheta\right)^p\;2^{\frac{p+1}{2}}\;\Gamma\left(\frac{p+1}{2}\right).$$

2.5 One-Step MLE

2.5.1 Parameter Estimation

We illustrate the construction of a one-step MLE on the same model of an ergodic diffusion process with the stochastic differential

$$\mathrm{d}X_t = S(\vartheta, X_t)\,\mathrm{d}t + \sigma(X_t)\,\mathrm{d}W_t, \quad X_0, \quad 0 \le t \le T. \tag{2.116}$$

As usual $\{W_t,\ 0 \le t \le T\}$ is a standard Wiener process, the initial value X_0 is a random variable independent on $\{W_t, t \ge 0\}$ and $\sigma(\cdot)$ is a known positive function. For simplicity of exposition we consider the one-dimensional case only $\vartheta \in \Theta = (\alpha, \beta)$ and the regularity conditions are also far from the minimum. The multidimensional version can be obtained in exactly the same way. We propose a method of construction of asymptotically efficient estimators which is in a certain sense simpler than the maximum likelihood. The process (2.116) is supposed to satisfy the condition $\mathcal{A}_0\,(\Theta)$; that is, it has ergodic properties with the density of invariant law

$$f(\vartheta, x) = G(\vartheta)^{-1}\,\sigma(x)^{-2} \exp\left\{2\int_0^x \frac{S(\vartheta, v)}{\sigma(v)^2}\,\mathrm{d}v\right\}$$

and this density has exponentially decreasing tails. If the conditions of Theorem 2.8 are fulfilled, then the MLE $\hat{\vartheta}_T$ is consistent, asymptotically normal

$$\mathcal{L}_\theta\left\{\sqrt{T}\left(\hat{\vartheta}_T - \vartheta\right)\right\} \Longrightarrow \mathcal{N}\left(0, \mathrm{I}(\vartheta)^{-1}\right),$$

and an asymptotically efficient estimator of the parameter ϑ. Here

$$\mathrm{I}(\vartheta) = \mathbf{E}_\vartheta\left(\frac{\dot{S}(\vartheta, \xi)}{\sigma(\xi)}\right)^2$$

is the Fisher information. Note that for calculation of this estimator in the nonlinear case we have to find the global maximum of the likelihood ratio function $L(\vartheta, \vartheta_1, X^T)$, $\vartheta \in \Theta$ calculating many times the Itô stochastic integral, and this procedure requires approximations of this integral which is not so easy to do for several reasons. That is why in this section we propose another approach based on the possibility of improving a consistent estimator up to the asymptotically efficient. This approach is due to Le Cam and can be presented as follows. Let $\bar{\vartheta}_T$ be a consistent estimator of the parameter ϑ such that $\eta_T = \sqrt{T}\left(\bar{\vartheta}_T - \vartheta\right)$ is bounded in probability, and suppose that the corresponding family of measures $\left\{\mathbf{P}_\vartheta^{(T)},\ \vartheta \in \Theta\right\}$ is LAN. That is, the likelihood ratio admits the representation

$$L\left(\vartheta + \frac{u}{\sqrt{T}}, \vartheta, X^T\right) = \exp\left\{u\Delta_T\left(\vartheta, X^T\right) - \frac{u^2}{2}\mathrm{I}(\vartheta) + r_T\left(\vartheta, u, X^T\right)\right\}, \tag{2.117}$$

where $u \in U_{\vartheta,T} = \left((\alpha-\vartheta)\sqrt{T}, (\beta-\vartheta)\sqrt{T}\right)$,

$$\Delta_T\left(\vartheta, X^T\right) = \frac{1}{\sqrt{T}}\int_0^T \frac{\dot S\left(\vartheta, X_t\right)}{\sigma\left(X_t\right)^2}\left[\mathrm{d}X_t - S\left(\vartheta, X_t\right)\,\mathrm{d}t\right] \tag{2.118}$$

and we have the convergences

$$\mathcal{L}_\vartheta\left\{\Delta_T\left(\vartheta, X^T\right)\right\} \Longrightarrow \mathcal{N}\left(0, \mathrm{I}(\vartheta)\right), \qquad r_T\left(\vartheta, u, X^T\right) \to 0. \tag{2.119}$$

Then we construct the estimator

$$\hat\vartheta^\circ_T = \bar\vartheta_T + \frac{\Delta_T\left(\bar\vartheta_T, X^T\right)}{\sqrt{T}\,\mathrm{I}(\bar\vartheta_T)} \tag{2.120}$$

and show that this estimator is asymptotically efficient. Forget for the moment that such a direct realization needs the definition of a stochastic Itô integral for the integrand $\dot S(\bar\vartheta_T, X_t)\ \sigma(X_t)^{-2}$ dependent on the "future" because the estimator $\bar\vartheta_T$ depends on the whole trajectory $\{X_t, 0 \le t \le T\}$.

To explain why this estimator can be asymptotically efficient we describe formally the asymptotics of this estimator as $T \to \infty$ and then will give the modification of the estimator, allowing us to construct and to study it in the framework of Itô calculus.

Using the convergence $\bar\vartheta_T \to \vartheta$ we can write

$$\begin{aligned}
\left(\hat\vartheta^\circ_T - \vartheta\right)\sqrt{T} &= \left(\bar\vartheta_T - \vartheta\right)\sqrt{T} + \frac{\Delta_T(\vartheta)}{\mathrm{I}(\vartheta)}\ (1+o(1)) \\
&\quad + \frac{1}{\mathrm{I}(\vartheta)\sqrt{T}}\int_0^T \frac{\dot S(\bar\vartheta_T, X_t)}{\sigma(X_t)^2}\left[S\left(\vartheta, X_t\right) - S\left(\bar\vartheta_T, X_t\right)\right]\ \mathrm{d}t\ (1+o(1)) \\
&= \eta_T + \frac{\Delta_T(\vartheta)}{\mathrm{I}(\vartheta)}(1+o(1)) - \frac{\eta_T}{\mathrm{I}(\vartheta)}\frac{1}{T}\int_0^T\left(\frac{\dot S(\vartheta, X_t)}{\sigma(X_t)}\right)^2 \mathrm{d}t\,(1+o(1)) \\
&= \frac{\Delta_T(\vartheta)}{\mathrm{I}(\vartheta)}\ (1+o(1)) + o(1) \Longrightarrow \mathcal{N}\left(0, \mathrm{I}(\vartheta)^{-1}\right),
\end{aligned}$$

where we used the law of large numbers

$$\frac{1}{T}\int_0^T\left(\frac{\dot S(\vartheta, X_t)}{\sigma(X_t)}\right)^2 \mathrm{d}t \to \mathrm{I}(\vartheta)$$

and smoothness of the function $S(\cdot)$. Therefore the estimator $\hat\vartheta^\circ_T$ can be asymptotically efficient, at least for the bounded loss functions.

We use as a first consistent estimator the estimator $\bar\vartheta_T$ of the method of moments described below and we suppose that the function $S(\vartheta, x)$ is sufficiently smooth. Remember that the dot means derivation over ϑ and the prime the derivation over x.

Regularity conditions $\mathcal{E}$.

$\mathcal{E}_1$. *The function $S(\theta,x)$ has three continuous derivatives on θ, the derivative $\dot S(\theta,x)$ and the function $\sigma(x)$ are continuously differentiable on x and*

$$\dot S(\theta,\cdot),\ \dot S'(\theta,\cdot),\ \ddot S(\theta,\cdot),\ \dddot S(\theta,\cdot),\ \sigma'(\cdot)\in\mathscr{P},$$

i.e., these partial derivatives have polynomial majorants.

$\mathcal{E}_2$. *The estimator $\bar\vartheta_T$ is uniformly consistent and there exists a constant $C>0$ such that*

$$\sup_{\vartheta\in\mathbb{K}}\mathbf{E}_\vartheta\left|\sqrt{T}\left(\bar\vartheta_T-\vartheta\right)\right|^p<C. \tag{2.121}$$

The Fisher information

$$\inf_{\vartheta\in\Theta}\mathrm{I}(\vartheta)>0. \tag{2.122}$$

We have to modify the definition of the one-step MLE by simply replacing the Itô integral in $\Delta_T\left(\theta,X^T\right)$ by the ordinary one. Note that ϑ in the statistic $\Delta_T\left(\vartheta,X^T\right)$ in (2.118) is the true value and now we consider $\Delta_T\left(\theta,X^T\right)$ as a random function of $\theta\in\Theta$. Let us put

$$\begin{aligned}\delta_T\left(\theta,X^T\right)&=\frac{1}{\sqrt{T}}\int_{X_0}^{X_T}\frac{\dot S(\theta,y)}{\sigma(y)^2}\,\mathrm{d}y-\frac{1}{2\sqrt{T}}\int_0^T\dot S'(\theta,X_t)\,\mathrm{d}t+\\&\quad+\frac{1}{\sqrt{T}}\int_0^T\dot S(\theta,X_t)\left(\frac{\sigma'(X_t)}{\sigma(X_t)}-\frac{S(\theta,X_t)}{\sigma(X_t)^2}\right)\,\mathrm{d}t,\end{aligned} \tag{2.123}$$

and define the *one-step maximum likelihood estimator* by the same formula

$$\hat\vartheta_T^\circ=\bar\vartheta_T+\frac{\delta_T\left(\bar\vartheta_T,X^T\right)}{\sqrt{T}\,\mathrm{I}(\bar\vartheta_T)}. \tag{2.124}$$

The difference between (2.120) and (2.124) is just in the representation of the random variable $\Delta_T\left(\cdot,X^T\right)$. It is easy to verify by the Itô formula that $\delta_T\left(\theta,X^T\right)=\Delta_T\left(\theta,X^T\right)$ for all $\theta\in\Theta$ with $\mathbf{P}_\vartheta^{(T)}$ probability 1 and in (2.123) we have no further stochastic integral.

Theorem 2.31. *Let the conditions $\mathcal{A}_0$ and $\mathcal{E}$ be fulfilled. Then the estimator $\hat\vartheta_T^\circ$ is uniformly consistent, asymptotically normal*

$$\mathcal{L}_\vartheta\left\{T^{1/2}\left(\hat\vartheta_T^\circ-\vartheta\right)\right\}\Longrightarrow\mathcal{N}\left(0,\mathrm{I}(\vartheta)^{-1}\right)$$

and asymptotically efficient for the loss functions $\ell(\cdot)\in\mathscr{W}_p$.

Proof. Let us write the statistic $\delta_T\left(\theta,X^T\right)$ as

$$\delta_T\left(\theta,X^T\right)=\delta_T^{(1)}\left(\theta,X^T\right)+\delta_T^{(2)}\left(\theta,X^T\right),$$

where

$$\delta_T^{(1)}\left(\theta, X^T\right) = \frac{1}{\sqrt{T}}\int_{X_0}^{X_T} \frac{\dot S(\theta,y)}{\sigma\left(y\right)^2}\,\mathrm{d}y - \frac{1}{2\sqrt{T}}\int_0^T \dot S'(\theta,X_t)\,\mathrm{d}t$$
$$+\frac{1}{\sqrt{T}}\int_0^T \dot S(\theta,X_t)\left(\frac{\sigma'(X_t)}{\sigma(X_t)}-\frac{S(\vartheta,X_t)}{\sigma(X_t)^2}\right)\,dt,$$
$$\delta_T^{(2)}\left(\theta, X^T\right) = \frac{1}{\sqrt{T}}\int_0^T \frac{\dot S(\theta,X_t)\ \left[S(\vartheta,X_t)-S(\theta,X_t)\right]}{\sigma(X_t)^2}\,\mathrm{d}t.$$

Here ϑ is the true value and θ is the variable. The one-step MLE can be written as

$$\sqrt{T}\left(\hat\vartheta_T^{\circ}-\vartheta\right)=\sqrt{T}\left(\bar\vartheta_T-\vartheta\right)+\frac{\delta_T^{(2)}\left(\bar\vartheta_T,X^T\right)}{\mathrm{I}(\bar\vartheta_T)}+\frac{\delta_T^{(1)}\left(\bar\vartheta_T,X^T\right)}{\mathrm{I}(\bar\vartheta_T)}.$$

The theorem will be proved if we verify that for any $p>0$

$$A_T=\sup_{\vartheta\in\mathbb{K}}\mathbf{E}_\vartheta\left|\sqrt{T}\left(\bar\vartheta_T-\vartheta\right)+\frac{\delta_T^{(2)}\left(\bar\vartheta_T,X^T\right)}{\mathrm{I}(\bar\vartheta_T)}\right|^p\longrightarrow 0, \tag{2.125}$$

$$B_T=\sup_{\vartheta\in\mathbb{K}}\mathbf{E}_\vartheta\left|\frac{\delta_T^{(1)}\left(\bar\vartheta_T,X^T\right)-\delta_T^{(1)}\left(\vartheta,X^T\right)}{\mathrm{I}(\bar\vartheta_T)}\right|^p\longrightarrow 0, \tag{2.126}$$

$$\mathcal{L}_\vartheta\left\{\frac{\delta_T^{(1)}\left(\vartheta,X^T\right)}{\mathrm{I}(\bar\vartheta_T)}\right\}\Longrightarrow\mathcal{N}\left(0,\mathrm{I}(\vartheta)^{-1}\right). \tag{2.127}$$

First note that the Fisher information by the conditions of the theorem is continuously differentiable over ϑ and

$$\dot{\mathrm{I}}\left(\vartheta\right)=2\ \mathbf{E}_\vartheta\left(\frac{\dot S\left(\vartheta,\xi\right)}{\sigma\left(\xi\right)^2}\left[\ddot S\left(\vartheta,\xi\right)+\dot S\left(\vartheta,\xi\right)\int_{\xi_0}^{\xi}\frac{\dot S\left(\vartheta,v\right)}{\sigma\left(v\right)^2}\mathrm{d}v\right]\right)$$

is bounded and separated from zero. Here ξ and ξ_0 are two independent random variables with the same stationary distribution. Hence we can put everywhere

$$\mathrm{I}\left(\bar\vartheta_T\right)^{-1}=\mathrm{I}\left(\vartheta\right)^{-1}\left(1+o(1)\right).$$

In particular (2.127) follows from this estimate, the equality

$$\delta_T^{(1)}\left(\vartheta,X^T\right)=\Delta_T\left(\vartheta,X^T\right)$$

and the central limit theorem.

Further, the Cauchy–Schwarz inequality allows us to write

$$
\begin{aligned}
&\mathbf{E}_\vartheta \left|\sqrt{T}\left(\bar\vartheta_T-\vartheta\right)+\frac{\delta_T^{(2)}\left(\bar\vartheta_T,X^T\right)}{\mathrm{I}(\vartheta)}\right|^p\\
&\quad=\mathbf{E}_\vartheta \left|\sqrt{T}\left(\bar\vartheta_T-\vartheta\right)\left(1-\frac{1}{\mathrm{I}(\vartheta)\,T}\int_0^T\frac{\dot S\left(\bar\vartheta_T,X_t\right)\,\dot S\left(\tilde\vartheta_T,X_t\right)}{\sigma\left(X_t\right)^2}\,\mathrm{d}t\right)\right|^p\\
&\quad\le\left(\mathbf{E}_\vartheta\left|\sqrt{T}\left(\bar\vartheta_T-\vartheta\right)\right|^{2p}\right)^{1/2}\\
&\qquad\qquad\left(\mathbf{E}_\vartheta\left|1-\frac{1}{\mathrm{I}(\vartheta)\,T}\int_0^T\frac{\dot S\left(\bar\vartheta_T,X_t\right)\,\dot S\left(\tilde\vartheta_T,X_t\right)}{\sigma\left(X_t\right)^2}\,\mathrm{d}t\right|^{2p}\right)^{1/2},
\end{aligned}
$$

where $\left|\tilde\vartheta_T-\vartheta\right|\le\left|\bar\vartheta_T-\vartheta\right|$. As $\mathbf{E}_\vartheta\left|\bar\vartheta_T-\vartheta\right|^{2p}\le C\,T^{-p}$ we have

$$
\begin{aligned}
&\mathbf{E}_\vartheta\left|1-\frac{1}{\mathrm{I}(\vartheta)\,T}\int_0^T\frac{\dot S\left(\bar\vartheta_T,X_t\right)\,\dot S\left(\tilde\vartheta_T,X_t\right)}{\sigma\left(X_t\right)^2}\,\mathrm{d}t\right|^{2p}\\
&\qquad\qquad=\mathbf{E}_\vartheta\left|1-\frac{1}{\mathrm{I}(\vartheta)\,T}\int_0^T\left(\frac{\dot S\left(\vartheta,X_t\right)}{\sigma\left(X_t\right)}\right)^2\mathrm{d}t\right|^{2p}+o(1).
\end{aligned}
$$

Now (2.125) follows from Proposition 1.18.

To finish the proof we have to show that

$$
\sup_{\vartheta\in\Theta}\mathbf{E}_\vartheta\left|\delta_T^{(1)}\left(\bar\vartheta_T,X^T\right)-\delta_T^{(1)}\left(\vartheta,X^T\right)\right|^p\longrightarrow 0. \tag{2.128}
$$

We began with the last convergence. For any $\varepsilon>0$ and $\delta>0$ we have

$$
\begin{aligned}
&\mathbf{E}_\vartheta\left|\delta_T^{(1)}\left(\bar\vartheta_T,X^T\right)-\delta_T^{(1)}\left(\vartheta,X^T\right)\right|^p\\
&\quad\le\mathbf{E}_\vartheta\left(\chi_{\left\{\left|\bar\vartheta_T-\vartheta\right|>\varepsilon\right\}}\left|\delta_T^{(1)}\left(\bar\vartheta_T,X^T\right)-\delta_T^{(1)}\left(\vartheta,X^T\right)\right|^p\right)\\
&\qquad+\mathbf{E}_\vartheta\left(\chi_{\left\{\left|\bar\vartheta_T-\vartheta\right|\le\varepsilon\right\}}\left|\delta_T^{(1)}\left(\bar\vartheta_T,X^T\right)-\delta_T^{(1)}\left(\vartheta,X^T\right)\right|^p\right)\\
&\quad\le 2^p\left(\mathbf{E}_\vartheta\sup_{\upsilon\in\Theta}\left|\eta_T\left(\upsilon,X^T\right)\right|^{2p}\right)^{1/2}\left(\mathbf{P}_\vartheta^{(T)}\left\{\left|\bar\vartheta_T-\vartheta\right|>\varepsilon\right\}\right)^{1/2}\\
&\qquad+\mathbf{E}_\vartheta\sup_{\left|\upsilon-\vartheta\right|\le\varepsilon}\left|\eta_T\left(\upsilon,X^T\right)-\eta_T\left(\vartheta,X^T\right)\right|^p.
\end{aligned}
$$

The stochastic process

$$
\eta_T\left(\upsilon,X^T\right)=\frac{1}{\sqrt{T}}\int_0^T\frac{S\left(\upsilon,X_t\right)}{\sigma\left(X_t\right)}\,\mathrm{d}W_t,\qquad \upsilon\in\Theta
$$

is differentiable with probability 1 because the process

$$\dot{\eta}_T\left(\upsilon,X^T\right)=\frac{1}{\sqrt{T}}\int_0^T\frac{\ddot{S}\left(\upsilon,X_t\right)}{\sigma\left(X_t\right)}\,\mathrm{d}W_t,\qquad \upsilon\in\Theta,$$

satisfies the estimate

$$\begin{aligned}\mathbf{E}_\vartheta\left|\dot{\eta}_T\left(\upsilon,X^T\right)-\dot{\eta}_T\left(\vartheta,X^T\right)\right|^2&=\mathbf{E}_\vartheta\left|\frac{1}{\sqrt{T}}\int_0^T\int_\vartheta^\upsilon\frac{\dddot{S}\left(y,X_t\right)}{\sigma\left(X_t\right)}\,\mathrm{d}y\,\mathrm{d}W_t\right|^2\\&\leq\left(\upsilon-\vartheta\right)\int_\vartheta^\upsilon\mathbf{E}_\vartheta\frac{1}{T}\int_0^T\left(\frac{\dddot{S}\left(y,X_t\right)}{\sigma\left(X_t\right)}\right)^2\mathrm{d}t\,\mathrm{d}y\\&=\left(\upsilon-\vartheta\right)\int_\vartheta^\upsilon\mathbf{E}_\vartheta\left(\frac{\dddot{S}\left(y,\xi\right)}{\sigma\left(\xi\right)}\right)^2\mathrm{d}y\leq C\left(\upsilon-\vartheta\right)^2.\end{aligned}$$

Hence the process $\dot{\eta}_T\left(\cdot,X^T\right)$ has a continuous modification and we can write with $\mathbf{P}_\vartheta^{(T)}$ probability 1

$$\begin{aligned}\sup_{|\upsilon-\vartheta|<\varepsilon}\left|\eta_T\left(\upsilon,X^T\right)-\eta_T\left(\vartheta,X^T\right)\right|&=\sup_{|\upsilon-\vartheta|<\varepsilon}\left|\int_\vartheta^\upsilon\dot{\eta}_T\left(y,X^T\right)\,\mathrm{d}y\right|\\&\leq\int_\vartheta^{\vartheta+\varepsilon}\left|\dot{\eta}_T\left(y,X^T\right)\right|\,\mathrm{d}y.\end{aligned}$$

This gives us the first estimate

$$\begin{aligned}&\sup_{\vartheta\in\Theta}\mathbf{E}_\vartheta\sup_{|\upsilon-\vartheta|\leq\varepsilon}\left|\eta_T\left(\upsilon,X^T\right)-\eta_T\left(\vartheta,X^T\right)\right|^p\\&\qquad\leq\varepsilon^{p-1}\sup_{\vartheta\in\Theta}\int_\vartheta^{\vartheta+\varepsilon}\mathbf{E}_\vartheta\left|\frac{1}{\sqrt{T}}\int_0^T\frac{\ddot{S}\left(\upsilon,X_t\right)}{\sigma\left(X_t\right)}\,\mathrm{d}W_t\right|^p\mathrm{d}\upsilon\\&\qquad\leq C'\,\varepsilon^{p-1}\sup_{\vartheta\in\Theta}\int_\vartheta^{\vartheta+\varepsilon}\mathbf{E}_\vartheta\left|\frac{\ddot{S}\left(\upsilon,\xi\right)}{\sigma\left(\xi\right)}\right|^p\mathrm{d}\upsilon\leq C\,\varepsilon^p.\end{aligned}$$

The second estimate

$$\sup_{\vartheta\in\Theta}\mathbf{E}_\vartheta\sup_{\upsilon\in\Theta}\left|\eta_T\left(\upsilon,X^T\right)\right|^{2p}\leq C$$

can be obtained in a similar way.

Note as well that

$$\sup_{\vartheta\in\Theta}\mathbf{P}_\vartheta^{(T)}\left\{\left|\bar{\vartheta}_T-\vartheta\right|>\varepsilon\right\}\leq\sup_{\vartheta\in\Theta}\frac{\mathbf{E}_\vartheta\left|\bar{\vartheta}_T-\vartheta\right|^2}{\varepsilon^2}\leq\frac{C}{\varepsilon^2T}.$$

Therefore if we put $\varepsilon=T^{-1/4}$, then

$$\sup_{\vartheta\in\Theta}\mathbf{E}_\vartheta\left|\delta_T\left(\bar{\vartheta}_T,X^T\right)-\delta_T\left(\vartheta,X^T\right)\right|^p\leq C\,T^{-p/2}+\tilde{C}\,T^{-1/4},$$

with corresponding constants C, $\tilde{C}$. So the convergence (2.126) is proved.

Example 2.32. (*Shift estimation*) Let us consider the problem of parameter estimation for the diffusion process

$$\mathrm{d}X_t = -\left(X_t - \vartheta\right)^3\,\mathrm{d}t + \sigma\,\mathrm{d}W_t, \quad X_0, \quad 0 \le t \le T.$$

The MLE cannot be written in explicit form, but the EMM

$$\bar{\vartheta}_T = \frac{1}{T}\int_0^T X_t\,\mathrm{d}t$$

is uniformly consistent, asymptotically normal and the moments converge too (see Example 2.30). As the other conditions of Theorem 2.31 for the trend coefficient $S\left(\vartheta, x\right) = \left(\vartheta - x\right)^3$ are fulfilled and Fisher information $\mathrm{I}\left(\vartheta\right) = \mathrm{I}$ does not depend on ϑ we can improve this estimator up to asymptotically efficient as follows:

$$\begin{aligned}\vartheta_T^\circ &= \bar{\vartheta}_T - \frac{1}{2\,T\,\mathrm{I}}\int_0^T \dot{S}'\left(\bar{\vartheta}_T, X_t\right)\,\mathrm{d}t \\ &\quad - \frac{1}{\sigma^2\,T\,\mathrm{I}}\int_0^T \dot{S}\left(\bar{\vartheta}_T, X_t\right)\,S\left(\bar{\vartheta}_T, X_t\right)\,\mathrm{d}t \\ &= \bar{\vartheta}_T - \frac{1}{\sigma^2 \mathrm{I}\,T}\int_0^T \left(\bar{\vartheta}_T - X_t\right)^5\,\mathrm{d}t.\end{aligned}$$

According to Theorem 2.31 this estimator is consistent, asymptotically normal

$$\mathcal{L}_\vartheta\left\{\sqrt{T}\left(\vartheta_T^\circ - \vartheta\right)\right\} \Longrightarrow \mathcal{N}\left(0, \mathrm{I}^{-1}\right)$$

and for any loss function $\ell\left(\cdot\right) \in \mathscr{W}_p$

$$\lim_{T\to\infty}\sup_{\left|\vartheta-\vartheta_0\right|<\delta}\mathbf{E}_\vartheta\,\ell\left(\sqrt{T}\left(\vartheta_T^\circ - \vartheta\right)\right) = \mathbf{E}\,\ell\left(\zeta\,\mathrm{I}^{-1/2}\right), \qquad \zeta \sim \mathcal{N}\left(0,1\right),$$

i.e., this estimator is asymptotically efficient.

2.5.2 Distribution Function Estimation

One-step MLE can be used in the problem of function estimation as well. To illustrate it we consider the problem of distribution function estimation by observations of the diffusion process

$$\mathrm{d}X_t = S(\vartheta, X_t)\,\mathrm{d}t + \sigma(X_t)\,\mathrm{d}W_t, \quad X_0, \quad 0 \le t \le T \tag{2.129}$$

where the trend coefficient and diffusion coefficient satisfy the usual conditions: for all $\vartheta \in \Theta = \left(\alpha, \beta\right)$ these are locally bounded functions, $\sigma\left(\cdot\right)$ is positive function and Equation (2.129) has a unique weak solution. We suppose as well that the observed process has ergodic properties with the invariant distribution function

$$F(\vartheta,x)=\frac{1}{G(\vartheta)}\int_{-\infty}^{x}\frac{1}{\sigma(v)^2}\exp\left\{2\int_0^v\frac{S(\vartheta,u)}{\sigma(u)^2}\,du\right\}\,dv.$$

We are interested in the asymptotically efficient estimation of the value $F(\vartheta,x)$, i.e., the value of the distribution function at a given point x. This is a parametric estimation problem because we estimate just a one-dimensional parameter — this value of the distribution function, when the underlying model is parameterized as well. Later in Section 4.1 we consider the problem of nonparametric estimation of the value $F(x)$ of the distribution function for an essentially wider class of ergodic diffusion processes.

First we propose a lower minimax bound on the risk of all estimators and then using one-step procedure we construct an asymptotically efficient estimator.

Proposition 2.33. *Suppose that the function $S(\vartheta,\cdot)\,\sigma(\cdot)^{-1}$ is continuously differentiable in the space $\mathscr{L}_2(f_\vartheta)$ (see (2.8)), the Fisher information*

$$\mathrm{I}(\vartheta)=\mathbf{E}_\vartheta\left(\frac{\dot S(\vartheta,\xi)}{\sigma(\xi)}\right)^2>0$$

and the function $F(\vartheta,x)$ is continuously differentiable over ϑ. Then

$$\varliminf_{\delta\to 0}\varliminf_{T\to\infty}\sup_{|\vartheta-\vartheta_0|<\delta}\mathbf{E}_\vartheta\,\ell\Big(T^{1/2}\left(\bar F_T(x)-F(\vartheta,x)\right)\Big)\geq\mathbf{E}\,\ell\left(\zeta\,d\left(\vartheta_0,x\right)\right),\quad(2.130)$$

where $\ell(\cdot)\in\mathscr{W}_p$, $\mathcal{L}\{\zeta\}=\mathcal{N}(0,1)$ and

$$d(\vartheta_0,x)=\dot F(\vartheta_0,x)\,\mathrm{I}(\vartheta_0)^{-1/2}.$$

Proof. The family of measures $\{\mathbf{P}_\vartheta,\ \vartheta\in\Theta\}$ is LAN (Proposition 2.2) and the slight modification of the proof of Theorem 2.4 in [109] gives us this lower bound.

The MLE $\hat\vartheta_T$ allows us to have an asymptotically efficient estimator by a simple *plug-in*. Indeed, if the conditions of Theorem 2.8 are fulfilled, then the estimator $F(\hat\vartheta_T,x)$ is asymptotically efficient, because

$$\sqrt{T}\left(F\left(\hat\vartheta_T,x\right)-F(\vartheta,x)\right)=\sqrt{T}\left(\hat\vartheta_T-\vartheta\right)\,\dot F\left(\tilde\vartheta_T,x\right)\Longrightarrow\mathcal{N}\left(0,d(\vartheta,x)^2\right).$$

But, as we wrote at the beginning of this section, the difficulty of its construction in nonlinear models suggests we seek another simpler asymptotically efficient estimator if possible.

Note that it is possible to use the one-step MLE ϑ_T° studied in the previous section for the plug-in estimator $F(\hat\vartheta_T^\circ,x)$ and we study such a procedure in the problem of density estimation below but here we prefer to consider an improvement of a consistent estimator of the distribution function based on the same *one-step idea.*

The empirical distribution function

$$\hat{F}_T(x) = \frac{1}{T}\int_0^T \chi_{\{X_t<x\}}\, dt$$

is a consistent (by the law of large numbers) and asymptotically normal

$$\mathcal{L}_\vartheta\left\{\sqrt{T}\left(\hat{F}_T(x) - F(\vartheta,x)\right)\right\} \Longrightarrow \mathcal{N}\left(0, d_F(\vartheta,x)^2\right),$$

estimator of the function $F(\vartheta, x)$ (see Proposition 1.51). Here

$$d_F(\vartheta,x)^2 = 4\,\mathbf{E}_\vartheta\left(\frac{F(\vartheta,\xi\wedge x)\left[1-F(\vartheta,\xi\vee x)\right]}{\sigma(\xi)\ f(\vartheta,\xi)}\right)^2 .$$

But this estimator is not asymptotically efficient because in general

$$d_F(\vartheta,x)^2 \geq d(\vartheta,x)^2 .$$

The construction (2.124) of asymptotically efficient estimator can be applied in this problem in the following way. Take $\hat{F}_T(x)$ as a starting consistent estimator and introduce a *one-step distribution function estimator* as

$$\hat{F}^\circ_T(x) = \hat{F}_T(x) + \frac{\dot{F}\left(\bar{\vartheta}_T(x),x\right)\delta_T\left(\bar{\vartheta}_T(x),X^T\right)}{\mathrm{I}\left(\bar{\vartheta}_T(x)\right)\sqrt{T}}, \tag{2.131}$$

where $\delta_T\left(\theta, X^T\right)$ is defined in (2.123) and $\bar{\vartheta}_T(x)$ is the estimator of the method of moments

$$\left|F\left(\bar{\vartheta}_T(x),x\right) - \hat{F}_T(x)\right| = \inf_{\theta\in\Theta}\left|F\left(\vartheta,x\right) - \hat{F}_T(x)\right| .$$

Indeed, it corresponds to the definition (2.107) with $q(y) = \chi_{\{y<x\}}$ and $m(\vartheta) = F(\vartheta,x)$. Remember, that here x is a fixed value and not a variable. If we denote by $\mathbb{M}\subset[0,1]$ the set

$$\mathbb{M} = \left\{F:\ F(\vartheta,x) = F,\ \ \vartheta\in\Theta\right\},$$

i.e., $\mathbb{M}$ is the set of all possible values of $F(\vartheta,x)$ when ϑ runs Θ and x is fixed. Then for increasing in ϑ function $F(\vartheta,x)$ this estimate can be written as

$$\bar{\vartheta}_T = \alpha\,\chi_{\left\{\hat{F}_T(x)\leq F(\alpha,x)\right\}} + \vartheta_T\chi_{\left\{\hat{F}_T(x)\in\mathbb{M}\right\}} + \beta\chi_{\left\{\hat{F}_T(x)\geq F(\beta,x)\right\}} \tag{2.132}$$

with evident modification for decreasing $F(\vartheta,x)$.

Note that the equation

$$\bar{\vartheta}(F) = \arg\inf_{\vartheta\in\Theta}\left|F(\vartheta,x) - F\right|, \qquad , \qquad F\in[0,1]$$

can be solved numerically before the experiment.

Proposition 2.34. *Let the conditions $\mathcal{A}_0(\Theta)$, $\mathcal{E}_1$ be fulfilled and*

$$\inf_{\vartheta\in\Theta} \mathrm{I}(\vartheta) > 0, \qquad \inf_{\vartheta\in\Theta} \left|\dot F(\vartheta,x)\right| > 0, \tag{2.133}$$

then the estimator $\hat F^\circ_T(x)$ is uniformly on compacts $\mathbb{K}\subset\Theta$ consistent, asymptotically normal

$$\mathcal{L}_\vartheta\left\{\sqrt{T}\left(\hat F^\circ_T(x) - F(\vartheta,x)\right)\right\} \Longrightarrow \mathcal{N}\left(0, d(\vartheta,x)^2\right) \tag{2.134}$$

and asymptotically efficient for the loss function $\ell(\cdot)\in\mathscr{W}_p$.

Proof. The proof is rather close to that given above, hence we just explain only the main steps here. Denote by $\vartheta(F,x)$ the function inverse to the function $F(\vartheta,x)$ (for a fixed value x) and remark that by condition (2.133) the equation

$$F(\vartheta,x) = F, \quad \vartheta\in\Theta$$

has a unique solution for any $F\in\mathbb{M}$. Further, for any compact $\mathbb{K}\in\Theta$ it is easy to show that

$$\sup_{\vartheta\in\mathbb{K}} \mathbf{P}_\vartheta^{(T)}\left\{\hat F_T(x)\in\mathbb{M}^c\right\} \le C T^{-p}$$

for any $p>0$ and some $C=C(p)>0$. Hence we can suppose that $\bar\vartheta_T(x)\in\Theta$. We have

$$\begin{aligned}\bar\vartheta_T(x) - \vartheta &= \vartheta\left(\hat F_T(x),x\right) - \vartheta\left(F(\vartheta,x),x\right)\\ &= \dot\vartheta\left(\tilde F_T(x),x\right)\left(\hat F_T(x) - F(\vartheta,x)\right) = \frac{\hat F_T(x) - F(\vartheta,x)}{\dot F\left(\tilde\vartheta_T,x\right)},\end{aligned}$$

where $\left|\tilde\vartheta_T - \vartheta\right| \le \left|\bar\vartheta_T - \vartheta\right|$. Therefore

$$\begin{aligned}\sqrt{T}\left(\hat F^\circ_T(x) - F(\vartheta,x)\right) &= \sqrt{T}\left(\hat F_T(x) - F(\vartheta,x)\right)\\ &\quad+\frac{\dot F\left(\bar\vartheta_T(x),x\right)\delta_T\left(\bar\vartheta_T(x),X^T\right)}{\mathrm{I}\left(\bar\vartheta_T(x)\right)}\\ &= \sqrt{T}\left(\hat F_T(x) - F(\vartheta,x)\right) + \frac{\dot F(\vartheta,x)}{\mathrm{I}(\vartheta)}\frac{1}{\sqrt{T}}\int_0^T \frac{\dot S(\vartheta,X_t)}{\sigma(X_t)}\mathrm{d}W_t\left(1+o(1)\right)\\ &\quad- \frac{\dot F\left(\bar\vartheta_T(x),x\right)\sqrt{T}\left(\bar\vartheta_T(x)-\vartheta\right)}{\mathrm{I}(\vartheta)}\frac{1}{T}\int_0^T\frac{\dot S(\vartheta,X_t)^2}{\sigma(X_t)^2}\mathrm{d}t\left(1+o(1)\right).\end{aligned}$$

Note that

$$\frac{\dot F\left(\bar\vartheta_T(x),x\right)}{\dot F(\tilde\vartheta_T(x),x)}\frac{1}{\mathrm{I}(\vartheta)\,T}\int_0^T\frac{\dot S(\vartheta,X_t)^2}{\sigma(X_t)^2}\mathrm{d}t \longrightarrow 1$$

as $T\longrightarrow\infty$. Hence we have the convergence (2.134). The convergence of moments follows from the same arguments as in Theorem 2.31.

2.5.3 Density Estimation

Let us consider the problem of invariant density $f(\vartheta,\cdot)$ estimation by the observations of the same process (2.129). The empirical density estimator

$$f_T^\circ(x) = \frac{1}{\sigma(x)^2\, T}\int_0^T \operatorname{sgn}(x - X_t)\, \mathrm{d}X_t + \frac{|X_T - x| - |X_0 - x|}{\sigma(x)^2\, T}$$

is unbiased, consistent and asymptotically normal (Proposition 1.25), but, of course, not asymptotically efficient in this parametric estimation problem. Hence we can try to improve it up to asymptotically efficient like (2.131) and to study the *one-step density estimator*

$$\hat{f}_T^\circ(x) = f_T^\circ(x) + \frac{\dot{f}\left(\bar\vartheta_T(x), x\right)\delta_T\left(\bar\vartheta_T(x), X^T\right)}{\mathrm{I}\left(\bar\vartheta_T(x)\right)\sqrt{T}},$$

with the help of the estimator

$$\bar\vartheta_T(x) = \arg\inf_{\vartheta\in\Theta}\left|f(\vartheta, x) - f_T^\circ(x)\right|.$$

Unfortunately, this program cannot be realized directly because the last equation can have many solutions or not to have it at all.

For example, in the case of the process

$$\mathrm{d}X_t = -\left(X_t - \vartheta\right)^3 \mathrm{d}t + \sigma\,\mathrm{d}W_t, \quad X_0$$

we have the invariant density

$$f(\vartheta, x) = \frac{1}{G\,\sqrt{\sigma}}\exp\left\{-\frac{(x-\vartheta)^4}{2\,\sigma^2}\right\}, \qquad G = \int_{\mathcal{R}} e^{-z^4/2}\,\mathrm{d}z$$

and the estimator

$$\bar\vartheta_T(x) = x \pm \left(-2\sigma^2\, \ln\left[G\,\sqrt{\sigma}\, f_T^\circ(x)\right]\right)_+^{1/4},$$

i.e., for $0 < f_T^\circ(x) < 1/\left(G\sqrt{\sigma}\right)$ there are two solutions, while for $f_T^\circ(x) \geq 1/\left(G\sqrt{\sigma}\right)$ there is one solution $\bar\vartheta_T(x) = x$. Hence the consistent estimation for wide sets Θ is impossible. That is why we consider here a different statement of the problem. Of course, in the case of distribution function estimation the condition (2.133) sometimes is not fulfilled as well.

If we estimate some function it is reasonable to compare the errors of different estimators not in one fixed point only, but in all points simultaneously. For example, for any estimator $\bar{f}_T(\cdot)$ we can calculate the risk

$$\mathcal{R}_T\left(\bar{f}_T, f_\vartheta\right) = \mathbf{E}_\vartheta\,\ell\left(\sqrt{T}\sup_x\left|\bar{f}_T(x) - f(\vartheta, x)\right|\right),$$

where $\ell(\cdot)$ is some loss function or

$$\mathcal{R}_T\left(\bar f_T, f_\vartheta\right) = \sqrt{T}\int_{\mathscr{R}} \mathbf{E}_\vartheta\left|\bar f_T(x) - f(\vartheta, x)\right|\,\mu(\mathrm{d}x)$$

where $\mu(\cdot)$ is some measure or to use any other risk based on some distance between two functions (estimator and true value) and then to try to find an estimator which minimizes (asymptotically) the corresponding risk.

In this section we consider the problem of asymptotically efficient density estimation for the *integral type risk*

$$\mathcal{R}_T\left(\tilde f_T, f_\vartheta\right) = \int_{\mathscr{R}} \mathbf{E}_\vartheta\,\ell\left(\sqrt{T}\left(\tilde f_T(x) - f(\vartheta, x)\right)\right)\,\mathrm{d}x,$$

where the loss function $\ell(\cdot) \in \mathscr{W}_p$.

Put

$$\mathcal{R}(\vartheta) = \int_{\mathscr{R}} \mathbf{E}\ \ell\left(\zeta\, d(\vartheta, x)\right)\,\mathrm{d}x,$$

where $\mathcal{L}\{\zeta\} = \mathcal{N}(0,1)$,

$$d(\vartheta, x) = 2\,\mathrm{I}(\vartheta)^{-1/2}\, f(\vartheta, x)\,\mathbf{E}_\vartheta \int_\xi^x \frac{\dot S(\vartheta, v)}{\sigma(v)^2}\mathrm{d}v$$

and $\mathrm{I}(\vartheta)$ is the Fisher information.

Proposition 2.35. *Suppose that the conditions $\mathcal{A}_0(\Theta)$, $\mathcal{E}$ be fulfilled. Then*

$$\varliminf_{\delta\to 0}\varliminf_{T\to\infty}\inf_{\tilde f_T}\sup_{|\vartheta-\vartheta_0|<\delta} \mathcal{R}_T\left(\tilde f_T, f_\vartheta\right) \geq \mathcal{R}(\vartheta_0) \tag{2.135}$$

and for the plug-in estimator $\tilde f_T(x) = f\left(\vartheta_T^\circ, x\right)$ we have the equality

$$\lim_{\delta\to 0}\lim_{T\to\infty}\sup_{|\vartheta-\vartheta_0|<\delta}\int_{\mathscr{R}} \mathbf{E}_\vartheta\,\ell\left(\sqrt{T}\left(f\left(\vartheta_T^\circ, x\right) - f(\vartheta, x)\right)\right)\,\mathrm{d}x = \mathcal{R}(\vartheta_0). \tag{2.136}$$

Proof. We follow here the proof of Theorem 12.1 in [109]. By conditions of the theorem the family of measures $\left\{\mathbf{P}_\vartheta^{(T)},\ \vartheta\in\Theta\right\}$ is LAN at the point ϑ_0, i.e., the normalized likelihood ratio

$$Z_T(u) = \frac{\mathrm{d}\mathbf{P}^{(T)}_{\vartheta_0+uT^{-1/2}}}{\mathrm{d}\mathbf{P}^{(T)}_{\vartheta_0}}(X^T), \qquad u\in\left(-\sqrt{T}(\alpha-\vartheta_0), (\beta-\vartheta_0)\sqrt{T}\right)$$

admits the representation (2.117)–(2.119). Moreover, using the stopping time τ_T and randomization, we can write the likelihood ratio $Z_T(u)$ as

$$Z_T(u) = \exp\left\{u\tilde{\Delta}_\tau\left(\vartheta_0, X^{(T)}\right) - \frac{u^2}{2}\mathrm{I}\left(\vartheta_0\right)) + r\left(\vartheta, u, X^T\right)\right\}$$

(2.10) with

$$\mathcal{L}_{\vartheta_0}\left\{\tilde{\Delta}_\tau\left(\vartheta_0, X^T\right)\right\} = \mathcal{N}\left(0, \mathrm{I}\left(\vartheta_0\right)\right)$$

for all T and $r\left(\vartheta, u, X^T\right) \to 0$ (see (2.9)). Let us put

$$Z_T^*(u) = \exp\left\{u\tilde{\Delta}_\tau\left(\vartheta_0, X^{(T)}\right) - \frac{u^2}{2}\mathrm{I}\left(\vartheta_0\right))\right\}$$

and note that for any $M > 0$

$$\sup_{|u|\leq M} \mathbf{E}_{\vartheta_0}\left|Z_T^*(u) - Z_T(u)\right| \to 0 \tag{2.137}$$

as $T \to \infty$, because $Z_T^*(u) > 0$, $Z_T(u) > 0$, $\mathbf{E}_{\vartheta_0} Z_T^*(u) = \mathbf{E}_{\vartheta_0} Z_T(u) = 1$, $Z_T^*(u) - Z_T(u) \to 0$ and we can apply the Lemma of Scheffe (see Billingsley [23], p. 51). Therefore for any $N > 0$ and $M \leq \delta\sqrt{T}$ we can write

$$\begin{aligned}
\sup_{|\vartheta-\vartheta_0|<\delta} \mathcal{R}_T\left(\tilde{f}_T, f_\vartheta\right) &= \sup_{|u|<\delta\sqrt{T}} \mathcal{R}_T\left(\tilde{f}_T, f_{\vartheta_0+u/\sqrt{T}}\right)\\
&\geq \sup_{|u|<M} \mathbf{E}_{\vartheta_0} Z_T\left(u\right)\int_{-N}^{N} \ell\left(\left(\bar{u}_T - u\right)\dot{f}\left(\vartheta_0, x\right)\left(1 + o\left(1\right)\right)\right)\,\mathrm{d}x\\
&\geq \frac{1}{2M}\int_{-N}^{N}\int_{-M}^{M} \mathbf{E}_{\vartheta_0} Z_T\left(u\right)\ell\left(\left(\bar{u}_T - u\right)\dot{f}\left(\vartheta_0, x\right)\right)\mathrm{d}u\,\mathrm{d}x\,\left(1 + o\left(1\right)\right)\\
&\geq \frac{1}{2M}\int_{-N}^{N}\int_{-M}^{M} \mathbf{E}_{\vartheta_0} Z_T\left(u\right)\ell^a\left(\left(\bar{u}_T - u\right)\dot{f}\left(\vartheta_0, x\right)\right)\mathrm{d}u\,\mathrm{d}x\,\left(1 + o\left(1\right)\right),
\end{aligned}$$

where $\ell^a = \max\left(\ell, a\right)$ (truncated loss function, $a > 0$), we used the expansion

$$f(\vartheta_0 + u/\sqrt{T}, x) = f\left(\vartheta_0, x\right) + u/\sqrt{T}\,\dot{f}\left(\vartheta_0, x\right) + o\left(\frac{u}{\sqrt{T}}\right),$$

and have denoted

$$\bar{u}_T(x) = \sqrt{T}\left(\bar{f}_T(x) - f\left(\vartheta_0, x\right)\right)\dot{f}\left(\vartheta_0, x\right)^{-1}.$$

If $\dot{f}\left(\vartheta_0, x\right) = 0$ we put $\bar{u}_T = 0$. Remember that by conditions $\boldsymbol{\mathcal{A}}_0\left(\Theta\right)$ and $\mathcal{E}_1$ the function $f(\vartheta, \cdot)$ is two times continuously differentiable and the derivative

$$\left|\dot{f}\left(\vartheta, x\right)\right| = 2\,f\left(\vartheta, x\right)\left|\mathbf{E}_\vartheta \int_\xi^x \frac{\dot{S}\left(\vartheta, v\right)}{\sigma\left(v\right)^2}\,\mathrm{d}v\right| \leq C\,\mathrm{e}^{-\gamma|x|}. \tag{2.138}$$

Further, using (2.137) we can write

$$
\begin{aligned}
&\frac{1}{2M}\int_{-N}^{N}\int_{-M}^{M}\mathbf{E}_{\vartheta_0} Z_T(u)\,\ell^a\Big(\big(\bar u_T(x)-u\big)\dot f(\vartheta_0,x)\Big)\,\mathrm{d}u\,\mathrm{d}x\\
&=\frac{1}{2M}\int_{-N}^{N}\int_{-M}^{M}\mathbf{E}_{\vartheta_0} Z_T^*(u)\,\ell^a\Big(\big(\bar u_T(x)-u\big)\dot f(\vartheta_0,x)\Big)\,\mathrm{d}u\,\mathrm{d}x\,(1+o(1))\\
&=\frac{1}{2M}\int_{-N}^{N}\mathbf{E}_{\vartheta_0}\int_{-M}^{M}\mathrm{e}^{u\Delta_\tau-\frac{u^2}{2}\mathrm{I}_0}\,\ell^a\Big(\big(\bar u_T(x)-u\big)\dot f(\vartheta_0,x)\Big)\,\mathrm{d}u\,\mathrm{d}x\,(1+o(1))
\end{aligned}
$$

where we denoted $\Delta_\tau=\tilde\Delta_\tau\left(\vartheta_0,X^T\right)$, $\mathrm{I}_0=\mathrm{I}(\vartheta_0)$ and for the last integral we have

$$
\begin{aligned}
&\mathbf{E}_{\vartheta_0}\mathrm{e}^{\frac{\Delta_\tau^2}{2\mathrm{I}_0}}\int_{-M}^{M}\mathrm{e}^{-\frac{1}{2\mathrm{I}_0}\left(u-\frac{\Delta_\tau}{\mathrm{I}_0}\right)^2}\,\ell^a\Big(\big(\bar u_T(x)-u\big)\dot f(\vartheta_0,x)\Big)\,\mathrm{d}u\\
&=\mathbf{E}_{\vartheta_0}\mathrm{e}^{\frac{\Delta_\tau^2}{2\mathrm{I}_0}}\int_{-M-\Delta_\tau}^{M-\Delta_\tau}\mathrm{e}^{-\frac{v^2}{2\mathrm{I}_0}}\,\ell^a\left(\left(\bar u_T(x)-\frac{\Delta_\tau}{\mathrm{I}_0}-v\right)\dot f(\vartheta_0,x)\right)\mathrm{d}u\\
&\geq\mathbf{E}_{\vartheta_0}\mathrm{e}^{\frac{\Delta_\tau^2}{2\mathrm{I}_0}}\chi_{\left\{\left|\frac{\Delta_\tau}{\mathrm{I}_0}\right|<M-\sqrt{M}\right\}}\int_{\mathbb{M}}\mathrm{e}^{-\frac{v^2}{2\mathrm{I}_0}}\,\ell^a\left(\left(\bar u_T(x)-\frac{\Delta_\tau}{\mathrm{I}_0}-v\right)\dot f(\vartheta_0,x)\right)\mathrm{d}u,
\end{aligned}
$$

where $\mathbb{M}=\left[-\sqrt{M},\sqrt{M}\right]$. The equality

$$
\inf_z\int_{-A}^{A}\mathrm{e}^{-\frac{v^2}{2\mathrm{I}_0}}\,\ell^a\left((z-v)\,b\right)\mathrm{d}v=\int_{-A}^{A}\mathrm{e}^{-\frac{v^2}{2\mathrm{I}_0}}\,\ell^a\left(vb\right)\mathrm{d}v
$$

for any b follows from Anderson's lemma (see [5] or [109], Section 2.10). Note that for $\ell(u)=|u|^p$ with $p\geq 1$ this can be shown by direct derivation. Therefore

$$
\begin{aligned}
&\frac{1}{2M}\mathbf{E}_{\vartheta_0}\mathrm{e}^{\frac{\Delta_\tau^2}{2\mathrm{I}_0}}\int_{-M}^{M}\mathrm{e}^{-\frac{1}{2\mathrm{I}_0}\left(u-\frac{\Delta_\tau}{\mathrm{I}_0}\right)^2}\,\ell^a\Big(\big(\bar u_T(x)-u\big)\dot f(\vartheta_0,x)\Big)\,\mathrm{d}u\\
&\qquad\geq\frac{1}{\sqrt{2\pi\mathrm{I}_0}}\int_{-\sqrt{M}}^{\sqrt{M}}\mathrm{e}^{-\frac{v^2}{2\mathrm{I}_0}}\,\ell^a\Big(v\dot f(\vartheta_0,x)\Big)\,\mathrm{d}v\,\frac{1}{2M}\int_{-M+\sqrt{M}}^{M-\sqrt{M}}\mathrm{e}^{\frac{y^2}{2\mathrm{I}_0}}\,\mathrm{e}^{-\frac{y^2}{2\mathrm{I}_0}}\,\mathrm{d}y\\
&\qquad=\frac{1}{\sqrt{2\pi\mathrm{I}_0}}\int_{-\sqrt{M}}^{\sqrt{M}}\mathrm{e}^{-\frac{v^2}{2\mathrm{I}_0}}\,\ell^a\Big(v\dot f(\vartheta_0,x)\Big)\,\mathrm{d}v\,\frac{2M-2\sqrt{M}}{2M}\longrightarrow\mathbf{E}\,\ell^a(\zeta_*)
\end{aligned}
$$

as $M\to\infty$. Here $\zeta_*=d(\vartheta_0,x)\ \zeta$, with $\zeta\sim\mathcal{N}(0,1)$. Finally, if $N\to\infty$ and $a\to\infty$ we obtain the lower bound (2.135).

To finish the proof we have to show (2.136) for a *plug-in* estimator

$$
f\left(\vartheta_T^\circ,x\right)=\frac{1}{G(\vartheta_T^\circ)\,\sigma(x)^2}\exp\left\{2\int_0^x\frac{S(\vartheta_T^\circ,v)}{\sigma(v)^2}\mathrm{d}v\right\}.
$$

Using continuous differentiability of the density and the asymptotic normality of the one-step MLE we can write

$$\sqrt{T}\left(f\left(\vartheta_T^\circ,x\right)-f\left(\vartheta,x\right)\right)=\sqrt{T}\left(\vartheta_T^\circ-\vartheta\right)\,\dot f\left(\bar\vartheta_T^\circ,x\right)\Longrightarrow\mathcal{N}\left(0,d\left(\vartheta,x\right)^2\right)$$

and this convergence is uniform on compacts $\vartheta\in\mathbb{K}$. Here $\left|\bar\vartheta_T^\circ-\vartheta\right|\leq\left|\vartheta_T^\circ-\vartheta\right|$. From the uniform convergence of moments and $\ell(\cdot)\in\mathscr{W}_p$ we have as well

$$\sup_{|\vartheta-\vartheta_0|<\delta}\mathbf{E}_\vartheta\,\ell\left(\sqrt{T}\left(f(\vartheta_T^\circ,x)-f\left(\vartheta,x\right)\right)\right)\longrightarrow\sup_{|\vartheta-\vartheta_0|<\delta}\mathbf{E}\;\ell\left(d\left(\vartheta,x\right)\zeta\right).$$

Now the equality (2.136) follows from the continuity of the functions $\dot f(\vartheta,x)$ and $\mathrm{I}(\vartheta)$ and the exponentially decreasing of $\dot f(\vartheta,x)$ for large values of x (2.138). Note that for $\ell(u)=u^2$ the optimal risk is

$$\mathcal{R}\left(\vartheta\right)=\frac{4}{\mathrm{I}\left(\vartheta\right)}\int_{\mathscr{R}}\left(\mathbf{E}_\vartheta\int_\xi^x\frac{\dot S\left(\vartheta,v\right)}{\sigma\left(v\right)^2}\,\mathrm{d}v\right)^2 f\left(\vartheta,x\right)^2\,\mathrm{d}x.$$

Suppose that ϑ is a shift parameter $f(\vartheta,x)=f(x-\vartheta)$ and the function $f(x)$ has one point x_* such that $f'(x_*)=0$, then in the case of a loss function $\ell(u)=|u|$ the integral can be calculated and we have

$$\begin{aligned}\mathcal{R}\left(\vartheta\right)&=\left(\frac{8}{\pi\mathrm{I}\left(\vartheta\right)}\right)^{1/2}\int_{\mathscr{R}}\left|\dot f\left(x-\vartheta\right)\right|\,f\left(x-\vartheta\right)\,\mathrm{d}x\\&=\left(\frac{8}{\pi\mathrm{I}\left(\vartheta\right)}\right)^{1/2}\int_{-\infty}^{x_*+\vartheta}f'\left(x-\vartheta\right)\,f\left(x-\vartheta\right)\,\mathrm{d}x\\&\quad-\left(\frac{8}{\pi\mathrm{I}\left(\vartheta\right)}\right)^{1/2}\int_{x_*+\vartheta}^{\infty}f'\left(x-\vartheta\right)\,f\left(x-\vartheta\right)\,\mathrm{d}x\\&=f\left(x_*\right)^2\left(\frac{8}{\pi\mathrm{I}\left(\vartheta\right)}\right)^{1/2}.\end{aligned}$$

2.6 Miscellaneous

As was shown in the preceding sections and is well known for other models as well (i.i.d., time series, point processes, other diffusion processes, etc.) the estimators MLE, BE, MDE, TFE, EMM and one-step MLE under regularity conditions (always including smoothness) are consistent and asymptotically normal. At the same time it is interesting to see what happens if some of these regularity conditions (except smoothness) are not fulfilled. This study allows for a better understanding of the real role of these conditions. Below we consider several such situations and we describe the properties of the estimators. We do not give detailed proofs because similar problems were already considered in [139], Sections 2.4–2.6 (diffusion with small noise) and in [145], Chapter 4 (Poisson process) and the proofs are quite similar. The properties of estimators for some nonsmooth trends are discussed in Chapter 3.

2.6.1 No True Model

Let us consider the problem of parameter estimation by observations of the ergodic diffusion process $X^T = \{X_t,\ 0 \leq t \leq T\}$, which is a solution of the stochastic differential equation

$$\mathrm{d}X_t = S_* (\vartheta_0, X_t)\ \mathrm{d}t + \sigma_* (X_t)\ \mathrm{d}W_t, \quad X_0 = x \tag{2.139}$$

with some unknown functions $S_* (\cdot, \cdot)$ and $\sigma_* (\cdot)$. We suppose that the observed process belongs to a family (w.r.t. $\vartheta \in \Theta$) of diffusion processes

$$\mathrm{d}X_t = S (\vartheta, X_t)\ \mathrm{d}t + \sigma (X_t)\ \mathrm{d}W_t, \quad X_0 = x, \tag{2.140}$$

with other functions $S (\cdot, \cdot)$ and $\sigma (\cdot)$. Note that we do not suppose that the true model (2.139) belongs to this family. This statement of the problem is quite important for applications, because there is always a gap between a mathematical model chosen by a statistician (here (2.140)) and the model of the real data (here (2.139)). Sometimes the difference is small and in the statistical literature the corresponding approach, studying the stability of estimators with respect to the small deviations in the model, is called *robust estimation* [102] (see as well Section 2.2.2 concerning asymptotic optimality of the second MDE). Here we do not suppose that the models are close and therefore the difference can be large.

Note that for the model (2.139) the diffusion coefficient $\sigma_* (x)^2$, $x \in \mathbb{I}_T$ can be estimated without error by observations $X^T = \{X_t, 0 \leq t \leq T\}$, where the interval $\mathbb{I}_T = (m_T, M_T)$ and $m_T = \inf_{0 \leq t \leq T} X_t$, $M_T = \sup_{0 \leq t \leq T} X_t$. Of course $m_T \to -\infty$ and $M_T \to \infty$. But we are in the situation when the model (2.140) is considered as a true one. Another motivation: we know well that the model (2.140) is wrong, but the true model is too cumbersome and we prefer to find in a chosen parametric family a member $S(\hat{\vartheta}_0, \cdot)$, which approximates the data in a certain good sense. Therefore we cannot use any information concerning the model (2.139) in the construction of the estimators, but nevertheless we suppose that it is possible to check some regularity conditions.

Hence we try to estimate the parameter ϑ_0 using the *wrong model* (2.140), i.e., we calculate the likelihood ratio on the base of the model (2.140)

$$L (\vartheta, X^T) = \exp \left\{ \int_0^T \frac{S (\vartheta, X_t)}{\sigma (X_t)^2}\ \mathrm{d}X_t - \frac{1}{2} \int_0^T \frac{S (\vartheta, X_t)^2}{\sigma (X_t)^2}\ \mathrm{d}t \right\}.$$

Then we substitute the observations (2.139) and calculate, say, the MLE as

$$\hat{\vartheta}_T = \arg \sup_{\vartheta \in \Theta} L (\vartheta, X^T),$$

or the Bayesian estimator

$$\tilde{\vartheta}_T = \frac{\int_\Theta y\, p(y)\, L\left(y, X^T\right)\, \mathrm{d}y}{\int_\Theta p(y)\, L\left(y, X^T\right)\, \mathrm{d}y}, \tag{2.141}$$

where $p(y),\ y \in \Theta$ is the density *a priori* of the random variable ϑ, or the MDE

$$\vartheta^*_T = \arg\inf_{\vartheta\in\Theta} \int_{-\infty}^{\infty} \left[\hat{F}_T(x) - F(\vartheta, x)\right]^2 \mu(\mathrm{d}x) \tag{2.142}$$

where $F(\vartheta, x)$ is the invariant distribution function of the model (2.140) and $\mu(\cdot)$ is some finite measure.

Such problems in the statistical literature are sometimes called parameter estimation for *misspecified* or *incorrect models* [181]. If we write $S(\vartheta, x) = S_*(\vartheta, x) + h(x)$ and $\sigma(x) = \sigma_*(x) + g(x)$, where $h(\cdot)$ and $g(\cdot)$ are some unknown functions, then we have parameter estimation for *contaminated models*, where $h(\cdot)$ and $g(\cdot)$ are *contaminations.* In Section 2.2.2 we already studied one such problem for the ergodic diffusion process (2.77) with small contamination in the trend coefficient.

We need several notations. Denote by $\mathbf{P}^{(T)}_{\vartheta_0}$ the measure of the process (2.139), $\mathbf{E}_{\vartheta_0}$ the expectation with respect to this measure, ξ_* - the stationary random variable of (2.139), $\mathbf{E}_{\vartheta_0}$ the expectation with the invariant density $f_*(\vartheta_0, \cdot)$ of the process (2.139) and put

$$D(\vartheta, \vartheta_0) = \mathbf{E}_{\vartheta_0}\left(\frac{S(\vartheta, \xi_*) - S_*(\vartheta_0, \xi_*)}{\sigma(\xi_*)}\right)^2,$$

$$H(\hat{\vartheta}_0, \vartheta, x) = 2\ \mathbf{E}_{\vartheta_0}\left(\chi_{\{\xi_* < x\}} \frac{\left[S(\hat{\vartheta}_0, \xi_*) - S_*(\vartheta_0, \xi_*)\right] \dot{S}(\hat{\vartheta}_0, \xi_*)}{\sigma(\xi_*)^2\ f_*(\vartheta_0, x)}\right),$$

$$K(\hat{\vartheta}_0, \vartheta) = \mathbf{E}_{\vartheta_0}\left(\frac{\dot{S}(\hat{\vartheta}_0, \xi_*)\ \sigma_*(\xi_*)}{\sigma(\xi_*)^2} + \frac{H(\hat{\vartheta}_0, \vartheta, \xi_*)}{\sigma_*(\xi_*)^2}\right)^2,$$

$$J(\hat{\vartheta}_0, \vartheta_0) = \mathbf{E}_{\vartheta_0}\left(\frac{\dot{S}(\hat{\vartheta}_0, \xi_*)^2 + \left(S(\hat{\vartheta}_0, \xi_*) - S(\vartheta_0, \xi_*)\right) \ddot{S}(\hat{\vartheta}_0, \xi_*)}{\sigma(\xi_*)^2}\right),$$

$$d(\hat{\vartheta}_0, \vartheta)^2 = \frac{K(\hat{\vartheta}_0, \vartheta_0)}{J(\hat{\vartheta}_0, \vartheta_0)^2},$$

where

$$\hat{\vartheta}_0 = \arg\inf_{\vartheta\in\Theta} D(\vartheta, \vartheta_0)$$

plays *the role of true value*, i.e., we show that the MLE converges to this value and is in this sense "*consistent*".

We suppose that the observed process satisfies the condition

$\mathcal{A}_*$. *The function $\sigma_*(\cdot)$ is positive, $\sigma_*(\cdot)^{-1} \in \mathcal{P}$, the function $S_*(\vartheta, \cdot)$ is locally bounded and*

$$\overline{\lim_{|x|\to\infty}}\ \operatorname{sgn}(x)\ \frac{S_*(\vartheta_0,x)}{\sigma_*(x)^2}<0. \tag{2.143}$$

The *regularity conditions* $\mathcal{F}$ in this problem will be

$\mathcal{F}_1$. *The function* $S(\vartheta,\cdot)$ *is three times continuously differentiable over* ϑ, *and the functions*

$$S(\vartheta,\cdot),\dot S(\vartheta,\cdot),\ddot S(\vartheta,\cdot),\dddot S(\vartheta,\cdot)\,S_*(\vartheta,\cdot),\sigma(\cdot),\sigma(\cdot)^{-1}\in\mathscr{P}.$$

$\mathscr{F}_2$. *For any* $\nu>0$ *the function*

$$g(\vartheta_0,\nu)\equiv\inf_{|\vartheta-\hat\vartheta_0|>\nu}\left[D(\vartheta,\vartheta_0)-D(\hat\vartheta_0,\vartheta_0)\right]>0$$

and $J(\hat\vartheta_0,\vartheta_0)>0$.

By this condition the function $D(\vartheta,\vartheta_0)$ has a unique minimum at the point $\vartheta=\hat\vartheta_0$.

Proposition 2.36. *Let the conditions* $\mathcal{A}_*$, $\mathcal{F}$ *be fulfilled. Then the MLE* $\hat\vartheta_T$ *converges to the value* $\hat\vartheta_0$, *is asymptotically normal:*

$$\mathcal{L}_{\vartheta_0}\left\{T^{1/2}\left(\hat\vartheta_T-\hat\vartheta_0\right)\right\}\Longrightarrow\mathcal{N}\left(0,d\left(\hat\vartheta_0,\vartheta_0\right)^2\right), \tag{2.144}$$

and (for any $p>0$*) the moments converge*

$$\lim_{T\to\infty}\mathbf{E}_{\vartheta_0}\left|T^{1/2}\left(\hat\vartheta_T-\hat\vartheta_0\right)\right|^p=d\left(\hat\vartheta_0,\vartheta_0\right)^p\mathbf{E}\left|\zeta\right|^p$$

Here $\zeta\sim\mathcal{N}(0,1)$.

Proof. First we establish the "consistency". For any $\nu>0$ we have

$$\begin{aligned}
&\mathbf{P}^{(T)}_{\vartheta_0}\left\{\left|\hat\vartheta_T-\hat\vartheta_0\right|>\nu\right\}\\
&\quad=\mathbf{P}^{(T)}_{\vartheta_0}\left\{\sup_{|\vartheta-\hat\vartheta_0|>\nu}L(\vartheta,X^T)>\sup_{|\vartheta-\hat\vartheta_0|\le\nu}L(\vartheta,X^T)\right\}\\
&\quad=\mathbf{P}^{(T)}_{\vartheta_0}\left\{\sup_{|\vartheta-\hat\vartheta_0|>\nu}L\left(\vartheta,\hat\vartheta_0,X^T\right)>\sup_{|\vartheta-\hat\vartheta_0|\le\nu}L\left(\vartheta,\hat\vartheta_0,X^T\right)\right\}\\
&\quad\le\mathbf{P}^{(T)}_{\vartheta_0}\left\{\sup_{|\vartheta-\hat\vartheta_0|>\nu}L\left(\vartheta,\hat\vartheta_0,X^T\right)>1\right\}
\end{aligned}$$

$$
\begin{aligned}
&\leq \mathbf{P}_{\vartheta_0}^{(T)} \left\{ \sup_{|\vartheta-\hat\vartheta_0|>\nu} \left[\int_0^T \frac{S(\vartheta,X_t)-S(\hat\vartheta_0,X_t)}{\sigma\left(X_t\right)^2}\, \sigma_*\left(X_t\right) \mathrm{d}W_t \right.\right.\\
&\qquad -\frac{1}{2}\int_0^T \frac{\left(S(\vartheta,X_t)-S_*(\vartheta_0,X_t)\right)^2}{\sigma\left(X_t\right)^2}\, \mathrm{d}t\\
&\qquad \left.\left. +\frac{1}{2}\int_0^T \frac{\left(S(\hat\vartheta_0,X_t)-S_*(\vartheta_0,X_t)\right)^2}{\sigma\left(X_t\right)^2}\, \mathrm{d}t \right] > 0 \right\}\\
&\leq \mathbf{P}_{\vartheta_0}^{(T)} \left\{ \sup_{\vartheta\in\Theta} \int_0^T \left[h\left(\vartheta,\hat\vartheta_0,X_t\right) - Q(\vartheta,\hat\vartheta_0,X_t)\right] \mathrm{d}W_t > \frac{T}{4}\, g\left(\vartheta_0,\nu\right)\right\}\\
&\qquad + \mathbf{P}_{\vartheta_0}^{(T)} \left\{ \sup_{\vartheta\in\Theta} \tilde Q(\vartheta,\hat\vartheta_0,X_0,X_T) > \frac{T}{4}\, g\left(\vartheta_0,\nu\right)\right\}, \qquad (2.145)
\end{aligned}
$$

where we put

$$
\begin{aligned}
h(\vartheta,\hat\vartheta_0,x) &= \frac{S(\vartheta,x)-S(\hat\vartheta_0,x)}{\sigma\left(x\right)^2}\,\sigma_*\left(x\right),\\
\delta\left(\vartheta,\vartheta_0,x\right) &= \frac{S(\vartheta,x)-S_*(\vartheta_0,x)}{\sigma\left(x\right)},\\
q(\vartheta,\hat\vartheta_0,x) &= \delta\left(\vartheta,\vartheta_0,x\right)^2 - \delta(\hat\vartheta_0,\vartheta_0,x)^2 - \mathbf{E}_{\vartheta_0}\left(\delta\left(\vartheta,\vartheta_0,\xi\right)^2 - \delta(\hat\vartheta_0,\vartheta_0,\xi)^2\right),\\
Q(\vartheta,\hat\vartheta_0,x) &= \frac{2}{\sigma_*\left(x\right) f_*\left(\vartheta_0,x\right)} \int_{-\infty}^{x} q(\vartheta,\hat\vartheta_0,v)\, f_*\left(\vartheta_0,v\right)\, \mathrm{d}v,\\
\tilde Q(\vartheta,\hat\vartheta_0,y,x) &= \int_y^x \sigma_*\left(v\right)^{-1}\, Q(\vartheta,\hat\vartheta_0,v)\, \mathrm{d}v.
\end{aligned}
$$

The functions $R(\vartheta,\hat\vartheta_0,x) = h(\vartheta,\hat\vartheta_0,x) - Q(\vartheta,\hat\vartheta_0,x)$ and $\tilde Q(\vartheta,\hat\vartheta_0,y,x)$ are continuously differentiable in ϑ and $R(\hat\vartheta_0,\hat\vartheta_0,x) = 0$, $\tilde Q(\hat\vartheta_0,\hat\vartheta_0,y,x) = 0$, hence

$$
R(\vartheta,\hat\vartheta_0,x) = \int_{\hat\vartheta_0}^{\vartheta} \dot R(v,\hat\vartheta_0,x)\, \mathrm{d}v,
$$

and by the Fubini theorem with probability 1 we have the estimate

$$
\sup_{\vartheta\in\Theta} \left| \int_0^T R(\vartheta,\hat\vartheta_0,X_t)\, \mathrm{d}W_t \right| \leq \int_\alpha^\beta \left| \int_0^T \dot R(v,\hat\vartheta_0,X_t)\, \mathrm{d}W_t \right| \mathrm{d}v.
$$

Therefore, for any $m \geq 1$

$$\mathbf{P}_{\vartheta_0}^{(T)}\left\{\sup_{\vartheta\in\Theta}\int_0^T R(\vartheta,\hat\vartheta_0,X_t)\,\mathrm{d}W_t>\frac{T}{4}\,g\left(\vartheta_0,\nu\right)\right\}$$
$$\leq \mathbf{P}_{\vartheta_0}^{(T)}\left\{\int_\alpha^\beta\left|\int_0^T \dot R(v,\hat\vartheta_0,X_t)\,\mathrm{d}W_t\right|\mathrm{d}v>\frac{T}{4}\,g\left(\vartheta_0,\nu\right)\right\}$$
$$\leq \frac{C\left(\beta-\alpha\right)^{2m-1}}{T^m g\left(\vartheta_0,\nu\right)^{2m}}\int_\alpha^\beta \mathbf{E}_{\vartheta_0}\left|\dot R(v,\hat\vartheta_0,\xi)\right|^{2m}\mathrm{d}v$$
$$\leq\frac{C}{T^m g\left(\vartheta_0,\nu\right)^{2m}}. \tag{2.146}$$

Recall that the function $\dot R(v,\hat\vartheta_0,x)\in\mathscr{P}$ and the invariant density has exponentially decreasing tails.

It can be shown like in (2.40) or (2.73) that

$$g\left(\vartheta_0,\nu\right)\geq\kappa\,\nu \tag{2.147}$$

with some $\kappa>0$. Hence, if we put $\nu=T^{-1/2+\mu}$, $\mu>0$ then, for any $m>1$ there exist a constant $C_m>0$, such that

$$\mathbf{P}_{\vartheta_0}^{(T)}\left\{\left|\hat\vartheta_T-\hat\vartheta_0\right|>T^{-1/2+\mu}\right\}\leq\frac{C_m}{T^\mu} \tag{2.148}$$

and we obtain the "consistency" with the estimate of the rate of convergence.

To prove asymptotic normality we use the *method of good sets.* Introduce the first *good set*

$$\mathbb{B}_1=\left\{\omega:\left|\hat\vartheta_T-\hat\vartheta_0\right|<T^{-1/4}\right\}$$

and suppose that $\omega\in\mathbb{B}_1$. Then MLE is one of the solutions of the maximum likelihood equation

$$\int_0^T\frac{\dot S\left(\vartheta,X_t\right)\,\sigma_*\left(X_t\right)}{\sigma\left(X_t\right)^2}\,\mathrm{d}W_t-\int_0^T\frac{S\left(\vartheta,X_t\right)-S_*\left(\vartheta,X_t\right)}{\sigma\left(X_t\right)^2}\dot S\left(\vartheta,X_t\right)\mathrm{d}t=0 \tag{2.149}$$

on the interval $\left(\hat\vartheta_0-T^{-1/4},\hat\vartheta_0+T^{-1/4}\right)$. Remember that with positive probability the MLE takes the values α and β. Let us denote

$$\lambda\left(\vartheta,\varepsilon,X^T\right)=\frac{\varepsilon}{\sqrt{T}}\int_0^T\frac{\dot S\left(\vartheta,X_t\right)\,\sigma_*\left(X_t\right)}{\sigma\left(X_t\right)^2}\,\mathrm{d}W_t$$
$$-\frac{1}{T}\int_0^T\frac{S\left(\vartheta,X_t\right)-S_*\left(\vartheta,X_t\right)}{\sigma\left(X_t\right)^2}\,\dot S\left(\vartheta,X_t\right)\,\mathrm{d}t,$$

where $\varepsilon>0$ is a *free parameter.* If we take $\varepsilon=T^{-1/2}$ then

$$\lambda\left(\vartheta,\varepsilon,X^T\right)=0,\qquad\vartheta\in\mathbb{A}_T=\left(\hat\vartheta_0-T^{-1/4},\hat\vartheta_0+T^{-1/4}\right) \tag{2.150}$$

is equivalent to (2.149). For its derivative we have

$$\frac{\partial\lambda\left(\vartheta,\varepsilon,X^T\right)}{\partial\vartheta}=\frac{\varepsilon}{\sqrt{T}}\int_0^T\frac{\ddot S\left(\vartheta,X_t\right)\,\sigma_*\left(X_t\right)}{\sigma\left(X_t\right)^2}\,\mathrm{d}W_t$$

$$-\frac{1}{T}\int_0^T\frac{S\left(\vartheta,X_t\right)-S_*\left(\vartheta,X_t\right)}{\sigma\left(X_t\right)^2}\ddot S\left(\vartheta,X_t\right)\,\mathrm{d}t-\frac{1}{T}\int_0^T\frac{\dot S\left(\vartheta,X_t\right)^2}{\sigma\left(X_t\right)^2}\,\mathrm{d}t$$

$$=\frac{\varepsilon}{\sqrt{T}}\int_0^T\ddot h\left(\vartheta,\hat\vartheta_0,X_t\right)\,\mathrm{d}W_t-\frac{1}{2T}\int_0^T\ddot q\left(\vartheta,\hat\vartheta_0,X_t\right)\,\mathrm{d}t-J\left(\vartheta,\vartheta_0\right),$$

where

$$J\left(\vartheta,\vartheta_0\right)=\frac{1}{2}\frac{\partial^2}{\partial\vartheta^2}D\left(\vartheta,\vartheta_0\right).$$

Introduce the second *good set*

$$\mathbb{B}_2=\left\{\omega:\begin{array}{l}\sup\limits_{\vartheta\in\mathbb{A}_T}\left|\varepsilon T^{-1/2}\int_0^T\ddot h(\vartheta,\hat\vartheta_0,X_t)\,\mathrm{d}W_t\right|<\frac{1}{4}\,J(\hat\vartheta_0,\vartheta_0)\\ \sup\limits_{\vartheta\in\mathbb{A}_T}\left|T^{-1}\int_0^T\ddot q(\vartheta,\hat\vartheta_0,X_t)\,\mathrm{d}t\right|<\frac{1}{2}\,J(\hat\vartheta_0,\vartheta_0)\\ \sup\limits_{\vartheta\in\mathbb{A}_T}\left|J\left(\vartheta,\vartheta_0\right)-J(\hat\vartheta_0,\vartheta_0)\right|<\frac{1}{4}\,J(\hat\vartheta_0,\vartheta_0).\end{array}\right.$$

Then for $\omega\in\mathbb{B}_1\cap\mathbb{B}_2$

$$\sup_{\vartheta\in\mathbb{A}_T}\frac{\partial\lambda\left(\vartheta,\varepsilon,X^T\right)}{\partial\vartheta}<-\frac{J(\hat\vartheta_0,\vartheta_0)}{4}$$

and Equation (2.150) has a unique solution $\hat\vartheta_T$.

This equation defines an implicit function $\hat\vartheta=\vartheta\left(\varepsilon\right)$ and we can obtain the first two terms of the expansion $\vartheta\left(\varepsilon\right)$ by the powers of ε (for $\omega\in\mathbb{B}_1\cap\mathbb{B}_2$):

$$\hat\vartheta_T=\hat\vartheta_0+\varepsilon\,\frac{\Delta_T(\hat\vartheta_0,\vartheta_0,X^T)}{J(\hat\vartheta_0,\vartheta_0)}\left(1+o(1)\right),\tag{2.151}$$

where the random variable

$$\Delta_T(\hat\vartheta_0,\vartheta_0,X^T)=\frac{1}{\sqrt{T}}\int_0^T\left[\frac{\dot S(\hat\vartheta_0,X_t)\,\sigma_*\left(X_t\right)}{\sigma\left(X_t\right)^2}+\frac{H(\hat\vartheta_0,\vartheta_0,X_t)}{\sigma_*\left(X_t\right)^2}\right]\,\mathrm{d}W_t$$

is asymptotically normal

$$\mathcal{L}_{\vartheta_0}\left\{\Delta_T(\hat\vartheta_0,\vartheta_0,X^T)\right\}\Longrightarrow\mathcal{N}\left(0,K(\hat\vartheta_0,\vartheta_0)\right).$$

The third set $\mathbb{B}_3$ is introduced to provide $o(1)$ in this expansion.

Then we obtain the estimates for the complements of these sets

$$\mathbf{P}_{\vartheta_0}^{(T)}\left\{\mathbb{B}_i^c\right\} \leq \frac{C_i}{T^\mu}, \qquad i=1,2,3,\ \mu>2.$$

Asymptotic normality of the MLE now follows from these estimates and the expansion (2.151).

The details of the proof can be found in [139], Theorem 2.13. We have just to add the verification of the uniform integrability of the random variable $\left|\hat{u}_T\right|^p$ where $\hat{u}_T=\sqrt{T}\left(\hat{\vartheta}_T-\hat{\vartheta}_0\right)$ for any $p>0$. From (2.145), (2.146) and (2.147) we obtain the estimate

$$\mathbf{P}_{\vartheta_0}^{(T)}\left\{\left|\hat{u}_T\right|>N\right\} \leq \frac{C}{T^m\, g\left(\vartheta_0, \frac{N}{\sqrt{T}}\right)^{2m}} \leq \frac{C}{\kappa\, N^{2m}}.$$

Hence

$$\mathbf{E}_{\vartheta_0}\left|\sqrt{T}\left(\hat{\vartheta}_T-\hat{\vartheta}_0\right)\right|^p \leq C$$

and we have the convergence of all moments.

Remark 2.37. For a fixed value u and $T \to \infty$ the likelihood ratio has the following representation:

$$\begin{aligned}
\ln L\left(\hat{\vartheta}_0+u/\sqrt{T}, \hat{\vartheta}_0, X^T\right) &= \frac{u}{\sqrt{T}}\int_0^T \frac{\dot{S}(\hat{\vartheta}_0, X_t)}{\sigma\left(X_t\right)^2}\,\sigma_*\left(X_t\right)\mathrm{d}W_t \\
&\quad -\frac{u^2}{2T}\int_0^T \frac{\dot{S}(\hat{\vartheta}_0, X_t)^2+\left[S(\hat{\vartheta}_0, X_t)-S_*(\vartheta_0, X_t)\right]\ddot{S}(\hat{\vartheta}_0, X_t)}{\sigma\left(X_t\right)^2}\,\mathrm{d}t \\
&\quad -\frac{u}{\sqrt{T}}\int_0^T \frac{S(\hat{\vartheta}_0, X_t)-S_*(\vartheta_0, X_t)}{\sigma\left(X_t\right)^2}\,\dot{S}(\hat{\vartheta}_0, X_t)\,\mathrm{d}t+r_T(u, X^T) \\
&= u\,\Delta_T(\hat{\vartheta}_0, \vartheta_0, X^T)-\frac{u^2}{2}\,J(\hat{\vartheta}_0, \vartheta_0)+\tilde{r}_T(u, X^T),
\end{aligned}$$

where the random variable $\tilde{r}_T(u, X^T) \to 0$ in probability.

Therefore the finite-dimensional distributions of the process $Z_T(u) = L(\hat{\vartheta}_0+u/\sqrt{T}, \hat{\vartheta}_0, X^T)$, $u \in U_T = \left(\sqrt{T}(\alpha-\hat{\vartheta}_0), \sqrt{T}(\beta-\hat{\vartheta}_0)\right)$, converge to the finite-dimensional distributions of the process

$$Z(u)=\exp\left\{u\,\Delta(\hat{\vartheta}_0, \vartheta_0)-\frac{u^2}{2}\,J(\hat{\vartheta}_0, \vartheta_0)\right\}, \qquad u \in \mathscr{R},$$

where $\Delta(\hat{\vartheta}_0, \vartheta_0) \sim \mathcal{N}\left(0, K(\hat{\vartheta}_0, \vartheta_0)\right)$. Hence if (following the Ibragimov–Khasminskii method) we prove the weak convergence of the process $Z_T(\cdot)$ to the process $Z(\cdot)$ then we obtain the asymptotic normality of the MLE as follows:

$$\mathbf{P}_{\vartheta_0}^{(T)}\left\{\sqrt{T}\left(\hat\vartheta_T-\hat\vartheta_0\right)<x\right\}=\mathbf{P}_{\vartheta_0}^{(T)}\left\{\sup_{u<x}Z_T(u)>\sup_{u\geq x}Z_T(u)\right\}$$

$$\longrightarrow \mathbf{P}_{\vartheta_0}\left\{\sup_{u<x}Z(u)>\sup_{u\geq x}Z(u)\right\}=\mathbf{P}\left\{\frac{\Delta(\hat\vartheta_0,\vartheta_0)}{J(\hat\vartheta_0,\vartheta_0)}<x\right\}.$$

The direct proof of this weak convergence can have some technical difficulties because, even for the limit process,

$$\mathbf{E}\ Z(u)=\exp\left\{\frac{u^2}{2}\left(K(\hat\vartheta_0,\vartheta_0)-J(\hat\vartheta_0,\vartheta_0)\right)\right\}\neq 1.$$

To avoid this inconvenience we can consider the process $Y_T(u)=Z_T(u)^q$, where $q>0$ is sufficiently small to provide the inequality $\mathbf{E}_{\vartheta_0}Y_T(u)\leq C$, where $C>0$ does not depend on T and u. The details can be found in [145], Section 4.1. Of course, we have

$$\mathbf{P}_{\vartheta_0}^{(T)}\left\{\sqrt{T}\left(\hat\vartheta_T-\hat\vartheta_0\right)<x\right\}=\mathbf{P}_{\vartheta_0}^{(T)}\left\{\sup_{u<x}Y_T(u)>\sup_{u\geq x}Y_T(u)\right\},$$

etc.

The Bayes estimator $\tilde\vartheta_T$ defined in (2.141)) has the same limit behavior, i.e., this estimator is "consistent":

$$\mathbf{P}_{\vartheta_0}-\lim_{T\to\infty}\tilde\vartheta_T=\hat\vartheta_0$$

and asymptotically normal

$$\mathcal{L}_{\vartheta_0}\left\{\sqrt{T}\left(\tilde\vartheta_T-\hat\vartheta_0\right)\right\}\Longrightarrow\mathcal{N}\left(0,d(\hat\vartheta_0,\vartheta_0)^2\right). \tag{2.152}$$

The proof can be carried out using the same *method of good sets* as in the proof of Theorem 2.17 in [139]. Note just that formally we can write

$$\tilde\vartheta_T=\hat\vartheta_0+T^{-1/2}\frac{\int_{U_T}u\,p\left(\hat\vartheta_0+u/\sqrt{T}\right)\ Z_T(u)\ du}{\int_{U_T}p\left(\hat\vartheta_0+v/\sqrt{T}\right)\ Z_T(v)\ dv}$$

and

$$\sqrt{T}\left(\tilde\vartheta_T-\hat\vartheta_0\right)\Longrightarrow\frac{\int_{\mathscr{R}}u\,Z(u)\ du}{\int_{\mathscr{R}}Z(v)\ dv}=\frac{\Delta\left(\hat\vartheta_0,\vartheta_0\right)}{J\left(\hat\vartheta_0,\vartheta_0\right)}\sim\mathcal{N}\left(0,d(\hat\vartheta_0,\vartheta_0)^2\right).$$

The MDE ϑ_T^* (2.142)) (under regularity conditions) converges to another value

$$\vartheta_0^* = \arg \inf_{\vartheta \in \Theta} \int_{-\infty}^{\infty} \left[F_* (\vartheta_0, x) - F(\vartheta, x)\right]^2 \mu(dx)$$

and is asymptotically normal

$$\mathcal{L}_{\vartheta_0} \left\{ \sqrt{T} \left(\vartheta_T^* - \vartheta_0^* \right) \right\} \Longrightarrow \mathcal{N}\left(0, d_*(\vartheta_0^*, \vartheta_0)^2\right). \tag{2.153}$$

Here

$$d_*(\vartheta_0^*, \vartheta_0)^2 = \frac{K_*(\vartheta_0^*, \vartheta_0)^2}{J_*(\vartheta_0^*, \vartheta_0)^2},$$

where

$$K_*(\vartheta_0^*, \vartheta_0)^2 = 4\,\mathbf{E}_{\vartheta_0} \left(\int_{-\infty}^{\infty} M(\vartheta_0, x, \xi)\, \dot F(\vartheta_0^*, x)\ \mu(\mathrm{d}x) \right)^2,$$

$$J_*(\vartheta_0^*, \vartheta_0) = \int_{\mathscr{R}} \left[\dot F(\vartheta_0^*, x)^2 - \left[F_*(\vartheta_0, x) - F(\vartheta_0^*, x)\right] \ddot F(\vartheta_0^*, x) \right] \mu(\mathrm{d}x) > 0,$$

$$M(\vartheta_0, x, y) = \frac{F_*(\vartheta_0, x \wedge y) - F_*(\vartheta_0, x)\ F_*(\vartheta_0, y)}{\sigma_*(y)\, f_*(\vartheta_0, y)}.$$

We give just a sketch of the proof. The detailed proof of these properties can be carried out in the same way as was done in Section 2.2.1, see as well [139], Theorems 7.12 and 7.14 (a diffusion-type process with small noise).

First, using the *identifiability condition*: for any $\nu > 0$

$$g_*(\nu) = \inf_{\left|\vartheta - \vartheta_0^*\right| > \nu} \left\| F(\vartheta, \cdot) - F_*(\vartheta_0, \cdot) \right\|_\mu^2 - \left\| F(\vartheta_0^*, \cdot) - F_*(\vartheta_0, \cdot) \right\|_\mu^2 > 0,$$

we obtain the consistency of the MDE:

$$\mathbf{P}_{\vartheta_0}^{(T)} \left\{ \left| \vartheta_T^* - \vartheta_0^* \right| > \nu \right\} \le \frac{C_m}{g_*(\nu)^m\, T^{m/2}} \le \frac{C_m^*}{\nu^{2m} T^{m/2}}, \qquad m > 2.$$

Here the norm

$$\left\| h(\cdot) \right\|_\mu^2 = \int_{-\infty}^{\infty} h(x)^2\ \mu(\mathrm{d}x),$$

and we used the estimate $g_*(\nu) \ge \kappa\, \nu$. So we localize the problem. Then we put $\nu = T^{-1/4}$ and consider the *minimum distance equation* (MDEq) on the interval $\left[\vartheta_0^* - T^{-1/4}, \vartheta_0^* + T^{-1/4}\right]$

$$\int_{-\infty}^{\infty} \left[\hat F_T(x) - F(\vartheta, x) \right]\ \dot F(\vartheta, x)\ \mu(\mathrm{d}x) = 0.$$

We write its solution as $\vartheta_T^* = \vartheta_0^* + T^{-1/2} u_T^*$ and rewrite the MDEq as

$$\int_{-\infty}^{\infty} \left[\hat F_T(x) - F_*(\vartheta_0, x) + F_*(\vartheta_0, x) \right.$$

$$\left. - F(\vartheta_0^*, x) - \frac{u_T^*}{\sqrt{T}} \dot F\left(\tilde\vartheta, x\right) \right]\ \dot F\left(\vartheta_T^*, x\right)\ \mu(\mathrm{d}x) = 0.$$

Further, we denote by $\eta_T(x) = \sqrt{T}\left(\hat{F}_T(x) - F_*(\vartheta_0, x)\right)$ and write the solution as

$$u_T^* = J_*(\vartheta_0^*, \vartheta_0)^{-1} \int_{-\infty}^{\infty} \eta_T(x)\ \dot{F}(\vartheta_0^*, x)\ \mu(\mathrm{d}x) + o(1).$$

For the process $\eta_T(x)$ we have the representation (2.65). Hence

$$\int_{-\infty}^{\infty} \eta_T(x)\ \dot{F}(\vartheta_0^*, x)\ \mu(\mathrm{d}x)$$
$$= \frac{1}{\sqrt{T}} \int_0^T \int_{-\infty}^{\infty} M(\vartheta_0, x, X_t)\ \dot{F}(\vartheta_0^*, x)\ \mu(\mathrm{d}x)\ \mathrm{d}W_t + o(1),$$

where we changed the order of integration. The asymptotic normality (2.153) now follows from the central limit theorem.

On Asymptotically Efficient Estimation

Note that if $S_*(\vartheta_0, x) = S(\vartheta_0, x),\ x \in \mathscr{R}$ and even $\sigma_*(x) \neq \sigma(x)$, then $\hat{\vartheta}_0 = \vartheta_0$, $H(\hat{\vartheta}_0, \vartheta, x) = 0$, and

$$K(\hat{\vartheta}_0, \vartheta) = \mathbf{E}_{\vartheta_0}\left(\frac{\dot{S}(\vartheta_0, \xi_*)\ \sigma_*(\xi_*)}{\sigma(\xi_*)^2}\right)^2,$$

$$J(\hat{\vartheta}_0, \vartheta) = \mathbf{E}_{\vartheta_0}\left(\frac{\dot{S}(\vartheta_0, \xi_*)}{\sigma(\xi_*)}\right)^2.$$

Hence using the Cauchy–Schwarz inequality it is easy to see that

$$d(\hat{\vartheta}_0, \vartheta)^2 \geq \mathrm{I}(\vartheta_0)^{-1} = \left(\mathbf{E}_{\vartheta_0}\left(\frac{\dot{S}(\vartheta_0, \xi_*)}{\sigma_*(\xi_*)}\right)^2\right)^{-1}$$

with equality if and only if

$$[\sigma(x) - \sigma_*(x)]\ \chi_{\{\dot{S}(\vartheta_0, x) \neq 0\}} = 0, \qquad x \in \mathscr{R}.$$

Therefore, the MLE is consistent even if we use the wrong diffusion coefficient $\sigma(\cdot)$ in the construction of the estimator. This interesting fact was noted by McKeague [181], who described as well the properties of the MLE for a misspecified ergodic diffusion process (under different regularity conditions and using another proof). In particular, he noted that the MLE $\hat{\vartheta}_T$ converges to the value $\hat{\vartheta}_0$ with probability 1 and is asymptotically normal.

The consistent estimation is possible in some situations, when we even do not know the trend coefficient. Indeed, the value $\hat{\vartheta}_0$ is a solution of the equation

$$\frac{\partial}{\partial \vartheta} D(\vartheta, \vartheta_0) = 2\,\mathbf{E}_{\vartheta_0} \left(\frac{[S(\vartheta, \xi_*) - S_*(\vartheta_0, \xi_*)]}{\sigma(\xi_*)^2} \dot{S}(\vartheta, \xi_*) \right) = 0 \qquad (2.154)$$

and if the function $[S(\vartheta, \cdot) - S_*(\vartheta_0, \cdot)]\,\sigma(\cdot)^{-2}$ for all $\vartheta \in \Theta$ is orthogonal to $\dot{S}(\vartheta, \cdot)$, then $\hat{\vartheta}_0 = \vartheta_0$.

Windows

To illustrate this possibility, let us consider a *contaminated* ergodic diffusion process

$$\mathrm{d}X_t = S(\vartheta, X_t)\ \mathrm{d}t + h(X_t)\ \mathrm{d}t + \sigma(X_t)\ \mathrm{d}W_t, \qquad (2.155)$$

i.e., $S_*(\vartheta, x) = S(\vartheta, x) + h(x)$, where $h(\cdot)$ is some unknown function but with known support. More precisely, we suppose that

$\mathcal{F}_0$. *The function $h(\cdot)$ belongs to the class $\mathscr{H}_{\mathbb{A}}$ of functions with compact support* supp $h(\cdot) \subset \mathbb{A}$, *where $\mathbb{A}$ is a known bounded set (window).*

$\mathcal{F}_2^*$. *For any $\nu > 0$*

$$\inf_{|\vartheta - \vartheta_0| > \nu} \mathbf{E}_{\vartheta_0} \left(\frac{S(\vartheta, \xi) - S(\vartheta_0, \xi)}{\sigma(\xi)} \right)^2 \chi_{\{\xi \in \mathbb{A}^c\}} > 0$$

and

$$\bar{\mathrm{I}}(\vartheta) = \mathbf{E}_{\vartheta} \left(\frac{\dot{S}(\vartheta, \xi)}{\sigma(\xi)} \right)^2 \chi_{\{\xi \in \mathbb{A}^c\}} > 0.$$

Note that this condition is given in terms of the *wrong model* $(h(x) \equiv 0)$ known to statisticians.

Of course, $\hat{\vartheta}_0 \neq \vartheta_0$ and MLE is not consistent. Our goal is to find an estimator which is consistent uniformly over the class $\mathscr{H}_{\mathbb{A}}$ and is asymptotically efficient.

Fix a value $\vartheta_0 \in \Theta$ and a function $h_0(\cdot) \in \mathscr{H}_{\mathbb{A}}$. By condition $\mathcal{F}_1$ the derivative $\dot{S}(\vartheta, \cdot)$ is bounded on $\mathbb{A}$. Denote by $K > 0$ a constant such that

$$\sup_{x \in \mathbb{A}} \left| \dot{S}(\vartheta_0, x) \right| \leq K$$

and define a *parametric-nonparametric* vicinity of the fixed model as

$$V_\delta = \left\{ \vartheta, h(\cdot) : \ |\vartheta - \vartheta_0| < \delta, \ \sup_{x \in \mathbb{A}} |h(x) - h_0(x)| < 2K\delta \right\}.$$

Introduce as well the estimator $\hat{\vartheta}_{T,\mathbb{A}^c}$ by the equation

$$\bar{L}\left(\hat{\vartheta}_{T,\mathbb{A}^c}, X^T\right) = \sup_{\vartheta \in \Theta} \bar{L}\left(\vartheta, X^T\right)$$

where $\bar{L}\left(\vartheta, X^T\right)$ is the likelihood ratio with excluded observations on the set $\mathbb{A}$:

$$\bar{L}\left(\vartheta, X^T\right) = \exp\left\{\int_0^T \frac{S\left(\vartheta, X_t\right)}{\sigma\left(X_t\right)^2}\,\chi_{\{X_t\in\mathbb{A}^c\}}\mathrm{d}X_t \right.$$
$$\left. -\frac{1}{2}\int_0^T \left(\frac{S\left(\vartheta, X_t\right)}{\sigma\left(X_t\right)}\right)^2 \chi_{\{X_t\in\mathbb{A}^c\}}\mathrm{d}t\right\}.$$

This is the likelihood ratio for the artificial process

$$\mathrm{d}X_t = S\left(\vartheta, X_t\right)\chi_{\{X_t\in\mathbb{A}^c\}}\,\mathrm{d}t + \sigma\left(X_t\right)\,\mathrm{d}W_t, \qquad X_0, \tag{2.156}$$

which is not really observed and $\mathcal{F}_2^*$ is the usual identifiability condition for this model.

Below $d\left(\vartheta, h\right)^2 = \bar{\mathrm{I}}\left(\vartheta, h\right)^{-1}$, where

$$\bar{\mathrm{I}}\left(\vartheta, h\right) = \mathbf{E}_{\vartheta,h}\left(\frac{\dot{S}\left(\vartheta, \xi_*\right)}{\sigma\left(\xi_*\right)}\right)^2 \chi_{\{\xi_*\in\mathbb{A}^c\}} > 0$$

is the Fisher information of the problem.

Proposition 2.38. (Höpfner–Kutoyants [99]) *Suppose that conditions $\mathcal{F}_0$, $\mathcal{F}_1$, $\mathcal{F}_2^*$ and $\boldsymbol{\mathcal{A}}_0\left(\Theta\right)$ are fulfilled, then for all estimators $\bar{\vartheta}_T$ and $\ell\left(\cdot\right)\in\mathscr{W}_p$*

$$\varliminf_{\delta\to 0}\varliminf_{T\to\infty}\sup_{\vartheta,h(\cdot)\in V_\delta}\mathbf{E}_{\vartheta,h}\,\ell\left(\sqrt{T}\left(\bar{\vartheta}_T-\vartheta\right)\right)\geq \mathbf{E}\;\ell\left(\zeta\, d\left(\vartheta_0, h_0\right)\right) \tag{2.157}$$

and the estimator $\hat{\vartheta}_{T,\mathbb{A}^c}$ is asymptotically efficient, i.e., we have the equality

$$\lim_{\delta\to 0}\lim_{T\to\infty}\sup_{\vartheta,h(\cdot)\in V_\delta}\mathbf{E}_{\vartheta,h}\,\ell\left(\sqrt{T}\left(\hat{\vartheta}_{T,\mathbb{A}^c}-\vartheta\right)\right) = \mathbf{E}\;\ell\left(\zeta\, d\left(\vartheta_0, h_0\right)\right). \tag{2.158}$$

Moreover this convergence is uniform over $h\left(\cdot\right)\in\mathscr{H}_{\mathbb{A}}$.

Proof. We start with the lower bound. To estimate the $\sup\limits_{\vartheta,h(\cdot)\in V_\delta}$ we take as the *least favorable function* $h\left(\cdot\right)\in V_\delta$ the function

$$h_*\left(\vartheta, x\right) = h_0\left(x\right) - \left[S\left(\vartheta, x\right) - S\left(\vartheta_0, x\right)\right]\,\chi_{\{x\in\mathbb{A}\}}.$$

It is easy to see that for sufficiently small δ

$$\sup_{x\in\mathbb{A}}\left|h_*\left(\vartheta, x\right) - h_0\left(x\right)\right| \leq \left|\vartheta-\vartheta_0\right|\sup_{x\in\mathbb{A}}\left|\dot{S}\left(\vartheta_0, x\right)\right|\,\left(1+o(1)\right)\leq 2K\delta,$$

the process (2.155) is transformed to the *least favorable process*

$$\mathrm{d}X_t = S(\vartheta, X_t)\,\chi_{\{X_t \in \mathbb{A}^c\}}\,\mathrm{d}t + \left[h_0(X_t) + S(\vartheta_0, X_t)\,\chi_{\{X_t \in \mathbb{A}\}}\right]\mathrm{d}t \\ + \sigma(X_t)\,\mathrm{d}W_t, \qquad X_0 \tag{2.159}$$

and of course

$$\sup_{\vartheta, h(\cdot) \in V_\delta} \mathbf{E}_{\vartheta,h}\,\ell\left(\sqrt{T}\left(\bar\vartheta_T - \vartheta\right)\right) \geq \sup_{|\vartheta - \vartheta_0| < \delta} \mathbf{E}_{\vartheta,h_*(\vartheta)}\,\ell\left(\sqrt{T}\left(\bar\vartheta_T - \vartheta\right)\right).$$

For the process (2.159) the corresponding family of measures

$$\left\{\mathbf{P}^{(T)}_{\vartheta,h_*(\vartheta)}, \vartheta \in \Theta\right\}$$

is LAN:

$$\bar Z_T(u) = \exp\left\{u\bar\Delta_T\left(\vartheta_0, X^T\right) - \frac{u^2}{2}\,\bar{\mathrm{I}}(\vartheta_0, h_0) + r_T\left(u, h_*, X^T\right)\right\},$$

where

$$\bar\Delta_T\left(\vartheta_0, X^T\right) = \frac{1}{\sqrt{T}}\int_0^T \frac{\dot S(\vartheta_0, X_t)}{\sigma(X_t)^2}\,\chi_{\{X_t \in \mathbb{A}^c\}}\left[\mathrm{d}X_t - S(\vartheta_0, X_t)\,\mathrm{d}t - h_0(X_t)\,\mathrm{d}t\right]$$

and

$$\mathcal{L}_{\vartheta_0,h_0}\left\{\bar\Delta_T\left(\vartheta_0, X^T\right)\right\} \Longrightarrow \mathcal{N}\left(0, \bar{\mathrm{I}}(\vartheta, h_0)\right), \qquad r_T\left(u, h_*, X^T\right) \to 0. \tag{2.160}$$

Therefore we can apply Theorem 2.4 to the model (2.159) and obtain the desired inequality (2.157).

As $h(\cdot)$ has bounded support, the convergences (2.160) are uniform on compacts $\mathbb{K} \subset \Theta$. To prove (2.158) we just need to check the condition $\mathscr{F}_2$ for the process of observations (2.155) and the wrong model (2.156). We have

$$D_h(\vartheta, \vartheta_0) = \mathbf{E}_{\vartheta_0,h}\left(\frac{S(\vartheta, \xi_*)\,\chi_{\{\xi_* \in \mathbb{A}^c\}} - S(\vartheta_0, \xi_*) - h(\xi_*)}{\sigma(\xi_*)}\right)^2$$

$$= \mathbf{E}_{\vartheta_0,h}\left(\frac{\left[S(\vartheta, \xi_*) - S(\vartheta_0, \xi_*)\right]\chi_{\{\xi_* \in \mathbb{A}^c\}} - S(\vartheta_0, \xi_*)\,\chi_{\{\xi_* \in \mathbb{A}\}} - h(\xi_*)}{\sigma(\xi_*)}\right)^2$$

$$= \mathbf{E}_{\vartheta_0,h}\left(\frac{S(\vartheta, \xi_*) - S(\vartheta_0, \xi_*)}{\sigma(\xi_*)}\right)^2 \chi_{\{\xi_* \in \mathbb{A}^c\}} \\ + \mathbf{E}_{\vartheta_0,h}\left(\frac{S(\vartheta_0, \xi_*) + h(\xi_*)}{\sigma(\xi_*)}\right)^2 \chi_{\{\xi_* \in \mathbb{A}\}}.$$

Now it is evident that for all $h(\cdot) \in \mathscr{H}_{\mathbb{A}}$ under condition $\mathscr{F}_2$ we have the equality

$$\hat\vartheta = \arg\inf_{\vartheta \in \Theta} D_h(\vartheta, \vartheta_0) = \vartheta_0$$

and the equality (2.151) is fulfilled too.

Let us denote by $\mathscr{H}_{\mathbb{A}}^{b}$ a class of functions $h(\cdot) \in \mathscr{H}_{\mathbb{A}}$ bounded by some constant. It can be shown that all convergence in Proposition 2.36 are uniform on compacts $\mathbb{K}$ and on $\mathscr{H}_{\mathbb{A}}^{b}$ and this proves (2.158). Note that the function $d(\vartheta, h)$ is a continuous functional of ϑ and $h(\cdot)$. Hence

$$\lim_{\delta \to 0} \sup_{\vartheta, h(\cdot) \in V_\delta} d(\vartheta, h) = d(\vartheta_0, h_0)$$

and this concludes the proof.

Remark 2.39. Another type of misspecification corresponds to the "noise process" $\{W_t, 0 \le t \le T\}$. Suppose that the observations $\{X_t^{(n)}, 0 \le t \le T\}$ are taken from the equation

$$\mathrm{d}X_t^{(n)} = \left(\vartheta, X_t^{(n)}\right) \,\mathrm{d}t + \sigma\left(X_t^{(n)}\right) \,\mathrm{d}W_t^{(n)}, \quad X_0,$$

where $W^{(n),T} = \left\{W_t^{(n)}, 0 \le t \le T\right\}$ is a differentiable with probability 1 process close in some sense to a Wiener process W^T ($W^{(n),T}$ converges to W^T as $n \to \infty$). Then to have an estimator close to the MLE we need to approximate the Itô integral. For example, if the system is linear

$$\mathrm{d}X_t^{(n)} = \vartheta h\left(X_t^{(n)}\right) \,\mathrm{d}t + \mathrm{d}W_t^{(n)}, \quad X_0,$$

then the estimator

$$\hat\vartheta_T^{(n)} = \left(\int_0^T h\left(X_t^{(n)}\right)^2 \mathrm{d}t\right)^{-1} \left[\int_0^T h\left(X_t^{(n)}\right) \,\mathrm{d}X_t^{(n)} - \frac{1}{2}\int_0^T h'\left(X_t^{(n)}\right) \mathrm{d}t\right]$$

is asymptotically ($n \to \infty$) equivalent to the MLE (see Afanas'ev [1]).

Note that this gives us an approximation of the MLE for the limit model, but we do not consider the problem of parameter estimation for the model with differentiable noise because this problem can be even singular, i.e., the parameter can be estimated without error.

2.6.2 Too Many True Models

Suppose that the observed diffusion process

$$\mathrm{d}X_t = S(\vartheta, X_t) \,\mathrm{d}t + \sigma(X_t) \,\mathrm{d}W_t, \quad X_0, \quad 0 \le t \le T \tag{2.161}$$

with $\vartheta \in \Theta = (\alpha, \beta)$ satisfies all the conditions $\boldsymbol{\mathcal{A}}$ of Theorem 2.8 except the condition of identifiability (2.18), which is replaced by the following one:

$\mathcal{G}$. *There exist $k > 1$ values $\vartheta_1, \ldots, \vartheta_k$ such that*

$$S(\vartheta_1, x) = S(\vartheta_i, x), \qquad i = 2, \ldots, k, \quad x \in \mathscr{R}.$$

Moreover, if we consider a subdivision $\Theta = \cup_{i=1}^k \Theta_i$, $\Theta_i \cap \Theta_l = \emptyset, i \neq l$, ϑ_i is an interior point of Θ_i, then for any $\nu > 0$ and any $i = 1, \ldots, k$

$$\inf_{|\vartheta - \vartheta_i| > \nu, \vartheta \in \Theta_i} \mathbf{E}_{\vartheta_1} \left| \frac{S(\vartheta, \xi) - S(\vartheta_i, \xi)}{\sigma(\xi)} \right|^2 > 0.$$

Put $\Theta_i = (\alpha_i, \beta_i], i = 1, \ldots, k-1$, where $\alpha_1 = \alpha$, $\alpha_{i+1} = \beta_i$ and $\Theta_k = (\alpha_k, \beta)$. Assume that ϑ_1 is the true value, but as the trends $S(\vartheta_i, x), i = 2, \ldots, k$ coincide with $S(\vartheta_1, x)$, then the true value cannot be estimated consistently and any one of $\vartheta_1, \ldots, \vartheta_k$ can be considered as a true value. We have too many true models. It is known that the MLE converges to this set of values (e.g., see Bagchi and Borkar [11]). Here we describe the probabilities of the convergence to each particular value and the limit distribution of the difference.

Let us introduce the Gaussian vector $\zeta = (\zeta_1, \ldots, \zeta_k)$ with zero mean and covariance

$$\rho_{il} = \mathbf{E}(\zeta_i \zeta_l) = \Big(\mathrm{I}(\vartheta_i)\,\mathrm{I}(\vartheta_l)\Big)^{-1/2} \mathbf{E}_{\vartheta_1} \left(\frac{\dot S(\vartheta_i, \xi)\, \dot S(\vartheta_l, \xi)}{\sigma(\xi)^2} \right),$$

where the Fisher information

$$\mathrm{I}(\vartheta_i) = \mathbf{E}_{\vartheta_1} \left(\frac{\dot S(\vartheta_i, \xi)}{\sigma(\xi)} \right)^2 > 0, \qquad i = 1, \ldots, k.$$

Define as well a discrete random variable

$$\hat\vartheta = \sum_{i=1}^k \vartheta_i \, \chi_{\{\mathbb{H}_i\}}$$

where the sets

$$\mathbb{H}_i = \left\{ \omega : |\zeta_i| > \max_{l \neq i} |\zeta_l| \right\}, \qquad i = 1, \ldots, k.$$

The MLE $\hat\vartheta_T$ is defined by the equality

$$\hat\vartheta_T = \arg\max_{\vartheta \in \Theta} L(\vartheta, X^T)$$

and we introduce k *local MLE's* as follows:

$$\hat\vartheta_T^{(i)} = \arg\max_{\vartheta \in \Theta_i} L(\vartheta, X^T), \quad i = 1, \ldots, k.$$

The (*global*) MLE can be written as

$$\hat{\vartheta}_T = \sum_{i=1}^{k} \hat{\vartheta}_T^{(i)} \chi_{\left\{\mathbb{H}_i^{(T)}\right\}},$$

where the sets

$$\mathbb{H}_i^{(T)} = \left\{\omega : \sup_{\vartheta \in \Theta_i} L\left(\vartheta, X^T\right) > \max_{l \neq i} \sup_{\vartheta \in \Theta_l} L\left(\vartheta, X^T\right)\right\}, \qquad i = 1, \ldots, k.$$

Define as well the random variable

$$\hat{\vartheta}^{(T)} = \sum_{i=1}^{k} \vartheta_i \, \chi_{\left\{\mathbb{H}_i^{(T)}\right\}}.$$

Note that this random variable is measurable with respect to observations.

Proposition 2.40. *Suppose that the conditions of Theorem 2.8 be fulfilled except the condition (1.72), which is replaced by the condition $\mathcal{G}$ and* $\det \rho \neq 0$. *Then the MLE $\hat{\vartheta}_T$ converges to $\hat{\vartheta}$ in distribution and*

$$\mathcal{L}_{\vartheta_1}\left\{\sqrt{T}\left(\hat{\vartheta}_T - \hat{\vartheta}^{(T)}\right)\right\} \Longrightarrow \mathcal{L}\left\{\hat{\zeta}\right\},$$

where the random variable

$$\hat{\zeta} = \sum_{i=1}^{k} \frac{\zeta_i}{\sqrt{\mathrm{I}\left(\vartheta_i\right)}} \chi_{\{\mathbb{H}_i\}}.$$

Proof. We give just a sketch of the proof, because it is quite close to the proofs of the similar results for diffusion processes with small noise asymptotics given in [151] and [139], Theorems 2.18 and 2.19. Formally we have k *normal* (identifiable) problems of parameter estimation, i.e., if $\Theta = \Theta_i$ then the MLE $\hat{\vartheta}_T^{(i)}$ is consistent $\hat{\vartheta}_T^{(i)} \to \vartheta_i$ and asymptotically normal.

Let us introduce the vector of likelihood ratios

$$\boldsymbol{Z}_T\left(\boldsymbol{u}\right) = \left(Z_T^{(1)}\left(u_1\right), \ldots, Z_T^{(k)}\left(u_k\right)\right), \qquad \boldsymbol{u} = \left(u_1, \ldots, u_k\right), \quad u_i \in \mathbb{U}_i$$

where

$$Z_T^{(1)}\left(u\right) = L\left(\vartheta_i + u/\sqrt{T}, \vartheta_1, X^T\right), \quad u \in \mathbb{U}_i = \left(\sqrt{T}\left(\alpha_i - \vartheta_i\right), \sqrt{T}\left(\beta_i - \vartheta_i\right)\right)$$

and the limit vector process

$$\boldsymbol{Z}\left(\boldsymbol{u}\right) = \left(Z^{(1)}\left(u_1\right), \ldots, Z^{(k)}\left(u_k\right)\right), \qquad \boldsymbol{u} = \left(u_1, \ldots, u_k\right) \in \mathscr{R}^k,$$

where

$$Z^{(i)}(u) = \exp\left\{u\,\zeta_i\,\mathrm{I}(\vartheta_i)^{1/2} - \frac{u^2}{2}\,\mathrm{I}(\vartheta_i)\right\}, \quad u \in \mathscr{R}$$

Suppose that we already proved the weak convergence of the process $\boldsymbol{Z}_T(\cdot)$ to the process $\boldsymbol{Z}(\cdot)$, then

$$\mathbf{P}_{\vartheta_1}^{(T)}\left\{\sqrt{T}\left(\hat\vartheta_T - \hat\vartheta^{(T)}\right) < x\right\} = \sum_{i=1}^k \mathbf{P}_{\vartheta_1}^{(T)}\left\{\sqrt{T}\left(\hat\vartheta_T^{(i)} - \vartheta_i\right) < x,\ \mathbb{H}_i^{(T)}\right\}$$

$$= \sum_{i=1}^k \mathbf{P}_{\vartheta_1}^{(T)}\left\{\sup_{u<x} Z_T^{(i)}(u) > \sup_{u\geq x} Z_T^{(i)}(u),\ \mathbb{H}_i^{(T)}\right\}$$

$$\longrightarrow \sum_{i=1}^k \mathbf{P}\left\{\sup_{u<x} Z^{(i)}(u) > \sup_{u\geq x} Z^{(i)}(u),\ \sup_{u\in\mathscr{R}} Z^{(i)}(u) > \max_{l\neq i}\sup_{u\in\mathscr{R}} Z^{(l)}(u)\right\}$$

$$= \sum_{i=1}^k \mathbf{P}\left\{\frac{\zeta_i}{\sqrt{\mathrm{I}(\vartheta_i)}} < x, |\zeta_i| > \max_{l\neq i}|\zeta_l|\right\}$$

$$= \sum_{i=1}^k \mathbf{P}\left\{\frac{\zeta_i}{\sqrt{\mathrm{I}(\vartheta_i)}} < x,\ \mathbb{H}_i\right\} = \mathbf{P}\left\{\hat\zeta < x\right\}.$$

To verify the weak convergence of the processes we have to check the convergence of finite-dimensional distributions and to establish two inequalities

$$\mathbf{E}_{\vartheta_1}\left|Z^{(i)}(u)^{1/2} - Z^{(i)}(v)^{1/2}\right|^2 \leq C\,|u-v|^2, \quad u,v \in \mathbb{U}_i$$

and

$$\mathbf{E}_{\vartheta_1} Z^{(i)}(u)^{1/2} \leq \frac{C}{|u|^N}, \quad u \in \mathbb{U}_i.$$

This can be done as in (2.38) and (2.39) (see as well [151], [139], Section 2.7).

It is interesting to note that the Bayes estimator

$$\tilde\vartheta_T = \left(\int_\Theta p(\vartheta)\,L\left(\vartheta, X^T\right)\,\mathrm{d}\vartheta\right)^{-1}\int_\Theta \vartheta\,p(\vartheta)\,L\left(\vartheta, X^T\right)\,\mathrm{d}\vartheta$$

has an entirely different asymptotic behavior. Below we change the variables $\vartheta = \vartheta_i + u/\sqrt{T}$:

$$\int_\Theta \vartheta\,p(\vartheta)\,L\left(\vartheta,\vartheta_1,X^T\right)\,\mathrm{d}\vartheta = \sum_{i=1}^k \int_{\Theta_i} \vartheta\,p(\vartheta)\,L\left(\vartheta,\vartheta_1,X^T\right)\,\mathrm{d}\vartheta$$

$$= \sum_{i=1}^k \int_{\mathbb{U}_i}\left(\vartheta_i + u/\sqrt{T}\right)\,p(\vartheta_i + u/\sqrt{T})\,L\left(\vartheta_i + u/\sqrt{T},\vartheta_i,X^T\right)\,\mathrm{d}u$$

$$= \sum_{i=1}^k \vartheta_i \int_{\mathbb{U}_i} p(\vartheta_i + u/\sqrt{T})\,Z_T^{(i)}(u)\,\mathrm{d}u$$

$$+T^{-1/2}\sum_{i=1}^{k}\int_{\mathbb{U}_i} u\,p(\vartheta_i+u/\sqrt{T})\;Z_T^{(i)}(u)\;\mathrm{d}u$$

$$\longrightarrow \sum_{i=1}^{k}\vartheta_i\,p(\vartheta_i)\int_{\mathscr{R}} Z^{(i)}(u)\;\mathrm{d}u=\sum_{i=1}^{k}\vartheta_i\,p(\vartheta_i)\,\sqrt{2\pi}\,\mathrm{I}(\vartheta_i)^{-1/2}\,\mathrm{e}^{\zeta_i^2/2}.$$

Hence it converges to the weighted mean value

$$\tilde{\vartheta}_T\Longrightarrow\tilde{\vartheta}=\sum_{i=1}^{k}\vartheta_i\,q_i,\qquad q_i=\frac{p(\vartheta_i)\;\mathrm{I}(\vartheta_i)^{-1/2}\,e^{\zeta_i^2/2}}{\sum_{l=1}^{k}p(\vartheta_l)\;\mathrm{I}(\vartheta_l)^{-1/2}\,e^{\zeta_l^2/2}}.$$

The asymptotic distribution of the random variables

$$\sqrt{T}\left(\tilde{\vartheta}_T-\tilde{\vartheta}^{(T)}\right)$$

with specially defined $\tilde{\vartheta}^{(T)}$ (close to $\tilde{\vartheta}$) can be described as well, but it is too cumbersome and we do not do it here. Note that we have to use $\tilde{\vartheta}^{(T)}$ and not $\tilde{\vartheta}$ because the last variable is not defined on the same probability space (as for MLE we used $\hat{\vartheta}^{(T)}$ and not $\hat{\vartheta}$). The similar problem for diffusion processes with small noise was considered in [139], Section 2.7, where the behavior of the MDE in such situations was studied too.

Note that MDE ϑ_T^* converges to another discrete random variable

$$\vartheta^*=\sum_{i=1}^{k}\vartheta_i\,\chi_{\{\mathbb{C}_i\}},$$

where the sets $\mathbb{C}_i$ are defined in the same way as the sets $\mathbb{H}_i$ but with the help of another Gaussian vector. The details can be found in [139], Theorems 7.11 and 7.12.

2.6.3 Null Fisher Information

The observed process is always ergodic diffusion

$$\mathrm{d}X_t=S(\vartheta,X_t)\;\mathrm{d}t+\sigma(X_t)\;\mathrm{d}W_t,\quad X_0,\quad 0\le t\le T$$

with a one-dimensional unknown parameter ϑ and the function $S(\vartheta,x)$ is smooth w.r.t. ϑ. Assume that all conditions of Theorem 2.8 are fulfilled except (2.17)

$$\inf_{\vartheta\in\Theta}\mathrm{I}(\vartheta)>0.$$

More precisely, the Fisher information is null at one point only and this point is the true value. Moreover, we suppose

$\mathcal{H}$. *The function* $S(\vartheta,\cdot)$ *is* $k+1$ *times continuously differentiable* w.r.t. ϑ, *the derivatives* $S^{(i)}(\vartheta,\cdot)\in\mathscr{P},\quad i=1,\ldots,k+1$, *the first* $k-1$ *derivatives at the point* ϑ_0 *are null:*

$$S^{(i)}(\vartheta_0, x) = 0, \quad i = 1, \ldots, k-1, \quad x \in \mathscr{R}$$

and

$$\mathrm{I}_k(\vartheta_0) = \frac{1}{(k!)^2} \int_{-\infty}^{\infty} \left(\frac{S^{(k)}(\vartheta_0, x)}{\sigma(x)} \right)^2 f(\vartheta_0, x)\, \mathrm{d}x > 0.$$

Introduce as well two random variables

$$\zeta_k \quad \sim \quad \mathcal{N}(0, \mathrm{I}_k(\vartheta_0)^{-1}), \qquad \zeta_k^+ = \max\{0, \zeta_k\}.$$

The properties of the MLE depend on the number k as follows.

Proposition 2.41. *Let the conditions of Theorem 2.8 be fulfilled except the condition (2.17), which is replaced by the condition $\mathcal{H}$. Then the MLE $\hat{\vartheta}_T$ is consistent and has the following limit distributions:*

- *if k is odd then*

$$\mathcal{L}_{\vartheta_0}\left\{T^{1/2k}\left(\hat{\vartheta}_T - \vartheta_0\right)\right\} \Longrightarrow \mathcal{L}\left\{(\zeta_k)^{1/k}\right\},$$

- *if k is even then*

$$\mathcal{L}_{\vartheta_0}\left\{T^{1/2k}\left(\hat{\vartheta}_T - \vartheta_0\right)\right\} \Longrightarrow \mathcal{L}\left\{(\zeta_k^+)^{1/k}\right\}.$$

Proof. We just explain how such limits arise. The detailed proof of a similar result can be found in [139], Section 2.4 (in the asymptotics of small noise). In the present situation the proof can be carried out in the same way.

The likelihood ratio

$$Z_T(u) = L\left(\vartheta_0 + uT^{-1/2k}, \vartheta_0, X^T\right)$$

admits the representation

$$Z_T(u) = \exp\left\{u^k\, \Delta_T(\vartheta_0, X^T) - \frac{u^{2k}}{2}\, \mathrm{I}_k(\vartheta_0) + r_T(\vartheta_0, u, X^T)\right\}$$

where

$$\Delta_T(\vartheta_0, X^T) = \frac{1}{k!\sqrt{T}} \int_0^T \frac{S^{(k)}(\vartheta_0, X_t)}{\sigma(X_t)^2}\, [\mathrm{d}X_t - S(\vartheta_0, X_t)\, \mathrm{d}t]$$

is asymptotically normal

$$\mathcal{L}_{\vartheta_0}\{\Delta_T(\vartheta_0, X^T)\} \Longrightarrow \mathcal{L}\{\Delta_k\} \sim \mathcal{N}(0, \mathrm{I}_k(\vartheta_0))$$

and $r_T(\vartheta_0, u, X^T) \to 0$ in probability.

The limit process

$$Z(u) = \exp\left\{u^k\, \Delta_k - \frac{u^{2k}}{2}\, \mathrm{I}_k(\vartheta_0)\right\}, \qquad u \in \mathscr{R}$$

takes its maximal value at the point $\hat{u}$ which is

- for k even, $\hat u = \left(\Delta_k/\mathrm{I}_k\left(\vartheta_0\right)\right)^{1/k}$ if $\Delta_k > 0$ and $\hat u = 0$ if $\Delta_k \leq 0$,
- for k odd, $\hat u = \left(\Delta_k/\mathrm{I}_k\left(\vartheta_0\right)\right)^{1/k}$.

Hence if we prove the weak convergence of the process $Z_T\left(\cdot\right)$ to the process $Z\left(\cdot\right)$, then we obtain the desired properties of the MLE. The convergence of finite-dimensional distributions follows from the above-mentioned representation of the likelihood ratio. The estimates

$$\mathbf{E}_{\vartheta_0}\left|Z_T^{1/2}\left(u\right)-Z_T^{1/2}\left(v\right)\right|^2 \leq C\,\left(u-v\right)^{2k}, \tag{2.162}$$

$$\mathbf{E}_{\vartheta_0}\,Z_T^{1/2}\left(u\right) < \frac{C_m}{\left|u\right|^m} \tag{2.163}$$

can be obtained by the same way as it was done in Theorem 2.8. In particular, to verify (2.162) we use the estimate (1.23) which gives us

$$\mathbf{E}_{\vartheta_0}\left|Z_T^{1/2}\left(u\right)-Z_T^{1/2}\left(v\right)\right|^2 \leq \mathbf{E}_{\vartheta_u}\int_0^T\left[\frac{S(\vartheta_u,X_t)-S(\vartheta_v,X_t)}{\sigma(X_t)}\right]^2\,\mathrm{d}t + \mathbf{E}_{\vartheta_v}\int_0^T\left[\frac{S(\vartheta_u,X_t)-S(\vartheta_v,X_t)}{\sigma(X_t)}\right]^2\,\mathrm{d}t,$$

where $\vartheta_v = \vartheta_0 + T^{-1/2k}v$. Further

$$\begin{aligned}
\mathbf{E}_{\vartheta_u}&\int_0^T\left[\frac{S(\vartheta_u,X_t)-S(\vartheta_v,X_t)}{\sigma(X_t)}\right]^2\,\mathrm{d}t\\
&=T\int_{-\infty}^{\infty}\left[\frac{S(\vartheta_u,x)-S(\vartheta_v,x)}{\sigma(x)}\right]^2\,f(\vartheta_u,x)\,\mathrm{d}x\\
&=T^{1-1/k}(u-v)^2\int_{-\infty}^{\infty}\left[\frac{\dot S(\vartheta+\gamma T^{-1/2k}(u-v),x)}{\sigma(x)}\right]^2\,f(\vartheta_u,x)\,\mathrm{d}x\\
&=T^{\frac{k-1}{k}}\left(u-v\right)^2\int_{-\infty}^{\infty}\left[\frac{\dot S(\vartheta_0,x)}{\sigma(x)}+\sum_{j=1}^{k-2}\gamma^j\left(u-v\right)^j\,\frac{S^{(j+1)}(\vartheta_0,x)}{\sigma(x)j!}\,T^{j/2k}\right.\\
&\qquad\left.+\gamma^{k-1}\left(u-v\right)^{k-1}\frac{S^{(k)}(\bar\vartheta_0,x)}{\sigma(x)(k-1)!}\,T^{1-(k-1)/2k}\right]^2\,f(\vartheta_u,x)\,\mathrm{d}x\\
&\leq\frac{\left(u-v\right)^{2k}}{\left((k-1)!\right)^2}\int_{-\infty}^{\infty}\left(\frac{S^{(k)}(\bar\vartheta_0,x)}{\sigma(x)}\right)^2\,f(\vartheta_u,x)\,\mathrm{d}x\leq C\,\left(u-v\right)^{2k}
\end{aligned}$$

because $S^{(k)}(\vartheta,\cdot),\sigma(\cdot)^{-1}\in\mathscr{P}$ and the function $f(\vartheta,x)$ has exponentially decreasing tails.

To verify (2.163) we note that there exists a vicinity $\left|\vartheta-\vartheta_0\right|<\nu$ of ϑ_0 such that

$$\inf_{\left|\vartheta-\vartheta_0\right|<\nu}\mathbf{E}_{\vartheta_0}\left(\frac{S(\vartheta,x)-S(\vartheta_0,x)}{\sigma(x)}\right)^2\geq\frac{\mathrm{I}_k(\vartheta_0)}{2}\,\left|\vartheta-\vartheta_0\right|^{2k}.$$

Hence we can use the proof of Lemma 2.11. Further details can be found in [139], Theorems 2.9 and 2.10 (in the asymptotic of small noise).

2.6.4 Optimal Observation Window

In the problems studied before it was supposed that an ergodic diffusion process

$$\mathrm{d}X_t = S(\vartheta, X_t)\ \mathrm{d}t + \sigma(X_t)\ \mathrm{d}W_t,\quad X_0,\quad 0 \le t \le T$$

can be observed on the whole line $\mathscr{R}$, i.e., all the values of X_t (even very large) can be observed. Now we consider the problem with a bounded *observation window* $\mathbb{A}$ and we suppose that the Borel set $\mathbb{A} \subset \mathscr{R}$ belongs to the class of sets $\mathscr{A}_\lambda$ of Lebesgue measure less or equal to some given value $\lambda > 0$.

The goal is to obtain an *optimal window in the class*, i.e., the window which provides the minimal asymptotic variance of the best estimator. The solution of this problem is almost evident and can be described heuristically as follows. Suppose we have already found such a window $\mathbb{A}^*$, then the MLE $\hat\vartheta_{T,\mathbb{A}^*}$, which is usually asymptotically efficient,can be constructed with the help of the likelihood ratio

$$L(\vartheta, X^T, \mathbb{A}^*) = \exp\left\{\int_0^T \frac{S(\vartheta, X_t)}{\sigma(X_t)^2}\,\chi_{\{X_t\in\mathbb{A}^*\}}\ \mathrm{d}X_t - \frac{1}{2}\int_0^T \left(\frac{S(\vartheta, X_t)}{\sigma(X_t)}\right)^2 \chi_{\{X_t\in\mathbb{A}^*\}}\mathrm{d}t\right\}$$

is asymptotically normal

$$\sqrt{T}\left(\hat\vartheta_{T,\mathbb{A}^*} - \vartheta\right) \Longrightarrow \mathcal{N}\left(0, \mathrm{I}(\vartheta, \mathbb{A}^*)^{-1}\right),$$

where the Fisher information

$$\mathrm{I}(\vartheta, \mathbb{A}^*) = \int_{\mathbb{A}^*}\left(\frac{\dot S(\vartheta, x)}{\sigma(x)}\right)^2 f(\vartheta, x)\ \mathrm{d}x.$$

The minimum of limit variance corresponds to the maximum of the Fisher information. Therefore we have the equality

$$\sup_{\mathbb{A}\in\mathscr{A}_\lambda}\int_{\mathbb{A}}\left(\frac{\dot S(\vartheta, x)}{\sigma(x)}\right)^2 f(\vartheta, x)\ \mathrm{d}x = \mathrm{I}(\vartheta, \mathbb{A}^*).$$

To solve this last maximization problem we consider the *level sets*

$$\mathbb{A}(\vartheta, r) = \left\{x : \left(\frac{\dot S(\vartheta, x)}{\sigma(x)}\right)^2 f(\vartheta, x) \ge r\right\},\qquad r > 0$$

and suppose that the equation

$$\text{meas}\left\{\mathbb{A}(\vartheta, r)\right\} = \lambda$$

has a unique solution $r = r_*$ (meas $\{\mathbb{A}\}$ means the Lebesgue measure of $\mathbb{A}$). Of course, $r_* = r_*(\vartheta)$. The class of sets $\{\mathbb{A}(\vartheta, r_*),\ \vartheta \in \Theta\}$ we denote by $\hat{\mathscr{A}}_\lambda$. The construction of the set $\mathbb{A}^* = \mathbb{A}(\vartheta, r_*)$ depends on the unknown parameter ϑ. Hence to choose the set $\mathbb{A}$ and estimate ϑ simultaneously is generally impossible.

Note that sometimes this problem can have trivial solutions. For example, suppose that $S(\vartheta, x) = h(\vartheta, x) + g(x)$, where the function $h(\vartheta, x)$ has compact support, i.e., there exists a bounded set $\mathbb{C} \subset \mathscr{R}$ such that $h(\vartheta, x) = 0$ for $x \in \mathbb{C}^c$. If meas$\{\mathbb{C}\} \leq \lambda$, then we can take $\mathbb{A}^* = \mathbb{C}$. In any case the set $\mathbb{N} = \left\{x : \dot{S}(\vartheta, x) = 0, \text{ for all } \vartheta \in \Theta\right\}$ can always be excluded from the observation window. We have another simplification if the integral in the Fisher information does not depend on ϑ (see the example below).

It is clear now that we have to estimate first the parameter ϑ and then use this estimate to construct the set $\mathbb{A}_T$, and having this estimated $\mathbb{A}_T$ we can finally estimate ϑ. Then if the preliminary estimator of the parameter is consistent, then the couple (set and estimator) can be asymptotically optimal. Note that all these are heuristical considerations only and any other choice of the windows (in the class $\mathscr{A}_\lambda$) and estimators are "admitted to the competition" in the lower bound.

We consider the following two-step procedures . First we have the *period of learning* on the interval $\left[0, \sqrt{T}\right]$, i.e., having observations on this interval $X^{\sqrt{T}} = \left\{X_t, 0 \leq t \leq \sqrt{T}\right\}$ we construct a set $\bar{\mathbb{A}}_T \in \mathscr{A}_\lambda$, and then we estimate the parameter by the observations hitting in the set $\bar{\mathbb{A}}_T$, i.e., by observations $Y_t = X_t\, \chi_{\{X_t \in \bar{\mathbb{A}}_T\}}$, $\sqrt{T} \leq t \leq T$. The couple $\left\{\bar{\mathbb{A}}_T, \bar{\vartheta}_T\right\}$ or $\bar{\vartheta}_{T,\bar{\mathbb{A}}_T}$ we call an *admissible strategy*.

Therefore we have two different problems. The first one is to construct a lower minimax bound on the risks of all admissible strategies and the second problem is to find an admissible strategy (windows + estimator) which achieves this bound.

This statement of the problem corresponds to the so-called *scheme of series*, when for each T we have a different problem.

Regularity conditions $\mathcal{I}$.

$\mathcal{I}_1$. *The function $S(\vartheta, \cdot)$ is two times continuously differentiable over ϑ, the derivatives $\dot{S}(\vartheta, \cdot), \ddot{S}(\vartheta, \cdot) \in \mathscr{P}$.*

$\mathcal{I}_2$. *The Fisher information $\mathrm{I}(\vartheta, \mathbb{A}^*)$ is a continuous function of θ, is uniformly positive:*

$$\inf_{\vartheta \in \Theta} \mathrm{I}(\vartheta, \mathbb{A}^*) > 0 \tag{2.164}$$

and for any $\vartheta_0 \in \Theta$, set $\mathbb{A}_0^* = \mathbb{A}(\vartheta_0, r_*(\vartheta_0))$ and $\nu > 0$

$$\inf_{|\vartheta-\vartheta_0|>\nu}\ \inf_{\mathbb{A}\in\hat{\mathscr{A}}} \mathbf{E}_{\vartheta_0}\left(\frac{S(\vartheta,\xi)-S(\vartheta_0,\xi)}{\sigma(\xi)}\right)^2 \chi_{\{\xi\in\mathbb{A}\}} > 0.$$

Remember that the set $\mathbb{A}^* = \mathbb{A}(\vartheta, r_*(\vartheta)) \in \hat{\mathscr{A}}_\lambda$.

Proposition 2.42. *Let the conditions $\mathcal{A}_0(\Theta)$, $\mathcal{I}_1$ and (2.164) be fulfilled, then*

$$\lim_{\delta\to 0}\lim_{T\to\infty}\ \inf_{\{\bar{\mathbb{A}}_T,\,\bar{\vartheta}_T\}}\ \sup_{|\vartheta-\vartheta_0|<\delta} T\,\mathbf{E}_\vartheta\left(\bar{\vartheta}_{T,\bar{\mathbb{A}}_T}-\vartheta\right)^2 \geq \mathrm{I}(\vartheta_0,\mathbb{A}_0^*)^{-1}. \quad (2.165)$$

Proof. Fix an admissible strategy $\{\bar{\mathbb{A}}_T,\ \bar{\vartheta}_T\}$ and consider the parameter estimation problem by observations $Y_t = X_t\,\chi_{\{X_t\in\bar{\mathbb{A}}_T\}}$, $\sqrt{T}\leq t\leq T$. Note that the contribution of the observations $X^{\sqrt{T}} = \left\{X_t, 0\leq t\leq \sqrt{T}\right\}$ even if we take $\mathbb{A}=\mathscr{R}$ and the asymptotically efficient MLE is negligible with respect to observations on the interval $[\sqrt{T},T]$. Therefore we consider observations on the interval $[\sqrt{T},T]$ only. The likelihood ratio can be written as

$$L(\vartheta, X_{\sqrt{T}}^T, \bar{\mathbb{A}}_T) = \tilde{f}\left(\vartheta, X_{\sqrt{T}}\right)\ \exp\left\{\int_{\sqrt{T}}^T \frac{S(\vartheta,X_t)}{\sigma(X_t)^2}\,\chi_{\{X_t\in\bar{\mathbb{A}}_T\}}\ \mathrm{d}X_t \right.$$
$$\left. -\frac{1}{2}\int_{\sqrt{T}}^T \left(\frac{S(\vartheta,X_t)}{\sigma(X_t)}\right)^2 \chi_{\{X_t\in\bar{\mathbb{A}}_T\}}\mathrm{d}t\right\}.$$

Here $\tilde{f}\left(\vartheta, X_{\sqrt{T}}\right) = f\left(\vartheta, X_{\sqrt{T}}\right)$ if $X_{\sqrt{T}}\in\bar{\mathbb{A}}_T$, otherwise $\tilde{f}\left(\vartheta, X_{\sqrt{T}}\right) = 1$. Hence the Fisher information is

$$\mathrm{I}_T\left(\vartheta,\bar{\mathbb{A}}_T\right) = \mathbf{E}_\vartheta\left(\frac{\partial}{\partial\vartheta}\ln L(\vartheta, X_{\sqrt{T}}^T, \bar{\mathbb{A}}_T)\right)^2$$
$$= \mathrm{I}_0(\vartheta) + \left(T-\sqrt{T}\right)\ \mathbf{E}_\vartheta\left(\frac{\dot{S}(\vartheta,\xi)}{\sigma(\xi)}\right)^2 \chi_{\{\xi\in\bar{\mathbb{A}}_T\}} = T\,\mathrm{I}\left(\vartheta,\bar{\mathbb{A}}_T\right)\ (1+o(1)).$$

Remember that the indicator $\chi_{\{X_t\in\bar{\mathbb{A}}_T\}}$ is $\mathfrak{F}_{\sqrt{T}}$ measurable.

Let $p(u), u\in\mathscr{R}$ be a continuously differentiable density with support $[-1,1]$ (hence $p(u)=0$ for $|u|=1$). Introduce as well the rescaled density

$$p_T(\vartheta) = \frac{H}{\sqrt{T}}\,p\left(\sqrt{T}(\vartheta-\vartheta_0)\right)$$

and put $\mathbb{B}_T = \{\vartheta : |\sqrt{T}(\vartheta-\vartheta_0)|<1\}$. Using the van Trees inequality (1.105) we obtain for any $\delta>0$ and $T>H^2\delta^{-2}$ the following estimate:

$$\sup_{|\vartheta-\vartheta_0|<\delta} \mathbf{E}_\vartheta \left(\bar\vartheta_{T,\bar{\mathbb{A}}_T}-\vartheta\right)^2 \geq \sup_{\vartheta\in\mathbb{B}_T} \mathbf{E}_\vartheta \left(\bar\vartheta_{T,\bar{\mathbb{A}}_T}-\vartheta\right)^2$$

$$\geq \int_{\mathbb{B}_T} \mathbf{E}_\vartheta \left(\bar\vartheta_{T,\bar{\mathbb{A}}_T}-\vartheta\right)^2 p_T(\vartheta)\,\mathrm{d}\vartheta$$

$$\geq \left\{T\int_{\mathbb{B}_T} \mathrm{I}\left(\vartheta,\bar{\mathbb{A}}_T\right)\, p_T(\vartheta)\,\mathrm{d}\vartheta\ (1+o(1))+T\,\mathrm{I}_p/H\right\}^{-1},$$

where

$$\mathrm{I}_p=\int_{-1}^{1}\frac{\dot p(u)^2}{p(u)}\,\mathrm{d}u$$

is the Fisher information corresponding to the density $p(\cdot)$. For $\vartheta\in\mathbb{B}_T$ we have as well the estimate

$$\mathbf{E}_\vartheta\left(\frac{\dot S(\vartheta,\xi)}{\sigma(\xi)}\right)^2\chi_{\{\xi\in\bar{\mathbb{A}}_T\}}\leq \mathbf{E}_{\vartheta_0}\left(\frac{\dot S(\vartheta_0,\xi)}{\sigma(\xi)}\right)^2\chi_{\{\xi\in\bar{\mathbb{A}}_T\}}+\varepsilon_T$$
$$\leq \mathrm{I}(\vartheta_0,\bar{\mathbb{A}}_T)+\varepsilon_T\leq \mathrm{I}(\vartheta_0,\mathbb{A}_0^*)+\varepsilon_T,$$

where $\varepsilon_T\to 0$ as $T\to\infty$ because the function $\mathrm{I}(\vartheta,\bar{\mathbb{A}}_T)$ is continuous at the point ϑ_0.

Hence

$$\sup_{|\vartheta-\vartheta_0|<\delta} T\,\mathbf{E}_\vartheta\left(\bar\vartheta_{T,\bar{\mathbb{A}}_T}-\vartheta\right)^2\geq\left(\mathrm{I}(\vartheta_0,\mathbb{A}_0^*)\ (1+o(1))+\varepsilon_T+\mathrm{I}(p)/H\right)^{-1}.$$

Letting now $T\to\infty$ and $H\to\infty$ we obtain (2.165).

This inequality allows us to introduce the following definition.

Definition 2.43. *Let the conditions $\mathcal{A}_0(\Theta)$, $\mathcal{I}_1$ and (2.164) be fulfilled then we call an admissible strategy $\{\bar{\mathbb{A}}_T,\bar\vartheta_T\}$ asymptotically optimal if for all $\vartheta_0\in\Theta$*

$$\lim_{\delta\to 0}\lim_{T\to\infty}\sup_{|\vartheta-\vartheta_0|<\delta} T\,\mathbf{E}_\vartheta\left(\bar\vartheta_{T,\bar{\mathbb{A}}_T}-\vartheta\right)^2=\mathrm{I}(\vartheta_0,\mathbb{A}_0^*)^{-1}.$$

We construct the asymptotically optimal strategy as follows. Fix some value $\vartheta_*\in\Theta$ and the corresponding set $\mathbb{A}_*=\mathbb{A}\left(\vartheta_*,r\left(\vartheta_*\right)\right)\in\mathscr{A}_\lambda$. Then we introduce the likelihood ratio

$$L(\vartheta,X^{\sqrt{T}},\mathbb{A}_*)=\exp\left\{\int_0^{\sqrt{T}}\frac{S(\vartheta,X_t)}{\sigma(X_t)^2}\,\chi_{\{X_t\in\mathbb{A}_*\}}\,\mathrm{d}X_t\right.$$
$$\left.-\frac{1}{2}\int_0^{\sqrt{T}}\left(\frac{S(\vartheta,X_t)}{\sigma(X_t)}\right)^2\chi_{\{X_t\in\mathbb{A}_*\}}\,\mathrm{d}t\right\}$$

and the MLE

$$\hat\vartheta_{\sqrt{T}}=\arg\sup_{\vartheta\in\Theta}L(\vartheta,X^{\sqrt{T}},\mathbb{A}^*).$$

Finally, we put $\mathbb{A}_T^* = \mathbb{A}\left(\hat{\vartheta}_{\sqrt{T}}, r\left(\hat{\vartheta}_{\sqrt{T}}\right)\right)$ and

$$\hat{\vartheta}_T = \arg\sup_{\vartheta\in\Theta} L(\vartheta, X^T_{\sqrt{T}}, \mathbb{A}_T^*).$$

We have the following

Proposition 2.44. *Let the conditions* $\boldsymbol{\mathcal{A}}_0(\Theta), \boldsymbol{\mathcal{I}}$ *be fulfilled, then* $\{\mathbb{A}_T^*, \hat{\vartheta}_T\}$ *the strategy is admissible and asymptotically optimal.*

Proof. First note that by conditions of the proposition the MLE $\hat{\vartheta}_{\sqrt{T}}$ is uniformly consistent. As the identifiability condition $\mathcal{I}_2$ is uniform over ϑ and $\mathbb{A} \in \mathscr{A}_\lambda$, then the MLE $\hat{\vartheta}_T$ is uniformly consistent and uniformly on compacts $\mathbb{K} \subset \Theta$ asymptotically normal. Moreover, we have the uniform convergence of moments

$$\sup_{\vartheta\in\mathbb{K}} T\ \mathbf{E}_\vartheta\left(\hat{\vartheta}_{T,\mathbb{A}_T^*} - \vartheta\right)^2 \to \sup_{\vartheta\in\mathbb{K}} \mathrm{I}\left(\vartheta, \mathbb{A}^*\right)^{-1}.$$

Hence

$$\sup_{|\vartheta-\vartheta_0|<\delta} T\ \mathbf{E}_\vartheta\left(\hat{\vartheta}_{T,\mathbb{A}_T^*} - \vartheta\right)^2 \to \sup_{|\vartheta-\vartheta_0|<\delta} \mathrm{I}\left(\vartheta, \mathbb{A}^*\right)^{-1} \to \mathrm{I}\left(\vartheta_0, \mathbb{A}_0^*\right)^{-1}$$

as $\delta \to 0$.

Example 2.45. (*Shift estimation*) Let us consider the problem of the optimal choice of window for the process

$$\mathrm{d}X_t = -\left(X_t - \vartheta\right)^3 \mathrm{d}t + \sigma\,\mathrm{d}W_t, \qquad X_0.$$

The invariant density is

$$f\left(x,\vartheta\right) = f\left(x-\vartheta\right) = G^{-1}\exp\left\{-\frac{\left(x-\vartheta\right)^4}{2\,\sigma^2}\right\}, \quad G = \int_{\mathscr{R}} e^{-z^4/2\sigma^2}\mathrm{d}z,$$

and the Fisher information

$$\mathrm{I}\left(\vartheta\right) = \frac{9}{\sigma^2\,G}\int_{\mathscr{R}}\left(x-\vartheta\right)^4\,\mathrm{e}^{-\frac{(x-\vartheta)^4}{2\,\sigma^2}}\,\mathrm{d}x = \frac{9}{\sigma^2\,G}\int_{\mathscr{R}} x^4\,\mathrm{e}^{-\frac{x^4}{2\sigma^2}}\,\mathrm{d}x$$

does not depend on ϑ. Therefore the optimal window is the level set of two intervals

$$\mathbb{A}^* = \left[-\left(x_* + \lambda/2\right), -x_*\right] \cup \left[x_*, \left(x_* + \lambda/2\right)\right],$$

where x_* is the unique positive solution of the equation

$$\frac{x_*}{x_* + \lambda/2} = \exp\left\{\frac{x_*^4 - \left(x_* + \lambda/2\right)^4}{8\sigma^2}\right\}. \tag{2.166}$$

As the Fisher information does not depend on ϑ we need not *learn* on the interval $\left[0, \sqrt{T}\right]$ and can immediately observe on the optimal window $\mathbb{A}^*$ and construct the corresponding MLE by the observations $Y^T = \left\{X_t \chi_{\{X_t \in \mathbb{A}^*\}}, \ 0 \leq t \leq T\right\}$.

Example 2.46. Let us suppose that the observed process is

$$\mathrm{d}X_t = -\vartheta \, X_t^3 \, \mathrm{d}t + \sigma \, \mathrm{d}W_t, \qquad X_0$$

where $\vartheta \in (\alpha, \beta)$, $\alpha > 0$. The invariant density is

$$f(x, \vartheta) = \frac{\vartheta^{1/4}}{G} \exp\left\{-\frac{\vartheta \, x^4}{2 \, \sigma^2}\right\}, \qquad x \in \mathscr{R}, \quad G = \int_{\mathscr{R}} \mathrm{e}^{-z^4/2\sigma^2} \mathrm{d}z,$$

and the Fisher information

$$\mathrm{I}(\vartheta) = \frac{\vartheta^{1/4}}{\sigma^2 \, G} \int_{\mathscr{R}} x^6 \, \mathrm{e}^{-\frac{\vartheta \, x^4}{2\sigma^2}} \, \mathrm{d}x$$

The optimal window is the same two-intervals set

$$\mathbb{A}_T^* = \left[-\left(x_* + \lambda/2\right), -x_*\right] \cup \left[x_*, \left(x_* + \lambda/2\right)\right],$$

but x_* is a solution of the equation similar to (2.166)

$$\frac{x_*}{x_* + \lambda/2} = \exp\left\{\hat{\vartheta}_{T,\mathbb{A}_{\sqrt{T}}} \, \frac{x_*^4 - \left(x_* + \lambda/2\right)^4}{12 \, \sigma^2}\right\}.$$

where the preliminary MLE is

$$\hat{\vartheta}_{T,\mathbb{A}_{\sqrt{T}}} = -\frac{\int_0^{\sqrt{T}} X_t^3 \, \chi_{\{|X_t| \leq \lambda/2\}} \, \mathrm{d}X_t}{\int_0^{\sqrt{T}} X_t^6 \, \chi_{\{|X_t| \leq \lambda/2\}} \, \mathrm{d}t}.$$

2.6.5 Asymptotic Expansions

In the regular case many estimators are consistent and asymptotically normal and even asymptotically efficient (MLE, BE, one-step MLE). Therefore many of them are asymptotically equivalent up to the first two terms in the following sense. Let the observed process be

$$\mathrm{d}X_t = S(\vartheta, X_t) \, \mathrm{d}t + \sigma(X_t) \, \mathrm{d}W_t, \quad X_0, \qquad 0 \leq t \leq T \qquad (2.167)$$

where $\vartheta \in \Theta = (\alpha, \beta)$ and the initial value X_0 has a density function $f_*(\vartheta, x)$. We suppose, as usual, that the coefficients $S(\cdot)$ and $\sigma(\cdot)$ are such that the conditions $\mathcal{ES}, \mathcal{EM}$ and $\mathcal{A}_0(\Theta)$ are fulfilled. Hence the process $X_t, t \geq 0$ has invariant measure with density function $f(\vartheta, x)$ and if $f_*(\vartheta, x) = f(\vartheta, x)$, then the process is stationary.

Let $\bar\vartheta_T$ be one of the above-mentioned estimators, then (under regularity conditions) we can write

$$\bar\vartheta_T = \vartheta + \bar u_T\, T^{-1/2} + o\left(T^{-1/2}\right), \tag{2.168}$$

where

$$\bar u_T = \frac{\Delta_T\left(\vartheta, X^T\right)}{\mathrm{I}(\vartheta)}, \qquad \Delta_T\left(\vartheta, X^T\right) = \int_0^T \frac{\dot S(\vartheta, X_t)}{\sigma(X_t)}\,\mathrm{d}W_t.$$

Remember, that

$$\Delta_T\left(\vartheta, X^T\right) \Longrightarrow \mathcal{N}(0, \mathrm{I}(\vartheta)), \qquad \mathrm{I}(\vartheta) = \mathbf{E}_\vartheta\left(\frac{\dot S(\vartheta, \xi)}{\sigma(\xi)}\right)^2.$$

Hence one way to compare these estimators is to calculate the next terms in the expansion of $\bar\vartheta_T$ by the powers of $T^{-1/2}$, i.e., to obtain the representation (*stochastic expansion*) like

$$\bar\vartheta_T = \vartheta + \sum_{l=1}^{k} \bar u_{l,T}\, T^{-l/2} + o\left(T^{-k/2}\right), \tag{2.169}$$

where $k \geq 1$ is some given number and $\bar u_{l,T}$ are bounded in probability random variables. Then if we have as well the corresponding expansion of the moments, then we can at least compare the expansions of the mean square errors

$$T\ \mathbf{E}_\vartheta\left(\bar\vartheta_T - \vartheta\right)^2 = \mathrm{I}(\vartheta) + \sum_{l=1}^{k} \bar A_l(\vartheta)\ T^{-l/2} + o\left(T^{-k/2}\right).$$

The more detailed analysis can allow us to obtain the asymptotic expansions of the distribution functions

$$\mathbf{P}_\vartheta^{(T)}\left\{T^{-1/2}\mathrm{I}(\vartheta)^{1/2}\left(\bar\vartheta_T - \vartheta\right) < y\right\} = \Phi(y) + \sum_{l=1}^{k} \bar F_l(\vartheta, y)\ T^{-l/2}$$
$$+ o\left(T^{-k/2}\right).$$

Here $\Phi(\cdot)$ is the distribution function of $\mathcal{N}(0,1)$.

For the MLE this program (stochastic expansion of the MLE and of its distribution function) was realized by Yoshida *et al.* [245], [212], [131], [213], [246].

Remember as well that another advantage of the asymptotic expansions is the possibility to know better the properties of the estimators for *moderate* values of T (because the error of the approximation is of the order of

$T^{-k/2}$). The asymptotic expansions of the estimators and of their distribution functions are well known in classical statistics (see Pfanzagl [197], [198] and references therein).

For diffusion processes (a signal in the white Gaussian noise model) such expansions (stochastic expansion, expansion of the moments and distribution functions) were first obtained by Burnashev [41], [42], who developed the *method (or approach) of good sets* (MGS) and obtained the stochastic expansions for the maximum likelihood and Bayesian estimators. A similar stochastic expansion of the MLE in the case of diffusion processes with small diffusion coefficients is given in [135] (see as well [139], Chapter 3), where we used the same MGS. Further it was Yoshida [243], [244], who obtained for this model the stochastic expansions of the MLE and BE and the expansions of their distribution functions in essentially a more general situation. Here we discuss the possibility of such expansions in the case of an ergodic diffusion process.

The representation (2.168) was not really proved in this chapter (except Section 2.6.1, where we used MGS) but it can be obtained quite directly using MGS. Below we illustrate this method by obtaining (2.169) with $k = 2$, i.e., we seek the next after asymptotically Gaussian term. Let $\bar\vartheta_T = \hat\vartheta_T$ (MLE).

Proposition 2.47. *Suppose that*

- *the conditions*

$$\lim_{|x|\to\infty} \operatorname{sgn}(x) \frac{S(\vartheta, x)}{\sigma(x)^2} < 0$$

  and

$$\inf_{|\theta-\vartheta|>\nu} \mathbf{E}_\vartheta \left(\frac{S(\theta,\xi) - S(\vartheta,\xi)}{\sigma(\xi)}\right)^2 > 0$$

  are fulfilled for any $\nu > 0$ and $\vartheta \in \Theta$,
- *the function $S(\theta, \cdot)$ is five times continuously differentiable w.r.t. ϑ, the derivatives $S^{(i)}(\theta, \cdot), l = 1, \ldots, 5$ and $\sigma(\cdot)^{-1}$ belong to the class $\mathcal{P}$.*

Then there exist a set $\mathbb{B}$ such that the MLE admits the representation

$$\sqrt{T}\left(\hat\vartheta_T - \vartheta\right) = \left\{\frac{\eta_1}{\mathrm{I}(\vartheta)} + \frac{\eta_1(\eta_2 - \eta_3)\,\mathrm{I}(\vartheta) - \eta_1^2\,\rho_{1,2}(\vartheta) - K\,\mathrm{I}(\vartheta)^2}{\mathrm{I}(\vartheta)^3\sqrt{T}} + \frac{\eta_T}{T^{3/4}}\right\}\chi_{\{\mathbb{B}\}} + \hat u_T\,\chi_{\{\mathbb{B}^c\}}, \tag{2.170}$$

where $\left|\eta_T\right| < 1$ and the other quantities are defined below in (2.172)–(2.175).

Moreover, for any $p > 0$ there exists a constant $C > 0$ such that

$$\mathbf{P}_\vartheta^{(T)}\{\mathbb{B}^c\} \le \frac{C}{T^p}. \tag{2.171}$$

Proof. We follow the same schema as in [135] (see as well [139], Section 3.1). Let us introduce the first *good set*

$$\mathbb{B}_1 = \left\{ \omega : \left| \hat{\vartheta}_T - \vartheta \right| < T^{-\gamma} \right\},$$

where $\gamma \in (0, 1/2)$ is some fixed value. Then for $\omega \in \mathbb{B}_1$ the estimator $\hat{\vartheta}_T$ is one of the solutions of the maximum likelihood equation

$$K(\theta, X_0) + \int_0^T \frac{\dot{S}(\theta, X_t)}{\sigma(X_t)^2} \left[\mathrm{d}X_t - S(\theta, X_t)\,\mathrm{d}t\right] = 0, \quad \theta \in \left(\vartheta - T^{-\gamma}, \vartheta + T^{-\gamma}\right),$$

where $K(\vartheta, x) = \partial \ln f_* (\vartheta, x)/\partial\vartheta$. If we put $v = \vartheta - \theta$, then we can write this equation as

$$\begin{aligned} &\varepsilon^2 K(\vartheta + v, X_0) + \frac{\varepsilon}{\sqrt{T}} \int_0^T \frac{\dot{S}(\vartheta + v, X_t)}{\sigma(X_t)} \,\mathrm{d}W_t \\ &\quad - \frac{1}{T} \int_0^T \frac{\dot{S}(\vartheta + v, X_t)\left[S(\vartheta + v, X_t) - S(\vartheta, X_t)\right]}{\sigma(X_t)^2} \,\mathrm{d}t = 0, \qquad |v| < T^{-\gamma}, \end{aligned}$$

where $\varepsilon = T^{-1/2}$. Remember, that the solutions of this equation depend on the whole trajectory X^T. Therefore it is preferable to use the form of the MLEq without the stochastic integral (1.48):

$$\begin{aligned} &\frac{\varepsilon}{\sqrt{T}} \int_{X_0}^{X_T} \frac{\dot{S}(\vartheta + v, y)}{\sigma(y)^2} \,\mathrm{d}y - \frac{1}{T} \int_0^T \frac{\dot{S}(\vartheta + v, X_t)\ S(\vartheta + v, X_t)}{\sigma(X_t)^2} \,dt \\ &\quad - \frac{\varepsilon}{2\sqrt{T}} \int_0^T \left(\frac{\dot{S}(\vartheta + v, X_t)}{\sigma(X_t)^2} \right)' \sigma(X_t)^2 \,\mathrm{d}t + \varepsilon^2 K(\vartheta + v, X_0) = 0. \end{aligned}$$

Here $h(X_t)' = h(x)'\big|_{x = X_t}$.

Let us denote the left hand side of this equation as $\lambda(v, \varepsilon)$, and note that the equation

$$\lambda(v, \varepsilon) = 0,$$

can have many solutions on the interval $\left[-T^{-\gamma}, T^{-\gamma}\right]$, but if we introduce the third *good set*

$$\mathbb{B}_2 = \left\{ \omega : \sup_{|v| \leq T^{-\gamma}} \frac{\partial \lambda(v, \varepsilon)}{\partial v} < -\frac{\mathrm{I}(\vartheta)}{2} \right\},$$

then this equation for $\omega \in \mathbb{B}_1 \cap \mathbb{B}_2$ has just one solution. Therefore, it defines an implicit function $v = v(\varepsilon)$. Using the rule of differentiating implicit functions we obtain the first terms of the expansion

$$v(\varepsilon) = v(0) + \varepsilon\, v'(0) + \frac{\varepsilon^2}{2}\, v''(0) + \frac{\varepsilon^3}{6} v'''(\tilde{\varepsilon}), \qquad \tilde{\varepsilon} \leq \varepsilon,$$

as

$$v(0) = 0, \qquad v'(0) = \frac{\eta_1}{\mathrm{I}_T}, \qquad v''(0) = \frac{2\eta_1\,\eta_2\,\mathrm{I}_T - 3\,\mathrm{J}_T\,\eta_1^2 + 2\,K\,\mathrm{I}_T^2}{\mathrm{I}_T^3},$$

where $K = K(\vartheta, X_0)$,

$$\eta_1 = \frac{1}{\sqrt{T}}\int_0^T \frac{\dot S(\vartheta, X_t)}{\sigma(X_t)}\,\mathrm{d}W_t, \qquad \eta_2 = \frac{1}{\sqrt{T}}\int_0^T \frac{\ddot S(\vartheta, X_t)}{\sigma(X_t)}\,\mathrm{d}W_t, \tag{2.172}$$

$$\mathrm{I}_T = \frac{1}{T}\int_0^T \left(\frac{\dot S(\vartheta, X_t)}{\sigma(X_t)}\right)^2 \mathrm{d}t, \qquad \mathrm{J}_T = \frac{1}{T}\int_0^T \frac{\dot S(\vartheta, X_t)\,\ddot S(\vartheta, X_t)}{\sigma(X_t)^2}\,\mathrm{d}t. \tag{2.173}$$

Note that by the Itô formula

$$\mathrm{I}_T = \mathrm{I}(\vartheta) - \frac{\varepsilon}{\sqrt{T}}\int_0^T \frac{g(\vartheta, X_t)}{\sigma(X_t)}\,\mathrm{d}W_t + \varepsilon^2\int_{X_0}^{X_T} \frac{g(\vartheta, v)}{\sigma(v)^2}\,\mathrm{d}v,$$

where

$$g(\vartheta, x) = \frac{2}{f(\vartheta, x)}\int_{-\infty}^{x}\left[\frac{\dot S(\vartheta, v)^2}{\sigma(v)^2} - \mathrm{I}(\vartheta)\right] f(\vartheta, v)\,\mathrm{d}v.$$

Further, by LLN

$$\mathrm{J}_T = \mathbf{E}_\vartheta\left(\frac{\dot S(\vartheta, \xi)\,\ddot S(\vartheta, \xi)}{\sigma(\xi)^2}\right) + O(\varepsilon) \equiv \rho_{1,2}(\vartheta) + O(\varepsilon). \tag{2.174}$$

Hence, we can finally write: for $\omega \in \mathbb{B}_1 \cap \mathbb{B}_2$

$$\sqrt{T}\left(\hat\vartheta_T - \vartheta\right) = \frac{\eta_1}{\mathrm{I}(\vartheta)} + \frac{\eta_1(\eta_2 - \eta_3)\,\mathrm{I}(\vartheta) - 1{,}5\,\eta_1^2\,\rho_{1,2}(\vartheta) - K\,\mathrm{I}(\vartheta)^2}{\mathrm{I}(\vartheta)^3\sqrt{T}} + \frac{R_T(\vartheta, v)}{T},$$

where

$$\eta_3 = \frac{1}{\sqrt{T}}\int_0^T \frac{g(\vartheta, X_t)}{\sigma(X_t)}\,\mathrm{d}W_t. \tag{2.175}$$

The last *good set* is

$$\mathbb{B}_3 = \left\{\omega : \sup_{|v|<T^{-\gamma}} T^{1/4}\left|R_T(\vartheta, v)\right| < 1\right\}.$$

Now we can write

$$
\begin{aligned}
&\sqrt{T}\left(\hat{\vartheta}_T-\vartheta\right)\\
&=\left\{\frac{\eta_1}{\mathrm{I}(\vartheta)}+\frac{\eta_1(\eta_2-\eta_3)\,\mathrm{I}(\vartheta)-1,5\,\eta_1^2\,\rho_{1,2}(\vartheta)-K\,\mathrm{I}(\vartheta)^2}{\mathrm{I}(\vartheta)^3\sqrt{T}}+\frac{\eta_T}{T^{3/4}}\right\}\chi_{\{\mathbb{B}\}}\\
&\quad+\hat{u}_T\,\chi_{\{\mathbb{B}^c\}},
\end{aligned}
$$

where $\left|\eta_T\right|<1$ and $\mathbb{B}=\cap_{l=1}^3\mathbb{B}_l$.

To verify (2.171) we first note that

$$
\mathbf{P}_\vartheta^{(T)}\{\mathbb{B}^c\}\leq\mathbf{P}_\vartheta^{(T)}\{\mathbb{B}_1^c\}+\mathbf{P}_\vartheta^{(T)}\{\mathbb{B}_2^c\}+\mathbf{P}_\vartheta^{(T)}\{\mathbb{B}_3^c\}
$$

and then estimate the probabilities $\mathbf{P}_\vartheta^{(T)}\{\mathbb{B}_l^c\}$ separately.

The first probability we estimate with the help of (2.27) as follows: for any $p_1>0$ there exists a constant $C_1>0$ such that

$$
\begin{aligned}
\mathbf{P}_\vartheta^{(T)}\{\mathbb{B}_1^c\}&=\mathbf{P}_\vartheta^{(T)}\left\{\sup_{|u|>T^{1/2-\gamma}}Z_T(u)>\sup_{|u|\leq T^{1/2-\gamma}}Z_T(u)\right\}\\
&\leq\mathbf{P}_\vartheta^{(T)}\left\{\sup_{|u|>T^{1/2-\gamma}}Z_T(u)>1\right\}\leq\frac{C_1}{T^{p_1}}.
\end{aligned}
$$

Here $Z_T(u)=L\left(\vartheta+\frac{u}{\sqrt{T}},\vartheta,X^T\right)$, we used the equality $Z_T(0)=1$ and we put $R=T^{1/2-\gamma}$.

To estimate the other probabilities we need the following elementary lemma.

Lemma 2.48. *Suppose that the function $h(\theta,x)$ is two times continuously differentiable w.r.t. θ and the derivatives $\dot{h}(\theta,\cdot),\ddot{h}(\theta,\cdot)\in\mathscr{P}$. Then the stochastic integral*

$$
\eta_T(\theta)=\int_0^T h(\theta,X_t)\;\mathrm{d}W_t
$$

is continuously differentiable (with probability 1) and for any $p>0$ there exist constants $C_1>0,C_2>0$ such that

$$
\mathbf{E}_\vartheta\sup_{a<\theta<b}\left|\eta_T(\theta)\right|^{2p}\leq C_1\,T^p+C_2\,(b-a)^{2p}\,T^p. \tag{2.176}
$$

Proof. Following [175] we first note that the random function

$$
\dot{\eta}_T(\theta)=\int_0^T\dot{h}(\theta,X_t)\;\mathrm{d}W_t,\qquad a\leq\theta\leq b
$$

is continuous with probability 1, because

$$
\mathbf{E}_\vartheta\left|\dot{\eta}_T(\theta_1)-\dot{\eta}_T(\theta_2)\right|^2\leq C\;(\theta_1-\theta_2)^2.
$$

Then using the Fubini theorem we obtain

$$\frac{\eta_T(\theta+\delta)-\eta_T(\theta)}{\delta}=\frac{1}{\delta}\int_\theta^{\theta+\delta}\dot\eta_T(\theta+v)\;\mathrm{d}v\rightarrow\dot\eta_T(\theta)$$

as $\delta\rightarrow 0$.

Further,

$$\sup_{a<\theta<b}\left|\eta_T(\theta)\right|\leq\left|\eta_T(a)\right|+\int_a^b\left|\dot\eta_T(\theta)\right|\,\mathrm{d}\theta.$$

Hence

$$\mathbf{E}_\vartheta\sup_{a<\theta<b}\left|\eta_T(\theta)\right|^{2p}\leq 2^{2p-1}\,\mathbf{E}_\vartheta\left|\eta_T(a)\right|^{2p}+$$
$$+2^{2p-1}\,(b-a)^{2p-1}\,\mathbf{E}_\vartheta\int_a^b\left|\dot\eta_T(\theta)\right|^{2p}\mathrm{d}\theta\leq C_1\,T^p+C_2\,(b-a)^{2p}\,T^p.$$

The probabilities $\mathbf{P}_\vartheta^{(T)}\left\{\mathbb{B}_2^c\right\}$ and $\mathbf{P}_\vartheta^{(T)}\left\{\mathbb{B}_3^c\right\}$ can be estimated with the help of this lemma in the same way as it was done in the proof of Theorem 3.1 in [139] (see as well (2.146) above).

This representation of the MLE allows us to obtain the expansion of the moments

The representation (2.170) of the MLE (in a different form with $K=0$) was obtained by Yoshida [245]. Moreover, using Malliavin calculus, Yoshida [245] presented the asymptotic expansion of the distribution function

$$\mathbf{P}_\vartheta^{(T)}\left\{\sqrt{T\mathrm{I}(\vartheta)}\left(\hat\vartheta_T-\vartheta\right)<x\right\}$$
$$=\Phi(x)+\frac{1}{\sqrt{T}}\left[A(\vartheta)-B(\vartheta)\,x^2\right]\,\phi(x)+o\left(\frac{1}{\sqrt{T}}\right)$$

where $\Phi(\cdot)$ and $\phi(\cdot)$ are the distribution function and density of $\mathcal{N}(0,1)$,

$$A(\vartheta)=-\frac{1}{2\,\mathrm{I}(\vartheta)^{3/2}}\,\mathbf{E}_\vartheta\left(\frac{\dot S(\vartheta,\xi)\,g(\vartheta,\xi)}{\sigma(\xi)^2}\right)$$

and

$$B(\vartheta)=\frac{1}{2\,\mathrm{I}(\vartheta)^{3/2}}\,\mathbf{E}_\vartheta\left(\frac{\dot S(\vartheta,\xi)\left[g(\vartheta,\xi)-\ddot S(\vartheta,\xi)\right]}{\sigma(\xi)^2}\right).$$

The convergence $o\left(\frac{1}{\sqrt{T}}\right)$ is uniform on $x\in\mathscr{R}$. This result (stochastic expansion plus expansion of the distribution function and the moments) was further extended on a large class of M estimators (including MLE) and multidimensional parameter by Sakamoto and Yoshida [212], [213]. In the last paper one can find the next after $T^{-1/2}$ terms of order T^{-1} of these expansions.

2.6.6 Recursive Estimation

We consider an opportunity to construct the approximate MLE in a recursive form. Note that recursive estimation has an essential advantage: it allows us to calculate estimators "on-line" (see, e.g., [188], [68] and references therein) and is attractive for many applications.

Let the observed process $X^t = \{X_s, 0 \leq s \leq t\}$ be ergodic diffusion

$$\mathrm{d}X_s = S(\vartheta, X_s)\ \mathrm{d}s + \sigma(X_s)\ \mathrm{d}W_s, \quad X_0, \quad 0 \leq s \leq t.$$

Then (under regularity conditions) the MLE $\hat{\vartheta}_t$ of the parameter $\vartheta \in \Theta = \mathscr{R}$ is one of the solutions of the MLEq

$$\int_0^t \frac{\dot{S}(\vartheta, X_s)}{\sigma(X_s)^2}\ [\mathrm{d}X_s - S(\vartheta, X_s)\ \mathrm{d}s] = 0, \qquad \vartheta \in \Theta,$$

where we suppose that the initial value X_0 does not depend on ϑ. We write this equation as

$$U_t(\vartheta) = 0, \qquad \vartheta \in \Theta$$

and note that $U_t(\vartheta), t \geq 0$ for each fixed ϑ is a continuous semimartingale. Assume that for some $t_0 > 0$ the MLE $\hat{\vartheta}_t, t \geq t_0$ is a continuous semimartingale with the stochastic differential

$$\mathrm{d}\hat{\vartheta}_t = a_t\ \mathrm{d}t + b_t\ \mathrm{d}X_t, \quad \hat{\vartheta}_{t_0}$$

and try to identify the coefficients a_t, b_t.

If the function $S(\vartheta, \cdot)$ is sufficiently smooth then the random function $U_t(\cdot) \in \mathscr{C}^2$ (a.s.) and, together with its derivatives, is jointly (ϑ, t) continuous. Then, one can apply Itô–Ventzell's formula ([129], Theorem 8.1) to $U_t(\hat{\vartheta}_t)$ and obtain the stochastic differential

$$\begin{aligned}\mathrm{d}U_t(\hat{\vartheta}_t) &= \dot{S}(\hat{\vartheta}_t, X_t)\ \left[\mathrm{d}X_t - S(\hat{\vartheta}_t, X_t)\ \mathrm{d}t\right] + H_t(\hat{\vartheta}_t)\mathrm{d}\hat{\vartheta}_t \\ &\quad + \left[\frac{1}{2}Q_t(\hat{\vartheta}_t)\ b_t^2 + \ddot{S}(\hat{\vartheta}_t, X_t)\ b_t\right]\ \mathrm{d}t, \end{aligned} \qquad (2.177)$$

where the random functions $H_t(\vartheta)$ and $Q_t(\vartheta)$ are

$$H_t(\vartheta) = \int_0^t \frac{\ddot{S}(\vartheta, X_s)}{\sigma(X_s)^2}\ [\mathrm{d}X_s - S(\vartheta, X_s)\ \mathrm{d}s] - \int_0^t \frac{\dot{S}(\vartheta, X_s)^2}{\sigma(X_s)^2}\ \mathrm{d}s,$$

and

$$Q_t(\vartheta) = \int_0^t \frac{\dddot{S}(\vartheta, X_s)}{\sigma(X_s)^2}\ [\mathrm{d}X_s - S(\vartheta, X_s)\ \mathrm{d}s] - 3\int_0^t \frac{\dot{S}(\vartheta, X_s)\,\ddot{S}(\vartheta, X_s)}{\sigma(X_s)^2}\ \mathrm{d}s,$$

i.e., $H_t(\vartheta)$ and $Q_t(\vartheta)$ are the second and third derivatives of the log-likelihood ratio $\ln L(\vartheta, \vartheta_0, X^t)$ with respect to ϑ.

The equality (2.177) can be solved (formally) w.r.t. $\mathrm{d}\hat{\vartheta}_t$ (remember that $U_t(\hat{\vartheta}_t) = 0$ and hence $\mathrm{d}U_t(\hat{\vartheta}_t) = 0$). In particular, we see that

$$b_t = -\frac{\dot{S}(\hat{\vartheta}_t, X_t)}{H_t(\hat{\vartheta}_t)}.$$

This provides us with the equation for the MLE

$$\mathrm{d}\hat{\vartheta}_t = \left[\frac{\dot{S}(\hat{\vartheta}_t, X_t)\ddot{S}(\hat{\vartheta}_t, X_t)}{H_t(\hat{\vartheta}_t)^2} - \frac{Q_t(\hat{\vartheta}_t)\,\dot{S}(\hat{\vartheta}_t, X_t)^2}{2\,H_t(\hat{\vartheta}_t)^3}\right]\mathrm{d}t$$
$$-\frac{\dot{S}(\hat{\vartheta}_t, X_t)}{H_t(\hat{\vartheta}_t)}\left[\mathrm{d}X_t - S(\hat{\vartheta}_t, X_t)\,\mathrm{d}t\right], \qquad (2.178)$$

This kind of representation for the MLE was obtained by many authors. Gerencsér *et al.* [82] used it for partially observed linear in state systems. Then Lazrieva and Toronjadze [155]-[157], Ljung *et al.* [176] and Levanony *et al.* [168], [83] presented this equation for inhomogeneous diffusion processes.

Below we follow the work by Levanony *et al.* [168]. We do not present here their regularity conditions $(A) = \{(A1) - (A8)\}$ and $(B) = \{(B1) - (B4)\}$, but just note that these conditions can be easily checked for an ergodic diffusion process if we suppose that the function $S(\vartheta, \cdot)$ is five times continuously differentiable w.r.t. ϑ, the derivatives belonging to $\mathscr{P}$, $\inf_x \sigma(x)^2 > 0$ and conditions $\boldsymbol{\mathcal{A}}_0$ and $\mathcal{A}_2$ are fulfilled.

Theorem 2.49. (Levanony *et al.* [168]) *Assume conditions $(A), (B)$ hold. If the MLE $\{\hat{\vartheta}_t\}, t \geq t_0$ has almost sure continuous trajectories, then* (2.178) *holds for $\hat{\vartheta}_t$ for all sufficiently large t. That is, for any $\varepsilon > 0$, there exists $t_0 = t_0(\varepsilon) < \infty$ such that* (2.178) *describes the MLE path on $[t_0, \infty)$ with probability $1 - \varepsilon$. If in addition, $\mathbf{P}_\vartheta^{(T)}\left\{H_t(\hat{\vartheta}_t) < 0,\ \forall t > 0\right\} = 1$, i.e., the log-likelihood ratio is strictly concave in some small neighborhood of the MLE for all $t > 0$, then* (2.178) *is the MLE evolution equation on $[t_0, \infty)$ a.s. for all $t_0 > 0$.*

The proof of this theorem as well as further references can be found in [168], where it is shown as well that the conditions (A) and (B) provide the consistency and asymptotic normality of the MLE.

Note that (2.178) is not suitable for recursive estimation because, the MLE $\{\hat{\vartheta}_t, t \geq 0\}$ can have discontinuous trajectories at the beginning and its trajectories $\{\hat{\vartheta}_t, t \geq t_0\}$ can be continuous for large values of t_0. The initial value $\hat{\vartheta}_{t_0}$ has to satisfy the condition $H_{t_0}(\hat{\vartheta}_{t_0}) < 0$ and for any ϑ we have $H_0(\vartheta) = 0$. Equation (2.178) has to be modified in such a way that *for any t_0* its solution $\{\vartheta_t,\ t > t_0\}$ "avoids no-solution situations" (e.g. when $H_t(\vartheta_t) = 0$). It is done with the help of switching as follows. The algorithm makes the estimator $\{\vartheta_t,\ t > t_0\}$ "follow the gradient" when $U_t(\vartheta_t) \neq 0$ until it enters some neighborhood of a local maximum and then keeps ϑ_t in this neighborhood as long as possible, i.e., as long, as singularity problems do not arise.

Let us define the set

$$\mathbb{A}_t = \left\{\vartheta : \ |U_t(\vartheta)| \leq \delta, \quad H_t(\vartheta) \leq -\varepsilon\right\},$$

where $\varepsilon > 0$ and $\delta > 0$ are some small numbers, and modify (2.178) as follows:

$$\begin{aligned}
\mathrm{d}\vartheta_t = &\left[\frac{\dot S(\vartheta_t, X_t)\ddot S(\vartheta_t, X_t)}{H_t(\vartheta_t)^2} - \frac{Q_t(\vartheta_t)\, \dot S(\vartheta_t, X_t)^2}{2\, H_t(\vartheta_t)^3} - \frac{\eta\, U_t(\vartheta_t)}{H_t(\vartheta_t)}\right] \chi_{\{\vartheta_t \in \mathbb{A}_t\}}\, \mathrm{d}t \\
&- \frac{\dot S(\vartheta_t, X_t)}{H_t(\vartheta_t)}\, \chi_{\{\vartheta_t \in \mathbb{A}_t\}}\, \left[\mathrm{d}X_t - S(\vartheta_t, X_t)\, \mathrm{d}t\right] + U_t(\vartheta_t)\, \chi_{\{\vartheta_t \in \mathbb{A}_t^c\}}\, \mathrm{d}t,
\end{aligned}$$

with an initial value $\vartheta_{t_0}, t_0 > 0$ and some constant $\eta > 0$. We see that if $\vartheta_t \in \mathbb{A}_t$ then the algorithm follows the likelihood ratio equation with decay term $\eta U/H$, whereas if $\vartheta_t \in \mathbb{A}_t^c$ it follows the gradient towards a local maximum. Note that (2.178) can have infinitely many switchings in bounded time intervals and this makes it difficult to implement. This problem can be solved using an appropriate smoothing. Choose continuous $0 < \delta_t \downarrow 0$, $0 < \varepsilon_t \downarrow 0$ where δ_t satisfies

$$\int_{t_0}^{\infty} \delta_t \, \mathrm{d}t = \infty, \qquad\qquad \frac{s}{t} < \frac{\delta_s}{\delta_t}, \quad \forall\, t_0 \leq s < t.$$

For example $\delta_t = t^{-\gamma}$, $\gamma < 1$. Redefine the set

$$\mathbb{A}_t = \left\{\vartheta : \ |U_t(\vartheta)| < t\, \delta_t, \quad H_t(\vartheta) \leq -\varepsilon_t\right\},$$

and let $\mathcal{A}_t$ be the collection of continuous functions $\phi^t = \{\phi_r, 0 \leq r \leq t\}$, $\phi^t \in \mathscr{C}[0,t]$ such that

$$\mathcal{A}_t = \left\{\phi^t : \ \exists\, s \leq t \text{ such that } H_s(\phi_s) \leq -2\varepsilon_s \text{ and } \phi_r \in \mathbb{A}_r \ \forall r \in [s,t]\right\}.$$

The proposed algorithm is

$$\begin{aligned}
\mathrm{d}\vartheta_t = &\left[\frac{\dot S(\vartheta_t, X_t)\ddot S(\vartheta_t, X_t)}{H_t(\vartheta_t)^2} - \frac{Q_t(\vartheta_t)\, \dot S(\vartheta_t, X_t)^2}{2\, H_t(\vartheta_t)^3} - \frac{\eta\, U_t(\vartheta_t)}{H_t(\vartheta_t)}\right] \chi_{\{\vartheta^t \in \mathcal{A}_t\}}\, \mathrm{d}t \\
&- \frac{\dot S(\vartheta_t, X_t)}{H_t(\vartheta_t)}\, \chi_{\{\vartheta^t \in \mathcal{A}_t\}}\, \left[\mathrm{d}X_t - S(\vartheta_t, X_t)\, \mathrm{d}t\right] + \frac{U_t(\vartheta_t)}{t}\, \chi_{\{\vartheta^t \in \mathcal{A}_t^c\}}\, \mathrm{d}t,
\end{aligned} \tag{2.179}$$

which holds in $[t_0, \infty)$ with any initial condition $\vartheta_{t_0}, t_0 > 0$.

The condition (B) is strengthened up to (B') describing the properties of $S(\vartheta, \cdot)$ as $|\vartheta| \to \infty$ and the following theorem is proved.

Theorem 2.50. *Under conditions (A) and (B′) the estimator ϑ_t defined by Equation (2.179) is consistent (a.s.) and asymptotically normal*

$$\sqrt{T}\left(\vartheta_t - \vartheta\right) \Longrightarrow \mathcal{N}\left(0, \mathrm{I}\left(\vartheta\right)^{-1}\right).$$

For the proof see [168], Theorem 5.1.

In the case of strict concaveness of the log-likelihood a simplified version of algorithm (2.179) was proposed in [169].

3

Special Models

We consider a partially observed two-dimensional diffusion process with coefficients depending on an unknown parameter and show the asymptotic normality of the MLE and BE of this parameter in regular case. Then we estimate the parameters of nonregular models of increasing singularity. The first is the point of cusp of the trend coefficient of the ergodic diffusion process. Next is the problem of delay parameter estimation of a Gaussian ergodic processes, and the last is the problem of parameter estimation of a diffusion process with a discontinuous trend coefficient. In all these nonregular problems the rates of convergence are better than in the regular case, the asymptotic distributions of the estimators are not Gaussian and properties of the estimators are quite similar. The last section is devoted to models for which the conditions of ergodicity are not fulfilled.

3.1 Partially Observed Systems

We are given a two-dimensional diffusion process $\{X_t, Y_t,\ t \geq 0\}$

$$\mathrm{d}Y_t = -\,a\,(\vartheta)\ Y_t\,\mathrm{d}t + b\,(\vartheta)\ \mathrm{d}V_t, \quad Y_0, \tag{3.1}$$

$$\mathrm{d}X_t = \quad c\,(\vartheta)\ Y_t\,\mathrm{d}t + \ \sigma\ \mathrm{d}W_t, \qquad X_0, \tag{3.2}$$

where $\{V_t, W_t,\ t \geq 0\}$ are two independent Wiener processes, the functions $a\,(\cdot), b\,(\cdot), c\,(\cdot)$ and the constant $\sigma \neq 0$ are known and ϑ is a finite-dimensional (unknown) parameter. The random variables X_0 and Y_0 are $\mathfrak{F}_0$-measurable, independent and Gaussian. The first equation (3.1) in the engineering literature is called the *state equation* and the second (3.2) is called the *observation equation*.

Suppose that only one component $X^T = \{X_t,\ 0 \leq t \leq T\}$ is observed and we have to estimate the parameter ϑ. Our goal is to construct the MLE and BE of this parameter, hence we need the likelihood ratio formula and the expression for this likelihood ratio depends on the conditional expectation of the nonobserved component Y_t. Therefore we first present the equations for this expectation (Kalman–Bucy filter) and then with their help we describe the estimators.

3.1.1 Kalman–Bucy Filter

The system (3.1)–(3.2) is well-known in the engineering literature but in a different context. It is supposed that the value of ϑ is known (the system is known) and the problem is to estimate the current value of the nonobserved component Y_t by observations $X^t = \{X_s,\ 0 \le s \le t\}$. Then the optimal in the mean square sense (Bayesian) estimator of Y_t is conditional mathematical expectation

$$\tilde{Y}_t = \mathbf{E}_\vartheta \left(Y_t | X^t\right), \qquad 0 \le t \le T,$$

and as was shown by Kalman and Bucy [117] (see as well [175], Chapter 10) the process $\left\{\tilde{Y}_t,\ 0 \le t \le T\right\}$ satisfies the linear equation

$$\mathrm{d}\tilde{Y}_t = -a(\vartheta)\ \tilde{Y}_t\,\mathrm{d}t + \frac{c(\vartheta)\ \gamma_t(\vartheta)}{\sigma^2}\left[\mathrm{d}X_t - c(\vartheta)\,\tilde{Y}_t\,\mathrm{d}t\right], \quad \tilde{Y}_0 \quad 0 \le t \le T, \tag{3.3}$$

where $\gamma_t(\vartheta) = \mathbf{E}_\vartheta\left(Y_t - \tilde{Y}_t\right)^2$ is a mean square error of estimation. The function $\gamma_t(\vartheta)$ is a solution of the *Riccati equation*

$$\frac{\partial \gamma_t(\vartheta)}{\partial t} = -2a(\vartheta)\ \gamma_t(\vartheta) - \frac{c(\vartheta)^2\,\gamma_t(\vartheta)^2}{\sigma^2} + b(\vartheta)^2 \quad 0 \le t \le T, \tag{3.4}$$

with initial value

$$\gamma_0(\vartheta) = \mathbf{E}_\vartheta\left(Y_0 - \mathbf{E}_\vartheta\left(Y_0 | X_0\right)\right)^2.$$

Equations (3.3) and (3.4) are called the *Kalman–Bucy filter* or *equations of the optimal linear filtration.* Therefore the solution of this system gives us the optimal estimator of the unobserved component. Recall that the calculation of $\tilde{Y}_t$ requires the knowledge of the parameter ϑ.

In the present section we are interested in parameter estimation for ergodic diffusion only. So we will study the asymptotic properties of MLE, BE and TFE of the parameter ϑ as $T \to \infty$. We simplify the system as follows. Suppose that the unknown parameter is one-dimensional $\vartheta \in \Theta = (\alpha, \beta)$ and the function $a(\vartheta)$ is strictly positive. Then the diffusion process $\{Y_t,\ t \ge 0\}$ is an Ornstein–Uhlenbeck process with ergodic properties and invariant Gaussian distribution $\mathcal{N}\left(0, \frac{b(\vartheta)^2}{2a(\vartheta)}\right)$. We put $\mathcal{L}(Y_0) = \mathcal{N}\left(0, \frac{b(\vartheta)^2}{2a(\vartheta)}\right)$, so the process $\{Y_t, t \ge 0\}$ will be stationary.

Put

$$\gamma_*(\vartheta) = \frac{a(\vartheta)\,\sigma^2}{c(\vartheta)^2}\left(\sqrt{1 + \frac{b(\vartheta)^2\,c(\vartheta)^2}{a(\vartheta)^2\,\sigma^2}} - 1\right)$$

and

$$r(\vartheta) = \sqrt{a(\vartheta)^2 + \frac{b(\vartheta)^2 c(\vartheta)^2}{\sigma^2}}.$$

If $\gamma_0 = \gamma_*(\vartheta)$, then we obtain $\frac{\partial \gamma_t(\vartheta)}{\partial t} = 0$ and $\gamma_t(\vartheta) = \gamma_*(\vartheta)$ for all $t \geq 0$, i.e., $\gamma_*(\vartheta)$ is a stationary solution of (3.4). Note that if $\gamma_0 \neq \gamma_*(\vartheta)$, then the solution $\gamma_t(\vartheta)$ of Equation (3.4) can be written explicitly [6]

$$\gamma_t(\vartheta) = \mathrm{e}^{-2r(\vartheta)t}\left[\frac{1}{\gamma_0 - \gamma_*(\vartheta)} + \frac{c(\vartheta)^2}{2r(\vartheta)\sigma^2}\left(1 - \mathrm{e}^{-2r(\vartheta)t}\right)\right]^{-1} + \gamma_*(\vartheta)$$

and it is easy to see that this function exponentially converges to the stationary solution.

We suppose that the system (3.3) and (3.4) is in a stationary regime: $\gamma_0(\vartheta) = \gamma_*(\vartheta)$. Further, it will be more convenient for us to introduce the function

$$\gamma(\vartheta) = \sqrt{a(\vartheta)^2 + \frac{b(\vartheta)^2 c(\vartheta)^2}{\sigma^2}} - a(\vartheta) = r(\vartheta) - a(\vartheta)$$

and rewrite Equation (3.3) for the estimator $m_t(\vartheta) = c(\vartheta)\ \tilde{Y}_t$:

$$\mathrm{d}m_t(\vartheta) = -a(\vartheta)\ m_t(\vartheta)\ \mathrm{d}t + \gamma(\vartheta)\ [\mathrm{d}X_t - m_t(\vartheta)\ \mathrm{d}t], \quad 0 \leq t \leq T. \quad (3.5)$$

As an initial value $m_0(\vartheta)$ we take a Gaussian random variable

$$m_0(\vartheta) \sim \mathcal{N}\left(0, \frac{\gamma(\vartheta)^2 \sigma^2}{2a(\vartheta)}\right). \quad (3.6)$$

Then the solution of Equation (3.5) is a stationary Gaussian process with the same invariant distribution.

The stochastic process $\left\{\bar{W}_t,\ 0 \leq t \leq T\right\}$ defined by the equation

$$\sigma\ \bar{W}_t = X_t - X_0 - \int_0^t m_s(\vartheta)\ \mathrm{d}s$$

is called the *innovation process* and is a standard Wiener process.

To construct the MLE of the parameter ϑ we have to calculate the function $\{m_t(y),\ 0 \leq t \leq T\},\ y \in \Theta$ defined by the equation

$$\mathrm{d}m_t(y) = -a(y)\ m_t(y)\ \mathrm{d}t + \gamma(y)\ [\mathrm{d}X_t - m_t(y)\ \mathrm{d}t], \quad m_0(y). \quad (3.7)$$

Note that $m_t(y)$ depends as well on the true value ϑ because the observed process X^T depends on ϑ and, if $y \neq \vartheta$, then $m_t(y)$ is not the conditional expectation of $a(\vartheta)Y_t$ and is just a stochastic process, a solution of the stochastic differential equation (3.7).

Let us denote $\mathbf{P}_\vartheta^{(T)}$ the measure induced in $(\mathscr{C}_T, \mathfrak{B}_T)$ by the process X^T. Fix some $\vartheta_1 \in \Theta$. The likelihood ratio

$$L\left(y, \vartheta_1, X^T\right) = \frac{\mathrm{d}\,\mathbf{P}_y^{(T)}}{\mathrm{d}\,\mathbf{P}_{\vartheta_1}^{(T)}}\left(X^T\right), \qquad y \in \Theta$$

has the following form:

$$L\left(y, \vartheta_1, X^T\right) = L_0\left(y, \vartheta_1, X_0\right) \exp\left\{\frac{1}{\sigma^2}\int_0^T \left[m_t\left(y\right) - m_t\left(\vartheta_1\right)\right]\mathrm{d}X_t - \right.$$
$$\left. -\frac{1}{2\sigma^2}\int_0^T \left[m_t\left(y\right)^2 - m_t\left(\vartheta_1\right)^2\right]\mathrm{d}t\right\},$$

where $L_0\left(y, \vartheta_1, X_0\right)$ is the ratio of two Gaussian densities of the laws (3.6) with $\vartheta = y$ and $\vartheta = \vartheta_1$.

3.1.2 Properties of Estimators

We are given the system (3.1) and (3.2) and by observations X^T we construct the MLE $\hat{\vartheta}_T$ defined by its equation

$$L\left(\hat{\vartheta}_T, \vartheta_1, X^T\right) = \sup_{y\in\Theta} L\left(y, \vartheta_1, X^T\right).$$

Introduce the quantity

$$\mathrm{I}\left(\vartheta\right) = \frac{\dot{a}\left(\vartheta\right)^2}{2\,a\left(\vartheta\right)} - \frac{2\dot{a}\left(\vartheta\right)\,\dot{r}\left(\vartheta\right)}{r\left(\vartheta\right) + a\left(\vartheta\right)} + \frac{\dot{r}\left(\vartheta\right)^2}{2r\left(\vartheta\right)}$$

which will play the role of Fisher information. Here and in the following the dot means the derivation w.r.t. ϑ.

The *regularity conditions* $\mathcal{J}$ in this problem are

$\mathcal{J}_1$. *The functions* $a\left(\vartheta\right)$, $b\left(\vartheta\right)$ *and* $c\left(\vartheta\right)$, $\vartheta \in [\alpha, \beta]$ *are continuously differentiable over* ϑ. *For all* $\vartheta \in [\alpha, \beta]$ *these functions satisfy the conditions:* $b\left(\vartheta\right) \neq 0$, $c\left(\vartheta\right) \neq 0$ *and* $a\left(\vartheta\right) > 0$.

$\mathcal{J}_2$. *For any compact* $\mathbb{K} \subset \Theta$ *any* $\nu > 0$ *we have*

$$\inf_{\vartheta\in\mathbb{K}}\inf_{|y-\vartheta|>\nu}\left(\left|a(y) - a\left(\vartheta\right)\right| + \left|r(y) - r\left(\vartheta\right)\right|\right) > 0, \tag{3.8}$$

and

$$\inf_{\vartheta\in\mathbb{K}}\left(\left|\dot{a}\left(\vartheta\right)\right| + \left|\dot{r}\left(\vartheta\right)\right|\right) > 0. \tag{3.9}$$

Theorem 3.1. *Let the conditions $\mathcal{J}$ be fulfilled, then the MLE $\hat{\vartheta}_T$ is uniformly consistent, asymptotically normal*

$$\mathcal{L}_\vartheta\left\{\sqrt{T}\left(\hat{\vartheta}_T-\vartheta\right)\right\}\Longrightarrow\mathcal{N}\left(0,\mathrm{I}\left(\vartheta\right)^{-1}\right) \tag{3.10}$$

and the moments of $\sqrt{T}\left(\hat{\vartheta}_T-\vartheta\right)$ converge. Moreover it is asymptotically efficient for polynomial loss functions.

Proof. The proof of this theorem is *parallel* to the proof of Theorem 2.8. We establish the same properties of the normalized likelihood ratio process

$$Z_T\left(u\right)=L\left(\vartheta+u/T,\vartheta,X^T\right),\quad u\in\mathbb{U}_T=\left(T\left(\alpha-\vartheta\right),T\left(\beta-\vartheta\right)\right)$$

where ϑ is the true value.

Remember that

$$\sqrt{T}\left(\hat{\vartheta}_T-\vartheta\right)=\arg\inf_{u\in\mathbb{U}_T}Z_T\left(u\right).$$

Introduce the stochastic process $\{\dot{m}_t\left(y\right),\ 0\leq t\leq T\}$ by the formal differentiating Equation (3.7) over y:

$$\begin{aligned}\mathrm{d}\dot{m}_t\left(y\right)&=-\dot{r}\left(y\right)\ m_t\left(y\right)\mathrm{d}t-r\left(y\right)\dot{m}_t\left(y\right)\mathrm{d}t\\&\quad+\dot{\gamma}\left(y\right)\ m_t\left(\vartheta\right)\ \mathrm{d}t+\dot{\gamma}\left(y\right)\sigma\ \mathrm{d}\bar{W}_t.\end{aligned} \tag{3.11}$$

It can be shown that $\{\dot{m}_t\left(y\right),\ 0\leq t\leq T\}$ is the mean-square derivative of $\{m_t\left(y\right),\ 0\leq t\leq T\}$. Indeed, using equations (3.7) and (3.11) and condition $\mathcal{J}_1$ we prove

$$\mathbf{E}_\vartheta\left[m_t\left(y+h\right)-m_t\left(y\right)-h\ \dot{m}_t\left(y\right)\right]^2=o\left(h^2\right).$$

Calculations are cumbersome but direct.

We have ($\vartheta_u=\vartheta+u/\sqrt{T}$)

$$\begin{aligned}\ln Z_T\left(u\right)=\ln Z_0\left(u\right)&+\int_0^T\frac{m_t\left(\vartheta_u\right)-m_t\left(\vartheta\right)}{\sigma}\mathrm{d}\bar{W}_t\\&-\frac{1}{2}\int_0^T\left(\frac{m_t\left(\vartheta_u\right)-m_t\left(\vartheta\right)}{\sigma}\right)^2\mathrm{d}t,\end{aligned}$$

where $Z_0\left(u\right)=L_0\left(\vartheta_u,\vartheta,X_0\right)$. Hence we can represent it as

$$\ln Z_T\left(u\right)=\frac{u}{\sigma\sqrt{T}}\int_0^T\dot{m}_t\left(\vartheta\right)\mathrm{d}\bar{W}_t-\frac{u^2}{2\,\sigma^2\,T}\int_0^T\dot{m}_t\left(\vartheta\right)^2\mathrm{d}t+o\left(1\right).$$

Note that the process $\dot{m}_t\left(\vartheta\right),\ 0\leq t\leq T$ satisfies Equation (3.11) with $y=\vartheta$

$$\mathrm{d}\dot{m}_t(\vartheta) = -\dot{a}(\vartheta)\; m_t(\vartheta)\;\mathrm{d}t - r(\vartheta)\,\dot{m}_t(\vartheta)\;\mathrm{d}t + \dot{\gamma}(\vartheta)\;\sigma\;\mathrm{d}\bar{W}_t, \quad \dot{m}_0(\vartheta).$$

The solution of this equation can be written as

$$\dot{m}_t(\vartheta) = \dot{m}_0(\vartheta)\;\mathrm{e}^{-r(\vartheta)\,t} - \dot{a}(\vartheta)\int_0^t \mathrm{e}^{-r(\vartheta)(t-s)} m_s(\vartheta)\;\mathrm{d}s$$
$$+\dot{\gamma}(\vartheta)\;\sigma\int_0^t \mathrm{e}^{-r(\vartheta)(t-s)}\;\mathrm{d}\bar{W}_s.$$

The process $\{m_t(\vartheta),\, 0 \le t \le T\}$ itself has the similar representation

$$m_t(\vartheta) = m_0(\vartheta)\;\mathrm{e}^{-a(\vartheta)\,t} + \gamma(\vartheta)\;\sigma\int_0^t \mathrm{e}^{-a(\vartheta)\,(t-s)}\;\mathrm{d}\bar{W}_s.$$

Therefore using the Fubini theorem we have

$$\dot{m}_t(\vartheta) = \sigma\int_0^t \mathrm{e}^{-r(\vartheta)(t-s)}\left[\dot{r}(\vartheta) - \dot{a}(\vartheta)\;\mathrm{e}^{\gamma(\vartheta)(t-s)}\right]\;\mathrm{d}\bar{W}_s + o(1).$$

Direct calculations give us

$$\mathbf{E}_\vartheta\,\dot{m}_t(\vartheta)^2 = \sigma^2\,\mathrm{I}(\vartheta)\;\left(1 + o(1)\right).$$

Hence by the law of large numbers we have the convergence

$$\mathbf{P}_\vartheta - \lim_{T\to\infty}\frac{1}{\sigma^2\,T}\int_0^T \dot{m}_t(\vartheta)^2\;\mathrm{d}t = \mathrm{I}(\vartheta)$$

which gives us the asymptotic normality

$$\mathcal{L}_\vartheta\left(\frac{1}{\sigma\sqrt{T}}\int_0^T \dot{m}_t(\vartheta)\,\mathrm{d}\bar{W}_t\right) \Longrightarrow \mathcal{N}\left(0, \mathrm{I}(\vartheta)\right).$$

Moreover, it can be shown that this convergence is uniform on compacts $\mathbb{K} \in \Theta$ and the moments converge too.

Note that condition (3.9) provides $\mathrm{I}(\vartheta) > 0$. Indeed, we have the equality

$$\mathrm{I}(\vartheta) = \left(\dot{a}(\vartheta) + \frac{a(\vartheta)}{r(\vartheta) + a(\vartheta)}\dot{r}(\vartheta)\right)^2 + \frac{\dot{r}(\vartheta)^2\, a(\vartheta)\left(r(\vartheta) - a(\vartheta)\right)}{r(\vartheta)\left(r(\vartheta) + a(\vartheta)\right)^2}$$

and $\mathrm{I}(\vartheta) = 0$ if $\dot{r}(\vartheta) = 0$ and $\dot{a}(\vartheta) = 0$, but these two equalities are impossible simultaneously due to condition (3.9). Remember that the equality $r(\vartheta) = a(\vartheta)$ is impossible too.

We have as well the estimate

$$\sup_{\vartheta\in\mathbb{K}}\mathbf{E}_\vartheta\left|\frac{1}{\sigma^2\,T}\int_0^T \dot{m}_t(\vartheta)^2\;\mathrm{d}t - \mathrm{I}(\vartheta)\right|^p \le \frac{C}{T^{p/2}}. \tag{3.12}$$

To prove this we replace the integral

$$\int_0^T \left[\frac{\dot m_t(\vartheta)^2}{\sigma^2} - \mathrm{I}(\vartheta)\right] \mathrm{d}t$$

by a stochastic integral (by the Itô formula) and then use the boundedness of all moments for a Gaussian process.

Therefore it can be shown the uniform on compacts $\mathbb{K}$ LAN of the family of measures $\left\{\mathbf{P}_\vartheta^{(T)},\ \vartheta\in\Theta\right\}$.

The second estimate

$$\sup_{\vartheta\in\mathbb{K}} \mathbf{E}_\vartheta \left|Z_T(u)^{1/2} - Z_T(v)^{1/2}\right|^2 \leq C\,\left(1+R^2\right)\,(u-v)^2, \tag{3.13}$$

is proved with the help of Lemma 1.13 (see (1.42)).

To check the identifiability of the model we have to verify that for any $\nu>0$

$$\inf_{\vartheta\in\mathbb{K}}\ \inf_{|y-\vartheta|>\nu}\ \mathbf{E}_\vartheta\left(m_t(y)-m_t(\vartheta)\right)^2>0.$$

Using Equations (3.5) and (3.7) we obtain for the difference $\delta_t = m_t(y) - m_t(\vartheta)$ the equation

$$\mathrm{d}\delta_t = -r(y)\,\delta_t\,\mathrm{d}t + \left[a(\vartheta)+\gamma(y)-r(y)\right] m_t(\vartheta)\,\mathrm{d}t + \sigma\left[\gamma(y)-\gamma(\vartheta)\right]\,\mathrm{d}\bar W_t \tag{3.14}$$

which can be solved as

$$\begin{aligned}\delta_t &= \sigma\int_0^t e^{-r(y)(t-s)}\left(a(\vartheta)-a(y)\right)m_s(\vartheta)\,\mathrm{d}s\\ &\quad+\sigma\left[\gamma(y)-\gamma(\vartheta)\right]\int_0^t e^{-r(y)(t-s)}\mathrm{d}\bar W_t + o(1).\end{aligned}$$

Suppose that $r(y)\neq a(\vartheta)$. Then using the solution of (3.5) and the Fubini theorem we can rewrite it as

$$\delta_t = \sigma\int_0^t \mathrm{e}^{-r(y)(t-s)}\left[A(y,\vartheta) - B(y,\vartheta)\,\mathrm{e}^{(r(y)-a(\vartheta)(t-s)}\right]\mathrm{d}\bar W_t + o(1),$$

where

$$\begin{aligned}A(y,\vartheta) &= \gamma(y)\left(1-\frac{\gamma(\vartheta)}{\gamma(y)+a(y)-a(\vartheta)}\right),\\ B(y,\vartheta) &= \gamma(\vartheta)\left(1-\frac{\gamma(y)}{\gamma(y)+a(y)-a(\vartheta)}\right).\end{aligned}$$

Direct calculation gives us

$$\sigma^{-2}\,\mathbf{E}_\vartheta\left(m_t\left(y\right)-m_t\left(\vartheta\right)\right)^2$$
$$=\frac{A(y,\vartheta)^2}{2r(y)}-\frac{2A(y,\vartheta)B(y,\vartheta)}{r(y)+a\left(\vartheta\right)}+\frac{B(y,\vartheta)^2}{2a\left(\vartheta\right)}+o\left(1\right)\qquad(3.15)$$

and

$$\frac{A(y,\vartheta)^2}{2r(y)}-\frac{2A(y,\vartheta)B(y,\vartheta)}{r(y)+a\left(\vartheta\right)}+\frac{B(y,\vartheta)^2}{2a\left(\vartheta\right)}$$
$$=\left(A(y,\vartheta)-\frac{2r(y)}{r(y)+a(\vartheta)}B(y,\vartheta)\right)^2+\frac{B(y,\vartheta)^2r(y)}{a(\vartheta)\left(r(y)+a(\vartheta)\right)^2}.$$

Hence (3.15) can be equal to zero if simultaneously $A(y,\vartheta)=0$ and $B(y,\vartheta)=0$. But these equalities imply $a\left(y\right)=a(\vartheta)$ and $r\left(y\right)=r\left(\vartheta\right)$, which is impossible by condition (3.8).

Suppose that $r(y)=a(\vartheta)$, then

$$\delta_t=\sigma\int_0^t \mathrm{e}^{-r(y)(t-s)}\left[\gamma\left(y\right)-\gamma\left(\vartheta\right)+\gamma\left(\theta\right)\left(t-s\right)\right]\mathrm{d}\bar W_t+o\left(1\right)$$

and

$$\sigma^{-2}\,\mathbf{E}_\vartheta\left(m_t\left(y\right)-m_t\left(\vartheta\right)\right)^2=\frac{\left(\gamma(y)-\gamma(\vartheta)\right)^2}{2r(y)}+\frac{\gamma(\vartheta)\left(\gamma(y)-\gamma(\vartheta)\right)}{2r(y)^2}$$
$$+\frac{\gamma(\vartheta)^2}{4r(y)^3}+o\left(1\right).$$

Note that

$$\left[\gamma(y)-\gamma(\vartheta)\right]^2+\left[\gamma(y)-\gamma(\vartheta)\right]\frac{\gamma(\vartheta)}{r(y)}+\frac{\gamma(\vartheta)^2}{2r(y)^2}$$
$$=\left(\gamma(y)-\gamma(\vartheta)+\frac{\gamma(\vartheta)}{2r(y)}\right)^2+\frac{\gamma(\vartheta)^2}{4r(y)^2}>0.$$

It is clear that by continuity we have a similar inequality in the vicinity of the points y defined by the equation $r(y)=a(\vartheta)$.

Therefore

$$\inf_{\vartheta\in\mathbb{K}}\inf_{|y-\vartheta|>\nu}\mathbf{E}_\vartheta\left(m_t\left(y\right)-m_t\left(\vartheta\right)\right)^2>0.$$

Remember that from the $\mathscr{L}_2$ differentiability of the function $m_t\left(y\right)$ we have the estimate

$$\mathbf{E}_\vartheta\left(m_t\left(y\right)-m_t\left(\vartheta\right)\right)^2\geq\frac{\sigma^2}{2}\mathrm{I}\left(\vartheta\right)\,\left(y-\vartheta\right)^2$$

for $|y-\vartheta|\leq\nu$, where $\nu>0$ is sufficiently small. Hence we can write

$$\mathbf{E}_\vartheta\left(m_t\left(y\right)-m_t\left(\vartheta\right)\right)^2\geq\kappa\left(y-\vartheta\right)^2\qquad(3.16)$$

with some $\kappa > 0$.

Having uniform on compacts LAN and two estimates (3.13) and (3.16) we can apply the same general theorem by Ibragimov and Khasminskii Theorem 2.6 and obtain the desired properties of the MLE described in Theorem 3.1. Note as well that LAN gives us the lower bound (2.11) on the risks of all estimators, and the convergence of moments together with the continuity of the Fisher information provide the asymptotical efficiency of the MLE for the loss functions $\ell(\cdot) \in \mathscr{W}_p$. Of course, the Bayes estimators with continuous positive prior density have the same asymptotic properties as the MLE.

Remark 3.2. The system (3.1) and (3.2) is entirely defined by the three parameters a, b, c but only two of them, say, $\vartheta_1 = a, \vartheta_2 = bc$, can be estimated due to the identifiability condition (3.8). This condition can be written as

$$\inf_{\vartheta \in \mathbb{K}} \inf_{|y-\vartheta|>\nu} \Big(|a(y) - a(\vartheta)| + |b(y)\, c(y) - b(\vartheta)\, c(\vartheta)| \Big) > 0.$$

It is evident that the parameter ϑ of the system

$$\begin{aligned} \mathrm{d}Y_t &= -a\, Y_t\, \mathrm{d}t + b\, \vartheta^{-1}\, \mathrm{d}V_t, & & Y_0, \\ \mathrm{d}X_t &= \quad c\, \vartheta\, Y_t\, \mathrm{d}t + \sigma\, \mathrm{d}W_t, & & X_0, \quad 0 \le t \le T, \end{aligned}$$

cannot be estimated consistently.

The two-dimensional case $\boldsymbol{\vartheta} = \left(\vartheta^{(1)}, \vartheta^{(2)}\right)$ for the system

$$\begin{aligned} \mathrm{d}Y_t &= -a(\boldsymbol{\vartheta})\; Y_t\, \mathrm{d}t + \; b(\boldsymbol{\vartheta})\; \mathrm{d}V_t, & & Y_0, \\ \mathrm{d}X_t &= \quad c(\boldsymbol{\vartheta})\; Y_t\, \mathrm{d}t + \sigma\; \mathrm{d}W_t, & & X_0, \quad 0 \le t \le T, \end{aligned}$$

can be done by a similar way. Cumbersome but direct calculations show the LAN of the underlying family of distributions and the estimates such as (3.13) and (3.16). The identifiability condition will be the same

$$\inf_{\boldsymbol{\vartheta} \in \mathbb{K}} \inf_{|\boldsymbol{y}-\boldsymbol{\vartheta}|>\nu} \Big(|a(\boldsymbol{y}) - a(\boldsymbol{\vartheta})| + |r(\boldsymbol{y}) - r(\boldsymbol{\vartheta})| \Big) > 0,$$

and we have to add the nondegeneracy of the Fisher information matrix

$$\mathbf{I}(\vartheta) = \frac{1}{\sigma^2} \begin{pmatrix} \mathbf{E}_{\boldsymbol{\vartheta}} \left(\frac{\partial m_t(\boldsymbol{\vartheta})}{\partial \vartheta^{(1)}} \right)^2 & , & \mathbf{E}_{\boldsymbol{\vartheta}} \left(\frac{\partial m_t(\boldsymbol{\vartheta})}{\partial \vartheta^{(1)}} \frac{\partial m_t(\boldsymbol{\vartheta})}{\partial \vartheta^{(2)}} \right) \\ \mathbf{E}_{\boldsymbol{\vartheta}} \left(\frac{\partial m_t(\boldsymbol{\vartheta})}{\partial \vartheta^{(1)}} \frac{\partial m_t(\boldsymbol{\vartheta})}{\partial \vartheta^{(2)}} \right) & , & \mathbf{E}_{\boldsymbol{\vartheta}} \left(\frac{\partial m_t(\boldsymbol{\vartheta})}{\partial \vartheta^{(2)}} \right)^2 \end{pmatrix}.$$

The elements of this matrix can be calculated directly from the corresponding stochastic differential equations.

3.1.3 Examples

We consider two examples with unknown parameters in the observation equation and unknown parameters in the state equation.

Example 3.3. Let the system be

$$\begin{aligned} dY_t &= -\vartheta\, Y_t\, dt + b\ dV_t, \qquad && Y_0, \\ dX_t &= \quad c\, Y_t\, dt + \sigma\ dW_t, \qquad && X_0, \quad 0 \le t \le T \end{aligned}$$

where $\vartheta \in (\alpha, \beta)$ and $\alpha > 0$ and we have to estimate ϑ by observations X^T. Then the conditions $\mathcal{J}$ are fulfilled and the MLE is asymptotically normal with

$$\mathrm{I}(\vartheta) = \frac{1}{2\vartheta} - \frac{2\vartheta}{r(\vartheta)\,(r(\vartheta) + a(\vartheta))} + \frac{\vartheta^2}{2r(\vartheta)^3},$$

where $r(\vartheta) = \sqrt{\vartheta^2 + b^2c^2/\sigma^2}$.

Example 3.4. Let the system be

$$\begin{aligned} dY_t &= -a\, Y_t\, dt + b\ dV_t, \qquad && Y_0, \\ dX_t &= \quad \vartheta\, Y_t\, dt + \sigma\ dW_t, \qquad && X_0, \quad 0 \le t \le T, \end{aligned}$$

where $\vartheta \in (\alpha, \beta)$ and/or $\alpha > 0$ or $\beta < 0$. Then the conditions $\mathcal{J}$ are fulfilled too and the MLE constructed by observations $\{X_t,\ 0 \le t \le T\}$ is asymptotically normal (3.10) with

$$\mathrm{I}(\vartheta) = \frac{\vartheta c^2}{2\,(a^2\sigma^2 + \vartheta^2 b^2)}.$$

Trajectory Fitting Estimator

The observed process $X_t,\ t \ge 0$, is not stationary and has no invariant distribution, so the direct calculation of the MDE ϑ_T^* or ϑ_T^{**} is impossible. At the same time it is possible to construct and to study the trajectory fitting estimator. Let us consider the following linear partially observed system

$$\begin{aligned} dY_t &= -\left[a(\vartheta)\, Y_t - A(\vartheta)\right]\, dt + b(\vartheta)\ dV_t, \qquad Y_0, \\ dX_t &= c(\vartheta)\ Y_t\ dt + \ dW_t, \quad X_0 = 0, \quad 0 \le t \le T, \end{aligned} \tag{3.17}$$

where $a(\vartheta) > 0, A(\vartheta)$ and $c(\vartheta) \neq 0$ are known functions and $\{V_t, t \ge 0\}$ and $\{W_t, t \ge 0\}$ are two independent Wiener processes.

By *innovation theorem* (see [175], Theorem 7.17) the process (3.17) admits the representation

$$\mathrm{d}X_t = c(\vartheta)\ \hat{Y}_t(\vartheta)\ \mathrm{d}t + \mathrm{d}\bar{W}_t, \quad X_0 = 0, \quad 0 \le t \le T,$$

where $\hat{Y}_t(\vartheta) = \mathbf{E}_\vartheta(Y_t|X^t), t \ge 0$ is a conditional expectation and the Wiener process $\bar{W}_t, t \ge 0$ is defined by the last equation. Therefore we can put $m_t(y) = c(y)\hat{Y}_t(y)$ and write the family of random functions

$$\{m_t(y),\ 0 \le t \le T\},\ y \in \Theta$$

as solutions of the stochastic differential equations

$$\mathrm{d}m_t(y) = -\left[a(y)\ m_t(y) - A(y)\right]\mathrm{d}t + \gamma(y)\left[\mathrm{d}X_t - m_t(y)\,\mathrm{d}t\right], \quad m_0(y).$$

Then with the help of the functions

$$M_t(y) = \int_0^t m_s(y)\ \mathrm{d}s$$

we define the trajectory fitting estimator $\vartheta_T^\star$ as

$$\vartheta_T^\star = \arg\inf_{y\in\vartheta}\int_0^T \left[X_t - M_t(y)\right]^2\ \mathrm{d}t.$$

This estimator was studied in a more general framework (multidimensional Y_t, X_t and ϑ) by Dietz and Kutoyants [22] and here we just present an example.

Example 3.5. Let the partially observed system be

$$\mathrm{d}Y_t = -a\left[Y_t - \theta\right]\ \mathrm{d}t + b\,\mathrm{d}V_t \qquad Y_0,$$
$$\mathrm{d}X_t = cY_t\ \mathrm{d}t + \ \mathrm{d}W_t \qquad X_0 = 0$$

with $a > 0, b \ne 0$ and $c \ne 0$ been known.

The Fisher information is

$$\mathrm{I}(\theta) = \frac{a^2c^2}{a^2 + b^2c^2} \ne 0,$$

and for the MLE $\hat{\theta}_T$ and the TFE $\theta_T^\star$ we have

$$\sqrt{T}\left(\hat{\theta}_T - \theta\right) \Longrightarrow \mathcal{N}\left(0, \mathrm{I}(\theta)^{-1}\right)$$

$$\sqrt{T}\left(\theta_T^\star - \theta\right) \Longrightarrow \mathcal{N}\left(0, \frac{6}{5}\ \mathrm{I}(\theta)^{-1}\right).$$

Note that if we replace in the definition of the TFE the Lebesgue measure dt by another measure ($(\frac{t}{T})^{T-1}$ dt or $\chi_{[T-1,T)}$ dt), then we obtain for it the same limit variance, i.e., without the coefficient $\frac{6}{5}$. Therefore the TFE has the same efficiency as the MLE.

3.2 Cusp Estimation

3.2.1 Model

Let us consider the problem of parameter estimation by the observations of the diffusion process

$$\mathrm{d}X_t = S\left(\vartheta, X_t\right)\,\mathrm{d}t + \sigma\left(X_t\right)\,\mathrm{d}W_t, \quad X_0, \quad 0 \leq t \leq T, \tag{3.18}$$

where $\vartheta \in \Theta = (\alpha, \beta)$ with $-\infty < \alpha < \beta < +\infty$. The trend coefficient $S(\vartheta, x) = s(x - \vartheta)$, where the function $s(\cdot)$ is regular everywhere except 0, and has a cusp in 0. More precisely, we suppose that

$\mathcal{K}$. *The function $\sigma(\cdot)$ is strictly positive and continuous, and the function $S(\vartheta, x)$ admits the representation*

$$S(\vartheta, x) = \begin{cases} a\left|x - \vartheta\right|^{\kappa} + h(x - \vartheta), & \text{if } x \leq \vartheta \\ b\left|x - \vartheta\right|^{\kappa} + h(x - \vartheta), & \text{if } x \geq \vartheta \end{cases},$$

where $\kappa \in (0, 1/2)$, $a \neq 0$, $b \neq 0$, and the function $h(\cdot)$ satisfies the Hölder condition of order $\mu > \kappa + 1/2$.

Therefore in this parameter estimation problem the usual regularity conditions are not fulfilled, the Fisher information is equal to infinity, and to describe the asymptotic ($T \to \infty$) proprieties of the maximum likelihood estimator and the Bayes estimators we need a special study.

We also suppose that the conditions $\mathcal{ES}$, $\mathcal{EM}$ and $\mathcal{A}_0(\Theta)$ are fulfilled. Remember that in this case the process (3.18) has ergodic properties with invariant density

$$f(\vartheta, x) = \frac{1}{G(\vartheta)\,\sigma(x)^2} \exp\left\{2\int_{\vartheta}^{x} \frac{S(\vartheta, v)}{\sigma(v)^2}\,\mathrm{d}v\right\}, \qquad x \in \mathscr{R},$$

where

$$G(\vartheta) = \int_{\mathscr{R}} \frac{1}{\sigma(x)^2} \exp\left\{2\int_{\vartheta}^{x} \frac{S(\vartheta, v)}{\sigma(v)^2}\,\mathrm{d}v\right\}\,\mathrm{d}x$$

is the normalizing constant.

3.2.2 Properties of Estimators

We consider the problem of estimation of ϑ and describe the properties of the MLE and BE as $T \to \infty$.

Remember that the likelihood ratio in this problem is

$$L\left(\theta,X^T\right)=\exp\left\{\int_0^T\frac{S\left(\theta,X_t\right)}{\sigma(X_t)^2}\,\mathrm{d}X_t-\frac{1}{2}\int_0^T\frac{S\left(\theta,X_t\right)^2}{\sigma(X_t)^2}\,\mathrm{d}t\right\}.$$

To simplify the exposition we have omitted the part $f\left(\theta,X_0\right)$ concerning the initial value in this formula.

The MLE $\hat{\vartheta}_T$ and BE (for quadratic loss function) $\tilde{\vartheta}_T$ are defined by the usual relations

$$L\left(\hat{\vartheta}_T,X^T\right)=\sup_{\theta\in\Theta}L\left(\theta,X^T\right)$$

and

$$\tilde{\vartheta}_T=\int_\Theta\theta\,p\left(\theta|X^T\right)\,\mathrm{d}\theta,\qquad p\left(\vartheta|X^T\right)=\frac{p(\theta)\,L\left(\theta,X^T\right)}{\int_\Theta p(v)\,L\left(v,X^T\right)\,\mathrm{d}v}.$$

We suppose that the prior density $p\left(\cdot\right)$ is a positive and continuous on Θ function. Let us put $H=\kappa+1/2$ (the *Hurst parameter*) and introduce the *fractional Brownian motion* $W^H\left(\cdot\right)$, i.e., the Gaussian random process with zero mean and the covariance function

$$\mathbf{E}\,W^H\left(u_1\right)W^H\left(u_2\right)=\frac{1}{2}\left[\left|u_1\right|^{2H}+\left|u_2\right|^{2H}-\left|u_1-u_2\right|^{2H}\right].$$

Let us introduce the random function

$$Z\left(u\right)=\exp\left\{W^H\left(u\right)-\frac{1}{2}\left|u\right|^{2H}\right\},\qquad u\in\mathscr{R},$$

and define two random variables $\hat{u}$ and $\tilde{u}$ by the relations

$$Z\left(\hat{u}\right)=\sup_{u\in R}Z\left(u\right),\qquad\tilde{u}=\frac{\int_{\mathscr{R}}u\,Z(u)\,\mathrm{d}u}{\int_{\mathscr{R}}Z(v)\,\mathrm{d}v}.$$

We introduce as well the function

$$\Gamma_\vartheta^2=\frac{1}{G\left(\vartheta\right)\,\sigma(\vartheta)^4}\int_{-\infty}^{+\infty}\left[d(x-1)\left|x-1\right|^\kappa-d(x)\left|x\right|^\kappa\right]^2\mathrm{d}x,\qquad(3.19)$$

where

$$d(x)=\begin{cases}a, & \text{if } x<0\\ b, & \text{if } x>0\end{cases}.$$

Note that $\Gamma_\vartheta^2<\infty$ since $\kappa<1/2$, and that it has the following representation (see [109], Section VI.4)

$$\Gamma_\vartheta^2=\frac{1}{G\left(\vartheta\right)\,\sigma\left(\vartheta\right)^4}\;\frac{\Gamma(1+\kappa)\Gamma(\frac{1}{2}-\kappa)}{2^{2\kappa}\sqrt{\pi}(2\kappa+1)}\left[a^2+b^2-2ab\cos(\pi\kappa)\right],$$

or equally

$$\Gamma_\vartheta^2 = \frac{\mathrm{B}(\kappa+1\,,\,\kappa+1)}{G(\vartheta)\;\sigma(\vartheta)^4}\;\left[\frac{a^2+b^2}{\cos(\pi\kappa)} - 2ab\right].$$

Here $\Gamma(\cdot)$ and $\mathrm{B}(\cdot,\cdot)$ are gamma and beta functions respectively. Finally, we put $\gamma_\vartheta = \Gamma_\vartheta^{1/H}$.

The first result concerns the lower minimax bound.

Proposition 3.6. *Suppose that the conditions* $\mathcal{A}_0(\Theta)$ *and* $\mathcal{K}$ *are fulfilled. Then, for any* $\vartheta_0 \in \Theta$,

$$\lim_{\delta\to 0}\varliminf_{T\to\infty}\inf_{\bar\vartheta_T}\sup_{|\vartheta-\vartheta_0|<\delta} T^{1/H}\,\mathbf{E}_\vartheta\left(\bar\vartheta_T - \vartheta\right)^2 \geq \frac{\mathbf{E}\,\tilde u^2}{\gamma_{\vartheta_0}^2},$$

where inf *is taken over all estimators* $\bar\vartheta_T$.

The proof of this proposition is based on the asymptotic behavior of the Bayesian estimators, so we discuss it a little later. The more general result can be found in [109], Section I.9.

This inequality allows us to define the asymptotically efficient estimators as follows.

Definition 3.7. *Let the conditions* $\mathcal{A}_0(\Theta)$ *and* $\mathcal{K}$ *be fulfilled. We call an estimator* $\bar\vartheta_T$ *asymptotically efficient if, for any* $\vartheta_0 \in \Theta$,

$$\lim_{\delta\to 0}\lim_{T\to\infty}\sup_{|\vartheta-\vartheta_0|<\delta} T^{1/H}\,\mathbf{E}_\vartheta\left(\bar\vartheta_T - \vartheta\right)^2 = \frac{\mathbf{E}\,\tilde u^2}{\gamma_{\vartheta_0}^2}.$$

The properties of the estimators are described in the following theorem.

Theorem 3.8. (Dachian and Kutoyants [50]) *Let the conditions* $\mathcal{A}_0(\Theta)$ *and* $\mathcal{K}$ *be fulfilled. Then the MLE and BE are, uniformly on compacts* $\mathbb{K}\subset\Theta$, *consistent, have the following limits in distribution*

$$\mathcal{L}_\vartheta\left\{T^{1/2H}\left(\hat\vartheta_T - \vartheta\right)\right\} \Longrightarrow \mathcal{L}_\vartheta\left\{\frac{\hat u}{\gamma_\vartheta}\right\},$$

$$\mathcal{L}_\vartheta\left\{T^{1/2H}\left(\tilde\vartheta_T - \vartheta\right)\right\} \Longrightarrow \mathcal{L}_\vartheta\left\{\frac{\tilde u}{\gamma_\vartheta}\right\},$$

and for any $p>0$ *we have the convergence of moments too:*

$$T^{p/2H}\,\mathbf{E}_\vartheta\left|\hat\vartheta_T - \vartheta\right|^p \longrightarrow \mathbf{E}_\vartheta\left|\frac{\hat u}{\gamma_\vartheta}\right|^p,$$

$$T^{p/2H}\,\mathbf{E}_\vartheta\left|\tilde\vartheta_T - \vartheta\right|^p \longrightarrow \mathbf{E}_\vartheta\left|\frac{\tilde u}{\gamma_\vartheta}\right|^p.$$

Moreover, the BE are asymptotically efficient.

Proof. For simplicity of exposition, the proof will be carried out in the case $a = b$ and for convenience of notation, as usual, C and c denote generic positive constants which can differ from formula to formula and even in the same formula.

As we are going to apply Theorems 2.6 and 2.12 by Ibragimov and Khasminskii, we have to establish several properties of the likelihood ratio process

$$Z_T(u) = L\left(\vartheta_u, \vartheta, X^T\right), \quad u \in \mathbb{U}_T = \left(T^\gamma(\alpha - \vartheta), T^\gamma(\beta - \vartheta)\right),$$

where $\gamma = 1/2H$ and $\vartheta_u = \vartheta + u/T^\gamma$. These properties will be described below in Lemmas 3.10–3.12. But before, let us establish the following result.

Lemma 3.9. *Let the conditions $\mathcal{A}_0(\Theta)$ and $\mathcal{K}$ be fulfilled. Then*

1. *For any $u_1, u_2 \in \mathbb{U}_T$, uniformly on compacts $\mathbb{K} \subset \Theta$, the limit of the integral*

$$T\,I = T\int_{\mathscr{R}} \frac{\left[S(\vartheta_{u_1}, x) - S(\vartheta, x)\right]\left[S(\vartheta_{u_2}, x) - S(\vartheta, x)\right]}{\sigma(x)^2} f(\vartheta, x)\,\mathrm{d}x$$

is equal to

$$\frac{1}{2}\Gamma_\vartheta^2\left[\left|u_1\right|^{2H} + \left|u_2\right|^{2H} - \left|u_1 - u_2\right|^{2H}\right].$$

Here $\vartheta_{u_i} = \vartheta + T^{-\gamma}u_i$. In particular,

$$\lim_{T\to\infty} T\int_{\mathscr{R}} \left(\frac{S(\vartheta_u, x) - S(\vartheta, x)}{\sigma(x)}\right)^2 f(\vartheta, x)\,\mathrm{d}x = \Gamma_\vartheta^2\left|u\right|^{2H}. \tag{3.20}$$

2. *There exists a constant $C > 0$, such that*

$$\sup_{\vartheta\in\mathbb{K}} T\int_{\mathscr{R}} \left(\frac{S(\vartheta_{u_1}, x) - S(\vartheta_{u_2}, x)}{\sigma(x)}\right)^2 f(\vartheta_{u_2}, x)\,\mathrm{d}x \le C\left|u_2 - u_1\right|^{2H} \tag{3.21}$$

for all $T > 1$ and $u_1, u_2 \in \mathbb{U}_T$ such that $\left|u_1 - u_2\right| < 1$.

3. *There exists a constant $c_* = c_*(\mathbb{K}) > 0$, such that*

$$\inf_{\vartheta\in\mathbb{K}} \int_{\mathscr{R}} \left(\frac{S(\vartheta + u, x) - S(\vartheta, x)}{\sigma(x)}\right)^2 f(\vartheta, x)\,\mathrm{d}x \ge c_*\left|u\right|^{2H} \tag{3.22}$$

for all $u \in (\alpha - \vartheta, \beta - \vartheta)$.

Proof. We start with 1. Let us write

$$T\,I = T\,I_1 + T\,I_2 + T\,I_3 + T\,I_4$$

with

$$I_1 = \int_{\mathscr{R}} \frac{\left[a\,|x-\vartheta_{u_1}|^\kappa - a\,|x-\vartheta|^\kappa\right]\left[a\,|x-\vartheta_{u_2}|^\kappa - a\,|x-\vartheta|^\kappa\right]}{\sigma(x)^2}\, f(\vartheta,x)\,\mathrm{d}x,$$

$$I_2 = \int_{\mathscr{R}} \frac{\left[h(x-\vartheta_{u_1}) - h(x-\vartheta)\right]\left[h(x-\vartheta_{u_2}) - h(x-\vartheta)\right]}{\sigma(x)^2}\, f(\vartheta,x)\,\mathrm{d}x,$$

$$I_3 = \int_{\mathscr{R}} \frac{\left[a\,|x-\vartheta_{u_1}|^\kappa - a\,|x-\vartheta|^\kappa\right]\left[h(x-\vartheta_{u_2}) - h(x-\vartheta)\right]}{\sigma(x)^2}\, f(\vartheta,x)\,\mathrm{d}x,$$

$$I_4 = \int_{\mathscr{R}} \frac{\left[a\,|x-\vartheta_{u_2}|^\kappa - a\,|x-\vartheta|^\kappa\right]\left[h(x-\vartheta_{u_1}) - h(x-\vartheta)\right]}{\sigma(x)^2}\, f(\vartheta,x)\,\mathrm{d}x.$$

In order to study I_1, let us fix a function $A(T)$ such that $A(T) \to +\infty$ and $A(T)/T^\gamma \to 0$ and write the integral I_1 as a sum of two: an integral J_1 over the interval $\mathbb{L} = (\vartheta - A(T)T^{-\gamma}, \vartheta + A(T)T^{-\gamma})$, and an integral J_2 over the set $\mathbb{M} = \mathscr{R} \setminus \mathbb{L}$.

For J_1 we have

$$\begin{aligned}
T\,J_1 &= T a^2 \int_{\vartheta - A(T)/T^\gamma}^{\vartheta + A(T)/T^\gamma} \frac{\left[|x-\vartheta_{u_1}|^\kappa - |x-\vartheta|^\kappa\right]\left[|x-\vartheta_{u_2}|^\kappa - |x-\vartheta|^\kappa\right]}{\sigma(x)^2} f(\vartheta,x)\mathrm{d}x \\
&= T a^2 \int_{-A(T)/T^\gamma}^{A(T)/T^\gamma} \frac{\left[|y-u_1/T^\gamma|^\kappa - |y|^\kappa\right]\left[|y-u_2/T^\gamma|^\kappa - |y|^\kappa\right]}{\sigma(y+\vartheta)^2} f(\vartheta,y+\vartheta)\mathrm{d}y \\
&\simeq T\,\frac{a^2\, f(\vartheta,\vartheta)}{\sigma(\vartheta)^2} \int_{-A(T)}^{A(T)} \left[|z-u_1|^\kappa - |z|^\kappa\right]\left[|z-u_2|^\kappa - |z|^\kappa\right] T^{-\gamma(2\kappa+1)}\,\mathrm{d}z \\
&= \frac{a^2\, f(\vartheta,\vartheta)}{2\,\sigma(\vartheta)^2} \left\{ \int_{-A(T)}^{A(T)} \left[|z-u_1|^\kappa - |z|^\kappa\right]^2 \mathrm{d}z + \int_{-A(T)}^{A(T)} \left[|z-u_2|^\kappa - |z|^\kappa\right]^2 \mathrm{d}z \right. \\
&\qquad \left. - \int_{-A(T)}^{A(T)} \left[|z-u_1|^\kappa - |z-u_2|^\kappa\right]^2 \mathrm{d}z \right\}
\end{aligned}$$

where the symbol "$\simeq$" means equality of limits, and is true since $A(T)/T^\gamma \to 0$ and the functions $f(\vartheta,\cdot)$ and $\sigma(\cdot)$ are continuous ϑ. It is easy to see that

$$\lim_{T\to\infty}\int\limits_{-A(T)}^{A(T)}\left[|z-u|^{\kappa}-|z|^{\kappa}\right]^2\,\mathrm{d}z=|u|^{2H}\int\limits_{-\infty}^{+\infty}\left[|x-1|^{\kappa}-|x|^{\kappa}\right]^2\,\mathrm{d}x.$$

Hence we have clearly

$$\lim_{T\to\infty}T\,J_1=\frac{1}{2}\,\Gamma_{\vartheta}^2\Big[|u_1|^{2H}+|u_2|^{2H}-|u_1-u_2|^{2H}\Big]. \tag{3.23}$$

To study J_2, let us at first note that

$$\begin{aligned}T\int\limits_{M}\left[|x-\vartheta_u|^{\kappa}-|x-\vartheta|^{\kappa}\right]^2\,\mathrm{d}x&\le 2\,T\int\limits_{A(T)/T^{\gamma}}^{+\infty}\left[|y-u/T^{\gamma}|^{\kappa}-|y|^{\kappa}\right]^2\,\mathrm{d}y\\&=2\int\limits_{A(T)}^{+\infty}\left[|z-u|^{\kappa}-|z|^{\kappa}\right]^2\,\mathrm{d}z\to 0,\end{aligned}$$

since $A(T)\to+\infty$ and the integral is finite. Hence, using the Cauchy–Schwarz inequality, we easily get

$$\begin{aligned}|T\,J_2|&=\left|T\,a^2\int\limits_{\mathbb{M}}\frac{\left[|x-\vartheta_{u_1}|^{\kappa}-|x-\vartheta|^{\kappa}\right]\left[|x-\vartheta_{u_2}|^{\kappa}-|x-\vartheta|^{\kappa}\right]}{\sigma(x)^2}f(\vartheta,x)\,\mathrm{d}x\right|\\&\le C\sqrt{T\int\limits_{\mathbb{M}}\left[|x-\vartheta_{u_1}|^{\kappa}-|x-\vartheta|^{\kappa}\right]^2\mathrm{d}x\times T\int\limits_{M}\left[|x-\vartheta_{u_2}|^{\kappa}-|x-\vartheta|^{\kappa}\right]^2\mathrm{d}x}.\end{aligned}$$

Therefore $\lim\limits_{T\to\infty}T\,J_2=0$, and combining it with (3.23),

$$\lim_{T\to\infty}T\,I_1=\frac{1}{2}\,\Gamma_{\vartheta}^2\Big[|u_1|^{2H}+|u_2|^{2H}-|u_1-u_2|^{2H}\Big].$$

So it remains to show that

$$\lim_{T\to\infty}T\,I_2=\lim_{T\to\infty}T\,I_3=\lim_{T\to\infty}T\,I_4=0.$$

For this, it is sufficient to remark that

$$\begin{aligned}T\int\limits_{\mathscr{R}}\frac{\left[h(x-\vartheta_u)-h(x-\vartheta)\right]^2}{\sigma(x)^2}f(\vartheta,x)\,\mathrm{d}x&\le T\,C\,|u/T^{\gamma}|^{2\mu}\int\limits_{\mathscr{R}}\frac{f(\vartheta,x)}{\sigma(x)^2}\,\mathrm{d}x\\&\le C\,|u|^{2\mu}\,T^{-(\mu-H)/H}\to 0,\end{aligned}$$

and apply the Cauchy–Schwarz inequality. So, part 1 is proved.

To verify part 2 we write

$$T\int_{\mathscr{R}}\left(\frac{S\left(\vartheta_{u_1},x\right)-S\left(\vartheta_{u_2},x\right)}{\sigma\left(x\right)}\right)^2 f\left(\vartheta_{u_2},x\right)\,\mathrm{d}x$$

$$\leq C\,T\int_{\mathscr{R}}\left[|x-\vartheta_{u_1}|^{\kappa}-|x-\vartheta_{u_2}|^{\kappa}\right]^2\mathrm{d}x$$

$$+C\,T\int_{\mathscr{R}}\left[h(x-\vartheta_{u_1})-h(x-\vartheta_{u_2})\right]^2\frac{f(\vartheta_{u_2},x)}{\sigma(x)^2}\,\mathrm{d}x$$

$$=C\,T\,I_1+C\,T\,I_2$$

with obvious notation.

For the first integral we have clearly

$$T\,I_1=T\int_{\mathscr{R}}\left[|y-u_1/T^{\gamma}|^{\kappa}-|y-u_2/T^{\gamma}|^{\kappa}\right]^2\mathrm{d}y$$

$$=\int_{\mathscr{R}}\left[|z-u_1|^{\kappa}-|z-u_2|^{\kappa}\right]^2\mathrm{d}z$$

$$=|u_1-u_2|^{2H}\int_{-\infty}^{+\infty}\left[|x-1|^{p}-|x|^{p}\right]^2\mathrm{d}x=C\,|u_1-u_2|^{2H}.$$

For the second one, taking into account that $|u_1-u_2|<1$ and $T>1$, we get

$$T\,I_2\leq T\,C\left|(u_1-u_2)/T^{\gamma}\right|^{2\mu}\int_{\mathscr{R}}\frac{f(\vartheta_{u_2},x)}{\sigma(x)^2}\,\mathrm{d}x$$

$$\leq T\,C\left|(u_1-u_2)/T^{\gamma}\right|^{2H}=C\,|u_1-u_2|^{2H}.$$

So, we get finally

$$T\int_{\mathscr{R}}\left(\frac{S\left(\vartheta_{u_1},x\right)-S\left(\vartheta_{u_2},x\right)}{\sigma\left(x\right)}\right)^2 f\left(\vartheta_{u_2},x\right)\,\mathrm{d}x$$

$$\leq C\,T\,I_1+C\,I_2\leq C\,T\,|u_1-u_2|^{2H}.$$

To prove part 3 we first write

$$
\begin{aligned}
J(u) &= \int_{\mathscr{R}} \left[S(\vartheta+u,x)-S(\vartheta,x)\right]^2 \frac{f(\vartheta,x)}{\sigma(x)^2}\,\mathrm{d}x \\
&= c\int_{\mathscr{R}} \left[\left|x-\vartheta-u\right|^\kappa - \left|x-\vartheta\right|^\kappa\right]^2 \frac{f(\vartheta,x)}{\sigma(x)^2}\,\mathrm{d}x \\
&\quad + c\int_{\mathscr{R}} \left[h(x-\vartheta-u)-h(x-\vartheta)\right]^2 \frac{f(\vartheta,x)}{\sigma(x)^2}\,\mathrm{d}x \\
&\quad \pm c\int_{\mathscr{R}} \left[\left|x-\vartheta-u\right|^\kappa - \left|x-\vartheta\right|^\kappa\right]\left[h(x-\vartheta-u)-h(x-\vartheta)\right] \frac{f(\vartheta,x)}{\sigma(x)^2}\,\mathrm{d}x \\
&= c\,I_1 + c\,I_2 \pm c\,I_3
\end{aligned}
$$

with obvious notation. For the first integral we have

$$
c\,I_1 \le C\int_{-\infty}^{+\infty} \left[\left|y-u\right|^\kappa - \left|y\right|^\kappa\right]^2 \mathrm{d}y = C\left|u\right|^{2H}
$$

and

$$
\begin{aligned}
c\,I_1 &\ge c\int_{\alpha}^{\beta} \left[\left|x-\vartheta-u\right|^\kappa - \left|x-\vartheta\right|^\kappa\right]^2 \frac{f(\vartheta,x)}{\sigma(x)^2}\,\mathrm{d}x \\
&\ge c\int_{\alpha}^{\beta} \left[\left|x-\vartheta-u\right|^\kappa - \left|x-\vartheta\right|^\kappa\right]^2 \mathrm{d}x \\
&= c\left|u\right|^{2H}\,\mathrm{sign}(u)\int_{(\alpha-\vartheta)/u}^{(\beta-\vartheta)/u} \left[\left|z-1\right|^\kappa - \left|z\right|^\kappa\right]^2 \mathrm{d}z \\
&\ge c\left|u\right|^{2H}\int_{0}^{1} \left[\left|z-1\right|^\kappa - \left|z\right|^\kappa\right]^2 \mathrm{d}z = c\left|u\right|^{2H},
\end{aligned}
$$

since for $u \in (0\,,\beta-\vartheta)$ we have $(\alpha-\vartheta)/u < 0$ and $(\beta-\vartheta)/u > 1$, and for $u \in (\alpha-\vartheta\,,0)$ we have $(\alpha-\vartheta)/u > 1$ and $(\beta-\vartheta)/u < 0$.

For the second integral we get clearly $c\,I_2 \le C\left|u\right|^{2\mu}$, and hence, using the Cauchy–Schwarz inequality, we obtain $\left|c\,I_3\right| \le C\left|u\right|^{H+\mu}$ for the last integral, and finally

$$
J(u) \ge c\left|u\right|^{2H} - C\left|u\right|^{H+\mu} = c\left|u\right|^{2H}\left(1-C\left|u\right|^{\mu-H}\right) \ge c_1\left|u\right|^{2H}
$$

for all u such that $\left|u\right| \le \delta$ where $\delta > 0$ is some fixed constant.

On the other hand, we have also

$$\inf_{|u|\geq\delta} J(u) = c_2 > 0,$$

since otherwise we should have $S(\vartheta + u^*, x) = S(\vartheta, x)$ for some fixed u^* and almost all $x \in \mathscr{R}$, which is impossible. Hence, for all $|u| \geq \delta$ we can write

$$J(u) \geq c_2 \geq c_2 \frac{|u|^{2H}}{(\beta - \alpha)^{2H}} = c_3 \, |u|^{2H}.$$

So, for all ϑ and $u \in (\alpha - \vartheta, \beta - \vartheta)$ we have

$$J(u) \geq c_* \, |u|^{2H}$$

with $c_* = \min(c_1, c_3)$. Therefore Lemma 3.9 is proved.

Now let us turn to the properties of the likelihood ratio process $Z_T(\cdot)$. Put $Z_\vartheta(u) = Z(\gamma_\vartheta u), u \in \mathscr{R}$, i.e.,

$$Z_\vartheta(u) = \exp\left\{\Gamma_\vartheta \, W^H(u) - \frac{1}{2}\Gamma_\vartheta^2 \, |u|^{2H}\right\}, \qquad u \in \mathscr{R}.$$

Lemma 3.10. *Let the conditions $\mathcal{A}_0(\Theta)$ and $\mathcal{K}$ be fulfilled. Then the marginal distributions of the likelihood ratio $Z_T(\cdot)$ converge to the marginal distributions of the stochastic process $Z_\vartheta(\cdot)$ and this convergence is uniform in ϑ on the compacts $\mathbb{K} \subset \Theta$.*

Proof. As before, we put $\vartheta_u = \vartheta + u/T^\gamma$. The function $Z_T(\cdot)$ can be written as

$$\ln Z_T(u) = \int_0^T \frac{S(\vartheta_u, X_t) - S(\vartheta, X_t)}{\sigma(X_t)} \, \mathrm{d}W_t$$
$$- \frac{1}{2}\int_0^T \left(\frac{S(\vartheta_u, X_t) - S(\vartheta, X_t)}{\sigma(X_t)}\right)^2 \mathrm{d}t.$$

Using local time $\Lambda_T(\vartheta, x)$ of this diffusion process, we can write the second integral as

$$\int_0^T \left(\frac{S(\vartheta_u, X_t) - S(\vartheta, X_t)}{\sigma(X_t)}\right)^2 \mathrm{d}t$$

$$= 2\int_{\mathscr{R}} \frac{\left[S(\vartheta_u, x) - S(\vartheta, x)\right]^2}{\sigma(x)^4} \Lambda_T(\vartheta, x)\ \mathrm{d}x$$

$$= T\int_{\mathscr{R}} \left(\frac{S(\vartheta_u, x) - S(\vartheta, x)}{\sigma(x)}\right)^2 f(\vartheta, x)\ \mathrm{d}x$$

$$+ T\int_{\mathscr{R}} \left(\frac{\left[S(\vartheta_u, x) - S(\vartheta, x)\right)}{\sigma(x)}\right)^2 \left(\frac{2\Lambda_T(\vartheta, x)}{T\sigma(x)^2} - f(\vartheta, x)\right) \mathrm{d}x.$$

For the random function

$$\eta_T(\vartheta, x) = \sqrt{T}\left(\frac{2\Lambda_T(\vartheta, x)}{T\sigma(x)^2} - f(\vartheta, x)\right)$$

and any $m \geq 2$, under condition $\mathcal{A}_0(\Theta)$ we have the estimate

$$\sup_{\vartheta\in\Theta} \mathbf{E}_\vartheta \left|\eta_T(\vartheta, x)\right|^p \leq C\, e^{-\gamma|x|} \tag{3.24}$$

with some positive constants C, γ (see Proposition 1.2). Hence we can write

$$\mathbf{E}_\vartheta \left|\int_{\mathscr{R}} \left(\frac{S(\vartheta_u, x) - S(\vartheta, x)}{\sigma(x)}\right)^2 \eta_T(\vartheta, x)\ \mathrm{d}x\right|$$

$$\leq \int_{\mathscr{R}} \left(\frac{S(\vartheta_u, x) - S(\vartheta, x)}{\sigma(x)}\right)^2 \mathbf{E}_\vartheta \left|\eta_T(\vartheta, x)\right|\ \mathrm{d}x$$

$$\leq C\int_{\mathscr{R}} \left(\frac{S(\vartheta_u, x) - S(\vartheta, x)}{\sigma(x)}\right)^2 \mathrm{e}^{-\gamma|x|}\ \mathrm{d}x.$$

For the last integral, according to (3.21) we have

$$T^{1/2}\int_{\mathscr{R}} \left(\frac{S(\vartheta_u, x) - S(\vartheta, x)}{\sigma(x)}\right)^2 \mathrm{e}^{-\frac{c_2|x|}{2}}\ \mathrm{d}x \leq C\, T^{-1/2}\ |u|^{2H} \to 0.$$

Hence (see (3.20))

$$\mathbf{P} - \lim_{T\to\infty} \int_0^T \left(\frac{S(\vartheta_u, X_t) - S(\vartheta, X_t)}{\sigma(X_t)}\right)^2 \mathrm{d}t = \Gamma_\vartheta^2\ |u|^{2H},$$

and by the central limit theorem (Proposition 1.21) the stochastic integral is asymptotically normal:

$$\int_0^T \frac{S(\vartheta_u, X_t) - S(\vartheta, X_t)}{\sigma(X_t)}\, dW_t \Longrightarrow \mathcal{N}\left(0, \Gamma_\vartheta^2\, |u|^{2H}\right).$$

Therefore we have the convergence of the one-dimensional distributions of $Z_T(u)$ to those of $Z_\vartheta(u)$. The proof of the convergence of the multidimensional distributions is based on a part 1 of Lemma 3.9 and the mentioned central limit theorem. It is quite similar to the one given, so we omit it.

Lemma 3.11. *Let the conditions $\mathcal{A}_0(\Theta)$ and $\mathcal{K}$ be fulfilled. Then, for any compact $\mathbb{K} \subset \Theta$, there exist some constant $C > 0$ such that*

$$\sup_{\vartheta \in \mathbb{K}} \mathbf{E}_\vartheta \left| Z_T^{1/2}(u_1) - Z_T^{1/2}(u_2) \right|^2 \leq C\, |u_1 - u_2|^{2H}$$

for all $T > 1$ and $u_1, u_2 \in \mathbb{U}_T$.

Proof. For $|u_1 - u_2| \geq 1$ the assertion is evident since for all ϑ and T we have

$$\mathbf{E}_\vartheta \left| Z_T^{1/2}(u_1) - Z_T^{1/2}(u_2) \right|^2 \leq 4 \leq 4\, |u_1 - u_2|^{2H}.$$

Suppose now that $|u_1 - u_2| < 1$. Remember that the stochastic process

$$V_t = \left(\frac{Z_t(u_2)}{Z_t(u_1)}\right)^{1/2}, \qquad 0 \leq t \leq T,$$

by the Itô formula admits the representation (with $\mathbf{P}_{\vartheta_{u_1}}$ probability 1)

$$V_T = 1 - \frac{1}{8}\int_0^T V_t\, \delta(X_t)^2\, dt - \frac{1}{4}\int_0^T V_t\, \delta(X_t)\, dW_t,$$

where

$$\delta(x) = \frac{S(\vartheta_{u_2}, x) - S(\vartheta_{u_1}, x)}{\sigma(x)}.$$

Hence (see (1.42))

$$\begin{aligned}\mathbf{E}_\vartheta \left| Z_T^{1/2}(u_1) - Z_T^{1/2}(u_2) \right|^2 &= 2 - 2\,\mathbf{E}_{\vartheta_{u_1}} V(T) \\ &\leq \frac{T}{8}\int_{\mathscr{R}} \delta(x)^2\, \left[f(\vartheta_{u_2}, x) + f(\vartheta_{u_1}, x)\right]\, dx \leq C\, |u_2 - u_1|^{2H},\end{aligned}$$

where we used the estimate (3.21). The lemma is proved.

Lemma 3.12. *Let the conditions $\mathcal{A}_0(\Theta)$ and $\mathcal{K}$ be fulfilled. Then, for any compact $\mathbb{K} \subset \Theta$, there exists some constant $\kappa > 0$ and some function $C(N)$ defined for all $N > 0$, such that*

$$\sup_{\vartheta \in \mathbb{K}} \mathbf{P}_\vartheta \left\{ Z_T(u) \geq \mathrm{e}^{-\kappa|u|^{2H}} \right\} \leq \frac{C(N)}{|u|^N} .$$

Proof. We follow the proof of Lemma 2.11. Below $0 < c_1 < 1$ and

$$\delta(u,x) = \frac{S(\vartheta_u, x) - S(\vartheta, x)}{\sigma(x)}$$

we have

$$\begin{aligned}
&\mathbf{P}_\vartheta \left\{ Z_T(u) \geq \mathrm{e}^{-\kappa|u|^{2H}} \right\} \\
&\quad = \mathbf{P}_\vartheta \left\{ c_1 \int_0^T \delta(u, X_t)\, \mathrm{d}W_t - \frac{c_1}{2} \int_0^T \delta(u, X_t)^2 \,\mathrm{d}t \geq -c_1 \kappa |u|^{2H} \right\} \\
&\quad \leq \mathbf{P}_\vartheta \left\{ c_1 \int_0^T \delta(u, X_t)\, \mathrm{d}W_t - \frac{c_1^2}{2} \int_0^T \delta(u, X_t)^2 \,\mathrm{d}t \geq c_1 \kappa |u|^{2H} \right\} \\
&\qquad + \mathbf{P}_\vartheta \left\{ \frac{c_1 - c_1^2}{2} \left[T J\left(\frac{u}{T^\gamma}\right) - \int_0^T \delta(u, X_t)^2 \,\mathrm{d}t \right] \right. \\
&\qquad\qquad \left. > \frac{c_1 - c_1^2}{2} T J\left(\frac{u}{T^\gamma}\right) - 2c_1 \kappa |u|^{2H} \right\} \\
&\quad \leq \mathrm{e}^{-c_1\kappa|u|^{2H}} + \mathbf{P}_\vartheta \left\{ \frac{c_1 - c_1^2}{2} \left[\int_0^T \left[\mathbf{E}_\vartheta\, \delta(u, X_t)^2 - \delta(u, X_t)^2 \right] \mathrm{d}t \right] \right. \\
&\qquad\qquad \left. \geq \left(\frac{c_1 - c_1^2}{2} c_* - 2c_1\kappa \right) |u|^{2H} \right\},
\end{aligned}$$

where we used the estimate (3.22). Let us denote

$$h(u,x) = \mathbf{E}_\vartheta\, \delta(u, X_t)^2 - \delta(u, x)^2 ,$$

and put

$$\kappa = \frac{c_1 - c_1^2}{8c_1}\, c_* .$$

Then, for any $M > 1$, the last probability can be estimated as follows:

$$\mathbf{P}_\vartheta \left\{ \int_0^T h(u, X_t)\ \mathrm{d}t > \frac{c_*}{2} |u|^{2H} \right\}$$

$$\leq \left(\frac{2}{c_* |u|^{2H}} \right)^{2M} \mathbf{E}_\vartheta \left(\int_0^T h(u, X_t)\ \mathrm{d}t \right)^{2M}$$

$$\leq C\ |u|^{-4MH} \left(\mathbf{E}_\vartheta \left(\int_{X_0}^{X_T} \frac{H(u,x)}{\sigma(x)} \mathrm{d}x \right)^{2M} + T^M \,\mathbf{E}_\vartheta\, H(u,\xi)^{2M} \right),$$

where ξ is a random variable with the density $f(\vartheta, \cdot)$ and

$$H(u,x) = \frac{2}{\sigma(x) f(\vartheta, x)} \int_{-\infty}^{x} h(u,v) f(\vartheta, v)\ \mathrm{d}v.$$

Remember that $T\,\mathbf{E}_\vartheta\, \delta(u,\xi)^2 \leq C\,|u|^{2H}$. The similar estimate is valid for the function

$$T^M \,\mathbf{E}_\vartheta\, H(u,\xi)^{2M} \leq C\, T^{-M}\ |u|^{4MH}.$$

Hence, using the estimate $T^\gamma > |u|\ (\beta - \alpha)^{-1}$, we finally obtain

$$\mathbf{P}_\vartheta \left\{ Z_T(u) \geq \mathrm{e}^{-\kappa |u|^{2H}} \right\} \leq \frac{C}{T^M} \leq \frac{C(M)}{|u|^{M/\gamma}}.$$

The properties of the likelihood ratio described in Lemmas 3.10–3.12 allow us to cite Theorems 2.6 and 2.12. Further, Theorem 3.1 now follows from the limit behavior of the Bayes estimators as follows. Fix some $\delta > 0$. Then

$$\inf_{\bar\vartheta_T} \sup_{|\vartheta - \vartheta_0| < \delta} T^{1/H}\, \mathbf{E}_\vartheta \left(\bar\vartheta_T - \vartheta \right)^2$$

$$\geq \inf_{\bar\vartheta_T} T^{1/H} \int_{\vartheta_0 - \delta}^{\vartheta_0 + \delta} \mathbf{E}_\vartheta \left(\bar\vartheta_T - \vartheta \right)^2 p_\delta(\vartheta)\ \mathrm{d}\vartheta$$

$$= T^{1/H} \int_{\vartheta_0 - \delta}^{\vartheta_0 + \delta} \mathbf{E}_\vartheta \left(\tilde\vartheta_T - \vartheta \right)^2 p_\delta(\vartheta)\ \mathrm{d}\vartheta \qquad (3.25)$$

where $p_\delta(\cdot)$ is a continuous positive function (density *a priori*) on the interval

$$\Theta_\delta = [\vartheta_0 - \delta, \vartheta_0 + \delta]$$

and $\tilde\vartheta_T$ is the corresponding Bayesian estimator. As the moments of Bayesian estimators converge uniformly on ϑ, we have the limit

$$\lim_{T\to\infty} T^{1/H} \int_{\vartheta_0 - \delta}^{\vartheta_0 + \delta} \mathbf{E}_\vartheta \left(\tilde\vartheta_T - \vartheta \right)^2 p_\delta(\vartheta)\ \mathrm{d}\vartheta = \int_{\vartheta_0 - \delta}^{\vartheta_0 + \delta} \frac{\mathbf{E}_\vartheta\, \tilde u^2}{\gamma_\vartheta^2}\, p_\delta(\vartheta)\ \mathrm{d}\vartheta.$$

Now as $\delta \to 0$ we obtain

$$\lim_{\delta\to 0} \int_{\vartheta_0 - \delta}^{\vartheta_0 + \delta} \frac{\mathbf{E}_\vartheta\, \tilde u^2}{\gamma_\vartheta^2}\, p_\delta(\vartheta)\ \mathrm{d}\vartheta = \frac{\mathbf{E}_{\vartheta_0}\, \tilde u^2}{\gamma_{\vartheta_0}^2}. \qquad (3.26)$$

3.2.3 Discussions

One can consider a situation when the trend coefficient has several cusps. For example, suppose that the trend coefficient $S(\vartheta, x) = s(x - \vartheta)$, where the function $s(\cdot)$ is regular everywhere except at points $x_1, \ldots, x_r$, and has cusps of order p in this points. More precisely, we suppose that

$$S(\vartheta, x) = \sum_{i=1}^{r} d_i(x - \vartheta - x_i) \left| x - \vartheta - x_i \right|^{\kappa} + h(x - \vartheta),$$

where

$$d_i(x) = \begin{cases} a_i, & \text{if } x < 0 \\ b_i, & \text{if } x > 0 \end{cases},$$

$\kappa \in (0, 1/2)$, $a_i \neq 0$, $b_i \neq 0$, and the function $h(\cdot)$ satisfies the Hölder condition of order $\mu > \kappa + 1/2$.

In this situation, we obtain the similar behavior of the estimators where the only difference is the constant Γ_ϑ^2, which is now given by

$$\Gamma_\vartheta^2 = \sum_{i=1}^{r} \Gamma_{\vartheta,i}^2$$

with $\Gamma_{\vartheta,i}^2$ defined as in (3.19), but using $d_i(\cdot)$ in place of $d(\cdot)$. Indeed, if we introduce r independent fractional Brownian motions W_i^H, $i = 1, \ldots, r$, then it is not difficult to establish that the likelihood ratio process $Z_T(\cdot)$ converges to the stochastic process

$$\begin{aligned} Z_\vartheta(u) &= \exp\left\{ \sum_{i=1}^{r} \Gamma_{\vartheta,i} \, W_i^H(u) - \frac{1}{2} \left| u \right|^{2H} \sum_{i=1}^{r} \Gamma_{\vartheta,i}^2 \right\} \\ &= \exp\left\{ \Gamma_\vartheta \, W^H(u) - \frac{1}{2} \Gamma_\vartheta^2 \left| u \right|^{2H} \right\}, \end{aligned}$$

as well as the analog of Lemmas 3.11 and 3.12.

Windows

The problem considered here belongs to the class of problems, where the observations X^T can be replaced by the observations $Y^T = \{Y_t, 0 \leq t \leq T\}$ with $Y_t = X_t \, \chi_{\{X_t \in [\alpha, \beta]\}}$. Indeed, the main contribution to the asymptotics (3.20) and to the estimate (3.22) is due to the integral

$$\begin{aligned} T \int_{\vartheta - A(T)/T^\gamma}^{\vartheta + A(T)/T^\gamma} &\left(\frac{S(\vartheta + u/T^\gamma, x) - S(\vartheta, x)}{\sigma(x)} \right)^2 f_T^\circ(x) \, dx \\ &= \int_0^T \left(\frac{S(\vartheta + u/T^\gamma, X_t) - S(\vartheta, X_t)}{\sigma(X_t)} \right)^2 \chi_{\{|X_t - \vartheta| \leq A(T)/T^\gamma\}} dt, \end{aligned}$$

where we have denoted $f_T^\circ(x) = 2\Lambda_T(\vartheta, x)/T\sigma(x)^2$. Remember that

$$A(T)/T^\gamma \to 0, \quad \text{and} \quad \alpha < \vartheta + A(T)/T^\gamma < \beta.$$

Hence, if we introduce the *window* $\mathbb{B} = [\alpha, \beta]$ and the *pseudo-likelihood ratio*

$$\bar{L}\left(\theta, X^T\right) = \exp\left\{\int_0^T \frac{S(\theta, X_t)}{\sigma(X_t)^2}\chi_{\{X_t\in\mathbb{B}\}}\,\mathrm{d}X_t - \frac{1}{2}\int_0^T \frac{S(\theta, X_t)^2}{\sigma(X_t)^2}\chi_{\{X_t\in\mathbb{B}\}}\,\mathrm{d}t\right\},$$

then, as it follows from the proofs given above, the corresponding stochastic process

$$\bar{Z}_T(u) = \frac{\bar{L}\left(\vartheta + u/T^\gamma, X^T\right)}{\bar{L}\left(\vartheta, X^T\right)}, \quad u \in \mathbb{U}_T$$

has the same asymptotic properties and estimates as $Z_T(\cdot)$. Therefore the MLE and BE constructed with the help of the *pseudo-likelihood ratio* will have the same asymptotic properties as if the whole observations X^T were used. This *pseudo-likelihood ratio* corresponds to the *misspecified model* of observations

$$\mathrm{d}X_t = S(\vartheta, X_t)\ \chi_{\{X_t\in\mathbb{B}\}}\,\mathrm{d}t + \sigma(X_t)\ \mathrm{d}W_t, \quad X_0, 0 \le t \le T. \tag{3.27}$$

Of course, we do not suppose that the observations come from Equation (3.27) and just use the *contrast function* $\bar{L}\left(\theta, X^T\right)$ to construct the estimators. As one of the consequences we need not remember all observations X^T and it is sufficient to save X_t which fits the *window* $\mathbb{B}$ only. Remember that the local time estimator of the density is sufficient statistics as well and in this problem we can save the values of $X_0, X_T,\ f_T^\circ(x), x \in [\alpha, \beta]$ only. Moreover, the detailed analysis shows that we can consider a two-step procedure with a preliminary consistent estimator $\bar{\vartheta}_{\sqrt{T}}$ constructed by the observations $Y^{\sqrt{T}} = \left\{X_t\,\chi_{\{X_t\in\mathbb{B}\}}, 0 \le t \le \sqrt{T}\right\}$. Then we introduce another window $\mathbb{B}_T = \left[\bar{\vartheta}_{\sqrt{T}} - h_T, \bar{\vartheta}_{\sqrt{T}} + h_T\right]$, where $h_T \to 0$ is a *slowly decreasing function* and construct the MLE and BE on the base of the *pseudo-likelihood ratio* $\bar{L}\left(\theta, X^T_{\sqrt{T}}\right)$ and the window $\mathbb{B}_T$. In this case the length of the window tends to zero (in the scheme of series), but nevertheless the MLE and the BE have the same asymptotic properties as those described in Theorem 3.8. Hence we have even an asymptotically efficient Bayes estimator constructed by the observations in the *vanishing window*. Note that the lower bound of Proposition 3.6 is based on the whole observations.

This reduction of the observations we consider a bit later in detail in Section 3.4.

3.3 Delay Estimation

In this section we deal with the linear stochastic differential equation

$$\mathrm{d}X_t = -\gamma\, X_{t-\tau}\, \mathrm{d}t + \sigma \mathrm{d}W_t, \quad X_0\,(s)\,, -\tau \leq s \leq 0, \quad 0 \leq t \leq T \qquad (3.28)$$

where $\vartheta = (\gamma, \tau) \in \Theta$ is an unknown two-dimensional parameter. We suppose that the function $\{X_0\,(s)\,, -\tau \leq s \leq 0\}$ is deterministic. We have to estimate ϑ by the observations $\{X_t, 0 \leq t \leq T\}$ and, as usual, describe the properties of the estimators (MLE and BE) as $T \to \infty$.

Note that the observed linear process (3.28) is not diffusion because its "trend coefficient" $S\,(\vartheta, X) = -\gamma\, X_{t-\tau}$ depends on the value of X_t at time $t - \tau$. This is a so-called *diffusion type* or *Itô process* (see [175] for a definition and the properties of the solutions of these equations). We just mention here that according to the general existence theorem ([175], Theorem 4.6) the strong solution of (3.28) exists and is unique. Moreover all the measures $\left\{\mathbf{P}_{\vartheta}^{(T)}, \vartheta \in \Theta\right\}$ induced in the measurable space of continuous on $[0, T]$ functions $(\mathscr{C}_T, \mathfrak{B}_T)$ by the solutions of (3.28) are equivalent.

The integral representation of the observed process

$$X_t = X_0(0) + \gamma \int_0^t X_{s-\tau}\, \mathrm{d}s + \sigma\, W_t,$$

shows that the random function

$$X_{t-\tau} = X_0(0) + \gamma \int_0^{t-\tau} X_{s-\tau}\, \mathrm{d}s + \sigma\, W_{t-\tau},$$

is as smooth with respect to parameter τ as the Wiener process W_t with respect to time t. Therefore the trend coefficient is not differentiable with respect to τ and the regularity conditions of Section 2.1 are not satisfied.

3.3.1 SDE with Delay

We are interested in the ergodic solutions of Equation (3.28), i.e., solutions having an ergodic property. Of course the process X_t with a deterministic initial value is not stationary, but it can have an invariant distribution and satisfy the law of large numbers with some invariant law.

Let us introduce the *fundamental solution* $\{x_0\,(t)\,, t \geq 0\}$ of the corresponding deterministic equation

$$\frac{\mathrm{d}x_0\,(t)}{\mathrm{d}t} = -\,\gamma\, x_0\,(t-\tau)\,, \quad x_0\,(s) = 0,\; s < 0, \quad x_0\,(0) = 1. \qquad (3.29)$$

This equation can be solved *step by step* (on the intervals $[k\tau, (k+1)\,\tau]$) and the function $x_0\,(t)\,, 0 \leq t \leq T$ can be written explicitly as

$$x_0\left(t\right)=\sum_{k=0}^{[t/\tau]}\left(-1\right)^k\frac{\gamma^k}{k!}\;\left(t-k\,\tau\right)^k$$

where $[A]$ is the integer part of A.

The properties of the solution of (3.28) are summarized in the following lemma.

Lemma 3.13. *The process* $\{X_t,\, t\geq 0\}$ *admits the representation*

$$X_t=X_0\,x_0\left(t\right)-\gamma\int_{-\tau}^{0}x_0\left(t-s-\tau\right)X_0\left(s\right)\mathrm{d}s+\sigma\int_0^t x_0\left(t-s\right)\mathrm{d}W_s. \quad (3.30)$$

A stationary solution of (3.28) exists if and only if $\gamma>0$ *and*

$$\tau\in\left(0,\frac{\pi}{2\gamma}\right).$$

Moreover the distribution of this stationary solution coincides with the distribution of the stochastic process

$$Y_t=\sigma\int_{-\infty}^{t}x_0\left(t-s\right)\mathrm{d}w_s,\quad t\in\mathscr{R} \quad (3.31)$$

where $\{w_s,\; s\in\mathscr{R}\}$ *is some Wiener processes on the line.*

Proof. For a proof see [125], Proposition 2.2, Theorems 2.5 and 2.8.

The stationary solution is a Gaussian process with mean zero and the covariance function

$$R(t-s,\boldsymbol{\vartheta})=\sigma^2\int_0^\infty x_0(|t-s|+v)\;x_0(v)\;\mathrm{d}v$$

because $x_0(t)=x_0(t,\boldsymbol{\vartheta})$. We put $r^2(\boldsymbol{\vartheta})=\sigma^{-2}R(0,\boldsymbol{\vartheta})$. It can be shown (see Hale *et al.* [96]) that for any compact set $\mathbb{K}\subset\boldsymbol{\Theta}$ there exist positive constants K_1 and c such that

$$\sup_{\boldsymbol{\vartheta}\in\mathbb{K}}|x_0(t)|\leq K_1\,\mathrm{e}^{-ct}. \quad (3.32)$$

This estimate and the representation (3.30) allow us to write

$$\sup_{\boldsymbol{\vartheta}\in\mathbb{K}}\left|\mathbf{E}_{\boldsymbol{\vartheta}}\left(X_tX_s\right)-R(t-s,\boldsymbol{\vartheta})\right|\leq K_2\,\mathrm{e}^{-c(t+s)}, \quad (3.33)$$

$$\sup_{\boldsymbol{\vartheta}\in\mathbb{K}}\left|R(t-s,\boldsymbol{\vartheta})\right|\leq K_3\,\mathrm{e}^{-|t-s|}. \quad (3.34)$$

Therefore, in our consideration we can replace the function $\mathbf{E}_{\boldsymbol{\vartheta}}\left(X_tX_s\right)$ by $R(t-s,\boldsymbol{\vartheta})$ with an asymptotically negligible difference.

3.3.2 Properties of MLE and BE

We observe a trajectory $X^T = \{X_t,\ 0 \le t \le T\}$ of the diffusion-type process given by the stochastic differential equation

$$\mathrm{d}X_t = -\,\gamma\, X_{t-\tau}\ \mathrm{d}t + \sigma\ \mathrm{d}W_t, \qquad t \ge 0, \tag{3.35}$$

and we suppose that the unknown parameter $\boldsymbol{\vartheta} \in \boldsymbol{\Theta}$. The only condition in this problem is

$\mathcal{L}$. *The set*

$$\boldsymbol{\Theta} = (\alpha_1, \beta_1) \times (\alpha_2, \beta_2)\,,$$

where the parameters $\alpha_1 > 0$, $\alpha_2 > 0$ *and* $\beta_2 \le \pi/2\beta_1$.

We assume that the deterministic *initial function* $\left\{X_0(s),\ \frac{\pi}{2\alpha_1} \le s \le 0\right\}$ is given too.

To construct the MLE and BE we introduce the likelihood ratio function

$$L(\boldsymbol{\vartheta}, X^T) = \frac{\mathrm{d}\,\mathbf{P}_{\boldsymbol{\vartheta}}^{(T)}}{\mathrm{d}\,\mathbf{P}^{(T)}}(X^T), \qquad \boldsymbol{\vartheta} \in \boldsymbol{\Theta},$$

(here $\mathbf{P}^{(T)} = \mathbf{P}_{0,0}^{(T)}$) by the formula

$$L(\boldsymbol{\vartheta}, X^T) = \exp\left\{-\frac{\gamma}{\sigma^2}\int_0^T X_{t-\tau}\ \mathrm{d}X_t - \frac{\gamma^2}{2\sigma^2}\int_0^T X_{t-\tau}^2\ \mathrm{d}t\right\}.$$

The maximum likelihood estimator (MLE) $\hat{\boldsymbol{\vartheta}}_T$ is defined by the equation

$$L(\hat{\boldsymbol{\vartheta}}_T, X^T) = \sup_{\boldsymbol{\vartheta} \in \boldsymbol{\Theta}} L(\boldsymbol{\vartheta}, X^T). \tag{3.36}$$

We will see later (Lemma 3.20) that the function $\left\{L(\boldsymbol{\vartheta}, X^T),\ \boldsymbol{\vartheta} \in \boldsymbol{\Theta}\right\}$ is continuous with probability 1. Hence the MLE $\hat{\boldsymbol{\vartheta}}_T$ exists and belongs to $\bar{\boldsymbol{\Theta}}$. If Equation (3.36) has more than one solution then we take any one as the MLE.

To introduce the Bayes estimator we suppose that $\boldsymbol{\vartheta}$ is a random vector with a prior density $\{p(\boldsymbol{y}), \boldsymbol{y} \in \boldsymbol{\Theta}\}$ and the loss function is quadratic. Then the Bayes estimator (BE) $\tilde{\boldsymbol{\vartheta}}_T$ is a posterior mean

$$\tilde{\boldsymbol{\vartheta}}_T = \int_{\boldsymbol{\Theta}} \boldsymbol{y}\, p(\boldsymbol{y}|X^T)\ \mathrm{d}\boldsymbol{y},$$

where the posterior density $p(\boldsymbol{y}|X^T)$ is given by

$$p(\boldsymbol{y}|X^T) = \frac{p(\boldsymbol{y})\ L(\boldsymbol{y}, X^T)}{\int_{\boldsymbol{\Theta}} p(\boldsymbol{v})\ L(\boldsymbol{v}, X^T)\ \mathrm{d}\boldsymbol{v}}.$$

The limit distributions of these estimators can be expressed with the help of the following stochastic process:

$$Z_\tau(u) = \exp\left\{\gamma\, W(u) - \frac{\gamma^2}{2}\,|u|\right\}, \qquad u \in \mathscr{R}, \tag{3.37}$$

where $W(\cdot)$ is a two-sided Wiener process, i.e., $W(u) = W_+(u)$ for $u \geq 0$ and $W(u) = W_-(-u)$ for $u < 0$. Here $W_+(u), W_-(u)$, $u \geq 0$ are two independent standard Wiener processes. Let us introduce two random variables $\hat{u}$ and $\tilde{u}$ by the equations

$$Z_\tau(\hat{u}) = \sup_{u \in \mathscr{R}} Z_\tau(u), \tag{3.38}$$

$$\tilde{u} = \frac{\int_{\mathscr{R}} u\, Z_\tau(u)\, \mathrm{d}u}{\int_{\mathscr{R}} Z_\tau(u)\, \mathrm{d}u}. \tag{3.39}$$

The process $Z_\tau(\cdot)$ arises as the limit of likelihood ratio in different nonregular estimation problems (see [109], Section 7.2, [139], Chapter 5 and references therein). Let us mention here three useful results concerning $\hat{u}$ and $\tilde{u}$. Below

$$\sigma_0^2 = 4\,\frac{\partial^2}{\partial z^2}\int_0^\infty x\, K_1(x)\, K_{\sqrt{1+8z}}(x) \int_0^x y\, K_1(y)\, I_{\sqrt{1+8z}}(y)\, \mathrm{d}y\, \mathrm{d}x\Bigg|_{z=0}, \tag{3.40}$$

where $K_{(\cdot)}(\cdot)$ and $I_{(\cdot)}(\cdot)$ are the modified Hankel and Bessel functions respectively.

Lemma 3.14. *Let $\gamma \neq 0$, then Equation (3.38) has a unique solution with probability 1 and the random variables $\hat{u}$ and $\tilde{u}$ have the following second moments*

$$\mathbf{E}\,\hat{u}^2 = \frac{26}{\gamma^2}, \tag{3.41}$$

$$\mathbf{E}\,\tilde{u}^2 = \frac{\sigma_0^2}{\gamma^2} \tag{3.42}$$

where $\sigma_0^2 \simeq 19.276 \pm 0.06$.

Proof. The uniqueness of the solution of Equation (3.38) follows from [109], Lemma VII.2.5, Equality (3.41) was obtained by Terent'yev [227] (see as well [109], Section 7.3) and the representation (3.40) and (3.42) was obtained by Golubev [88].

Introduce the normalizing matrix

$$\varphi_T = \begin{pmatrix} T^{-1/2}, & 0 \\ 0 & , T^{-1} \end{pmatrix}.$$

Proposition 3.15 and Theorems 3.17, 3.18, presented below, are generalizations to the two-dimensional case ($\boldsymbol{\vartheta} = (\gamma, \tau)$) of our work (Küchler and

Kutoyants [126]) devoted to the estimation of the one-dimensional parameter $\vartheta = \tau$.

The first result allows us to define the asymptotically efficient estimators in this problem.

Proposition 3.15. *Suppose that the condition $\mathcal{L}$ is fulfilled. Then for any $\vartheta_0 \in \Theta$*

$$\lim_{\delta\to 0}\lim_{T\to\infty}\inf_{\bar\vartheta_T}\sup_{|\vartheta-\vartheta_0|<\delta} \mathbf{E}_\vartheta \left|\varphi_T^{-1}\left(\bar\vartheta_T-\vartheta\right)\right|^2 \geq r\left(\vartheta_0\right)^{-2} + \mathbf{E}_{\vartheta_0}\left(\tilde u\right)^2. \quad (3.43)$$

The proof of this proposition, which is a particular case of Theorem I.9.1 in [109], is based on the asymptotic properties of the BE so we will discuss it after the proof of Theorem 3.18 below. Similar arguments were used in Section 3.2 (see (3.25) and (3.26)).

Therefore we can give the following

Definition 3.16. *We call an estimator ϑ^*_T asymptotically efficient if for all $\vartheta_0 \in \Theta$*

$$\lim_{\delta\to 0}\lim_{T\to\infty}\sup_{|\vartheta-\vartheta_0|<\delta} \mathbf{E}_\vartheta \left|\varphi_T^{-1}\left(\vartheta^*_T-\vartheta\right)\right|^2 = r\left(\vartheta_0\right)^{-2} + \mathbf{E}_{\vartheta_0}\left(\tilde u\right)^2. \quad (3.44)$$

Let $\mathbb{K}$ below be an arbitrary compact set in Θ and $\hat v = \zeta\, r\left(\vartheta\right)^{-1}$ be a Gaussian random variable, $\zeta \sim \mathcal{N}\left(0,1\right)$.

The asymptotic properties of the MLE $\hat\vartheta_T = \left(\hat\gamma_T, \hat\tau_T\right)$ and BE $\tilde\vartheta_T$ are described in the following theorems.

Theorem 3.17. *Let the condition $\mathcal{L}$ be fulfilled. Then the MLE $\hat\vartheta_T$ of the parameter ϑ is uniformly in $\vartheta \in \mathbb{K}$ consistent,*

$$\mathcal{L}_\vartheta\left\{\varphi_T^{-1}\left(\hat\vartheta_T-\vartheta\right)\right\} \Longrightarrow \mathcal{L}_\vartheta\left(\hat w\right), \quad (3.45)$$

where the random vector $\hat w = \left(\hat v, \hat u\right)$ and this convergence is uniform on $\vartheta \in \mathbb{K}$. Moreover, for any $p > 0$ it holds

$$\lim_{T\to\infty} \mathbf{E}_\vartheta \left|\varphi_T^{-1}\left(\hat\vartheta_T-\vartheta\right)\right|^p = \mathbf{E}_\vartheta\left|\hat w\right|^p.$$

Theorem 3.18. *Let the condition $\mathcal{L}$ be fulfilled and the prior density $p(\cdot)$ on Θ, be a positive, continuous function. Then the BE $\tilde\vartheta_T$ is uniformly in $\vartheta \in \mathbb{K}$ consistent,*

$$\mathcal{L}_\vartheta\left\{\varphi_T^{-1}\left(\tilde\vartheta_T-\vartheta\right)\right\} \Longrightarrow \mathcal{L}_\vartheta\left(\tilde w\right),$$

where the random vector $\tilde w = \left(\hat v, \tilde u\right)$, this convergence is uniform on $\vartheta \in \mathbb{K}$ and for any $p > 0$ it holds that

$$\lim_{T\to\infty} \mathbf{E}_\vartheta \left|\varphi_T^{-1}\left(\tilde\vartheta_T-\vartheta\right)\right|^p = \mathbf{E}_\vartheta\left|\tilde w\right|^p.$$

Moreover the BE is asymptotically efficient.

Proof. The proof is based on the results of Theorems 2.6 and 2.12, so we check the conditions of these theorems with the help of the three lemmas presented below. That is, we show the convergence of marginal distributions of the normalized likelihood ratio $Z_T(\cdot)$ and establish two estimates on the increments $Z_T(\cdot)$ and on its decrease.

Let us denote $\boldsymbol{w} = (v, u)$ and introduce the normalized likelihood ratio

$$Z_T(v,u) = \frac{\mathrm{d}\,\mathbf{P}^{(T)}_{\vartheta+\varphi_T \boldsymbol{w}}}{\mathrm{d}\,\mathbf{P}^{(T)}_{\vartheta}}(X^T), \qquad \boldsymbol{w} \in \mathbb{W}_T,$$

where the set

$$\mathbb{W}_T = \left((\alpha_1-\gamma)\sqrt{T}, (\beta_1-\gamma)\sqrt{T}\right) \times \left((\alpha_2-\tau)\,T, (\beta_2-\tau)\,T\right).$$

Outside of $\mathbb{W}_T$ we define $Z_T(\boldsymbol{w})$ linearly decreasing to zero in the band of width 1 and put $Z_T(\boldsymbol{w}) = 0$ on the rest of the plane $\mathscr{R}^2$. So the function $Z_T(\boldsymbol{w})$ is defined for all $\boldsymbol{w} \in \mathscr{R}^2$ and it can be shown with the help of Lemmas 3.19 and 3.20 that it belongs to the space $\mathscr{C}_0 = \mathscr{C}_0\left(\mathscr{R}^2\right)$ of continuous on $\mathscr{R}^2$ functions decreasing to zero at infinity. Denote by $\mathfrak{B}_0 = \mathfrak{B}_0\left(\mathscr{R}^2\right)$ the corresponding Borel σ-algebra.

Let us introduce the limit likelihood ratio

$$Z_\vartheta\left(\boldsymbol{w}\right) = Z_\gamma\left(v\right)\, Z_\tau\left(u\right), \qquad \boldsymbol{w} = (v,u) \in \mathscr{R}^2,$$

where

$$Z_\gamma\left(v\right) = \exp\left\{v\,\Delta - \frac{v^2}{2}\,r\left(\vartheta\right)^2\right\}, \quad \mathcal{L}_\vartheta\left(\Delta\right) = \mathcal{N}\left(0, r\left(\vartheta\right)^2\right)$$

and $Z_\tau\left(\cdot\right)$ is defined in (3.37).

We have the following presentations for the MLE and BE:

$$\left(\sqrt{T}(\hat{\gamma}_T-\gamma), T\left(\hat{\tau}_T-\tau\right)\right) \equiv \hat{\boldsymbol{w}}_T = \arg\sup_{\boldsymbol{w}\in\mathbb{W}_T} Z_T(\boldsymbol{w})$$

and

$$\left(\sqrt{T}(\tilde{\gamma}_T-\gamma), T\left(\tilde{\tau}_T-\tau\right)\right) \equiv \tilde{\boldsymbol{w}}_T = \frac{\int_{\mathbb{W}_T} \boldsymbol{w}\, p\left(\vartheta+\varphi_T \boldsymbol{w}\right)\, Z_T(\boldsymbol{w})\,\mathrm{d}\boldsymbol{w}}{\int_{\mathbb{W}_T} p\left(\vartheta+\varphi_T \boldsymbol{w}\right)\, Z_T(\boldsymbol{w})\,\mathrm{d}\boldsymbol{w}}.$$

Lemma 3.19. *Let the condition* $\boldsymbol{\mathcal{L}}$ *be fulfilled, then the marginal distributions of* $Z_T(\cdot)$ *converge to the marginal distributions of* $Z_T(\cdot)$ *and this convergence is uniform over the compacts* $\mathbb{K} \subset \boldsymbol{\Theta}$.

Proof. The likelihood ratio $Z_T(\cdot)$ with $\mathbf{P}_\vartheta^{(T)}$ probability 1 admits the representation

$$
\begin{aligned}
\ln Z_T(v,u) &= -\frac{v}{\sigma\sqrt{T}}\int_0^T X_{t-\tau-\frac{u}{T}}\,\mathrm{d}W_t - \frac{v^2}{2\sigma^2 T}\int_0^T X_{t-\tau-\frac{u}{T}}^2\,\mathrm{d}t\\
&\quad -\frac{\gamma}{\sigma}\int_0^T \left(X_{t-\tau-\frac{u}{T}} - X_{t-\tau}\right)\mathrm{d}W_t - \frac{\gamma^2}{2\sigma^2}\int_0^T \left(X_{t-\tau-\frac{u}{T}} - X_{t-\tau}\right)^2\mathrm{d}t\\
&\quad -\frac{v\gamma}{\sigma^2\sqrt{T}}\int_0^T X_{t-\tau-\frac{u}{T}}\left(X_{t-\tau-\frac{u}{T}} - X_{t-\tau}\right)\mathrm{d}t\\
&= -v\, I_1(u) - \frac{v^2}{2}\, I_2(u) - \frac{\gamma}{\sigma}\, I_3(u) - \frac{\gamma^2}{2\sigma^2}\, I_4(u) - \frac{v\gamma}{\sigma^2}\, I_5(u)
\end{aligned}
\tag{3.46}
$$

with obvious notation. We show that for a fixed u

$$
\begin{aligned}
&\mathbf{P}_\vartheta - \lim_{T\to\infty} I_2(u) = r(\vartheta)^2, \quad \mathbf{P}_\vartheta - \lim_{T\to\infty} I_4(u) = \sigma^2\,|u|,\\
&\mathbf{P}_\vartheta - \lim_{T\to\infty} I_5(u) = 0,
\end{aligned}
$$

uniformly on $\vartheta\in\mathbb{K}$, which imply the uniform on $\vartheta\in\mathbb{K}$ convergences

$$
\mathcal{L}_\vartheta\left\{I_1(u)\right\} \Longrightarrow \mathcal{L}_\vartheta(\Delta) = \mathcal{N}\Big(0, r(\vartheta)^2\Big), \quad \mathcal{L}_\vartheta\left\{I_3(u)\right\} \Longrightarrow \mathcal{L}\Big(\sigma\, W(u)\Big),
$$

where the random variable Δ and the two-sided Wiener process $W(\cdot)$ are independent. These provide the convergence of one-dimensional distributions only. Then we show how the general case can be done.

At first we show that uniformly in $\vartheta\in\mathbb{K}$

$$
\mathbf{P}_\vartheta - \lim_{T\to\infty}\frac{1}{T\sigma^2}\int_0^T X_{t-\tau-\frac{u}{T}}^2\,\mathrm{d}t = r(\vartheta)^2.
\tag{3.47}
$$

To simplify the exposition we suppose that the process X_t is stationary. The estimates (3.32) and (3.34) show that the difference is asymptotically negligible. Therefore we can use the representation

$$
X_t = \sigma\int_{-\infty}^t x_0(t-s)\;\mathrm{d}W_s, \quad \mathbf{E}_\vartheta\, X_t^2 = R(0,\vartheta)^2.
$$

The process $\{X_t, t\geq 0\}$ is Gaussian so the direct calculation gives us the estimate (below $t_u = t-\tau-\frac{u}{T}$ $s_u = s-\tau-\frac{u}{T}$)

$$
\begin{aligned}
&\mathbf{E}_\vartheta\left(\frac{1}{T}\int_0^T X_{t_u}^2\,\mathrm{d}t - \mathbf{E}_\vartheta\, X_{t_u}^2\right)^2\\
&\qquad = \frac{1}{T^2}\int_0^T\int_0^T \mathbf{E}_\vartheta\left(X_{t_u}^2 - \mathbf{E}_\vartheta\, X_{t_u}^2\right)\left(X_{s_u}^2 - \mathbf{E}_\vartheta\, X_{s_u}^2\right)\;\mathrm{d}t\;\mathrm{d}s\\
&\qquad = \frac{2}{T^2}\int_0^T\int_0^T \left(\int_{-\infty}^{t_u\wedge s_u} x_0(t_u-q)\, x_0(s_u-q)\,\mathrm{d}q\right)^2\;\mathrm{d}t\;\mathrm{d}s\\
&\qquad = \frac{2}{T^2}\int_0^T\int_0^T R(t-s,\vartheta)\;\mathrm{d}t\;\mathrm{d}s \leq \frac{C}{T}
\end{aligned}
$$

because we have the estimate (3.33) as well. Hence for any $\delta > 0$

$$\sup_{\vartheta\in\mathbb{K}} \mathbf{P}_\vartheta^{(T)} \left\{ \left| \frac{1}{T}\int_0^T X_{t_u}^2\,\mathrm{d}t - \mathbf{E}_\vartheta\, X_{t_u}^2 \right| > \delta \right\}$$
$$\leq \frac{1}{\delta^2} \sup_{\vartheta\in\mathbb{K}} \mathbf{E}_\vartheta \left(\frac{1}{T}\int_0^T X_{t_u}^2\,\mathrm{d}t - \mathbf{E}_\vartheta\, X_{t_u}^2 \right)^2 \leq \frac{C}{\delta^2 T} \to 0$$

and we have the uniform on $\vartheta \in \mathbb{K}$ convergence of I_2, which implies the uniform asymptotic normality of the stochastic integral I_1.

For the integral I_4 we show that for any $\delta > 0$

$$\lim_{T\to\infty} \sup_{\vartheta\in\mathbb{K}} \mathbf{P}_\vartheta^{(T)} \left\{ \left| \int_0^T \left(X_{t-\tau-\frac{u}{T}} - X_{t-\tau}\right)^2 \mathrm{d}t - \sigma^2\, |u| \right| > \delta \right\} = 0.$$

Let $u > 0$. The difference $Y_t(u) = X_{t-\tau-\frac{u}{T}} - X_{t-\tau}$ can be written as

$$\begin{aligned} Y_t(u) &= -\gamma \int_{t-\tau}^{t-\tau-\frac{u}{T}} X_{s-\tau}\,\mathrm{d}s + \sigma\,\left(W_{t-\tau-\frac{u}{T}} - W_{t-\tau}\right) \\ &= \frac{\gamma\, u}{T}\, X_{t-2\tau}\,(1+o(1)) + \frac{\sigma}{\sqrt{T}} W_t(u). \end{aligned}$$

Here

$$W_t(u) = T^{1/2}\left(W_{t-\tau-\frac{u}{T}} - W_{t-\tau}\right)$$

is (for each fixed t) a Wiener process with respect to $u \geq 0$: it is continuous with probability 1, $W_t(0) = 0$ and

$$\mathbf{E}\, W_t(u) = 0, \quad \mathbf{E}\, W_t(u)^2 = u, \quad \mathbf{E}\, W_t(u_1)\, W_t(u_2) = \min(u_1, u_2).$$

Therefore

$$Y_t(u) = \frac{\sigma}{\sqrt{T}}\, W_t(u)\,(1+o(1)).$$

The process X_t is Gaussian and stationary. Hence we have $T^{-1/2} X_t \to 0$ as well as $T^{-1/2}\,\mathbf{E}_\vartheta\, |X_t|^p \to 0$ for any $p > 0$ as $T \to \infty$. We can write

$$\begin{aligned} &\sup_{\vartheta\in\mathbb{K}} \mathbf{E}_\vartheta \left(\int_0^T Y_t(u)^2\,\mathrm{d}t - |u|\sigma^2 \right)^2 \\ &\quad = \sup_{\vartheta\in\mathbb{K}} \mathbf{E}_\vartheta \left(\frac{\sigma^2}{T}\int_0^T \left[W_t(u)^2 - |u|\right]\,\mathrm{d}t \right)^2 (1+o(1)) \\ &\quad \leq \frac{\sigma^4}{T^2} \int_0^T\int_0^T \sup_{\vartheta\in\mathbb{K}} \mathbf{E}_\vartheta \left(W_t(u)^2\, W_s(u)^2 - u^2 \right)(1+o(1))\,\mathrm{d}s\,\mathrm{d}t. \end{aligned}$$

For the values s, t such that $|t-s| > |u|/T$ these increments of the Wiener process are independent and we have

$$\mathbf{E}_\vartheta\, W_t(u)^2\, W_s(u)^2 = u^2.$$

In the case $|t-s| \leq |u|/T$, say, $t < s$ we have $s - u/T < t$ and using the independence of the increments $W_s - W_t$, $W_{t-u/T} - W_t$, $W_{t-u/T} - W_{s-u/T}$ we can write

$$\left|\mathbf{E}_\vartheta\, W_t(u)^2\, W_s(u)^2 - u^2\right| \leq 6\, u^2.$$

Therefore

$$\sup_{\vartheta\in\mathbb{K}} \int_0^T \int_0^T \mathbf{E}_\vartheta \left(\left[Y_t(u)^2 - \frac{|u|\sigma^2}{T}\right]\left[Y_s(u)^2 - \frac{|u|\sigma^2}{T}\right]\right) \mathrm{d}t\,\mathrm{d}s \leq C\frac{u^2}{T}.$$

Similar arguments allow us to prove the uniform convergence

$$\mathbf{P}_\vartheta - \lim_{T\to\infty} \int_0^T Y_t(u_1)\, Y_t(u_2)\, \mathrm{d}t = |u_1| \wedge |u_2|\, \sigma^2. \tag{3.48}$$

The central limit theorem for stochastic integrals and convergence (3.46) provide the uniform asymptotic normality

$$\mathcal{L}_\vartheta \left\{\int_0^T Y_t(u)\, \mathrm{d}W_t\right\} \Longrightarrow \mathcal{N}\left(0, |u|\sigma^2\right).$$

Note that the integrals I_1 and I_3 are asymptotically normal and independent because

$$\lim_{T\to\infty} \mathbf{E}_\vartheta\,(I_1\, I_3) = 0.$$

The same central limit theorem and (3.48) give us the convergence of the finite-dimensional distributions of the process

$$z_T(u) = \int_0^T Y_t(u)\, \mathrm{d}W_t, \quad u \in \mathscr{R}$$

to the finite dimensional distributions of the process $\{\sigma\, W(u),\, u \in \mathscr{R}\}$.

Direct but cumbersome calculation gives us

$$\mathbf{E}_\vartheta \left(\frac{1}{\sqrt{T}} \int_0^T X_{t-\tau-\frac{u}{T}}\left(X_{t-\tau-\frac{u}{T}} - X_{t-\tau}\right) \mathrm{d}t\right)^2 \leq \frac{C\,|u|}{T}.$$

Therefore $I_5 \to 0$.

Lemma 3.20. *Let the condition $\mathcal{L}$ be fulfilled. Then there exists a constant C_0 such that for any $R > 0$*

$$\sup_{\vartheta\in\mathbb{K}} \mathbf{E}_\vartheta \left|Z_T^{\frac{1}{4}}(v_1,u_1) - Z_T^{\frac{1}{4}}(v_2,u_2)\right|^4 \leq C_0\,(1+R^6)\left(|v_1-v_2|^4 + |u_1-u_2|^2\right).$$

Proof. With the help of Lemma 1.13 ($m = 2$) we obtain for the random function

$$V_T = \left(\frac{\mathrm{d}\,\mathbf{P}_{\vartheta_1}^{(T)}}{\mathrm{d}\,\mathbf{P}_{\vartheta_2}^{(T)}}(X_T) \right)^{\frac{1}{4}},$$

(here $\vartheta_i = \left(\gamma + v_i/\sqrt{T}, \tau + u_i/T\right), i = 1,2$) the following estimate (with $\mathbf{P}_{\vartheta_1}^{(T)}$ probability 1)

$$\begin{aligned}
|V_T - 1| &\leq \frac{3}{33\sigma^2} \int_0^T V_t \left[\left(\gamma + \frac{v_1}{\sqrt{T}}\right) X_{t_{u_1}} - \left(\gamma + \frac{v_2}{\sqrt{T}}\right) X_{t_{u_2}} \right]^2 \mathrm{d}t \\
&\quad + \frac{1}{4\sigma} \left| \int_0^T V_t \left[\left(\gamma + \frac{v_1}{\sqrt{T}}\right) X_{t_{u_1}} - \left(\gamma + \frac{v_2}{\sqrt{T}}\right) X_{t_{u_2}} \right] \mathrm{d}W_t \right| \\
&\leq \frac{3\left(v_1 - v_2\right)^2}{16\sigma^2 T} \int_0^T V_t\, X_{t_{u_1}}^2 \,\mathrm{d}t + \frac{3\gamma_2^2}{16\sigma^2} \int_0^T V_t\, \left(X_{t_{u_1}} - X_{t_{u_2}}\right)^2 \mathrm{d}t \\
&\quad + \frac{|v_1 - v_2|}{4\sigma\sqrt{T}} \left| \int_0^T V_t\, X_{t_{u_1}} \,\mathrm{d}W_t \right| + \frac{\gamma_2}{4\sigma} \left| \int_0^T V_t\, \left(X_{t_{u_1}} - X_{t_{u_1}}\right) \mathrm{d}W_t \right|,
\end{aligned}$$

where $t_{u_i} = t - \tau - u_i/T$ and $\gamma_2 = \gamma + v_2/\sqrt{T}$. Therefore

$$\begin{aligned}
\mathbf{E}_\vartheta \left| Z_T(v_1,u_1)^{1/4} - Z_T(v_2,u_2)^{1/4} \right|^4 &= \mathbf{E}_\vartheta\; Z_T(v_2,u_2) |V(T) - 1|^4 \\
&\leq C_1\, \frac{\left(v_1 - v_2\right)^8}{T} \int_0^T \mathbf{E}_{\vartheta_2} V_t^4\, X_{t_{u_1}}^8 \;dt \\
&\quad + C_2\, T^3 \int_0^T \mathbf{E}_{\vartheta_2} V_t^4\, \left(X_{t_{u_1}} - X_{t_{u_2}}\right)^8 \mathrm{d}t \\
&\quad + C_3\, \frac{\left(v_1 - v_2\right)^4}{T} \int_0^T \mathbf{E}_{\vartheta_2} V_t^4\, X_{t_{u_1}}^4 \;\mathrm{d}t \\
&\quad + C_4\, T^3 \int_0^T \mathbf{E}_{\vartheta_2} V_t^4\, \left(X_{t_{u_1}} - X_{t_{u_2}}\right)^4 \mathrm{d}t.
\end{aligned}$$

Note that

$$\mathbf{E}_{\vartheta_2} V_t^4\, X_{t_{u_1}}^8 = \mathbf{E}_{\vartheta_1}\; X_{t_{u_1}}^8 \leq C.$$

The moments of $X_{t_{u_1}} - X_{t_{u_2}}$ can be estimated as follows:

$$\begin{aligned}
&\mathbf{E}_{\vartheta_1} \left(X_{t_{u_1}} - X_{t_{u_2}}\right)^8 \\
&\quad = \mathbf{E}_{\vartheta_1} \left(\gamma_1 \int_{t_{u_1}}^{t_{u_2}} X_{s-\tau-u_1/T} \,\mathrm{d}s + \sigma\left(W_{t_{u_1}} - W_{t_{u_2}}\right) \right)^8 \\
&\quad \leq C \left| \frac{u_1 - u_2}{T} \right|^7 \int_{t_{u_1}}^{t_{u_2}} \mathbf{E}_{\vartheta_1} X_{s-\tau-u_1/T}^8 \,\mathrm{d}s + C \left| \frac{u_1 - u_2}{T} \right|^4 \\
&\quad \leq C \left(\left| \frac{u_1 - u_2}{T} \right|^8 + \left| \frac{u_1 - u_2}{T} \right|^4 \right).
\end{aligned}$$

Similar arguments provide the estimate

$$\mathbf{E}_{\vartheta_1}\left(X_{t_{u_1}}-X_{t_{u_2}}\right)^4\leq C\left(\left|\frac{u_1-u_2}{T}\right|^4+\left|\frac{u_1-u_2}{T}\right|^2\right).$$

Therefore we have

$$\begin{aligned}&\mathbf{E}_{\vartheta}\left|Z_T(v_1,u_1)^{1/4}-Z_T(v_2,u_2)^{1/4}\right|^4\\&\leq C\left(\left|v_1-v_2\right|^8+\left|v_1-v_2\right|^4+\left|u_1-u_2\right|^8+\left|u_1-u_2\right|^4+\left|u_1-u_2\right|^2\right)\\&\leq C_0\left(1+R^6\right)\left(\left|v_1-v_2\right|^4+\left|u_1-u_2\right|^2\right),\end{aligned}$$

where the constant $C_0>0$ can be chosen not depending on R and ϑ.

The last estimate is given in the following lemma.

Lemma 3.21. *Let the condition $\mathcal{L}$ be fulfilled, then for any $N>0$ there exist constants $C=C(N)>0$ and $\kappa>0$ such that*

$$\sup_{\vartheta\in\mathbb{K}}\mathbf{P}_{\vartheta}^{(T)}\left\{Z_T(v,u)>\mathrm{e}^{-\kappa\left(v^2+|u|\right)}\right\}\leq\frac{C}{\left(v^2+|u|\right)^N}.\tag{3.49}$$

Proof. The proof is similar to the proof of Lemma 2.11. Let us denote

$$q(t,\vartheta_1,\vartheta)=\frac{\gamma_1\,X_{t-\tau_1}-\gamma\,X_{t-\tau}}{\sigma},\qquad\vartheta_1=(\gamma_1,\tau_1),$$

where $\gamma_1=\gamma+\frac{v}{\sqrt{T}}$, $\tau_1=\tau+\frac{u}{T}$ and introduce the set

$$\mathbb{A}=\left\{\omega:\ \int_0^T q(t,\vartheta_1,\vartheta)^2\,\mathrm{d}t\geq 8\kappa\left(v^2+|u|\right)\right\},$$

where the number $\kappa>0$ will be chosen later. Then we can write

$$\begin{aligned}&\mathbf{P}_{\vartheta}^{(T)}\left\{Z_T(v,u)>\mathrm{e}^{-\kappa\left(v^2+|u|\right)}\right\}\\&\quad=\mathbf{P}_{\vartheta}^{(T)}\left\{-\int_0^T q(t,\vartheta_1,\vartheta)\,\mathrm{d}W_t\right.\\&\qquad\qquad\left.-\frac{1}{2}\int_0^T q(t,\vartheta_1,\vartheta)^2\,\mathrm{d}t\geq-\kappa\left(v^2+|u|\right)\right\}\\&\quad\leq\mathbf{P}_{\vartheta}^{(T)}\left\{-\int_0^T\frac{q(t,\vartheta_1,\vartheta)}{2}\,\mathrm{d}W_t\right.\\&\qquad\qquad\left.-\int_0^T\frac{q(t,\vartheta_1,\vartheta)^2}{8}\,\mathrm{d}t\geq\frac{\kappa}{2}\left(v^2+|u|\right),\mathbb{A}\right\}\\&\qquad\qquad+\mathbf{P}_{\vartheta}^{(T)}\left\{\mathbb{A}^c\right\}\leq\mathrm{e}^{-\frac{\kappa}{2}\left(v^2+|u|\right)}+\mathbf{P}_{\vartheta}^{(T)}\left\{\mathbb{A}^c\right\},\end{aligned}$$

where we used the Markov inequality and the equality

$$\mathbf{E}_{\vartheta} \exp\left\{-\int_0^T \frac{q(t,\vartheta_1,\vartheta)}{2}\mathrm{d}W_t - \int_0^T \frac{q(t,\vartheta_1,\vartheta)^2}{8}\mathrm{d}t\right\} = 1.$$

Further, put $Y_t = q(t,\vartheta_1,\vartheta)^2 - \mathbf{E}_{\vartheta}\, q(t,\vartheta_1,\vartheta)^2$. Then we have

$$\begin{aligned}\mathbf{P}_{\vartheta}^{(T)}\{\mathbb{A}^c\} &= \mathbf{P}_{\vartheta}^{(T)}\left\{\int_0^T q(t,\vartheta_1,\vartheta)^2\,\mathrm{d}t < 8\kappa\left(v^2+|u|\right)\right\}\\ &= \mathbf{P}_{\vartheta}^{(T)}\left\{\left|\int_0^T Y_t\,\mathrm{d}t\right| > \int_0^T \mathbf{E}_{\vartheta}\, q(t,\vartheta_1,\vartheta)^2\,\mathrm{d}t - 8\kappa\left(v^2+|u|\right)\right\}. \qquad (3.50)\end{aligned}$$

For the last mathematical expectation we can write

$$\begin{aligned}\mathbf{E}_{\vartheta}\, q(t,\vartheta_1,\vartheta)^2 &= \sigma^{-2}\,\mathbf{E}_{\vartheta}\left[\left(\gamma+\frac{v}{\sqrt{T}}\right)X_{t-\tau-\frac{u}{T}} - \gamma\, X_{t-\tau}\right]^2\\ &= \frac{v^2}{T\sigma^2}R(0,\vartheta) + 2\,\frac{\gamma_1\gamma}{\sigma^2}\left[R(0,\vartheta) - R\left(\frac{u}{T},\vartheta\right)\right].\end{aligned}$$

The function $R(\cdot)$ has explicit representation on the interval $[0,\tau]$

$$R(s,\vartheta) = \frac{1}{2\gamma\sigma^2}\left[\frac{1+\sin(\gamma\tau)}{\cos(\gamma\tau)}\cos(\gamma s) - \sin(\gamma s)\right]$$

and is symmetric $R(-s,\vartheta) = R(s,\vartheta)$ (see [125], Proposition 2.13). We can see that the right derivative of this function at zero is equal to $\dot R(0+) = -1/2\sigma^2$. Hence there exist numbers $\nu > 0$ and κ_1 such that for $|s| < \nu$

$$\inf_{\vartheta\in\mathbb{K}}\left[R(0,\vartheta) - R(s,\vartheta)\right] \geq \kappa_1\,|s|.$$

Note as well that for the other values of s we have

$$\inf_{\vartheta\in\mathbb{K}}\inf_{\nu\leq|s|<\tau}\left[R(0,\vartheta) - R(s,\vartheta)\right] = g(\nu) > 0.$$

Therefore for $|s| < \beta_2 - \alpha_2$

$$\inf_{\vartheta\in\mathbb{K}}\left[R(0,\vartheta) - R(s,\vartheta)\right] \geq g(\nu) \geq g(\nu)\,\frac{|s|}{\beta_2-\alpha_2} = \kappa_2\,|s|. \qquad (3.51)$$

We have as well the obvious estimate

$$\inf_{\vartheta\in\mathbb{K}} R(0,\vartheta) = \kappa_3 > 0.$$

Finally we obtain

$$\inf_{\vartheta\in\mathbb{K}} \mathbf{E}_\vartheta \int_0^T q\left(t,\vartheta_1,\vartheta\right)^2 \mathrm{d}t \geq \kappa_3\, v^2 + 2\alpha_1^2\, \kappa_2\, \left|u\right| \geq \kappa_*\left(v^2+\left|u\right|\right), \qquad (3.52)$$

where $\kappa_* = \min\left(\kappa_3, 2\alpha_1^2\, \kappa_2\right)$.

Now we choose $\kappa = \kappa_*/16$ and (3.50) becomes

$$\mathbf{P}_\vartheta^{(T)}\left\{\mathbb{A}^c\right\} \leq \mathbf{P}_\vartheta^{(T)}\left\{\left|\int_0^T Y_t\,\mathrm{d}t\right| > \frac{\kappa_*}{2}\left(v^2+\left|u\right|\right)\right\}.$$

For the last step we use the estimate which follows from the following lemma

Lemma 3.22. *Let $\{y_t, t>0\}$ be a stochastic process with mean zero and for some $m>2$*

$$\mathbf{E}\left|y_t\right|^{m(2k-1)} < C_1, \qquad \int_0^\infty t^{k-1}\left[\alpha\left(t\right)\right]^{(m-2)/m}\,\mathrm{d}t < C_2, \qquad (3.53)$$

where $\alpha(t)$ is the coefficient of the strong mixing. Then

$$\mathbf{E}\left|\int_0^T y_t\,\mathrm{d}t\right|^{2k} \leq C_3\,T^k. \qquad (3.54)$$

Proof. For the proof see Lemma 2.1 in [123].

Remember that $\{X_t, t\geq 0\}$ is Gaussian process with covariance function majorized by an exponentially decreasing function. Therefore the conditions (3.53) are fulfilled.

We write the integral of the process

$$Y_t = q\left(t,\vartheta_1,\vartheta\right)^2 - \mathbf{E}_\vartheta\, q\left(t,\vartheta_1,\vartheta\right)^2,\ t\geq 0$$

as a sum of three terms:

$$\int_0^T Y_t\,\mathrm{d}t = \frac{v^2}{\sigma^2 T}\int_0^T \left[X_{t-\tau-u/T}^2 - \mathbf{E}_\vartheta\, X_{t-\tau-u/T}^2\right]\,\mathrm{d}t$$
$$+\frac{2v\,\gamma}{\sigma^2\sqrt{T}}\int_0^T \left[X_{t-\tau-u/T}\left(X_{t-\tau-u/T} - X_{t-\tau}\right) - R\left(0,\vartheta\right) + R\left(u/T,\vartheta\right)\right]\,\mathrm{d}t$$
$$+\frac{\gamma^2}{\sigma^2}\int_0^T \left[\left(X_{t-\tau-u/T} - X_{t-\tau}\right)^2 - \mathbf{E}_\vartheta\left(X_{t-\tau-u/T} - X_{t-\tau}\right)^2\right]\,\mathrm{d}t.$$

For the first term by Lemma 3.22 we have

$$\mathbf{P}_\vartheta^{(T)}\left\{\frac{v^2}{\sigma^2 T}\left|\int_0^T \left[X_{t-\tau-u/T}^2 - \mathbf{E}_\vartheta\, X_{t-\tau-u/T}^2\right]\,\mathrm{d}t\right| > c\left(v^2+\left|u\right|\right)\right\}$$
$$\leq \frac{C\,v^{4k}}{\left(v^2+\left|u\right|\right)^{2k}}\quad T^{-k}.$$

Remember that

$$v^2+|u|\leq\left[(\beta_1-\alpha_1)^2+\beta_2-\alpha_2\right]\,T. \tag{3.55}$$

Hence the last expression can be estimated as

$$\frac{C\,v^{4k}}{\left(v^2+|u|\right)^{2k}}\quad T^{-k}\leq\frac{C'}{\left(v^2+|u|\right)^{k}}.$$

For the second term using the same lemma we obtain ($t_u=t-\tau-u/T$)

$$\begin{aligned}\mathbf{P}_{\vartheta}^{(T)}\Biggl\{\Bigl|\frac{v}{\sqrt{T}}\int_0^T\Bigl[X_{t_u}\left(X_{t_u}-X_{t-\tau}\right)-R\left(0,\vartheta\right)\\ +R\left(u/T,\vartheta\right)\Bigr]\mathrm{d}t\Bigr|>c\left(v^2+|u|\right)\Biggr\}\\ \leq\frac{C\,v^{2k}}{\left(v^2+|u|\right)^{2k}}\leq\frac{C}{\left(v^2+|u|\right)^{k}}.\end{aligned}$$

For the third term we consider separately two sets. The first one is

$$\left\{v,u:\quad T^{3/4}<v^2+|u|<\left[(\beta_1-\alpha_1)^2+\beta_2-\alpha_2\right]\,T\right\}$$

and the estimate is

$$\begin{aligned}\mathbf{P}_{\vartheta}^{(T)}\Biggl\{\Bigl|\int_0^T\Bigl[\left(X_{t_u}-X_{t-\tau}\right)^2-2R\left(0,\vartheta\right)\\ +2R\left(u/T,\vartheta\right)\Bigr]\,\mathrm{d}t\Bigr|>c\left(v^2+|u|\right)\Biggr\}\\ \leq\frac{C}{\left(v^2+|u|\right)^{2k}}\quad T^{k}\leq\frac{C}{\left(v^2+|u|\right)^{2k/3}},\end{aligned}$$

where we used the inequality $T^k<\left(v^2+|u|\right)^{4k/3}$. On the set

$$\left\{v,u:v^2+|u|\leq T^{3/4}\right\}$$

we at first note that

$$2\,\left[R\left(0,\vartheta\right)-R\left(u/T,\vartheta\right)\right]=\frac{|u|}{T}\,\left(1+O\left(T^{-1/4}\right)\right).$$

Therefore we consider the main term $\frac{|u|}{T}$ only. We can write

$$\begin{aligned}\sigma^{-2}\left(X_{t_u}-X_{t-\tau}\right)^2-\frac{|u|}{T}&=\left(-\frac{\gamma}{\sigma}\int_{t-\tau}^{t_u}X_{s-\tau}\,\mathrm{d}s+\left(W_{t_u}-W_{t-\tau}\right)\right)^2-\frac{|u|}{T}\\ &=\left(\frac{\gamma}{\sigma}\int_{t-\tau}^{t_u}X_{s-\tau}\,\mathrm{d}s\right)^2-2\frac{\gamma}{\sigma}\int_{t-\tau}^{t_u}X_{s-\tau}\,\mathrm{d}s\,\left(W_{t_u}-W_{t-\tau}\right)\\ &\quad+\left(W_{t_u}-W_{t-\tau}\right)^2-\frac{|u|}{T}.\end{aligned}$$

We have as previously three terms. For the first one

$$\mathbf{P}_{\vartheta}^{(T)}\left\{\int_0^T\left(\int_{t-\tau}^{t_u} X_{s-\tau}\,\mathrm{d}s\right)^2\mathrm{d}t > c\left(v^2+|u|\right)\right\}$$
$$\leq \frac{C}{\left(v^2+|u|\right)^{2k}}\,\mathbf{E}_{\vartheta}\left(\int_0^T\left(\int_{t-\tau}^{t_u} X_{s-\tau}\,\mathrm{d}s\right)^2\mathrm{d}t\right)^{2k}$$
$$\leq \frac{C}{\left(v^2+|u|\right)^{2k}}\;\frac{u^{4k}}{T^{2k}} \leq \frac{C\,u^{2k}}{\left(v^2+|u|\right)^{2k}}\;T^{-k/2} \leq \frac{C}{\left(v^2+|u|\right)^{k/2}}$$

where we have used again the estimate (3.55). Using the similar arguments we obtain the second estimate

$$\mathbf{P}_{\vartheta}^{(T)}\left\{\int_0^T\int_{t-\tau}^{t_u} X_{s-\tau}\,\mathrm{d}s\;\left(W_{t_u}-W_{t-\tau}\right)\mathrm{d}t > c\left(v^2+|u|\right)\right\}$$
$$\leq \frac{C}{\left(v^2+|u|\right)^{2k}}\;\frac{|u|^{3k}}{T^{k}} \leq \frac{C\,u^{2k}}{\left(v^2+|u|\right)^{2k}}\;T^{-k/4} \leq \frac{C}{\left(v^2+|u|\right)^{k/4}}.$$

And for the last term we have

$$\mathbf{P}_{\vartheta}^{(T)}\left\{\left|\frac{1}{T}\int_0^T\left[T\left(W_{t_u}-W_{t-\tau}\right)^2-|u|\right]\,\mathrm{d}t\right| > c\left(v^2+|u|\right)\right\}$$
$$\leq \frac{C\,T^{-2k}}{\left(v^2+|u|\right)^{2k}}\,\mathbf{E}_{\vartheta}\left|\int_0^T\left[T\left(W_{t_u}-W_{t-\tau}\right)^2-|u|\right]\,\mathrm{d}t\right|^{2k}$$
$$\leq \frac{C}{\left(v^2+|u|\right)^{2k}}\;\frac{u^{2k}}{T^{k}} \leq \frac{C}{\left(v^2+|u|\right)^{k}}.$$

Therefore we obtain (3.49) with $N=k/4$.

The properties of the likelihood ratio $Z_T(\cdot)$ established in Lemmas 3.19–3.21 allow us to apply Theorems 2.6 and 2.12 and so to obtain the desired properties of the MLE and BE.

To prove the lower minimax bound (3.43) we (following [109], Theorem 1.9.1) fix some $\delta>0$ and $\vartheta_0\in\Theta$. Then we introduce a prior distribution $\{p_\delta(y),\, y\in\Theta_\delta\}$, where the set

$$\Theta_\delta=\{\vartheta:\;|\vartheta-\vartheta_0|\leq\delta\}.$$

We take the function $p_\delta(\cdot)$ positive on Θ_δ and continuous. Denote as $\tilde{\vartheta}_T^{\delta}$ the Bayes estimator of the parameter $\vartheta\in\Theta_\delta$. For any estimator $\bar{\vartheta}_T$ we have the obvious inequalities

$$\sup_{|\vartheta-\vartheta_0|<\delta} \mathbf{E}_\vartheta \left|\varphi_T^{-1}\left(\bar{\vartheta}_T-\vartheta\right)\right|^2 \geq \int_{\Theta_\delta} \mathbf{E}_\vartheta \left|\varphi_T^{-1}\left(\bar{\vartheta}_T-\vartheta\right)\right|^2 p_\delta\left(\vartheta\right)\,\mathrm{d}\vartheta$$
$$\geq \int_{\Theta_\delta} \mathbf{E}_\vartheta \left|\varphi_T^{-1}\left(\tilde{\vartheta}_T^\delta-\vartheta\right)\right|^2 p_\delta\left(\vartheta\right)\,\mathrm{d}\vartheta.$$

By Theorem 3.18 we have the uniform on ϑ convergence

$$\mathbf{E}_\vartheta \left|\varphi_T^{-1}\left(\tilde{\vartheta}_T^\delta-\vartheta\right)\right|^2 \longrightarrow \mathbf{E}_\vartheta \left|\tilde{w}\right|^2.$$

Hence for any $\bar{\vartheta}_T$

$$\varliminf_{T\to\infty} \sup_{|\vartheta-\vartheta_0|<\delta} \mathbf{E}_\vartheta \left|\varphi_T^{-1}\left(\bar{\vartheta}_T-\vartheta\right)\right|^2 \geq \int_{\Theta_\delta} \mathbf{E}_\vartheta \left|\tilde{w}\right|^2 p_\delta\left(\vartheta\right)\,\mathrm{d}\vartheta.$$

Note that $\left\{\mathbf{E}_\vartheta \left|\tilde{w}\right|^2,\ \vartheta\in\Theta_\delta\right\}$ is a continuous function of ϑ. Therefore

$$\varliminf_{\delta\to 0} \varliminf_{T\to\infty} \sup_{|\vartheta-\vartheta_0|<\delta} \mathbf{E}_\vartheta \left|\varphi_T^{-1}\left(\bar{\vartheta}_T-\vartheta\right)\right|^2 \geq \mathbf{E}_{\vartheta_0} \left|w\right|^2 = r\left(\vartheta_0\right)^{-2} + \mathbf{E}_{\vartheta_0}\left(\tilde{u}\right)^2,$$

which proves Proposition 3.15.

3.3.3 Discussion

Similar problems of the estimation of parameter τ were considered in [137] (see also [139], Section 2.4) for the models of observation

$$\mathrm{d}X_t = \gamma\, X_{t-\tau}\,\mathrm{d}t + \varepsilon\,\mathrm{d}W(t), \quad X_0 = x_0,\ 0\leq t\leq T, \quad \tau>0, \tag{3.56}$$

and

$$\mathrm{d}X_t = \gamma\, X_{\tau t}\,\mathrm{d}t + \varepsilon\,\mathrm{d}W(t), \quad X_0 = x_0,\ 0\leq t\leq T \quad 0<\tau<1, \tag{3.57}$$

but in asymptotics of *small noise*: $\varepsilon\to 0$ with T fixed. In that problem the maximum likelihood $\hat{\tau}_\varepsilon$ and the Bayesian estimators $\tilde{\tau}_\varepsilon$ are consistent and asymptotically normal with the same limit variance

$$\mathscr{L}\left\{\varepsilon^{-1}\left(\hat{\tau}_\varepsilon-\tau\right)\right\} \Longrightarrow \mathcal{N}\left(0,\mathrm{I}_1\left(\tau\right)^{-1}\right), \qquad \mathrm{I}_1\left(\tau\right) = \gamma^4\int_0^T x_0\left(t-2\tau\right)^2\,\mathrm{d}t$$

for the model (3.56) and

$$\mathscr{L}\left\{\varepsilon^{-1}\left(\hat{\tau}_\varepsilon-\tau\right)\right\} \Longrightarrow \mathcal{N}\left(0,\mathrm{I}_2\left(\tau\right)^{-1}\right), \qquad \mathrm{I}_1\left(\tau\right) = \gamma^4\int_0^T x_0\left(\tau^2\,t\right)^2\,\mathrm{d}t$$

for the model (3.57) with the *usual "regular" rate* ε^{-1} and both are asymptotically efficient. Here $\left\{x_0\left(t\right),0\leq t\leq T\right\}$ are the solutions of the equations

(3.56) and (3.57) with $\varepsilon = 0$ respectively. Multidimensional generalization can be found in [148]. We see that the same mathematical model of observations but with two different asymptotics is at the same time *regular* ($\varepsilon \to 0$) and *nonregular* ($T \to \infty$), depending on the type of limit.

Note that in the ergodic case of the process

$$\mathrm{d}X_t = [\vartheta_1 X_t + \vartheta_2 X_{t-\tau}]\, \mathrm{d}t + \mathrm{d}W_t, \quad X_0(s), -\tau \leq s \leq 0, \quad 0 \leq t \leq T$$

the properties of the MLE and BE of the parameter $\boldsymbol{\vartheta} = (\vartheta_1, \vartheta_2, \tau)$ can be studied as well (using the approach applied in this section).

3.4 Change-Point Estimation

Suppose that the observed diffusion process

$$\mathrm{d}X_t = S(\boldsymbol{\vartheta}, X_t)\, \mathrm{d}t + \sigma(X_t)\, \mathrm{d}W_t, \quad X_0, \quad 0 \leq t \leq T, \qquad (3.58)$$

has a discontinuous trend coefficient, i.e., there exists a point $x_* = x_*(\boldsymbol{\vartheta})$ such that the limits from the left $S(\boldsymbol{\vartheta}, x_*-)$ and from the right $S(\boldsymbol{\vartheta}, x_*+)$ are different. Therefore the trend coefficient is *switching* when $X_t = x_*(\boldsymbol{\vartheta})$. Such a kind of model corresponds to the *space change-point problem*. As usual, we do not know the true value $\boldsymbol{\vartheta} \in \boldsymbol{\Theta}$ and have to estimate it by the observations $X^T = \{X_t, 0 < t \leq T\}$. We suppose that the conditions $\mathcal{ES}$ and $\mathcal{EM}$ are fulfilled, hence there exists a weak solution and the corresponding measures are equivalent. The regularity conditions of Section 2.1 are not fulfilled because we suppose that the point of discontinuity depends on the unknown parameter, hence the Fisher information is equal to infinity. We consider several statements of the problem.

We start with the simplest model of the diffusion process with trend coefficient taking two values only and describe the properties of the Bayesian estimator. In this case the proof is a bit simpler than in the general case considered further and this model is used later in studying the role of contamination (misspecification) in the change-point problems. Then we consider several generalizations of the model (two and more jumps, simultaneous estimation of the *smooth and discontinuous parameters*, multidimensional case) and describe the properties of the MLE and BE.

Note that the asymptotics of these estimators is similar to that of the same estimators in the delay estimation problem. The difference is in the particular forms of the constants in the limiting likelihood ratios.

3.4.1 Simple Switching

Let us consider the simplest model of diffusion process with the trend coefficient taking just two values $+1$ and -1:

$$\mathrm{d}X_t = -\ \mathrm{sgn}\,(X_t - \vartheta)\ \mathrm{d}t + \mathrm{d}W_t, \quad X_0, \quad 0 \le t \le T. \tag{3.59}$$

It is easy to see that the condition $\mathcal{A}_0\,(\boldsymbol{\Theta})$ is fulfilled and (3.59) is an ergodic diffusion process with stationary two-sided exponential density

$$f\,(\vartheta, x) = \mathrm{e}^{-2\,|x-\vartheta|}, \qquad x \in \mathscr{R}.$$

The unknown parameter $\vartheta \in \Theta = (\alpha, \beta)$ with finite α and β. The measures $\left\{\mathbf{P}_\vartheta^{(T)}, \vartheta \in \Theta\right\}$ are equivalent and the likelihood ratio is

$$L\left(\theta, \theta_1, X^T\right) = \exp\left\{2\,|X_0 - \theta_1| - 2\,|X_0 - \theta| - 2\int_0^T \chi_{\{\theta\wedge\theta_1 \le X_t \le \theta\vee\theta_1\}}\,\mathrm{d}X_t\right\},$$

where $\theta_1 \in \Theta$ is some fixed value. The MLE $\hat{\vartheta}_T$ is defined by the equation

$$L\left(\hat{\vartheta}_T, \theta_1, X^T\right) = \sup_{\theta\in\Theta} L\left(\theta, \theta_1, X^T\right), \tag{3.60}$$

and the Bayes estimator $\tilde{\vartheta}_T$ for quadratic loss function and density *a priori* $p\,(y)\,,\ y \in \Theta$ is defined as the ratio of two integrals:

$$\tilde{\vartheta}_T = \left(\int_\alpha^\beta p\,(z)\ L(z, \theta_1\, X^T)\ \mathrm{d}z\right)^{-1} \int_\alpha^\beta y\,p\,(y)\ L(y, \theta_1\, X^T)\ \mathrm{d}y. \tag{3.61}$$

We suppose that the prior density is a continuous positive function on $[\alpha, \beta]$.
The limit process for the normalized likelihood ratio

$$L\left(\vartheta + u/T, \vartheta; X^T\right) = Z_T\,(u)$$

will be like $Z_T(\cdot)$ in the preceding section

$$Z(u) = \exp\left\{2\,W(u) - 2|u|\right\}, \qquad u \in \mathscr{R},$$

where $W(\cdot)$ is a two-sided Wiener process. Define the corresponding random variables $\hat{u}$ and $\tilde{u}$ by the equations

$$Z\,(\hat{u}) = \sup_{u\in\mathscr{R}} Z\,(u)\,, \qquad \tilde{u} = \left(\int Z(u)\,\mathrm{d}u\right)^{-1} \int u\,Z(u)\,\mathrm{d}u. \tag{3.62}$$

We have the following

Proposition 3.23. *The Bayes estimator $\tilde{\vartheta}_T$ is uniformly consistent, converges in distribution to the random variable $\tilde{u}$:*

$$\mathcal{L}_\vartheta\left\{T\left(\tilde{\vartheta}_T - \vartheta\right)\right\} \Longrightarrow \mathcal{L}_\vartheta\left\{\tilde{u}\right\}, \tag{3.63}$$

and the moments converge. In particular

$$\mathbf{E}_\vartheta\left|T\left(\tilde{\vartheta}_T - \vartheta\right)\right|^2 \longrightarrow \frac{\sigma_0^2}{4} \simeq 4.819 \pm 0.015.$$

Proof. The value of the σ_0^2 is given in (3.40). Below we show that the normalized likelihood ratio ($u > 0$)

$$Z_T(u) = \exp\left\{2\left|X_0-\vartheta\right| - 2\left|X_0-\vartheta-u/T\right| - 2\int_0^T \chi_{\{\vartheta \le X_t \le \vartheta+u/T\}} \mathrm{d}W_t \right.$$
$$\left. - 2\int_0^T \chi_{\{\vartheta \le X_t \le \vartheta+u/T\}} \mathrm{d}t \right\}$$

satisfies the conditions of Theorem 2.12.

First we obtain an estimate for the ordinary integral. We have ($u > 0$)

$$\begin{aligned}
&\mathbf{E}_\vartheta \left(\int_0^T \chi_{\{\vartheta \le X_t \le \vartheta+u/T\}} \mathrm{d}t - u\right)^2 \\
&= \mathbf{E}_\vartheta \left(\int_\vartheta^{\vartheta+u/T} 2\Lambda_T(\vartheta,x)\,\mathrm{d}x - u\,f(\vartheta,\vartheta)\right)^2 \\
&\le 2\,\mathbf{E}_\vartheta \left(\int_\vartheta^{\vartheta+u/T} \left[2\Lambda_T(\vartheta,x) - T\,f(\vartheta,x)\right]\mathrm{d}x\right)^2 \\
&\qquad + 2T^2\left(\int_\vartheta^{\vartheta+u/T} \left[f(\vartheta,x) - f(\vartheta,\vartheta)\right]\mathrm{d}x\right)^2 \\
&\le \frac{2u^2}{T} \sup_{x\in(\alpha,\beta)} \mathbf{E}_\vartheta \left(\sqrt{T}\left[\frac{2\Lambda_T(\vartheta,x)}{T} - f(\vartheta,x)\right]\right)^2 + \frac{2\,u^4}{T^2} \\
&\le \frac{C\,u^2}{T} + \frac{2\,u^4}{T^2},
\end{aligned} \tag{3.64}$$

where we used the estimates

$$\sup_{\vartheta\in\Theta}\sup_{x\in(\alpha,\beta)} \mathbf{E}_\vartheta \left(\sqrt{T}\left[\frac{2\Lambda_T(\vartheta,x)}{T} - f(\vartheta,x)\right]\right)^2 < \infty$$

and

$$\left|f(\vartheta,x) - f(\vartheta,\vartheta)\right| \le 2\left|x-\vartheta\right|.$$

The first one follows from the representation (1.25)

$$2\Lambda_T(x) = \left|X_T - x\right| - \left|X_0 - x\right| - \int_0^T \operatorname{sgn}(X_t - x)\,\mathrm{d}X_t$$

and $\mathbf{E}_\vartheta\,\Lambda_T(x) = T\,f(\vartheta,x)$ and the second is trivial.

Therefore, for any compact $\mathbb{K} \subset \Theta$

$$\lim_{T\to\infty}\sup_{\vartheta\in\mathbb{K}} \mathbf{E}_\vartheta \left(\int_0^T \chi_{\{\vartheta \le X_t \le \vartheta+u/T\}} \mathrm{d}t - u\right)^2 = 0. \tag{3.65}$$

This " law of large numbers" allows us to prove the uniform in $\vartheta \in \mathbb{K}$ convergence of marginal distributions

$$\mathcal{L}_\vartheta \left\{ Z_T(u_1), \ldots, Z_T(u_k) \right\} \Longrightarrow \mathcal{L}_\vartheta \left\{ Z(u_1), \ldots, Z(u_k) \right\}. \tag{3.66}$$

To do it we have to apply Proposition 1.21 to the stochastic integral

$$\sum_{i=1}^{k} \lambda_i \int_0^T \chi_{\{\vartheta \le X_t \le \vartheta + u_i/T\}} \mathrm{d}W_t = \int_0^T \sum_{i=1}^{k} \lambda_i \, \chi_{\{\vartheta \le X_t \le \vartheta + u_i/T\}} \mathrm{d}W_t$$

and to use the convergence (3.65). The estimate

$$\sup_{\vartheta \in \mathbb{K}} \sup_{|u|+|v|<R} |u-v|^{-1} \mathbf{E}_\vartheta \left| Z_T^{1/2}(u) - Z_T^{1/2}(v) \right|^2 \le C \left(1 + R^2\right) \tag{3.67}$$

can be derived quite similar to that of Lemma 3.20.

We obtain the last estimate

$$\sup_{\vartheta \in \mathbb{K}} \mathbf{P}_\vartheta^{(T)} \left\{ Z_T(u) > \mathrm{e}^{-\kappa |u|} \right\} \le \frac{C_N}{|u|^N} \tag{3.68}$$

for any $N > 0$ as follows.

Let $|u| \le T^{1/2}$ and introduce the sets $(u > 0)$

$$\mathbb{B}_t = \{\omega : \vartheta \le X_t \le \vartheta + u/T\}$$

and

$$\mathbb{C} = \left\{ \omega : \int_0^T \chi_{\{\vartheta \le X_t \le \vartheta + u/T\}} \mathrm{d}t - \frac{u}{2} > \frac{u}{4} \right\}.$$

Then we have

$$\begin{aligned}
&\mathbf{P}_\vartheta^{(T)} \left\{ Z_T(u) > \mathrm{e}^{-\frac{1}{4}|u|} \right\} \\
&\quad = \mathbf{P}_\vartheta^{(T)} \left\{ 2 \int_0^T \chi_{\{\mathbb{B}_t\}} \mathrm{d}W_t - 2 \int_0^T \chi_{\{\mathbb{B}_t\}} \mathrm{d}t > -\frac{u}{4} \right\} \\
&\quad \le \mathbf{P}_\vartheta^{(T)} \left\{ \int_0^T \chi_{\{\mathbb{B}_t\}} \mathrm{d}W_t - \frac{1}{2} \int_0^T \chi_{\{\mathbb{B}_t\}} \mathrm{d}t > \frac{u}{8}, \mathbb{C} \right\} + \mathbf{P}_\vartheta^{(T)} \{\mathbb{C}^c\} \\
&\quad \le \mathrm{e}^{-u/8} + \mathbf{P}_\vartheta^{(T)} \{\mathbb{C}^c\}.
\end{aligned}$$

Here we omitted $f(\vartheta, X_0)$ to simplify the exposition. The last probability can be estimated as follows

$$\begin{aligned}
\mathbf{P}_\vartheta^{(T)} \{\mathbb{C}^c\} &= \mathbf{P}_\vartheta^{(T)} \left\{ \int_0^T \chi_{\{\mathbb{B}_t\}} \mathrm{d}t - u \le -\frac{u}{4} \right\} \\
&\le \mathbf{P}_\vartheta^{(T)} \left\{ \left| \int_0^T \chi_{\{\mathbb{B}_t\}} \mathrm{d}t - u \right| > \frac{u}{4} \right\} \le \left(\frac{4}{u}\right)^{2N} \mathbf{E}_\vartheta \left| \int_0^T \chi_{\{\mathbb{B}_t\}} \mathrm{d}t - u \right|^{2N} \\
&\le C \left(\frac{1}{T^N} + \frac{u^{2N}}{T^N} \right) \le \frac{C_1}{T^N}
\end{aligned}$$

for all $|u| < T^{1/2}$. The last estimates were obtained using the same arguments as that for obtaining (3.64).

Let $|u| \geq T^{1/2}$. Put $\kappa_* = \mathrm{e}^{-2|\beta-\alpha|}$. Then we can write (for $u > 0$)

$$T \int_{\vartheta}^{\vartheta+u/T} f(\vartheta, x)\,\mathrm{d}x \geq \kappa_* |u|$$

and as above we have

$$\begin{aligned}
&\mathbf{P}_{\vartheta}^{(T)} \left\{ Z_T(u) > \mathrm{e}^{-\kappa_*|u|/4} \right\} \\
&\quad \leq \mathrm{e}^{-\kappa_*|u|/8} + \mathbf{P}_{\vartheta}^{(T)} \left\{ \int_{\vartheta}^{\vartheta+u/T} \left[\Lambda_T(x) - T f(\vartheta, x) \right] \mathrm{d}x < -\frac{1}{4}\kappa_*|u| \right\} \\
&\quad \leq \mathrm{e}^{-\kappa_*|u|/8} \\
&\qquad + \left(\frac{4\sqrt{T}}{\kappa_* |u|} \right)^{2N} \mathbf{E}_{\vartheta} \left| \int_{\vartheta}^{\vartheta+u/T} \sqrt{T} \left[\frac{\Lambda_T(x)}{T} - f(\vartheta, x) \right] \mathrm{d}x \right|^{2N} \\
&\quad \leq \mathrm{e}^{-\kappa_*|u|/8} + \frac{C}{T^N} \leq \frac{C_2}{|u|^N}
\end{aligned} \tag{3.69}$$

because $T \geq |u| / (\beta - \alpha)$.

Hence choosing

$$\kappa = \min\left(\frac{1}{8}, \frac{\kappa_*}{8} \right), \qquad C_N = \min\left(C_1, C_2 \right)$$

we obtain the estimate (3.68).

The properties (3.66)–(3.68) of the likelihood ratio allow us to apply Theorem 2.12 and to obtain all the desired properties of the Bayes estimator.

Other Estimators

It will follow from Theorem 3.26 below that the MLE $\hat{\vartheta}_T$ is consistent and has the following limit distribution:

$$\mathcal{L}_{\vartheta} \left\{ T \left(\hat{\vartheta}_T - \vartheta \right) \right\} \Longrightarrow \mathcal{L}_{\vartheta} \left\{ \hat{u} \right\} \tag{3.70}$$

but is not asymptotically efficient. In particular,

$$\mathbf{E}_{\vartheta} \left| T \left(\hat{\vartheta}_T - \vartheta \right) \right|^2 \to 6.5$$

(see (3.41) with $\gamma = 2$).

We remember the properties of EMM and describe MDE for this model. Note that

$$\mathbf{E}_{\vartheta}\ \xi = \vartheta.$$

Hence EMM

$$\bar{\vartheta}_T = \frac{1}{T}\int_0^T X_t\,\mathrm{d}t$$

is a consistent estimator of ϑ and it is easy to verify that it is asymptotically normal

$$\mathcal{L}_\vartheta\left\{\sqrt{T}\left(\bar{\vartheta}_T - \vartheta\right)\right\} \Longrightarrow \mathcal{N}\left(0, \frac{5}{4}\right).$$

Remember that we have

$$\sqrt{T}\left(\bar{\vartheta}_T - \vartheta\right) = \frac{1}{\sqrt{T}}\int_0^T \left(X_t - \vartheta\right)\,\mathrm{d}t = \frac{H(X_T) - H(X_0)}{\sqrt{T}}$$
$$-\frac{2}{\sqrt{T}}\int_0^T \mathrm{e}^{2|\vartheta - X_t|}\int_{-\infty}^{X_t}\left(v - \vartheta\right)\,\mathrm{e}^{-2|\vartheta - v|}\mathrm{d}v\,\mathrm{d}W_t$$

with corresponding function $H(\cdot)$. Hence the limit variance of EMM is

$$4\,\mathbf{E}_\vartheta\left(\mathrm{e}^{2|\vartheta-\xi|}\int_{-\infty}^{\xi}\left(v - \vartheta\right)\,\mathrm{e}^{-2|\vartheta - v|}\mathrm{d}v\right)^2 = \frac{5}{4}.$$

The main difference between the MLE and EMM is the rate of convergence. With the EMM we have just $\sqrt{T}$ consistent estimation.

We have the same rate in the case of minimum distance estimation. The distribution function is

$$F(\vartheta, x) = \chi_{\{x>\vartheta\}} + \frac{1}{2}\,\mathrm{sgn}\left(\vartheta - x\right)\,\mathrm{e}^{-2|x-\vartheta|}, \quad x \in \mathscr{R}$$

and MDE

$$\vartheta^*_T = \arg\inf_{\theta\in\Theta}\int_{-\infty}^{\infty}\left[\hat{F}_T(x) - F(\theta, x)\right]^2\mathrm{d}x$$

by Theorem 2.19 is consistent and asymptotically normal

$$\mathcal{L}_\vartheta\left\{\sqrt{T}\left(\vartheta^*_T - \vartheta\right)\right\} \Longrightarrow \mathcal{N}\left(0, \mathrm{d}\left(\vartheta\right)^2\right)$$

with

$$\mathrm{d}\left(\vartheta\right)^2 = \left(\int_{\mathscr{R}}\dot{F}\left(\vartheta, x\right)^2\,\mathrm{d}x\right)^{-2}\int_{\mathscr{R}}\int_{\mathscr{R}}\dot{F}\left(\vartheta, x\right)\,\dot{F}\left(\vartheta, y\right)\,D_\vartheta\left(x, y\right)\,\mathrm{d}x\,\mathrm{d}y.$$

The function $\mathrm{d}\left(\vartheta\right)^2$ can be calculated directly (see (2.63) for $D_\vartheta\left(x, y\right)$).

3.4.2 General Case

Consider the ergodic diffusion process

$$\mathrm{d}X_t = S\left(\vartheta, X_t\right)\,\mathrm{d}t + \sigma\left(X_t\right)\,\mathrm{d}W_t, \quad X_0,\ t \geq 0 \tag{3.71}$$

with invariant density $f(\theta, x)$ and the parameter $\vartheta \in \Theta = (\alpha, \beta)$. We suppose that the diffusion coefficient $\sigma(x)^2$ is a known positive continuous function and the trend coefficient $S(\theta, x)$ is a discontinuous function along the two curves

$$\left\{x_*^{(1)}(\theta),\ \theta \in [\alpha, \beta]\right\}, \qquad \text{and} \qquad \left\{x_*^{(2)}(\theta),\ \theta \in [\alpha, \beta]\right\}.$$

Let us denote

$$\Gamma_\vartheta^2 = \sum_{i=1}^{2} \left|\dot{x}_*^{(i)}(\vartheta)\right| \left(\frac{S\left(\vartheta, x_*^{(i)}(\vartheta)+\right) - S\left(\vartheta, x_*^{(i)}(\vartheta)-\right)}{\sigma\left(x_*^{(i)}(\vartheta)\right)}\right)^2 f\left(\vartheta, x_*^{(i)}(\vartheta)\right),$$

where

$$\dot{x}_*^{(i)}(\vartheta) = \frac{\mathrm{d}x_*^{(i)}}{\mathrm{d}\vartheta}(\vartheta), \quad i = 1, 2, \qquad \vartheta \in [\alpha, \beta]$$

and introduce the processes

$$Z(u) = \exp\left\{W(u) - \frac{|u|}{2}\right\}, \qquad u \in \mathscr{R},$$

and $Z_\vartheta(u) = Z\left(\Gamma_\vartheta^2 u\right)$. Here $W(\cdot)$ is a *two-sided Wiener process*. The random variables $\hat{u}$ and $\tilde{u}$ are defined by the same formulae

$$Z(\hat{u}) = \sup_u Z(u), \qquad \tilde{u} = \frac{\int_{\mathscr{R}} u\, Z(u)\ \mathrm{d}u}{\int_{\mathscr{R}} Z(u)\ \mathrm{d}u}.$$

The *regularity conditions* $\boldsymbol{\mathcal{M}}$ in this problem are the following.

$\mathcal{M}_1$. *The function* $S(\theta, x)$ *is continuously differentiable on* θ *for all* $x < x_*^{(1)}(\theta)$, $x_*^{(1)}(\theta) < x < x_*^{(2)}(\theta)$ *and* $x > x_*^{(2)}(\theta)$ *with derivative* $\dot{S}(\theta, \cdot) \in \mathscr{P}$ *and is discontinuous along the curves* $x_*^{(i)}(\cdot)$; *i.e., for* $i = 1, 2$ *and all* $\theta \in [\alpha, \beta]$

$$\lim_{\varepsilon \to 0+} \left[S\left(\theta, x_*^{(i)}(\theta) + \varepsilon\right) - S\left(\theta, x_*^{(i)}(\theta) - \varepsilon\right)\right] = r_i(\theta)$$

and

$$|r_1(\theta)| + |r_2(\theta)| > 0.$$

$\mathcal{M}_2$. *The functions* $\left\{x_*^{(i)}(\theta),\ \theta \in [\alpha, \beta]\right\}$ *are continuous and continuously differentiable with derivatives*

$$\inf_{\vartheta \in \Theta} \left|\dot{x}_*^{(i)}(\theta)\right| > 0$$

and

$$x_*^{(1)}(\theta) < x_*^{(2)}(\theta), \qquad \theta \in [\alpha, \beta].$$

The first result is the lower minimax bound.

Proposition 3.24. *Let the conditions $\mathcal{A}_0(\Theta)$ and $\mathcal{M}$ be fulfilled. Then for any $\vartheta_0 \in \Theta$*

$$\lim_{\delta\to 0}\lim_{T\to\infty}\inf_{\bar\vartheta_T}\sup_{|\vartheta-\vartheta_0|<\delta} T\,\mathbf{E}_\vartheta\left(\bar\vartheta_T-\vartheta\right)^2 \geq \frac{\sigma_0^2}{\Gamma_\vartheta^4}, \tag{3.72}$$

where σ_0^2 is defined in (3.40).

The proof is based on the asymptotics of the Bayesian estimators and, as we have already discussed it two times (see Propositions 3.6 and 3.15), we omit it here. As usual, we define the asymptotically efficient estimators with the help of this bound.

Definition 3.25. *Let the conditions $\mathcal{A}_0(\Theta)$ and $\mathcal{M}$ be fulfilled. We call an estimator $\bar\vartheta_T$ asymptotically efficient if, for any $\vartheta_0 \in \Theta$,*

$$\lim_{\delta\to 0}\lim_{T\to\infty}\sup_{|\vartheta-\vartheta_0|<\delta} T\,\mathbf{E}_\vartheta\left(\bar\vartheta_T-\vartheta\right)^2 = \frac{\sigma_0^2}{\Gamma_\vartheta^4}.$$

The MLE $\hat\vartheta_T$ and the BE $\tilde\vartheta_T$ are defined as usual by (3.60) and (3.61) respectively.

Theorem 3.26. *Let the conditions $\mathcal{A}_0(\Theta)$ and $\mathcal{M}$ be fulfilled. Then the MLE $\hat\vartheta_T$ and the BE $\tilde\vartheta_T$ are uniformly on compacts $\mathbb{K}\subset\Theta$ consistent, their limit distributions are*

$$\mathcal{L}_\vartheta\left\{T\left(\hat\vartheta_T-\vartheta\right)\right\}\Longrightarrow\mathcal{L}_\vartheta\left\{\frac{\hat u}{\Gamma_\vartheta^2}\right\},\qquad \mathcal{L}_\vartheta\left\{T\left(\tilde\vartheta_T-\vartheta\right)\right\}\Longrightarrow\mathcal{L}_\vartheta\left\{\frac{\tilde u}{\Gamma_\vartheta^2}\right\}, \tag{3.73}$$

the moments converge and the estimator $\tilde\vartheta_T$ is asymptotically efficient.

Proof. The proof contains all the elements of the proof of Proposition 3.23. We check the conditions of Theorems 2.6 and 2.12 for the normalized likelihood ratio

$$Z_T(u) = L\left(\vartheta+u/T,\vartheta,X^T\right),\quad u\in U_T=\left(T\left(\alpha-\vartheta\right),T\left(\beta-\vartheta\right)\right)$$

with the help of the three lemmas.

Lemma 3.27. *Let the conditions $\mathcal{A}_0(\Theta)$ and $\mathcal{M}$ be fulfilled. Then the marginal distributions of $Z_T(\cdot)$ converge to the marginal distributions of $Z_\vartheta(\cdot)$ and this convergence is uniform over compacts $\mathbb{K}\subset\Theta$.*

Proof. At the beginning we show that

$$\mathbf{P}_\vartheta - \lim_{T\to\infty}\int_0^T\left(\frac{S\left(\vartheta+u/T,X_t\right)-S\left(\vartheta,X_t\right)}{\sigma\left(X_t\right)}\right)^2 \mathrm{d}t = |u|\;\Gamma_\vartheta^2 \tag{3.74}$$

and this convergence is uniform on compacts $\mathbb{K}\subset\Theta$. Put

$$\delta_\vartheta\left(y,x\right)=\frac{S\left(\vartheta+y,x\right)-S\left(\vartheta,x\right)}{\sigma\left(x\right)}.$$

We have

$$\int_0^T \delta_\vartheta\left(u/T,X_t\right)^2\,\mathrm{d}t=\int_{-\infty}^{\infty}\delta_\vartheta\left(u/T,x\right)^2\,\frac{2\,\eta_T\left(\vartheta,x\right)}{\sigma\left(x\right)^2}\,\mathrm{d}x$$
$$=T\int_{-\infty}^{\infty}\delta_\vartheta\left(u/T,x\right)^2\,f\left(\vartheta,x\right)\,\mathrm{d}x+\frac{1}{\sqrt{T}}\int_{-\infty}^{\infty}\delta_\vartheta\left(u/T,x\right)^2\,W_T\left(\vartheta,x\right)\,\mathrm{d}x$$

where $\Lambda_T\left(\vartheta,x\right)$ is the local time of the diffusion process and

$$\eta_T\left(\vartheta,x\right)=\sqrt{T}\left(\frac{2\,\Lambda_T\left(\vartheta,x\right)}{T\,\sigma\left(x\right)^2}-f\left(\vartheta,x\right)\right).$$

Remember that under condition $\mathcal{A}_0\left(\Theta\right)$ for any $p>0$

$$\sup_{\vartheta\in\mathbb{K}}\sup_x\mathbf{E}_\vartheta\left|\eta_T\left(\vartheta,x\right)\right|^p<\infty \tag{3.75}$$

(see (1.35) and moreover as $\dot S(\theta,\cdot),\sigma\left(\cdot\right)^{-1}\in\mathscr{P}$ we have as well

$$\sup_{\vartheta\in\mathbb{K}}\sup_{\theta\in\Theta}\mathbf{E}_\vartheta\left|\int_{-\infty}^{\infty}\delta_\vartheta\left(\theta-\vartheta,x\right)^2\,\eta_T\left(\vartheta,x\right)\,\mathrm{d}x\right|^p<\infty. \tag{3.76}$$

Therefore

$$\int_0^T\delta_\vartheta\left(u/T,X_t\right)^2\,\mathrm{d}t=T\int_{-\infty}^{\infty}\delta_\vartheta\left(u/T,x\right)^2\,f\left(\vartheta,x\right)\,\mathrm{d}x\ \left(1+o(1)\right).$$

Suppose (without loss of generality) that $\dot x_*^{(1)}\left(\vartheta\right)<0$ and $\dot x_*^{(2)}\left(\vartheta\right)>0$. Let $u>0$, then we can write

$$\int_{-\infty}^{\infty}\delta_\vartheta\left(u/T,x\right)^2\,f\left(\vartheta,x\right)\,\mathrm{d}x=\int_{-\infty}^{x_*^{(1)}(\vartheta+u/T)}\delta_\vartheta\left(u/T,x\right)^2\,f\left(\vartheta,x\right)\,\mathrm{d}x$$
$$+\int_{x_*^{(1)}(\vartheta+u/T)}^{x_*^{(1)}(\vartheta)}\delta_\vartheta\left(u/T,x\right)^2\,f\left(\vartheta,x\right)\,\mathrm{d}x$$
$$+\int_{x_*^{(1)}(\vartheta)}^{x_*^{(2)}(\vartheta)}\delta_\vartheta\left(u/T,x\right)^2\,f\left(\vartheta,x\right)\,\mathrm{d}x$$
$$+\int_{x_*^{(2)}(\vartheta)}^{x_*^{(2)}(\vartheta+u/T)}\delta_\vartheta\left(u/T,x\right)^2\,f\left(\vartheta,x\right)\,\mathrm{d}x$$
$$+\int_{x_*^{(2)}(\vartheta+u/T)}^{\infty}\delta_\vartheta\left(u/T,x\right)^2\,f\left(\vartheta,x\right)\,\mathrm{d}x$$
$$=I_1+I_2+I_3+I_4+I_5 \tag{3.77}$$

with obvious notation. The function $\delta_\vartheta(u/T,x)$ in the integrals I_1, I_3 and I_5 is differentiable on u, hence (using the condition $\mathcal{A}_0(\Theta)$) we obtain the estimates

$$I_1 + I_3 + I_5 \leq C\,\frac{u^2}{T^2}, \tag{3.78}$$

where the constant $C > 0$ does not depend on ϑ, u, and T.

Further, for fixed u and $T \to \infty$ we have the expansion

$$\begin{aligned} T\,I_2 &= T\int_{x_*^{(1)}(\vartheta+u/T)}^{x_*^{(1)}(\vartheta)} \delta_\vartheta\left(u/T,x\right)^2\, f\left(\vartheta,x\right)\,\mathrm{d}x \\ &= -u\,\dot{x}_*^{(1)}\left(\vartheta\right)\left(\frac{r_1\left(\vartheta\right)}{\sigma\left(x_*^{(1)}\left(\vartheta\right)\right)}\right)^2 f\left(\vartheta, x_*^{(1)}\left(\vartheta\right)\right)\,\left(1+o(1)\right) \\ &= u\left|\dot{x}_*^{(1)}\left(\vartheta\right)\right| \frac{r_1\left(\vartheta\right)^2 f\left(\vartheta, x_*^{(1)}\left(\vartheta\right)\right)}{\sigma\left(x_*^{(1)}\left(\vartheta\right)\right)^2}\left(1+o(1)\right). \end{aligned} \tag{3.79}$$

We have the similar equality for the fourth integral as well:

$$T\,I_4 = u\left|\dot{x}_*^{(2)}\left(\vartheta\right)\right| \frac{r_2\left(\vartheta\right)^2 f\left(\vartheta, x_*^{(2)}\left(\vartheta\right)\right)}{\sigma\left(x_*^{(2)}\left(\vartheta\right)\right)^2}\left(1+o(1)\right). \tag{3.80}$$

For the negative values of u we obtain the same estimates. Hence the convergence (3.74) is established.

Note that the estimate (3.76) allows us to prove the convergence of moments as well:

$$\sup_{\vartheta\in\mathbb{K}} \mathbf{E}_\vartheta \left|\int_0^T \left(\frac{S\left(\vartheta+u/T,X_t\right)-S\left(\vartheta,X_t\right)}{\sigma\left(X_t\right)}\right)^2\,\mathrm{d}t - |u|\;\Gamma_\vartheta^2\right|^p \leq \frac{C}{T^{p/2}}$$

for any $p > 0$.

It can be shown by a similar argument that for any $u_l,\ u_m \in \mathscr{R}$

$$\mathbf{P}_\vartheta - \lim_{T\to\infty}\int_0^T q_\vartheta\left(u_l/T,X_t\right)\,\delta_\vartheta\left(u_m/T,X_t\right)\mathrm{d}t = \chi_{\{u_l u_m>0\}}\;|u_l|\wedge|u_m|\;\Gamma_\vartheta^2. \tag{3.81}$$

Further, the convergence (3.81) provides the joint asymptotic normality

$$\mathcal{L}_\vartheta\left\{I_T(u_1,\vartheta),\ldots,I_T(u_k,\vartheta)\right\} \Longrightarrow \mathcal{L}\left\{\Gamma_\vartheta\,W\left(u_1\right),\ldots,\Gamma_\vartheta\,W\left(u_k\right)\right\}$$

of the following stochastic integrals:

$$I_T(u_l,\vartheta) = \int_0^T \delta_\vartheta\left(u_l/T,X_t\right)\,\mathrm{d}W_t,\quad l=1,\ldots,k$$

for any $k = 1, 2, \ldots$. Note that all these convergences are uniform on $\vartheta \in \mathbb{K}$.

Finally, we obtain uniform on $\vartheta \in \mathbb{K}$ convergence of the finite dimensional distributions

$$\mathcal{L}_\vartheta \left\{ Z_T(u_1), \ldots, Z_T(u_k) \right\} \Longrightarrow \mathcal{L} \left\{ Z_\vartheta(u_1), \ldots, Z_\vartheta(u_k) \right\} \tag{3.82}$$

because

$$Z_T(u_m) = Z_0(u_m) \exp\left\{ I_T(u_l, \vartheta) - \frac{1}{2} \int_0^T \delta_\vartheta \left(u_m/T, X_t\right)^2 \mathrm{d}t \right\},$$

where

$$Z_0(u_m) = \frac{G\left(\vartheta + u_m/T\right)}{G\left(\vartheta\right)} \exp\left\{ 2 \int_0^{X_0} \frac{\delta_\vartheta \left(u_m/T, v\right)}{\sigma\left(v\right)} \mathrm{d}v \right\} \to 1.$$

Lemma 3.28. *Let the conditions $\mathcal{A}_0\left(\Theta\right)$ and $\mathcal{M}$ be fulfilled. Then there exist constants C_1 and C_2 such that for any $R > 0$ and $|u| + |v| < R$ we have*

$$\sup_{\vartheta \in \mathbb{K}} \mathbf{E}_\vartheta \left| Z_T^{1/2}\left(u\right) - Z_T^{1/2}\left(v\right) \right|^2 \le C_2 \left(1 + R\right) \left|u - v\right| \tag{3.83}$$

$$\sup_{\vartheta \in \mathbb{K}} \mathbf{E}_\vartheta \left| Z_T^{1/8}\left(u\right) - Z_T^{1/8}\left(v\right) \right|^4 \le C_1 \left(1 + R^2\right) \left|u - v\right|^2. \tag{3.84}$$

Proof. Let us denote

$$V_T = \left(\frac{Z_T\left(u\right)}{Z_T\left(v\right)} \right)^{1/8}.$$

To prove the first inequality (3.83) we write ($\vartheta_v = \vartheta + v/T$)

$$\begin{aligned} \mathbf{E}_\vartheta \left| Z_T^{1/2}\left(u\right) - Z_T^{1/2}\left(v\right) \right|^2 &= 2 - 2\, \mathbf{E}_\vartheta \left(Z_T\left(u\right)\, Z_T\left(v\right) \right)^{1/2} \\ &= 2 - 2\, \mathbf{E}_{\vartheta_v} V_T^4 = 2\,\mathbf{E}_* \left(1 - \exp\left\{ -\frac{1}{8} \int_0^T \delta_{\vartheta_v} \left(\frac{u - v}{T}, X_t \right)^2 \mathrm{d}t \right\} \right) \\ &\le \frac{1}{4} \mathbf{E}_* \int_0^T \delta_{\vartheta_v} \left(\frac{u - v}{T}, X_t \right)^2 \mathrm{d}t, \end{aligned}$$

where we used the inequality $1 - \mathrm{e}^x \le x$ and denoted as $\mathbf{E}_*$ the mathematical expectation with respect to measure $\mathbf{P}_*^{(T)}$ of the process

$$\mathrm{d}X_t = \frac{1}{2} \left[S\left(\vartheta_u, X_t\right) + S\left(\vartheta_v, X_t\right) \right] \mathrm{d}t + \sigma\left(X_t\right) \mathrm{d}W_t. \tag{3.85}$$

The change of the measure is done in the following way.

$$\begin{aligned}
\mathbf{E}_{\vartheta_v} V_T^4 &= \mathbf{E}_{\vartheta_v} \exp\left\{\int_0^T \frac{\delta_{\vartheta_v}\left(\frac{u-v}{T}, X_t\right)}{2} \mathrm{d}W_t - \frac{1}{2}\int_0^T \left(\frac{\delta_{\vartheta_v}\left(\frac{u-v}{T}, X_t\right)}{2}\right)^2 \mathrm{d}t\right\} \\
&\qquad \times \exp\left\{-\frac{1}{8}\int_0^T \delta_{\vartheta_v}\left(\frac{u-v}{T}, X_t\right)^2 \mathrm{d}t\right\} \\
&= \mathbf{E}_{\vartheta_v} \frac{\mathrm{d}\mathbf{P}_*^{(T)}}{\mathrm{d}\mathbf{P}_{\vartheta_v}^{(T)}} \exp\left\{-\frac{1}{8}\int_0^T \delta_{\vartheta_v}\left(\frac{u-v}{T}, X_t\right)^2 \mathrm{d}t\right\} \\
&= \mathbf{E}_* \exp\left\{-\frac{1}{8}\int_0^T \delta_{\vartheta_v}\left(\frac{u-v}{T}, X_t\right)^2 \mathrm{d}t\right\}
\end{aligned}$$

Further, denote by $\Lambda_T(x)$ and $f_*(x)$ the local time and the invariant density of the process (3.85). Then as in (3.77) we have

$$\begin{aligned}
\mathbf{E}_* \int_0^T \delta_{\vartheta_v}\left(\frac{u-v}{T}, X_t\right)^2 \mathrm{d}t &= \int_{-\infty}^{\infty} \delta_{\vartheta_v}\left(\frac{u-v}{T}, x\right)^2 \mathbf{E}_* \frac{2\,\Lambda_T(x)}{\sigma(x)^2} \mathrm{d}x \\
&= \int_{-\infty}^{\infty} \delta_{\vartheta_v}\left(\frac{u-v}{T}, x\right)^2 f_*(x)\,\mathrm{d}x = I_1 + I_2 + I_3 + I_4 + I_5
\end{aligned}$$

where we use notation similar to that of (3.77). Remember that the function $\delta_{\vartheta_v}\left(\frac{u-v}{T}, \cdot\right)$ is continuously differentiable on u, v in the integrals $I_1,\ I_3,\ I_5$, and hence we have the estimate (3.78). The integrands in the integrals I_2 and I_3 are bounded, therefore

$$I_2 + I_3 \le C\,T\left|x_*^{(1)}(\vartheta_v) - x_*^{(1)}(\vartheta_u)\right| + C\,T\left|x_*^{(2)}(\vartheta_u) - x_*^{(2)}(\vartheta_v)\right| \le C\,|u-v|$$

because the derivatives $\dot{x}_*^{(1)}(\cdot)$, $\dot{x}_*^{(2)}(\cdot)$ are bounded functions.

Finally we have the estimate (3.83)

$$\begin{aligned}
\mathbf{E}_\vartheta \left|Z_T^{1/2}(u) - Z_T^{1/2}(v)\right|^2 &\le C\,T^{-1}\,|u-v|^2 + C\,|u-v| \\
&\le C\,(1+R)\,|u-v|
\end{aligned}$$

To verify the second inequality (3.84) we proceed as in the proof of Lemma 1.13. We can write

$$\begin{aligned}
\mathbf{E}_\vartheta \left|Z_T^{1/8}(u) - Z_T^{1/8}(v)\right|^4 &\le \mathbf{E}_\vartheta\, Z_T^{1/2}(v)\left|V_T - 1\right|^4 \\
&\le C\,\mathbf{E}_\vartheta\, Z_T^{1/2}(v)\left|\int_0^T V_t\,\delta_{\vartheta_v}\left(\frac{u-v}{T}, X_t\right)^2 \mathrm{d}t\right|^4 \\
&\le C\,\mathbf{E}_\vartheta\, Z_T^{1/2}(v)\left|\int_0^T V_t^2\,\delta_{\vartheta_v}\left(\frac{u-v}{T}, X_t\right)^2 \mathrm{d}t\right|^2
\end{aligned}$$

$$\leq C\left(\mathbf{E}_{\vartheta_v}\left(\sup_{0\leq t\leq T}V_t^4\right)\right)^{1/2}\left(\mathbf{E}_{\vartheta}\left(\int_0^T\delta_{\vartheta_v}\left(\frac{u-v}{T},X_t\right)^2\mathrm{d}t\right)^4\right)^{1/2}.$$

The first expectation is bounded (see (1.4))

$$\mathbf{E}_{\vartheta_v}\left(\sup_{0\leq t\leq T}V_t^4\right)\leq 1.$$

Indeed,

$$0<V_t^4\leq 1-\frac{1}{8}\int_0^t V_s^4\delta_{\vartheta_v}\left(\frac{u-v}{T},X_s\right)\mathrm{d}W_s$$

and for the second integral we have

$$\int_0^T\delta_{\vartheta_v}\left(\frac{u-v}{T},X_t\right)^2\mathrm{d}t=\int_{-\infty}^{\infty}\delta_{\vartheta_v}\left(\frac{u-v}{T},x\right)^2\frac{2\,\Lambda_T\left(\vartheta,x\right)}{\sigma\left(x\right)^2}\,\mathrm{d}x.$$

For the last integral we have the same estimate as above and this gives us

$$\mathbf{E}_{\vartheta}\left(\int_0^T\delta_{\vartheta_v}\left(\frac{u-v}{T},X_t\right)^2\mathrm{d}t\right)^4\leq C\left(\left|u-v\right|^8+\left|u-v\right|^4\right).$$

Hence the inequality (3.84) is proved.

Lemma 3.29. *Let the conditions $\mathcal{A}_0(\Theta)$ and $\mathcal{M}$ be fulfilled. Then for any $N>1$ there exist constants $\kappa>0$ and $C(N)>0$ such that*

$$\sup_{\vartheta\in\mathbb{K}}\mathbf{P}_{\vartheta}\left\{Z_T\left(u\right)\geq \mathrm{e}^{-\frac{\kappa}{4}\left|u\right|}\right\}\leq\frac{C(N)}{\left|u\right|^N}. \tag{3.86}$$

Proof. The proof of this lemma is similar to the proof of Lemma 2.11. Using the estimates (3.77)–(3.80) we can write for the small values of h

$$\mathbf{E}_{\vartheta}\left(\frac{S\left(\vartheta+h,\xi\right)-S\left(\vartheta,\xi\right)}{\sigma\left(\xi\right)}\right)^2\geq\kappa_1\left|h\right|-C\left|h\right|^2$$

with some $\kappa_1>0$. Hence for sufficiently small $\nu>0$ and $|h|<\nu$ we have the estimate

$$\mathbf{E}_{\vartheta}\left(\frac{S\left(\vartheta+h,\xi\right)-S\left(\vartheta,\xi\right)}{\sigma\left(\xi\right)}\right)^2\geq\frac{\kappa}{2}\left|h\right|. \tag{3.87}$$

For $|h|\geq\nu$ we use the inequality

$$g\left(\nu\right)=\inf_{\vartheta\in\mathbb{K}}\inf_{|h|>\nu}\mathbf{E}_{\vartheta}\left(\frac{S\left(\vartheta+h,\xi\right)-S\left(\vartheta,\xi\right)}{\sigma\left(\xi\right)}\right)^2>0,$$

which can be proved as follows. Suppose that this inequality is not fulfilled. Then there exist ϑ^* and $h^* \neq 0$ such that

$$S(\vartheta^* + h^*, x) = S(\vartheta^*, x)$$

for almost all x, but this is impossible because the functions $S(\vartheta^* + h^*, x)$ and $S(\vartheta^*, x)$ have discontinuities in different points.

Therefore

$$\inf_{\vartheta \in \mathbb{K}} \mathbf{E}_\vartheta \left(\frac{S(\vartheta + h, \xi) - S(\vartheta, \xi)}{\sigma(\xi)} \right)^2 \geq g(\nu) \geq g(\nu) \frac{|h|}{\beta - \alpha} = \kappa_2 |h| .$$

Finally we can write

$$\inf_{\vartheta \in \mathbb{K}} T\, \mathbf{E}_\vartheta \left(\frac{S(\vartheta + u/T, \xi) - S(\vartheta, \xi)}{\sigma(\xi)} \right)^2 \geq \kappa |u|$$

with $\kappa = \min(\kappa_1/2, \kappa_2)$.

The rest of the proof of this lemma is quite close to the proof of Lemma 2.11.

The properties of the likelihood ratio $Z_T(\cdot)$ established in these three lemmas allow us to apply Theorems 2.6 and 2.12 and therefore to finish the proof of Theorem 3.26.

3.4.3 Examples

We consider several examples of SDE with discontinuous trend coefficients.

Example 3.30. The first example is the switching process (3.59). It is even simpler because the trend coefficient has one jump only. The conditions $\mathcal{A}_0(\Theta)$ and a version of $\mathcal{M}$ corresponding to one discontinuity with $x_*(\vartheta) = \vartheta$ for this process are fulfilled and the MLE has the properties described in the Theorem 3.26.

Example 3.31. Another example of simple switching is the diffusion process

$$\mathrm{d}X_t = -X_t \left(a + b\, \chi_{\{\vartheta < X_t < c + \vartheta\}} \right) \mathrm{d}t + \mathrm{d}W_t, \quad X_0, 0 \leq t \leq T,$$

where $a > 0$, $b \neq 0$, $c > 0$, $\vartheta \in (\alpha, \beta)$. The conditions $\mathcal{A}_0(\Theta)$ and $\mathcal{M}$ are fulfilled with $x_*^{(1)}(\vartheta) = \vartheta$, $x_*^{(2)}(\vartheta) = c + \vartheta$ and the MLE and BE have the corresponding asymptotic properties.

Example 3.32. The next example is the problem of the estimation parameter $\vartheta \in (\alpha, \beta)$ by observations of an ergodic diffusion process

$$\mathrm{d}X_t = -X_t \,\mathrm{sgn}\Big(X_t^2 - 2\,\vartheta\, X_t + 1\Big)\,\mathrm{d}t + \sigma\,\mathrm{d}W_t, \quad X_0,\ 0 \le t \le T.$$

We suppose that $\alpha > 1$. Then the trend coefficient

$$S(\vartheta, x) = -\,x\ \mathrm{sgn}\Big(x^2 - 2\,\vartheta\, x + 1\Big)$$

is discontinuous along the functions

$$x_*^{(1)}(\vartheta) = \vartheta - \sqrt{\vartheta^2 - 1}, \quad \vartheta \in (\alpha, \beta),$$
$$x_*^{(2)}(\vartheta) = \vartheta + \sqrt{\vartheta^2 - 1}, \quad \vartheta \in (\alpha, \beta),$$

we have $x_*^{(1)}(\vartheta) < x_*^{(2)}(\vartheta)$ for all $\vartheta \in (\alpha, \beta)$, and

$$\dot{x}_*^{(1)}(\vartheta) = 1 - \frac{\vartheta}{\sqrt{\vartheta^2 - 1}} < 0,$$
$$\dot{x}_*^{(2)}(\vartheta) = 1 + \frac{\vartheta}{\sqrt{\vartheta^2 - 1}} > 0.$$

Denote $A(x) = \mathrm{e}^{x^2/\sigma^2}$, and put $x_i = x_*^{(i)}(\vartheta)$. Then the invariant density can be written as

$$f(\vartheta, x) = \frac{A(x_1)}{G(\vartheta)\,A(x)}\,\chi_{\{x \le x_1\}} + \frac{A(x)}{G(\vartheta)\,A(x_1)}\,\chi_{\{x_1 < x < x_2\}}$$
$$+ \frac{A(x_2)^2}{G(\vartheta)\,A(x_1)\,A(x)}\,\chi_{\{x \ge x_2\}},$$

and the function

$$\Gamma_\vartheta^2 = 4\left(\frac{\vartheta}{\sqrt{\vartheta^2 - 1}} - 1\right)\frac{\left(\vartheta - \sqrt{\vartheta^2 - 1}\right)^2}{\sigma^2\,G(\vartheta)} +$$
$$+4\left(\frac{\vartheta}{\sqrt{\vartheta^2 - 1}} + 1\right)\frac{\left(\vartheta + \sqrt{\vartheta^2 - 1}\right)^2}{\sigma^2\,G(\vartheta)}\,\frac{A\left(\vartheta + \sqrt{\vartheta^2 - 1}\right)}{A\left(\vartheta - \sqrt{\vartheta^2 - 1}\right)} > 0.$$

Hence the conditions $\boldsymbol{\mathcal{A}}_0(\Theta)$ and $\boldsymbol{\mathcal{M}}$ are fulfilled and the estimators $\hat{\vartheta}_T$ and $\tilde{\vartheta}_T$ have all the properties mentioned in Theorem 3.26.

3.4.4 Discussion

Theorem 3.26 can be generalized in several directions. The first one corresponds to the diffusion process with the trend coefficient $S(\vartheta, x)$ having $k > 2$ discontinuities at points $x_*^{(1)}(\vartheta), \ldots, x_*^{(k)}(\vartheta)$ and continuously differentiable on ϑ between these points. The properties of estimators will be the same as described in Theorem 3.26 with the function

$$\Gamma_\vartheta^2=\sum_{i=1}^{k}\left|\dot x_*^{(i)}\left(\vartheta\right)\right|\left(\frac{S\left(\vartheta,x_*^{(i)}\left(\vartheta\right)+\right)-S\left(\vartheta,x_*^{(i)}\left(\vartheta\right)-\right)}{\sigma\left(x_*^{(i)}\left(\vartheta\right)\right)}\right)^2 f\left(\vartheta,x_*^{(i)}\left(\vartheta\right)\right).$$

The proof of this is quite close to that given above.

The second problem is the simultaneous estimation of the two-dimensional parameter $\boldsymbol{\vartheta}=\left(\vartheta^{(1)},\vartheta^{(2)}\right)$ with *smooth* $\vartheta^{(1)}$ and *discontinuous* $\vartheta^{(2)}$ components. In particular we can take an ergodic diffusion process

$$\mathrm{d}X_t=S(\vartheta^{(1)},X_t-\vartheta^{(2)})\,\mathrm{d}t+\sigma\left(X_t\right)\,\mathrm{d}W_t,\qquad X_0,\ 0\leq t\leq T,\tag{3.88}$$

and we suppose that the function $S(y,x)$ is continuously differentiable on y and is continuously differentiable on x except for one point x_*, where it has a jump

$$S(\vartheta^{(1)},x_*+)-S(\vartheta^{(1)},x_*-)\neq 0.$$

The identifiability condition will be

$$\inf_{\boldsymbol{\vartheta}\in\mathbb{K}}\ \inf_{\left|y-\vartheta^{(1)}\right|>\nu}\mathbf{E}_{\boldsymbol{\vartheta}}\left(\frac{S(y,\xi-\vartheta^{(2)})-S(\vartheta^{(1)},\xi-\vartheta^{(2)})}{\sigma\left(\xi\right)}\right)^2>0.\tag{3.89}$$

It can be easily shown that the limit process for the normalized likelihood ratio

$$Z_T\left(u,v\right)=L\left(\vartheta^{(1)}+u/\sqrt{T},\vartheta_2+v/T,\vartheta^{(1)},\vartheta^{(2)};X^T\right)$$

is

$$Z_{\boldsymbol{\vartheta}}\left(u,v\right)=\exp\left\{u\,\Delta\left(\boldsymbol{\vartheta}\right)-\frac{u^2}{2}\,\mathrm{I}\left(\boldsymbol{\vartheta}\right)\right\}\,\exp\left\{\Gamma_{\boldsymbol{\vartheta}}\,W\left(v\right)-\frac{\left|v\right|}{2}\,\Gamma_{\boldsymbol{\vartheta}}^2\right\},$$

where

$$\mathcal{L}_{\boldsymbol{\vartheta}}\left\{\Delta\left(\boldsymbol{\vartheta}\right)\right\}=\mathcal{N}\Big(0,\mathrm{I}\left(\boldsymbol{\vartheta}\right)\Big),\quad \mathrm{I}\left(\boldsymbol{\vartheta}\right)=\int_{-\infty}^{\infty}\left(\frac{\dot S\left(\vartheta^{(1)},x-\vartheta^{(2)}\right)}{\sigma\left(x\right)}\right)^2 f\left(\boldsymbol{\vartheta},x\right)\,\mathrm{d}x.$$

Here $\dot S\left(y,x\right)=\frac{\partial S(y,x)}{\partial y}$. The function

$$\Gamma_{\boldsymbol{\vartheta}}^2=\left(\frac{S(\vartheta^{(1)},x_*+)-S(\vartheta^{(1)},x_*-)}{\sigma\left(x_*+\vartheta^{(2)}\right)}\right)^2 f\left(\boldsymbol{\vartheta},x_*+\vartheta^{(2)}\right)$$

and the random variable $\Delta\left(\boldsymbol{\vartheta}\right)$ and two-sided Wiener process $W\left(\cdot\right)$ are independent.

The random variables $\sqrt{T}\left(\hat{\vartheta}_{1,T}-\vartheta^{(1)}\right)$ and $T\left(\hat{\vartheta}_{2,T}-\vartheta^{(2)}\right)$ are asymptotically independent and have the following limit distributions:

$$\mathcal{L}_{\vartheta}\left\{\sqrt{T}\left(\hat{\vartheta}_{1,T}-\vartheta^{(1)}\right)\right\} \Longrightarrow \mathcal{N}\left(0,\mathrm{I}\left(\vartheta\right)^{-1}\right),$$
$$\mathcal{L}_{\vartheta}\left\{T\left(\hat{\vartheta}_{2,T}-\vartheta^{(2)}\right)\right\} \Longrightarrow \mathcal{L}_{\vartheta}\left\{\frac{\hat{u}}{\Gamma_{\vartheta}^{2}}\right\}.$$

The proof is similar to that given above.

The third problem concerns the multidimensional parameter $\boldsymbol{\vartheta}=(\vartheta_1,\ldots,\vartheta^{(d)})$ estimation by the observations of an ergodic diffusion process

$$\mathrm{d}X_t = S\left(\boldsymbol{\vartheta},X_t\right)\,\mathrm{d}t+\sigma\left(X_t\right)\mathrm{d}W_t,\quad X_0,\quad 0\leq t\leq T$$

with the trend coefficient

$$S\left(\boldsymbol{\vartheta},x\right)=g\left(x-\vartheta^{(1)},\ldots,x-\vartheta^{(d)}\right),\qquad x\in\mathscr{R},$$

where the function $g\left(x_1,\ldots,x_d\right)$ is continuously differentiable with respect to $x_1,\ldots,x_d$ except for the points $x_1^*,\ldots,x_d^*$. At these points the function $g\left(\cdot\right)$ has jumps, i.e., if we denote $g_i\left(x,y\right)=g\left(x-\vartheta^{(1)},\ldots,y,\ldots,x-\vartheta^{(d)}\right)$, where y is the i-th variable, then for any $i=1,\ldots,d$

$$\lim_{\varepsilon\rightarrow 0+} g\left(x_1^*+\vartheta^{(i)},x_i^*+\varepsilon\right)-g\left(x_1^*+\vartheta^{(i)},x_i^*-\varepsilon\right)=r_i\left(\vartheta^{(i)}\right)\neq 0.$$

It can be shown that

$$\begin{aligned}\mathbf{P}_{\vartheta}-\lim_{T\Rightarrow\infty}&\int_0^T\delta\left(\frac{\boldsymbol{u}}{T},X_t\right)\,\delta\left(\frac{\boldsymbol{u}'}{T},X_t\right)\mathrm{d}t\\&=\sum_{i=1}^{d}\chi_{\{u_iu_i'>0\}}\;\left(\left|u_i\right|\wedge\left|u_i'\right|\right)\left(\frac{r_i\left(\vartheta^{(i)}\right)}{\sigma\left(x_i^*+\vartheta^{(i)}\right)}\right)^2\,f\left(\boldsymbol{\vartheta},x_i^*+\vartheta^{(i)}\right)\\&=\sum_{i=1}^{d}\chi_{\{u_iu_i'>0\}}\;\left(\left|u_i\right|\wedge\left|u_i'\right|\right)\,\Gamma_{i,\boldsymbol{\vartheta}}^2,\end{aligned}\tag{3.90}$$

where

$$\delta\left(\boldsymbol{h},x\right)=\frac{S\left(\boldsymbol{\vartheta}+\boldsymbol{h},x\right)-S\left(\boldsymbol{\vartheta},x\right)}{\sigma\left(x\right)}.$$

The limit process for the likelihood ratio is

$$Z_{\boldsymbol{\vartheta}}\left(\boldsymbol{u}\right)=\prod_{i=1}^{d}\exp\left\{\Gamma_{i,\boldsymbol{\vartheta}}\,W_i\left(u_i\right)-\frac{\left|u_i\right|}{2}\Gamma_{i,\boldsymbol{\vartheta}}^2\right\},\qquad \boldsymbol{u}=\left(u_1,\ldots,u_d\right)\in\mathscr{R}^d,$$

where $W_i\left(\cdot\right),\ i=1,\ldots,d$ are independent two-sided Winer processes.

Therefore the components of the vector $T\left(\hat{\boldsymbol{\vartheta}}_T-\boldsymbol{\vartheta}\right)$ are asymptotically independent and converge in distribution to $\hat{\boldsymbol{u}}=(\hat{u}_1,\ldots,\hat{u}_d)$, where

$$\hat{u}_i=\arg\sup_{u\in\mathscr{R}}\left[W_i\left(u_i\right)-\frac{|u_i|}{2}\Gamma_{i,\vartheta}\right].$$

The asymptotic behavior of the Bayes estimates can be easily described as well. The proof are close to the proof of Theorem 3.26 but there are several differences. To prove (3.87) we subdivide the integral

$$\begin{aligned}\int_{-\infty}^{\infty}\delta\left(\frac{\boldsymbol{u}}{T},x\right)\,\delta\left(\frac{\boldsymbol{u}'}{T},x\right)\,f\left(\boldsymbol{\vartheta},x\right)\,\mathrm{d}x\\ =\sum_{i=0}^{d}\int_{\mathbb{C}_i}\delta\left(\frac{\boldsymbol{u}}{T},x\right)\,\delta\left(\frac{\boldsymbol{u}'}{T},x\right)\,f\left(\boldsymbol{\vartheta},x\right)\,\mathrm{d}x\\ +\sum_{i=1}^{d}\int_{\mathbb{D}_i}\delta\left(\frac{\boldsymbol{u}}{T},x\right)\,\delta\left(\frac{\boldsymbol{u}'}{T},x\right)\,f\left(\boldsymbol{\vartheta},x\right)\,\mathrm{d}x,\end{aligned}$$

where the sets

$$\mathbb{C}_0=\left(-\infty,x_1^{(1)}\right),\quad \mathbb{C}_i=\left(x_i^{(2)},x_{i+1}^{(1)}\right),\quad \mathbb{C}_d=\left(x_d^{(2)},\infty\right),$$
$$\mathbb{D}_i=\left(x_i^{(1)},x_i^{(2)}\right),$$

and

$$x_i^{(1)}=\min\left(\vartheta^{(i)},\vartheta^{(i)}+\frac{u_i\wedge u_i'}{T}\right),\qquad x_i^{(2)}=\max\left(\vartheta^{(i)},\vartheta^{(i)}+\frac{u_i\vee u_i'}{T}\right).$$

The functions $\delta\left(\frac{\boldsymbol{u}}{T},x\right)$ and $\delta\left(\frac{\boldsymbol{u}'}{T},x\right)$ are continuously differentiable on ϑ on the intervals $\mathbb{C}_i$. Therefore their asymptotic contributions in the limit are zero. We have two different limits of the integrals $\int_{\mathbb{D}_i}$ depending on the sign of the product u_iu_i'. If $u_iu_i'>0$ then both functions $\delta\left(\frac{\boldsymbol{u}}{T},x\right)$ and $\delta\left(\frac{\boldsymbol{u}'}{T},x\right)$ have jumps on $\mathbb{D}_i$ and

$$T\int_{\mathbb{D}_i}\delta\left(\frac{\boldsymbol{u}}{T},x\right)\,\delta\left(\frac{\boldsymbol{u}'}{T},x\right)\,f\left(\boldsymbol{\vartheta},x\right)\,\mathrm{d}x=\left|u_i-u_i'\right|\,\Gamma_{i,\boldsymbol{\vartheta}}^2\,\left(1+o\left(1\right)\right).$$

If $u_iu_i'<0$ (say, $u_i<0$, $u_i'>0$) then

$$\begin{aligned}T\int_{\mathbb{D}_i}\delta\left(\frac{\boldsymbol{u}}{T},x\right)\,\delta\left(\frac{\boldsymbol{u}'}{T},x\right)\,f\left(\boldsymbol{\vartheta},x\right)\,\mathrm{d}x\\ =T\int_{x_i^{(1)}}^{\vartheta^{(i)}}\delta\left(\frac{\boldsymbol{u}}{T},x\right)\,\delta\left(\frac{\boldsymbol{u}'}{T},x\right)\,f\left(\boldsymbol{\vartheta},x\right)\,\mathrm{d}x\\ +T\int_{\vartheta^{(i)}}^{x_i^{(2)}}\delta\left(\frac{\boldsymbol{u}}{T},x\right)\,\delta\left(\frac{\boldsymbol{u}'}{T},x\right)\,f\left(\boldsymbol{\vartheta},x\right)\,\mathrm{d}x\end{aligned}$$

and on each interval one of the functions $\delta(\cdot)$ is continuously differentiable and so both integrals are asymptotically negligible.

Another difference: to study the MLE we need the estimate

$$\sup_{\vartheta\in\mathbb{K}} \mathbf{E}_\vartheta \left|Z_T^{1/2n}(\boldsymbol{u}) - Z_T^{1/2n}(\boldsymbol{v})\right|^{2m} \leq C_1\left(1+R^r\right)\left|\boldsymbol{u}-\boldsymbol{v}\right|^r, \qquad (3.91)$$

where $r > d$. This estimate can be obtained like (3.84), choosing m and n sufficiently large. The identifiability condition providing (3.86) can be checked as it was done in Lemma 3.29.

Example 3.33. Suppose that the observed diffusion process is

$$\mathrm{d}X_t = -X_t\left(a + b\,\chi_{\left\{\vartheta^{(1)} < X_t < \vartheta^{(2)}\right\}}\right)\mathrm{d}t + \sigma\,\mathrm{d}W_t, \qquad X_0, \quad 0 \leq t \leq T,$$

where

$$a > 0,\ b \neq 0, \quad \vartheta^{(1)} \in (\alpha_1, \beta_1), \quad \vartheta^{(2)} \in (\alpha_2, \beta_2), \quad \alpha_2 > \beta_1.$$

Then the conditions $\mathcal{A}_0(\Theta)$ and two-dimensional version of $\mathcal{M}$ are fulfilled and the MLE and BE have the corresponding asymptotical properties.

Example 3.34. The second process is

$$\mathrm{d}X_t = \left[\chi_{\left\{X_t < \vartheta^{(1)}\right\}} - \chi_{\left\{X_t > \vartheta^{(2)}\right\}}\right]\mathrm{d}t + \sigma\,\mathrm{d}W_t, \quad X_0, \quad t \geq 0, \qquad (3.92)$$

and its invariant density

$$f(\vartheta, x) = G(\vartheta)^{-1}\begin{cases} \mathrm{e}^{x-\underline{\vartheta}}, & \text{if } x < \underline{\vartheta} \\ 1, & \text{if } \underline{\vartheta} \leq x \leq \overline{\vartheta} \\ \mathrm{e}^{\overline{\vartheta}-x}, & \text{if } x > \overline{\vartheta}, \end{cases} \qquad (3.93)$$

where $G(\vartheta) = 2 + \left|\vartheta^{(2)} - \vartheta^{(1)}\right|$, $\underline{\vartheta} = \min\left(\vartheta^{(1)}, \vartheta^{(2)}\right)$ and $\overline{\vartheta} = \max\left(\vartheta^{(1)}, \vartheta^{(2)}\right)$.

Here once more the conditions $\mathcal{A}_0(\Theta)$ and two-dimensional version of $\mathcal{M}$ are fulfilled and the MLE and BE have the corresponding asymptotical properties.

Windows

As follows from the proofs the main contribution to the likelihood ratio is due to the observations near the jump point. This suggests restricting the observations by the window including all the possible jump points of the model. Indeed, if the observed process is simple switching (3.59), then the likelihood ratio is the following function:

$$L\left(\vartheta, X^T\right) = \exp\left\{-2\left|X_0-\vartheta\right| - \int_0^T \operatorname{sgn}\left(X_t-\vartheta\right)\,\mathrm{d}X_t - \frac{T}{2}\right\}, \qquad \vartheta\in\Theta$$

and the MLE $\hat{\vartheta}_T$ can be written as

$$\begin{aligned}\hat{\vartheta}_T &= \arg\sup_{\vartheta\in\Theta} L\left(\vartheta, X^T\right) = \arg\sup_{\vartheta\in\Theta}\frac{1}{T}\int_0^T \operatorname{sgn}\left(\vartheta-X_t\right)\,\mathrm{d}X_t\ \left(1+o\left(1\right)\right)\\ &= \arg\sup_{\vartheta\in\Theta} f_T^\circ\left(\vartheta\right)\ \left(1+o\left(1\right)\right).\end{aligned} \tag{3.94}$$

Therefore, the MLE is asymptotically equivalent (has the same limit distribution) to the *maximum empirical density estimator*

$$\hat{\theta}_T = \arg\sup_{\vartheta\in\Theta} f_T^\circ\left(\vartheta\right).$$

Its construction is based on the local time $\Lambda_T\left(\Theta\right) = \left\{\Lambda_T\left(\vartheta\right), \vartheta\in\Theta\right\}$ only. Therefore, $\Lambda_T\left(\Theta\right)$ is sufficient statistics and, as it uses the observations of the process X_t fitting in the window $[\alpha,\beta]$ only, all the other observations can be cancelled.

Note that for the process (3.59) as it follows from the proof given above, the random function

$$\eta_T\left(u\right) = \Lambda_T\left(\vartheta+\frac{u}{T}\right) - \Lambda_T\left(\vartheta\right), \qquad u\in\mathscr{R}$$

converges in distribution on compacts $|u|\leq K$ to the double-sided Wiener process with shift

$$\eta\left(u\right) = W\left(u\right) - \frac{|u|}{2}, u\in\mathscr{R}.$$

Here ϑ is the true value.

In the general case of observations (3.71) we have the similar effect. For example, if $\vartheta\in(\alpha,\beta)$ and the trend coefficient $S\left(\vartheta,x\right)$ is discontinuous along the strictly increasing curve $x_*\left(\vartheta\right), \vartheta\in\left(\alpha,\beta\right)$, then we can introduce the set $\mathbb{B} = \left[x_*\left(\alpha\right), x_*\left(\beta\right)\right]$ (*window*) and to restrict ourselves by the observations $X_t\in\mathbb{B}$. This means, that we can introduce the *pseudo-likelihood ratio*

$$\begin{aligned}\bar{L}\left(\theta, X^T\right) = \exp\Bigg\{&\int_0^T \frac{S\left(\theta,X_t\right)}{\sigma\left(X_t\right)^2}\chi_{\{X_t\in\mathbb{B}\}}\,\mathrm{d}X_t\\ &-\frac{1}{2}\int_0^T\left(\frac{S\left(\theta,X_t\right)}{\sigma\left(X_t\right)}\right)^2\chi_{\{X_t\in\mathbb{B}\}}\,\mathrm{d}t\Bigg\}\end{aligned} \tag{3.95}$$

and then with the help of this function to construct the MLE and BE. These estimators will have the same asymptotic properties as *the true MLE and BE*, i.e., the same limit distributions and convergence of moments.

Obviously, if the function $S(\vartheta, x)$ has two jumps along the increasing curves $x_*^{(i)}(\cdot), i = 1, 2$ then the window is

$$\mathbb{B} = \left[x_*^{(1)}(\alpha), x_*^{(1)}(\beta)\right] \cup \left[x_*^{(2)}(\alpha), x_*^{(2)}(\beta)\right].$$

If the set $\mathbb{B}$ nevertheless seems to be quite large, it is possible to seek a narrower one as was done in Section 2.6.4 in two steps. First we estimate consistently the parameter ϑ by observations $\left\{X_t, 0 \le t \le \sqrt{T}\right\}$ and then having the estimator $\bar{\vartheta}_{\sqrt{T}}$ we chose the window

$$\mathbb{B}_T = \left[x_*\left(\bar{\vartheta}_{\sqrt{T}}\right) - b_T, x_*\left(\bar{\vartheta}_{\sqrt{T}}\right) + b_T\right]$$

with $b_T \to 0$ slowly, say, $b_T = T^{-1/4}$. Then we use the "likelihood ratio" (3.95) with $\mathbb{B} = \mathbb{B}_T$ starting at time $\sqrt{T}$ and construct the estimators. The *first properties* of the MLE and BE (consistency and limit distributions) can be the same.

For example, let the observed process be

$$\mathrm{d}X_t = S(X_t - \vartheta)\, \mathrm{d}t + \sigma(X_t)\, \mathrm{d}W_t,$$

where the function $S(x)$ has a jump at the point $x = 0$ and $\bar{\vartheta}_{\sqrt{T}}$ be an estimator of the method of moments (say, for the simple switching it can be the empirical mean $\bar{X}_T$). This estimator is consistent and asymptotically normal

$$T^{1/4}\left(\bar{\vartheta}_{\sqrt{T}} - \vartheta\right) \Longrightarrow \mathcal{N}\left(0, \sigma_*^2\right).$$

Then

$$\mathbf{P}_\vartheta^{(T)}\left\{\left|\bar{\vartheta}_{\sqrt{T}} - \vartheta\right| > T^{-1/8}\right\} \to 0$$

as $T \to \infty$. Hence we can take the window $\mathbb{B}_T = \left[\bar{\vartheta}_{\sqrt{T}} - T^{-1/4}, \bar{\vartheta}_{\sqrt{T}} + T^{-1/4}\right]$ and the pseudo-likelihood ratio

$$\bar{L}\left(\theta, X^T\right) = \exp\left\{\int_{\sqrt{T}}^{T} \frac{S(\theta, X_t)}{\sigma(X_t)^2} \chi_{\{X_t \in \mathbb{B}_T\}}\, \mathrm{d}X_t - \frac{1}{2}\int_{\sqrt{T}}^{T}\left(\frac{S(\theta, X_t)}{\sigma(X_t)}\right)^2 \chi_{\{X_t \in \mathbb{B}_T\}}\, \mathrm{d}t\right\}.$$

It is easy to show that this function provides the same properties of the normalized likelihood ratio $Z_T(\cdot)$ as the true likelihood ratio. Hence the MLE and BE constructed with the help of this function have the same asymptotic properties.

3.4.5 Contaminated Switching

It is interesting to see what happens if the observed process is *contaminated* in the following sense. The statistician supposes that he observes a diffusion process

$$\mathrm{d}X_t = S(\boldsymbol{\vartheta}, X_t)\,\mathrm{d}t + \sigma(X_t)\,\mathrm{d}W_t, \quad X_0, \quad 0 \le t \le T \tag{3.96}$$

from the family of processes defined by the finite-dimensional parameter $\boldsymbol{\vartheta}$ but in reality he observes another process

$$\mathrm{d}X_t = S_*(\boldsymbol{\vartheta}, X_t)\,\mathrm{d}t + \sigma(X_t)\,\mathrm{d}W_t, \quad X_0, \quad 0 \le t \le T, \tag{3.97}$$

where in general $S_*(\boldsymbol{\vartheta}, x) \ne S(\boldsymbol{\vartheta}, x)$. Nevertheless he constructs the MLE $\hat{\boldsymbol{\vartheta}}_T$ and a Bayes estimator $\tilde{\boldsymbol{\vartheta}}_T$ on the base of the *theoretical* model (3.96) and substitutes the observed process (3.97).

We already discussed such statements in Section 2.6.1 (no true model) and in particular, we noted that in general these estimators converge to the value $\boldsymbol{\vartheta}_*$ which minimizes the Kullback–Leibler distance between the measures $\mathbf{P}_{\boldsymbol{\theta}}^{(T)}$ and $\mathbf{P}_{\boldsymbol{\theta}}^{*(T)}$, i.e.,

$$\boldsymbol{\vartheta}_* = \arg\inf_{\boldsymbol{\theta}\in\Theta} \mathbf{E}_{\vartheta}^* \left(\frac{S_*(\boldsymbol{\theta}, \xi) - S(\boldsymbol{\vartheta}, \xi)}{\sigma(\xi)}\right)^2 \tag{3.98}$$

and usually $\boldsymbol{\vartheta}_* \ne \boldsymbol{\vartheta}$. Here $\mathbf{E}_{\vartheta}^*$ is the expectation with respect to the invariant measure of the observed process (3.97).

In the change-type problems it is not so and the consistent estimation is possible even if the contamination is not small (see, for example, [139], Section 5.3: estimation for diffusion process with small diffusion coefficient and [43]: hypotheses testing for the same model). It is possible to prove a similar result for an ergodic diffusion process, but for the sake of simplicity we consider a simple switching model (3.58) with contamination, i.e., the observed process is

$$\mathrm{d}X_t = -\ \mathrm{sgn}(X_t - \vartheta)\,\mathrm{d}t + h(X_t)\,\mathrm{d}t + \mathrm{d}W_t, \quad X_0, \quad 0 \le t \le T, \tag{3.99}$$

where $h(\cdot)$ is unknown to the observer function and $\vartheta \in \Theta = (\alpha, \beta)$. We suppose that

$$h(\cdot) \in \mathcal{H}_\gamma = \left\{g(\cdot) : \sup_{x\in\mathscr{R}} |g(x)| \le \gamma\right\},$$

where $\gamma < 1$.

The invariant density is

$$f_h(\vartheta, x) = G(h, \vartheta)^{-1} \exp\left\{-2|x - \vartheta| + 2\int_\vartheta^x h(v)\,\mathrm{d}v\right\}$$

where $G(h, \vartheta)$ is the normalizing constant. Note that condition $\gamma < 1$ provides

$$\sup_{\vartheta\in\Theta}\sup_{h(\cdot)\in\mathcal{H}_\gamma} G(h,\vartheta) < \infty.$$

The limit process for the likelihood ratio is: for $u > 0$

$$Z_*(u) = \exp\left\{\Gamma(h,\vartheta)\,W(u) - \frac{|u|}{2}\Gamma_+(h,\vartheta)^2\right\}$$

and for $u \leq 0$

$$Z_*(u) = \exp\left\{\Gamma(h,\vartheta)\,W(u) - \frac{|u|}{2}\Gamma_-(h,\vartheta)^2\right\}$$

where

$$\Gamma(h,\vartheta)^2 = \frac{4}{G(h,\vartheta)}, \qquad \Gamma_\pm(h,\vartheta)^2 = \frac{4(1\pm h(\vartheta))}{G(h,\vartheta)}.$$

Define the random variable

$$u^* = \arg\sup_{u\in\mathscr{R}} Z_*(u).$$

This variable is well defined.

Proposition 3.35. *The MLE $\hat\vartheta_T$ is uniformly in $h(\cdot)\in\mathcal{H}_\gamma$ consistent,*

$$\mathbf{P}_{h,\vartheta} - \lim_{T\to\infty}\hat\vartheta_T = \vartheta,$$

the normalized difference converges in distribution

$$\mathcal{L}_{h,\vartheta}\left\{T\left(\hat\vartheta_T - \vartheta\right)\right\} \Longrightarrow \mathcal{L}_{h,\vartheta}\{u^*\},$$

and the moments converge too

$$\lim_{T\to\infty}\mathbf{E}_{h,\vartheta}\left|T\left(\hat\vartheta_T - \vartheta\right)\right|^p = \mathbf{E}_{h,\vartheta}|u^*|^p$$

for any $p > 0$.

Proof. At first we verify that ϑ_* defined by (3.98) is equal to ϑ for any $h(\cdot)\in\mathcal{H}_\gamma$. We have for $y > \vartheta$

$$\begin{aligned}\Phi(y,\vartheta) &= \mathbf{E}_{h,\vartheta}\Big(\operatorname{sgn}(\xi - y) - \operatorname{sgn}(\xi-\vartheta) + h(\xi)\Big)^2\\ &= \int_{-\infty}^{\vartheta} h(x)^2\, f_h(\vartheta,x)\;\mathrm{d}x + \int_y^{\infty} h(x)^2\, f_h(\vartheta,x)\;\mathrm{d}x\\ &\quad + \int_\vartheta^y [2-h(x)]^2\, f_h(\vartheta,x)\;\mathrm{d}x.\end{aligned}$$

Hence for $y > \vartheta$

$$\begin{aligned}\frac{\partial \Phi(y,\vartheta)}{\partial y} &= -h(y)^2\ f_h(\vartheta,y) + [2-h(y)]^2\ f_h(\vartheta,y)\\ &= 4\ [1-h(y)]\ f_h(\vartheta,y) > 0,\end{aligned}$$

and similarly for $y < \vartheta$ we have

$$\begin{aligned}\frac{\partial \Phi(y,\vartheta)}{\partial y} &= h(y)^2\ f_h(\vartheta,y) - [2+h(y)]^2\ f_h(\vartheta,y)\\ &= -4\ [1+h(y)]\ f_h(\vartheta,y) < 0.\end{aligned}$$

Hence

$$\vartheta = \arg\inf_{y\in\Theta} \mathbf{E}_{h,\vartheta}\Big(\operatorname{sgn}(\xi-y) - \operatorname{sgn}(\xi-\vartheta) + h(\xi)\Big)^2.$$

Remember that in such misspecified problems

$$\mathbf{E}_{h,\vartheta}\,\frac{\mathrm{d}\,\mathbf{P}_y^{(T)}}{\mathrm{d}\,\mathbf{P}_\vartheta^{(T)}}\left(X^T\right) \neq 1,$$

and we study the MLE defined as

$$\hat\vartheta_T = \arg\sup_{y\in\Theta} L\left(y,\theta_1;X^T\right) = \arg\sup_{y\in\Theta} L\left(y,\theta_1;X^T\right)^{1/q}.$$

Put

$$Z_T(u) = L\left(\vartheta+u/T,\vartheta;X^T\right)^{1/q},\quad u\in\mathbb{U}_T = \left(T(\alpha-\vartheta),T(\beta-\vartheta)\right).$$

Then for $u > 0$ we have the representation

$$\begin{aligned}Z_T(u)^q = \exp\Big\{&2\left|X_0-\vartheta\right| - 2\left|X_0-\vartheta-u/T\right|\\ &+2\int_0^T \chi_{\{\vartheta<X_t<\vartheta+u/T\}}\,\mathrm{d}W_t - 2\int_0^T \chi_{\{\vartheta<X_t<\vartheta+u/T\}}\,\mathrm{d}t\\ &+2\int_0^T \chi_{\{\vartheta<X_t<\vartheta+u/T\}}\,h(X_t)\ \mathrm{d}t\Big\}.\end{aligned}$$

Using the equalities

$$\begin{aligned}\int_0^T \chi_{\{\vartheta<X_t<\vartheta+u/T\}}\,h(X_t)\ \mathrm{d}t &= \int_\vartheta^{\vartheta+u/T} h(x)\ 2\,\Lambda_T(h,\vartheta,x)\ \mathrm{d}x\\ &= T\int_\vartheta^{\vartheta+u/T} h(x)\ f_h(\vartheta,x)\ \mathrm{d}x\ (1+o(1))\end{aligned}$$

we obtain

$$\int_0^T \chi_{\{\vartheta<X_t<\vartheta+u/T\}}\, h\left(X_t\right)\,\mathrm{d}t = u\, h\left(\vartheta\right)\; f_h\left(\vartheta,\vartheta\right)\;\left(1+o(1)\right),$$

and the convergence

$$Z_T\left(u\right) \Longrightarrow Z\left(u\right),$$

for $u > 0$. Here $Z\left(u\right) = Z_*\left(u\right)^{1/q}$.

Further, the inequality

$$\mathbf{E}_{h,\vartheta}\left|Z_T\left(u\right)^{1/8} - Z_T\left(v\right)^{1/8}\right|^4 \leq C\left(1+R^2\right)\,\left|u-v\right|^2$$

can be derived like (3.91) and to obtain the last estimate

$$\mathbf{P}_{h,\vartheta}^{(T)}\left\{Z_T\left(u\right) \geq \mathrm{e}^{-\kappa\left|u\right|}\right\} \leq \frac{C_N}{\left|u\right|^N}$$

we repeat the proof of (3.69), where we replace κ_* by $\left(1-\gamma\right)\kappa_*$.

Having all these properties of the "likelihood ratio" $Z_T\left(\cdot\right)$ we have the weak convergence

$$\mathbf{Q}_{h,\vartheta}^{(T)} \Longrightarrow \mathbf{Q}_{h,\vartheta},$$

in the space $\mathscr{C}_0\left(\mathscr{R}\right)$ and the estimate (2.27) for the tails of $Z_T\left(\cdot\right)$, which give us the desired properties of the MLE (for details see Theorem A21 in [109]).

Remark 3.36. Note that condition $\gamma < 1$ is essential. If $\gamma = 1$ then the function $h(x) = \operatorname{sgn}(x-\vartheta)$ belongs to $\mathcal{H}_1$ and we observe a Wiener process only, hence the consistent estimation of ϑ is of course impossible.

It is almost evident that we can have the similar effect in the general case of observations

$$\mathrm{d}X_t = S\left(\boldsymbol{\vartheta}, X_t\right)\,\mathrm{d}t + h(X_t)\,\mathrm{d}t + \sigma\left(X_t\right)\,\mathrm{d}W_t,\quad X_0,\quad 0 \leq t \leq T,$$

with contamination $h\left(\cdot\right)$. We have just to suppose that the function $h\left(\cdot\right)$ can not compensate entirely the jump of the function $S\left(\boldsymbol{\vartheta},\cdot\right)$.

3.5 Non Ergodic Processes

In this section we consider three examples of homogeneous diffusion processes of the type

$$\mathrm{d}X_t = \vartheta\, S\left(X_t\right)\,\mathrm{d}t + \sigma\,\mathrm{d}W_t,\quad X_0 = x_0,\quad 0 \leq t \leq T \tag{3.100}$$

where $S\left(\cdot\right)$ is a known function and $\vartheta \in \left(\alpha,\beta\right)$ is the parameter unknown to the observer. We choose situations when at least one of the conditions of ergodicity

$$V(\vartheta,x)=\int_0^x \exp\left\{-\frac{2\vartheta}{\sigma^2}\int_0^y S(v)\,\mathrm{d}v\right\}\,\mathrm{d}y\to\pm\infty,\quad \text{as}\quad x\to\pm\infty,\tag{3.101}$$

$$G(\vartheta)=\int_{-\infty}^{\infty} \exp\left\{\frac{2\vartheta}{\sigma^2}\int_0^x S(v)\,\mathrm{d}v\right\}\,\mathrm{d}x<\infty\tag{3.102}$$

is not fulfilled. Therefore the process $\{X_t,\ t\geq 0\}$ has no finite invariant measure and has no ergodic properties like the large numbers law. We start with a diffusion process which satisfies (3.101) and does not satisfy (3.102) (null recurrent) and then consider two examples with diffusion processes (going to infinity with time) with both conditions nonfulfilled.

The MLE $\hat{\vartheta}_T$ in this linear problem is the ratio of two integrals

$$\hat{\vartheta}_T=\frac{\int_0^T S(X_t)\,\mathrm{d}X_t}{\int_0^T S(X_t)^2\,\mathrm{d}t}=\vartheta+\sigma\,\frac{\int_0^T S(X_t)\,\mathrm{d}W_t}{\int_0^T S(X_t)^2\,\mathrm{d}t}$$

and its asymptotic behavior is entirely defined by the asymptotics of these integrals. Of course, we have to write the expression like (1.115) but for simplicity of exposition we prefer to write it in this form. Remember that usually the asymptotic contribution of the events $\left\{\hat{\vartheta}_T\leq\alpha\right\}$ and $\left\{\hat{\vartheta}_T\geq\beta\right\}$ to the variance of estimators is negligible.

Note as well that in all examples the integral

$$\int_0^T S(X_t)^2\,\mathrm{d}t\longrightarrow\infty$$

and we have already the consistency of the estimator (3.102)

$$\mathbf{P}_\vartheta-\lim_{T\to\infty}\hat{\vartheta}_T=\vartheta.$$

Therefore we need to describe the asymptotics of the difference $\hat{\vartheta}_T-\vartheta$.

3.5.1 Null Recurrent Process

The diffusion process is null-recurrent, if the condition (3.101) is fulfilled and $G(\vartheta)=\infty$. Hence the time to return to any bounded region is finite with probability 1 but mathematical expectation of this time is equal to infinity.

For example, any diffusion process

$$\mathrm{d}X_t=S(X_t)\,\mathrm{d}t+\sigma(X_t)\,\mathrm{d}W_t,\quad X_0=x_0,\ t\geq 0,\tag{3.103}$$

with trend coefficient $S(\cdot)$ having a compact support and bounded diffusion coefficient $\sigma(\cdot)^2$ is null recurrent.

Suppose for simplicity that the trend function $S(x)\equiv 0$ and we have

$$\mathrm{d}Y_t=\bar{\sigma}(Y_t)\,\mathrm{d}W_t,\quad Y_0=y_0,\ t\geq 0.\tag{3.104}$$

Remember that the general case (3.103) can be reduced to this one if we put $Y_t = V(X_t)$, where

$$V(x) = \int_0^x \exp\left\{-2\int_0^y \frac{S(v)}{\sigma(v)^2}\mathrm{d}v\right\}\,\mathrm{d}y.$$

Indeed, in this case the process Y_t by the Itô formula admits the differential (3.104) with $\bar\sigma(y) = V'\left(V^{-1}(y)\right)\,\sigma\left(V^{-1}(y)\right)$, where $V^{-1}(y)$ is the function inverse to $V(x) = y$.

This class of diffusion processes was first studied by Khasminskii [97], who obtained the following result. Suppose that the function $\bar\sigma(\cdot)$ is locally Lipschitz and

$$0 < \bar\sigma(x)^2 \le C\,\left(1 + x^2\right). \tag{3.105}$$

Suppose that the diffusion coefficient $\sigma(\cdot)^2$ is such that the function

$$f(y) = \frac{2}{\bar\sigma(y)^2}$$

has the following asymptotics:

$$f(y) = B\,|y|^b\,(1 + o(1)), \quad \text{as} \quad |y| \to \infty, \tag{3.106}$$

where $b > -1$. Then the process $\{Y_t, t \ge 0\}$ is *null recurrent* with invariant density $f(\cdot)$. Note that

$$\int_{-\infty}^{\infty} f(y)\,\mathrm{d}y = \infty.$$

Put $\gamma = (2+b)^{-1}$ and denote by η a random variable with stable distribution function having the Laplace transform

$$\mathbf{E}\,\mathrm{e}^{-p\eta} = \mathrm{e}^{-p^\gamma}.$$

Then we have the following result.

Lemma 3.37. *Let $h(\cdot)$ be a measurable function such that the integral*

$$\bar h = \int_{-\infty}^{\infty} h(y)\,f(y)\,\mathrm{d}y$$

converges absolutely. Then the following weak convergence holds:

$$\frac{1}{T^\gamma}\int_0^T h(Y_t)\,\mathrm{d}t \Longrightarrow \frac{K_*(B,\gamma)\,\bar h}{\eta^\gamma}.$$

Proof. See [97], Theorems 4.11.1, 4.11.2 and the Corollary. The constant

$$K_*(B,\gamma) = \frac{\Gamma(1+\gamma)}{2\,(\gamma^2 B)^\gamma\,\Gamma(1-\gamma)}. \tag{3.107}$$

Here $\Gamma(\cdot)$ is the Gamma function.

Note that this convergence replaces the law of large numbers for the ergodic diffusion processes ($G < \infty, \gamma = 1$). The central limit theorem for stochastic integral in the case of null-recurrent processes was obtained recently by Höpfner and Löcherbach [101]. We formulate it in the form of the following lemma.

Lemma 3.38. *Suppose that the function $\bar{\sigma}(\cdot)$ is locally Lipschitz and satisfies (3.105), (3.106). Then for any measurable function $h(\cdot)$ such that*

$$\bar{h^2} = \int_{-\infty}^{\infty} h(y)^2 \; f(y)\,\mathrm{d}y < \infty$$

the convergence

$$\left(\frac{1}{T^{\gamma/2}}\int_0^T h(X_t)\;\mathrm{d}W_t,\ \frac{1}{T^{\gamma}}\int_0^T h(X_t)^2\;\mathrm{d}t\right) \Longrightarrow \left(\frac{J_*^{1/2}\,\zeta}{\eta^{\gamma/2}},\ \frac{J_*}{\eta^{\gamma}}\right) \tag{3.108}$$

holds. Here $J_* = K_*(B,\gamma)\;\bar{h^2}$ and $\zeta \sim \mathcal{N}(0,1)$ is independent of η.

Below we consider just an example of parameter estimation problem for null-recurrent diffusion process.

Suppose that the observed process is

$$\mathrm{d}X_t = -\,\vartheta\,\frac{X_t}{1+X_t^2}\,\mathrm{d}t + \sigma\,\mathrm{d}W_t, \quad X_0 = x_0,\ 0 \le t \le T, \tag{3.109}$$

and we have to estimate the parameter ϑ by the observations X^T. To describe the properties of the MLE

$$\hat{\vartheta}_T = -\left(\int_0^T \frac{X_t^2}{(1+X_t^2)^2}\,\mathrm{d}t\right)^{-1} \int_0^T \frac{X_t}{1+X_t^2}\,\mathrm{d}X_t$$

we need the following notations:

$$G(\vartheta) = \int_{-\infty}^{\infty} \frac{\mathrm{d}x}{(1+x^2)^{\vartheta/\sigma^2}}, \quad f(\vartheta,x) = \frac{1}{G(\vartheta)\,(1+x^2)^{\vartheta/\sigma^2}},$$

$$\bar{h^2} = \frac{2}{\sigma^2}\int_{-\infty}^{\infty} \frac{x^2}{(1+x^2)^{2+\vartheta/\sigma^2}}\,\mathrm{d}x, \quad \mathrm{I}(\vartheta) = \frac{\bar{h^2}}{2\,G(\vartheta)},$$

$$\gamma = 1/2 + \vartheta/\sigma^2, \qquad J_*(\vartheta) = K_*(B,\gamma)\;\bar{h^2},$$

where the constant $K_*(B,\gamma)$ is the same as in (3.107) with

$$B = \frac{2}{\sigma^2}\left(1+\frac{2\vartheta}{\sigma^2}\right)^{-\frac{4\vartheta}{\sigma^2+2\vartheta}}. \tag{3.110}$$

The properties of the MLE are given in the following proposition.

Proposition 3.39. (Höpfner and Kutoyants [99]) *If $\vartheta > \sigma^2/2$, then the MLE is asymptotically normal*

$$\mathcal{L}_\vartheta\left\{\sqrt{T}\left(\hat{\vartheta}_T - \vartheta\right)\right\} \Longrightarrow \mathcal{N}\left(0, \mathrm{I}\left(\vartheta\right)^{-1}\right). \tag{3.111}$$

If $-\sigma^2/2 < \vartheta < \sigma^2/2$, then the MLE is asymptotically mixing normal, i.e.,

$$\mathcal{L}_\vartheta\left\{T^{\gamma/2}\left(\hat{\vartheta}_T - \vartheta\right)\right\} \Longrightarrow \frac{\zeta\,\eta^{\gamma/2}}{J_*\left(\vartheta\right)^{1/2}}, \tag{3.112}$$

where η and $\zeta \sim \mathcal{N}(0,1)$ are independent random variables.

Proof. Note that

$$\hat{\vartheta}_T - \vartheta = -\sigma\left(\int_0^T \frac{X_t^2}{(1+X_t^2)^2}\,\mathrm{d}t\right)^{-1}\int_0^T \frac{X_t}{1+X_t^2}\,\mathrm{d}W_t,$$

therefore we have to study the asymptotic behavior of these two integrals.

Let $\vartheta/\sigma^2 > 1/2$, then the process (3.109) has ergodic properties. Indeed, the conditions (3.101) and (3.102) are fulfilled and the invariant density is $f(\vartheta,\cdot)$ (defined above) is integrable over $\mathscr{R}$. By the law of large numbers

$$\mathbf{P}_\vartheta - \lim_{T\longrightarrow\infty}\frac{1}{T}\int_0^T \frac{X_t^2}{\sigma^2\left(1+X_t^2\right)^2}\,\mathrm{d}t = \mathrm{I}\left(\vartheta\right)$$

and the stochastic integral is (by the central limit theorem) asymptotically normal:

$$\mathcal{L}_\vartheta\left\{\frac{1}{\sqrt{T}}\int_0^T \frac{X_t}{\sigma\left(1+X_t^2\right)}\,\mathrm{d}W_t\right\} \Longrightarrow \mathcal{N}\left(0, \mathrm{I}\left(\vartheta\right)\right).$$

Hence we obtain the asymptotic normality (3.111) of the MLE.

Consider now the case $\vartheta < \sigma^2/2$, when $G(\vartheta) = \infty$. To study the asymptotic behavior of the integrals

$$\int_0^T \frac{X_t}{1+X_t^2}\,\mathrm{d}W_t, \qquad \int_0^T \frac{X_t^2}{(1+X_t^2)^2}\,\mathrm{d}t$$

we transform the process $\{X_t, t \geq 0\}$ to the form (3.104) without trend and then we use the convergence (3.108). It is easy to see that

$$V\left(\vartheta, x\right) = \int_0^x \left(1+z^2\right)^{\vartheta/\sigma^2}\,\mathrm{d}z.$$

Hence we can put

$$Y_t = \int_0^{X_t} \left(1+z^2\right)^{\vartheta/\sigma^2} \, dz$$

and the process $\{Y_t, t \geq 0\}$ by the Itô formula satisfies Equation (3.104) with $y_0 = V(\vartheta, x_0)$. The function

$$\bar{\sigma}(y) = \left(1 + V^{-1}(\vartheta, y)^2\right)^{\vartheta/\sigma^2} \sigma,$$

where $V^{-1}(\vartheta, y)$ is the inverse to the $y = V(\vartheta, x)$ function. The condition (3.105) is fulfilled because $\vartheta > -\sigma^2/2$. Elementary calculus gives us the asymptotics (3.106) with B defined in (3.110) and $b = -\frac{4\vartheta}{\sigma^2+2\vartheta}$.

Note that the condition $\vartheta < \sigma^2/2$ is equivalent to $b > -1$.

We have

$$\frac{1}{T^\gamma}\int_0^T \frac{X_t^2}{\sigma^2\left(1+X_t^2\right)^2}\, dt = \frac{1}{T^\gamma}\int_0^T \frac{V^{-1}(\vartheta, Y_t)^2}{\sigma^2\left[1+V^{-1}(\vartheta, Y_t)^2\right]^2}\, dt$$
$$\Longrightarrow K_*(\vartheta)\, \bar{h}^2\, \eta^{-\gamma},$$

where $K_*(\vartheta) = K_*(B, \gamma)$. Now the convergence (3.112) follows from Lemma 3.38.

Remark 3.40. Note that the rate of convergence in the null-recurrent case is unknown to the observer. Hence it is interesting to estimate the parameter $\gamma \in (0,1)$ as well. The MLE

$$\hat{\gamma}_T = \frac{1}{2} + \frac{\hat{\vartheta}_T}{\sigma^2}$$

is consistent and is asymptotically mixing normal

$$\mathcal{L}_\vartheta\left\{T^{\gamma/2}\left(\hat{\gamma}_T - \gamma\right)\right\} \Longrightarrow \sigma^{-2}\, J_*(\vartheta)^{-1/2}\, \zeta\, \eta^{\gamma/2}.$$

This convergence follows immediately from (3.112).

Remark 3.41. Note that in 1983 I gave a talk on the seminar by D. Dacunha-Castelle in Ecole Normale Superieure (Paris, rue d'Ulm), where I presented the properties of the MLE in the regular case of an ergodic diffusion process and L. Le Cam asked me if I know the properties of this estimator in the null recurrent case. My answer at that time was negative.

Remark 3.42. The family of measures $\left\{\mathbf{P}_\vartheta^{(T)}, \vartheta \in \left(-\sigma^2/2, \sigma^2/2\right)\right\}$, as is usual in such situations (see Höpfner and Löcherbach [101]) is *locally asymptotically mixing normal* (LAMN). In particular, for any ϑ_0 the normalized likelihood ratio $Z_T(u) = L\left(\vartheta_0 + T^{-\gamma_0/2}u, \vartheta_0, X^T\right)$ with $\mathbf{P}_{\vartheta_0}^{(T)}$ probability 1 is equal to

$$Z_T(u) = \exp\left\{u\,\Delta_T\left(\vartheta_0, X^T\right) - \frac{u^2}{2}\,\mathrm{I}_T\left(\vartheta_0, X^T\right)\right\},$$

where $\gamma_0 = \gamma(\vartheta_0)$,

$$\Delta_T\left(\vartheta_0, X^T\right) = \frac{1}{\sigma^2 T^{\gamma_0/2}} \int_0^T \frac{X_t}{1+X_t^2}\left[\mathrm{d}X_t + \frac{\vartheta_0 X_t}{1+X_t^2}\,\mathrm{d}t\right],$$
$$\mathrm{I}_T\left(\vartheta_0, X^T\right) = \frac{1}{\sigma^2 T^{\gamma_0}} \int_0^T \left(\frac{X_t}{1+X_t^2}\right)^2 \mathrm{d}t$$

and

$$\mathcal{L}_{\vartheta_0}\left\{\Delta_T\left(\vartheta_0, X^T\right), \mathrm{I}_T\left(\vartheta_0, X^T\right)\right\} \Longrightarrow \mathcal{L}\left\{\frac{J_*(\vartheta_0)^{1/2}\,\zeta}{\eta^{\gamma_0/2}}, \frac{J_*(\vartheta_0)}{\eta^{\gamma_0}}\right\}. \tag{3.113}$$

It is possible to construct a lower minimax bound on the risks of all estimators like the Hajek–Le Cam bound but there are two particularities of this model. The first one is the *random limit Fisher information*

$$\mathrm{I}_T\left(\vartheta_0, X^T\right) \to \mathrm{I}(\vartheta_0) = \frac{J_*(\vartheta_0)}{\eta^{\gamma_0}},$$

and the second: the normalizing function $T^{-\gamma/2}$ essentially depends on ϑ ($\gamma = \gamma(\vartheta)$) and this provides the limits

$$\frac{T^{\frac{\gamma(\vartheta)}{2}}}{T^{\frac{\gamma(\vartheta_0)}{2}}} \longrightarrow 0 \quad \text{or} \quad \infty$$

if $\vartheta \neq \vartheta_0$. The case of random limit information was studied by Jeganathan [115] and the necessary modification due to the different rates of the normalizing functions was proposed by Davies [58]. Here we present a version of Davies' result applied to our model.

Proposition 3.43. *Let $\vartheta \in \Theta = \left(-\sigma^2/2, \sigma^2/2\right)$. Then for any bounded loss function $\ell(\cdot) \in \mathscr{W}$ and any $\vartheta_0 \in \Theta$*

$$\varliminf_{M\to\infty} \varliminf_{T\to\infty} \inf_{\bar\vartheta_T} \sup_{|\vartheta-\vartheta_0|\leq MT^{-\gamma_0/2}} \mathbf{E}_\vartheta\, \ell\left(T^{\gamma_0/2}\left(\bar\vartheta_T - \vartheta\right)\right) \geq \mathbf{E}\,\ell\left(\zeta\,\mathrm{I}(\vartheta_0)^{-1/2}\right), \tag{3.114}$$

where inf *is taken over all estimators of ϑ and $\zeta \sim \mathcal{N}(0,1)$.*

Proof. The main lines below follow the proof of Theorem II.12.1 in [109]. Let us put $\vartheta = \vartheta_0 + T^{-\gamma_0/2}\,u$, then we have

$$\begin{aligned}
\mathrm{J}_T &= \sup_{|\vartheta-\vartheta_0|\leq MT^{-\gamma_0/2}} \mathbf{E}_\vartheta \, \ell\left(T^{\gamma_0/2}\left(\bar\vartheta_T-\vartheta\right)\right) \\
&= \sup_{|u|\leq M} \mathbf{E}_{\vartheta_0} \, Z_T(u)\,\ell\left(T^{\gamma_0/2}\left(\bar\vartheta_T-\vartheta_0\right)-u\right) \\
&\geq \frac{1}{2M}\int_{-M}^{M} \mathbf{E}_{\vartheta_0} \, Z_T(u)\,\ell\left(\bar u_T-u\right)\,\mathrm{d}u \\
&= \frac{1}{2M}\int_{-M}^{M} \mathbf{E}_{\vartheta_0} \, \mathrm{e}^{u\Delta_T-\frac{u^2}{2}\mathrm{I}_T}\,\ell\left(\bar u_T-u\right)\,\mathrm{d}u,
\end{aligned}$$

where we put

$$\bar u_T = T^{\gamma_0/2}\left(\bar\vartheta_T-\vartheta_0\right), \qquad \Delta_T=\Delta_T\left(\vartheta_0,X^T\right), \qquad \mathrm{I}_T=\mathrm{I}_T\left(\vartheta_0,X^T\right).$$

Further, we write the last integral as

$$\begin{aligned}
&\frac{1}{2M}\mathbf{E}_{\vartheta_0}\,\mathrm{e}^{\frac{\Delta_T^2}{2\mathrm{I}_T}}\int_{-M}^{M}\mathrm{e}^{-\frac{\mathrm{I}_T}{2}\left(u-\frac{\Delta_T}{\mathrm{I}_T}\right)^2}\,\ell\left(\bar u_T-u\right)\,\mathrm{d}u \\
&\quad= \frac{1}{2M}\mathbf{E}_{\vartheta_0}\,\mathrm{e}^{\frac{\Delta_T^2}{2\mathrm{I}_T}}\int_{-M-\frac{\Delta_T}{\mathrm{I}_T}}^{M-\frac{\Delta_T}{\mathrm{I}_T}}\mathrm{e}^{-\frac{\mathrm{I}_T}{2}v^2}\,\ell\left(\bar v_T-v\right)\,\mathrm{d}v \\
&\quad\geq \frac{1}{2M}\mathbf{E}_{\vartheta_0}\,\mathrm{e}^{\frac{\Delta_T^2}{2\mathrm{I}_T}}\,\chi_{\left\{\left|\frac{\Delta_T}{\mathrm{I}_T}\right|\leq M-\sqrt{M}\right\}}\int_{-\sqrt{M}}^{\sqrt{M}}\mathrm{e}^{-\frac{\mathrm{I}_T}{2}v^2}\,\ell\left(\bar v_T-v\right)\,\mathrm{d}v.
\end{aligned}$$

Here we changed the variables $v=u-\Delta_T/\mathrm{I}_T$ and put $\bar v_T=\bar u_T-\Delta_T/\mathrm{I}_T$. The last inequality is evident. The loss function $\ell(\cdot)$ is symmetric and nondecreasing, hence we can apply Anderson's lemma [5] (see as well [109], section II.10) and write

$$\int_{-\sqrt{M}}^{\sqrt{M}}\mathrm{e}^{-\frac{\mathrm{I}_T}{2}u^2}\,\ell\left(\bar v_T-v\right)\,\mathrm{d}v \geq \int_{-\sqrt{M}}^{\sqrt{M}}\mathrm{e}^{-\frac{\mathrm{I}_T}{2}u^2}\,\ell\left(v\right)\,\mathrm{d}v.$$

Therefore using the convergence (3.113) we can write

$$\begin{aligned}
\mathrm{J}_T &\geq \frac{1}{2M}\mathbf{E}_{\vartheta_0}\,\mathrm{e}^{\frac{\Delta_T^2}{2\mathrm{I}_T}}\,\chi_{\left\{\left|\frac{\Delta_T}{\mathrm{I}_T}\right|\leq M-\sqrt{M},\ |\mathrm{I}_T|\leq N\right\}}\int_{-\sqrt{M}}^{\sqrt{M}}\mathrm{e}^{-\frac{\mathrm{I}_T}{2}v^2}\,\ell\left(v\right)\,\mathrm{d}v \\
&\longrightarrow \frac{1}{2M}\mathbf{E}\,\mathrm{e}^{\frac{\zeta^2}{2}}\,\chi_{\left\{\left|\zeta\,\mathrm{I}_0^{-1/2}\right|\leq M-\sqrt{M},\ |\mathrm{I}_0|\leq N\right\}}\int_{-\sqrt{M}}^{\sqrt{M}}\mathrm{e}^{-\frac{\mathrm{I}_0}{2}v^2}\,\ell\left(v\right)\,\mathrm{d}v.
\end{aligned}$$

Here $\mathrm{I}_0=\mathrm{I}\left(\vartheta_0\right)$. Below in the conditional expectation we use the independence of ζ and η. Remember that ζ is $\mathcal{N}(0,1)$, hence with probability 1

$$\begin{aligned}
&\frac{1}{2M}\mathbf{E}\left(\mathrm{e}^{\frac{\zeta^2}{2}}\,\chi_{\left\{|\zeta|\leq\left(M-\sqrt{M}\right)\mathrm{I}_0^{1/2}\right\}}\middle|\,\eta\right) \\
&\quad= \frac{1}{2M}\int_{-\left(M-\sqrt{M}\right)\mathrm{I}_0^{1/2}}^{\left(M-\sqrt{M}\right)\mathrm{I}_0^{1/2}}\mathrm{e}^{z^2/2}\,\frac{1}{\sqrt{2\pi}}\mathrm{e}^{-z^2/2}\,\mathrm{d}z = \frac{\left(M-\sqrt{M}\right)\,\mathrm{I}\left(\vartheta_0\right)^{1/2}}{M\sqrt{2\pi}}.
\end{aligned}$$

Therefore

$$\varliminf_{T\to\infty} \mathrm{J}_T \geq \left(1 - M^{-1/2}\right) \int_{-\sqrt{M}}^{\sqrt{M}} \mathbf{E}\ \chi_{\{|\mathrm{I}_0|\leq N\}}\ e^{-\frac{\mathrm{I}_0}{2} v^2}\ \ell(v) \sqrt{\frac{\mathrm{I}(\vartheta_0)}{2\pi}}\ \mathrm{d}v.$$

This relation holds for any $M > 0$ and $N > 0$, hence it holds as well as $M \to \infty$ and $N \to \infty$ and we finally obtain

$$\varliminf_{M\to\infty} \varliminf_{T\to\infty} \mathrm{J}_T \geq \frac{1}{\sqrt{2\pi}} \int_{\mathscr{R}} e^{-\frac{y^2}{2}}\ \mathbf{E}\ \ell\left(y\,\mathrm{I}(\vartheta_0)^{-1/2}\right)\ \mathrm{d}y = \mathbf{E}\ \ell\left(\frac{\zeta\ \eta^{\gamma_0/2}}{J_*(\vartheta_0)^{\gamma_0/2}}\right).$$

We can define now asymptotically efficient estimator as an estimator $\bar{\vartheta}_T$ satisfying the equality

$$\lim_{M\to\infty} \lim_{T\to\infty} \sup_{|\vartheta-\vartheta_0|\leq MT^{-\gamma_0/2}} \mathbf{E}_\vartheta\, \ell\left(T^{\gamma_0/2}\left(\bar{\vartheta}_T - \vartheta\right)\right) = \mathbf{E}\ \ell\left(\zeta\, \mathrm{I}(\vartheta_0)^{-1/2}\right),$$

for all $\vartheta_0 \in \left(-\sigma^2/2, \sigma^2/2\right)$. Note that for fixed M and $|u| \leq M$ the ratio

$$\frac{T^{\gamma_0}}{T^{\gamma_u}} \to 1.$$

Here $\gamma_u = \gamma\left(\vartheta_0 + T^{-\gamma_0/2}\, u\right)$, therefore

$$T^{\frac{\gamma_0}{2}}\left(\hat{\vartheta}_T - \vartheta\right) = T^{\frac{\gamma}{2}}\left(\hat{\vartheta}_T - \vartheta\right)\ (1 + o(1))$$

and the MLE is an asymptotically efficient estimator in this problem too.

Remark 3.44. In the null recurrent case we have $V(\vartheta, \pm\infty) = \pm\infty$ and $G(\vartheta) = \infty$. Kulinich [128] studied the properties of the MLE $\hat{\vartheta}_T$ of the parameter ϑ constructed by the observations

$$\mathrm{d}X_t = [a(X_t) + \vartheta b(X_t)]\ \mathrm{d}t + \sigma(X_t)\ \mathrm{d}W_t$$

in the situation when the ratio

$$\frac{V(\vartheta, x)}{F_*(\vartheta, x)}$$

has finite limits as $x \to \pm\infty$. Here

$$F_*(\vartheta, x) = \int_0^x \frac{1}{\sigma(y)^2} \exp\left\{2\int_0^y \frac{a(v) + \vartheta b(v)}{\sigma(v)^2}\ \mathrm{d}v\right\}\ \mathrm{d}y.$$

In particular, if

$$\mathrm{d}X_t = \vartheta\cos(X_t)\ \mathrm{d}t + \mathrm{d}W_t, \quad X_0, \quad 0 \leq t \leq T,$$

then

$$\mathcal{L}_\vartheta\left\{\sqrt{T}\left(\hat{\vartheta}_T - \vartheta\right)\right\} \Longrightarrow \mathcal{N}(0, 2).$$

The proof and another example can be found in [128].

3.5.2 Polynomial Growth Process

We consider a homogeneous diffusion process having an exact rate of increasing on infinity studied by Gikhman and Skorokhod [84], section 1.17 (see as well Keller *et al.* [120]). In particular, for a process

$$\mathrm{d}X_t = S\left(X_t\right)\,\mathrm{d}t + \sigma\left(X_t\right)\,\mathrm{d}W_t, \quad X_0 = x_0, \quad t \geq 0$$

they propose conditions on $S\left(\cdot\right)$ and $\sigma\left(\cdot\right)$ such that

$$\mathbf{P}\left\{\lim_{t\to\infty}\frac{X_t}{x_t} = 1\right\} = 1$$

where $\{x_t,\ t \geq 0\}$ is a solution of the ordinary differential equation

$$\frac{\mathrm{d}x_t}{\mathrm{d}t} = S\left(x_t\right), \qquad x_0, \quad t \geq 0$$

and $x_t \to \infty$ as $t \to \infty$ (see [84], Section 1.17, Theorem 4).

Below we consider the following example:

$$\mathrm{d}X_t = \vartheta\ \left|X_t\right|^{\kappa}\mathrm{d}t + \sigma\ \mathrm{d}W_t, \quad X_0 = x_0, \quad t \geq 0 \tag{3.115}$$

of the homogeneous diffusion process, where the parameter $\vartheta \in (\alpha, \beta)\,,\ \alpha > 0$. The value $\kappa \in (0,1)$ (or $\kappa \in (-1,0)$) is known and we have to estimate ϑ by observations $X^T = \{X_t, 0 \leq t \leq T\}$. For the process $\{X_t,\ \geq 0\}$ we have the following property:

$$\mathbf{P}_\vartheta\left\{\lim_{t\to\infty}\frac{X_t}{t^{\frac{1}{1-\kappa}}} = \left[(1-\kappa)\ \vartheta\right]^{\frac{1}{1-\kappa}}\right\} = 1 \tag{3.116}$$

(see [84], Section 1.17, Theorems 2 and 5). This means that the solution X_t of the stochastic differential equation (3.115) is asymptotically equivalent (in this sense) to the solution x_t of the ordinary differential equation

$$\frac{\mathrm{d}x_t}{\mathrm{d}t} = \vartheta\ \left|x_t\right|^{\kappa}, \qquad x_0, \quad t \geq 0.$$

The MLE of the parameter ϑ is

$$\begin{aligned}\hat\vartheta_T &= \left(\int_0^T \left|X_t\right|^{2\kappa}\mathrm{d}t\right)^{-1}\int_0^T \left|X_t\right|^{\kappa}\mathrm{d}X_t\\ &= \vartheta + \sigma\left(\int_0^T \left|X_t\right|^{2\kappa}\mathrm{d}t\right)^{-1}\int_0^T \left|X_t\right|^{\kappa}\mathrm{d}W_t.\end{aligned}$$

Put

$$\gamma = \frac{1+\kappa}{1-\kappa}.$$

As follows from the proposition below, the behavior of the MLE for this model is similar to that of the MLE for ergodic process.

Proposition 3.45. *The MLE $\hat{\vartheta}_T$ as $T \to \infty$ is consistent and asymptotically normal*

$$\mathcal{L}_\vartheta\left\{T^{\gamma/2}\left(\hat{\vartheta}_T - \vartheta\right)\right\} \Longrightarrow \mathcal{N}\left(0, \mathrm{I}(\vartheta)^{-1}\right), \tag{3.117}$$

where

$$\mathrm{I}(\vartheta) = \frac{(1-\kappa)^\gamma}{\sigma^2(1+\kappa)}\,\vartheta^{\frac{2\kappa}{1-\kappa}}$$

Proof. The convergence (3.116) provides us the limit

$$\frac{1}{\sigma^2 T^\gamma}\int_0^T |X_t|^{2\kappa}\,\mathrm{d}t \to \mathrm{I}(\vartheta).$$

Now (3.117) follows immediately from the central limit theorem.

The lower bound in this problem is almost exactly the same as in the ergodic case, because the corresponding family of measures is LAN. Indeed, the likelihood ratio $Z_T(u) = L\left(\vartheta + T^{-\gamma/2}u, \vartheta, X^T\right)$ admits the representation

$$Z_T(u) = \exp\left\{u\,\Delta_T\left(\vartheta, X^T\right) - \frac{u^2}{2}\mathrm{I}(\vartheta) + r_T\right\},$$

where $r_T \to 0$ in probability and

$$\Delta_T\left(\vartheta, X^T\right) = \frac{1}{\sigma\,T^{\gamma/2}}\int_0^T |X_t|^\kappa\,\mathrm{d}W_t \Longrightarrow \mathcal{N}\left(0, \mathrm{I}(\vartheta)\right).$$

Hence we have the Hajek–Le Cam bound

$$\varliminf_{\delta\to 0}\ \varliminf_{T\to\infty}\ \inf_{\bar{\vartheta}_T}\ \sup_{|\vartheta-\vartheta_0|\le\delta} \mathbf{E}_\vartheta\,\ell\left(T^{\gamma/2}\left(\bar{\vartheta}_T - \vartheta\right)\right) \ge \mathbf{E}\,\ell\left(\zeta\,\mathrm{I}(\vartheta_0)^{-1/2}\right),$$

and it is easy to show that the MLE is asymptotically efficient at least for bounded loss functions $\ell(\cdot)$.

3.5.3 Exponential Growth Process

Here we present another example of a stochastic process with exact (exponential) rate in infinity.

Let us consider the problem of parameter estimation by the observations X^T of the Ornstein–Uhlenbeck process

$$\mathrm{d}X_t = \vartheta\,X_t\,\mathrm{d}t + \sigma\,\mathrm{d}W_t, \quad X_0, \quad 0 \le t \le T, \tag{3.118}$$

where the unknown parameter $\vartheta \in \mathscr{R}$. The MLE is

$$\hat{\vartheta}_T = \frac{\int_0^T X_t\,\mathrm{d}X_t}{\int_0^T X_t^2\,\mathrm{d}t}.$$

The asymptotic behavior of this estimator depends on the value of ϑ. Let us put

$$\eta_1 = \frac{w_1^2 - 1}{2\int_0^1 w_s^2\,\mathrm{d}s},$$

where $w_s, 0 \leq s \leq 1$ is a Wiener process and denote by η_2 a random variable (Cauchy) with the density function

$$f(x) = \frac{1}{\pi}\,\frac{1}{1+x^2}.$$

The properties of the MLE are described in the following proposition.

Proposition 3.46. *If $\vartheta < 0$, then*

$$\mathcal{L}_\vartheta\left\{\sqrt{T}\left(\hat{\vartheta}_T - \vartheta\right)\right\} \Longrightarrow \mathcal{N}\left(0, -2\vartheta\right), \tag{3.119}$$

if $\vartheta = 0$, then

$$\mathcal{L}_\vartheta\left\{T\left(\hat{\vartheta}_T - 0\right)\right\} \Longrightarrow \mathcal{L}\left\{\eta_1\right\} \tag{3.120}$$

and if $\vartheta > 0$, then

$$\mathcal{L}_\vartheta\left\{\sigma^{-1}\mathrm{e}^{\vartheta T}\left(\hat{\vartheta}_T - \vartheta\right)\right\} \Longrightarrow \mathcal{L}\left\{2\vartheta\frac{\zeta_1}{\tilde{X}_0 + \zeta_2}\right\}, \tag{3.121}$$

where ζ_1, ζ_2 are two independent Gaussian $\mathcal{N}(0,1)$ variables, $\tilde{X}_0 = X_0\sqrt{2\vartheta}/\sigma$. In particular, if $X_0 = 0$ then

$$\mathcal{L}_\vartheta\left\{\sigma^{-1}\mathrm{e}^{\vartheta T}\left(\hat{\vartheta}_T - \vartheta\right)\right\} \Longrightarrow \mathcal{L}\left\{\eta_2\right\}.$$

Proof. The first convergence (3.119) follows immediately, because the process (3.114) for these values of ϑ is ergodic.

If $\vartheta = 0$, then the observed process is $X_t = X_0 + \sigma\, W_t$ and we have (see Feigin [72])

$$\begin{aligned} T\left(\hat{\vartheta}_T - 0\right) &= T\,\frac{2\sigma\, X_0\, W_T + \sigma^2\left(W_T^2 - T\right)}{2\int_0^T \left(X_0 + \sigma\, W_t\right)^2 \mathrm{d}t} \\ &= \frac{2\sigma X_0\,\frac{W_T}{T} + \sigma^2\left(\tilde{w}_1^2 - 1\right)}{2\int_0^1\left(\frac{X_0}{\sqrt{T}} + \sigma\,\tilde{w}_s\right)^2 \mathrm{d}s} \Longrightarrow \eta_1, \end{aligned}$$

where we put $\tilde{w}_s = T^{-1/2}\, W_{sT},\ 0 \leq s \leq 1$.

If $X_0 = 0$, then

$$\mathcal{L}_\vartheta\left\{T\left(\hat{\vartheta}_T - 0\right)\right\} = \mathcal{L}\left\{\eta_1\right\}.$$

If $\vartheta > 0$ then the process $\{X_t,\ t \geq 0\}$ goes to $\pm$ infinity as $t \to \infty$. Indeed, the solution of (3.118) can be written as

$$X_t = X_0\, e^{\vartheta t} + \sigma\, e^{\vartheta t} \int_0^t e^{-\vartheta s}\, dW_s = (X_0 + Z_t)\, e^{\vartheta t},$$

where

$$Z_t = \sigma \int_0^t e^{-\vartheta s}\, dW_s \to Z \sim \mathcal{N}\left(0, \frac{\sigma^2}{2\vartheta}\right).$$

Hence

$$\lim_{t\to\infty} \frac{X_t}{e^{\vartheta t}} = X_0 + Z = X_0 + \frac{\sigma}{\sqrt{2\vartheta}}\, \zeta_2, \quad \zeta_2 \sim \mathcal{N}(0,1)$$

i.e., we have

$$\mathbf{P}_\vartheta\left\{\lim_{t\to\infty} X_t = +\infty\right\} = \mathbf{P}\left\{X_0 + \frac{\sigma}{\sqrt{2\vartheta}}\, \zeta_1 > 0\right\},$$

$$\mathbf{P}_\vartheta\left\{\lim_{t\to\infty} X_t = -\infty\right\} = \mathbf{P}\left\{X_0 + \frac{\sigma}{\sqrt{2\vartheta}}\, \zeta_1, < 0\right\}.$$

We have to study two integrals

$$\int_0^T (X_0 + Z_t)\, e^{\vartheta t}\, dW_t, \qquad \text{and} \qquad \int_0^T (X_0 + Z_t)^2\, e^{2\vartheta t}\, dt.$$

Introduce the stochastic process

$$Y_t = \int_0^t e^{\vartheta s} dW_s, \qquad t \geq 0.$$

We have by the Itô formula

$$e^{-\vartheta T} \int_0^T X_t\, dW_t = e^{-\vartheta T} \int_0^T (X_0 + Z_t)\, dY_t = (X_0 + Z_T)\, e^{-\vartheta T} Y_T$$
$$- \frac{T\sigma}{2}\, e^{-\vartheta T} - \sigma e^{-\vartheta T} \int_0^T e^{-\vartheta t} Y_t\, dW_t \Longrightarrow \frac{X_0 + Z}{\sqrt{2\vartheta}}\, \zeta_1, \qquad \zeta_1 \sim \mathcal{N}(0,1),$$

as $T \to \infty$ and with probability 1

$$e^{-2\vartheta T} \int_0^T X_t^2\, dt = \frac{(X_0 + Z_T)^2}{2\vartheta} - o(1) \longrightarrow \frac{(X_0 + Z)^2}{2\vartheta}.$$

Now the convergence (3.121) follows directly from these representations.

Recall that X_0 is independent of the Wiener process, hence it is independent of Z and ζ. Note as well that the random variables Z and ζ are asymptotically independent because they are jointly asymptotically normal and

$$\sqrt{2\vartheta} \lim_{t\to\infty} \mathbf{E}_\vartheta \left(Z_t\, e^{-\vartheta t}\, Y_t\right) = \mathbf{E}_\vartheta \left(Z\, \zeta\right) = 0.$$

Remark 3.47. It is easy to see that the family of measures $\left\{\mathbf{P}_\vartheta^{(n)}, \vartheta \in \mathscr{R}\right\}$ is

- LAN for $\vartheta < 0$, i.e.,

$$L\left(\vartheta + u/\sqrt{T}, \vartheta, X^T\right) = \exp\left\{u\,\zeta\,\sqrt{\frac{2\vartheta}{\sigma^2}} - \frac{u^2\vartheta}{\sigma^2} + r_T\right\}, \quad r_T \to 0,$$

- for $\vartheta = 0$ it is

$$L\left(\vartheta + u/T, \vartheta, X^T\right) = \exp\left\{\frac{u}{2}\left(w_1^2 - 1\right) - \frac{u^2}{2}\int_0^1 w_s^2 \mathrm{d}s + r_T\right\}, \quad r_T \to 0,$$

- and for $\vartheta < 0$ it is locally asymptotically mixing normal

$$L\left(\vartheta + \mathrm{e}^{-\vartheta T} u, \vartheta, X^T\right) = \exp\left\{u\,\mathrm{I}_0\left(\vartheta\right)^{1/2}\,\zeta - \frac{u^2}{2}\,\mathrm{I}_0\left(\vartheta\right) + r_T\right\}, \quad r_T \to 0,$$

  where the random variable

$$\mathrm{I}_0\left(\vartheta\right) = \frac{\left(X_0 + Z\right)^2}{2\vartheta\sigma^2}.$$

In the first case (ergodic) the lower bound on the risks of all estimators is already given in Section 2.1 and in the last situation $\vartheta > 0$ we can have the lower bound similar to (3.114). In particular for a bounded loss function $\ell\left(\cdot\right)$ it is

$$\varliminf_{M\to\infty} \varliminf_{T\to\infty} \inf_{\bar\vartheta_T} \sup_{|\vartheta-\vartheta_0|\le M\mathrm{e}^{-\vartheta_0 T}} \mathbf{E}_\vartheta\, \ell\left(\mathrm{e}^{\vartheta_0 T}\left(\bar\vartheta_T - \vartheta\right)\right) \ge \mathbf{E}\;\ell\left(\zeta\,\mathrm{I}\left(\vartheta_0\right)^{-1/2}\right). \tag{3.122}$$

The proof is quite close to that given above. Note that for any $M > 0$, $|u| \le M$ and $\vartheta_u = \vartheta_0 + \mathrm{e}^{-\vartheta_0 T}\,u$

$$\mathrm{e}^{(\vartheta_0-\vartheta_u)T} = \mathrm{e}^{uT\,\mathrm{e}^{-\vartheta_0 T}} \to 1$$

as $T \to \infty$. Therefore we have for $\vartheta = \vartheta_0 + \mathrm{e}^{-\vartheta_0 T}\,u$ and the MLE

$$\mathrm{e}^{\vartheta_0 T}\left(\hat\vartheta_T - \vartheta\right) = \mathrm{e}^{\vartheta T}\left(\hat\vartheta_T - \vartheta\right)\,\left(1 + o(1)\right).$$

Hence it is asymptotically efficient.

Remark 3.48. The model (3.118) can be easily generalized up to

$$\mathrm{d}X_t = \vartheta\,S\left(X_t\right)\,dt + \mathrm{d}W_t, \qquad X_0, \quad 0 \le t \le T, \tag{3.123}$$

where $\vartheta > 0$, the function $S\left(\cdot\right)$ admits the representation

$$S(x) = cx + r(x), \qquad \text{with} \quad |r(x)| \leq K\left(1 + |x|^{\gamma}\right),$$

and the constants $c > 0$, $K > 0$ and $\gamma \in [0, 1)$. It can be shown that the process (3.123) has the following asymptotics: with probability 1

$$\lim_{t \to \infty} e^{\vartheta\, ct} X_t = X_0 + Y + Z,$$

where the random variables X_0 and (Y, Z) are independent,

$$Y = \int_0^{\infty} e^{-c\vartheta t}\, dW_t, \qquad Z = \vartheta \int_0^{\infty} e^{-c\vartheta t}\, r(X_t)\, dt.$$

This convergence allows us to study the MLE $\hat{\vartheta}_T$ and to show that

$$\mathcal{L}_{\vartheta}\left\{e^{c\vartheta T}\left(\hat{\vartheta}_T - \vartheta\right)\right\} \Longrightarrow \mathcal{L}\left\{\frac{\zeta\,\sqrt{2\vartheta}}{X_0 + Y + Z}\right\}.$$

Here the random variable $\zeta \sim \mathcal{N}(0, 1)$ is independent of X_0, Y, Z. The proof can be found in our work (Dietz and Kutoyants [64]), where the properties of the TFE for this model are described as well.

4

Nonparametric Estimation

We consider the problems of invariant distribution function, density and trend coefficient estimation in the situations when the trend coefficient is an unknown function. In every problem we propose a lower minimax bound on the risk of all estimators and then construct asymptotically efficient estimators.

4.1 Distribution Function Estimation

We consider the problem of estimation of the invariant distribution function $F(\cdot)$ at a given point x. At first we introduce a lower (minimax) bound on the risk of all estimators and then we show that the risk of the empirical distribution function $\hat{F}_T(x)$ attains this bound. Hence this estimator is asymptotically efficient in the sense of the given bound. We recall as well the class of unbiased estimators of the distribution function introduced in Section 1.3 (Proposition 1.51) and show that these estimators can be asymptotically efficient too.

Let $X^T = \{X_t,\ 0 \le t \le T\}$ be a solution of the stochastic differential equation

$$\mathrm{d}X_t = S(X_t)\,\mathrm{d}t + \sigma(X_t)\,\mathrm{d}W_t, \quad X_0, \quad 0 \le t \le T, \tag{4.1}$$

where $\{W_t, 0 \le t \le T\}$ is the standard Wiener process. The trend coefficient $S(\cdot)$ is unknown to the observer and the diffusion coefficient $\sigma(\cdot)^2$ is a known continuous positive function. Recall that the diffusion coefficient can be estimated without error.

Remember that $\mathscr{S}_\sigma$ is the class of functions $S(\cdot)$ such that for the given function $\sigma(\cdot)$ the conditions $\mathcal{ES}$, $\mathcal{EM}$ and $\mathcal{RP}$ are fulfilled (see (1.142)). Of course, this is as well a condition on $\sigma(\cdot)$.

Remember that by the condition $\mathcal{RP}$

$$V(S,x) = \int_0^x \exp\left\{-2\int_0^y \frac{S(v)}{\sigma(v)^2}\,\mathrm{d}v\right\}\mathrm{d}y \to \pm\infty, \qquad \text{as} \quad x \to \pm\infty \tag{4.2}$$

and

$$G(S)=\int_{-\infty}^{\infty}\sigma(y)^{-2}\exp\left\{2\int_0^y\frac{S(v)}{\sigma(v)^2}\,\mathrm{d}v\right\}\,\mathrm{d}y<\infty. \qquad (4.3)$$

Hence the process $\{X_t, t\geq 0\}$ has ergodic properties with the invariant distribution function

$$F_S(x)=G(S)^{-1}\int_{-\infty}^{x}\sigma(y)^{-2}\exp\left\{2\int_0^y\frac{S(v)}{\sigma(v)^2}\,\mathrm{d}v\right\}\,\mathrm{d}y$$

and the corresponding density $f_S(x)$. We suppose that the initial value X_0 has a *stationary* distribution function $F_S(\cdot)$. Therefore the process $\{X_t,\ t\geq 0\}$ is stationary.

Further, by condition $\mathcal{EM}$ the measures $\left\{\mathbf{P}_S^{(T)},\ S(\cdot)\in\mathscr{S}_\sigma\right\}$ (induced in the measurable space $(\mathscr{C}_T,\ \mathfrak{B}_T)$ of continuous on $[0,T]$ functions) are equivalent and the likelihood ratio is given by (1.35) with the corresponding notation.

To estimate $F_S(x)$ we use, as in classical statistics, the *empirical distribution function* (EDF)

$$\hat{F}_T(x)=\frac{1}{T}\int_0^T\chi_{\{X_t<x\}}\,\mathrm{d}t. \qquad (4.4)$$

The uniform (in x) consistency and asymptotic normality of this estimator are established in Proposition 1.51 and Theorem 1.52. Below we show that as in classical statistics the EDF is an asymptotically efficient estimator.

Let us denote

$$\mathrm{I}_F(S,x)=\left(4\ \mathbf{E}_S\left(\frac{F_S(\xi\wedge x)\left[1-F_S(\xi\vee x)\right]}{\sigma(\xi)f_S(\xi)}\right)^2\right)^{-1},$$

where $\xi\wedge x=\min(\xi,x)$ and $\xi\vee x=\max(\xi,x)$. This quantity plays the role of Fisher information in our problem. Here and in the following ξ is a random variable with the distribution function $F_S(\cdot)$.

4.1.1 Lower Bound

Let us introduce the set of functions $\mathscr{S}_\sigma^*\subset\mathscr{S}_\sigma$ with the following property: for any $S_*(\cdot)\in\mathscr{S}_\sigma^*$ there exists a $\delta>0$-vicinity

$$V_\delta=\left\{S(\cdot)\ :\sup_{y\in\mathscr{R}}|S(y)-S_*(y)|\leq\delta,\quad S(\cdot)\in\mathscr{S}_\sigma^*\right\} \qquad (4.5)$$

such that

$$\sup_{S(\cdot)\in V_\delta}G(S)<\infty. \qquad (4.6)$$

This condition guarantee the existence of the unknown distribution function $F_S(\cdot)$ for all $S(\cdot)\in V_\delta$. If it is not fulfilled and for all $\delta>0$ this sup is equal to infinity, then the family of ergodic diffusion processes "touch" the null recurrent process and we have another type of statistical problem.

Remark 4.1. An example of such a set can be constructed as follows. Let us put $\sigma(\cdot) \equiv 1$ and define

$$\mathscr{S}_1^* = \left\{ S(\cdot) : \varlimsup_{|y|\to\infty} y\, S(y) < -\frac{1}{2} \right\}. \tag{4.7}$$

Then for any $S_*(\cdot) \in \mathscr{S}_1^*$ there exist numbers $\gamma > 1/2$ and $A > 0$ such that for all $|y| > A$ the estimate holds

$$y\, S_*(y) < -\gamma.$$

Hence we can take any $\delta < \gamma - 1/2$ and the condition (4.6) will be fulfilled.

The mathematical expectation with respect to the measure $\mathbf{P}_S^{(T)}$ is denoted by $\mathbf{E}_S$ and if $S(\cdot) = S_*(\cdot)$ we write $\mathbf{E}$.

Theorem 4.2. *Let $S_*(\cdot) \in \mathscr{S}_\sigma^*$ and $\mathrm{I}_F(S_*, x) > 0$. Then for any loss function $\ell(\cdot) \in \mathscr{W}_p$*

$$\lim_{\delta\to 0} \lim_{T\to\infty} \inf_{\bar F_T(x)} \sup_{S(\cdot)\in V_\delta} \mathbf{E}_S \ell(T^{1/2} \left(\bar F_T(x) - F_S(x)\right)) \geq \mathbf{E}\,\ell\left(\zeta\, \mathrm{I}_F(S_*, x)^{-1/2}\right), \tag{4.8}$$

where $\zeta \sim \mathcal{N}(0,1)$ and inf *is taken over all possible estimators $\bar F_T(x)$.*

Proof. The lower bound is constructed as follows. The supremum on V_δ is estimated from below by the supremum on some parametric subfamily with a special parameterization passing through the *central model* with $S(\cdot) = S_*(\cdot)$. For this parametric model we apply the inequality of Hajek–Le Cam, then maximize the right-hand side of the last inequality and find the *least favorable parametric family* (with minimal Fisher information). This last quantity (risk) is just the right-hand side of (4.8). We follow the approach initiated by Levit [170] concerning the asymptotically efficient estimation of the smooth functionals.

Let us introduce the parametric family of functions

$$S_\vartheta(y) = S_*(y) + (\vartheta - \vartheta_*)\, \psi(y)\, \sigma(y)^2,$$

where the value of ϑ_* will be chosen later,

$$\vartheta \in \Theta = (\vartheta_* - \gamma, \vartheta_* + \gamma), \quad \gamma > 0$$

and the function $\psi(\cdot)$ is continuous with compact support. Then for the functions $S_\vartheta(\cdot), \vartheta \in \Theta$ the conditions (4.2) and (4.3) are fulfilled too and for the small values of γ

$$S_*(y) + (\vartheta - \vartheta_*)\,\psi(y)\,\sigma(y)^2 \in V_\delta.$$

The corresponding family (parameterized by $\vartheta \in \Theta$) of stochastic differential equations is

$$\mathrm{d}X_t = [S_*(X_t) + (\vartheta - \vartheta_*)\psi(X_t)\,\sigma(X_t)^2]\,\mathrm{d}t + \sigma(X_t)\,\mathrm{d}W_t, \quad X_0, \quad 0 \le t \le T. \tag{4.9}$$

We denote the invariant distribution function and density as $F_\vartheta(\cdot)$ and $f_\vartheta(\cdot)$ and for $\vartheta = \vartheta_*$ we put $F_{\vartheta_*}(\cdot) = F(\cdot)$ and $f_{\vartheta_*}(\cdot) = f(\cdot)$ respectively. We can consider now the problem of estimation of the parameter ϑ by the observations (4.9).

In our assumptions the stochastic process (4.9) has ergodic properties at the point $\vartheta = \vartheta_*$ hence the family of measures $\{\mathbf{P}_\vartheta^{(T)}, \vartheta \in \Theta\}$ is *locally asymptotically normal* at the point $\vartheta = \vartheta_*$, i.e., the likelihood ratio

$$Z_T(u) \equiv \frac{\mathrm{d}\mathbf{P}^{(T)}_{\vartheta_*+T^{-1/2}u}}{\mathrm{d}\mathbf{P}^{(T)}_{\vartheta_*}}(X^{(T)}) = \frac{G(S_*)}{G\left(S_{\vartheta_*+T^{-1/2}u}\right)}\exp\left\{\frac{2u}{\sqrt{T}}\int_0^{X_0}\psi(v)\,\mathrm{d}v\right\}$$

$$\times\exp\left\{\frac{u}{\sqrt{T}}\int_0^T\psi(X_t)\,[\mathrm{d}X_t - S_*(X_t)\,\mathrm{d}t] - \frac{u^2}{2T}\int_0^T\psi(X_t)^2\,\sigma(X_t)^2\,\mathrm{d}t\right\}$$

admits the representation

$$Z_T(u) = \exp\left\{u\,\Delta_T - \frac{u^2}{2}\,\mathrm{I}_\psi + r_T(u, X^T)\right\}, \tag{4.10}$$

where $u \in \mathbb{U}_\gamma = (-\gamma T^{1/2}, \gamma T^{1/2})$,

$$\Delta_T = T^{-1/2}\int_0^T\psi(X_t)\,\sigma(X_t)\;\mathrm{d}W_t, \qquad \mathcal{L}_{\vartheta_*}\{\Delta_T\} \Longrightarrow \mathcal{N}(0, \mathrm{I}_\psi),$$

$$\mathrm{I}_\psi = \int_{-\infty}^{\infty}\psi(y)^2\,\sigma(y)^2\,f(y)\,\mathrm{d}y, \qquad \mathbf{P}_{\vartheta_*} - \lim_{T\to\infty} r_T(u, X^T) = 0.$$

Thus by Hajek–Le Cam's inequality (Theorem 2.4) for any $\ell(\cdot) \in \mathscr{W}_p$ we have

$$\varliminf_{\delta\to 0}\varliminf_{T\to\infty}\inf_{\bar\vartheta_T}\sup_{|\vartheta-\vartheta_*|<\delta}\mathbf{E}_\vartheta\,\ell\left(T^{1/2}(\bar\vartheta - \vartheta)\right) \ge \mathbf{E}\,\ell(\zeta\,\mathrm{I}_\psi^{-1/2}) \tag{4.11}$$

with $\mathcal{L}\{\zeta\} = \mathcal{N}(0,1)$.

We put

$$F_\vartheta(x) = \int_{-\infty}^{x}\frac{1}{G(S_\vartheta)\sigma(y)^2}\exp\left\{2\int_0^y\frac{S_*(v)}{\sigma(v)^2}\,\mathrm{d}v + 2(\vartheta - \vartheta_*)\int_0^y\psi(v)\mathrm{d}v\right\}\mathrm{d}y.$$

The function $\psi(\cdot)$ has a compact support, hence we can expand $F_\vartheta(\cdot)$ by the powers of $\vartheta - \vartheta_*$ at the vicinity of the point ϑ_* and obtain

$$
\begin{aligned}
F_\vartheta(x) &= F(x) + 2\,(\vartheta - \vartheta_*) \left\{ \int_{-\infty}^{x} \int_0^y \psi(v)\,\mathrm{d}v\, f(y)\,\mathrm{d}y \right. \\
&\quad \left. - F(x) \int_{-\infty}^{\infty} \int_0^y \psi(v)\,\mathrm{d}v\, f(y)\,\mathrm{d}y \right\} + o(\vartheta - \vartheta_*) \\
&= F(x) + 2\,(\vartheta - \vartheta_*)\, \mathbf{E} \left\{ [\chi_{\{\xi<x\}} - F(x)] \Psi(\xi) \right\} + o(\vartheta - \vartheta_*),
\end{aligned}
$$

where ξ as before has a distribution function $F(\cdot)$ and we denote

$$
\Psi(\xi) = \int_0^\xi \psi(y)\,\mathrm{d}y.
$$

Let us put $\vartheta_* = F(x)$ and introduce the class of functions $\psi(\cdot)$ (with compact supports) as

$$
\mathscr{K} = \left\{ \psi(\cdot) : \mathbf{E} \left\{ [\chi_{\{\xi<x\}} - F(x)] \Psi(\xi) \right\} = \frac{1}{2} \right\}.
$$

Then for $\psi(\cdot) \in \mathscr{K}$ we have the expansion

$$
\begin{aligned}
F_\vartheta(x) &= F(x) + 2\,(\vartheta - \vartheta_*)\, \mathbf{E} \left\{ [\chi_{\{\xi<x\}} - F(x)] \Psi(\xi) \right\} + o(\vartheta - \vartheta_*) \\
&= \vartheta + o(\vartheta - \vartheta_*).
\end{aligned}
$$

For the function $\psi(\cdot) \in \mathscr{K}$ the following inequalities hold:

$$
\begin{aligned}
&\sup_{S(\cdot) \in V_\delta} \mathbf{E}_S \ell \left(T^{1/2} (\bar{F}_T(x) - F_S(x)) \right) \geq \\
&\quad \geq \sup_{|\vartheta - \vartheta_*| < \gamma} \mathbf{E}_\vartheta \ell \left(T^{1/2} (\bar{F}_T(x) - F_\vartheta(x)) \right) \\
&\quad = \sup_{|\vartheta - \vartheta_*| < \gamma} \mathbf{E}_\vartheta \ell \left(T^{1/2} (\bar{\vartheta}_T - \vartheta + o(\vartheta - \vartheta_*)) \right),
\end{aligned}
$$

where we consider $\bar{F}_T(x)$ as an arbitrary estimator of ϑ and put $\bar{\vartheta}_T = \bar{F}_T(x)$. So we can choose $\gamma = \gamma(\delta) \to 0$ as $\delta \to 0$ in such a way that for any estimator $\bar{F}_T(x)$

$$
\begin{aligned}
&\underline{\lim}_{\delta \to 0}\, \underline{\lim}_{T \to \infty} \sup_{S(\cdot) \in V_\delta} \mathbf{E}_S \ell \left(T^{1/2} (\bar{F}_T(x) - F_S(x)) \right) \\
&\quad \geq \underline{\lim}_{\gamma \to 0}\, \underline{\lim}_{T \to \infty} \sup_{|\vartheta - \vartheta_*| < \gamma} \mathbf{E}_\vartheta \ell \left(T^{1/2} (\bar{\vartheta}_T - \vartheta + o(\vartheta - \vartheta_*)) \right) \\
&\quad \geq \frac{1}{\sqrt{2\pi}} \int_{-\infty}^{\infty} \ell \left(y\, \mathrm{I}_\psi^{-1/2} \right) \mathrm{e}^{-\frac{y^2}{2}}\, \mathrm{d}y.
\end{aligned}
$$

The last integral is a monotonically decreasing function of $\mathrm{I}_\psi = \mathbf{E}\, \psi(\xi)^2 \sigma(\xi)^2$, hence to find the least favorable parametric family (defined by $\psi(\cdot)$) we have to minimize I_ψ over the class $\mathscr{K}$.

The Fubini theorem allows us to write

$$
\begin{aligned}
&\mathbf{E}\left\{[\chi_{\{\xi<x\}} - F(x)]\Psi(\xi)\right\} \\
&\quad = \int_{-\infty}^{x} (1-F(x))\,\Psi(y)\,f(y)\,\mathrm{d}y - F(x)\int_{x}^{\infty}\Psi(y)\,f(y)\,\mathrm{d}y \\
&\quad = (F(x)-1)\int_{-\infty}^{0}\int_{-\infty}^{0}\chi_{\{v\geq y\}}\psi(v)f(y)\,\mathrm{d}v\,\mathrm{d}y \\
&\qquad +(1-F(x))\int_{0}^{x}\int_{0}^{x}\chi_{\{v<y\}}\psi(v)f(y)\,\mathrm{d}v\,\mathrm{d}y \\
&\qquad -F(x)(1-F(x))\Psi(x) - F(x)\int_{x}^{\infty}\int_{x}^{\infty}\chi_{\{v<y\}}\psi(v)f(y)\,\mathrm{d}v\,\mathrm{d}y \\
&\quad = (F(x)-1)\int_{-\infty}^{x}\psi(v)F(v)\,\mathrm{d}v - F(x)\int_{x}^{\infty}\psi(v)(1-F(v))\,\mathrm{d}v \\
&\quad = \int_{-\infty}^{\infty}\psi(v)\left([F(x)-1]F(v)\chi_{\{v<x\}} + [F(v)-1]F(x)\chi_{\{v\geq x\}}\right)\mathrm{d}v.
\end{aligned}
$$

The function $\psi(\cdot)$ belongs to the class $\mathscr{K}$. Hence by the Cauchy–Schwarz inequality

$$
\begin{aligned}
\frac{1}{4} &= \left(\mathbf{E}[\chi_{\{\xi<x\}} - F(x)]\Psi(\xi)\right)^2 \\
&= \left(\int_{-\infty}^{\infty}\psi(v)[F(v\vee x)-1]F(v\wedge x)\,\mathrm{d}v\right)^2 \\
&\leq \int_{-\infty}^{\infty}\psi(v)^2\,\sigma(v)^2\,f(v)\,\mathrm{d}v\int_{-\infty}^{\infty}\frac{([F(v\vee x)-1]F(v\wedge x))^2}{\sigma(v)^2\,f(v)}\,\mathrm{d}v \\
&= \mathbf{E}\psi(\xi)^2\sigma(\xi)^2\;\mathbf{E}\left(\frac{[F(\xi\vee x)-1]F(\xi\wedge x)}{\sigma(\xi)\,f(\xi)}\right)^2.
\end{aligned}
$$

Therefore

$$
\mathrm{I}_{\psi} \geq \left\{4\,\mathbf{E}\left(\frac{[1-F(\xi\vee x)]F(\xi\wedge x)}{\sigma(\xi)\,f(\xi)}\right)^2\right\}^{-1} = \mathrm{I}_F\left(S_*,x\right)
$$

for all $\psi(\cdot)\in\mathscr{K}$ and we obtain the equality for the function

$$
\psi_*(v) = \frac{2\,\mathrm{I}_F\left(S_*,x\right)}{\sigma(v)^2\,f(v)}\,[1-F(v\vee x)]\;F(v\wedge x).
$$

This function has no compact support but we can introduce a sequence of functions $\psi_*^n(\cdot)$, $n=1,2,\ldots$ having compact supports (with the help of a truncation $\psi_*(v)\,\chi_{\{|v|<n\}}$), and smoothed at the vicinities of points $v=n, v=-n$ in such a way that $\psi_*^n(\cdot)\in\mathscr{K}$ and $\psi_*^n(\cdot)$ approximates $\psi_*(\cdot)$ as $n\to\infty$ in the following sense:

$$\lim_{n\to\infty} \mathbf{E}\ \sigma(\xi)^2\ \left(\psi_*^n(\xi) - \psi_*(\xi)\right)^2 = 0.$$

Then we can write

$$\inf_{\psi(\cdot)\in\mathscr{K}} \mathrm{I}_\psi = \lim_{n\to\infty} \mathrm{I}_{\psi_*^n} = \mathrm{I}_F\left(S_*, x\right),$$

which provides the desired inequality (4.8).

The inequality (4.8) suggests we introduce the following definition of an asymptotically efficient estimator:

Definition 4.3. *Let the conditions of Theorem 4.2 be fulfilled, then we say that the estimator* $\bar{F}_T(x)$ *is asymptotically efficient for the loss function* $\ell(\cdot)$ *if for any* $x \in \mathscr{R}$ *and any* $S_*(\cdot) \in \mathscr{S}_\sigma^*$

$$\lim_{\delta\to 0}\lim_{T\to\infty}\sup_{S(\cdot)\in V_\delta} \mathbf{E}_S\, \ell\left(T^{1/2}(\bar{F}_T(x) - F_S(x))\right) = \mathbf{E}\, \ell(\zeta\, \mathrm{I}_F\left(S_*, x\right)^{-1/2}),$$

where $\mathcal{L}(\zeta) = \mathcal{N}(0,1)$.

Remark 4.4. Remember that it is possible to describe a set of functions satisfying the condition $\mathrm{I}_F(S,x) > 0$ as follows. Let us put $\sigma(\cdot) \equiv 1$ and define

$$\mathscr{S}_1^* = \left\{ S(\cdot)\ :\ \overline{\lim_{|y|\to\infty}}\ y\, S(y) < -\frac{3}{2} \right\}. \tag{4.12}$$

Then for any $S(\cdot) \in \mathscr{S}_1^*$ there exist numbers $\gamma > 3/2$ and $A > 0$ such that for all $|y| > A$ the estimate holds:

$$y\, S(y) < -\gamma.$$

Hence we can take any $\delta < \gamma - 3/2$ and for all $S(\cdot) \in \mathscr{S}_1^*$ we have

$$\inf_{S(\cdot)\in V_\delta} \mathrm{I}_F\left(S, x\right) > 0.$$

We show below that for the polynomial loss functions the EDF is asymptotically efficient.

4.1.2 EDF

We already have the asymptotic normality of the EDF with the limit variance equal to $\mathrm{I}_F(S,x)^{-1}$ (Proposition 1.51). Hence if this convergence is uniform over V_δ, then the estimator is asymptotically efficient for the bounded loss functions. To extend this efficiency to loss functions $\ell(\cdot) \in \mathscr{W}_p$ we have to strengthen the conditions.

$\mathcal{N}$. *There exists a number* $p_* \geq 2$ *such that for any* $S_*(\cdot) \in \mathscr{S}_\sigma^*$

$$\sup_{S(\cdot)\in V_\delta} \mathbf{E}_S \left\{ \left| \int_x^\xi \frac{F_S(v\wedge x)(1-F_S(v\vee x))}{\sigma(v)^2 f_S(v)} dv \right|^{p_*} + \left| \frac{F_S(\xi\wedge x)(1-F_S(\xi\vee x))}{\sigma(\xi) f_S(\xi)} \right|^{p_*} \right\} < \infty. \tag{4.13}$$

Moreover, the law of large numbers

$$\mathbf{P}_S - \lim_{T\to\infty} \frac{4}{T} \int_0^T \left(\frac{F_S(X_t\wedge x) - F_S(x)F_S(X_t)}{\sigma(X_t) f_S(X_t)} \right)^2 dt = \mathrm{I}_F(S,x)^{-1}$$

is uniform on $S(\cdot)\in V_\delta$.

Of course, this is an additional condition on the set of functions $\mathscr{S}_\sigma^*$. Remember, that the uniform on $S(\cdot)\in V_\delta$ LLN for a function

$$h_S(y,x) = 2\frac{F_S(y\wedge x) - F_S(x)F_S(y)]}{\sigma(y) f_S(y)}$$

is undestood as follows: for any $\varepsilon > 0$

$$\sup_{S(\cdot)\in V_\delta} \mathbf{P}_S^{(T)} \left\{ \left| \frac{4}{T} \int_0^T h_S(X-t,x)^2\, dt - \mathrm{I}_F(S,x)^{-1} \right| > \varepsilon \right\} \to 0.$$

as $T\to\infty$.

Remark 4.5. Let $\sigma(\cdot)\equiv 1$, then it is easy to see that the expectation of the first term in condition (4.13) is finite for the functions $S(\cdot)$ satisfying the condition $\overline{\lim}_{|y|\to\infty}\, y\, S(y) < -p^* - \frac{1}{2}$ and the expectation of the second term is finite for the functions $S(\cdot)$ satisfying the condition

$$\overline{\lim_{|y|\to\infty}}\; y\, S(y) < -\frac{p^*+1}{2}.$$

Hence we can take

$$\mathscr{S}_1^* = \left\{ S(\cdot) \;:\; \overline{\lim_{|y|\to\infty}}\; y\, S(y) < -p^* - \frac{1}{2} \right\}$$

and for any $S_*(\cdot)\in\mathscr{S}_1^*$ there exist numbers $\gamma > p^* + \frac{1}{2}$ and $A > 0$ such that for all $|y| > A$ the estimate holds:

$$y\, S_*(y) < -\gamma.$$

Therefore we can take any $\delta < \gamma - p^* - \frac{1}{2}$ and for all $S_*(\cdot)\in\mathscr{S}_1^*$ we have (4.13).

Theorem 4.6. *Let the conditions $\mathscr{N}$ be fulfilled and $\mathrm{I}_F(S,x)$ be continuous at the point $S(\cdot) = S_*(\cdot)$ in the uniform metric. Then the EDF $\hat{F}_T(x)$ is uniformly consistent in $S(\cdot) \in V_\delta$, asymptotically normal and asymptotically efficient for the loss functions $\ell(\cdot) \in \mathscr{W}_p$ with $p < p_*$.*

Proof. According to Proposition 1.51 the normalized difference

$$\eta_T(x) = \sqrt{T}\left(\hat{F}_T(x) - F_S(x)\right) = \frac{1}{\sqrt{T}} \int_0^T \left[\chi_{\{X_t<x\}} - F_S(x)\right] \mathrm{d}t$$

is asymptotically normal with the limit variance $I_F(S,x)^{-1}$.

Moreover this convergence is uniform. Indeed, we have the representation

$$\sqrt{T}\left(\hat{F}_T(x) - F_S(x)\right) = \frac{H_x(S,X_T) - H_x(S,X_0)}{\sqrt{T}} - \frac{2}{\sqrt{T}} \int_0^T \frac{F_S(X_t \wedge x) - F_S(X_t)\, F(x)}{\sigma(X_t)\, f_S(X_t)}\, \mathrm{d}W_t,$$

where

$$H_x(S,y) = 2\int_0^y \frac{F_S(v \wedge x) - F_S(v)\, F(x)}{\sigma(v)^2\, f_S(v)}\, \mathrm{d}v.$$

By the uniform central limit theorem (Proposition 1.20) and condition $\mathscr{N}$ the stochastic integral

$$\frac{2}{\sqrt{T}} \int_0^T \frac{F_S(X_t \wedge x) - F_S(X_t) F_S(x)}{\sigma(X_t)\, f_S(X_t)}\, \mathrm{d}W_t$$

is uniformly asymptotically normal with the parameters $\left(0, \mathrm{I}_F(S,x)^{-1}\right)$. Further, by the same condition $\mathscr{N}$ for any $\varepsilon > 0$

$$\sup_{S(\cdot)\in V_\delta} \mathbf{P}_S^{(T)}\left\{|H_x(S,\xi)| > \varepsilon\sqrt{T}\right\} \leq \frac{1}{\left(\varepsilon\sqrt{T}\right)^{p_*}} \sup_{S(\cdot)\in V_\delta} \mathbf{E}_S\, |H_x(S,\xi)|^{p_*} \to 0$$

as $T \to \infty$. Therefore the random variables $\eta_T(x)$ are uniformly asymptotically normal and we have to check the uniform integrability of the variables

$$\ell_T(x) = \ell\left(\sqrt{T}\left(\hat{F}_T(x) - F_S(x)\right)\right).$$

Elementary inequality gives us the estimates: for any $p < p_*$:

$$\sup_{T,S(\cdot)\in V_\delta} \mathbf{E}_S \left|\ell_T(x)\right|^{p_*/p} \leq C_1 + C_2 \sup_{T,S(\cdot)\in V_\delta} \mathbf{E}_S \left|\eta_T(x)\right|^{p_*}$$

$$\leq C_1 + C_2 T^{-p_*/2} \sup_{S(\cdot)\in V_\delta} \mathbf{E}_S \left|H_x(S, X_T) - H_x(S, X_0)\right|^{p_*}$$

$$+C_3\, T^{-1} \sup_{S(\cdot)\in V_\delta} \mathbf{E}_S \int_0^T \left(\frac{[F_S(X_t\wedge x) - F_S(X_t)F_H(x)]}{\sigma(X_t)\, f_S(X_t)}\right)^{p_*} \mathrm{d}t$$

$$\leq C_1 + C_2 T^{-p_*/2}\, 2^{p_*} \sup_{S(\cdot)\in V_\delta} \mathbf{E}_S \left|H_x(S,\xi)\right|^{p_*}$$

$$+C_3 \sup_{S(\cdot)\in V_\delta} \mathbf{E}_S \left(\frac{[F_S(\xi\wedge x) - F_S(\xi)F_S(x)]}{\sigma(\xi)\, f_S(\xi)}\right)^{p_*} \leq C. \qquad (4.14)$$

Hence

$$\sup_{S(\cdot)\in V_\delta} \mathbf{E}_S \ell\left(\sqrt{T}(\hat{F}_T(x) - F_S(x)\right) \longrightarrow \sup_{S(\cdot)\in V_\delta} \mathbf{E}\ell\left(\zeta\, \mathrm{I}_F(S,x)^{-1/2}\right) \qquad (4.15)$$

and the asymptotic efficiency of the EDF now follows from the continuity of $\mathrm{I}_F(S,x)$ at the point $S(\cdot) = S_*(\cdot)$. We have

$$\lim_{\delta\to 0} \sup_{S(\cdot)\in V_\delta} \mathbf{E}\,\ell\left(\zeta\, \mathrm{I}_F(S,x)^{-1/2}\right) = \mathbf{E}\,\ell\left(\zeta\, \mathrm{I}_F(S_*,x)^{-1/2}\right),$$

where $\mathcal{L}\{\zeta\} = \mathcal{N}(0,1)$.

4.1.3 Example

Suppose (for simplicity) that $\sigma(\cdot) \equiv 1$. Then the observed diffusion process is

$$\mathrm{d}X_t = S(X_t)\,\mathrm{d}t + \,\mathrm{d}W_t, \qquad X_0, \quad 0 \leq t \leq T, \qquad (4.16)$$

where $S(\cdot)$ is an unknown function.

Let us define the class $\mathscr{S}_0 \subset \mathscr{S}_1^*$ of locally integrable functions $S(\cdot)$ satisfying the relation

$$\overline{\lim_{|y|\to\infty}}\ \mathrm{sgn}\,(y)\ S\,(y) < 0. \qquad (4.17)$$

It is easy to see that for $S(\cdot) \in \mathscr{S}_0$ Equation (4.16) has a unique weak solution and the process $\{X_t, t \geq 0\}$ has ergodic properties.

Proposition 4.7. *Let $S(\cdot) \in \mathscr{S}_0$, then the EDF $\hat{F}_T(x)$ is asymptotically efficient for the loss functions $\ell(\cdot) \in \mathscr{W}_p$ with any $p \geq 2$.*

Proof. We check the condition $\boldsymbol{\mathcal{N}}$ and then this proposition will follow from Theorem 4.6. We use the same arguments as in checking (1.35).

It is easy to see that for $S_*(\cdot) \in \mathscr{S}_0$ there exist positive constants A and γ such that for all $|y| > A$ the inequality holds

$$\operatorname{sgn}(y)\ S_*(y) \le -\gamma. \tag{4.18}$$

Hence, for all $S(\cdot) \in V_\delta$ with $\delta < \gamma$ and all $y > A$

$$\exp\left\{2\int_0^y S(v)\,\mathrm{d}v\right\} \le \exp\left\{2\int_0^A |S(v|)\,\mathrm{d}v - 2(\gamma-\delta)(y-A)\right\}$$
$$\le C\,\mathrm{e}^{-2(\gamma-\delta)y}.$$

We have the similar estimate for $y < -A$. Therefore

$$\sup_{S(\cdot)\in V_\delta} G(S) \le \sup_{S(\cdot)\in V_\delta} \int_{-A}^{A} \exp\left\{2\int_{-A}^{A} |S(v)|\ \mathrm{d}v\right\}\,\mathrm{d}y$$
$$+\,2C\int_A^\infty \exp\{-2(\gamma-\delta)\,|y|\}\,\mathrm{d}y \le C.$$

Hence condition (4.6) is fulfilled. As the function $S_*(\cdot)$ is locally integrable, we have the second estimate as well:

$$\inf_{S(\cdot)\in V_\delta} G(S) \ge \kappa > 0$$

with some $\kappa > 0$.

Further, we note that for $y \ge A$ we have for any $S(\cdot) \in V_\delta$ the estimate

$$\frac{1-F_S(y)}{f_S(y)} = \int_y^\infty \mathrm{e}^{2\int_y^v S(u)\ \mathrm{d}u}\mathrm{d}v \le \int_y^\infty \mathrm{e}^{-2(\gamma-\delta)(v-y)}\mathrm{d}v = \frac{1}{2(\gamma-\delta)}$$

and similarly for $y \le -A$

$$\frac{F_S(y)}{f_S(y)} = \int_{-\infty}^y \mathrm{e}^{2\int_y^v S(u)\ \mathrm{d}u}\mathrm{d}v \le \int_{-\infty}^y \mathrm{e}^{2(\gamma-\delta)(v-y)}\mathrm{d}v = \frac{1}{2(\gamma-\delta)}.$$

Hence we can write

$$\sup_{S(\cdot)\in V_\delta} \mathbf{E}_S\left|\int_x^\xi \frac{F_S(v\wedge x)(1-F_S(v\vee x))}{f_S(v)}\mathrm{d}v\right|^{p_*} \le \sup_{S(\cdot)\in V_\delta}\left|\int_{-A}^{A}\frac{\mathrm{d}v}{f_S(v)}\right|^{p_*}$$
$$+\sup_{S(\cdot)\in V_\delta} \mathbf{E}_S\chi_{\{\xi\ge A\}}\left|\int_A^\xi \frac{(1-F_S(v))}{f_S(v)}\mathrm{d}v\right|^{p_*}$$
$$+\sup_{S(\cdot)\in V_\delta} \mathbf{E}_S\chi_{\{\xi\le -A\}}\left|\int_\xi^{-A} \frac{F_S(v)}{f_S(v)}\mathrm{d}v\right|^{p_*}$$
$$\le C + \frac{1}{(2(\gamma-\delta))^{p_*}}\sup_{S(\cdot)\in V_\delta} \mathbf{E}_S\chi_{\{|\xi|\ge A\}}\,|\xi - A|^{p_*}$$
$$+\frac{1}{(2(\gamma-\delta))^{p_*}}\sup_{S(\cdot)\in V_\delta} \mathbf{E}_S\chi_{\{|\xi|\le -A\}}\,|\xi + A|^{p_*}.$$

The mathematical expectation can be easily estimated as follows:

$$\sup_{S(\cdot)\in V_\delta} \mathbf{E}_S \chi_{\{|\xi|\ge A\}} |\xi|^{p_*} \le \frac{C}{\kappa} \int_A^\infty y^{p_*} \,\mathrm{e}^{-2(\gamma-\delta)(y-A)}\, \mathrm{d}y < \infty. \tag{4.19}$$

The same arguments can be used to prove the estimate

$$\sup_{S(\cdot)\in V_\delta} \mathbf{E}_S \left| \frac{F_S(\xi\wedge x)(1-F_S(\xi\vee x))}{f_S(\xi)} \right|^{p_*} < \infty.$$

The continuity of $\mathrm{I}_F(S,x)$ in a uniform metric can be checked as follows. For any $\varepsilon > 0$ we can choose such a number $A_\varepsilon \ge A$ that

$$\int_{-\infty}^{-A_\varepsilon} \frac{F_S(y)^2}{f_S(y)}\, \mathrm{d}y + \int_{A_\varepsilon}^{\infty} \frac{(1-F_S(y))^2}{f_S(y)}\, \mathrm{d}y < \frac{\varepsilon}{3}$$

for all $S(\cdot)\in V_\delta$. Then we take such a $\delta = \delta(\varepsilon)$ that

$$\sup_{S(\cdot)\in V_\delta} \int_{-A_\varepsilon}^{A_\varepsilon} \left| \frac{F_S(y\wedge x)^2(1-F_S(y\vee x))^2}{f_S(y)} \right.$$
$$\left. - \frac{F_{S_*}(y\wedge x)^2(1-F_{S_*}(y\vee x))^2}{f_{S_*}(y)} \right| \mathrm{d}y \le \frac{\varepsilon}{3}.$$

Therefore we have the estimate

$$\sup_{S(\cdot)\in V_\delta} \left| \mathrm{I}_F(S,x)^{-1} - \mathrm{I}_F(S_*,x)^{-1} \right| \le \varepsilon$$

which shows the continuity of the function $\mathrm{I}_F(S,x)$ at the point $S(\cdot) = S_*(\cdot)$.

It remains to check the uniform law of large numbers. Let us denote

$$m_S(y,x) = 4\left(\frac{F_S(y\wedge x) - F_S(x)F_S(y)}{f_S(y)}\right)^2 - \mathrm{I}_F(S,x)^{-1},$$

and

$$M_S(y,x) = \int_x^y \frac{2}{f_S(z)} \int_{-\infty}^{z} m_S(v,x)\, f_S(v)\, \mathrm{d}v\, \mathrm{d}z.$$

Note that $\mathbf{E}_S\, m_S(\xi,x) = 0$. Using the Itô formula we write

$$\frac{4}{T}\int_0^T \left(\frac{F_S(X_t\wedge x) - F_S(x)F_S(X_t)}{\sigma(X_t) f_S(X_t)}\right)^2 \mathrm{d}t - \mathrm{I}_F(S,x)^{-1}$$
$$= \frac{M_S(X_T,x) - M_S(X_0,x)}{T}$$
$$-\frac{2}{T}\int_0^T \frac{1}{f_S(X_t)} \int_{-\infty}^{X_t} m_S(v,x)\, f_S(v)\, \mathrm{d}v\, \mathrm{d}W_t.$$

Therefore for any $\varepsilon > 0$ by Tchebychev inequality we have

$$\sup_{S(\cdot)\in V_\delta} \mathbf{P}_S^{(T)} \left\{ \left| \frac{1}{T} \int_0^T m_S(X_t, x)\, \mathrm{d}t \right| > \varepsilon \right\}$$

$$\leq \frac{2}{\varepsilon^2 T^2} \sup_{S(\cdot)\in V_\delta} \mathbf{E}_S \left(M_S(X_T, x) - M_S(X_0, x) \right)^2$$

$$+ \frac{8}{\varepsilon^2 T} \sup_{S(\cdot)\in V_\delta} \mathbf{E}_S \left(\frac{1}{f_S(\xi)} \int_{-\infty}^{\xi} m_S(v, x)\, f_S(v)\, \mathrm{d}v \right)^2 .$$

Now the uniform law of large numbers follows from the finiteness of these mathematical expectations, which can be proved with the help of arguments like those given above.

4.1.4 Other Metrics

It is possible to show the asymptotic efficiency of the EDF for the other metrics too. The case of a $\mathscr{L}_2$ risk function was studied by Kutoyants and Negri [149]. It is shown that under mild regularity conditions the following equality holds:

$$\lim_{\delta\to 0} \lim_{T\to\infty} \inf_{\bar F_T} \sup_{S(\cdot)\in V_\delta} T\, \mathbf{E}_S \int_{\mathscr{R}} \left| \bar F_T(x) - F_S(x) \right|^2 \mu(\mathrm{d}x)$$

$$= \int_{\mathscr{R}} \mathrm{I}_F(S_*, x)^{-1}\, \mu(\mathrm{d}x). \qquad (4.20)$$

Note that the EDF is asymptotically efficient in this sense too (here $\mu(\cdot)$ is some finite measure). In Section 4.3.4 we consider a slightly more general problem of function estimation with $\mathscr{L}_2$ type risk and obtain (4.20) as a particular case.

Similar problems (lower bound and asymptotic efficiency of the EDF) for the uniform metric were considered by Negri [185], [186]. In particular, it is shown that for the loss functions $\ell(\cdot) \in \mathscr{W}_{e,2}$ we have the inequality

$$\underline{\lim}_{\delta\to 0}\, \underline{\lim}_{T\to\infty} \inf_{\bar F_T} \sup_{S(\cdot)\in V_\delta} \mathbf{E}_S\, \ell\left(\sup_x \sqrt{T} \left| \bar F_T(x) - F_S(x) \right| \right) \geq \mathbf{E}_{S_*}\, \ell\left(\sup_x |\eta(x)| \right),$$

where $\eta(\cdot)$ is a Gaussian process with the covariance function

$$R_{S_*}(x, y) = 4\, \mathbf{E}_{S_*} \left(\frac{F_{S_*}(\xi\wedge x) F_{S_*}(\xi\wedge y) \left[1 - F_{S_*}(\xi\vee x)\right] \left[1 - F_{S_*}(\xi\vee y)\right]}{\sigma(\xi)^2 f_{S_*}(\xi)^2} \right)$$

and for bounded loss functions and $S(\cdot) \in \mathscr{S}_0$ the EDF is asymptotically efficient in this sense too.

4.2 Density Estimation

Now we consider the problem of stationary density $f_S(x)$ estimation by the observations $X^T = \{X_t,\ 0 \le t \le T\}$ of the same diffusion process

$$\mathrm{d}X_t = S(X_t)\,\mathrm{d}t + \sigma(X_t)\,\mathrm{d}W_t, \quad X_0, \quad t \ge 0 \tag{4.21}$$

where the functions $S(\cdot)$ and $\sigma(\cdot)$ are such that the conditions $\mathcal{ES}$, $\mathcal{EM}$ and $\mathcal{RP}$ are fulfilled, i.e., the unknown trend coefficient $S(\cdot) \in \mathscr{S}_\sigma$ and the diffusion coefficient $\sigma(\cdot)^2$ is a known, continuous positive function.

The condition $\mathcal{RP}$ provides the existence of the invariant density function

$$f_S(x) = G(S)^{-1}\,\sigma(x)^{-2}\exp\left\{2\int_0^x \frac{S(y)}{\sigma(y)^2}\,\mathrm{d}y\right\}$$

and we consider the problem of its estimation. As usual, we suppose for simplicity of exposition that the initial value X_0 has the *invariant density* $f_S(\cdot)$ and therefore the process $\{X_t,\, t \ge 0\}$ is stationary. This problem is quite close to the problem of distribution function estimation because both functions $F_S(x)$ and $f_S(x)$ can be represented as mathematical expectations of some random variables

$$F_S(x) = \mathbf{E}_S\,\chi_{\{\xi<x\}}, \qquad f_S(x) = \mathbf{E}_S\left(\frac{2\,S(\xi)}{\sigma(x)^2}\,\chi_{\{\xi<x\}}\right).$$

In Section 4.3 we consider a more general statement, where the problems of distribution and density function estimation are just two particular cases.

We begin with the problem of density estimation at a given point x and propose the lower bound on the risks of all estimators, which is similar to (4.8). Then we verify that the class of asymptotically efficient (in the sense of this bound) estimators is quite large (local time estimators, class of unbiased estimators, kernel-type estimators). Special attention is paid to the local time estimator (LTE), because any other asymptotically efficient estimator can be represented as a sum of the LTE and an asymptotically vanishing term. It is shown that the normalized LTE converges in distribution to the Gaussian process on the whole line. Having too many asymptotically efficient estimators, we consider a later (Section 4.6) the problem of second order asymptotically efficient estimation and show that under regularity conditions a special kernel-type estimator is asymptotically efficient in the sense of this bound too.

4.2.1 Lower Bound

We construct the lower bound with the help of the same approach as in Section 4.1. In particular, the nonparametric vicinity

$$V_\delta = \left\{S(\cdot):\ \sup_{y\in\mathscr{R}} |S(y) - S_*(y)| \le \delta, \quad S(\cdot) \in \mathscr{S}_\sigma^*\right\}$$

is defined with the help of the same set $\mathscr{S}_\sigma^*$ of functions which guarantee (4.6). Let us denote

$$\mathrm{I}_f(S,x) = \left\{ 4\, f_S(x)^2\, \mathbf{E}_S \left(\frac{\chi_{\{\xi > x\}} - F_S(\xi)}{\sigma(\xi) f_S(\xi)} \right)^2 \right\}^{-1}.$$

This quantity will play the role of Fisher information in our problem.

Theorem 4.8. *Let $S_*(\cdot) \in \mathscr{S}_\sigma^*$ and $\mathrm{I}_f(S,x) > 0$. Then for any loss function $\ell(\cdot) \in \mathscr{W}_p$,*

$$\lim_{\delta \to 0} \varliminf_{T \to \infty} \inf_{\bar f_T} \sup_{S(\cdot) \in V_\delta} \mathbf{E}_S\, \ell\left(T^{1/2} (\bar f_T(x) - f_S(x)) \right)$$
$$\geq \mathbf{E}\, \ell\left(\zeta\, \mathrm{I}_f(S_*, x)^{-1/2} \right), \tag{4.22}$$

where $\zeta \sim \mathscr{N}(0,1)$ and inf *is taken over all possible estimators $\bar f_T(x)$.*

Proof. We follow exactly the same way as in the preceding section and the difference is just in the choice of the parameterization.

As before we introduce the parametric family of functions

$$S_\vartheta(x) = S_*(x) + (\vartheta - \vartheta_*)\, \psi(x) \sigma(x)^2$$

where $\vartheta \subset \Theta = (\vartheta_* - \gamma, \vartheta_* + \gamma)$, $\gamma > 0$ and the function $\psi(\cdot)$ is continuous and has a compact support. The value $\gamma = \gamma(\delta) > 0$ is sufficiently small to provide $S_*(\cdot) + (\vartheta - \vartheta_*)\psi(\cdot)\sigma(\cdot)^2 \in V_\delta$ for all $\vartheta \in \Theta$.

The corresponding family of stochastic differential equations will be given by Equation (4.9).

Consider now the problem of the parameter ϑ estimation by the observations (4.9). Set

$$f_\vartheta(x) = G(\vartheta)^{-1} \sigma(x)^{-2} \exp\left\{ 2 \int_0^x \frac{S_*(y)}{\sigma(y)^2}\, \mathrm{d}y + 2(\vartheta - \vartheta_*) \int_0^x \psi(y)\, \mathrm{d}y \right\}.$$

The function $\psi(\cdot)$ has compact support, hence we can expand $f_\vartheta(\cdot)$ by the powers of $\vartheta - \vartheta_*$ at the vicinity of ϑ_* and obtain

$$f_\vartheta(x) = f_{S_*}(x) + 2\,(\vartheta - \vartheta_*) f_{S_*}(x)\; \mathbf{E}_{S_*} \Psi(\xi) + o(\vartheta - \vartheta_*),$$

where ξ has the distribution function $F_{S_*}(\cdot)$ and

$$\Psi(y) = \int_y^x \psi(v)\, \mathrm{d}v.$$

Set $\vartheta_* = f_{S_*}(x)$ and introduce the following class of functions:

$$\mathscr{K} = \left\{ \psi(\cdot) :\ \mathbf{E}_{S_*} \Psi(\xi) = (2\, f_{S_*}(x))^{-1} \right\}.$$

Then for $\psi(\cdot) \in \mathscr{K}$ we have the expansion

$$f_\vartheta(x) = \vartheta + o(\vartheta - \vartheta_*)$$

and the estimates

$$\begin{aligned}\sup_{S(\cdot)\in V_\delta} \mathbf{E}_S\, \ell\left(T^{1/2}(\bar f_T(x) - f_S(x))\right) &\\ \geq \sup_{|\vartheta-\vartheta_*|<\gamma} \mathbf{E}_\vartheta\, \ell\left(T^{1/2}(\bar f_T(x) - f_\vartheta(x))\right) &\\ = \sup_{|\vartheta-\vartheta_*|<\gamma} \mathbf{E}_\vartheta\, \ell\left(T^{1/2}(\bar\vartheta_T - \vartheta + o(\vartheta - \vartheta_*))\right), &\end{aligned}$$

where $\bar\vartheta_T = \bar f_T(x)$ is an arbitrary estimator of ϑ.

Repeating the same arguments as in Section 4.1 we obtain the Hajek–Le Cam inequality

$$\varliminf_{\delta\to 0} \varliminf_{T\to\infty} \inf_{\bar\vartheta_T} \sup_{|\vartheta-\vartheta_*|<\delta} \mathbf{E}_\vartheta\, \ell\left(T^{1/2}(\bar\vartheta_T - \vartheta)\right) \geq \mathbf{E}\,\ell(\zeta\, \mathrm{I}_\psi^{-1/2}). \tag{4.23}$$

So we can choose $\gamma = \gamma(\delta) \to 0$ as $\delta \to 0$ in such a way that for any estimator $\bar f_T(x)$ of $f_S(x)$

$$\varliminf_{\delta\to 0} \varliminf_{T\to\infty} \sup_{S(\cdot)\in V_\delta} \mathbf{E}_S\, \ell\left(T^{1/2}(\bar f_T(x) - f_S(x))\right) \geq \mathbf{E}\,\ell\left(\zeta\, \mathrm{I}_\psi^{-1/2}\right).$$

To maximize the right hand side on the class $\mathscr{K}$ we use once more Fubini's theorem (below $f(\cdot) = f_{S_*}(\cdot)$ and $F(\cdot) = F_{S_*}(\cdot)$)

$$\begin{aligned}\mathbf{E}\,\Psi(\xi) &= \int_{-\infty}^{\infty} f(y) \int_y^x \psi(v)\,\mathrm{d}v\,\mathrm{d}y\\ &= \int_{-\infty}^{x}\int_y^x f(y)\,\psi(v)\ \mathrm{d}y\,\mathrm{d}v + \int_x^{\infty}\int_y^x f(y)\,\psi(v)\,\mathrm{d}v\,\mathrm{d}y\\ &= \int_{-\infty}^{x}\int_{-\infty}^{x} f(y)\,\psi(v)\,\chi_{\{v>y\}}\,\mathrm{d}y\,\mathrm{d}v\\ &\qquad - \int_x^{\infty}\int_x^{\infty} f(y)\,\psi(v)\,\chi_{\{v\leq y\}}\,\mathrm{d}v\,\mathrm{d}y\\ &= \int_{-\infty}^{x} \psi(v)\,F(v)\,\mathrm{d}v - \int_x^{\infty} \psi(v)\,(1 - F(v))\,\mathrm{d}v\\ &= \int_{-\infty}^{\infty} \psi(v)[F(v) - \chi_{\{v>x\}}]\,\mathrm{d}v.\end{aligned}$$

Thus by the Cauchy–Schwarz inequality

$$(2\,f(x))^{-2} = (\mathbf{E}\Psi(\xi))^2 = \left(\int_{-\infty}^{\infty} \psi(v)[F(v) - \chi_{\{v>x\}}]\,\mathrm{d}v\right)^2$$
$$\leq \int_{-\infty}^{\infty} \psi(v)^2\,\sigma(v)^2\,f(v)\,\mathrm{d}v \int_{-\infty}^{\infty} \frac{\left(F(v) - \chi_{\{v>x\}}\right)^2}{\sigma(v)^2\,f(v)}\,\mathrm{d}v$$
$$= \mathbf{E}\,\psi(\xi)^2\sigma(\xi)^2\ \mathbf{E}\left(\frac{F(\xi) - \chi_{\{\xi>x\}}}{\sigma(\xi)\,f(\xi)}\right)^2.$$

Therefore

$$\mathrm{I}_\psi \geq \left\{4\,f(x)^2\ \mathbf{E}\left(\frac{\chi_{\{\xi>x\}} - F(\xi)}{\sigma(\xi)\,f(\xi)}\right)^2\right\}^{-1} = \mathrm{I}_f(S_*, x)$$

for all $\psi(\cdot) \in \mathscr{K}$.

We have equality in the Cauchy–Schwarz inequality if we put

$$\psi_*(v) = C(x)\,\sigma(v)^{-2}\,f(v)^{-1}\left[\chi_{\{v>x\}} - F(v)\right]$$

with the constant

$$C(x) = 2\,f(x)\,\mathrm{I}_f(S, x).$$

This function has a discontinuity at the point $v = x$ and its support is not compact, hence it does not belong to the class $\mathscr{K}$. Therefore we consider the sequence of problems with the functions $\psi_n(\cdot)$ defined by the equalities

$$m_n\,\psi_n(v) = \begin{cases} 0, & |v| \geq n+1, \\ \psi_*(-n)(v+n+1), & -n-1 < v < -n, \\ \psi_*(n)(v-n-1), & n < v < n+1, \\ \psi_*(v), & -n \leq v < x - n^{-1}, x + n^{-1} < v \leq n, \end{cases} \tag{4.24}$$

and

$$\psi_n(v) = \psi_*(x - n^{-1})$$
$$+2n^{-1}\left(\psi_*(x+n^{-1}) + 2n^{-1} - \psi_*(x-n^{-1}) + 2n^{-1}\right)(v - x + n^{-1}),$$

for $x - n^{-1} \leq v \leq x + n^{-1}$. Here m_n is such that

$$\mathbf{E}\int_\xi^x \psi_n(y)\,\mathrm{d}y = \frac{1}{2\,f(x)}.$$

Then $\psi_n(\cdot) \in \mathscr{K}$ for all n and from the convergence

$$\lim_{n\to\infty} \mathbf{E}\,\sigma(\xi)^2\left(\psi_n(\xi) - \psi_*(\xi)\right)^2 = 0$$

we have $I_{\psi_n} \geq \mathrm{I}_f(S, x)$ and

$$\lim_{n\to\infty} \mathrm{I}_{\psi_n} = \mathrm{I}_f(S_*, x).$$

Therefore we obtain the required estimate (4.22).

Definition 4.9. *Let the conditions of Theorem 4.8 be fulfilled. Then we say that the estimator $\bar{f}_T(x)$ is asymptotically efficient for the loss function $\ell(\cdot)$ if*

$$\lim_{\delta\to 0}\lim_{T\to\infty}\sup_{S(\cdot)\in V_\delta}\mathbf{E}_S\,\ell\left(T^{1/2}(f_T(x)-f_S(x))\right)=\mathbf{E}\,\ell\left(\zeta\,\mathrm{I}_f(S_*,x)^{-1/2}\right). \quad (4.25)$$

Such estimators are also called *locally asymptotically minimax*. Below we construct several asymptotically efficient estimators.

4.2.2 Local-Time Estimator

Remember that the *local time estimator* (LTE) (or *empirical density function*) (derivative of empirical distribution function) has the form

$$f_T^{\circ}(x)=\frac{1}{\sigma(x)^2\,T}\int_0^T \mathrm{sgn}\,(x-X_t)\;\mathrm{d}X_t+\frac{|X_T-x|-|X_0-x|}{\sigma(x)^2\,T} \quad (4.26)$$

and is consistent and asymptotically normal (Proposition 1.57). Its asymptotic variance $\mathrm{I}_f(S,x)^{-1}$ coincides with the right hand side of (4.22) for $\ell(u)=u^2$. Therefore to show its optimality we have to strengthen the regularity conditions to provide the uniform convergence of moments.

Let us introduce the following
regularity conditions $\mathcal{O}$.

$\mathcal{O}_1$. *The function $S(\cdot)\in\mathscr{S}_\sigma^*$ and for some $p_*\geq 2$*

$$\sup_{S(\cdot)\in V_\delta}\mathbf{E}_S\left\{\left|\frac{\chi_{\{\xi>x\}}-F_S(\xi)}{\sigma(\xi)\,f_S(\xi)}\right|^{p_*}+\left|\int_0^{\xi}\frac{\chi_{\{v>x\}}-F_S(v)}{\sigma(v)^2 f_S(v)}\mathrm{d}v\right|^{p_*}\right\}<\infty.$$

$\mathcal{O}_2$. *The law of large numbers*

$$\mathbf{P}_S-\lim_{T\to\infty}\frac{4\,f_S(x)^2}{T}\int_0^T\left(\frac{\chi_{\{X_t>x\}}-F_S(X_t)}{\sigma(X_t)\,f_S(X_t)}\right)^2 dt=\mathrm{I}_f(S,x)^{-1} \quad (4.27)$$

is uniform on $S(\cdot)\in V_\delta$ for some $\delta>0$.

Remark 4.10. Remember that if $\sigma(\cdot)\equiv 1$ and $S_*(\cdot)\in\mathscr{S}_1^*$ where

$$\mathscr{S}_1^*=\left\{S(\cdot)\;:\;\overline{\lim_{|y|\to\infty}}\;y\,S(y)<-p^*-\frac{1}{2}\right\},$$

then there exist numbers $\gamma>p^*+\frac{1}{2}$ and $A>0$ such that for all $|y|>A$ the estimate $y\,S_*(y)<-\gamma$ holds. Therefore for any $\delta<\gamma-p^*-\frac{1}{2}$ the condition $\mathcal{O}_1$ is fulfilled.

Theorem 4.11. *Let the condition* $\mathcal{O}$ *be fulfilled and* $\mathrm{I}_f(S,x)$ *be a continuous function of* $S(\cdot)$ *with respect to the uniform convergence on* V_δ*. Then the estimator* $f_T^\circ(x)$ *is uniformly in* $S(\cdot) \in V_\delta$ *consistent, asymptotically normal*

$$\mathcal{L}_S\left\{T^{1/2}\left(f_T^\circ(x) - f_S(x)\right)\right\} \Longrightarrow \mathcal{N}\left(0, \mathrm{I}_f(S,x)^{-1}\right) \tag{4.28}$$

and asymptotically efficient for the loss functions $\ell(\cdot) \in \mathscr{W}_p$ *with* $p < p_*$.

Proof. According to Propositions 1.11 and 1.57 (see (1.32)) we have the following representation for the estimator:

$$\begin{aligned}\sqrt{T}\left(f_T^\circ(x) - f_S(x)\right) =& 2\,\frac{f_S(x)}{\sqrt{T}}\int_{X_0}^{X_T}\left(\frac{\chi_{\{v>x\}} - F_S(v)}{\sigma(v)^2\, f_S(v)}\right)\mathrm{d}v\\ &- 2\,\frac{f_S(x)}{\sqrt{T}}\int_0^T\left(\frac{\chi_{\{X_t>x\}} - F_S(X_t)}{\sigma(X_t)\, f_S(X_t)}\right)\mathrm{d}W_t.\end{aligned} \tag{4.29}$$

Using calculus similar to (4.14), (4.15) and the condition $\mathcal{O}_1$ we obtain the uniform integrability of the random variables

$$\left|\eta_T(x)\right|^p = \left|T^{1/2}\left(f_T^\circ(x) - f_S(x)\right)\right|^p,$$

which provides us (together with the continuity of the Fisher information) the desired asymptotical efficiency of the LTE for the loss functions $\ell(\cdot) \in \mathscr{W}_p$.

Example

We consider the same model as in Section 4.1.3. Therefore we have the diffusion process

$$\mathrm{d}X_t = S(X_t)\,\mathrm{d}t + \mathrm{d}W_t, \qquad X_0, \quad 0 \le t \le T,$$

where the unknown function $S(\cdot) \in \mathscr{S}_0$.

Proposition 4.12. *Let* $S(\cdot) \in \mathscr{S}_0$*. Then the LTE* $f_T^\circ(x)$ *is asymptotically efficient for the loss functions* $\ell(\cdot) \in \mathcal{W}_p$ *with any* $p \ge 2$.

Proof. We have to check the conditions of Theorems 4.8 and 4.11. Using the estimate (4.18) we obtain immediately for all $S(\cdot) \in V_\delta$

$$\int_A^\infty\left(\frac{1 - F_S(y)}{f_S(y)}\right)^p f_S(y)\,\mathrm{d}y \le \frac{1 - F_S(A)}{2^p\,(\gamma-\delta)^p} < \frac{1}{2^p\,(\gamma-\delta)^p}$$

and

$$\int_{-\infty}^{-A}\left(\frac{F_S(y)}{f_S(y)}\right)^p f_S(y)\,\mathrm{d}y \le \frac{F_S(-A)}{2^p\,(\gamma-\delta)^p} < \frac{1}{2^p\,(\gamma-\delta)^p}.$$

Hence

$$\sup_{S(\cdot)\in V_\delta} \mathbf{E}_S \left|\frac{\chi_{\{\xi>x\}} - F_S(\xi)}{f_S(\xi)}\right|^{p_*} \leq C_1(\gamma) < \infty.$$

Further

$$\int_A^\infty \left(\int_0^A + \int_A^y \left(\frac{1-F_S(v)}{f_S(v)}\right) \mathrm{d}v\right)^p f_S(y)\,\mathrm{d}y$$
$$\leq 2^{p-1}\left(\int_0^A \left(\frac{1-F_S(v)}{f_S(v)}\right) \mathrm{d}v\right)^p (1-F_S(A))$$
$$+2^{p-1}\int_A^\infty \left(\frac{y-A}{2(\gamma-\delta)}\right)^p f_S(y)\,\mathrm{d}y \leq C_2(\gamma)$$

and the condition

$$\sup_{S(\cdot)\in V_\delta} \mathbf{E}_S \left|\int_0^\xi \frac{\chi_{\{v>x\}} - F_S(v)}{f_S(v)} \mathrm{d}v\right|^{p_*} < \infty$$

is fulfilled as well.

The continuity of $\mathrm{I}_f(S,x)$ and the uniform law of large numbers can be checked in the same way as was done in the example of Section 4.1.3.

Weak Convergence of the LTE

The next step is to prove for this model the weak convergence of the corresponding stochastic processes $\eta_T(x) = \sqrt{T}\left(f_T^\circ(x) - f(x)\right),\ x \in \mathscr{R}$. Remember that in the i.i.d. case the similar limit is a collection of independent random variables and such weak convergence can not be established.

The random variables $\eta_T(x),\ x \in \mathscr{R}$, by Proposition 1.57 are asymptotically normal with parameters $\left(0, \mathrm{I}_f(S,x)\right),\ x \in \mathscr{R}$. Let us denote by $\eta(\cdot) = \{\eta(x),\ x \in \mathscr{R}\}$ the Gaussian process with mean 0 and the covariance function

$$R_S(x,y) = 4 f_S(x)\, f_S(y)\ \mathbf{E}_S \left(\frac{[\chi_{\{\xi>x\}} - F_S(\xi)][\chi_{\{\xi>y\}} - F_S(\xi)]}{\sigma(\xi)^2 f_S(\xi)^2}\right).$$

Introduce the space $\mathscr{C}_0 = \mathscr{C}_0(\mathscr{R})$ of continuous functions

$$z(\cdot) = \{z(x),\ x \in \mathscr{R}\}$$

tending to zero at infinity, i.e., $\lim_{|x|\to\infty} z(x) = 0$. We define the distance between two functions $z_1(\cdot),\ z_2(\cdot) \in \mathscr{C}_0(\mathscr{R})$ as

$$\rho(z_1, z_2) = \sup_{x\in\mathscr{R}} |z_1(x) - z_2(x)|\,.$$

We denote the corresponding Borel σ-algebra as $\mathfrak{B}_0$. Let $\Phi(z)$ be a continuous functional defined on $z(\cdot) \in \mathscr{C}_0(\mathscr{R})$. For example,

$$\Phi(z) = \sup_x |z(x)|, \qquad \Phi(z) = \inf_x z(x), \qquad \Phi(z) = \int_{\mathscr{R}} h(x)\, z(x)\, \mathrm{d}x$$

(for some classes of functions $h(\cdot)$) or any other. We would like to show that

$$\Phi\left(\eta_T\right) \Longrightarrow \Phi\left(\eta\right). \tag{4.30}$$

If we denote by $\mathbf{Q}_S^{(T)}$ and $\mathbf{Q}_S$ the measures induced by the processes $\eta_T(\cdot)$ and $\eta(\cdot)$ in the measurable space $(\mathscr{C}_0, \mathfrak{B}_0)$ respectively, then the convergence (4.30) will be equivalent to the weak convergence

$$\mathbf{Q}_S^{(T)} \Longrightarrow \mathbf{Q}_S. \tag{4.31}$$

We suppose for simplicity that $\sigma(y) \equiv 1$, i.e., the observed process is

$$\mathrm{d}X_t = S\left(X_t\right)\, \mathrm{d}t + \mathrm{d}W_t, \qquad X_0, \qquad 0 \le t \le T$$

and the unknown function $S(\cdot)$ is such that this equation has a unique weak solution.

Theorem 4.13. *Let $S(\cdot) \in \mathscr{S}_\sigma$ and $S(\cdot) \in \mathscr{P}$. Then the stochastic process $\eta_T(\cdot)$ converges weakly to $\eta(\cdot)$.*

Proof. To verify (4.30) we show the convergence of marginal distributions of $\eta_T(\cdot)$ and check the tightness of this family of measures. We use Theorem A21 in [109]. As the local time $\Lambda_T(\cdot)$ is continuous with a probability 1 process, the stochastic process $\eta_T(\cdot)$ is continuous with probability 1 too. Note that the function $\eta_T(x)$, $x \in \mathscr{R}$ is random on the random interval $\left[X_m^{(T)}, X_M^{(T)}\right]$ and is equal to $-\sqrt{T} f_S(x)$ outside of this interval. Here $X_m^{(T)}$ and $X_M^{(T)}$ are the minimal and the maximal values of the process $\{X_t,\ 0 \le t \le T\}$. The convergence of finite-dimensional distributions follows directly from the representation (4.29)

$$\begin{aligned}\eta_T(x_i) &= \frac{2 f_S(x_i)}{\sqrt{T}} \int_{X_0}^{X_T} \left(\frac{\chi_{\{v > x_i\}} - F_S(v)}{f_S(v)}\right) \mathrm{d}v \\ &\quad -2\, \frac{f_S(x_i)}{\sqrt{T}} \int_0^T \left(\frac{\chi_{\{X_t > x_i\}} - F_S(X_t)}{f_S(X_t)}\right) \mathrm{d}W_t, \qquad i = 1, \ldots, k\end{aligned}$$

and from the central limit theorem because for any $\boldsymbol{x} = \{x_1, \ldots, x_k\}$ and $\boldsymbol{\lambda} = \{\lambda_1, \ldots, \lambda_k\}$ we have

$$\begin{aligned}\sum_{i=1}^{k}\lambda_i\frac{f_S(x_i)}{\sqrt{T}}&\int_0^T\left(\frac{\chi_{\{X_t>x_i\}}-F_S(X_t)}{f_S(X_t)}\right)\mathrm{d}W_t\\&=\frac{1}{\sqrt{T}}\int_0^T\sum_{i=1}^{k}\lambda_i\,f_S(x_i)\left(\frac{\chi_{\{X_t>x_i\}}-F_S(X_t)}{f_S(X_t)}\right)\mathrm{d}W_t\\&=\frac{1}{\sqrt{T}}\int_0^T h_S(\boldsymbol{\lambda},\boldsymbol{x},X_t)\,\mathrm{d}W_t,\end{aligned}$$

where $h_S(\boldsymbol{\lambda},\boldsymbol{x},X_t)$ is a square-integrable function. Therefore

$$\mathcal{L}_S\left\{\sum_{i=1}^{k}\lambda_i\,\eta_T(x_i)\right\}\Longrightarrow\mathcal{N}\left(0,R_S\right)=\mathcal{L}_S\left\{\sum_{i=1}^{k}\lambda_i\,\eta(x_i)\right\},$$

where

$$R_S=\mathbf{E}_S\;h_S(\boldsymbol{\lambda},\boldsymbol{x},\xi)^2=\sum_{i,j=1}^{k}\lambda_i\,\lambda_j\;R_S(x_i,x_j).$$

Further, by Proposition 1.11 (see (1.35))

$$\mathbf{E}_S\,\eta_T\left(x\right)^2\le C\,\mathrm{e}^{-\gamma|x|}. \tag{4.32}$$

We write the process $\eta_T(x)$ as the sum

$$\eta_T(x)=Y_T(x)+Z_T(x),\qquad Y_T(x)=\frac{2f_S(x)}{\sqrt{T}}\int_{X_0}^{X_T}\left(\frac{\chi_{\{v>x\}}-F_S(v)}{f_S(v)}\right)\mathrm{d}v.$$

For the process $Y_T(\cdot)$ we have the estimate

$$\begin{aligned}|Y_T(x)|\le\frac{2f_S(x)}{\sqrt{T}}&\left|\int_0^{X_T}\frac{1-F_S(v)}{f_S(v)}\,\mathrm{d}v\right|+\frac{2f_S(x)}{\sqrt{T}}\left|\int_0^{X_T}\frac{F_S(v)}{f_S(v)}\,\mathrm{d}v\right|\\&+\frac{2f_S(x)}{\sqrt{T}}\left|\int_0^{X_0}\frac{1-F_S(v)}{f_S(v)}\,\mathrm{d}v\right|+\frac{2f_S(x)}{\sqrt{T}}\left|\int_0^{X_0}\frac{F_S(v)}{f_S(v)}\,\mathrm{d}v\right|.\end{aligned}$$

Note as well that the density $f_S(\cdot)$ is a bounded function. Hence for any $\varepsilon>0$

$$\begin{aligned}\mathbf{P}_S^{(T)}\left\{\sup_x|Y_T(x)|>\varepsilon\right\}&\le\mathbf{P}_S^{(T)}\left\{\left|\int_0^{\xi}\frac{1-F_S(v)}{f_S(v)}\mathrm{d}v\right|>\frac{\varepsilon\sqrt{T}}{C}\right\}\\&\quad+\mathbf{P}_S^{(T)}\left\{\left|\int_0^{\xi}\frac{F_S(v)}{f_S(v)}\mathrm{d}v\right|>\frac{\varepsilon\sqrt{T}}{C}\right\}\to0\end{aligned}$$

as $T\to\infty$. Therefore it is sufficient to study the process

$$Z_T(x)=\frac{2f_S(x)}{\sqrt{T}}\int_0^T\frac{\chi_{\{X_t>x\}}-F_S\left(X_t\right)}{f_S(X_t)}\,\mathrm{d}W_t,\qquad x\in\mathscr{R}.$$

Hence if we show that

$$\mathbf{E}_S \left(Z_T(x) - Z_T(y)\right)^4 \leq C \ \left(1+y^2\right)|x-y|^2, \tag{4.33}$$

then we can apply Theorem A21 in [109], which provides (4.31).

Let $y > x$ and put

$$\Psi(x,v) = f_S(x)\left[\chi_{\{v>x\}} - F_S(v)\right] f_S(v)^{-1}.$$

Then we can write

$$\begin{aligned}
\mathbf{E}_S \left(Z_T(x) - Z_T(y)\right)^4 &\leq C\ \mathbf{E}_S \left(\int_{\mathscr{R}} \left[\Psi(x,v) - \Psi(y,v)\right]^2 f_T^\circ(v)\ \mathrm{d}v\right)^2 \\
&\leq 2\,\mathbf{E}_S \left(\int_{\mathscr{R}} \left[\Psi(x,v) - \Psi(y,v)\right]^2 \left(f_T^\circ(v) - f_S(v)\right)\ \mathrm{d}v\right)^2 \\
&\quad +2\left(\int_{\mathscr{R}} \left[\Psi(x,v) - \Psi(y,v)\right]^2 f_S(v)\ \mathrm{d}v\right)^2,
\end{aligned}$$

where $f_T^\circ(v)$ is the empirical density (LTE).

Let us write

$$\int_{\mathscr{R}} \left[\Psi(x,v) - \Psi(y,v)\right]^2 f_S(v)\ \mathrm{d}v = I^y + I_y^x + I_x,$$

and

$$\int_{\mathscr{R}} \left[\Psi(x,v) - \Psi(y,v)\right]^2 \eta_T(v)\ \mathrm{d}v = J^y + J_y^x + J_x,$$

where

$$I^y = \left(f_S(x) - f_S(y)\right)^2 \int_{-\infty}^{y} \frac{F_S(v)^2}{f_S(v)}\ \mathrm{d}v,$$

$$I_y^x \leq 2 f_S(x)^2 \int_y^x \frac{F_S(v)^2}{f_S(v)}\ \mathrm{d}v + 2 f_S(y)^2 \int_y^x \frac{\left[1-F_S(v)\right]^2}{f_S(v)}\ \mathrm{d}v,$$

$$I_x = \left(f_S(x) - f_S(y)\right)^2 \int_x^{\infty} \frac{\left[1-F_S(v)\right]^2}{f_S(v)}\ \mathrm{d}v,$$

$$J^y = \left(f_S(x) - f_S(y)\right)^2 \int_{-\infty}^{y} \frac{F_S(v)^2}{f_S(v)^2}\ \eta_T(v)\ \mathrm{d}v,$$

$$J_y^x \leq 2 f_S(x)^2 \int_y^x \frac{F_S(v)^2}{f_S(v)^2}\ \eta_T(v)\ \mathrm{d}v + 2 f_S(y)^2 \int_y^x \frac{\left[1-F_S(v)\right]^2}{f_S(v)^2}\ \eta_T(v)\ \mathrm{d}v,$$

$$J_x = \left(f_S(x) - f_S(y)\right)^2 \int_x^{\infty} \frac{\left[1-F_S(v)\right]^2}{f_S(v)^2}\ \eta_T(v)\ \mathrm{d}v.$$

Further, note that the function $S(v)\, f_S(v), v \in \mathscr{R}$ by condition $S(\cdot)\,\mathscr{S}_0$ is bounded, hence

$$\left(f_S(x) - f_S(y)\right)^2 = 4 \left(\int_y^x S(v)\, f_S(v)\, \mathrm{d}v \right)^2 \leq C\, (x-y)^2 .$$

Therefore if x, y belongs to a bounded set, then

$$I^y + I_x \leq C\,(x-y)^2, \qquad I_y^x \leq C\,(x-y)\,. \tag{4.34}$$

Hence we have to study these quantities as $x, y \to \pm\infty$. Let $A < y < x \to \infty$ and $x - y \leq 1$. Then for any $v > y$ we have $S\,(v) \leq -\gamma$ (see (4.18)) and

$$\begin{aligned}\left|f_S\,(y) - f_S\,(x)\right| &\leq 2 \int_y^x |S(v)|\; f_S(v)\, \mathrm{d}v = 2 \int_y^x |S(v)|\; \frac{f_S(v)}{f_S(y)}\, \mathrm{d}v\, f_S(y)\\ &\leq 2 \int_y^x |S(v)|\; \exp\left\{-2\gamma\,(v-y)\right\} \mathrm{d}v\; f_S(y)\\ &\leq C\, \left(|x-y| + |x-y|^p\right)\, f_S(y) \leq C\, \left(1 + |y|^{p-1}\right)\, |x-y|\; f_S(y)\end{aligned}$$

because the function $S\,(\cdot) \in \mathscr{P}$.

Further

$$\begin{aligned}f_S\,(y)^2 \int_{-\infty}^y \frac{F_S(v)^2}{f_S(v)}\, \mathrm{d}v &\leq f_S\,(y)^2 \int_{-\infty}^A \frac{F_S(v)^2}{f_S(v)}\, \mathrm{d}v + f_S\,(y) \int_A^y \frac{f_S(y)}{f_S(v)}\, \mathrm{d}v\\ &\leq C\, f_S\,(y)^2 + f_S\,(y) \int_A^y \mathrm{e}^{-2\gamma(y-v)}\, \mathrm{d}v \leq C\, f_S(y).\end{aligned}$$

Hence

$$I^y \leq C\, |x-y|^2$$

with some constant $C > 0$. Similar estimates allow us to establish

$$I_x \leq C\, |x-y|^2, \qquad I_x^y \leq C\, |x-y|\,.$$

These estimates together with (4.34) provide us the inequality

$$\int_{\mathscr{R}} \left[\Psi\,(x,v) - \Psi\,(y,v)\right]^2 f_S\,(v)\, \mathrm{d}v \leq C\, |x-y| \tag{4.35}$$

which is valid for all $x, y \in \mathscr{R}$.

Now we estimate $\mathbf{E}_S\left(J^y + J_y^x + J_x\right)^2$. We have

$$\begin{aligned}&\mathbf{E}_S\,(J^y)^2\\ &= \left(f_S\,(x) - f_S\,(y)\right)^4 \mathbf{E}_S \left(\int_{-\infty}^y \frac{F_S(v)^2}{f_S(v)^2} \sqrt{1+v^2}\,\eta_T\,(v)\, \frac{\mathrm{d}v}{\sqrt{1+v^2}} \right)^2\\ &\leq C\,(1+|y|^p)\,|x-y|^4\; f_S\,(y)^4 \int_{-\infty}^y \left(\frac{F_S(v)}{f_S(v)}\right)^4 (1+v^2)\,\mathbf{E}_S\,\eta_T\,(v)^2\, \mathrm{d}v\\ &\leq C\,(1+|y|^p)\,|x-y|^4\; f_S\,(y)^4 \left(\int_{-\infty}^A \left(\frac{F_S(v)}{f_S(v)}\right)^4 (1+v^2)\,\mathbf{E}_S\,\eta_T\,(v)^2\, \mathrm{d}v \right.\\ &\qquad \left. + \int_A^y \frac{(1+v^2)}{f_S(v)^4}\; \mathbf{E}_S\,\eta_T\,(v)^2\, \mathrm{d}v \right).\end{aligned}$$

In the last integral we apply the elementary inequality

$$\mathbf{E}_S\, \eta_T(v)^2 \le 2\,\mathbf{E}_S\, Y_T(v)^2 + 2\,\mathbf{E}_S\, Z_T(v)^2$$

and write

$$\begin{aligned}
&f_S(y)^4 \int_A^y \frac{(1+v^2)}{f_S(v)^4}\, \mathbf{E}_S\, Z_T(v)^2\, \mathrm{d}v \\
&\quad = 4 f_S(y)^4 \int_A^y \frac{(1+v^2)}{f_S(v)^2}\, \mathbf{E}_S \left(\frac{\chi_{\{\xi>v\}} - F_S(\xi)}{f_S(\xi)}\right)^2 \mathrm{d}v \\
&\quad = 4 f_S(y)^4 \int_A^y \frac{(1+v^2)}{f_S(v)^2} \left(\int_{-\infty}^v \frac{F_S(u)^2}{f_S(u)}\mathrm{d}u + \int_v^\infty \frac{[1-F_S(u)]^2}{f_S(u)}\mathrm{d}u\right).
\end{aligned} \tag{4.36}$$

Remember that

$$\begin{aligned}
\int_{-\infty}^v \frac{F_S(u)^2}{f_S(u)}\mathrm{d}u &= \int_{-\infty}^{-A} \frac{F_S(u)^2}{f_S(u)}\mathrm{d}u + \int_{-A}^{A} \frac{F_S(u)^2}{f_S(u)}\mathrm{d}u + \int_A^v \frac{F_S(u)^2}{f_S(u)}\mathrm{d}u \\
&\le \int_{-\infty}^{-A} \left(\int_{-\infty}^u \frac{f_S(z)}{f_S(u)}\right)^2 f_S(u)\ \mathrm{d}u + C + \int_A^v \frac{1}{f_S(u)}\mathrm{d}u \\
&\le C + \int_A^v \frac{1}{f_S(u)}\mathrm{d}u
\end{aligned}$$

because $f_S(z)/f_S(u) \le \exp\{-2\gamma(u-z)\}$ and

$$\int_v^\infty \frac{[1-F_S(u)]^2}{f_S(u)}\mathrm{d}u = \int_v^\infty \left(\int_u^\infty \frac{f_S(z)}{f_S(u)}\mathrm{d}z\right)^2 f_S(u)\ \mathrm{d}u \le C\,\mathrm{e}^{-2\gamma v}.$$

Hence the main contribution in (4.36) is due to the term

$$\begin{aligned}
&f_S(y)^4 \int_A^y \frac{(1+v^2)}{f_S(v)^2} \int_A^v \frac{1}{f_S(u)}\mathrm{d}u\,\mathrm{d}v \\
&\quad = f_S(y) \int_A^y (1+v^2) \left(\frac{f_S(y)}{f_S(v)}\right)^3 \int_A^v \frac{f_S(v)}{f_S(u)}\mathrm{d}u\,\mathrm{d}v \\
&\quad \le f_S(y)\,\mathbf{E}_S\,(J^y)^2 \int_A^y (1+v^2)\,\mathrm{e}^{-6\gamma(y-v)} \int_A^v \mathrm{e}^{-2\gamma(v-u)}\mathrm{d}u\,\mathrm{d}v \\
&\quad \le C\left(1+y^2\right)\, f_S(y).
\end{aligned}$$

The other terms in $\mathbf{E}_S\,(J^y)^2$ as well as $\mathbf{E}_S\,(J_x)^2$ can be estimated in a similar way and this gives us the inequalities

$$\mathbf{E}_S\,(J^y)^2 + \mathbf{E}_S\,(J_x)^2 \le C\ (1+y^p)\ |x-y|^4\ f_S(y). \tag{4.37}$$

Further

$$\mathbf{E}_S \left(J_x^y\right)^2 \leq 32 f_S(x)^2 (x-y) \int_y^x \left(\frac{f_S(x)}{f_S(v)}\right)^2 \mathbf{E}_S \left(\frac{\chi_{\{\xi>v\}} - F_S(\xi)}{f_S(\xi)}\right)^2 dv$$
$$+32 f_S(x)^4 (x-y) \int_y^x \frac{\left[1-F_S(v)\right]^4}{f_S(v)^2} \mathbf{E}_S \left(\frac{\chi_{\{\xi>v\}} - F_S(\xi)}{f_S(\xi)}\right)^2 dv.$$

These integrals can be estimated exactly as was done above and this provides us the estimate

$$\mathbf{E}_S \left(J_x^y\right)^2 \leq C\,(x-y)^2. \tag{4.38}$$

Now the required relation (4.33) follows from (4.34)–(4.38) and the elementary inequalities. The convergence of marginal distribution together with the estimate (4.32) allows us to cite Theorem A21 in [109].

As a corollary of Theorem 4.13 we obtain the uniform consistency of the LTE: for any $\varepsilon > 0$

$$\lim_{T\to\infty} \mathbf{P}_S^{(T)} \left\{ \sup_x \left| f_T^\circ(x) - f_S(x) \right| > \varepsilon \right\} = 0. \tag{4.39}$$

The proof follows directly from the weak convergence of the process $\eta_T(\cdot)$. Note, that a similar consistency of the LTE was established by van Zanten under less restrictive conditions [231] and Negri proved the asymptotic efficiency of the LTE in the case of loss function with the distance defined by the uniform metric [187].

Remark 4.14. This weak convergence of LTE is equivalent to the weak convergence of the local time process

$$\sqrt{T}\left(\frac{2\Lambda_T(\cdot)}{T} - f_S(\cdot)\right) \Longrightarrow \eta(\cdot) \tag{4.40}$$

which can be interesting itself. Moreover, if the diffusion coefficient $\sigma(x)^2 \neq 1$, then we suppose that the function $\sigma(\cdot), \sigma(\cdot)^{-1} \in \mathscr{P}$, is continuously differentiable with derivative $\sigma'(\cdot) \in \mathscr{P}$. In this case the obvious modification of the given proof will give us the weak convergence of $\eta_T(\cdot)$.

4.2.3 Other Estimators

Unbiased Estimators

The class of unbiased estimators given in Section 1.3 (see (1.164)) was introduced with the help of the functions

$$R_x(y) = \frac{2\,\chi_{\{y\le x\}}\,h(y)}{\sigma(x)^2\,h(x)}, \qquad N_x(y) = \frac{\chi_{\{y\le x\}}\,h'(y)\,\sigma(y)^2}{\sigma(x)^2\,h(x)}. \tag{4.41}$$

as follows:

$$\tilde f_T(x) = \frac{1}{T}\int_0^T R_x(X_t)\,\mathrm{d}X_t + \frac{1}{T}\int_0^T N_x(X_t)\,\mathrm{d}t.$$

Here $h(\cdot)$ is an arbitrary function from $\mathscr{C}^1(\mathscr{R})$ with $h(x)\neq 0$ at the point x.

Remember that these estimators are unbiased, because

$$\mathbf{E}_S\left[R_x(\xi)\,S(\xi) + N_x(\xi)\right] = f_S(x),$$

consistent and asymptotically normal

$$\mathcal{L}_S\left\{\sqrt{T}\left(\tilde f_T(x) - f_S(x)\right)\right\} \Longrightarrow \mathcal{N}\left(0, \mathrm{I}_f(S,x)^{-1}\right).$$

Therefore to show their asymptotic efficiency we have to check the uniform integrability of the random variables

$$\eta_T^*(x) = T^{1/2}\left(\tilde f_T(x) - f_S(x)\right) = \frac{M_S(x,X_T) - M_S(x,X_0)}{\sqrt{T}} - \frac{2 f_S(x)}{\sqrt{T}}\int_0^T \frac{\chi_{\{X_t>x\}} - F_S(X_t)}{\sigma(X_t)\; f_S(X_t)}\,\mathrm{d}W_t,$$

where (see (1.166))

$$M_S(x,z) = 2 f_S(x)\,\chi_{\{z\ge x\}}\int_x^z \frac{1-F_S(y)}{\sigma(y)^2\, f_S(y)}\mathrm{d}y + 2\,\chi_{\{z<x\}}\int_x^z \left[\frac{h(y)}{\sigma(x)^2\,h(x)} - \frac{f_S(x)}{\sigma(y)^2\, f_S(y)}\,F_S(y)\right]\,\mathrm{d}y.$$

Hence if we suppose that the condition $\mathcal{K}_2$ is fulfilled and for some $p_*\ge 2$

$$\sup_{S(\cdot)\in V_\delta} \mathbf{E}_S\left\{\left|M_S(x,\xi)\right|^{p_*} + \left|\frac{\chi_{\{\xi>x\}} - F_S(\xi)}{\sigma(\xi)\; f_S(\xi)}\right|^{p_*}\right\} < \infty,$$

then for the loss functions $\ell(\cdot)\in\mathscr{W}_p$ with $p<p_*$ we have the convergence

$$\lim_{T\to\infty}\sup_{S(\cdot)\in V_\delta}\mathbf{E}_S\,\ell\left(\eta_T^*(x)\right) = \sup_{S(\cdot)\in V_\delta}\mathbf{E}\,\ell\left(\zeta\,\mathrm{I}_f(S,x)^{-1/2}\right).$$

Therefore for continuous w.r.t. uniform metric Fisher information $\mathrm{I}_f(S,x)$ these unbiased estimators are asymptotically efficient in the sense of Definition 4.9. It can be shown that if the function $h(\cdot)\in\mathscr{P}$ and $S(\cdot)\in\mathscr{S}_0$, then the estimator $\tilde f_T(x)$ is asymptotically efficient for the loss functions $\ell(\cdot)\in\mathscr{W}_p$.

Kernel-type Estimators

Remember that the *kernel-type estimator* (1.151) is

$$\hat{f}_T(x) = \frac{1}{\sqrt{T}} \int_0^T Q\left(\sqrt{T}\,(X_t - x)\right)\, dt,$$

where the kernel $Q(\cdot)$ is a nonnegative function with compact support $[A, B]$ and it satisfies the usual conditions:

$$\int_A^B Q(u)\, du = 1, \qquad \int_A^B u\, Q(u)\, du = 0.$$

Of course, we can take other bandwidths as well (see Section 4.5). This estimator admits the representation

$$\begin{aligned}\sqrt{T}\left(\hat{f}_T(x) - f_S(x)\right) &= \sqrt{T}\left(f_T^{\circ}(x) - f_S(x)\right)\\ &= -\frac{2 f_S(x)}{\sqrt{T}} \int_0^T \frac{\chi_{\{X_t > x\}} - F_S(X_t)}{\sigma(X_t)\; f_S(X_t)}\, dW_t + \delta_T(x),\end{aligned}$$

where $\delta_T(x) \to 0$. Therefore to prove the asymptotic efficiency of the kernel-type estimators we need just to add to the conditions of Theorem 4.11 providing the (uniform) convergence to zero of the moments of $\delta_T(x)$.

Interval Estimation

We see that all the estimators considered above have the same limit variance

$$\mathcal{L}_S\left\{\sqrt{T}\left(\bar{f}_T(x) - f_S(x)\right)\right\} \Longrightarrow \mathcal{N}\Big(0, A(S,x)\Big), \qquad A(S,x) = \mathrm{I}_f(S,x)^{-1}.$$

Therefore, if we have to construct a *confidence interval* of the asymptotic level $1 - \varepsilon$ ($\varepsilon \in (0,1)$)

$$\left[\bar{f}_T(x) - \frac{z_\varepsilon\, \bar{A}_T(x)^{1/2}}{\sqrt{T}},\; \bar{f}_T(x) + \frac{z_\varepsilon\, \bar{A}_T(x)^{1/2}}{\sqrt{T}}\right], \tag{4.42}$$

we need a consistent estimator $\bar{A}_T(x)$ of $A(S,x)$. Here z_ε is a $(1-\varepsilon)$-quantile of the standard Gaussian law.

Let us introduce the following *empirical estimator*:

$$\hat{A}_T(x) = \frac{4\, f_T^{\circ}(x)}{T} \int_0^T \left(\frac{\chi_{\{X_t > x\}} - \hat{F}_T(X_t)}{\sigma(X_t)\left[\hat{f}_T(X_t) + \delta_T\right]}\right)^2 dt,$$

where $\hat{F}_T(\cdot)$ is the EDF, $\delta_T = T^{-1/4}$ and $\hat{f}_T(\cdot)$ is the kernel-type estimator of the density.

Proposition 4.15. (Dehay and Kutoyants [59]) *Let the condition* $\mathcal{A}_0$ *be fulfilled. Then the estimator* $A_T(x)$ *is consistent, i.e.,*

$$\mathbf{P}_S - \lim_{T\to\infty} A_T(x) = A(S,x)$$

and the confidence interval (4.42) is of asymptotic level $1-\varepsilon$.

Note that other estimators of the quantity $A(S,x)$ were studied by Guillou and Merlevède [92] and Blanke and Merlevède, [28].

Similar confidence interval

$$\left[\hat F_T(x) - \frac{z_\varepsilon\, \bar B_T(x)^{1/2}}{\sqrt{T}},\ \hat F_T(x) + \frac{z_\varepsilon\, \bar B_T(x)^{1/2}}{\sqrt{T}}\right]$$

can be constructed in the problem of distribution function estimation [59]. Here

$$\hat B_T(x) = \frac{4}{T}\int_0^T \left(\frac{\hat F_T(X_t\wedge x) - \hat F_T(X_t)\,\hat F_T(x)}{\sigma(X_t)\,\left[\hat f_T(X_t)+\delta_T\right]}\right)^2 \mathrm{d}t$$

is the empirical estimator of the limit variance

$$B(S,x) = 4\ \mathbf{E}\left(\frac{F_S(\xi\wedge x) - F_S(\xi)\,\hat F_S(x)}{\sigma(\xi)\ f_S(\xi)}\right)^2$$

of EDF:

$$\mathcal{L}_S\left\{\sqrt{T}\left(\hat F_T(x) - F_S(x)\right)\right\} \Longrightarrow \mathcal{N}\Big(0, B(S,x)\Big), \qquad B(S,x) = \mathrm{I}_F(S,x)^{-1}.$$

It is shown that under the same condition $\mathcal{A}_0$ this estimator is consistent.

Of course, these two estimators can be used in the problems of Fisher information estimation as follows:

$$\hat{\mathrm{I}}_{f,T}(x) = A_T(x)^{-1} \longrightarrow \mathrm{I}_f(S,x), \qquad \hat{\mathrm{I}}_{F,T}(x) = B_T(x)^{-1} \longrightarrow \mathrm{I}_F(S,x).$$

4.3 Semiparametric Estimation

Let $R(\cdot)$ and $N(\cdot)$ be some (known) functions and we have to estimate the value of the parameter

$$\vartheta_S = \mathbf{E}_S\Big(R(\xi)\,S(\xi) + N(\xi)\Big)$$

by the observations $X^T = \{X_t,\ 0\le t\le T\}$ of the diffusion process

$$\mathrm{d}X_t = S(X_t)\,\mathrm{d}t + \sigma(X_t)\,\mathrm{d}W_t, \quad X_0, \quad 0\le t\le T. \tag{4.43}$$

We suppose as before that the diffusion coefficient $\sigma(\cdot)^2$ is a known positive function and the trend coefficient is unknown to the observer, but is such that the conditions $\mathcal{ES}$, $\mathcal{EM}$ and $\mathcal{RP}$ are fulfilled. Therefore the model is known up to the function but our goal is to estimate a one-dimensional parameter ϑ_S. Such type of estimation problems sometimes are called *semiparametric* [36].

For example, if we put $R(y) \equiv 0$ and $N(y) = y^p,\ p > 0$, then we have the problem of moment estimation

$$\vartheta_S = \int_{-\infty}^{\infty} y^p\ f_S(y)\ \mathrm{d}y.$$

As in the preceding two sections we derive the lower bound on the risks of all estimators and then we show that the *empirical estimator*

$$\hat\vartheta_T = \frac{1}{T}\int_0^T R\left(X_t\right)\,\mathrm{d}X_t + \frac{1}{T}\int_0^T N\left(X_t\right)\,\mathrm{d}t \tag{4.44}$$

is asymptotically (as $T \to \infty$) efficient in the sense of this bound.

The special choices of the functions $R\left(\cdot\right)$ and $N\left(\cdot\right)$ will give us the problems of distribution function and density estimation as well.

4.3.1 Lower Bound

We define the nonparametric vicinity of a fixed model exactly in the same way as was done above. We fix some $S_*(\cdot) \in \mathscr{S}_\sigma^*$ and $\delta > 0$ and introduce the set

$$V_\delta = \left\{S(\cdot):\ \sup_{x\in\mathscr{R}} \left|S(x) - S_*(x)\right| < \delta, \qquad S(\cdot) \in \mathscr{S}_\sigma^*\right\}.$$

Remember that

$$\sup_{S(\cdot)\in V_\delta} G(S) < \infty.$$

We suppose that the estimated parameter ϑ_S is finite, i.e.,

$\mathcal{P}_1$. $$\sup\nolimits_{S(\cdot)\in V_\delta} \mathbf{E}_S\Big(\left|R(\xi)\ S(\xi)\right| + \left|N(\xi)\right|\Big) < \infty.$$

We denote as $f(\cdot)$ the invariant density for the process (4.43) with $S(x) = S_*(x)$ and $\mathbf{E}$ will be the mathematical expectation with respect to the measure $\mathbf{P}_{S_*}^{(T)}$. The distribution function will be $F(\cdot)$ and the corresponding value of the parameter is denoted as

$$\vartheta = \mathbf{E}\,\Big(R(\xi)\ S_*(\xi_0) + N(\xi)\Big) \equiv \int_{-\infty}^{\infty}\Big[R(y)\ S_*(y) + N(y)\Big]\ f(y)\ \mathrm{d}y.$$

The role of Fisher information in this (semiparametric) problem will be played by the quantity

$$\mathrm{I}_\vartheta\left(S\right)=\left\{\mathbf{E}_S\left(\frac{R(\xi)\,\sigma(\xi)^2\,f_S(\xi)+2\,M_S(\xi)}{\sigma(\xi)\,f_S(\xi)}\right)^2\right\}^{-1}$$

with

$$M_S(y)=\mathbf{E}_S\left(\left[F_S(y)-\chi_{\{\xi<y\}}\right]\left(R(\xi)S(\xi)+N(\xi)\right)\right),\qquad M(y)=M_{S_*}(y).$$

Theorem 4.16. *Let $S_*\left(\cdot\right)\in\mathscr{S}_\sigma^*$, the condition $\mathcal{P}_1$ be fulfilled and $\mathrm{I}_\vartheta(S_*)>0$. Then for any loss function $\ell(\cdot)\in\mathscr{W}_p$,*

$$\varliminf_{\delta\to 0}\varliminf_{T\to\infty}\inf_{\bar\vartheta_T}\sup_{S(\cdot)\in V_\delta}\mathbf{E}_S\,\ell\left(T^{1/2}(\bar\vartheta_T-\vartheta_S)\right)\geq\mathbf{E}\,\ell\left(\zeta\,\mathrm{I}_\vartheta\left(S_*\right)^{-1/2}\right)\tag{4.45}$$

where $\mathcal{L}(\zeta)=\mathcal{N}(0,1)$ and inf *is taken over all possible estimators $\bar\vartheta_T$ of the unknown parameter.*

Proof. This proof is quite close to the proof of Theorems 4.2 and 4.8 with the only difference in the choice of the parameterization in the parametric sub-family $S_h(\cdot)=S_*(\cdot)+(h-\vartheta)\,\psi(\cdot)\,\sigma(\cdot)^2$. Recall that $\psi(\cdot)$ is a function with compact support and such $S_h(\cdot)\in V_\delta$.

Note that the corresponding family of measures

$$\left\{\mathbf{P}_h^{(T)},\;h\in(\vartheta-\gamma,\vartheta+\gamma)\right\}$$

is LAN (see (4.10)) , and we have the obvious inequality

$$\varliminf_{\delta\to 0}\varliminf_{T\to\infty}\inf_{\bar\vartheta_T}\sup_{S(\cdot)\in V_\delta}\mathbf{E}_S\,\ell\left(T^{1/2}(\bar\vartheta_T-\vartheta_S)\right)$$
$$\geq\varliminf_{\gamma\to 0}\varliminf_{T\to\infty}\inf_{\bar\vartheta_T}\sup_{|h-\vartheta|<\gamma}\mathbf{E}_h\,\ell\left(T^{1/2}(\bar\vartheta_T-\vartheta_h)\right)$$

where $\mathbf{E}_h$ is the mathematical expectation with respect to the measure $\mathbf{P}_h^{(T)}$ and

$$\vartheta_h=\mathbf{E}_h\Big(R\left(\xi\right)S_h\left(\xi\right)+N\left(\xi\right)\Big).$$

The random variable ξ here has $f_h(\cdot)\equiv f_{S_h}(\cdot)$ as a probability density function. For the small values of γ we can expand the variable ϑ_h by the powers of $h-\vartheta$. This expansion is given by the formula

$$\begin{aligned}f_h(x)&=f(x)+2\,(h-\vartheta)f(x)\,\left\{\int_0^x\psi(y)\,\mathrm{d}y\right.\\&\qquad\left.-\int_{-\infty}^{\infty}f(y)\int_0^y\psi(v)\,\mathrm{d}v\,\mathrm{d}y\right\}+o(h-\vartheta)\\&=f(x)+2\,(h-\vartheta)\;f(x)\;\mathbf{E}\Psi(\xi,x)+o(h-\vartheta),\end{aligned}$$

where

$$\Psi(\xi,x)=\int_{\xi}^{x}\psi(y)\,\mathrm{d}y.$$

Therefore denoting $Q\,(y)=R\,(y)\,S_*\,(y)+N\,(y)$ we can write

$$\begin{aligned}\vartheta_h-\vartheta&=(h-\vartheta)\ \mathbf{E}_h\left(R(\xi)\,\psi(\xi)\,\sigma\,(\xi)^2\right)+\ \mathbf{E}_h\ Q(\xi)-\vartheta\\&=(h-\vartheta)\left\{\mathbf{E}\left(R(\xi)\,\psi(\xi)\,\sigma\,(\xi)^2\right)+2\,\mathbf{E}\left(\frac{M(\xi)\,\psi(\xi)}{f(\xi)}\right)\right\}+o(h-\vartheta)\\&=(h-\vartheta)\int\left[R\,(y)\ \sigma(y)^2f(y)+2M(y)\right]\psi(y)\mathrm{d}y+o(h-\vartheta)\\&=(h-\vartheta)\int\left[R\,(y)\ \sigma(y)+\frac{2}{\sigma(y)f(y)}M(y)\right]\ \psi(y)\,\sigma(y)\ f(y)\,\mathrm{d}y\\&\quad+o(h-\vartheta).\end{aligned}\tag{4.46}$$

Let us introduce the class of functions $\psi\,(\cdot)$ such that the derivative $\partial\vartheta_h/\partial h$ at the point $h=\vartheta$ is equal to 1. This class is given as follows:

$$\mathscr{K}=\left\{\psi(\cdot):\int_{\mathscr{R}}\left[R\,(y)\ \sigma(y)^2\,f(y)+2M(y)\right]\ \psi(y)\,\mathrm{d}y=1\right\}.$$

Then for all functions $\psi(\cdot)\in\mathscr{K}$ we have

$$\vartheta_h=h+o(h-\vartheta).$$

For the LAN families of measures we can apply the inequality of Hajek–Le Cam

$$\varliminf_{\gamma\to 0}\ \varliminf_{T\to\infty}\ \inf_{\bar h_T}\ \sup_{|h-\vartheta|<\gamma}\ \mathbf{E}_h\,\ell\left(T^{1/2}(\bar h_T-h)\right)\geq\mathbf{E}\,\ell\left(\zeta\,\mathrm{I}_\psi^{-1/2}\right).$$

Therefore

$$\begin{aligned}&\varliminf_{\delta\to 0}\ \varliminf_{T\to\infty}\ \inf_{\bar\vartheta_T}\ \sup_{S(\cdot)\in V_\delta}\ \mathbf{E}_S\,\ell\left(T^{1/2}(\bar\vartheta_T-\vartheta_S)\right)\\&\quad\geq\varliminf_{\gamma\to 0}\ \varliminf_{T\to\infty}\ \inf_{\bar h_T}\ \sup_{|h-\vartheta|<\gamma}\ \mathbf{E}_h\,\ell\Big(T^{1/2}\left(\bar h_T-h+o(h-\vartheta)\right)\Big)\geq\mathbf{E}\,\ell\left(\zeta\,\mathrm{I}_\psi^{-1/2}\right)\end{aligned}$$

too. Now we have a family of lower bounds defined by the functions $\psi\,(\cdot)\in\mathscr{K}$ and we will find the *least favorable* parametric family (with the highest lower bound) in the class $\mathscr{K}$ by the same way as was done in Sections 4.1.1 and 4.2.1, i.e., we use the Cauchy–Schwarz inequality

$$\begin{aligned}1&=\left(\int\left[R\,(y)\ \sigma(y)^2\,f(y)+2M(y)\right]\ \psi(y)\,\mathrm{d}y\right)^2\\&\leq\mathrm{I}_\psi\ \mathbf{E}\left(\frac{R(\xi)\,\sigma(\xi)^2\,f(\xi)+2\,M(\xi)}{\sigma(\xi)\,f(\xi)}\right)^2.\end{aligned}$$

Therefore

$$\mathrm{I}_{\psi} \geq \left\{ \mathbf{E} \left(\frac{R(\xi)\,\sigma(\xi)^2\, f(\xi) + 2\,M(\xi)}{\sigma(\xi)\, f(\xi)} \right)^2 \right\}^{-1}$$

and the worst $\psi_*(\cdot)$ corresponds to the Fisher information $\mathrm{I}_{\psi_*} = \mathrm{I}_{\vartheta}\,(S_*)$.

This function $\psi_*\,(\cdot)$ will be

$$\psi_*(v) = \frac{C}{\sigma(v)} \left(R(y)\,\sigma(y) + \frac{2}{\sigma(y)\,f(y)}\,M(y) \right)$$

with the normalizing constant $C > 0$. Note that this function has no compact support and therefore cannot belong to $\mathscr{K}$. As is usual in such situations, we introduce a sequence of functions $\psi_n(\cdot) \in \mathscr{K}$, $n \to \infty$ such that

$$\lim_{n\to\infty} \mathbf{E}\ \sigma(\xi)^2 \left(\psi_n(\xi) - \psi_*\,(\xi)\right)^2 = 0.$$

Then

$$\inf_{\psi(\cdot)\in\mathscr{K}} \mathrm{I}_{\psi} = \lim_{n\to\infty} \mathrm{I}_{\psi_n} = \mathrm{I}_{\vartheta}\,(S_*)\,.$$

Recall that $\mathbf{E}\ \ell\left(\zeta\,\mathrm{I}_{\psi}^{-1/2}\right)$ is a monotonic function of I_{ψ}. Therefore the inequality (4.45) is proved.

Definition 4.17. *Let the conditions of Theorem 4.16 be fulfilled. Then we call an estimator $\bar{\vartheta}_T$ asymptotically efficient for a loss function $\ell\,(\cdot)$ if*

$$\lim_{\delta\to 0}\lim_{T\to\infty}\sup_{S(\cdot)\in V_\delta} \mathbf{E}_S\ \ell\left(T^{1/2}(\bar{\vartheta}_T - \vartheta_S)\right) = \mathbf{E}\ \ell\left(\zeta\,\mathrm{I}_{\vartheta}\,(S_*)^{-1/2}\right). \tag{4.47}$$

4.3.2 Empirical Estimator

Let us study the asymptotic behavior of the *empirical estimator*

$$\begin{aligned}\hat{\vartheta}_T &= \frac{1}{T}\int_0^T R(X_t)\,\mathrm{d}X_t + \frac{1}{T}\int_0^T N(X_t)\,\mathrm{d}t\\ &= \frac{1}{T}\int_0^T R(X_t)\,\sigma(X_t)\,\mathrm{d}W_t + \frac{1}{T}\int_0^T Q_S(X_t)\,\mathrm{d}t,\end{aligned}$$

where $Q_S\,(y) = R\,(y)\,S\,(y) + N\,(y)$. Note that $\vartheta_S = \mathbf{E}_S\ Q_S(\xi)$ and

$$M_S(y) = \int_{-\infty}^{y} \left[\mathbf{E}_S\ Q_S(\xi) - Q_S(y)\right]\, f_S(y)\,\mathrm{d}y.$$

Introduce the function

$$H_S(y) = \int_0^y \frac{2\,M_S(v)}{\sigma\,(v)^2\ f_S\,(v)}\,\mathrm{d}v.$$

Our *regularity conditions* $\mathcal{P}$ are

$\mathcal{P}_2$. *The condition* $\mathcal{P}_1$ *is fulfilled and there exist* $p_* \geq 2$ *such that*

$$\sup_{S(\cdot)\in V_\delta} \mathbf{E}_S \left|H_S(\xi)\right|^{p_*} < \infty, \qquad \sup_{S(\cdot)\in V_\delta} \mathbf{E}_S \left|\frac{R(\xi)\,\sigma(\xi)^2\,f_S(\xi) + 2\,M_S(\xi)}{\sigma(\xi)\,f_S(\xi)}\right|^{p_*} < \infty.$$

$\mathcal{P}_2$. *The law of large numbers*

$$\mathbf{P}_S - \lim_{T\to\infty} \frac{1}{T}\int_0^T \left[\frac{R(\xi)\,\sigma(\xi)^2\,f_S(\xi) + 2\,M_S(\xi)}{\sigma(\xi)\,f_S(\xi)}\right]^2 \mathrm{d}t = \mathrm{I}_\vartheta\,(S)^{-1}$$

is uniform on the set V_δ.

The asymptotic efficiency of empirical estimator is established in the following theorem.

Theorem 4.18. *Let the conditions* $\mathcal{P}$ *be fulfilled and the Fisher information* $\mathrm{I}_\vartheta\,(S)$ *be a continuous function at the point* $S(\cdot) = S_*(\cdot)$ *(in uniform metric). Then the estimator* $\hat\vartheta_T$ *is uniformly consistent, asymptotically normal*

$$\mathcal{L}_S\left\{T^{1/2}(\hat\vartheta_T - \vartheta_S)\right\} \Longrightarrow \mathcal{N}\left(0, \mathrm{I}_\vartheta\,(S)^{-1}\right) \tag{4.48}$$

and asymptotically efficient for the loss functions $\ell(\cdot) \in \mathscr{W}_p$ *with* $p < p_*$.

Proof. This estimator is unbiased because $\mathbf{E}_S\,\hat\vartheta_T = \mathbf{E}_S\,Q_S\,(\xi) = \vartheta_S$. Further, the difference $\eta_T = T^{1/2}(\hat\vartheta_T - \vartheta_S)$ has the representation

$$\eta_T = \frac{1}{\sqrt{T}}\int_0^T R(X_t)\,\sigma(X_t)\,\mathrm{d}W_t + \frac{1}{\sqrt{T}}\int_0^T \left[Q(X_t) - \mathbf{E}_S\;Q(\xi)\right]\,\mathrm{d}t.$$

As is usual in such problems we transform the ordinary integral in stochastic integral with the help of the Itô formula in the following way:

$$\eta_T = \frac{H_S(X_T) - H_S(X_0)}{\sqrt{T}} + \frac{1}{\sqrt{T}}\int_0^T \left[R(X_t)\,\sigma(X_t) + \frac{2M_S(X_t)}{\sigma(X_t)\,f_S(X_t)}\right]\,\mathrm{d}W_t. \tag{4.49}$$

We have for the first term

$$\sup_{S(\cdot)\in V_\delta} \mathbf{E}_S \left(\frac{H_S(X_T) - H_S(X_0)}{\sqrt{T}}\right)^2 \leq \frac{4}{T}\sup_{S(\cdot)\in V_\delta} \mathbf{E}_S\,H_S(\xi)^2 \to 0.$$

The uniform version of the central limit theorem for stochastic integrals allows us to check the uniform in $S(\cdot) \in V_\delta$ weak convergence

$$\mathcal{L}_S\left\{\frac{1}{\sqrt{T}}\int_0^T\left[R(X_t)\,\sigma(X_t)+\frac{2M_S(X_t)}{\sigma(X_t)\,f_S(X_t)}\right]\,\mathrm{d}W_t\right\}\Longrightarrow\mathcal{N}\left(0,\mathrm{I}_\vartheta(S)^{-1}\right).$$

Therefore we have the convergence (4.48).

To verify the uniform integrability of the family of random variables $\ell\left(\eta_T\right)$ we proceed exactly as was done in Section 4.2.3. Now the asymptotic efficiency follows from the convergence

$$\lim_{T\to\infty}\sup_{S(\cdot)\in V_\delta}\mathbf{E}_S\,\ell\left(\eta_T\right)=\sup_{S(\cdot)\in V_\delta}\mathbf{E}_S\,\ell\left(\zeta\,\mathrm{I}_\vartheta(S)^{-1/2}\right)$$

and the continuity of the Fisher information.

4.3.3 Remarks

The problems of distribution function, density and moments estimation can be considered as particular cases of the parameter estimation problem presented here as follows.

Distribution Function Estimation

Put $R(y)\equiv 0$ and $N(y)=\chi_{\{y<x\}}$. Then

$$\vartheta_S=\mathbf{E}_S\,\chi_{\{\xi<x\}}=F_S(x)$$

and the problem of estimation ϑ_S is equivalent to the problem of estimation of the distribution function at point x. The direct calculation provides the value

$$\mathrm{I}_F\left(S,x\right)=\left\{4\,\mathbf{E}_S\left(\frac{F_S(\xi\wedge x)[1-F_S(\xi\vee x)]}{\sigma(\xi)\,f_S(\xi)}\right)^2\right\}^{-1}.$$

This is the Fisher information in this problem and the empirical estimator is equal to the empirical distribution function

$$\hat\vartheta_T\equiv\hat F_T(x)=\frac{1}{T}\int_0^T\chi_{\{X_t<x\}}\,\mathrm{d}t.$$

Therefore, it is a consistent, asymptotically normal

$$\mathcal{L}_S\left\{\sqrt{T}\left(\hat F_T(x)-F_S(x)\right)\right\}\Longrightarrow\mathcal{N}\left(0,\mathrm{I}_F(S,x)^{-1}\right)$$

and asymptotically efficient estimator of the function $F_S(x)$ (see Section 4.1 for details).

Density Estimation

Note at first that for ergodic diffusion processes the invariant density $f_S(x)$ satisfies the equality

$$f_S(x) = \sigma(x)^{-2} \mathbf{E}_S \left(\operatorname{sgn}(x-\xi)\, S(\xi)\right).$$

Therefore if we put $N(y) \equiv 0$ and $R(y) = \sigma(x)^{-2} \operatorname{sgn}(x-y)$ then

$$\vartheta_S = \mathbf{E}_S\ R(\xi)\, S(\xi) = f_S(x)$$

and the estimation of ϑ_S corresponds to the estimation of the density function $f_S(x)$.

The Fisher information will be

$$\mathrm{I}_f(S,x) = \left\{4\, f_S(x)^2\, \mathbf{E}_S \left(\frac{\chi_{\{\xi>x\}} - F_S(\xi)}{\sigma(\xi) f_S(\xi)}\right)^2\right\}^{-1}$$

and under regularity conditions $\boldsymbol{\mathcal{P}}$ the estimator

$$\hat{\vartheta}_T = \frac{1}{\sigma(x)^2 T} \int_0^T \operatorname{sgn}(x - X_t)\ \mathrm{d}X_t$$

is unbiased, consistent, uniformly asymptotically normal

$$\mathcal{L}_S\left\{\sqrt{T}\left(\hat{\vartheta}_T - f_S(x)\right)\right\} \Longrightarrow \mathcal{N}\left(0, \mathrm{I}_f(S,x)^{-1}\right)$$

and asymptotically efficient (see Section 4.2.2 for details).

The density function $f_S(x)$ can also be written as

$$f_S(x) = 2\sigma(x)^{-2}\ \mathbf{E}_S\left(\chi_{\{\xi<x\}} S(\xi)\right)$$

and this equality suggests introducing the corresponding estimator

$$\hat{\vartheta}_T = \frac{2}{\sigma(x)^2 T} \int_0^T \chi_{\{X_t<x\}}\ \mathrm{d}X_t$$

which has the same asymptotic properties as the LTE.

It was shown in Sections 1.3 and 4.2 that the equality

$$f_S(x) = \mathbf{E}_S\ \left(R_x(\xi) S(\xi) + N_x(\xi)\right)$$

can be realized for a wide class of functions $R_x(\cdot)$, $N_x(\cdot)$ (see (4.41)). Therefore, we can put $\vartheta_S = \mathbf{E}_S\ (R_x(\xi)S(\xi) + N_x(\xi))$ and the empirical estimator $\hat{\vartheta}_T$ by Theorem 4.18 is unbiased, consistent, asymptotically normal and asymptotically efficient.

Moments Estimation

Let $R(y) \equiv 0$ and $N(y) = |y|^p$. Then $\vartheta_S = \mathbf{E}_S |\xi|^p$, $p \geq 2$ is a moment of invariant distribution. The semiparametric estimation of such moments were studied in [152]. The Fisher information is

$$\mathrm{I}_\vartheta(S) = \left\{ 4\, \mathbf{E}_S \left(\frac{A_S(\xi)}{\sigma(\xi)\, f_S(\xi)} \right)^2 \right\}^{-1},$$

where

$$A_S(y) = \mathbf{E}_S \left([F_S(y) - \vartheta_S]\, \chi_{\{\xi<y\}} N(\xi) \right).$$

It can be shown that the empirical estimator

$$\hat{\vartheta}_T = \frac{1}{T} \int_0^T N(X_t)\, \mathrm{d}t$$

is unbiased, consistent, asymptotically normal and asymptotically efficient.

4.3.4 Integral Type Risk

We consider the problem of function estimation

$$\vartheta_S(x) = \mathbf{E}_S \Big(R(\xi, x)\, S(\xi) + N(\xi, x) \Big), \qquad x \in \mathscr{R},$$

for the same model of observations as in Section 4.3; that is, the observed process $X^T = \{X_t,\ 0 \leq t \leq T\}$ is diffusion with the stochastic differential

$$\mathrm{d}X_t = S(X_t)\, \mathrm{d}t + \sigma(X_t)\, \mathrm{d}W_t, \quad X_0, \quad 0 \leq t \leq T, \tag{4.50}$$

where $S(\cdot)$ is an unknown function and the functions $R(\cdot,\cdot)$, $N(\cdot,\cdot)$ and $\sigma(\cdot)^2 > 0$ are known to the observer. The risk of any estimator $\left\{\bar{\vartheta}_T(x),\ x \in \mathscr{R}\right\}$ will be measured by the risk function

$$\mathcal{R}_T(\bar{\vartheta}_T, S) = \mathbf{E}_S \int_{\mathscr{R}} \left|\bar{\vartheta}_T(x) - \vartheta_S(x)\right|^2 \mu(\mathrm{d}x), \tag{4.51}$$

where $\mu(\cdot)$ is some finite measure on $\mathscr{R}$. Remember that the problems of distribution function and density estimation can be obtained as particular cases of this one, choosing the appropriate functions $R(\cdot, x)$ and $N(\cdot, x)$. For example, if $N(\cdot, x) = 0$ and $R(y, x) = \sigma(x)^{-2} \operatorname{sgn}(x - y)$, then we have the problem of density $f_S(\cdot)$ estimation and the risk of an estimator $\left\{\bar{f}_T(x),\ x \in \mathscr{R}\right\}$ is

$$\mathcal{R}_T(\bar{f}_T, S) = T\ \mathbf{E}_S \int_{\mathscr{R}} \left|\bar{f}_T(x) - f_S(x)\right|^2 \mu(\mathrm{d}x).$$

We follow the same schema and begin with the lower bound on the integral-type risk of all estimators and then we show that the empirical estimator

$$\hat{\vartheta}_T(x) = \frac{1}{T}\int_0^T R(X_t, x)\,\mathrm{d}X_t + \frac{1}{T}\int_0^T N(X_t, x)\,\mathrm{d}t \tag{4.52}$$

is asymptotically efficient. We take the same nonparametric vicinity V_δ of a fixed model $S_*(\cdot)$ and the role of Fisher information will be played by the same quantity

$$\mathrm{I}_\vartheta(S, x) = \left\{ \mathbf{E}_S\left(\frac{R(\xi,x)\,\sigma(\xi)^2\,f_S(\xi) + 2\,M_S(\xi,x)}{\sigma(\xi)\,f_S(\xi)}\right)^2\right\}^{-1}$$

with

$$M_S(y,x) = \mathbf{E}_S\left(\left[F_S(y) - \chi_{\{\xi<y\}}\right]\left(R(\xi,x)S(\xi) + N(\xi,x)\right)\right).$$

Put

$$\begin{aligned}\mathcal{R}_*(S) &= \int_{\mathscr{R}} \mathbf{E}_S\Big(\frac{R(\xi,x)\,\sigma(\xi)^2\,f_S(\xi) + 2\,M_S(\xi,x)}{\sigma(\xi)\,f_S(\xi)}\Big)^2 \mu(\mathrm{d}x)\\ &= \int_{\mathscr{R}} \mathrm{I}_\vartheta(S,x)^{-1}\,\mu(\mathrm{d}x),\end{aligned}$$

and introduce the first *regularity condition*

$\mathcal{Q}_1$. *The function $S_*(\cdot) \in \mathscr{S}_\sigma^*$ and for some $\delta > 0$*

$$\sup_{S(\cdot)\in V_\delta} \mathbf{E}_S\left(|R(\xi,x)\,S(\xi)| + |N(\xi,x)|\right) < \infty$$

and

$$\sup_{S(\cdot)\in V_\delta} \int_{\mathscr{R}} \mathbf{E}_S\Big(\frac{R(\xi,x)\,\sigma(\xi)^2\,f_S(\xi) + 2\,M_S(\xi,x)}{\sigma(\xi)\,f_S(\xi)}\Big)^2 \mu(\mathrm{d}x) < \infty.$$

The lower bound is given by the following theorem.

Theorem 4.19. *Let condition $\mathcal{Q}_1$ be fulfilled then*

$$\varliminf_{\delta\to 0}\ \varliminf_{T\to\infty}\ \inf_{\bar{\vartheta}_T}\ \sup_{S(\cdot)\in V_\delta} \mathcal{R}_T(\bar{\vartheta}_T, S) \geq \mathcal{R}_*(S_*). \tag{4.53}$$

Proof. We follow the proof of a similar result obtained in the problem of distribution function estimation for i.i.d. observations by Gill and Levit [86]. Introduce a parametric family of diffusion processes

$$\mathrm{d}X_t = \left[S_*(X_t) + \sum_{i=1}^k \theta_i\,\psi_i(X_t)\,\sigma(X_t)^2\right]\mathrm{d}t + \sigma(X_t)\,\mathrm{d}W_t, \quad X_0, \quad 0 \le t \le T, \tag{4.54}$$

where $\psi_i(\cdot)$, $i = 1, \ldots, k$ are continuously differentiable functions with compact support, and parameter $\boldsymbol{\theta} = (\theta_1, \ldots, \theta_k) \in \Theta_\delta = [-\varepsilon, \varepsilon]^k$, $\varepsilon = \varepsilon(\delta)$ is such that

$$S(\boldsymbol{\theta}, \cdot) = S_*(\cdot) + \sum_{i=1}^k \theta_i\, \psi_i(\cdot)\; \sigma(\cdot)^2 \in V_\delta.$$

Please note the difference in the notation: $\vartheta_S(\cdot)$ is the estimated function and $\boldsymbol{\theta}$ is a parameter of the introduced family of processes. The stationary density function is

$$f(\theta, x) = \frac{1}{G(S(\boldsymbol{\theta})\,\sigma(x)^2} \exp\left\{2\int_0^x \frac{S_*(v)}{\sigma(v)^2}\,\mathrm{d}v + 2\sum_{i=1}^k \theta_i \int_0^x \psi_i(v)\,\mathrm{d}v\right\},$$

and the initial value X_0 has this density function. Therefore the process $\{X_t, t \geq 0\}$ is stationary and ergodic. We denote by $\mathbf{P}_{\boldsymbol{\theta}}^{(T)}$ the measure induced by the process (4.54) and by $\mathbf{E}_{\boldsymbol{\theta}}^{(T)}$ the expectation with respect to this measure.

Suppose that $\boldsymbol{\theta}$ is a random vector independent of the Wiener process $\{W_t,\ t \geq 0\}$ with independent components and its density function $q(\boldsymbol{\theta}) = \prod_{i=1}^k q_i(\theta_i)$, be such that the trace of the Fisher information of the prior distribution

$$\operatorname{Tr} \mathrm{I}(q) = \sum_{i=1}^k \mathrm{I}_{ii}(q) = \sum_{i=1}^k \int_{-\varepsilon}^{\varepsilon} \left(\frac{\partial q_i(\theta_i)}{\partial \theta_i}\right)^2 q_i(\theta_i)^{-1}\,\mathrm{d}\theta_i < \infty$$

and $q_i(\pm\varepsilon) = 0$. Then for any absolutely continuous vector function

$$\Psi(S(\boldsymbol{\theta})) = \Big(\Psi_1(S(\boldsymbol{\theta})), \ldots, \Psi_k(S(\boldsymbol{\theta}))\Big)$$

and any estimator $\bar{\Psi}_T = \left(\bar{\Psi}_{T,1}, \ldots, \bar{\Psi}_{T,k}\right)$ of this function we have the (multidimensional, $\mathscr{L}_2$-norm type) van Trees inequality (1.108)

$$\mathbb{E}\sum_{i=1}^k \left|\bar{\Psi}_{T,i} - \Psi_i(S(\boldsymbol{\theta}))\right|^2 \geq \frac{\left(\mathrm{E}_Q \sum_{i=1}^k \frac{\partial \Psi_i(S(\boldsymbol{\theta}))}{\partial \theta_i}\right)^2}{\mathrm{E}_Q \operatorname{Tr} \mathrm{I}^{(T)}(\boldsymbol{\theta}) + \operatorname{Tr}\mathrm{I}(q)}. \tag{4.55}$$

Here $\mathbb{E}$ and E_Q are expectations with respect to measures $\mathrm{d}\mathbb{P} = q(\boldsymbol{\theta})\ \mathrm{d}\boldsymbol{\theta} \times \mathrm{d}\,\mathbf{P}_{\boldsymbol{\theta}}^{(T)}$ and $q(\boldsymbol{\theta})\ \mathrm{d}\boldsymbol{\theta}$ respectively, the trace of the Fisher information

$$\begin{aligned}
\operatorname{Tr} \mathrm{I}^{(T)}(\boldsymbol{\theta}) &= \sum_{i=1}^k \mathrm{I}_{ii}^{(T)}(\boldsymbol{\theta}) \\
&= \sum_{i=1}^k \mathbf{E}_{\boldsymbol{\theta}}^{(T)} \left(2\int_{-\infty}^{X_0} \mathbf{E}_{\boldsymbol{\theta}}\left[\chi_{\{v>\xi\}}\psi_i(v)\right]\mathrm{d}v + \int_0^T \psi_i(X_t)\,\sigma(X_t)\,\mathrm{d}W_t\right)^2 \\
&= T\sum_{i=1}^k \mathbf{E}_{\boldsymbol{\theta}}\Big(\psi_i(\xi)\,\sigma(\xi)\Big)^2\Big(1 + o(1)\Big) = T\sum_{i=1}^k \mathrm{I}_{ii}(\boldsymbol{\theta})\,\Big(1 + o(1)\Big) < \infty.
\end{aligned}$$

Introduce a complete orthonormal system of functions $\{\phi_i(\cdot),\ i=1,2,\ldots\}$ in the space $\mathscr{L}_2(\mu)$ and define the Fourier coefficients $\bar{\boldsymbol{\Phi}}_T = \left(\bar{\Phi}_{T,1},\ldots,\bar{\Phi}_{T,k}\right)$ and $\boldsymbol{\Phi}(S) = \left(\Phi_1(S),\ldots,\Phi_k(S)\right)$ of the functions $\bar{\vartheta}_T(\cdot)$, $\vartheta_S(\cdot)$ as

$$\bar{\Phi}_{T,i} = \int_{\mathscr{R}} \bar{\vartheta}_T(x)\,\phi_i(x)\,\mu(\mathrm{d}x), \qquad \Phi_i(S) = \int_{\mathscr{R}} \vartheta_S(x)\,\phi_i(x)\,\mu(\mathrm{d}x).$$

We have the obvious inequalities

$$\begin{aligned}\sup_{S(\cdot)\in V_\delta} \mathcal{R}_T(\bar{\vartheta}_T, S) &\geq \sup_{\boldsymbol{\theta}\in\Theta_\delta} \mathcal{R}_T\left(\bar{\vartheta}_T, S(\boldsymbol{\theta})\right) \geq \int_{\Theta_\delta} \mathcal{R}_T\left(\bar{\vartheta}_T, S(\boldsymbol{\theta})\right) q(\boldsymbol{\theta})\,\mathrm{d}\boldsymbol{\theta} \\ &= T\,\mathbb{E}\sum_{i=1}^{\infty}\left|\bar{\Phi}_{T,i} - \Phi_i(S)\right|^2 \geq T\,\mathbb{E}\sum_{i=1}^{k}\left|\bar{\Phi}_{T,i} - \Phi_i(S)\right|^2,\end{aligned}$$

where we used Parseval's identity. We can now apply the van Trees inequality (4.55) to the last sum and obtain the estimate

$$T\,\mathbb{E}\sum_{i=1}^{k}\left|\bar{\Phi}_{T,i} - \Phi_i(S)\right|^2 \geq \frac{\left(\mathrm{E}_Q \operatorname{div}\boldsymbol{\Phi}(S(\boldsymbol{\theta}))\right)^2}{\mathrm{E}_Q \operatorname{Tr}\mathrm{I}(\boldsymbol{\theta})\,(1+o(1)) + T^{-1}\operatorname{Tr}\mathrm{I}(q)}.$$

The direct calculation yields

$$\begin{aligned}\frac{\partial\Phi_i(S(\boldsymbol{\theta}))}{\partial\theta_i} &= \int_{\mathscr{R}}\int_{\mathscr{R}} \Bigg[R(y,x)\ \sigma(y) \\ &\quad + \frac{2M_{S(\boldsymbol{\theta})}(y,x)}{\sigma(y) f(\boldsymbol{\theta},y)}\Bigg]\ \psi_i(y)\,\sigma(y)\ f(\boldsymbol{\theta},y)\phi_i(x)\mathrm{d}y\,\mu(\mathrm{d}x) \\ &= \int_{\mathscr{R}} \psi_i(y)\,\sigma(y)\ f(\boldsymbol{\theta},y)\Bigg(\int_{\mathscr{R}}\Bigg[R(y,x)\ \sigma(y) \\ &\quad + \frac{2M_{S(\boldsymbol{\theta})}(y,x)}{\sigma(y) f(\boldsymbol{\theta},y)}\Bigg]\ \phi_i(x)\mu(\mathrm{d}x)\Bigg)\,\mathrm{d}y.\end{aligned}$$

Choosing T sufficiently large and ε sufficiently small we can obtain the following estimate:

$$\frac{\left(\mathrm{E}_Q \operatorname{div}\boldsymbol{\Phi}(S(\boldsymbol{\theta}))\right)^2}{\mathrm{E}_Q \operatorname{Tr}\mathrm{I}(\boldsymbol{\theta})\,(1+o(1)) + T^{-1}\operatorname{Tr}\mathrm{I}(q)} = \frac{\left(\operatorname{div}\boldsymbol{\Phi}(S_*)\right)^2}{\sum_{i=1}^{k}\mathbf{E}_{S_*}\left(\psi_i(\xi)\,\sigma(\xi)\right)^2}\left(1+o(1)\right).$$

Let us introduce the functions

$$\psi_i^*(y) = \frac{1}{\sigma(y)}\int_{\mathscr{R}}\left[R(y,x)\ \sigma(y) + \frac{2M_{S_*}(y,x)}{\sigma(y) f(y)}\right]\ \phi_i(x)\,\mu(\mathrm{d}x), \quad i=1,\ldots,k,$$

where $f(y) = f_{S_*}(y)$. We have

$$\mathbf{E}_{S_*}\left(\psi_i^*(\xi)\,\sigma(\xi)\right)^2 \leq \int_{\mathscr{R}} \mathrm{I}_{S_*}(x)^{-1}\,\mu(\mathrm{d}x) < \infty.$$

Therefore we can introduce a sequence $\psi_i^{(n)}(\cdot)$, $n = 1, 2, \ldots, i = 1, \ldots, k$, of continuously differentiable functions with compact supports and such that for all i

$$\lim_{n\to\infty} \mathbf{E}_{S_*}\left(\left[\psi_i^{(n)}(\xi) - \psi_i^*(\xi)\right]\sigma(\xi)\right)^2 = 0.$$

Finally we put $\psi_i(\cdot) = \psi_i^{(n)}(\cdot)$; that is, the parametric family of diffusion processes (4.54) is constructed with such functions $\psi_i^{(n)}(\cdot)$. Then for the large values of n and k we have the desired estimate

$$\frac{\left(\operatorname{div}\boldsymbol{\Phi}(S_*)\right)^2}{\sum_{i=1}^k \mathbf{E}_{S_*}\left(\psi_i(\xi)\,\sigma(\xi)\right)^2} = \int_{\mathscr{R}} \mathrm{I}_{S_*}(x)^{-1}\,\mu(\mathrm{d}x)\,\left(1 + o(1)\right),$$

because by Parseval's identity

$$\mathbf{E}_{S_*}\sum_{i=1}^{\infty}\left(\psi_i(\xi)\,\sigma(\xi)\right)^2 = \int_{\mathscr{R}} \mathrm{I}_{S_*}(x)^{-1}\,\mu(\mathrm{d}x).$$

Definition 4.20. *Let the condition $\mathcal{Q}_1$ be fulfilled. Then an estimator $\bar\vartheta_T(\cdot)$ is asymptotically efficient if for any $S_*(\cdot) \in \mathscr{S}_\sigma^*$ we have*

$$\lim_{\delta\to 0}\lim_{T\to\infty}\sup_{S(\cdot)\in V_\delta} \mathcal{R}_T(\bar\vartheta_T, S) = \mathcal{R}_*(S_*). \tag{4.56}$$

Let us introduce the function

$$H_S(y,x) = \int_0^y \frac{2\,M_S(v,x)}{\sigma(v)^2\,f_S(v)}\,\mathrm{d}v$$

and the condition $\boldsymbol{\mathcal{Q}}$

$\mathcal{Q}_2$. *The condition $\mathcal{Q}_1$ is fulfilled and for some $\delta > 0$*

$$\sup_{S(\cdot)\in V_\delta}\int_{\mathscr{R}} \mathbf{E}_S\, H_S(\xi,x)^2\,\mu(\mathrm{d}x) < \infty.$$

Theorem 4.21. *Let condition $\boldsymbol{\mathcal{Q}}$ be fulfilled, and the functional $\mathcal{R}_*(S)$ be continuous in uniform topology at the point $S_*(\cdot)$. Then the empirical estimator $\hat\vartheta_T(\cdot)$ (4.52) is asymptotically efficient.*

Proof. This follows from the representation (4.49) and elementary estimates.

The problems of distribution function and density estimation are particular cases of the one considered. We present here these two results as well.

In the first problem we introduce the condition.

$\mathcal{Q}_F$. *Let $S_*(\cdot) \in \mathscr{S}_\sigma^*$ and for some $\delta > 0$*

$$\sup_{S(\cdot)\in V_\delta} \left\{ \int_{\mathscr{R}} \mathbf{E}_S \left| \int_x^\xi \frac{F_S(v\wedge x)(1-F_S(v\vee x))}{\sigma(v)^2 f_S(v)} \mathrm{d}v \right|^2 \mu(\mathrm{d}x) \right.$$
$$\left. + \int_{\mathscr{R}} \mathbf{E}_S \left| \frac{F_S(\xi\wedge x)(1-F_S(\xi\vee x))}{\sigma(\xi) f_S(\xi)} \right|^2 \mu(\mathrm{d}x) \right\} < \infty.$$

Put

$$\mathcal{R}_F(S) = 4 \int_{\mathscr{R}} \mathbf{E} \left(\frac{F_S(\xi\wedge x)(1-F_S(\xi\vee x))}{\sigma(\xi) f_S(\xi)} \right)^2 \mu(\mathrm{d}x).$$

Therefore we have the following proposition.

Proposition 4.22. *Let the condition $\mathcal{Q}_F$ be fulfilled and the functional $\mathcal{R}_F(S)$ be continuous in the uniform topology at the point $S_*(\cdot)$. Then we have*

$$\lim_{\delta\to 0} \lim_{T\to\infty} \inf_{\bar{F}_T} \sup_{S(\cdot)\in V_\delta} T\, \mathbf{E}_S \int_{\mathscr{R}} \left| \bar{F}_T(x) - F_S(x) \right|^2 \mu(\mathrm{d}x) = \mathcal{R}_F(S_*).$$

Proof. This result follows from Theorems 4.19 and 4.21. To prove the equality we use the empirical distribution function.

In the second problem we need a similar condition.

$\mathcal{Q}_f$. *For any $S_*(\cdot) \in \mathscr{S}_\sigma^*$ and some $\delta > 0$*

$$\sup_{S(\cdot)\in V_\delta} \left\{ \int_{\mathscr{R}} \mathbf{E}_S \left| \frac{\chi_{\{\xi>x\}} - F_S(\xi)}{\sigma(\xi)\, f_S(\xi)} \right|^2 \mu(\mathrm{d}x) \right.$$
$$\left. + \int_{\mathscr{R}} \mathbf{E}_S \left| \int_0^\xi \frac{\chi_{\{v>x\}} - F_S(v)}{\sigma(v)^2 f_S(v)} \mathrm{d}v \right|^2 \mu(\mathrm{d}x) \right\} < \infty.$$

Denote

$$\mathcal{R}_f(S) = 4 \int_{\mathscr{R}} f_S(x)^2\, \mathbf{E}_S \left(\frac{(\chi_{\{\xi>x\}} - F_S(\xi))}{\sigma(\xi) f_S(\xi)} \right)^2 \mu(\mathrm{d}x).$$

Proposition 4.23. *Let the condition $\mathcal{Q}_f$ be fulfilled and the functional $\mathcal{R}_f(S)$ be continuous in the uniform topology at the point $S(\cdot)$. Then we have*

$$\lim_{\delta\to 0}\lim_{T\to\infty}\inf_{\bar f_T}\sup_{S(\cdot)\in V_\delta} T\,\mathbf{E}_S\int_{\mathscr{R}}\left|\bar f_T(x)-f_S(x)\right|^2\mu(\mathrm{d}x)=\mathcal{R}_f(S_*)\,.$$

Proof. This is obviously another particular case of Theorems 4.19 and 4.21. To prove the equality we use the empirical density function (LTE).

4.4 Density Derivative Estimation

We observe a trajectory $X^T=\{X_t, 0\le t\le T\}$ of the diffusion process

$$\mathrm{d}X_t=S(X_t)\,\mathrm{d}t+\sigma(X_t)\,\mathrm{d}W_t,\quad X_0,\quad 0\le t\le T, \tag{4.57}$$

where $S(\cdot)\in\mathscr{S}_\sigma$ is an unknown function. For simplicity of exposition, we suppose that the diffusion coefficient $\sigma(\cdot)^2\equiv 1$. Remember that if $\sigma(\cdot)^2>0$ and is continuously differentiable, then the process

$$Y_t=H(X_t),\qquad\qquad H(x)=\int_0^x\frac{\mathrm{d}v}{\sigma(v)} \tag{4.58}$$

has by the Itô formula the stochastic differential

$$\mathrm{d}Y_t=\left[\frac{S(X_t)}{\sigma(X_t)}-\frac{1}{2}\sigma'(X_t)\right]\,\mathrm{d}t+\mathrm{d}W_t,\quad Y_0,\quad 0\le t\le T,$$

i.e., we obtain the SDE

$$\mathrm{d}Y_t=\tilde S(Y_t)\,\mathrm{d}t+\mathrm{d}W_t,\quad Y_0,\quad 0\le t\le T, \tag{4.59}$$

for diffusion process $Y_t, t\ge 0$ with

$$\tilde S(y)=S(h(y))\sigma(h(y))-1/2\sigma'(h(y)),\qquad \sigma(\cdot)^2\equiv 1, \tag{4.60}$$

where $h(y)$ is the function inverse to $H(x)$. Remember that $\sigma(\cdot)$ is continuous and $\neq 0$. Hence the function $H(\cdot)$ is strictly monotonic.

As the trend coefficient $S(\cdot)$ is supposed to be unknown, the invariant density

$$f_S(y)=G(S)^{-1}\exp\left\{2\int_0^y S(v)\,\mathrm{d}v\right\}$$

is an unknown function too. Here we consider the problem of the asymptotically efficient estimation of its first derivative

$$f_S'(x)=\frac{\partial f_S(x)}{\partial x}=2\,S(x)\;f_S(x)\,.$$

This problem of estimation first can be interesting itself, because sometimes we need to know the first (or some higher) derivatives of the unknown density function. Second, it allows us to understand better what happens in the problem of the estimation of the trend coefficient $S(x)$ because

$$S(x) = \frac{f_S'(x)}{2\, f_S(x)},$$

where the function $f_S(x)$ can be estimated with the rate $\sqrt{T}$ (see Section 4.2). Therefore for the estimation of $S(x)$ we need a good estimator of the derivative $f_S'(x)$. It can be easily shown that this function can be estimated with the rate of convergence depending on the smoothness of $S(\cdot)$, i.e., if the function $S(\cdot)$ is k times continuously differentiable then the rate of convergence of any estimator cannot be better than $T^{\frac{k}{2k+1}}$. Below we propose a lower bound on the rate of convegence of estimators in the similar problem of trend estimation (Theorem 4.38).

As usual, we consider two problems. The first one is to construct a lower minimax bound on the risks of all estimators and the second is to find an estimator, which is asymptotically efficient in the sense of this bound. We finish with concluding remarks where we propose some possible generalizations of these results.

Let us fix an integer $k \geq 2$ and suppose that the function $S(\cdot)$ is k-times differentiable. Introduce the set

$$\Sigma(k,R) = \left\{ S(\cdot) \in \mathscr{S}_\sigma \;:\; \int_{\mathscr{R}} \left[f_S^{(k+1)}(x) \right]^2 \,\mathrm{d}x \leq 4\,R \right\},$$

where $R > 0$ is some known constant (we write $4R$ and not R because in such a form the results of this section are more convenient for application in the next section in the problem of trend estimation) and define the risk of an estimator $\bar\vartheta_T(x)$ of the function $f_S'(x)$ as

$$\mathcal{R}_T\left(\bar\vartheta_T, f_S'\right) = \mathbf{E}_S \int_{\mathscr{R}} \left(\bar\vartheta_T(x) - f_S'(x)\right)^2 \,\mathrm{d}x.$$

We consider two types of lower bounds for this risk: *global* and *local*. In the first case we measure the quality of estimation by the risk

$$\sup_{S\in\Sigma_D} \mathcal{R}_T\left(\bar\vartheta_T, f_S'\right)$$

where $\Sigma_D \subset \Sigma(k,R)$ and in the second (local) case we study the risk

$$\sup_{S\in\Sigma_\delta} \mathcal{R}_T\left(\bar\vartheta_T, f_S'\right),$$

where Σ_δ is some δ-neighborhood of a fixed function $S_*(\cdot) \in \mathscr{S}$.

In both cases the asymptotic behavior of the minimax risks is quite similar, i.e., we have as $T \to \infty$

$$\inf_{\bar\vartheta_T}\sup_{S\in\Sigma_D}\mathcal{R}_T\left(\bar\vartheta_T,f'_S\right)\sim\inf_{\bar\vartheta_T}\sup_{S\in\Sigma_\delta}\mathcal{R}_T\left(\bar\vartheta_T,f'_S\right)\sim 4\,\Pi(k,R)\,T^{-\frac{2k}{2k+1}},$$

where $\Pi(k,R)$ is the *Pinsker's constant* defined by the equality

$$\Pi(k,R)=(2k+1)\left(\frac{k}{\pi\,(k+1)\,(2k+1)}\right)^{\frac{2k}{2k+1}}R^{\frac{1}{2k+1}}\,. \qquad (4.61)$$

In the first case we need some additional conditions (see below the conditions $\mathcal{S}_D$) on the *tails of the family of invariant densities*, because the least favorable parametric family of invariant densities approaches the uniform over $\mathscr{R}$ and the corresponding diffusion process is close to the null recurrent diffusion process. Remember that in the null recurrent case we have no law of large numbers and the rate of convergence is essentially the worst. Note that the conditions $\mathscr{Q}_D$ below are used in the study of the asymptotically efficient estimator only and are not used in the construction of the lower bound.

The statement of the problem and the results obtained in this and the next two sections follow *Pinsker's approach* [201] (see [192] for discussion and references). At the same time the proofs follow the work by Golubev and Levit [91] (see as well [214]).

4.4.1 Global Risk

Fix some constant and $B_0 > 0$ and introduce for any $D > 0$ the sets

$$\Sigma_1(D)=\left\{S(\cdot)\;:\;\sup_{B>B_0}B^2\int_{|x|\geq B}S(x)^2\,f_S(x)^2\,dx\leq D\right\}, \qquad (4.62)$$

$$\Sigma_2(D)=\left\{S(\cdot)\;:\;\sup_{B>B_0}B^{-2}\int_{|x|\leq B}S(x)^2\,f_S(x)^2\,\mathbf{E}_S\,H_S(\xi,x)^2\,dx\leq D\right\}, \qquad (4.63)$$

$$\Sigma_3(D)=\left\{S(\cdot)\;:\;\int_{\mathscr{R}}S(x)^2\,f_S(x)^2\,\mathbf{E}_S\left[\frac{\chi_{\{\xi>x\}}-F_S(\xi)}{f_S(\xi)}\right]^2dx\leq D\right\}, \qquad (4.64)$$

where

$$H_S(z,x)=\int_0^z\frac{\chi_{\{y>x\}}-F_S(y)}{f_S(y)}\,dy.$$

Of course

$$\Sigma_i(D_1)\subset\Sigma_i(D_2)\,,\qquad i=1,2,3$$

for any $D_1\leq D_2$. Let us denote

$$\Sigma(D) = \Sigma_1(D) \cap \Sigma_2(D) \cap \Sigma_3(D).$$

Condition $\boldsymbol{S}_D$.

$\mathcal{S}_1$. *The function* $S(\cdot) \in \mathscr{S}$ *is* $k >$ *times continuously differentiable and*

$$S(\cdot) \in \Sigma(k, R).$$

$\mathcal{S}_2$. *There exists a constant* $B_0 > 0$ *such that*

$$S(\cdot) \in \Sigma(D)$$

for some $D > 0$.

Put

$$\Sigma_D = \Sigma(k, R) \cap \Sigma(D).$$

Of course, Σ_D depends on k, R, B_0 and D. The (global) lower bound on the minimax risk of any estimator of the derivative is given in the following theorem.

Theorem 4.24. (Dalalyan and Kutoyants [54]) *Let the condition* $\boldsymbol{S}_D$ *be fulfilled. Then*

$$\underline{\lim}_{D\to\infty} \underline{\lim}_{T\to\infty} \inf_{\bar{\theta}_T} \sup_{S\in\Sigma_D} T^{\frac{2k}{2k+1}}\, \mathcal{R}_T(\bar{\theta}_T, f_S') \geq 4\,\Pi(k, R), \tag{4.65}$$

where inf *is taken over all estimators* $\bar{\theta}_T(\cdot)$ *of the derivative* $f_S'(\cdot)$.

Proof. To derive this inequality we proceed, roughly speaking, as before (see, e.g., the proof of Theorems 4.2, 4.8 and 4.19), i.e., we introduce a parametric subfamily governed by a trend coefficient $S(\vartheta, x), \vartheta \in \Theta_T$ belonging to the set $\Sigma(k, R)$ and satisfying the inequalities (4.62)–(4.64) for all $\vartheta \in \Theta_T$. Then in the problem of parameter estimation we construct a lower bound on the risks of all estimators and for the final point we choose the least favorable parametric subfamily (with highest lower bound). The limit value of the risk function in this last problem will be equal to $\Pi_*(k, R)$. The essential difference with the preceding proofs is in the choice of the parametric subfamily. In particular, the set Θ_T is a subset of $\mathscr{R}^{2L+1}$, where $L = L_T \to \infty$ with the rate $c\ T^{\frac{1}{2k+1}}$ and a specially chosen constant $c > 0$. The parameter ϑ is random with some prior distribution. Finally, the *least favorable parametric subfamily* means the least favorable trend coefficient $S(\vartheta, x)$ and the least favorable prior distribution. This program was realized by Golubev and Levit [91] in the problem of second order asymptotically efficient estimation of the distribution function in the i.i.d. case and our work was inspired by their result.

Let us introduce the trend coefficient

$$S(\boldsymbol{\vartheta}, x) = S_0(x) + \sum_{|l| \le L} \vartheta_l \, e_l(x) \, U(A - |x|),$$

where $\boldsymbol{\vartheta} = (\vartheta_{-L}, \ldots, \vartheta_L) \in \boldsymbol{\Theta}_T$,

$$S_0(x) = -\operatorname{sgn}(x)(k+2)\big(|x| - A\big)^{k+1} \chi_{\{|x|>A\}},$$

$A = \ln(1+T)$ and $U(x)$ is a $(k+1)$ times differentiable increasing function vanishing for $x \le 0$ and equal to one for $x \ge 1$. The functions $\{e_l(\cdot), l \in \mathbb{Z}\}$ are the elements of the trigonometric basis in $\mathscr{L}_2[-A, A]$, i.e.,

$$e_l(x) = \frac{1}{\sqrt{A}} \begin{cases} \sin\left(\dfrac{\pi l x}{A}\right), & \text{if } l > 0, \\ 1/\sqrt{2}, & \text{if } l = 0, \\ \cos\left(\dfrac{\pi l x}{A}\right), & \text{if } l < 0. \end{cases}$$

The set $\boldsymbol{\Theta}_T$ and the positive number $L = L_T \to \infty$ will be chosen later.

It is easy to see that $S(\boldsymbol{\vartheta}, \cdot) \in \mathscr{S}$ and the corresponding diffusion process

$$\mathrm{d}X_t = S(\boldsymbol{\vartheta}, X_t)\,\mathrm{d}t + \mathrm{d}W_t, \quad X_0,\ t \ge 0 \tag{4.66}$$

is recurrent positive with the invariant density

$$f(\boldsymbol{\vartheta}, x) = \frac{1}{G(\boldsymbol{\vartheta})} \exp\left\{2 \int_0^x S(\boldsymbol{\vartheta}, y)\,\mathrm{d}y\right\}.$$

The function $S(\boldsymbol{\vartheta}, x)$ is k-times differentiable w.r.t. x and the kth derivative is continuous. Hence the density $f(\boldsymbol{\vartheta}, x)$ is $(k+1)$ times continuously differentiable too.

We have to choose the set $\boldsymbol{\Theta}_T$ in such a way that the function $S(\boldsymbol{\vartheta}, \cdot) \in \Sigma_D$. Then we can write

$$\sup_{S(\cdot) \in \Sigma(k,R,D)} \mathcal{R}_T\left(\bar{\vartheta}_T, f_S'\right) \ge \sup_{\boldsymbol{\vartheta} \in \boldsymbol{\Theta}_T} \mathcal{R}_T\left(\bar{\vartheta}_T, f_{\boldsymbol{\vartheta}}'\right)$$

and to construct the lower bound in this parametric problem.

Let us put

$$\boldsymbol{\Theta}_T = \prod_{|l| \le L} \left[-K\sqrt{\sigma_l}, K\sqrt{\sigma_l}\right],$$

where

$$\sigma_l = \sigma_{l,T} = \frac{2A}{T}\left(\left|\frac{A}{l}\right|^k - 1\right)_+, \qquad l \ne 0, \tag{4.67}$$

with

$$\Lambda = \Lambda_T = A \left(\frac{R\,(k+1)\,(2k+1)}{k\,\pi^{2k}} \right)^{1/(2k+1)} T^{1/(2k+1)} \tag{4.68}$$

and $\sigma_0 = 0$. The integer L is chosen to be equal to $[\Lambda]$ (integer part) and the constant $K > 0$ will be described later. Here $(a)_+ = \max(a, 0)$. The reason for this choice of σ_l will be clear a bit later. Note that $\boldsymbol{\vartheta} \in \boldsymbol{\Theta}_T$ is equivalent to

$$\vartheta_0 = 0, \qquad |\vartheta_l| \le K\,\sqrt{\sigma_l}, \quad l = \pm 1, \ldots, \pm L.$$

The invariant density of this parametric family is close to the uniform over $[-A, A]$ density and to show this we need the following technical lemma.

Lemma 4.25. *For any $m = 0, 1, \ldots, k$, we have*

$$\sup_{\boldsymbol{\vartheta} \in \boldsymbol{\Theta}_T} \sup_{x \in [-A,A]} \left| \left(\int_0^x S(\boldsymbol{\vartheta}, y)\mathrm{d}y \right)^{(m)} \right| \le C\, A^{k+2}\, T^{-\frac{1}{4k+2}}. \tag{4.69}$$

Proof. For $m = 0$, using the inequality $|\vartheta_l| \le K\sqrt{\sigma_l}$ and the definitions (4.67) and (4.68), we obtain

$$\sup_{|x| \le A} \left| \int_0^x S(\boldsymbol{\vartheta}, v)\mathrm{d}v \right| \le A \sup_{|x| \le A} \left| S(\boldsymbol{\vartheta}, x) \right| \le A \sup_{|x| \le A} \sum_{l \ne 0} \left| \vartheta_l\, e_l(x) \right|$$

$$\le \sqrt{A} \sum_{l \ne 0} |\vartheta_l| = \frac{2^{3/2} K A}{\sqrt{T}} \sum_{l=1}^{L} \left(\left| \frac{\Lambda}{l} \right|^k - 1 \right)_+^{1/2}$$

$$\le \frac{4 K A \Lambda^{(k/2)+1}}{\sqrt{T}} \le \frac{C A^{k+2}}{T^{(k-1)/(4k+2)}} \le \frac{C A^{k+2}}{T^{1/(4k+2)}} .$$

Note that this chain of inequalities proves the bound (4.69) for $m = 1$ as well. Suppose that this bound holds for $1, 2, \ldots, m$ and we prove it for $m + 1$. Since all the derivatives of the function $U(\cdot)$ up to the order m are bounded, we have (for $|x| \le A$)

$$\left| S^{(m)}(\boldsymbol{\vartheta}, x) \right| \le \frac{1}{\sqrt{A}} \sum_{|l| \le L} |\vartheta_l| \left| \frac{\pi l}{A} \right|^m + \frac{C A^{k+2}}{T^{1/(4k+2)}}$$

$$\le \frac{C}{A^m} \sum_{l=1}^{L} l^m \sqrt{\sigma_l} + \frac{C A^{k+2}}{T^{1/(4k+2)}}$$

$$\le \frac{C}{A^{m-1}\sqrt{T}} \sum_{l=1}^{L} l^{k-1} \left(\frac{\Lambda}{l} \right)^{k/2} + \frac{C A^{k+2}}{T^{1/(4k+2)}}$$

$$\le \frac{C \Lambda^k}{A^{m-1}\sqrt{T}} + \frac{C A^{k+2}}{T^{1/(4k+2)}} \le \frac{C A^{k+2}}{T^{1/(4k+2)}} .$$

The estimate (4.69) allows us to write

$$f(\boldsymbol{\vartheta},x)=\frac{1+o_T(1)}{2A}\,\exp\left\{-2\left(|x|-A\right)^{k+2}\,\chi_{\{|x|>A\}}\right\}. \qquad (4.70)$$

Therefore the invariant density is close in uniform metric to the density

$$f_0\left(x\right)=\frac{1}{2A}\,\chi_{\{|x|\leq A\}}$$

and has exponentially decreasing tails outside the interval $[-A,A]$. Note as well that $o_T(1)$ in (4.70) is uniform in $\boldsymbol{\vartheta}\in\boldsymbol{\Theta}_T$.

We are going to use the van Trees inequality to diminish the risk. Hence we suppose that $\boldsymbol{\vartheta}$ is a random vector. Its components ϑ_l are independent random variables, which can be represented as

$$\vartheta_l=\sqrt{\sigma_l\left(\varepsilon\right)}\,\eta_l,\qquad l=\pm 1,\ldots,\pm L,\qquad \vartheta_0=0, \qquad (4.71)$$

where

$$\sigma_l(\varepsilon)=\frac{2A\left(1+\varepsilon\right)}{T}\left(\left|\frac{A(1-\varepsilon)}{l}\right|^k-1\right)_+,\quad l\neq 0 \qquad (4.72)$$

and η_l are i.i.d. random variables with continuously differentiable density function $p\left(v\right)=p_K\left(v\right),v\in[-K,K]$. We suppose that $p\left(\cdot\right)$ has the following properties:

$$|\eta_l|<K,\quad \mathbf{E}\,\eta_l=0,\quad \mathbf{E}\,\eta_l^2=1,\quad \mathrm{I}=\int\frac{[p'(x)]^2}{p(x)}\,\mathrm{d}x=1+\varepsilon, \qquad (4.73)$$

where $\varepsilon\rightarrow 0$ when $K\rightarrow\infty$. By continuity of $p\left(\cdot\right)$ we have $p\left(\pm K\right)=0$.

The equality (4.71) defines the prior distribution $\mathrm{Q}\left(\cdot\right)$ of $\boldsymbol{\vartheta}$ on $\boldsymbol{\Theta}_T$ with the density

$$q\left(\boldsymbol{\theta}\right)=\prod_{l=-L,l\neq 0}^{L}\sigma_l\left(\varepsilon\right)^{-1/2}\,p\left(\theta_l\,\sigma_l\left(\varepsilon\right)^{-1/2}\right),\qquad \boldsymbol{\theta}\in\boldsymbol{\Theta}_T,$$

where the vector $\boldsymbol{\theta}=\left(\theta_{-L},\ldots,\theta_{-1},\theta_1\ldots,\theta_L\right)$.

The Fisher information I_l of this prior distribution with respect to ϑ_l (for $l\neq 0$) is then equal to $(1+\varepsilon)/\sigma_l(\varepsilon)$.

We have an elementary inequality

$$\sup_{\boldsymbol{\vartheta}\in\boldsymbol{\Theta}_T}\mathcal{R}_T\left(\bar{\vartheta}_T,f'_{\boldsymbol{\vartheta}}\right)\geq\mathbb{E}\int_{\mathscr{R}}\left(\bar{\vartheta}_T(x)-f'(\boldsymbol{\vartheta},x)\right)^2\mathrm{d}x,$$

where $\mathbb{E}$ is the mathematical expectation with respect to the probability measure $\mathrm{d}\mathbf{P}_{\boldsymbol{\vartheta}}^{(T)}\times\mathrm{Q}(\mathrm{d}\boldsymbol{\vartheta})$. If the underlying expression does not depend on X^T and the expectation is taken w.r.t. the measure Q only we write E_Q.

Let us introduce the Bayesian risk

$$\mathcal{R}_T(\mathrm{Q})=\inf_{\bar{\vartheta}_T}\mathbb{E}\int_{\mathscr{R}}\left(\bar{\vartheta}_T(x)-f'(\boldsymbol{\vartheta},x)\right)^2\mathrm{d}x,$$

and we try now to find a maximal lower bound corresponding to the *least favorable* prior distribution. In particular, we show that the choice (4.71) and (4.72) provides an approximation of a distribution *a priori* with asymptotically maximal lower bound.

Let $\psi_{l,\vartheta}$ and $\psi_{l,T}$ be the Fourier coefficients on $[-A, A]$ of $f'(\vartheta, \cdot)$ and $\bar\vartheta_T(\cdot)$ respectively, i.e.,

$$\psi_{l,\vartheta} = \int_{-A}^{A} f'(\vartheta, x)\, e_l(x)\, \mathrm{d}x,$$

$$\psi_{l,T} = \int_{-A}^{A} \bar\vartheta_T(x)\, e_l(x)\, \mathrm{d}x.$$

Since $\{e_l(\cdot)\}$ is an orthonormal sequence and $f'(\vartheta, \cdot)$ belongs to the Hilbert space generated by this sequence, using the Parseval identity, we can write

$$\mathcal{R}_T(\mathrm{Q}) \geq \inf_{\bar\vartheta_T} \mathbb{E} \int_{-A}^{A} \left(\bar\vartheta_T(x) - f'(\vartheta, x)\right)^2 \mathrm{d}x \geq \inf_{\bar\vartheta_T} \sum_{|l|<L} \mathbb{E}\left(\psi_{l,T} - \psi_{l,\vartheta}\right)^2 .$$

By the two-dimensional van Trees inequality

$$\mathcal{R}_T(\mathrm{Q}) \geq (1 + o_T(1)) \sum_{0<l\leq L} \frac{\left(\mathrm{E}_Q\left[\partial\psi_{l,\vartheta}/\partial\vartheta_l + \partial\psi_{-l,\vartheta}/\partial\vartheta_{-l}\right]\right)^2}{\mathrm{E}_Q\left[\mathrm{I}_{l,T}(\vartheta) + \mathrm{I}_{-l,T}(\vartheta)\right] + \mathrm{I}_l + \mathrm{I}_{-l}}, \tag{4.74}$$

where

$$\mathrm{E}_Q\, \mathrm{I}_{l,T}(\vartheta) = \int_{\Theta_T} \mathrm{I}_{l,T}(\vartheta)\, Q\,(\mathrm{d}\vartheta)$$

and $\mathrm{I}_{l,T}(\vartheta)$ is the Fisher information corresponding to the value ϑ_l. By definition

$$\mathrm{I}_{l,T}(\vartheta) = \mathbf{E}_\vartheta \left(\frac{\partial f(\vartheta, X_0)}{\partial\vartheta_l} + \int_0^T \frac{\partial S(\vartheta, X_t)}{\partial\vartheta_l}\, \mathrm{d}W_t\right)^2$$

$$= \mathbf{E}_\vartheta \left(\frac{\partial f(\vartheta, X_0)}{\partial\vartheta_l}\right)^2 + T\, \mathbf{E}_\vartheta \left(\frac{\partial S(\vartheta, X_0)}{\partial\vartheta_l}\right)^2 .$$

For the first term it is easy to show that

$$\left|\frac{\partial f(\vartheta, x)}{\partial\vartheta_l}\right| \leq 2A$$

and for the second we have

$$\mathbf{E}_\vartheta \left(\frac{\partial S(\vartheta, X_0)}{\partial\vartheta_l}\right)^2 = \int_{-A}^{A} e_l(x)^2\, U(A - |x|)^2\, f(\vartheta, x)\, \mathrm{d}x \leq \frac{1}{2A}.$$

Therefore

$$\mathrm{I}_{l,T}(\boldsymbol{\vartheta}) = \frac{T}{2A}\,\left(1 + o_T\left(1\right)\right).$$

We have the equality

$$\frac{\partial f'\left(\boldsymbol{\vartheta},x\right)}{\partial \vartheta_l} = 2\, f(\boldsymbol{\vartheta},x)\,\frac{\partial S\left(\boldsymbol{\vartheta},x\right)}{\partial \vartheta_l} + 4f'(\boldsymbol{\vartheta},x)\,\mathbf{E}_{\boldsymbol{\vartheta}}\int_{\xi}^{x}\frac{\partial S\left(\boldsymbol{\vartheta},v\right)}{\partial \vartheta_l}\,\mathrm{d}v.$$

Using Lemma 4.25, we obtain the estimate

$$\sup_{|x|\leq A}\left|f'(\boldsymbol{\vartheta},x)\right|\,\mathbf{E}_{\boldsymbol{\vartheta}}\left|\int_{\xi}^{x}\frac{\partial S\left(\boldsymbol{\vartheta},v\right)}{\partial \vartheta_l}\,\mathrm{d}v\right| = o_T(A^{-2}), \tag{4.75}$$

which gives us

$$\frac{\partial f'\left(\boldsymbol{\vartheta},x\right)}{\partial \vartheta_l} = 2f(\boldsymbol{\vartheta},x)\,e_l(x)\,U(A-|x|) + o_T(A^{-2}). \tag{4.76}$$

So the partial derivative of $\psi_{\boldsymbol{\vartheta},l}$ with respect to ϑ_l can be evaluated in the following way:

$$\begin{aligned}\frac{\partial \psi_{l,\boldsymbol{\vartheta}}}{\partial \vartheta_l} &= \int_{-A}^{A} e_l(x)\,\frac{\partial f'\left(\boldsymbol{\vartheta},x\right)}{\partial \vartheta_l}\,\mathrm{d}x \\ &= 2\int_{-A}^{A} \mathrm{e}_l^2(x)\,U(A-|x|)f(\boldsymbol{\vartheta},x)\,\mathrm{d}x + o_T(A^{-1}) \\ &= 2\int_{-A}^{A} \mathrm{e}_l^2(x)f(\boldsymbol{\vartheta},x)\,\mathrm{d}x + o_T(A^{-1}).\end{aligned}$$

This equality and the elementary identity $\mathrm{e}_l^2(x) + \mathrm{e}_{-l}^2(x) = 1/A$ imply

$$\frac{\partial \psi_{l,\boldsymbol{\vartheta}}}{\partial \vartheta_l} + \frac{\partial \psi_{-l,\boldsymbol{\vartheta}}}{\partial \vartheta_{-l}} = \frac{2}{A}\,\left(1 + o_T(1)\right). \tag{4.77}$$

As we already have

$$\mathrm{I}_{l,T}(\boldsymbol{\vartheta}) + \mathrm{I}_{-l,T}(\boldsymbol{\vartheta}) = \frac{T}{A}\,\left(1 + o_T(1)\right),$$

the inequality (4.74) can be rewritten as

$$\mathcal{R}_T(\mathrm{Q}) \geq 4\,A^{-1}\left(1 + o_T(1)\right)\sum_{0<l\leq L}\frac{\sigma_l(\varepsilon)}{T\,\sigma_l(\varepsilon) + 2A\,(1+\varepsilon)}. \tag{4.78}$$

Further we can write

$$\sum_{0<l\leq L} \frac{\sigma_l(\varepsilon)}{T\,\sigma_l(\varepsilon)+2A\,(1+\varepsilon)}$$
$$= \frac{1}{T}\sum_{0<l\leq L}\left(\frac{l}{\Lambda\,(1-\varepsilon)}\right)^k\left(\left(\frac{\Lambda\,(1-\varepsilon)}{l}\right)^k-1\right)$$
$$= \frac{\Lambda}{T}(1-\varepsilon)\int_0^1\left(1-x^k\right)\,\mathrm{d}x\,\left(1+o_{_T}(1)\right)$$
$$= \frac{\Lambda\,(1-\varepsilon)\,k}{T\,(k+1)}\,\left(1+o_{_T}(1)\right).$$

Now replacing Λ by its expression (4.68) we obtain the following inequality

$$\mathcal{R}_T(Q)\geq\frac{4\,k\,\Lambda\,(1-\varepsilon)^2}{T\,A\,(k+1)}\,(1+o_T(1))$$
$$=T^{-2k/(2k+1)}\,4\,\Pi(k,R)(1+o_T(1))\,(1-\varepsilon)^2.$$

Suppose for instant that the values $\sigma_l\,(\varepsilon)$ describing the prior distribution are not yet defined. Let us write the principal part of the right-hand sum in (4.78) as

$$\Psi(\boldsymbol{y})=\frac{4}{A}\sum_{l>0}\frac{y_l}{T\,y_l+2A}.$$

Then if we maximize this functional (using Lagrange multipliers method) over the set

$$\mathcal{E}(k,R)=\left\{\boldsymbol{y}=(y_l)_{l>0}\;\Big|\;2A^{-2}\sum_{l>0}y_l\left(\frac{\pi l}{A}\right)^{2k}\leq 4\,R\right\},$$

then we obtain the equality

$$\sup_{\boldsymbol{y}\in\mathcal{E}(k,R)}\;\Psi(\boldsymbol{y})=\Psi(\boldsymbol{\sigma})=T^{-2k/(2k+1)}4\Pi(k,R)$$

with

$$\boldsymbol{\sigma}=(\sigma_l,\;l=\pm1,\ldots,\pm L)$$

defined in (4.67), (4.68). Note that the set $\mathcal{E}(k,R)$ corresponds to the condition $S\,(\boldsymbol{\vartheta},\cdot)\in\Sigma\,(k,R)$ (see below).

Now we verify that for this parametric family the trend coefficient is such that the conditions (4.62)–(4.64) are satisfied with some $D>0$.

Lemma 4.26. *If T is sufficiently large then the functions $\left\{S(\boldsymbol{\vartheta},\cdot),\;\boldsymbol{\vartheta}\in\boldsymbol{\Theta}_T\right\}$ satisfy the conditions (4.62)–(4.64).*

Proof. We show firstly that there exists a constant D such that

$$\sup_{\vartheta\in\Theta_T}\int_{-\infty}^{\infty}\left[f'(\vartheta,u)\right]^2\mathbf{E}_\vartheta\left[\frac{\chi_{\{\xi>u\}}-F(\vartheta,\xi)}{f(\vartheta,\xi)}\right]^2\mathrm{d}u\le D \tag{4.79}$$

for all $T>T_0$. Note that for $z>y>A$ the following inequality is true:

$$(z-A)^{k+2}-(y-A)^{k+2}\ge(z-y)(y-A)^{k+1}.$$

Hence it follows that

$$\frac{1-F(\vartheta,y)}{f(\vartheta,y)}\le\int_y^{\infty}\exp\left\{-(z-y)(y-A)^{k+1}\right\}\mathrm{d}z=\frac{1}{(y-A)^{k+1}}\,.$$

This leads us to the estimate

$$\begin{aligned}\int_A^{\infty} f'(\vartheta,u)^2\,\mathbf{E}_\vartheta&\left[\left(\frac{1-F(\vartheta,\xi)}{f(\vartheta,\xi)}\right)^2\chi_{\{\xi>u\}}\right]\mathrm{d}u\\&\le\int_A^{\infty} f'(\vartheta,u)^2\,\mathbf{E}_\vartheta\left[\frac{\chi_{\{\xi>u\}}}{(\xi-A)^{2k+2}}\right]\mathrm{d}u\\&\le\int_A^{\infty} f'(\vartheta,u)^2\,(u-A)^{-2k}\,\mathrm{d}u<1\end{aligned} \tag{4.80}$$

for T large enough. In the same way, the inequality

$$(y-A)^{k+2}\le(x-A)^{k+1}(y-A),\qquad \text{for}\ \ A\le y\le x,$$

implies

$$\begin{aligned}f(\vartheta,u)\int_A^u\frac{F(\vartheta,y)^2}{f(\vartheta,y)}\,\mathrm{d}y&\le\int_A^u\frac{f(\vartheta,u)}{f(\vartheta,y)}\,\mathrm{d}y\\&\le\int_A^u \mathrm{e}^{(u-A)^{k+1}(y-u)}\mathrm{d}y\le\frac{1}{(u-A)^{k+1}}\,.\end{aligned}$$

Hence

$$\int_A^{\infty} f'(\vartheta,u)^2\,\mathbf{E}_\vartheta\left[\left(\frac{F(\vartheta,\xi)}{f(\vartheta,\xi)}\right)^2\chi_{\{A<\xi<u\}}\right]\mathrm{d}u \tag{4.81}$$

$$\begin{aligned}&\le C\int_A^{\infty} f'(\vartheta,u)^2\,|f'(\vartheta,u)|^{-1}\mathrm{d}u\\&=C\int_A^{\infty}|f'(\vartheta,u)|\,\mathrm{d}u=Cf(\vartheta,A)<1\end{aligned} \tag{4.82}$$

if T is large enough. Then, we have

$$\int_{-A-1}^{A} \frac{F(\vartheta,y)^2}{f(\vartheta,y)}\,\mathrm{d}y \le \int_{-A-1}^{A} \frac{1}{f(\vartheta,y)}\,\mathrm{d}y = 4A^2(1+o_T(1))$$

and

$$\int_A^\infty f'(\vartheta,u)^2\,\mathbf{E}_\vartheta\left[\left(\frac{F(\vartheta,\xi)}{f(\vartheta,\xi)}\right)^2 \chi_{\{-A-1<\xi<A\}}\right]\mathrm{d}u \le 8A^2\int_A^\infty f'(\vartheta,u)^2\mathrm{d}u < C. \tag{4.83}$$

Proceeding as for (4.80), one can show that

$$\int_{-\infty}^{-A-1} \frac{F(\vartheta,y)^2}{f(\vartheta,y)}\,\mathrm{d}y \le \sup_{y\le -A-1}\frac{F(\vartheta,y)^2}{f(\vartheta,y)^2} \le \frac{1}{(A+1-A)^{2k+2}} = 1.$$

Thus,

$$\int_A^\infty f'(\vartheta,u)^2\,\mathbf{E}_\vartheta\left[\left(\frac{F(\vartheta,\xi)}{f(\vartheta,\xi)}\right)^2 \chi_{\{\xi<-A\}}\right]\mathrm{d}u \le 1, \tag{4.84}$$

for T large enough. Combining the inequalities (4.80)–(4.83), we obtain

$$\int_A^\infty f'(\vartheta,u)^2\,\mathbf{E}_\vartheta\left[\left(\frac{\chi_{\{\xi>u\}}-F(\vartheta,\xi)}{f(\vartheta,\xi)}\right)^2\right]\mathrm{d}u \le C. \tag{4.85}$$

In the same way it can be shown that

$$\int_{-\infty}^{-A} f'(\vartheta,u)^2\,\mathbf{E}_\vartheta\left[\left(\frac{\chi_{\{\xi>u\}}-F(\vartheta,\xi)}{f(\vartheta,\xi)}\right)^2\right]\mathrm{d}u \le C. \tag{4.86}$$

For $u\in[-A,A]$, using Lemma 4.25 we have

$$f'(\vartheta,u) = \frac{o_T(1)}{2A}$$

and

$$\begin{aligned}\mathbf{E}_\vartheta\left[\left(\frac{\chi_{\{\xi>u\}}-F(\vartheta,\xi)}{f(\vartheta,\xi)}\right)^2\right] &\le \mathbf{E}_\vartheta\left[\left(\frac{1-F(\vartheta,\xi)}{f(\vartheta,\xi)}\right)^2\chi_{\{\xi>A+1\}}\right]\\ &+\mathbf{E}_\vartheta\left[\left(\frac{1}{f(\vartheta,\xi)}\right)^2\chi_{\{-A-1\le\xi\le A+1\}}\right]+\mathbf{E}_\vartheta\left[\left(\frac{F(\vartheta,\xi)}{f(\vartheta,\xi)}\right)^2\chi_{\{\xi<-A-1\}}\right]\le C.\end{aligned}$$

Hence

$$\int_{-A}^{A} f'(\vartheta,u)^2 \, \mathbf{E}_\vartheta \left[\left(\frac{\chi_{\{\xi>u\}} - F(\vartheta,\xi)}{f(\vartheta,\xi)} \right)^2 \right] \mathrm{d}u \le C$$

and the inequality (4.64) is proved.

We pass now to the proof of the following estimate:

$$\int_A^\infty \left[f'(\vartheta,u)\right]^2 \mathbf{E}_\vartheta \left[\int_0^\xi \frac{\chi_{\{y>u\}} - F(\vartheta,y)}{f(\vartheta,y)} \, \mathrm{d}y \right]^2 \mathrm{d}u \le D\, A^2. \tag{4.87}$$

To prove it, one has to consider the cases $\xi < -A$, $\xi \in [-A, A]$, $\xi \in [A, u]$ and $\xi > u$ separately. In the first, third and fourth cases we obtain (4.87) proceeding as in (4.64). For the second one, we have

$$\begin{aligned}
&\int_A^\infty \left[f'(\vartheta,u)\right]^2 \mathbf{E}_\vartheta \left[\int_0^\xi \frac{\chi_{\{y>u\}} - F(\vartheta,y)}{f(\vartheta,y)} \, \mathrm{d}y \; \chi_{\{-A\le\xi\le A\}} \right]^2 \mathrm{d}u \\
&\qquad \le \int_A^\infty \left[f'(\vartheta,u)\right]^2 \left[\int_{-A}^{A} \frac{1}{f(\vartheta,y)} \, \mathrm{d}y \right]^2 \mathrm{d}u \\
&\qquad \le C A^4 \int_A^\infty \left[f'(\vartheta,u)\right]^2 \mathrm{d}u \le D\, A^2.
\end{aligned} \tag{4.88}$$

So we proved (4.87) which implies the estimate (4.63).

It remains to check the condition (4.62). We suppose that $A > 2$ and $B > 2$. If $A > B/2$, then

$$\int_B^\infty \left[f'(\vartheta,x)\right]^2 \mathrm{d}x \le \int_0^\infty \left[f'(\vartheta,x)\right]^2 \mathrm{d}x = \frac{C}{A^2} \le \frac{D}{B^2}\,.$$

For $A \le B/2$, we have

$$\begin{aligned}
\int_B^\infty \left[f'(\vartheta,x)\right]^2 \mathrm{d}x &\le C \int_B^\infty (x-A)^{2k+2} \, \mathrm{e}^{-2(x-A)^{k+2}} \mathrm{d}x \\
&\le C \int_B^\infty (x-A)^{2k+3} \, \mathrm{e}^{-2(x-A)^{k+2}} \mathrm{d}x \\
&= D \int_{(B-A)^{k+2}}^\infty y \, \mathrm{e}^{-2y} \mathrm{d}y \le D \int_{(B-A)}^\infty \mathrm{e}^{-2y} \mathrm{d}y \\
&\le D \, \mathrm{e}^{-2(B-A)} < D \, \mathrm{e}^{-B} < \frac{D}{B^2}.
\end{aligned} \tag{4.89}$$

This inequality completes the proof of Lemma 4.26.

Lemma 4.27. *The following relation holds:*

$$\varGamma_T = \left\{ \vartheta \in \varTheta_T \; : \frac{1}{A^2} \sum_{l\in\mathbb{Z}} \left(\frac{\pi l}{A} \right)^{2k} \vartheta_l^2 < 4\, R\,(1-\varepsilon) \right\} \subseteq \left\{ \vartheta \; : \; S(\vartheta,\cdot) \in \varSigma_D \right\}, \tag{4.90}$$

if T is large enough.

Proof. Since $\boldsymbol{\Gamma}_T$ is a subset of $\boldsymbol{\Theta}_T$, the conditions (4.62) – (4.64) by Lemma 4.26 are satisfied for any $\boldsymbol{\vartheta} \in \boldsymbol{\Gamma}_T$. So, only the condition

$$\sup_{\boldsymbol{\vartheta}\in\boldsymbol{\Gamma}_T} \int_{-\infty}^{\infty} \left[f_{\boldsymbol{\vartheta}}^{(k+1)}(x)\right]^2 \mathrm{d}x \le 4\,R$$

needs to be checked. Note that

$$f^{(k+1)}(\boldsymbol{\vartheta},x) = \left[2S^{(k)}(\boldsymbol{\vartheta},x) + P\big(S^{(k-1)}(\boldsymbol{\vartheta},x),\ldots,S(\boldsymbol{\vartheta},x)\big)\right] f(\boldsymbol{\vartheta},x),$$

where $P(z_1,\ldots,z_k)$ is a polynomial. By Lemma 4.25

$$S^{(m)}(\boldsymbol{\vartheta},x) = O\left(\frac{A^{k+2}}{T^{\frac{1}{4k+2}}}\right) \qquad m = 0,1,\ldots,k-1.$$

So, on the one hand,

$$\left[P\big(S^{(k-1)}(\boldsymbol{\vartheta},x),\ldots,S(\boldsymbol{\vartheta},x)\big)\right]^2 = o\left(\frac{1}{T^{\frac{1}{4k+1}}}\right), \quad \text{for } \ x \in [-A,A].$$

On the other hand, using the orthonormality of the trigonometric basis $\{e_l\}$ and the definition of $\boldsymbol{\Gamma}_T$, we obtain

$$\begin{aligned}
4\int_{-A}^{A} \left[S^{(k)}(\boldsymbol{\vartheta},x) f(\boldsymbol{\vartheta},x)\right]^2 \mathrm{d}x &= \frac{1+o_T(1)}{A^2}\int_{-A}^{A} \left[S^{(k)}(\boldsymbol{\vartheta},x)\right]^2 \mathrm{d}x \\
&= \frac{1+o_T(1)}{A^2}\sum_{l\in\mathbb{Z}} \left(\frac{\pi l}{A}\right)^{2k} \vartheta_l^2 < 4\,R\,(1-\varepsilon)\,\big(1+o_T(1)\big) \quad (4.91)
\end{aligned}$$

for any $\boldsymbol{\vartheta} \in \boldsymbol{\Gamma}_T$. Therefore,

$$\begin{aligned}
\int_{-A}^{A} \left[f^{(k+1)}(\boldsymbol{\vartheta},x)\right]^2 \mathrm{d}x &= 4\int_{-A}^{A} \left[S^{(k)}(\boldsymbol{\vartheta},x) f_0(x)\right]^2 \mathrm{d}x\, \big(1+o_T(1)\big) \\
&\le 4\,R(1-\varepsilon)\big(1+o_T(1)\big). \qquad (4.92)
\end{aligned}$$

It can be easily checked that

$$\int_{|x|>A} \left[f^{(k+1)}(\boldsymbol{\vartheta},x)\right]^2 \mathrm{d}x \le C A^{-2}. \qquad (4.93)$$

Combining (4.91) and (4.92) we obtain

$$\int_{-\infty}^{\infty} \left[f_{\boldsymbol{\vartheta}}^{(k+1)}(x)\right]^2 \mathrm{d}x \le 4\,R\,(1-\varepsilon) + o_T(1).$$

This completes the proof of Lemma 4.27.

Lemma 4.28. *The probability of the event* $S\,(\boldsymbol{\vartheta},\cdot) \notin \Sigma_D$ *is exponentially small and consequently*

$$\mathbf{Q}\Big(S\,(\boldsymbol{\vartheta},\cdot) \notin \Sigma_D\Big) = o(T^{-1}).$$

Proof. The relation (4.90) implies

$$\mathrm{Q}\Big(S(\vartheta,\cdot)\notin\Sigma_D\Big)\le\mathrm{Q}\Big(\vartheta\notin\boldsymbol{\Gamma}_T\Big).$$

Below $\Lambda_\varepsilon=\Lambda(1-\varepsilon)$

$$\begin{aligned}\frac{1}{A^2}\,\mathrm{E}_Q\Bigg[\sum_{|l|\le L}\left(\frac{\pi l}{A}\right)^{2k}\vartheta_l^2\Bigg]&=\frac{1}{A^2}\sum_{|l|\le L}\left(\frac{\pi l}{A}\right)^{2k}\sigma_l(\varepsilon)\\&=\frac{2(1+\varepsilon)}{A^2}\sum_{l=1}^{L}\left(\frac{\pi l}{A}\right)^{2k}\frac{2A}{T}\left(\left|\frac{\Lambda(1-\varepsilon)}{l}\right|^k-1\right)\\&=\frac{4(1+\varepsilon)\,\pi^{2k}\,\Lambda_\varepsilon^{2k+1}}{T\,A^{2k+1}}\sum_{l=1}^{L}\frac{1}{\Lambda_\varepsilon}\left(\frac{l}{\Lambda_\varepsilon}\right)^{2k}\left(\left|\frac{\Lambda_\varepsilon}{l}\right|^k-1\right)\\&=\frac{4(1+\varepsilon)\,\pi^{2k}\,\Lambda_\varepsilon^{2k+1}}{T\,A^{2k+1}}\int_0^1\left(x^k-x^{2k}\right)\,\mathrm{d}x\;(1+o_T(1))\\&=4\,R(1-\varepsilon)^{2k+1}\big(1+o_T(1)\big)\,(1+\varepsilon)\,.\end{aligned}$$

Hence, for T sufficiently large

$$\frac{1}{A^2}\,\mathrm{E}_Q\Bigg[\sum_{|l|\le L}\left(\frac{\pi l}{A}\right)^{2k}\vartheta_l^2\Bigg]\le 4\,R\,(1-\varepsilon)^2.$$

Hoeffding's inequality gives us the following upper bound:

$$\begin{aligned}\mathrm{Q}\Big(\vartheta\notin\boldsymbol{\Gamma}_T\Big)&=\mathrm{Q}\Big\{\frac{1}{A^2}\sum_{|l|\le L}\left(\frac{\pi l}{A}\right)^{2k}\vartheta_l^2\ge 4\,R(1-\varepsilon)\Big\}\\&\le\mathrm{Q}\Big\{\frac{1}{A^2}\sum_{|l|\le L}\left(\frac{\pi l}{A}\right)^{2k}(\vartheta_l^2-\mathrm{E}_Q\vartheta_l^2)\ge 4\,R\varepsilon(1-\varepsilon)\Big\}\\&\le\exp\Big\{-\frac{8\,R^2\,\varepsilon^2\,(1-\varepsilon)^2}{N}\Big\},\end{aligned}$$

where

$$\begin{aligned}N=\frac{K^4}{A^4}\sum_{|l|\le L}\left(\frac{\pi l}{A}\right)^{4k}\sigma_l\,(\varepsilon)^2&\le\frac{C}{T^2}\sum_{|l|<\Lambda}l^{4k}\left(\left|\frac{\Lambda}{l}\right|^k-1\right)^2\\&\le\frac{C}{T^2}\sum_{|l|<\Lambda}l^{2k}\Lambda^{2k}\le CT^{-2}\Lambda^{4k+1}=C\,T^{-1/(2k+1)}.\end{aligned}$$

So we have

$$\mathrm{Q}\Big(S(\vartheta,\cdot)\notin\Sigma_D\Big)\le\mathrm{Q}\Big(\vartheta\notin\boldsymbol{\Gamma}_T\Big)\le\exp\{-c\,T^{1/(2k+1)}\}.$$

The Lemma 4.28 is proved.

Now everything is ready to finish the proof of the Theorem 4.24.

Let us introduce the following class of estimators

$$\mathcal{W}_T = \left\{ \bar\vartheta_T(\cdot) \;:\; \mathcal{R}_T(\bar\vartheta_T, f'_{S_0}) = \mathbf{E_0} \int_{-\infty}^{\infty} \left(\bar\vartheta_T(x) - f'(\mathbf{0},x)\right)^2 \mathrm{d}x < 1 \right\}, \tag{4.94}$$

with $S(\boldsymbol{\vartheta},\cdot) = S_0(\cdot)$ corresponding to the value $\boldsymbol{\vartheta} = \mathbf{0}$. Note that without loss of generality we can consider only estimators $\bar\vartheta_T(\cdot) \in \mathcal{W}_T$. For the other estimators the result is evident.

We have the following obvious inequalities:

$$\begin{aligned}
\inf_{\bar\vartheta_T\in\mathcal{W}_T} \sup_{S\in\Sigma_D} \mathcal{R}_T(\bar\vartheta_T, f'_S) &\geq \inf_{\bar\vartheta_T\in\mathcal{W}_T} \sup_{S\in\Sigma_D\cap\boldsymbol{\Theta}_T} \mathcal{R}_T(\bar\vartheta_T, f'_S) \\
&\geq \inf_{\bar\vartheta_T\in\mathcal{W}_T} \int_{\boldsymbol{\Theta}_T} \mathcal{R}_T(\bar\vartheta_T, f'_{\boldsymbol{\vartheta}})\, \mathrm{Q}(\mathrm{d}\boldsymbol{\vartheta}) - \sup_{\bar\vartheta_T\in\mathcal{W}_T} \int_{\boldsymbol{\Theta}_T\setminus\Sigma_D} \mathcal{R}_T(\bar\vartheta_T, f'_{\boldsymbol{\vartheta}})\, \mathrm{Q}(\mathrm{d}\boldsymbol{\vartheta}) \\
&\geq \mathcal{R}_T(\mathrm{Q}) - \sup_{\bar\vartheta_T\in\mathcal{W}_T} \int_{\boldsymbol{\Theta}_T\setminus\Sigma_D} \mathcal{R}_T(\bar\vartheta_T, f'_{\boldsymbol{\vartheta}})\, \mathrm{Q}(\mathrm{d}\boldsymbol{\vartheta}).
\end{aligned} \tag{4.95}$$

We have already found a lower bound for the first term. For the second term we have

$$\begin{aligned}
&\sup_{\bar\vartheta_T\in\mathcal{W}_T} \int_{\boldsymbol{\Theta}_T\setminus\Sigma_D} \mathcal{R}_T(\bar\vartheta_T, f'_{\boldsymbol{\vartheta}})\, \mathrm{Q}(\mathrm{d}\boldsymbol{\vartheta}) \\
&\qquad\qquad \leq (8 \sup_{\boldsymbol{\vartheta}\in\boldsymbol{\Theta}_T} \| f'(\boldsymbol{\vartheta},\cdot) \|_2^2 + 2)\, \mathrm{Q}\Big(S(\boldsymbol{\vartheta},\cdot) \notin \Sigma_D\Big).
\end{aligned} \tag{4.96}$$

It follows from Lemma 4.25 that the $\mathscr{L}^2$ norm of $f'(\boldsymbol{\vartheta},\cdot)$ is bounded uniformly on $\boldsymbol{\vartheta} \in \boldsymbol{\Theta}_T$. Hence,

$$\inf_{\bar\vartheta_T\in\mathcal{W}_T} \sup_{S\in\Sigma_D} \mathcal{R}_T(\bar\vartheta_T, f'_S) \geq \mathcal{R}_T(\mathrm{Q}) - o(T^{-1}).$$

Therefore

$$\underline{\lim}_{T\to\infty} \inf_{\bar\vartheta_T} \sup_{S\in\Sigma_D} T^{\frac{2k}{2k+1}} \mathcal{R}_T(\bar\vartheta_T, f'_S) \geq 4\, \Pi(k,R)(1-\varepsilon).$$

This proves the inequality (4.65), since ε can be taken as small as we want.

Now we have a global lower bound for the minimax risk, which allows us to introduce the following definition.

Definition 4.29. *We call an estimator $\vartheta^*_T(\cdot)$ asymptotically efficient if*

$$\lim_{D\to\infty} \lim_{T\to\infty} \sup_{S(\cdot)\in\Sigma_D} T^{\frac{2k}{2k+1}}\, \mathbf{E}_S \int_{-\infty}^{\infty} \left(\vartheta^*_T(x) - f'_S(x)\right)^2 \mathrm{d}x = 4\, \Pi(k,R). \tag{4.97}$$

To construct an asymptotically efficient estimator we consider a family of kernel-type estimators

$$\bar{\vartheta}_{K,T}(x) = \frac{2}{T}\int_0^T K_T(x - X_t)\,\chi_{\{|X_t|<B_T\}}\,\mathrm{d}X_t, \tag{4.98}$$

where $K_T(\cdot)$ is a kernel-type function and $B_T \to \infty$ and then show that the estimator $\vartheta_T^*(\cdot) = \bar{\vartheta}_{K^*,T}(\cdot)$ with $B_T = \sqrt{T}$ and the kernel

$$K_T^*(x) = \nu_T\, K^*(\nu_T\, x) \tag{4.99}$$

where

$$K^*(x) = \frac{1}{\pi}\int_0^1 (1 - u^k)\cos(ux)\,\mathrm{d}u$$

and

$$\nu_T = \left(\frac{\pi\, R\,(k+1)\,(2k+1)}{k}\right)^{\frac{1}{2k+1}} T^{\frac{1}{2k+1}} \to \infty \tag{4.100}$$

is asymptotically efficient. This means that (4.98) gives the optimal kernel in the problem of the estimation of the first derivative of the invariant density.

In particular, if $k = 2$, then the optimal kernel has the following form:

$$K^*(x) = \frac{2\,(\sin x - x\cos x)}{\pi\, x^3}$$

and the estimator is

$$\vartheta_T^*(x)$$
$$= 4\int_0^T \frac{\sin\Big((x - X_t)\nu_T\Big) - (x - X_t)\nu_T\cos\Big((x - X_t)\nu_T\Big)}{T\nu_T^2\,\pi\,(x - X_t)^3}\,\chi_{\{|X_t|<\sqrt{T}\}}\mathrm{d}X_t,$$

where $\nu_T = (10.5\,\pi)^{1/5}\, R^{1/5}\, T^{1/5}$. At the point $x = 0$ the function $K^*(\cdot)$ by continuity is equal to 2/3.

The form (4.98) of the estimator $\bar{\vartheta}_T(\cdot)$ for the derivative is quite a natural one because, say, its expectation can be written as follows:

$$\begin{aligned}
\mathbf{E}_S\,\bar{\vartheta}_T(x) &= 2\,\mathbf{E}_S\, K_T(x - \xi)\; S(\xi)\,\chi_{\{|\xi|<B_T\}} \\
&= 2\int_{-B_T}^{B_T} K_T(x - y)\; S(y)\; f_S(y)\,\mathrm{d}y \\
&= \nu_T\int_{-\sqrt{T}}^{\sqrt{T}} K\left(\nu_T(x - y)\right)\; f_S'(y)\,\mathrm{d}y \\
&= \int_{(x-\sqrt{T})\nu_T}^{(x+\sqrt{T})\nu_T} K(u)\; f_S'(x - u/\nu_T)\,\mathrm{d}u \longrightarrow f_S'(x)
\end{aligned}$$

for the "usual kernels" $K_T(u) = \nu_T\, K(\nu_T\, u)$, $\nu_T \to \infty$ satisfying (4.32).

Theorem 4.30. (Dalalyan and Kutoyants [54]) *Let the condition $\boldsymbol{S}_D$ be fulfilled. Then the estimator $\vartheta_{K^*,T}(\cdot)$ is asymptotically efficient, i.e., for this choice of the kernel (4.99) and $B_T = \sqrt{T}$ we have equality in (4.97).*

Proof. First we show that

$$\overline{\lim_{T\to\infty}} \inf_{\bar\vartheta_T} \sup_{S\in\Sigma_D} T^{\frac{2k}{2k+1}}\, \mathbf{E}_S \int_{-\infty}^{\infty} \left(\bar\vartheta_T(x) - f_S'(x)\right)^2 \mathrm{d}x$$
$$\le \overline{\lim_{T\to\infty}} \sup_{S\in\Sigma_D} T^{\frac{2k}{2k+1}}\, \mathbf{E}_S \int_{-\infty}^{\infty} \left(\vartheta_T^*(x) - f_S'(x)\right)^2 \mathrm{d}x \le 4\, \Pi(k,R).$$

Then, this upper bound together with (4.65) yields the equality

$$\lim_{T\to\infty} \inf_{\bar\vartheta_T} \sup_{S\in\Sigma_D} T^{\frac{2k}{2k+1}}\, \mathbf{E}_S \int_{-\infty}^{\infty} \left(\bar\vartheta_T(x) - f_S'(x)\right)^2 \mathrm{d}x = 4\, \Pi(k,R).$$

In the proof of the upper bound we find the form $K^*(\cdot)$ of the optimal kernel and the normalization rate N_T.

It is evident that

$$\inf_{\bar\vartheta_T} \sup_{S(\cdot)\in\Sigma_D} \mathbf{E}_S \int_{-\infty}^{\infty} \left(\bar\vartheta_T(x) - f_S'(x)\right)^2 \mathrm{d}x$$
$$\le \inf_{K} \sup_{S(\cdot)\in\Sigma_D} \mathbf{E}_S \int_{-\infty}^{\infty} \left(\bar\vartheta_{K,T}(x) - f_S'(x)\right)^2 \mathrm{d}x.$$

Hence, it is sufficient to study the risk

$$\mathcal{R}_T(K, f_S') = \mathbf{E}_S \int_{-\infty}^{\infty} \left(\bar\vartheta_{K,T}(x) - f_S'(x)\right)^2 \mathrm{d}x,$$

where $S(\cdot) \in \Sigma_D$ is the unknown trend coefficient.

To evaluate this risk we will use the Fourier transformations. Let us denote

$$\varphi_S(\lambda) = \int_{-\infty}^{\infty} \mathrm{e}^{i\lambda x}\, f_S'(x)\,\mathrm{d}x,$$
$$\varphi_{K,T}(\lambda) = \int_{-\infty}^{\infty} \mathrm{e}^{i\lambda x}\, \bar\vartheta_{K,T}(x)\,\mathrm{d}x,$$
$$\varphi_K(\lambda) = \int_{-\infty}^{\infty} \mathrm{e}^{i\lambda x}\, K_T(x)\,\mathrm{d}x,$$
$$\varphi_T(\lambda) = \frac{1}{T}\int_0^T \mathrm{e}^{i\lambda X_t}\, \chi_{\{|X_t|<B_T\}}\, \mathrm{d}X_t.$$

By Parseval's identity

$$\mathcal{R}_T(K, f_S') = \frac{1}{2\pi}\, \mathbf{E}_S \int_{-\infty}^{\infty} \left|\varphi_{K,T}(\lambda) - \varphi_S(\lambda)\right|^2 \mathrm{d}\lambda.$$

As the estimator $\vartheta_{T,K}$ is a convolution, its Fourier transform is a product of two Fourier transforms. Indeed

$$\begin{aligned}\varphi_{K,T}(\lambda) &= \frac{2}{T}\int_{-\infty}^{\infty} \mathrm{e}^{i\lambda x}\int_0^T K_T(x-X_t)\,\chi_{\{|X_t|<B_T\}}\,\mathrm{d}X_t\,\mathrm{d}x\\ &= \frac{2}{T}\int_0^T\int_{-\infty}^{\infty}\mathrm{e}^{i\lambda x}K_T(x-X_t)\,\mathrm{d}x\,\chi_{\{|X_t|<B_T\}}\,\mathrm{d}X_t\\ &= \frac{2}{T}\int_0^T \mathrm{e}^{i\lambda X_t}\varphi_K(\lambda)\,\chi_{\{|X_t|<B_T\}}\,\mathrm{d}X_t = 2\varphi_K(\lambda)\,\varphi_T(\lambda).\end{aligned} \quad (4.101)$$

So the quadratic risk can be rewritten as

$$\begin{aligned}\mathcal{R}_T(K,f_S') &= \frac{1}{2\pi}\,\mathbf{E}_S\int_{-\infty}^{\infty}|2\varphi_K(\lambda)\varphi_T(\lambda)-\varphi_S(\lambda)|^2\,\mathrm{d}\lambda\\ &= \frac{1}{2\pi}\int_{-\infty}^{\infty}\mathbf{E}_S\left|2\varphi_K(\lambda)\varphi_T(\lambda)-\varphi_S(\lambda)\right|^2\mathrm{d}\lambda\\ &= \frac{2}{\pi}\int_{-\infty}^{\infty}|\varphi_K(\lambda)|^2\,\mathbf{Var}_S\,\varphi_T(\lambda)\,\mathrm{d}\lambda\\ &\quad+\frac{1}{2\pi}\int_{-\infty}^{\infty}\left|2\varphi_K(\lambda)\,\mathbf{E}_S\,\varphi_T(\lambda)-\varphi_S(\lambda)\right|^2\mathrm{d}\lambda.\end{aligned}$$

For the mathematical expectation of $\varphi_T(\lambda)$, we have

$$\begin{aligned}\mathbf{E}_S\,\varphi_T(\lambda) &= \frac{1}{T}\,\mathbf{E}_S\int_0^T \mathrm{e}^{i\lambda X_t}S(X_t)\chi_{\{|X_t|<B_T\}}\,\mathrm{d}t = \mathbf{E}_S\left[\mathrm{e}^{i\lambda\xi}S(\xi)\chi_{\{|\xi|<B_T\}}\right]\\ &= \frac{1}{2}\,\varphi_S(\lambda)-\frac{1}{2}\int_{|x|>B_T}\mathrm{e}^{i\lambda x}\,f_S'(x)\,\mathrm{d}x.\end{aligned}$$

The following lemma describes the behavior of the bias and variance terms and tells us how to choose B_T.

Lemma 4.31. *Let the conditions $\mathcal{S}_D$ be fulfilled and the function $K_T(\cdot)$ be such that $|\varphi_K(\lambda)|\le 1$ for all $\lambda\in\mathscr{R}$. Then there exists a constant $C>0$ such that*

$$\int_{\mathscr{R}}|\varphi_K(\lambda)|^2\,\mathbf{Var}_S\left[\varphi_T(\lambda)\right]\,\mathrm{d}\lambda \le \left[\frac{1}{\sqrt{T}}\left(\int_{\mathscr{R}}|\varphi_K(\lambda)|^2\mathrm{d}\lambda\right)^{1/2}+\frac{C}{\sqrt{T}}+\frac{CB_T}{T}\right]^2,$$

$$\int_{\mathscr{R}}\left|2\varphi_K(\lambda)\,\mathbf{E}_S\,\varphi_T(\lambda)-\varphi_S(\lambda)\right|^2\mathrm{d}\lambda \le \left[\left(\int_{\mathscr{R}}|\varphi_K(\lambda)-1|^2|\varphi_S(\lambda)|^2\mathrm{d}\lambda\right)^{1/2}+\frac{C}{B_T}\right]^2$$

for any $S(\cdot) \in \Sigma_D$.

Proof. We prove here only the first inequality; the second can be proved in the same way. Note that the Itô formula gives us the following representation:

$$\varphi_T(\lambda) - \mathbf{E}_S\, \varphi_T(\lambda) = H_S(\lambda, X_T) - H_S(\lambda, X_0) + \int_0^T \left[e^{i\lambda X_t} - h_S(\lambda, X_t) \right] dW_t$$

with

$$h_S(\lambda, y) = 2 \int_{-B_T}^{B_T} e^{i\lambda x} f_S'(x)\, \frac{\chi_{\{x<y\}} - F_S(y)}{f_S(y)}\, dx,$$

and $H_S(\lambda, x) = \int_0^x h_S(\lambda, y)\, dy$. Consequently

$$\begin{aligned} \mathbf{Var}_S\left[\varphi_T(\lambda)\right] = T^{-2}\, \mathbf{E}_S \Big| H_S(\lambda, X_T) - H_S(\lambda, X_0) \\ + \int_0^T \left(e^{i\lambda X_t} \chi_{\{|X_t|<B_T\}} - h_S(\lambda, X_t) \right) dW_t \Big|^2. \end{aligned}$$

Using the triangular inequality we have

$$\int_{\mathcal{R}} |\varphi_K(\lambda)|^2\, \mathbf{Var}_S\left[\varphi_T(\lambda)\right]\, d\lambda \le \left(\sqrt{A_1} + 2\sqrt{A_2} + \sqrt{A_3}\right)^2, \tag{4.102}$$

with

$$\begin{aligned} A_1 &= \frac{1}{T^2} \int_{\mathcal{R}} |\varphi_K(\lambda)|^2\, \mathbf{E}_S \left| \int_0^T e^{i\lambda X_t} \chi_{\{|X_t|<B_T\}}\, dW_t \right|^2 d\lambda \le \frac{1}{T} \int_{-\infty}^{\infty} |\varphi_K(\lambda)|^2\, d\lambda, \\ A_2 &= \frac{1}{T^2} \int_{\mathcal{R}} |\varphi_K(\lambda)|^2\, \mathbf{E}_S \left| H_S(\lambda, \xi) \right|^2 d\lambda, \\ A_3 &= \frac{1}{T^2} \int_{\mathcal{R}} |\varphi_K(\lambda)|^2\, \mathbf{E}_S \left| \int_0^T h_S(\lambda, X_t)\, dW_t \right|^2 d\lambda. \end{aligned}$$

To bound the term A_2 we use the fact that $|\varphi_K(\lambda)|$ is less than 1. Thus according to Parseval's identity and condition (4.63) we have

$$\begin{aligned} A_2 &\le \frac{4}{T^2}\, \mathbf{E}_S \int_{-\infty}^{\infty} \left| \int_{-B_T}^{B_T} e^{i\lambda x} f_S'(x) \int_0^{\xi} \frac{\chi_{\{y>x\}} - F_S(y)}{f_S(y)}\, dy\, dx \right|^2 d\lambda \\ &= \frac{8\pi}{T^2} \int_{-B_T}^{B_T} \left[f_S'(x) \right]^2 \mathbf{E}_S \left[\int_0^{\xi} \frac{\chi_{\{y>x\}} - F_S(y)}{f_S(y)}\, dy \right]^2 dx \le \frac{8\pi D\, B_T^2}{T^2}. \end{aligned}$$

Repeating exactly the same arguments one can check that

$$A_3 \le \frac{8\pi D}{T^2}.$$

Thus Lemma 4.31 is proved.

We choose the kernel-type function $K_T(\cdot)$ in the following way:

$$\varphi_K(\lambda) = \varphi_\nu(\lambda) = \left(1 - \left|\frac{\lambda}{\nu}\right|^k\right)_+ , \tag{4.103}$$

where $\nu = \nu_T$ is a positive number. As we show below, the integrals of the right hand sides in Lemma 4.31 converge both to zero with the rate $T^{-k/(2k+1)}$. Therefore the choice $B_T = \sqrt{T}$ gives us the following upper estimate for quadratic risk:

$$\mathcal{R}_T(K, f_S') \le L_T(\varphi_\nu, \varphi_S)\big(1 + o_T(1)\big),$$

where $o_T(1)$ tends to zero uniformly on $S(\cdot) \in \Sigma_D$ and

$$L_T(\varphi_\nu, \varphi_S) = \frac{1}{2\pi T}\int_{-\infty}^{\infty} \left(4|\varphi_\nu(\lambda)|^2 + T|\varphi_\nu(\lambda) - 1|^2 |\varphi_S(\lambda)|^2\right) \mathrm{d}\lambda. \tag{4.104}$$

Since the function $f_S'(x)$ is in the ellipsoid defined by $\Sigma(k, R)$, its Fourier transform should belong to the following set

$$\Phi = \left\{\varphi(\cdot) \;:\; \frac{1}{2\pi}\int_{\mathcal{R}} \lambda^{2k}\, |\varphi(\lambda)|^2 \,\mathrm{d}\lambda \le 4R\right\}.$$

Replacing in L_T the function φ_ν by its explicit expression, we obtain

$$L_T(\varphi_\nu, \varphi_S) = \frac{2}{\pi T}\int_{-\nu}^{\nu} \left(1 - \left|\frac{\lambda}{\nu}\right|^k\right)^2 \mathrm{d}\lambda + \frac{1}{2\pi\nu^{2k}}\int_{\mathcal{R}} |\lambda|^{2k}|\varphi_S(\lambda)|^2 \,\mathrm{d}\lambda.$$

Since $\varphi_S(\cdot)$ belongs to Φ, the second term of the right hand side is less than $4R/\nu^{2k}$ and the first term can be calculated explicitly:

$$\int_{-\nu}^{\nu} \left(1 - \left|\frac{\lambda}{\nu}\right|^k\right)^2 \mathrm{d}\lambda = \frac{4\,\nu\,k^2}{(k+1)\,(2k+1)}.$$

It leads us to the following inequality:

$$\begin{aligned}&\inf_{\nu>0}\ \sup_{S(\cdot)\in\Sigma_D} L_T(\varphi_\nu, \varphi_S)\\ &\quad\le \inf_{\nu>0}\left\{\frac{8\,k^2\,\nu}{\pi T\,(k+1)\,(2k+1)} + 4R\,\nu^{-2k}\right\} = \inf_{\nu>0} U(\nu).\end{aligned} \tag{4.105}$$

The function $U(\nu)$ is continuously differentiable and strictly convex. Consequently it attains the minimum at the point ν_* satisfying the following equation:

$$\frac{8\,k^2}{\pi\,T\,(k+1)\,(2k+1)} = \frac{8\,k\,R}{\nu_*^{2k+1}},$$

which leads to

$$\nu_* = \left(\frac{R\pi\, T\,(k+1)\,(2k+1)}{k}\right)^{\frac{1}{2k+1}}$$

and

$$\inf_{\nu>0} U(\nu) = U(\nu_*) = \frac{4(2k+1)\,R}{\nu_*^{2k}} = 4\,\Pi(k,R)\,T^{-\frac{2k}{2k+1}}\,. \tag{4.106}$$

Theorem 4.30 is proved.

Note that the inverse Fourier transformation of the optimal $\varphi_{\nu_*}(\cdot)$ gives

$$K_T^*(x) = \frac{1}{2\pi}\int_{\mathscr{R}} \mathrm{e}^{-i\lambda x}\,\varphi_{\nu_*}(\lambda)\,\mathrm{d}\lambda = \frac{1}{2\pi}\int_{-\nu_*}^{\nu_*} \mathrm{e}^{-i\lambda x}\left(1-\left|\frac{\lambda}{\nu_*}\right|^k\right)\mathrm{d}\lambda$$
$$= \frac{\nu_*}{\pi}\int_0^1 \left(1-u^k\right)\,\cos\left(\nu_*\,ux\right)\,\mathrm{d}u = \nu_T\,K^*\left(\nu_T\,x\right)$$

which corresponds to (4.99).

Remark 4.32. One can use similar arguments to find Pinsker's constant in the problem of the estimation of higher derivatives. Say, for the rth derivative $f_S^{(r)}(\cdot)$ with $r\geq 1$ the optimal rate of convergence is $T^{-\frac{k}{2k+2r-1}}$ and Pinsker's constant is

$$\Pi(r,k,R)$$
$$= \left(\frac{2k+2r-1}{2r-1}\right)\left(\frac{\pi\,(k+2r-1)\,(2k+2r-1)}{k}\right)^{-\frac{2k}{2k+2r-1}} R^{\frac{2r-1}{2k+2r-1}}.$$

The proof is close to that given above and the differences are in the choice of $\Lambda, K^*(\cdot)$ and ν_T:

$$\Lambda = A\left(\frac{R\,(k+2r-1)\,(2k+2r-1)}{k\,\pi^{2k+2r-2}}\right)^{\frac{1}{2k+2r-1}} T^{\frac{1}{2k+2r-1}},$$
$$K^*(x) = \frac{1}{\pi}\int_0^1 u^{r-1}\,\left(1-u^k\right)\,\phi_r(ux)\,\mathrm{d}u,$$
$$\nu_T = \left(\frac{R\,\pi\,(k+2r-1)\,(2k+2r-1)}{r}\right)^{\frac{1}{2k+2r-1}} T^{\frac{1}{2k+2r-1}},$$

where $\phi_r(x)$ is the $(r-1)$th derivative of the function $\cos(x)$. The asymptotically efficient estimator is

$$\vartheta_T^* = \frac{2}{T}\,\nu_T^r\int_0^T K^*\Big(\nu_T\,(x-X_t)\Big)\,\chi_{\{|X_t|\leq\sqrt{T}\}}\,\mathrm{d}X_t.$$

Of course $\Pi(1,k,R) = \Pi(k,R)$.

Remark 4.33. If the observed process is

$$\mathrm{d}X_t = S(X_t)\,\mathrm{d}t + \sigma(X_t)\,\mathrm{d}W_t, \quad X_0, \quad 0 \le t \le T,$$

then one way is to transform it using (4.58) to the process (4.59) and then to use Theorems 4.24 and 4.30, where in the conditions Q the function $S(\cdot)$ is replaced by $\tilde{S}(\cdot)$ (see (4.60)). Another way is to modify the proofs of these theorems in the following way. Suppose that the function $\sigma(\cdot)$ is $(k+1)$-times differentiable, separated from zero, i.e.,

$$\inf_{x\in\mathscr{R}} |\sigma(x)| > 0,$$

and $S(\cdot)$ is an unknown k times differentiable function satisfying corresponding (modified) conditions. In particular in the conditions $\boldsymbol{\mathcal{S}}_D$ the quantity

$$\frac{\chi_{\{y>x\}} - F_S(y)}{f_S(y)}$$

should be replaced by

$$\frac{\chi_{\{y>x\}} - F_S(y)}{\sigma(y)^2 f_S(y)}$$

and the parametric family is defined by the trend coefficient

$$S(\vartheta, x) = S_0(x)\,\sigma(x)^2 + \sum_{|l|\le L} \vartheta_l\, e_l(x)\, U(A - |x|)\,\sigma(x)^2,$$

where $S_0(\cdot)$, L, ϑ_l and $e_l(\cdot)$ are the same as in the given proof. The asymptotically efficient estimator is then

$$\vartheta_T^* = \frac{2\,\nu_T}{T}\int_0^T K^*\left(\nu_T\,(x - X_t)\right)\frac{\chi_{\{|X_t|\le\sqrt{T}\}}}{\sigma(X_t)^2}\,\mathrm{d}X_t.$$

4.4.2 Local Risk

The lower bound (4.65) and the corresponding criteria of optimality (4.98) have a *global nature* in the sense that the sup is calculated over the set $\Sigma(k, R)$ as is usually done in this kind of problems (see [192] and the references therein). Remember that the function $S(\cdot) \in \mathscr{S}_\sigma$ and so it satisfies the conditions $\mathcal{RP}$. Therefore each diffusion process (4.57) with $S(\cdot) \in \Sigma(k, R)$ is ergodic, but when we calculate the $\sup_{S\in\Sigma(k,R)}$ we "touch" the null recurrent process. In particular, the parametric family (4.66) has the invariant density (4.70), which is close to the uniform on $[-A, A]$ density $f_0(\cdot)$ as $A \to \infty$. The interval $[-A, A] \to \mathscr{R}$ sufficiently slowly, so the Fisher information $\mathrm{I}_{l,T} = T/2A\,(1 + o_T(1))$ and we can construct the lower bound (4.65), but when we study the asymptotically efficient estimator we need to control the asymptotic behavior of the *empirical characteristic function* $\varphi_T(\cdot)$. We do it with the help of the conditions (4.62)–(4.64). Note that this choice of parametric family does not allow us to use the *uniform on* $S \in \Sigma(k,R)$ condition of ergodicity like

$$\overline{\lim_{|x|\to\infty}} \sup_{S\in\Sigma(k,R)} \operatorname{sgn}(x)\; S(x) < 0,$$

which immediately gives us the uniform convergence in (4.62) and the uniform boundedness of the integrals in (4.63) and (4.64). The parametric family (4.66) does not satisfy it. At the same time the condition $\boldsymbol{\mathcal{S}}_D$ is quite cumbersome and we would like to have Pinsker's type bound and asymptotically efficient estimator without them. That is why we consider below a somewhat different construction of the *local* lower bound inspired by the work by Golubev [89]. Moreover, in the next section we consider the problem of trend coefficient estimation in the same *local* statement.

The statement of the problem is the following. The observed process as before is

$$\mathrm{d}X_t = S(X_t)\,\mathrm{d}t + \mathrm{d}W_t, \quad X_0,\ 0 \le t \le T,$$

where the function $S(\cdot) \in \mathscr{S}_\sigma$ is unknown to the observer.

Let us fix a *central function* function $S_*(\cdot) \in \mathscr{S}_\sigma$ and the nonparametric vicinity

$$V_\delta = \left\{ S(\cdot)\ :\ \sup_x |S(x) - S_*(x)| \le \delta \right\}.$$

Introduce the

Conditions $\boldsymbol{\mathcal{S}}_\delta$.

$\mathcal{S}_3$. *The function* $S(\cdot) \in \mathscr{P}$ *is such that*

$$\overline{\lim_{|x|\to\infty}} \operatorname{sgn}(x)\; S(x) < 0. \tag{4.107}$$

$\mathcal{S}_4$. *The function* $S(\cdot)$ *is* $k \ge 1$ *times continuously differentiable and belongs to the set*

$$\Sigma_\delta = \left\{ S(\cdot) \in V_\delta\ :\ \int_{\mathscr{R}} \left[f_S^{(k+1)}(x) - f_{S_*}^{(k+1)}(x) \right]^2 \mathrm{d}x \le 4\,R \right\}.$$

$\mathcal{S}_5$. *The Fourier transform* $\varphi_*(\cdot)$ *of the function* $f'_{S_*}(\cdot)$ *satisfies the conditions*

$$\int_{\mathscr{R}} |\lambda|^{2k+\tau} \left|\varphi_*(\lambda)\right|^2 \mathrm{d}\lambda < \infty \tag{4.108}$$

with some positive constant τ.

The condition (4.108) means that the function $S_*(\cdot)$ is a little bit smoother than the other functions $S(\cdot) \in \Sigma_\delta$. For example, if the function $S_*(\cdot)$ is

$(k+1)$-times continuously differentiable and $S_*(\cdot) \in \mathscr{P}$, then this condition is satisfied with $\tau = 2$. To simplify the notation we will write $f_*(\cdot)$ instead of $f_{S_*}(\cdot)$.

The first result is the local minimax bound on the $\mathscr{L}_2$ risk

$$\mathcal{R}_T\left(\bar\vartheta_T, f_S'\right) = \mathbf{E}_S \int_{\mathscr{R}} \left(\bar\vartheta_T(x) - f_S'(x)\right)^2 \mathrm{d}x.$$

Local means that the this risk is estimated over the shrinking set Σ_δ.

Theorem 4.34. (Dalalyan and Kutoyants [55]) *Let the conditions $\boldsymbol{S}_\delta$ be fulfilled. Then*

$$\lim_{\delta\to 0}\,\lim_{T\to\infty}\,\inf_{\bar\vartheta_T}\,\sup_{S\in\Sigma_\delta} T^{\frac{2k}{2k+1}}\,\mathcal{R}_T\left(\bar\vartheta_T, f_S'\right) \geq 4\,\Pi(k,R).$$

Proof. The proof of this theorem is quite close to the proof of Theorem 4.24, i.e., we diminish the risk of the nonparametric problem by the risk of a specially constructed parametric family. The difference is just in the choice of this parametric family.

Let us denote

$$\overline{\lim_{|x|\to\infty}}\ \mathrm{sgn}\,(x)\ S(x) = -\delta_0.$$

Note that as $\delta \to 0$ we can consider $\delta < \delta_0$. So the condition (4.107) is fulfilled for all $S(\cdot) \in V_\delta$. Therefore the functions $f_S(\cdot), S(\cdot) \in V_\delta$, are well defined.

We fix an increasing interval $[-A, A]$ with $A = A_T = \ln(T+1)$ and a sequence of its sub-intervals $\mathbb{I}_m = \left[a_m - AT^{-\beta}, a_m + AT^{-\beta}\right]$ where $\beta = (2k+1)^{-1}$ and

$$a_m = 2m\,A\,T^{-\beta}, \qquad m = 0, \pm 1, \pm 2, \ldots, \pm M.$$

Here $M = M_T$ is the greatest integer such that $\mathbb{I}_M \subseteq [-A, A]$. Let us introduce now the parameterization

$$S(\boldsymbol{\vartheta}, x) = S_*(x) + \sum_{|m|<M} \sqrt{\frac{2A}{T^\beta f_*(a_m)}}\ \sum_{|l|<L} \vartheta_{l,m}\ \phi_{l,m}(x), \tag{4.109}$$

where

$$f_*(x) = \frac{1}{G(S_*)} \exp\left\{2\int_0^x S_*(v)\ \mathrm{d}v\right\}$$

and

$$\phi_{l,m}(x) = \sqrt{\frac{T^\beta}{A}}\ e_l\big(T^\beta A^{-1}(x - a_m)\big)\,U(A - |x - a_m|T^\beta). \tag{4.110}$$

Here $e_l(\cdot)$ is the trigonometric basis in $\mathscr{L}_2[-1,1]$, i.e.,

$$e_l(x) = \begin{cases} \sin(\pi l x) \ , & \text{if } l > 0, \\ 1/\sqrt{2} \quad , & \text{if } l = 0, \\ \cos(\pi l x) \ , & \text{if } l < 0, \end{cases}$$

the function $U(x)$ is $(k+1)$-times differentiable, increasing, vanishing for $x \leq 0$ and equal to one for $x \geq 1$. The parameter $\boldsymbol{\vartheta} = (\vartheta_{l,m})_{|l| \leq L, |m| \leq M}$ and the integer $L = L_T$ will be chosen later.

It is easy to show that the functions $S(\boldsymbol{\vartheta}, \cdot)$ are k-times continuously differentiable and coincide with $S_*(\cdot)$ outside of the interval $[-A, A]$.

The parametric space $\boldsymbol{\Theta}_T$ that we consider is the set of all finite sequences $\{\vartheta_{l,m}\}_{|l| \leq L, |m| \leq M}$ such that $|\vartheta_{l,m}| \leq K\sqrt{\sigma_l(\varepsilon)}$ for all l and m. Here

$$\sigma_l(\varepsilon) = \sigma_{l,T}(\varepsilon) = \frac{1}{2\,A\,T^{2k\beta}} \left(\left| \frac{\Lambda(1-\varepsilon)}{l} \right|^k - 1 \right)_+ , \qquad l \neq 0, \tag{4.111}$$

with

$$\Lambda = \Lambda_T = A \left(\frac{R\,(k+1)\,(2k+1)}{k\,\pi^{2k}} \right)^{1/(2k+1)} \tag{4.112}$$

and $\sigma_0(\varepsilon) = 0$. The integer L is chosen to be equal to $[\Lambda]$. Note that the dimension of the space $\boldsymbol{\Theta}_T$ increases like $T^{\beta} \Lambda_T$, as $T \to \infty$.

Let $\{\eta_{l,m}\}_{l,m \in \mathbb{Z}}$ be i.i.d. random variables with common probability density $p(\cdot)$ satisfying the conditions (4.73), where $\varepsilon \to 0$, as $K \to \infty$.

We introduce distribution *a prior* $Q(\cdot)$ on $\boldsymbol{\Theta}_T$ putting

$$\vartheta_{l,m} = \sqrt{\sigma_l(\varepsilon)}\,\eta_{l,m}$$

for all integers l different from 0. The coefficients $\vartheta_{0,m}$ will be deterministic and equal to 0. The Fisher information $\mathrm{I}_{l,m}$ of this prior distribution with respect to $\vartheta_{l,m}$ (for $l \neq 0$) is then equal to $(1+\varepsilon)/\sigma_l(\varepsilon)$.

Since the minimax risk is bounded below by the Bayesian one, we are looking for a lower bound of the Bayesian risk defined by

$$\mathcal{R}_T(Q) = \inf_{\bar{\vartheta}_T} \mathbb{E} \int_{-\infty}^{\infty} \left(\bar{\vartheta}_T(x) - f'(\boldsymbol{\vartheta}, x) \right)^2 \mathrm{d}x,$$

where $\mathbb{E}$ is the mathematical expectation with respect to the probability measure $\mathrm{d}\,\mathbf{P}_{\boldsymbol{\vartheta}}^{(T)} \times Q(\mathrm{d}\boldsymbol{\vartheta})$. Here $f(\boldsymbol{\vartheta}, x) = f_{S(\boldsymbol{\vartheta},\cdot)}(x)$.

Further, let us define the class $\mathcal{W}_T$ (see (4.94)) of estimators. Then according to (4.95) we have the estimate

$$\inf_{\bar{\vartheta}_T \in \mathcal{W}_T} \sup_{S \in \Sigma_\delta} \mathcal{R}_T(\bar{\vartheta}_T, f_S') \geq \mathcal{R}_T(\mathrm{Q}) - \sup_{\bar{\vartheta}_T \in \mathcal{W}_T} \int_{\boldsymbol{\Theta}_T \setminus \Sigma_\delta} \mathcal{R}_T(\bar{\vartheta}_T, f_{\boldsymbol{\vartheta}}')\,\mathrm{Q}(\mathrm{d}\boldsymbol{\vartheta}). \tag{4.113}$$

We show below that the prior distribution Q is the *least favorable*, i.e., it maximizes asymptotically the Bayesian risk $\mathcal{R}_T(\mathrm{Q})$. To evaluate $\mathcal{R}_T(\mathrm{Q})$ we denote by $\psi_{l,m,\vartheta}$ and $\psi_{l,m,T}$ the Fourier coefficients on $[-A, A]$ of $f'(\vartheta, \cdot)$ and $\bar{\vartheta}_T$ with respect to the orthonormal sequence

$$e_{l,m}(x) = \sqrt{T^{\beta}/A}\, e_l\big(T^{\beta} A^{-1}(x - a_m)\big)\chi_{\{x \in \mathbb{I}_m\}}.$$

This means that

$$\psi_{l,m,\vartheta} = \int_{-A}^{A} f'(\vartheta, x)\, e_{l,m}(x)\, \mathrm{d}x,$$
$$\psi_{l,m,T} = \int_{-A}^{A} \bar{\vartheta}_T(x)\, e_{l,m}(x)\, \mathrm{d}x.$$

Since the $\mathscr{L}_2[-A, A]$ norm of the function $\bar{\vartheta}_T(\cdot) - f'(\vartheta, \cdot)$ is greater than the norm of its projection on the subspace $\mathscr{L}_2[-A, A]$ generated by the orthonormal sequence $\{e_{l,m}\}_{1 \le |l| \le L,\, |m| \le M}$, we have

$$\mathcal{R}_T(Q) \ge \inf_{\bar{\vartheta}_T} \mathbb{E} \int_{-A}^{A} \left(\bar{\vartheta}_T(x) - f'(\vartheta, x)\right)^2 \mathrm{d}x$$
$$\ge \inf_{\bar{\vartheta}_T} \sum_{|m| \le M} \sum_{1 \le |l| \le L} \mathbb{E}\left(\psi_{l,m,T} - \psi_{l,m,\vartheta}\right)^2.$$

According to the van Trees inequality, we have

$$\mathcal{R}_T(Q) \ge \sum_{|m| \le M} \sum_{1 \le |l| \le L} \frac{\left(\mathrm{E}_Q\left[\partial \psi_{l,m,\vartheta} / \partial \vartheta_{l,m}\right]\right)^2}{\mathrm{E}_Q\, \mathrm{I}_{l,m,T}(\vartheta) + \mathrm{I}_{l,m}}, \tag{4.114}$$

where the Fisher information $\mathrm{I}_{l,m,T}(\vartheta)$ is defined by the formula

$$\mathrm{I}_{l,m,T}(\vartheta) = \mathbf{E}_{\vartheta}\left[\frac{\partial \ln f(\vartheta, X_0)}{\partial \vartheta_{l,m}}\right]^2 + T\, \mathbf{E}_{\vartheta}\left[\frac{\partial S(\vartheta, \xi)}{\partial \vartheta_{l,m}}\right]^2.$$

Since $|a_m| \le A$ and $S_0(\cdot)$ has a polynomial majorant, using the Taylor formula one can check that for any $x \in \mathbb{I}_m$

$$f_*(x) = f_*(a_m)\big(1 + o_T(1)\big), \qquad x \in \mathbb{I}_m.$$

Therefore as in the global case we have the estimate

$$\mathrm{I}_{l,m,T}(\vartheta) = T\,\left(1 + o_T(1)\right)\, \mathbf{E}_{\vartheta}\left[\frac{\partial S(\vartheta, \xi)}{\partial \vartheta_{l,m}}\right]^2 \le \frac{2AT}{T^{\beta}}\,(1 + \varepsilon)$$

for sufficiently large T.

Using the fact that only CT^{β} elements of the set $\boldsymbol{\Theta}_T$ are nonzero and each one is less than $CT^{-2k\beta}$, one can easily show that

$$f(\vartheta, x) = f_*(x)\big(1 + o_T(1)\big), \tag{4.115}$$

where $o_T(1)$ is uniform on $x \in \mathscr{R}$ and $\vartheta \in \Theta_T$. Hence

$$\begin{aligned}
\frac{\partial \psi_{l,m,\boldsymbol{\vartheta}}}{\partial \vartheta_{l,m}} &= 2\int_{I_m} \frac{\partial S(\boldsymbol{\vartheta},x)}{\partial \vartheta_{l,m}}\, f(\boldsymbol{\vartheta},x)\, e_{l,m}(x)\,\mathrm{d}x\,\left(1+o_T(1)\right) \\
&= 2\int_{\mathbb{I}_m} \sqrt{\frac{2A}{T^{\beta} f_*(a_m)}}\, f_*(x)\, e_{l,m}(x)^2\,\mathrm{d}x\,\left(1+o_T(1)\right) \\
&= 2\int_{\mathbb{I}_m} \sqrt{\frac{2A\, f_*(a_m)}{T^{\beta}}}\, e_{l,m}(x)^2\,\mathrm{d}x\,\left(1+o_T(1)\right). \qquad (4.116)
\end{aligned}$$

Hence

$$\frac{\partial \psi_{l,m,\boldsymbol{\vartheta}}}{\partial \vartheta_{l,m}} = 2\sqrt{2Af_*(a_m)\,T^{-\beta}}\,\left(1+o_T(1)\right),$$

where $o_T(1)$ tends to zero uniformly on l,m and $\boldsymbol{\vartheta} \in \Theta_T$. Now, the inequality (4.114) can be rewritten as follows:

$$\mathcal{R}_T(Q) \geq \left(1+o_T(1)\right) \sum_{|m|\leq M}\sum_{|l|\leq L} \frac{8\,A\,T^{-\beta}\, f_*(a_m)}{2A\,T^{1-\beta}\,(1+\varepsilon)+\mathrm{I}_{l,m}}\,. \qquad (4.117)$$

Using the convergence of Riemann's sums, it is easy to show that

$$\sum_{|m|\leq M} 2AT^{-\beta} f_*(a_m) = \int_{-A}^{A} f_*(x)\,\mathrm{d}x\,\left(1+o_T(1)\right) = 1+o_T(1). \qquad (4.118)$$

Hence, using once more the convergence of Riemann's sums, we get

$$\begin{aligned}
\mathcal{R}_T(Q) &\geq \frac{\left(1+o_T(1)\right)}{(1+\varepsilon)} \sum_{|l|\leq L} \frac{4\,\sigma_l\,(\varepsilon)}{2A\,T^{k\beta}\,\sigma_l\,(\varepsilon)+1} \\
&= \frac{4\,(1+o_T\,(1))}{AT^{2k\beta}\,(1+\varepsilon)} \sum_{l=1}^{L}\left(1-\left|\frac{l}{\Lambda\,(1-\varepsilon)}\right|^k\right) \\
&= \frac{4\Lambda\,(1-\varepsilon)\,(1+o_T\,(1))}{AT^{2k\beta}\,(1+\varepsilon)} \int_0^1 \left(1-x^k\right)\,\mathrm{d}x \\
&\geq 4\,(1-\varepsilon)^2\; T^{-\frac{2k}{2k+1}}\; \Pi\,(k,R)\;\left(1+o_T\,(1)\right).
\end{aligned}$$

Proceeding exactly as in Lemma 4.27, it can be shown that

$$\begin{aligned}
&\int_{-\infty}^{\infty} \left[\left(2S(\boldsymbol{\vartheta},x)f(\boldsymbol{\vartheta},x)-2S_*(x)f_*(x)\right)^{(k)}\right]^2\mathrm{d}x \\
&\qquad = 4\int_{-\infty}^{\infty} \left[S^{(k)}(\boldsymbol{\vartheta},x)-S_*^{(k)}(x)\right]^2 f_0^2(x)\,\mathrm{d}x\,\left(1+o_T(1)\right) \\
&\qquad = \frac{4A}{T^{\beta}} \sum_{|m|\leq M}\sum_{|l|\leq L} f_*(a_m)\,\vartheta_{l,m}^2\left(\frac{\pi l\,T^{\beta}}{A}\right)^{2k}\,\left(1+o_T(1)\right).
\end{aligned}$$

Let us denote this last sum by $B_T(\boldsymbol{\vartheta})$ and introduce a subset of $\mathscr{R}^{\mathbb{Z}\times\mathbb{Z}}$ defined by

$$\boldsymbol{\Gamma}_T = \left\{\boldsymbol{\vartheta} = \{\vartheta_{l,m}\} \;\Big|\; B_T(\boldsymbol{\vartheta}) \le 4\,R\,(1-\varepsilon)\right\}.$$

For T sufficiently large, the set $\{S(\boldsymbol{\vartheta},\cdot)\,|\,\boldsymbol{\vartheta}\in\boldsymbol{\Gamma}_T\}$ is included in the set Σ_δ. Moreover, using (4.118), we can write

$$\begin{aligned}\mathbb{E}\big[B_T(\boldsymbol{\vartheta})\big] &= 4AT^{(2k-1)\beta}\sum_{|m|\le M} f_*(a_m)\sum_{|l|\le L}\sigma_l(\varepsilon)\left(\frac{\pi l}{A}\right)^{2k}(1+o_T(1))\\ &= \sum_{|l|\le L} 2T^{2k\beta}\,\sigma_l(\varepsilon)\left(\frac{\pi l}{A}\right)^{2k}(1+o_T(1)) \le 4\,R(1-\varepsilon)^{2k}(1+o_T(1)).\end{aligned}$$

We can now apply Hoeffding's inequality to show that the Q-measure of the complement set $\boldsymbol{\Gamma}_T^c$ decreases exponentially (see Lemma 4.28)

$$Q(S(\boldsymbol{\vartheta},\cdot)\notin\Sigma_\delta) \le Q(\boldsymbol{\Gamma}_T^c) = o(T^{-1}).$$

Now the second term of (4.113) can be evaluated as follows:

$$\begin{aligned}&\sup_{\bar\vartheta_T\in\mathcal{W}_T}\int_{\boldsymbol{\Theta}_T\setminus\Sigma_\delta}\mathcal{R}_T(\bar\vartheta_T, f'_{\boldsymbol{\vartheta}})\,\mathrm{Q}(\mathrm{d}\boldsymbol{\vartheta})\\ &\qquad\le\left(8\sup_{\boldsymbol{\vartheta}\in\boldsymbol{\Theta}_T}\left\|f'(\boldsymbol{\vartheta},\cdot)\right\|^2+2\right)Q(S(\boldsymbol{\vartheta},\cdot)\notin\Sigma_\delta)\end{aligned}$$

Remember that the $\mathscr{L}_2$ norm of $f'(\boldsymbol{\vartheta},\cdot)$ is bounded uniformly on $\boldsymbol{\vartheta}\in\boldsymbol{\Theta}_T$. Therefore

$$\underline{\lim}_{\delta\to 0}\,\underline{\lim}_{T\to\infty}\,\inf_{\bar\vartheta_T}\,\sup_{S(\cdot)\in\Sigma_\delta} T^{\frac{2k}{2k+1}}\,\mathcal{R}\left(\bar\vartheta_T, f'_S\right)\ge 4\,\Pi(k,R)\,(1-\varepsilon).$$

This completes the proof of the theorem, since ε can be taken as small as we want.

Now we can define an asymptotically efficient estimator of the derivative of invariant density.

Definition 4.35. *Let the conditions $\boldsymbol{S}_\delta$ be fulfilled. An estimator $\vartheta_T^*(\cdot)$ is called asymptotically efficient if*

$$\lim_{\delta\to 0}\lim_{T\to\infty}\sup_{S(\cdot)\in\Sigma_\delta} T^{\frac{2k}{2k+1}}\,\mathcal{R}\left(\vartheta_T^*, f'_S\right) = 4\,\Pi(k,R) \tag{4.119}$$

for any function $S_*(\cdot)$ satisfying (4.108).

Let us introduce the kernel

$$K^*(x) = \frac{1}{\pi}\int_0^1 \left(1 - u^{k+\rho}\right)\,\cos(ux)\,\mathrm{d}u, \tag{4.120}$$

where $\rho = \rho_T = 1/\ln\ln(T+1)$ and the corresponding kernel type estimator

$$\vartheta_T^*(x) = \frac{2\nu_T}{T}\int_0^T K^*\Big(\nu_T(x - X_t)\Big)\,\mathrm{d}X_t, \tag{4.121}$$

where ν_T is

$$\nu_T = \left(\frac{\pi\,R\,(k+1)\,(2k+1)}{4k}\right)^{\frac{1}{2k+1}}\,T^{\frac{1}{2k+1}}. \tag{4.122}$$

Theorem 4.36. (Dalalyan and Kutoyants [55]) *Let the condition $\boldsymbol{S}_\delta$ be fulfilled, then the kernel-type estimator $\theta_T^*(\cdot)$ is asymptotically efficient.*

Proof. As in the proof of Theorem 4.30 we consider the class of kernel type estimators

$$\theta_{K,T}(x) = \frac{2}{T}\int_0^T K_T(x - X_t)\,\mathrm{d}X_t,$$

where $K_T(\cdot)$ is an arbitrary square integrable function and find an upper bound for the linear risk

$$\mathcal{R}_T(K, f_S') = \mathbf{E}_S\int_{\mathscr{R}} \left(\theta_{K,T}(x) - f_S'(x)\right)^2 \mathrm{d}x,$$

where $S(\cdot) \in \Sigma_\delta$ is the unknown trend coefficient. To evaluate this risk we will use the following Fourier transforms. Let us denote

$$\varphi_S(\lambda) = \int_{-\infty}^{\infty} \mathrm{e}^{i\lambda x} f_S'(x)\,\mathrm{d}x, \qquad \varphi_{K,T}(\lambda) = \int_{-\infty}^{\infty} \mathrm{e}^{i\lambda x}\theta_{K,T}(x)\,\mathrm{d}x,$$

$$\varphi_K(\lambda) = \int_{-\infty}^{\infty} \mathrm{e}^{i\lambda x} K_T(x)\,\mathrm{d}x, \qquad \varphi_T(\lambda) = \frac{1}{T}\int_0^T \mathrm{e}^{i\lambda X_t}\mathrm{d}X_t.$$

By Parseval's identity

$$\mathcal{R}_T(K, f_S') = \frac{1}{2\pi}\,\mathbf{E}_S\int_{\mathscr{R}} \left|\varphi_{K,T}(\lambda) - \varphi_S(\lambda)\right|^2 \mathrm{d}\lambda.$$

Remember that $\varphi_{K,T}(\lambda) = 2\varphi_K(\lambda)\,\varphi_T(\lambda)$ (see (4.101)) and $2\,\mathbf{E}_S\,\varphi_T(\lambda) = \varphi_S(\lambda)$. Therefore the quadratic risk can be rewritten as

$$\begin{aligned}
\mathcal{R}_T(K, f_S') &= \frac{1}{2\pi}\,\mathbf{E}_S\int_{\mathscr{R}} \left|2\varphi_K(\lambda)\varphi_T(\lambda) - \varphi_S(\lambda)\right|^2 \mathrm{d}\lambda \\
&= \frac{1}{2\pi}\int_{\mathscr{R}}\Big(4\left|\varphi_K(\lambda)\right|^2 \mathbf{Var}_S\,\varphi_T(\lambda) + |\varphi_K(\lambda) - 1|^2|\varphi_S(\lambda)|^2\Big)\,\mathrm{d}\lambda.
\end{aligned}$$

We choose the kernel-type function $K_T(\cdot)$ in the following way:

$$\varphi_K(\lambda) = \varphi_\nu(\lambda) = \left(1 - \left|\frac{\lambda}{\nu}\right|^{k+\rho}\right)_+,$$

where $\nu = \nu_T$ is a positive number tending to infinity and $\rho = \rho_T > 0$ tends to zero such that

$$\lim_{T\to\infty} \rho_T \log \nu_T = \infty.$$

The following lemma describes the asymptotic behavior of the variance of the empirical characteristic function.

Lemma 4.37. *For any $S(\cdot) \in \Sigma_\delta$, we have*

$$\int_{\mathcal{R}} |\varphi_\nu(\lambda)|^2 \, \mathbf{Var}_S\, \varphi_T(\lambda)\, \mathrm{d}\lambda \le \frac{1}{T} \int_{\mathcal{R}} |\varphi_\nu(\lambda)|^2 \,\mathrm{d}\lambda \,\left(1 + o_T(1)\right).$$

Proof. By the definition of $\varphi_T(\cdot)$ we have

$$\varphi_T(\lambda) = \frac{1}{T}\int_0^T \mathrm{e}^{i\lambda X_t}\, S(X_t)\,\mathrm{d}t + \frac{1}{T}\int_0^T \mathrm{e}^{i\lambda X_t}\,\mathrm{d}W_t.$$

The variance of the last term is equal to T^{-1}. Thus, if we show that the quantity

$$R_T = \int_{\mathcal{R}} |\varphi_\nu(\lambda)|^2 \, \mathbf{Var}_S \left[\frac{1}{T}\int_0^T \mathrm{e}^{i\lambda X_t}\, S(X_t)\,\mathrm{d}t\right] \mathrm{d}\lambda$$

converges to zero faster than

$$\frac{1}{T}\int_{\mathcal{R}} |\varphi_\nu(\lambda)|^2 \,\mathrm{d}\lambda = \frac{C\nu}{T},$$

then the lemma will be proved. We show below that

$$R_T \le \frac{C}{T}.$$

Indeed, the appropriate use of the Itô formula gives

$$\int_0^T \mathrm{e}^{i\lambda X_t}\, S(X_t)\,\mathrm{d}t - T\,\mathbf{E}_S\left[\mathrm{e}^{i\lambda\xi} S(\xi)\right] = \hat{H}_S(\lambda, X_T) - \hat{H}_S(\lambda, X_0) - \int_0^T \hat{g}_S(\lambda, X_t)\,\mathrm{d}W_t,$$

where $\hat{H}_S(\cdot, X_t)$ and $\hat{g}_S(\cdot, X_t)$ are respectively the Fourier transforms of the functions $S(\cdot) H_S(\cdot, X_t)$ and $S(\cdot) g_S(\cdot, X_t)$ defined by

$$H_S(x,u) = 2f_S(x) \int_0^u \frac{\chi_{\{y>x\}} - F_S(y)}{f_S(y)}\, dy,$$

$$g_S(x,u) = 2f_S(x) \frac{\chi_{\{u>x\}} - F_S(u)}{f_S(u)}.$$

Using once more Parseval's identity one can show that

$$\begin{aligned} R_T &\leq \int_{\mathscr{R}} \mathbf{Var}_S\left[\frac{1}{T}\int_0^T e^{i\lambda X_t}\, S(X_t)\, dt\right] d\lambda \\ &\leq \frac{8}{T^2}\int_{\mathscr{R}} \mathbf{E}_S \left|\hat{H}_S(\lambda,\xi)\right|^2 d\lambda + \frac{2}{T}\int_{\mathscr{R}} \mathbf{E}_S \left|\hat{g}_S(\lambda,\xi)\right|^2 d\lambda \\ &= \frac{8}{T^2}\int_{\mathscr{R}} S^2(x)\, \mathbf{E}_S\left[H_S(x,\xi)^2\right] dx + \frac{2}{T}\int_{\mathscr{R}} S(x)^2\, \mathbf{E}_S\left[g_S(x,\xi)^2\right] dx, \end{aligned}$$

where the first inequality is due to the fact that $|\varphi_\nu(\cdot)| \leq 1$. These two integrals are bounded uniformly on $S(\cdot) \in \Sigma_\delta$ by Proposition 1.22 and the fact that the trend coefficient $S(\cdot)$ has a polynomial majorant. The proof of Lemma 4.37 is finished.

We return to the proof of Theorem 4.36. By virtue of the last lemma, we have the following upper estimate for the quadratic risk:

$$\mathcal{R}_T(K, f_S') \leq L_T(\varphi_\nu, \varphi_S)\, \left(1 + o_T(1)\right),$$

where $o_T(1)$ tends to zero uniformly on $S(\cdot) \in \Sigma_\delta$ and

$$L_T(\varphi_\nu, \varphi_S) = \frac{1}{2\pi T}\int_{\mathscr{R}} \left(4|\varphi_\nu(\lambda)|^2 + T|\varphi_\nu(\lambda) - 1|^2 |\varphi_S(\lambda)|^2\right) d\lambda.$$

Since the function $S(\cdot)$ is in the set Σ_δ, the Fourier transform $\varphi_S(\cdot)$ should belong to the set

$$\Phi = \left\{\varphi(\cdot) \;\middle|\; \frac{1}{2\pi}\int_{\mathscr{R}} \lambda^{2k} |\varphi(\lambda) - \varphi_*(\lambda)|^2\, d\lambda \leq 4\, R\right\},$$

where $\varphi_*(\cdot) = \varphi_{S_*}(\cdot)$. Replacing in L_T the function φ_ν by its explicit expression, we obtain

$$\begin{aligned} L_T(\varphi_\nu, \varphi_S) = \frac{2}{\pi T}\int_{-\nu}^{\nu} \left(1 - \left|\frac{\lambda}{\nu}\right|^{k+\rho}\right)^2 d\lambda + \frac{1}{2\pi}\int_{-\nu}^{\nu} \left|\frac{\lambda}{\nu}\right|^{2k+2\rho} |\varphi_S(\lambda)|^2\, d\lambda \\ + \frac{1}{2\pi}\int_{|\lambda|>\nu} \left|\frac{\lambda}{\nu}\right|^{2k+2\rho} |\varphi_S(\lambda)|^2\, d\lambda. \qquad (4.123) \end{aligned}$$

Since $\varphi_S(\cdot)$ belongs to Φ we have

$$\int_{-\nu}^{\nu} \left|\frac{\lambda}{\nu}\right|^{2k+2\rho} |\varphi_S(\lambda) - \varphi_*(\lambda)|^2\, d\lambda + \int_{|\lambda|>\nu} |\varphi_S(\lambda) - \varphi_*(\lambda)|^2\, d\lambda \leq \frac{8\pi R}{\nu^{2k}}$$

and taking into account (4.108) for T sufficiently large, we obtain

$$\int_{-\nu}^{\nu}\left|\frac{\lambda}{\nu}\right|^{2k+2\rho}|\varphi_*(\lambda)|^2\,\mathrm{d}\lambda+\int_{|\lambda|>\nu}|\varphi_*(\lambda)|^2\,\mathrm{d}\lambda$$
$$\leq\int_{\mathscr{R}}\left|\frac{\lambda}{\nu}\right|^{2k+2\rho}|\varphi_*(\lambda)|^2\,\mathrm{d}\lambda\leq\int_{\mathscr{R}}\frac{|\lambda|^{2k}+|\lambda|^{2k+\tau}}{\nu^{2k+2\rho}}|\varphi_*(\lambda)|^2\,\mathrm{d}\lambda\leq\frac{C}{\nu^{2k+2\rho}}.$$

Thus the sum of the second and the third terms of (4.123) is bounded by

$$\frac{4R}{\nu^{2k}}+\frac{C}{\nu^{2k+2\rho}}=\frac{4R}{\nu^{2k}}\,(1+o_T(1))$$

and the first term can be calculated explicitly:

$$\frac{2}{\pi T}\int_{-\nu}^{\nu}\left(1-\left|\frac{\lambda}{\nu}\right|^{k+\rho}\right)^2\mathrm{d}\lambda=\frac{8\nu\,k^2\,(1+o_T(1))}{\pi T\,(k+1)(2k+1)}\,.$$

So we can repeat the calculations (4.105)–(4.106) and obtain the desired relation

$$\inf_{\nu>0}\sup_{S\in\Sigma_\delta}L_T(\varphi_\nu,\varphi_S)\leq 4\,\Pi(k,R)\,T^{-\frac{2k}{2k+1}}\,.$$

4.5 Trend Coefficient Estimation

We consider now the problem of estimation of the function $S(\cdot)$ itself by the observations of ergodic diffusion process

$$\mathrm{d}X_t=S(X_t)\,\mathrm{d}t+\sigma(X_t)\,,\mathrm{d}W_t,\quad X_0,\quad 0\leq t\leq T, \tag{4.124}$$

and we are interested by the asymptotically efficient estimation. As before, we suppose that the function $S(\cdot)\in\mathscr{S}_\sigma$, so this equation has a unique weak solution, the measures are equivalent and the process has ergodic properties with the invariant density $f_S(\cdot)$. The initial value X_0 is supposed to be a random variable independent of the Wiener process $\{W_t, 0\leq t\leq T\}$ and having the density function $f_S(\cdot)$.

We suppose as well that the function $S(\cdot)$ is $k\geq 1$ times continuously differentiable and k is known. This problem is similar to the problem of density estimation in the i.i.d. case and to the derivative estimation considered in the preceding section. Remember that in this kind of problem the rate of convergence of the estimators depends on the smoothness of the unknown function. We first show that for a wide class of loss functions including polynomials the optimal rate is $T^{\frac{k}{2k+1}}$ and then we verify that the wide class of kernel-type estimators is *asymptotically efficient in order*. To have an estimator asymptotically efficient *up to the constant* we restrict ourself to the quadratic loss functions and consider the $\mathscr{L}_2$-type integral risk function. Then we propose a lower minimax bound for this risk and finally construct a kernel-type estimator, which is asymptotically efficient *up to the constant* (Pinsker's approach).

4.5.1 Optimal Rate

Introduce the *regularity condition* $\boldsymbol{\mathcal{T}}$.

$\mathcal{T}_1$. *The functions* $\sigma(\cdot)^2$ and $S(\cdot)$ *are* $k \geq 2$*-times differentiable and the* k*th derivative of the function* $S(x)\, f_S(x)$ *is bounded by some constant* $R > 0$.

$\mathcal{T}_2$. *The functions* $S(\cdot)\,,\ \sigma(\cdot)\,,\ \sigma(\cdot)^{-1} \in \mathscr{P}$ and

$$\overline{\lim_{|x|\to\infty}}\ \operatorname{sgn}(x)\ \frac{S(x)}{\sigma(x)^2} < 0. \tag{4.125}$$

Let us denote by $\Sigma^*(k,R)$ the set of functions satisfying the condition $\boldsymbol{\mathcal{T}}$.

Theorem 4.38. *Let the condition* $\boldsymbol{\mathcal{T}}$ *be fulfilled. Then for any* x_* *and any loss function* $\ell(\cdot) \in \mathscr{W}_p$*, we have*

$$\varliminf_{T\to\infty}\ \inf_{\bar S_T}\ \sup_{S\in\Sigma^*(k,R)}\ \mathbf{E}_S\, \ell\Big(T^{\frac{k}{2k+1}}\left(\bar S_T(x_*) - S(x_*)\right)\Big) \geq \Pi, \tag{4.126}$$

where inf *is taken over all estimators* $\bar S_T(x_*)$ *of the function* $S(\cdot)$ *at the point* x_* *and* Π *is some positive constant.*

Proof. We follow the proof of Theorem IV.5.1, [109] of the similar result in the i.i.d. case (see as well [70], [139], Theorem 4.3 and [78]).

To prove (4.126) we construct a parametric family $\{S(\vartheta,\cdot)\,,\ \vartheta \in \Theta\}$ belonging to the set $\Sigma^*(k,R)$ and will diminish the global risk of estimation $S(x_*)$ by the risk of parameter estimation with special parameterization. This parametric family of measures will be *AN* (*asymptotically normal*) and we use a version of the Hajek–Le Cam lower bound.

This parametric family is constructed as follows. Let $S_*(\cdot) \in \Sigma^*(k,R/2)$ and $f_*(\cdot) = f_{S_*}(\cdot)$ be the corresponding invariant density. Introduce a function $\psi(\cdot) \in \mathscr{C}^\infty$ with compact support $[-1,1]$ and $\psi(0) \neq 0$.

Further, put

$$S(\vartheta,x) = S_*(x) + \vartheta\, T^{-\frac{k}{2k+1}}\, \psi\left(\gamma\,(x-x_*)\,T^{\frac{1}{2k+1}}\right), \qquad \vartheta \in \Theta = (-\delta,\delta)\,,$$

where

$$\gamma = \frac{f_*(x_*)}{\sigma(x_*)^2} \int_{-1}^{1} \psi(y)^2\ \mathrm{d}y$$

and $\delta = \gamma^{-k}$.

We have for the kth derivative w.r.t. x

$$S^{(k)}(\vartheta,x) = S_*^{(k)}(x) + \vartheta\gamma^k\, \psi^{(k)}\left(\gamma\,(x-x_*)\,T^{\frac{1}{2k+1}}\right).$$

Suppose as well that the function $\psi(\cdot)$ satisfies the condition $\left|\psi^{(k)}(y)\right| < R/2$, then $S(\vartheta,\cdot) \in \Sigma^*(k,R)$ for all $\vartheta \in \Theta$.

Let us consider the problem of parameter estimation by the observations

$$\mathrm{d}X_t = S(\vartheta, X_t)\ \mathrm{d}t + \sigma(X_t)\ \mathrm{d}W_t, \quad X_0,\ 0 \le t \le T, \tag{4.127}$$

where $S_*(\cdot)$ and $\psi(\cdot)$ are supposed to be known functions. This is the so-called *scheme of series*, i.e., for each $T > 0$ we have an ergodic diffusion process (4.127) and as $T \to \infty$ we have a sequence of problems indexed by T $(S(\vartheta,x) = S_T(\vartheta,x))$. Therefore we can write

$$\begin{aligned}
&\sup_{S\in\Sigma^*(k,R)} \mathbf{E}_S\, \ell\Big(T^{\frac{k}{2k+1}}\left(\bar S_T(x_*) - S(x_*)\right)\Big)\\
&\qquad \ge \sup_{\vartheta\in\Theta} \mathbf{E}_\vartheta\, \ell\Big(T^{\frac{k}{2k+1}}\left(\bar S_T(x_*) - S(\vartheta, x_*)\right)\Big)\\
&\qquad = \sup_{\vartheta\in\Theta} \mathbf{E}_\vartheta\, \ell\Big(T^{\frac{k}{2k+1}}\left(\bar S_T(x_*) - S_*(x_*) - \vartheta\, T^{\frac{k}{2k+1}}\psi(0)\right)\Big)\\
&\qquad = \sup_{|\vartheta|<\delta} \mathbf{E}_\vartheta\, \ell_*\left(\bar\vartheta_T - \vartheta\right),
\end{aligned}$$

where we have denoted $\bar\vartheta_T = T^{\frac{k}{2k+1}}\left(\bar S_T(x_*) - S_*(x_*)\right)$ and put $\ell_*(u) = \ell(u\,\psi(0))$.

The likelihood ratio $Z_T(\vartheta) = L\left(\vartheta, 0, X^T\right)$ has the form

$$\begin{aligned}
Z_T(\vartheta) = \frac{f(\vartheta, X_0)}{f_*(X_0)} \exp\Bigg\{&\vartheta\, T^{\frac{k}{2k+1}} \int_0^T \frac{\psi\left(\gamma(X_t - x_*)T^{\frac{1}{2k+1}}\right)}{\sigma(X_t)}\,\mathrm{d}W_t\\
&-\frac{\vartheta^2}{2}T^{-\frac{2k}{2k+1}} \int_0^T \left(\frac{\psi\left(\gamma(X_t - x_*)T^{\frac{1}{2k+1}}\right)}{\sigma(X_t)}\right)^2 \mathrm{d}t\Bigg\}, \quad \vartheta\in\Theta.
\end{aligned}$$

For the second integral we have

$$\begin{aligned}
&\mathbf{E}_* \int_0^T \frac{\psi\left(\gamma(X_t - x_*)T^{\frac{1}{2k+1}}\right)^2}{\sigma(X_t)^2}\,\mathrm{d}t\\
&\qquad = T\int_{\mathscr{R}} \frac{\psi\left(\gamma(x - x_*)T^{\frac{1}{2k+1}}\right)^2}{\sigma(x)^2}\, f_*(x)\,\mathrm{d}x\\
&\qquad = \gamma^{-1}T^{\frac{2k}{2k+1}} \int_{-1}^1 \frac{\psi(y)^2 f_*\left(x_* + y\gamma^{-1}T^{-\frac{1}{2k+1}}\right)}{\sigma\left(x_* + y\gamma^{-1}T^{-\frac{k}{2k+1}}\right)^2}\,\mathrm{d}y\\
&\qquad = T^{\frac{2k}{2k+1}}\left(1 + o_T(1)\right)
\end{aligned}$$

As the function $S_*(\cdot)$ satisfies the condition (4.125) we have the convergence

$$T^{-\frac{2k}{2k+1}}\int_0^T\left(\frac{\psi\left(\gamma\left(X_t-x_*\right)T^{\frac{1}{2k+1}}\right)}{\sigma\left(X_t\right)}\right)^2\,\mathrm{d}t\longrightarrow 1.$$

Therefore the first integral (by CLT) is asymptotically normal

$$\mathcal{L}_*\left\{T^{-\frac{k}{2k+1}}\int_0^T\frac{\psi\left(\gamma\left(X_t-x_*\right)T^{\frac{1}{2k+1}}\right)}{\sigma\left(X_t\right)}\,\mathrm{d}W_t\right\}\Longrightarrow\mathcal{N}(0,1)$$

and the likelihood ratio is *AN*, i.e., it admits the representation

$$Z_T(\vartheta)=\exp\left\{\vartheta\,\Delta_T-\frac{\vartheta^2}{2}+r_T\right\},\qquad \vartheta\in[-\delta,\delta],$$

where $\Delta_T\Rightarrow\mathcal{N}(0,1)$ and $r_T\to 0$. We say *AN* and not LAN because we have no reparameterization $\vartheta=\vartheta_0+\varphi_T u$ with $\varphi_T\to 0$ and so this asymptotic normality of the likelihood ratio is not *local* in the parameter space. Nevertheless it follows from the proof of the Hajek–Le Cam inequality, that for the minimax risk in such problems the following lower bound holds:

$$\underline{\lim}_{T\to\infty}\inf_{\bar\vartheta_T}\sup_{\vartheta\in\Theta}\mathbf{E}_\vartheta\,\ell_*\left(\bar\vartheta_T-\vartheta\right)\geq\frac{1}{2}\,\mathbf{E}\,\ell\big(\psi(0)\,\zeta\big)\,\chi_{\left\{|\zeta|\leq\gamma^{-k/2}\right\}}\equiv\Pi.\qquad(4.128)$$

Here $\zeta\sim\mathcal{N}(0,1)$.

To prove it (following Remark II.12.2 in [109]) we use the representation

$$Z_T(\vartheta)=\exp\left\{\vartheta\,\tilde\Delta_\tau-\frac{\vartheta^2}{2}+r_\tau\right\},\qquad \vartheta\in[-\delta,\delta],$$

where $\mathcal{L}_*\left\{\tilde\Delta_\tau\right\}=\mathcal{N}(0,1)$ and $r_\tau\to 0$. The Gaussian variable $\tilde\Delta_\tau$ is defined like (2.9). We introduced the process

$$Z_\tau(\vartheta)=\exp\left\{\vartheta\,\tilde\Delta_\tau-\frac{\vartheta^2}{2}\right\},\qquad \vartheta\in[-\delta,\delta],$$

and note that

$$\mathbf{E}_*\left|Z_T(\vartheta)-Z_\tau(\vartheta)\right|\to 0\qquad(4.129)$$

because $\mathbf{E}_*\,Z_T(\vartheta)=\mathbf{E}_*\,Z_\tau(\vartheta)=1,\ Z_T(\vartheta)>0, Z_\tau(\vartheta)>0$ and $Z_T(\vartheta)-Z_\tau(\vartheta)\to 0$ (Scheffe's lemma). This convergence allows us to write

$$\begin{aligned}\sup_{|\vartheta|\leq\delta}\mathbf{E}_\vartheta\,\ell_*\left(\bar\vartheta_T-\vartheta\right)&\geq\frac{1}{2\delta}\int_{-\delta}^{\delta}\mathbf{E}_\vartheta\,\ell_*\left(\bar\vartheta_T-\vartheta\right)\,\mathrm{d}\vartheta\\&=\frac{1}{2\delta}\int_{-\delta}^{\delta}\mathbf{E}_*\;Z_T(\vartheta)\;\ell_*\left(\bar\vartheta_T-\vartheta\right)\,\mathrm{d}\vartheta\\&=\frac{1}{2\delta}\int_{-\delta}^{\delta}\mathbf{E}_*\;Z_\tau(\vartheta)\;\ell_*\left(\bar\vartheta_T-\vartheta\right)\,\mathrm{d}\vartheta\;(1+o(1)).\end{aligned}$$

Below we suppose that $\delta > 1$, change the variable $\vartheta = \tilde{\Delta}_\tau + u$ and put $\bar{u}_T = \bar{\vartheta}_T - \vartheta$

$$
\begin{aligned}
\frac{1}{2\delta} & \int_{-\delta}^{\delta} \mathbf{E}_* \; Z_\tau(\vartheta) \; \ell_*\left(\bar{\vartheta}_T - \vartheta\right) \, \mathrm{d}\vartheta \\
&= \frac{1}{2\delta} \mathbf{E}_* \; \mathrm{e}^{\tilde{\Delta}_\tau^2/2} \int_{-\delta}^{\delta} \mathrm{e}^{-(\vartheta - \tilde{\Delta}_\tau)^2/2} \ell_*\left(\bar{\vartheta}_T - \vartheta\right) \, \mathrm{d}\vartheta \\
&\geq \frac{1}{2\delta} \mathbf{E}_* \; \mathrm{e}^{\tilde{\Delta}_\tau^2/2} \, \chi_{\{|\tilde{\Delta}_\tau| \leq \delta - \sqrt{\delta}\}} \int_{-\sqrt{\delta}}^{\sqrt{\delta}} \mathrm{e}^{-u^2/2} \ell_*\left(\bar{u}_T - u\right) \, \mathrm{d}u \\
&\geq \frac{1}{2\delta} \int_{-\sqrt{\delta}}^{\sqrt{\delta}} \mathrm{e}^{-u^2/2} \ell_*(u) \, \mathrm{d}u \; \mathbf{E}_* \; \mathrm{e}^{\tilde{\Delta}_\tau^2/2} \, \chi_{\{|\tilde{\Delta}_\tau| \leq \delta - \sqrt{\delta}\}} \\
&= \frac{1}{2\delta} \int_{-\sqrt{\delta}}^{\sqrt{\delta}} \mathrm{e}^{-u^2/2} \ell_*(u) \, \mathrm{d}u \int_{-\delta + \sqrt{\delta}}^{\delta - \sqrt{\delta}} \mathrm{e}^{v^2/2} \frac{1}{\sqrt{2\pi}} \mathrm{e}^{-v^2/2} \, \mathrm{d}v \\
&= \frac{2\delta - 2\sqrt{\delta}}{2\delta} \frac{1}{\sqrt{2\pi}} \int_{-\sqrt{\delta}}^{\sqrt{\delta}} \mathrm{e}^{-u^2/2} \ell_*(u) \, \mathrm{d}u \\
&= \left(1 - \delta^{-1/2}\right) \mathbf{E}\, \ell_*(\zeta) \, \chi_{\{|\zeta| \leq \gamma^{-k/2}\}},
\end{aligned}
$$

where we used the inequality

$$
\int_{-\sqrt{\delta}}^{\sqrt{\delta}} \mathrm{e}^{-u^2/2} \ell_*\left(\bar{u}_T - u\right) \, \mathrm{d}u \geq \int_{-\sqrt{\delta}}^{\sqrt{\delta}} \mathrm{e}^{-u^2/2} \ell_*(u) \, \mathrm{d}u
$$

which is valid for any symmetric, nondecreasing function $\ell(\cdot)$ with $\ell(0) = 0$. This inequality follows from Anderson's lemma (see, e.g. [109], Section II.10), but in this one-dimensional case it can be proved directly (see, e.g. [136], p. 181).

Therefore the inequality (4.128) is proved because we can always take γ sufficiently small.

This theorem shows that an estimator of the trend coefficient converging to the value $S(x_*)$ uniformly over the set $\Sigma^*(k, R)$ faster than $T^{-\frac{k}{2k+1}}$ does not exist. Indeed, let $\ell(u) = |u|^p$, $p > 0$. Then for such estimators we will have

$$
\lim_{T \to \infty} \sup_{\vartheta \in \Theta} T^{\frac{pk}{2k+1}} \, \mathbf{E}_S \left|\bar{S}_T(x_*) - S(x_*)\right|^p = 0,
$$

which contradicts (4.126).

The kernel type estimator

$$
\hat{S}_T(x) = \frac{\bar{\vartheta}_T^*(x)}{f_T^\circ(x)}, \qquad \bar{\vartheta}_T^*(x) = T^{-\frac{2k}{2k+1}} \int_0^T K\left(T^{\frac{1}{2k+1}}\left(X_t - x\right)\right) \mathrm{d}X_t \tag{4.130}
$$

with the kernel

$$\int_A^B K(u)\,\mathrm{d}u = 1, \qquad \int_A^B u^j\, K(u)\,\mathrm{d}u = 0, \quad j = 1, \ldots, k \tag{4.131}$$

studied in Section 1.3 is asymptotically normal (Proposition 1.65)

$$\mathcal{L}_S\left\{T^{\frac{k}{2k+1}}\left(\hat{S}_T(x_*) - S(x_*)\right)\right\} \Longrightarrow \mathcal{N}\left(0, d_S(x_*)^2\right). \tag{4.132}$$

Therefore it can be shown that it is *asymptotically efficient in the order of convergence* at least for the bounded loss functions $\ell(\cdot)$ (we need just to verify the uniform over $\Sigma^*(k,R)$ convergence (4.132)). Note that the regularity conditions $\boldsymbol{\mathcal{T}}$ are stronger than that of Proposition 1.65 because here we need the uniform over $S(\cdot) \in \Sigma^*(k,R)$ convergence. A slight modification of this estimator like

$$\hat{S}_T(x) = \frac{\bar{\vartheta}^*_T(x)}{f^\circ_T(x) + \varepsilon_T},$$

with $\varepsilon_T \to 0$ sufficiently slowly, will provide its asymptotical efficiency for polynomial loss functions as well (see below (4.136)). Remember that $f^\circ_T(x)$ is the local-time estimator of the invariant density and $\bar{\vartheta}^*_T(x)$ is a consistent estimator of the function $S(x)\, f_S(x) = \frac{1}{2}\, f'_S(x)$.

Of course, it is interesting to have an asymptotically efficient estimator of the trend up to the constant. Unfortunately, the constant Π here is not *exact* (or sometimes it is said that the bound (4.126) is not *sharp*). This means that we cannot construct an estimator for which the inequality (4.126) is transformed in equality. To have an estimator asymptotically efficient up to the constant we follow *Pinsker's approach*, as was already done in the preceding section. Therefore we take the loss function $\ell(u) = u^2$ and we calculate the risk of an estimator $\bar{S}_T(\cdot)$ as a mean of the weighted $\mathscr{L}^2(\mathscr{R})$ norm:

$$\mathcal{R}_T\left(\bar{S}_T, S\right) = \mathbf{E}_S \int_{\mathscr{R}} \left(\bar{S}_T(x) - S(x)\right)^2 f_S(x)^2\, \mathrm{d}x.$$

Having the estimate (1.35) it is easy to show that the same result is true for the risk

$$\mathcal{R}_T\left(\bar{S}_T, S\right) = \mathbf{E}_S \int_{\mathscr{R}} \left(\bar{S}_T(x) - S(x)\right)^2 f^\circ_T(x)^2\, \mathrm{d}x.$$

For the other weights the proof can be more complicated. Note that as $S(\cdot) \notin \mathscr{L}^2(\mathscr{R})$ the weighting is necessary.

Below we introduce the lower minimax bound on the risks $\mathcal{R}_T\left(\bar{S}_T, S\right)$ of any estimator $\bar{S}_T(\cdot)$ and then we propose an estimator which is asymptotically efficient (up to the constant) in the sense of this bound. Remember that this problem is closely related to the problem of estimation of the derivative $f'_S(\cdot)$ (w.r.t. x) of the invariant density because

$$S(x) = \frac{\left(f_S(x)\,\sigma(x)^2\right)'}{2\,f_S(x)}.$$

The invariant density $f_S(x)$ can be estimated with the rate $\sqrt{T}$. Therefore its contribution to the asymptotic variance of the estimator can be negligible and the main problem is to have an asymptotically efficient estimator of the derivative. We already constructed such an estimator in the preceding section and we use it here.

To simplify the exposition we suppose that the diffusion coefficient $\sigma(x)^2 \equiv 1$ and so the observed process is

$$\mathrm{d}X_t = S(X_t)\,\mathrm{d}t + \mathrm{d}W_t, \quad X_0, \quad 0 \le t \le T. \tag{4.133}$$

We use the same regularity condition $\boldsymbol{S}_\delta$ as in Section 4.4.

Therefore we fix a *central function* function $S_*(\cdot) \in \mathscr{S}_\sigma$ and its non-parametric δ-vicinity V_δ. Further, we choose the central function $S_*(\cdot)$ a bit smoother than the others in V_δ and put

$$\Sigma_\delta = \left\{ S(\cdot) \in V_\delta : \int_{\mathscr{R}} \left[f_S^{(k+1)}(x) - f_{S_*}^{(k+1)}(x) \right]^2 \mathrm{d}x \le 4\,R \right\}.$$

4.5.2 Lower Bound

We start with a lower bound for the minimax risk. In order to do it, we use the lower bound established in the previous section and some results concerning the behavior of the local time estimator.

Theorem 4.39. (Dalalyan and Kutoyants [55]) *Let the condition $\boldsymbol{S}_\delta$ be fulfilled. Then*

$$\underline{\lim}_{\delta\to 0}\ \underline{\lim}_{T\to\infty}\ \inf_{\bar S_T}\ \sup_{S\in\Sigma_\delta} T^{\frac{2k}{2k+1}}\, \mathcal{R}_T\left(\bar S_T, S\right) \ge \Pi(k,R),$$

where

$$\Pi(k,R) = (2k+1)\left(\frac{k}{\pi\,(k+1)\,(2k+1)}\right)^{\frac{2k}{2k+1}} R^{\frac{1}{2k+1}}.$$

Proof. If we denote $\bar\theta_T(x) = \bar S_T(x)\ f_T^\circ(x)$, then using the triangle inequality we obtain

$$\mathbf{E}_S \int_{\mathscr{R}} \left(\bar S_T(x) - S(x)\right)^2 f_S(x)^2\,\mathrm{d}x \ge \left(\sqrt{\mathcal{R}_2} - \sqrt{\mathcal{R}_1}\right)^2,$$

where we denote

$$\mathcal{R}_1 = \mathbf{E}_S \int_{\mathscr{R}} \bar S_T(x)^2 \left(f_S(x) - f_T^\circ(x)\right)^2 \mathrm{d}x,$$

$$\mathcal{R}_2 = \mathbf{E}_S \int_{\mathscr{R}} \left(\bar\theta_T(x) - S(x)\,f_S(x)\right)^2 \mathrm{d}x.$$

In order to estimate the first term, we remark that we can consider only those estimators $\bar{S}_T(\cdot)$ which satisfy the condition

$$\left|\bar{S}_T(x)\right| \leq b_T\, \mathrm{e}^{x/\ln b_T}, \tag{4.134}$$

where $b_T = \ln(T+1)$. Indeed, for T large enough, we have

$$\sup_{S \in V_\delta} \left|S(x)\right| \leq C(1+|x|^m) \leq b_T\, \mathrm{e}^{x/\ln b_T}.$$

hence the risk of the truncated estimator

$$\tilde{S}_T(x) = \begin{cases} b_T\, \mathrm{e}^{x/\ln b_T} &, \quad \text{if} \quad \bar{S}_T(x) > b_T\, \mathrm{e}^{x/\ln b_T}, \\ \bar{S}_T(x) &, \quad \text{if} \quad \left|\bar{S}_T(x)\right| \leq b_T\, \mathrm{e}^{x/\ln b_T}, \\ -b_T\, \mathrm{e}^{x/\ln b_T} &, \quad \text{if} \quad \bar{S}_T(x) < -b_T\, \mathrm{e}^{x/\ln b_T} \end{cases}$$

is less than the risk of $\bar{S}_T(\cdot)$. This means that in the proof of the lower bound we can consider only the estimators satisfying this condition.

Remember that for $S(\cdot) \in V_\delta$ under condition (4.107) we have the estimate (1.35)

$$\sup_{S \in V_\delta} T\ \mathbf{E}_S \left|f_T^\circ(x) - f_S(x)\right|^2 \leq C\, \mathrm{e}^{-\gamma|x|}. \tag{4.135}$$

This inequality together with (4.134) give us the estimate

$$\mathcal{R}_1 \leq \frac{C\, b_T^2}{T}$$

for any $S(\cdot) \in \Sigma_\delta$ and T large enough.

Let us denote

$$\mathcal{R}_T(\Sigma_\delta(k,R)) = \inf_{\bar{S}_T} \sup_{S \in \Sigma_\delta} T^{\frac{2k}{2k+1}}\, \mathcal{R}_T\left(\bar{S}_T, S\right).$$

Then we have the obvious inequalities

$$\begin{aligned} \underline{\lim}_{\delta \to 0}\ \underline{\lim}_{T \to \infty} \sqrt{\mathcal{R}_T(\Sigma_\delta(k,R))} &\geq \underline{\lim}_{\delta \to 0}\ \underline{\lim}_{T \to \infty} \inf_{\bar{S}_T} \sup_{S \in \Sigma_\delta} T^{\frac{k}{2k+1}} \left(\sqrt{\mathcal{R}_2} - \sqrt{\mathcal{R}_1}\right) \\ &\geq \underline{\lim}_{\delta \to 0}\ \underline{\lim}_{T \to \infty} \inf_{\bar{S}_T} \sup_{S \in \Sigma_\delta} T^{\frac{k}{2k+1}} \left(\sqrt{\mathcal{R}_2} - \frac{C\, b_T}{\sqrt{T}}\right) \geq \sqrt{\Pi(k,R)}. \end{aligned}$$

This completes the proof of the theorem.

4.5.3 Efficient Estimator

To construct an asymptotically efficient estimator of the trend coefficient we use the asymptotically efficient estimator of the derivative of the invariant density studied in the preceding section (local bound). First we give the corresponding definition.

Definition 4.40. *Let the condition $\boldsymbol{S}_\delta$ be fulfilled. Then we call an estimator $S_T^*(\cdot)$ asymptotically efficient, if*

$$\lim_{\delta\to 0}\lim_{T\to\infty}\sup_{S\in\Sigma_\delta} T^{\frac{2k}{2k+1}}\,\mathbf{E}_S\int_{-\infty}^{\infty}\left(S_T^*(x)-S(x)\right)^2 f_S^2(x)\,\mathrm{d}x = \Pi(k,R).$$

Remember that $S(x) = f_S'(x)/2f_S(x)$. Therefore to estimate $S(x)$ we need two *good estimators* of $f_S'(x)$ and $f_S(x)$.

We have already found an asymptotically efficient estimator $\theta_T^*(\cdot)$ of the derivative $f_S'(\cdot)$. The invariant density $f_S(\cdot)$ is well estimated by the local time estimator $f_T^\circ(\cdot)$. The main property of the estimator $f_T^\circ(\cdot)$ that we need is the convergence to $f_S(\cdot)$ with a rate faster than $T^{k/(2k+1)}$. So we can replace $f_T^\circ(\cdot)$ by another estimator which converges slower than $\sqrt{T}$ but faster than $T^{k/(2k+1)}$.

These heuristics lead us to investigate the behavior of the estimator

$$\hat{S}_T(x) = \frac{\theta_T^*(x)}{2f_T^\circ(x)+\varepsilon_T \mathrm{e}^{-l_T|x|}} \tag{4.136}$$

where $\bar\theta_T(\cdot)$ is the asymptotically efficient estimator of the derivative of the invariant density studied in the previous section, $f_T^\circ(\cdot)$ is the local time estimator, $\varepsilon_T = T^{-(1-\kappa)/2}$ and $l_T = \left[\ln(T+1)\right]^{-1}$. The positive constant κ is chosen to be strictly smaller than $\beta = 1/(2k+1)$.

Theorem 4.41. (Dalalyan and Kutoyants [55]) *The estimator $\ddot{S}_T(\cdot)$ is asymptotically efficient in the problem of trend coefficient estimation.*

Proof. Let us denote $\bar f_T(x) - f_T^\circ(x) + \varepsilon_T \mathrm{e}^{-l_T|x|}$ and

$$\mathbb{B}_T(x) = \left\{\omega \;\middle|\; f_S(x) - f_T^\circ(x) < \varepsilon_T\,\mathrm{e}^{-l_T|x|}\right\}.$$

According to (4.135) and Tchebychev's inequality, we have

$$\mathbf{P}_S^{(T)}\left[\mathbb{B}_T^c(x)\right] \leq CT^{-\kappa p}\mathrm{e}^{-\gamma_*|x|},$$

where p can be chosen as large as we want and $\gamma_* < \gamma$.

We need the following estimate.

Lemma 4.42. *Let the conditions (4.107) and $S_*(\cdot)\in\mathscr{P}$ be fulfilled. Then there exists a constant C such that*

$$\mathbf{E}_S\left[\bar\theta_T^4(x)\right] \leq C\,T^{4\beta} \tag{4.137}$$

for any $x\in\mathscr{R}$ and $S(\cdot)\in\Sigma_\delta$.

Proof. Using an elementary inequality $(a+b)^4 \le 8a^4 + 8b^4$ and Hölder's inequalities one can show that

$$\begin{aligned}\mathbf{E}_S\left[\bar{\theta}_T^4(x)\right] \le \frac{C}{T^4}\,\mathbf{E}_S\left[\int_0^T K_T^*(x-X_t)\,S(X_t)\,\mathrm{d}t\right]^4 \\ +\frac{C}{T^4}\,\mathbf{E}_S\left[\int_0^T K_T^*(x-X_t)\,\mathrm{d}W_t\right]^4 \\ \le C\,\mathbf{E}_S\left[K_T^*(x-\xi)^4 S(\xi)^4\right] + \frac{C}{T^2}\,\mathbf{E}_S\left[K_T^*(x-\xi)^4\right].\end{aligned}$$

It follows from (4.120), that the supremum norm of the optimal kernel $K_T^*(x) = N_T K^*(xN_T)$ is bounded by CT^β, i.e.,

$$\sup_{x\in\mathscr{R}}\left|K_T^*(x)\right| \le CT^\beta.$$

These estimates imply the inequality (4.137).

Note that proceeding as in the proof of Lemma 4.25, one can show that

$$\mathbf{E}_S\left|\bar{\theta}_T(x) - f_S'(x)\right|^4 \le CT^{2\beta-2}.$$

This estimate is much better than (4.137), but for the proof of the efficiency of $\hat{S}_T(\cdot)$, the inequality of Lemma 4.42 is sufficient.

Using the Cauchy–Schwarz inequality and Lemma 4.42, together with the estimate

$$\inf_{S\in V_\delta}\inf_{|x|<B} f_S(x) > 0$$

one can show that

$$\begin{aligned}&\mathbf{E}_S\left[\left(\hat{S}_T(x)-S(x)\right)^2 f_S(x)^2\,\chi_{\{\mathbb{B}_T^c(x)\}}\right] \\ &\le 2\,\mathbf{E}_S\left[\hat{S}_T(x)^2 f_S(x)^2\,\chi_{\{\mathbb{B}_T^c(x)\}}\right] + 2S(x)^2\,f_S(x)^2\,\mathbf{P}_S^{(T)}\left\{\mathbb{B}_T^c\right\} \\ &\le \frac{C\,\varepsilon_T^{-2}\,\mathrm{e}^{-2\gamma|x|}}{T^{\kappa p-2\beta}} + C\,S(x)^2\,f_S(x)^2\,T^{-\kappa p} \le C\,T^{2-\kappa p}\,\mathrm{e}^{-2\gamma|x|}.\end{aligned}$$

We choose p such that $\kappa p > 3$. Then we have

$$\int_{\mathscr{R}}\mathbf{E}_S\left[\left(\hat{S}_T(x)-S(x)\right)^2 f_S(x)^2\,\chi_{\{\mathbb{B}_T^c(x)\}}\right]\mathrm{d}x \le \frac{C}{T}.$$

To evaluate the risk over the set $\mathbb{B}_T(x)$, we use the triangular inequality

$$\int_{\mathscr{R}}\mathbf{E}_S\left[\left(\hat{S}_T(x)-S(x)\right)^2 f_S(x)^2\chi_{\{\mathbb{B}_T(x)\}}\right]\mathrm{d}x \le \left(\sqrt{\mathcal{A}_1}+\sqrt{\mathcal{A}_2}\right)^2,$$

where

$$\mathcal{A}_1 = \int_{\mathscr{R}} \mathbf{E}_S \left[[\hat{S}_T(x) - \frac{S(x) f_S(x)}{2\bar{f}_T(x)}]^2 \, f_S(x)^2 \, \chi_{\{\mathbb{B}_T(x)\}} \right] \mathrm{d}x,$$
$$\mathcal{A}_2 = \int_{\mathscr{R}} \mathbf{E}_S \left[[\frac{S(x) f_S(x)}{2\bar{f}_T(x)} - S(x)]^2 \, f_S(x)^2 \, \chi_{\{\mathbb{B}_T(x)\}} \right] \mathrm{d}x.$$

Since $f_S(x) < \bar{f}_T(x)$ for any $\omega \in \mathbb{B}_T(x)$, and $2S(x)f_S(x) = f_S'(x)$, we have the following obvious inequalities:

$$4\mathcal{A}_1 \leq \int_{\mathscr{R}} \mathbf{E}_S \left(\bar{\theta}_{K^*,T}(x) - f_S'(x) \right)^2 \mathrm{d}x,$$
$$4\mathcal{A}_2 \leq \int_{\mathscr{R}} S(x)^2 \, \mathbf{E}_S \left(\bar{f}_T(x) - f_S(x) \right)^2 \mathrm{d}x.$$

The condition $S^{(k)}(\cdot) \in \mathscr{P}$ and the estimate (4.135) imply that the term $\mathcal{A}_2$ is of the order of $\varepsilon_T^2 \, l_T^{-2m}$, which is smaller than $T^{-2k/(2k+1)}$ since $\kappa < 1/(2k+1)$. Finally, combining the results of Theorem 4.30 and those obtained above we come to

$$\sup_{S \in \Sigma_\delta^*} T^{\frac{2k}{2k+1}} \mathcal{R}_T \left(\hat{S}_T, S \right) \leq o_T(1) + \sup_{S \in \Sigma_\delta^*} \left(T^{\frac{2k}{2k+1}} \sqrt{\mathcal{A}_1} + o_T(1) \right)^2$$
$$\leq o_T(1) + \sup_{S \in \Sigma_\delta^*(k, 4R + r_\delta)} \left(T^{\frac{k}{2k+1}} \sqrt{\mathcal{R}_T \left(\bar{\vartheta}_{K^*,T}, f_S' \right)} + o_T(1) \right)^2$$
$$\leq \Pi(k, R + r_\delta) \, (1 + o_\delta(1)) \, (1 + o_T(1))$$

where the term $o_T(1)$ may depend on δ and $r_\delta \to 0$ as $\delta \to 0$. This last estimate completes the proof of Theorem 4.41.

4.5.4 Adaptive Estimator

The construction of the adaptive estimator proposed below was done by Dalalyan [53] and relies heavily on the papers by Golubev [90], and Cavalier *et al.* [46].

First note that the asymptotically efficient estimator (4.136) can be rewritten (in an asymptotically equivalent form) as

$$\bar{S}_T(x) = \frac{\frac{1}{\alpha^2} \int_0^T K^* \left(\frac{x - X_t}{\alpha} \right)' \mathrm{d}t}{2\sqrt{T} \int_0^T Q \left(\sqrt{T} (x - X_t) \right) \mathrm{d}t + 2\varepsilon \sqrt{T} \, \mathrm{e}^{-\ell_T |x|}},$$

where $\alpha = \alpha_T^* = \nu_T^{-1}$, $\varepsilon = \mathrm{e}^{\sqrt{\ln T}}$, $l_T = [\ln(T+1)]^{-1}$ and $K^*(\cdot)'$ is the derivative of the kernel $K^*(\cdot)$

$$\alpha_T^* = \nu_T^{-1} = \left(\frac{4k}{\pi R T (k+1)(2k+1)} \right)^{\frac{1}{2k+1}},$$
$$K^*(x) = \frac{1}{\pi} \int_0^1 (1 - u^{k + \rho_T}) \cos(ux) \, \mathrm{d}u$$

(see (4.120), (4.121) and (4.122)). We replaced in the definition (4.136) the estimator of the derivative (4.120) by the derivative of the kernel-type estimator

$$\hat{f}_T(x) = \frac{1}{T\alpha}\int_0^T K^*\left(\frac{x - X_t}{\alpha}\right) \mathrm{d}t$$

of the invariant density. It is shown (Theorem 4.41) that if $S(\cdot) \in \Sigma_\delta = \Sigma_\delta(S_*, k, R)$, then this estimator is asymptotically efficient *up to the constant*. Here we take as the kernel $Q(\cdot)$ any positive, differentiable, symmetric density with support in $[-1, 1]$.

Therefore, to construct an asymptotically efficient estimator we have to know that $S(\cdot) \in \Sigma_\delta(S_*, k, R)$ with known values of k and R. Now we consider the same problem of trend coefficient $S(\cdot) \in \Sigma_\delta$ estimation, but in the situation when the values $k \geq 2$ and $R > 0$ are unknown.

The Fourier transform of the *optimal kernel* $K^*(\cdot)$ is (see (4.103))

$$\varphi_{K^*}(\lambda) = \left(1 - |\lambda|^{k+p}\right)_+$$

and depends on k. Hence, we can seek this kernel in the class of functions with Fourier transform

$$h_\beta(\lambda) = \left(1 - |\lambda|^\beta\right)_+.$$

Remember as well that the integral risk of estimates

$$\mathcal{R}_T(S_T, S) = \mathbf{E}_S \int_{\mathscr{R}} \left(S_T(x) - S(x)\right)^2 f_S^2(x)\,\mathrm{d}x$$

satisfies the inequality

$$\mathcal{R}_T(S_T^*, S) \leq L_T(\alpha, \varphi_{K^*}, \varphi_S)(1 + o_T(1)),$$

where

$$L_T(\alpha, h_\beta, \varphi_S) = \frac{1}{2\pi T}\int_{\mathscr{R}} \left(4|h_\beta(\alpha\lambda)|^2 + T|\lambda(h_\beta(\alpha\lambda) - 1)\,\varphi_S(\lambda)|^2\right) \mathrm{d}\lambda$$

and $o_T(1)$ is a term tending uniformly to zero (see (4.104)). Here $\varphi_S(\cdot)$ is a Fourier transform of the invariant density.

Thus it suffices to give a good adaptive choice of real parameters α and β in order to obtain an adaptive estimator of $S(\cdot)$. The values of these parameters that are of interest for us are those minimizing the risk $\mathcal{R}_T(S_T, S)$ or equivalently $L_T(\alpha, h_\beta, \varphi_S)$.

The minimizers of $L_T(\cdot)$ depend obviously on the unknown function $S(\cdot)$, so they cannot be used in an estimation procedure. A standard method for overcoming this difficulty is to estimate $L_T(\alpha, h, \varphi_S)$ by a data dependent functional $l_T(\alpha, h)$ that does not involve the function $S(\cdot)$. Then the minimizers of the latter functional could be chosen as parameters for the adaptive procedure. Perhaps the most straightforward idea for estimating $L_T(\alpha, h_\beta, \varphi_S)$ is to utilize the plug-in estimator $L_T(\alpha, h_\beta, \hat{\varphi}_T)$, $\hat{\varphi}_T(\cdot)$ being the empirical characteristic function. But it is well known that the plug-in estimators of quadratic functionals are asymptotically inefficient. That is why a smarter solution would be to apply the plug-in method to $L_T(\cdot)$ considered as a linear functional of $|\varphi_S(\cdot)|^2$. By some arguments relying on the martingale representation of the local time estimator and the integration by parts formula, one can show that a good estimator of $|\varphi_S(\lambda)|^2$ is $|\hat{\varphi}_T(\lambda)|^2 - 4/(T\lambda^2)$. On the other hand, the minimization of $L_T(\alpha, h_\beta, \varphi_S)$ w.r.t. parameters α and β obviously is equivalent to the minimization of

$$\tilde{L}_T(\alpha, h_\beta, \varphi_S) = T\int_{\mathcal{R}} \lambda^2\left(h_\beta^2(\alpha\lambda) - 2h_\beta(\alpha\lambda)\right)\left|\varphi_S(\lambda)\right|^2 \mathrm{d}\lambda + 4\int_{\mathcal{R}} |h_\beta(\alpha\lambda)|^2 \mathrm{d}\lambda,$$

since it is just $L_T(\alpha, h_\beta, \varphi_S) - T\int_{\mathcal{R}} \lambda^2 |\varphi_S(\lambda)|^2 \mathrm{d}\lambda$. For this reason, we define the functional

$$l_T(h) = T\int_{\mathcal{R}} \lambda^2\left(h_\beta^2(\lambda) - 2h(\lambda)\right)\left|\hat{\varphi}_T(\lambda)\right|^2 \mathrm{d}\lambda + 8\int_{\mathcal{R}} h_\beta(\lambda)\mathrm{d}\lambda. \qquad (4.138)$$

This functional depends on the observed path via the empirical characteristic function $\hat{\varphi}_T$. To obtain the adaptive kernel K_β and the adaptive bandwidth α, one should minimize the expression $l_T(h)$ over a suitably chosen subset $\mathbb{H}_T^N$ of the set

$$\mathbb{H}_T = \left\{h : x \mapsto h_\beta(\alpha x) = (1 - |\alpha x|^\beta)_+ \,\middle|\, \alpha \in [T^{-1/3}, (\ln T)^{-1}], \beta \geq 1\right\}$$

such that $\#\mathbb{H}_T^N = N$. For any positive integers i and j let us denote

$$\alpha_i = \left(1 + \frac{1}{\ln T}\right)^{-i} \quad \text{and} \quad \beta_j = \left(1 - \frac{j}{\ln T}\right)^{-1}. \qquad (4.139)$$

The finite subset $\mathbb{H}_T^N$ of $\mathbb{H}_T$ is defined as follows:

$$\mathbb{H}_T^N = \left\{h : x \mapsto (1 - |\alpha_i x|^{\beta_j})_+ \,\middle|\, \alpha_i \in [T^{-1/3}, (\ln T)^{-1}],\ j = 1, \ldots, \lfloor \ln T \rfloor\right\},$$

where $\lfloor a \rfloor$ stands for the largest integer strictly smaller than the real number a. It is evident that the cardinality of $\mathbb{H}_T^N$ is less than $(\ln T)^3$. From now on, we denote the N elements of this set by $h_1, h_2, \ldots, h_N$. Thus for constructing the adaptive estimator it suffices to maximize the functional $l_T(\cdot)$ over a set of cardinality $(\ln T)^3$.

Let us now summarize the method. We start by computing the values α_i and β_j according to (4.139). Then we determine the function $\tilde{h}_T \in \mathbb{H}_T^N$ such that

$$l_T(\tilde{h}_T) = \min_{h \in \mathbb{H}_T^N} l_T(h).$$

If the function satisfying the latter equality is not unique, we denote one of them by $\tilde{h}_T$. Next we apply to $\tilde{h}_T$ the inverse Fourrier transform in order to define the kernel

$$\tilde{K}_T(x) = \frac{1}{2\pi}\int_{\mathscr{R}} \tilde{h}_T(\lambda)\cos(\lambda x)\,\mathrm{d}\lambda.$$

This form of the kernel already comprises the bandwidth since $\tilde{h}_T(\lambda) = h_{\tilde{\beta}_T}(\tilde{\alpha}_T\lambda)$, where $\tilde{\alpha}_T$ and $\tilde{\beta}_T$ are the values of α_i and β_j corresponding to $\tilde{h}_T$. Further, we choose another kernel function $Q(\cdot)$, which is positive, differentiable, symmetric, supported by $[-1,1]$ and with integral equal to one. Finally, we set $\varepsilon_T = \mathrm{e}^{\sqrt{\ln T}}$, $\ell_T = 1/\ln T$ and define the estimator

$$\tilde{S}_T(x) = \frac{\int_0^T \tilde{K}_T(x-X_t)'\,\mathrm{d}t}{2\sqrt{T}\int_0^T Q\big(\sqrt{T}(x-X_t)\big)\,\mathrm{d}t + 2\sqrt{T}\,\varepsilon_T \mathrm{e}^{-\ell_T|x|}}.$$

Note that the function $\tilde{K}_T(\cdot)$ is differentiable, since $\min\beta_j > 1$.

We need to impose an additional assumption on $S_*(\cdot)$. We denote by $P_t(S,x,\mathbb{A})$ the transition probability corresponding to the instant t, that is $P_t(S,x,\mathbb{A}) = \mathbf{P}_S(X_t \in \mathbb{A}\,|\,X_0 = x)$, for all $x \in \mathscr{R}$, and all $\mathbb{A} \in \mathfrak{B}(\mathscr{R})$. The density of the measure $P_t(S,x,\cdot)$ will be denoted by $p_t(S,x,y)$.

Theorem 4.43. (Dalalyan [53]) *Let the condition $\boldsymbol{S}_\delta$ be satisfied and the density $p_t(S_*,x,y)$ be finite for a $t>0$ uniformly in $x \in \mathscr{R}$ and $y \in \mathscr{R}$. Then*

$$\lim_{\delta\to 0}\lim_{T\to\infty}\sup_{S\in\Sigma_\delta} T^{\frac{2k}{2k+1}}\,\mathcal{R}_T(\tilde{S}_T,S) = \Pi(k,R).$$

4.6 Second Order Efficiency

4.6.1 Problem

The class of asymptotically efficient estimators is quite large. Remember that in the problem of invariant density estimation the local time estimator (Section 4.2.2), the class of unbiased estimators (Section 4.2.4) and all kernel-type estimators (Section 4.2.5) are asymptotically normal with the same limit variance, which coincides with the variance of the asymptotically efficient estimator. This means that under corresponding regularity conditions we have for any of these estimators, say, $\hat{f}_T(x)$ and any point $x \in \mathscr{R}$ the convergence

$$T\,\mathbf{E}_S\left(\hat{f}_T(x) - f_S(x)\right)^2 \longrightarrow 4f_S(x)^2\,\mathbf{E}_S\left(\frac{\chi_{\{\xi>x\}} - F_S(\xi)}{\sigma(\xi)\,f_S(\xi)}\right)^2.$$

In Section 4.3 we show that the corresponding equality holds for $\mathcal{L}_2$-type risk too, i.e.,

$$T\int_{\mathscr{R}} \mathbf{E}_S\left(\hat f_T(x)-f_S(x)\right)^2 \mathrm{d}x \longrightarrow \mathcal{R}(S)$$
$$= 4\int_{\mathscr{R}} f_S(x)^2\,\mathbf{E}_S\left(\frac{\chi_{\{\xi>x\}}-F_S(\xi)}{\sigma(\xi)\,f_S(\xi)}\right)^2 \mathrm{d}x.$$

Hence it is reasonable to seek the second order asymptotically efficient estimators of the invariant density in the following sense. Suppose that the function $f_S(\cdot)$ is k times differentiable and belongs to some set Σ_* defined below. Consider the quantity

$$\mathscr{R}\left(\bar f_T, f_S\right) = \int_{\mathscr{R}} \mathbf{E}_S\left(\bar f_T(x)-f_S(x)\right)^2 \mathrm{d}x - \frac{\mathcal{R}(S)}{T},$$

where $\bar f_T(x)$ is an arbitrary estimator of the density. Obviously for the asymptotically efficient estimators mentioned above we have $T\mathscr{R}(\bar f_T, f_S)\to 0$. Moreover, using the representation (1.32) for the LTE $f_T^\circ(\cdot)$, we can easily see that $T^{3/2}\mathscr{R}(f_T^\circ, f_S)$ has a nondegenerate limit. A more detailed analysis shows that if the unknown density is $k\geq 2$ differentiable, then there exists a nondegenerate limit for $T^{\frac{2k}{2k-1}}\mathscr{R}\left(\hat f_T, f_S\right)$. Therefore we can compare these estimators according to the limits of this quantity.

As usual, we consider two problems. The first one is to find a constant $\hat\Pi(k,R)>0$ such that the following inequality holds:

$$\lim_{T\to\infty}\inf_{\bar f_T}\sup_{S(\cdot)\in\Sigma_*} T^{\frac{2k}{2k-1}}\mathscr{R}\left(\bar f_T, f_S\right)\geq -\hat\Pi(k,R),$$

where inf is taken over all estimators $\bar f_T(\cdot)$. This inequality gives us the lower bound for the risks of all estimators. The second problem is to construct an estimator $\hat f_T(\cdot)$ which attains this bound, i.e., such that

$$\lim_{T\to\infty}\sup_{S(\cdot)\in\Sigma_*} T^{\frac{2k}{2k-1}}\mathscr{R}\left(\hat f_T, f_S\right) = -\hat\Pi(k,R).$$

This estimator we call *second order asymptotically efficient.*

In the present section we show that

$$\hat\Pi(k,R) = 2(2k-1)\left(\frac{4k}{\pi(k-1)(2k-1)}\right)^{\frac{2k}{2k-1}} R^{\frac{1}{2k-1}}$$

and then we construct the second order efficient estimator.

Similar statements can be considered in the problem of invariant distribution function estimation too.

The approach taken in this section is inspired by the work by Golubev and Levit [91], who considered the second order efficient estimation of the distribution function in the i.i.d. case and we apply here the method developed in their work. This problem belongs to the class of problems initiated by Pinsker [201] and already treated in Sections 4.4 and 4.5.

To solve the first problem we deal, roughly speaking, in the way similar to that of the preceding sections, i.e., we construct a parametric family of trend coefficients $\left\{S(\vartheta,\cdot),\ \vartheta\in\Theta_T\subset\mathscr{R}^{d_T}\right\}$, which is a subfamily of the original nonparametric family Σ_*. This allows us to obtain the first lower bound for the risk and hence to reduce the problem to the calculation of the risk in the parameter estimation problem. Then we choose the least favorable parametric family and the risk of this last problem will give us the constant $\mathrm{P}(k,R)$. As in Sections 4.4 and 4.5, the parameter ϑ is of growing dimension $d_T=2L_T\to\infty$, and this parameter is supposed to be random with some prior distribution. The least favorable family corresponds to the parametric family with minimal Fisher information and a special prior distribution of the parameter. Note that the choice of Fisher information and prior distribution can be considered separately and this simplifies the problem.

To simplify the exposition we put $\sigma(x)\equiv 1$, so the observations are

$$\mathrm{d}X_t=S(X_t)\ \mathrm{d}t+\mathrm{d}W_t,\quad X_0,\quad 0\le t\le T, \tag{4.140}$$

and we have to estimate the invariant density

$$f_S(x)=\frac{1}{G(S)}\exp\left\{2\int_0^x S(v)\,\mathrm{d}v\right\}$$

by the observations $X^T=\{X_T,\ 0\le t\le T\}$.

Let us define the class Σ^* of functions $S(\cdot)$ such that the conditions $\mathcal{A}_0$ and $S(\cdot)\in\mathscr{P}$ are fulfilled uniformly on $S(\cdot)\in\Sigma^*$. This means that

$$\overline{\lim_{|x|\to\infty}}\ \sup_{S(\cdot)\in\Sigma^*}\ \operatorname{sgn}(x)\ S(x)<0 \tag{4.141}$$

and

$$\sup_{S(\cdot)\in\Sigma^*}|S(x)|\le C\left(1+|x|^\nu\right), \tag{4.142}$$

where the constants $C>0$ and $\nu>0$ do not depend on $S(\cdot)$. In particular, for the functions $S(\cdot)$ satisfying (4.141) there exist constants $\gamma_*>0$ and $A_*>0$ such that the following estimate:

$$\operatorname{sgn}(x)\ S(x)<-\gamma_*,\qquad\qquad |x|\ge A_* \tag{4.143}$$

holds.

The inequalities (4.142) and (4.143) for fixed values C,ν,γ_* and A_* define an example of the set $\Sigma^*=\Sigma^*(C,\nu,\gamma_*,A_*)$.

Fix some integer $k>1$. The function $S(\cdot)$ is supposed to be $(k-1)$-times differentiable and to belong to the set

$$\Sigma(k-1,R,S_*)=\left\{S(\cdot)\in\Sigma^*:\ \int_{\mathscr{R}}[f_S^{(k)}(x)-f_{S_*}^{(k)}(x)]^2\,\mathrm{d}x\le R\right\},$$

where $R > 0$ is some constant and $f_S^{(k)}(x)$ is the kth derivative of the function $f_S(x)$ w.r.t. x. The set $\Sigma(k-1, R, S_*)$ is a Sobolev ball of smoothness k and radius R centered at $f_*(\cdot)$. The choice of the central function is not arbitrary, it should be smoother than the other functions of the class. For simplicity we consider

$$S_*(x) = -x.$$

In this case the corresponding diffusion is an Ornstein–Uhlenbeck process and the invariant density is infinitely differentiable. Finally we define the parameter set

$$\Sigma_* = \Sigma_*(k-1, R, S_*) \cap \Sigma^*.$$

4.6.2 Lower Bound

We start with the lower bound on the risks of all estimators given in the following theorem.

Theorem 4.44. (Dalalyan and Kutoyants [56]) *Let the integer $k > 1$. Then*

$$\varliminf_{T\to\infty} \inf_{\bar f_T} \sup_{S\in\Sigma_*} T^{\frac{2k}{2k-1}} \mathscr{R}_T(\bar f_T, f_S) \geq -\hat\Pi(k, R), \tag{4.144}$$

where

$$\hat\Pi(k, R) = 2(2k-1)\left(\frac{4k}{\pi(k-1)(2k-1)}\right)^{\frac{2k}{2k-1}} R^{-\frac{1}{2k-1}}.$$

Proof. The main steps of the proof are similar to those of the Theorem 4.34. Hence in the beginning we repeat the construction of the parametric family.

The minimax risk over the nonparametric set Σ_* is evaluated by the Bayesian risk over a parametric set of increasing dimension. Then the van Trees inequality is applied.

Let us introduce the same parametric family as in Section 4.4 of diffusion processes

$$\mathrm{d}X_t = S(\vartheta, X_t)\,\mathrm{d}t + \mathrm{d}W_t, \quad X_0, \quad 0 \leq t \leq T, \tag{4.145}$$

where $S(\vartheta, \cdot)$ is chosen in the following way (see (4.109) and (4.110)). We fix an increasing interval

$$\mathbb{A} = [-A, A]$$

with $A = A_T = \gamma_*^{-1} \log(T+1)$ and a sequence of its sub-intervals

$$\mathbb{I}_m = \left[a_m - AT^{-\beta}, a_m + AT^{-\beta}\right],$$

where $\beta = (2k-1)^{-1}$ and

$$a_m = 2mAT^{-\beta}, \qquad m = 0, \pm 1, \pm 2, \ldots, \pm M.$$

Here $M = M_T$ is the greatest integer such that the interval I_M is entirely included in $[-A, A]$. Let us denote

$$S(\boldsymbol{\vartheta}, x) = S_*(x) + \sum_{|m|\leq M} \sqrt{\frac{2A}{T^\beta f_*(a_m)}} \sum_{|l|\leq L} \vartheta_{l,m}\, \phi_{l,m}(x),$$

where $\boldsymbol{\vartheta}$ is the $(2M+1)\times(2L+1)$ matrix of coefficients $\vartheta_{l,m}$ and

$$\phi_{l,m}(x) = \sqrt{\frac{T^\beta}{A}}\, e_l(T^\beta A^{-1}(x - a_m))\, U(A - |x - a_m| T^\beta).$$

Here $e_l(\cdot)$ is the trigonometric basis on $[-1,1]$, that is

$$e_l(x) = \begin{cases} \sin(\pi l x)\ , & \text{if } l > 0, \\ 1/\sqrt{2} \quad , & \text{if } l = 0, \\ \cos(\pi l x)\ , & \text{if } l < 0, \end{cases}$$

the function $U(x)$ is $(k+1)$-times differentiable, increasing, vanishing for $x \leq 0$ and equal to one for $x \geq 1$. The integer $L = L_T$ will be chosen later.

The parameter $\boldsymbol{\vartheta} \in \boldsymbol{\Theta}_T$ is of increasing dimension and

$$|\vartheta_{l,m}| \leq K\sqrt{\sigma_l(\varepsilon)}, \qquad \sigma_l(\varepsilon) = \frac{1}{2AT^{1-\beta}}\left(\left|\frac{\Lambda(1-\varepsilon)}{l}\right|^k - 1\right)_+, \tag{4.146}$$

for $l \neq 0$, and $\sigma_0 = T^{-\beta}$, for $l = 0$. Above we used the standard notation $B_+ = \max(0, B)$; ε is a positive number and

$$\Lambda = \Lambda_T = A\left(\frac{R\,(k-1)\,(2k-1)}{4\,k\,\pi^{2k-2}}\right)^{\frac{1}{2k-1}}.$$

The number $L = L_T$ is now the integer part of Λ. The choice of the σ_l and Λ will be clarified a little later.

Note that for $S(\cdot) \in \Sigma_*$ we have the estimate (see (1.35) and (4.135))

$$\sup_{\boldsymbol{\vartheta}\in\boldsymbol{\Theta}_T} \int_{|x|>A} f(\boldsymbol{\vartheta}, x)^2\, \mathbf{E}_{\boldsymbol{\vartheta}} \left(\frac{\chi_{\{\xi > x\}} - F_{\boldsymbol{\vartheta}}(\xi)}{f(\boldsymbol{\vartheta}, \xi)}\right)^2 \mathrm{d}x \leq C\,\mathrm{e}^{-\gamma_* A},$$

which implies that the choice $A = \gamma_*^{-1}\log T$ makes this term asymptotically negligible. Therefore it is sufficient to study the lower bound for the risk

$$\hat{\mathscr{R}}_T(\bar{f}_T, f_S) = \mathbf{E}_S \int_{-A}^{A} [\bar{f}_T(x) - f_S(x)]^2 \mathrm{d}x - \frac{1}{T}\int_{-A}^{A} \mathrm{I}_f^{-1}(S, x)\mathrm{d}x$$

because

$$|\mathscr{R}_T(\bar{f}_T, f_S) - \hat{\mathscr{R}}_T(\bar{f}_T, f_S)| \leq CT^{-2},$$

and the order of the term $\hat{\mathscr{R}}_T$ will be shown to be less than T^{-2}.

We consider now only the estimators belonging to the set

$$\mathcal{W}_T = \{\bar{f}_T : \hat{\mathscr{R}}_T(\bar{f}_T, f_*) < 1\}.$$

It is evident that in the proof of the lower bound one can drop all the estimators that do not belong to $\mathcal{W}_T$. Moreover, it is clear from that for T large enough this set is not empty (since there exist consistent estimators). Further suppose that the parameter $\boldsymbol{\vartheta}$ is a random matrix with a prior distribution $\mathrm{Q}(\mathrm{d}\boldsymbol{\vartheta})$ on the set $\boldsymbol{\Theta}_T$. Then we have the obvious inequality

$$\sup_{\boldsymbol{\vartheta}\in\boldsymbol{\Theta}_T} \hat{\mathscr{R}}_T(\bar{f}_T, f_{\boldsymbol{\vartheta}}) \geq \mathbf{E}_Q\, \hat{\mathscr{R}}_T(\bar{f}_T, f_{\boldsymbol{\vartheta}}) = \int_{\boldsymbol{\Theta}_T} \hat{\mathscr{R}}_T(\bar{f}_T, f_{\boldsymbol{\vartheta}})\, \mathrm{Q}(\mathrm{d}\boldsymbol{\vartheta}),$$

where the expectation $\mathbf{E}_Q$ is defined by the last equality. We can write now the following sequence of inequalities:

$$\begin{aligned}
\inf_{\bar{f}_T\in\mathcal{W}_T} \sup_{S\in\Sigma_*} \hat{\mathscr{R}}_T(\bar{f}_T, f_S) &\geq \inf_{\bar{f}_T\in\mathcal{W}_T} \sup_{S\in\Sigma_*\cap\boldsymbol{\Theta}_T} \hat{\mathscr{R}}_T(\bar{f}_T, f_S) \\
&\geq \inf_{\bar{f}_T\in\mathcal{W}_T} \int_{\boldsymbol{\Theta}_T} \hat{\mathscr{R}}_T(\bar{f}_T, f_{\boldsymbol{\vartheta}})\, \mathrm{Q}(\mathrm{d}\boldsymbol{\vartheta}) - \sup_{\bar{f}_T\in\mathcal{W}_T} \int_{\boldsymbol{\Theta}_T\setminus\Sigma_*} \hat{\mathscr{R}}_T(\bar{f}_T, f_{\boldsymbol{\vartheta}})\, \mathrm{Q}(\mathrm{d}\boldsymbol{\vartheta}) \\
&\geq \hat{\mathscr{R}}_T(\mathrm{Q}) - \sup_{\bar{f}_T\in\mathcal{W}_T} \int_{\boldsymbol{\Theta}_T\setminus\Sigma_*} \hat{\mathscr{R}}_T(\bar{f}_T, f_{\boldsymbol{\vartheta}})\, \mathrm{Q}(\mathrm{d}\boldsymbol{\vartheta}). \qquad (4.147)
\end{aligned}$$

Here $\hat{\mathscr{R}}_T(\mathrm{Q})$ denotes the second order Bayesian risk with respect to the prior Q, that is

$$\hat{\mathscr{R}}_T(\mathrm{Q}) = \mathbf{E}_{\mathrm{Q}}\left[\hat{\mathscr{R}}_T(\hat{f}_T, f_{\boldsymbol{\vartheta}})\right].$$

The next step is to choose a prior distribution that maximises the second order Bayesian risk and, at the same time, is essentially concentrated on the set Σ_*, that is the probability of the set $\boldsymbol{\Theta}_T\setminus\Sigma_*$ is sufficiently small. This distribution Q is usually called the asymptotically least favorable prior distribution. In our case, it is defined in the following way: let $\eta_{l,m}$, $l = 0, \pm1, \pm2, \ldots, \pm L$, $m = 0, \pm1, \ldots, \pm M$, be i.i.d. random variables with common density $p(\cdot)$, such that

$$|\eta_{l,m}| \leq K, \quad \mathbf{E}[\eta_{l,m}] = 0, \quad \mathbf{E}[\eta_{l,m}^2] = 1, \quad J = \int_{-K}^{K} \frac{\dot{p}(x)^2}{p(x)}\, \mathrm{d}x = 1 + \varepsilon,$$

where $\varepsilon \to 0$ as $K \to \infty$ (for example, one can set $K = \varepsilon^{-1}$). Then Q is the distribution of the random array $\vartheta_{l,m} = \sqrt{\sigma_l(\varepsilon)}\,\eta_{l,m}$. In other words, it is the product measure

$$Q(\mathrm{d}\boldsymbol{\vartheta}) = \prod_{|m|\leq M} \prod_{|l|\leq L} \frac{1}{\sqrt{\sigma_l(\varepsilon)}}\, p\left(\frac{\vartheta_{l,m}}{\sqrt{\sigma_l(\varepsilon)}}\right) \mathrm{d}\vartheta_{l,m}. \qquad (4.148)$$

Therefore the Fisher information of one component of this prior distribution Q is

$$\mathrm{I}_{l,m} = \mathrm{I}_l = \frac{1+\varepsilon}{\sigma_l(\varepsilon)}.$$

We will evaluate now the Bayesian risk, which will give us the main part of the minimax risk. Remark firstly that the set of functions

$$e_{l,m}(x) = \sqrt{A^{-1}T^{\beta}}\, e_l\big(A^{-1}T^{\beta}(x-a_m)\big)$$

forms an orthonormal basis of the Hilbert space $\mathscr{L}^2[-A,A]$ and hence, using the well known Parseval's identity, we can write

$$\begin{aligned}
\mathbf{E}_Q \hat{\mathscr{R}}_T(\bar{f}_T, f_{\vartheta}) &= \mathbb{E}\int_{-A}^{A} [\bar{f}_T(x) - f(\vartheta,x)]^2\,\mathrm{d}x - \frac{1}{T}\mathbb{E}\int_{-A}^{A} I_f^{-1}(S,x)\mathrm{d}x \\
&\geq \sum_{|m|\leq M}\sum_{|l|\leq L}\left[\mathbb{E}\,|\psi_{l,m,T} - \psi_{l,m,\vartheta}|^2 - \frac{4}{T}\mathbb{E}\,|\Psi_{l,m,\vartheta}(\xi)|^2\right] \\
&\quad - \sum_{|m|\leq M}\sum_{|l|> L}\frac{4}{T}\mathbb{E}\,|\Psi_{l,m,\vartheta}(\xi)|^2 := \mathcal{A}_{1T} + \mathcal{A}_{2T}, \qquad (4.149)
\end{aligned}$$

where $\mathbb{E}$ is the expectation with respect to the probability measure $\mathrm{d}\mathbf{P}_{\vartheta}^{(T)} \times Q(\mathrm{d}\vartheta)$ and $\psi_{l,m,T}$, $\psi_{l,m,\vartheta}$ and $\Psi_{l,m,\vartheta}(y)$ are the Fourier coefficients

$$\begin{aligned}
\psi_{l,m,T} &= \int_{\mathbb{I}_m} \bar{f}_T(x)\, e_{l,m}(x)\,\mathrm{d}x, \\
\psi_{l,m,\vartheta} &= \int_{\mathbb{I}_m} f(\vartheta,x)\, e_{l,m}(x)\,\mathrm{d}x, \\
\Psi_{l,m,\vartheta}(y) &= \int_{\mathbb{I}_m} f(\vartheta,x)\left(\frac{\chi_{\{y>x\}} - F_{\vartheta}(y)}{f(\vartheta,y)}\right) e_{l,m}(x)\,\mathrm{d}x,
\end{aligned}$$

of the corresponding functions.

Direct calculation provides the equality

$$\mathrm{I}_{l,m}(\vartheta) = \mathbf{E}_{\vartheta}\left(\frac{\partial S(\vartheta,\xi)}{\partial\vartheta_{l,m}}\right)^2 = \frac{2A}{T^{\beta} f_*(a_m)}\;\mathbf{E}_{\vartheta}\left[\phi_{l,m}^2(\xi)\right] < \infty.$$

This quantity plays the role of Fisher information concerning the parameter $\vartheta_{l,m}$ in the van Trees inequality. That is, the following inequality holds:

$$\mathbb{E}\,|\psi_{l,m,T} - \psi_{l,m,\vartheta}|^2 \geq \frac{\left(\mathbf{E}_Q \frac{\partial\psi_{l,m,\vartheta}}{\partial\vartheta_{l,m}}\right)^2}{T\,\mathbf{E}_Q\,\mathrm{I}_{l,m}(\vartheta) + \mathrm{I}_{l,m}}.$$

From this inequality one can deduce that

$$\begin{aligned}
&\mathbb{E}|\psi_{l,m,T} - \psi_{l,m,\vartheta}|^2 - \frac{4}{T}\mathbb{E}[\Psi_{l,m,\vartheta}^2(\xi)] \\
&\qquad \geq -\frac{4\mathrm{I}_l\,\mathbb{E}[\Psi_{l,m,\vartheta}^2(\xi)]}{T(T\,\mathbf{E}_Q\,\mathrm{I}_{l,m}(\vartheta) + \mathrm{I}_l)} + \mu_{l,m,T}, \qquad (4.150)
\end{aligned}$$

where

$$\mu_{l,m,T} = \frac{(\mathbf{E}_Q \frac{\partial \psi_{l,m,\vartheta}}{\partial \vartheta_{l,m}})^2 - 4\,\mathbf{E}_Q\left[\mathrm{I}_{l,m}(\vartheta)\right]\mathbb{E}\left|\Psi_{l,m,\vartheta}(\xi)\right|^2}{T\,\mathbf{E}_Q\left[\mathrm{I}_{l,m}(\vartheta)\right] + \mathrm{I}_l}\,. \tag{4.151}$$

We now need the following auxiliary results.

Lemma 4.45. *For any integer $l \neq 0$, we have*

$$\mathbb{E}[\Psi^2_{l,m,\vartheta}(\xi)\chi_{\{\xi \notin \mathbb{I}_m\}}] \leq \frac{C\,A^3}{l^2\,T^{3\beta}}\,.$$

Proof. We denote $\delta = A/T^\beta$ and divide the indicator in two parts

$$\mathbb{E}[\Psi^2_{l,m,\vartheta}(\xi)\chi_{\{\xi \notin \mathbb{I}_m\}}] = \mathbb{E}[\Psi^2_{l,m,\vartheta}(\xi)\chi_{\{\xi > a_m+\delta\}}] + \mathbb{E}[\Psi^2_{l,m,\vartheta}(\xi)\chi_{\{\xi < a_m-\delta\}}].$$

The first term can be evaluated as follows:

$$\begin{aligned}
&\mathbb{E}[\Psi^2_{l,m,\vartheta}(\xi)\chi_{\{\xi > a_m+\delta\}}] \\
&\quad = \mathbb{E}\left(\int_{\mathbb{I}_m} f(\vartheta,x)e_{l,m}(x)\,\mathrm{d}x\right)^2 \left(\frac{1-F_\vartheta(\xi)}{f(\vartheta,\xi)}\,\chi_{\{\xi > a_m+\delta\}}\right)^2 \\
&\quad \leq C\,\mathbf{E}_Q\left(\int_{\mathbb{I}_m} f(\vartheta,x)\,e_{l,m}(x)\,\mathrm{d}x\right)^2,
\end{aligned}$$

since one can prove exactly like (4.115), that $f(\vartheta,x) = f_*(x)\big(1+o_T(1)\big)$ $(f_*(x) = f_{S_*}(x))$ and

$$\int_{a_m+\delta}^{\infty} \left(\frac{1-F_*(y)}{f_*(y)}\right)^2 f_*(y)\,\mathrm{d}y < CAf_*(a_m+\delta)^{-1}.$$

The first multiplier will be evaluated using the integration by parts formula:

$$\begin{aligned}
\int_{\mathbb{I}_m} f(\vartheta,x)\,e_{l,m}(x)\,\mathrm{d}x = \frac{\delta}{\pi l}\Bigg[&\big(f(\vartheta,a_m+\delta) - f(\vartheta,a_m-\delta)\big)e_{-l,m}(a_m-\delta) \\
&- \int_{\mathbb{I}_m} f'(\vartheta,x)\,e_{-l,m}(x)\,\mathrm{d}x\Bigg]
\end{aligned}$$

Using the fact that $f(\vartheta,x)$ is differentiable, its derivative is $2S(\vartheta,x)f(\vartheta,x)$ and that $S(\vartheta,\cdot)$ is bounded by $2A$ on the interval $[-A,A]$ uniformly bounded on $\vartheta \in \Theta_T$, we get

$$\left|\int_{\mathbb{I}_m} f(\vartheta,x)e_{l,m}(x)\,\mathrm{d}x\right| \leq \frac{CA\delta^{\frac{3}{2}}}{l^2}\,f_*(a_m).$$

Combining these estimates we obtain the desired result for the first term. The term involving the indicator of the event $\xi < a_m - \delta$ can be evaluated analogously. This completes the proof of the lemma.

Lemma 4.46. *If $l \neq 0$ and $y \in \mathbb{I}_m$, then the following representation holds:*

$$\Psi_{l,m,\vartheta}(y) = \frac{A(1+o_T(1))}{\pi\, l\, T^\beta}\left(e_{-l,m}(y) - e_{-l,m}(a_m - AT^{-\beta})\right) + \frac{A\Phi_{l,m,\vartheta}(y)}{\pi l T^\beta},$$

where the sequence $\Phi_{l,m,\vartheta}$ is defined as follows:

$$\Phi_{l,m,\vartheta}(y) = \int_{\mathbb{I}_m} f'(\vartheta,x)\left(\frac{\chi_{\{y>x\}} - F(\vartheta,y)}{f(\vartheta,y)}\right)e_{-l,m}(x)\,\mathrm{d}x.$$

Proof. The proof of this lemma is quite close to the proof of the preceding one. Using the integration by parts formula one gets

$$\begin{aligned}\Psi_{l,m,\vartheta}(y) &= \frac{\delta}{\pi l f(\vartheta,y)}\left(f(\vartheta,y)e_{-l,m}(y) - f(\vartheta,a_m-\delta)e_{-l,m}(a_m-\delta)\right)\\ &\quad - \frac{\delta F(\vartheta,y)}{\pi l f(\vartheta,y)}\left(f(\vartheta,a_m+\delta) - f(\vartheta,a_m)\right)e_{-l,m}(a_m-\delta)\\ &\quad - \frac{\delta}{\pi l}\int_{\mathbb{I}_m} f'(\vartheta,x)\left(\frac{\chi_{\{y>x\}} - F(\vartheta,y)}{f(\vartheta,y)}\right)e_{-l,m}(x)\,\mathrm{d}x.\end{aligned}$$

To finish the proof, it remains to remark that on the interval $\mathbb{I}_m$, one has

$$|f(\vartheta,y) - f(\vartheta,a_m-\delta)| \le C\delta|f'(\vartheta,y)| \le C\,A\,\delta\,|f(\vartheta,y)|,$$

since $f'(\vartheta,x) = 2S(\vartheta,x)f'(\vartheta,x)$ and $S(\vartheta,x)$ is bounded by $1+|x|$.

Corollary

For any $l \neq 0$, we define the number c_l to be equal to 1, if $l<0$, and to 3, if $l>0$. Then we have

$$\mathbf{E}\left[\Psi^2_{l,m,\vartheta}(\xi)\right] \le \frac{A^2 f_*(a_m)(1+o_T(1))}{\pi^2 l^2 T^{2\beta}}\left(\sqrt{c_l} + \sqrt{\mathbf{E}\left[\Phi^2_{l,m,\vartheta}(\xi)\right]}\right)^2 + \frac{CA^3}{l^2 T^{3\beta}},$$

and

$$\sum_{l,m\in\mathbb{Z}} \mathbf{E}\left[\Phi^2_{l,m,\vartheta}(\xi)\right] \le C. \tag{4.152}$$

Lemma 4.47. *The residual terms $\mu_{l,m,T}$ defined in (4.151) satisfy the relation*

$$|\mu_{l,m,T}| \le \frac{C\delta_T^3}{T} \tag{4.153}$$

where C is a constant.

Proof. The proof of this lemma is rather technical. We give below the main ideas of the proof and leave the technical details to the reader. First of all, due to Lemma 4.45, it is enough to prove this inequality for

$$\bar{\mu}_{l,m,T} = \frac{\left(\mathbf{E}_Q \frac{\partial \psi_{l,m,\vartheta}}{\partial \vartheta_{l,m}}\right)^2 - 4\, \mathbf{E}_Q\left[\mathrm{I}_{l,m}(\vartheta)\right] \mathbb{E}\left[\Psi^2_{l,m,\vartheta}(\xi)\chi_{\{\xi\in \mathbb{I}_m\}}\right]}{T\, \mathbf{E}_Q\left[\mathrm{I}_{l,m}(\vartheta)\right] + \mathrm{I}_l}.$$

Next, it can be shown (see (4.116)) that

$$\frac{\partial \psi_{l,m,\vartheta}}{\partial \vartheta_{l,m}} = 2\sqrt{\frac{2A}{f_*(a_m)T^\beta}} \int_{\mathbb{I}_m} \phi_{l,m}(x)\Psi_{l,m,\vartheta}(x)\, f(\vartheta, x)\, \mathrm{d}x. \qquad (4.154)$$

Since all the functions under the integral are differentiable with bounded derivatives and the length of the interval $\mathbb{I}_m$ is $2\delta_T$, we have

$$\frac{\partial \psi_{l,m,\vartheta}}{\partial \vartheta_{l,m}} = \sqrt{\frac{32\delta_T^3}{f_*(a_m)}}\, \phi_{l,m}(a_m)\Psi_{l,m,\vartheta}(a_m)\, f(\vartheta, a_m) + \delta_T^{\frac{5}{2}} O_T(1)$$

Using the same arguments, one can check that

$$\begin{aligned} &\mathrm{I}_{l,m}(\vartheta)\, \mathbf{E}_\vartheta\left[\Psi_{l,m,\vartheta}(\xi)^2\chi_{\{\xi\in\mathbb{I}_m\}}\right] \\ &\quad = \frac{2\delta_T}{f_*(a_m)}\, \mathbf{E}_\vartheta[\phi_{l,m}(\xi)^2]\, \mathbf{E}_\vartheta\left[\Psi^2_{l,m,\vartheta}(\xi)\chi_{\{\xi\in\mathbb{I}_m\}}\right] \\ &\quad = \frac{8\delta_T^3}{f_*(a_m)}\, \phi_{l,m}(a_m)^2 f(\vartheta, a_m)\Psi^2_{l,m,\vartheta}(a_m)\, f(\vartheta, a_m) + \delta_T^4 O_T(1). \end{aligned} \qquad (4.155)$$

It is clear now that the difference of the terms (4.154) and (4.155) is of the order of δ_T^4. Since the denominator of $\bar{\mu}_{l,m,T}$ is larger than $CT\delta_T$, we get the estimate (4.153).

According to (4.153) the sum of the terms $\mu_{l,m,T}$ is less than a power of A divided by $T^{1+2\beta}$, which is clearly of the order of $o_T(1)/T^{\frac{2k}{2k-1}}$. This implies that the main part of the convergence of the second order Bayesian risk is given by the first terms of inequality (4.150).

Now we will evaluate the sum of the first terms in the inequality (4.150). Since the function $\phi_{l,m}(\cdot)$ is close in supremum norm to the normalised function $e_{l,m}(\cdot)$, we have

$$\mathbf{E}_Q\left[\mathrm{I}_{l,m}(\vartheta)\right] \geq \frac{2A(1-\varepsilon)}{T^\beta}\,.$$

Remark that the sum over m of the first terms in the inequality (4.150) corresponding to the case $l = 0$ is smaller in order than $T^{-\frac{2k}{2k-1}}$, since the value I_0 is chosen significantly smaller than the other I_l's. Further, using (4.152), together with the inequality of Minkovsky, we get

$$\mathcal{A}_{1T} \geq -\frac{4}{T} \sum_{m=-M}^{M} \sum_{l=\pm 1}^{\pm L} \frac{\mathrm{I}_l \, \mathbf{E}\, |\Psi_{l,m,\vartheta}(\xi)|^2}{T\,\mathbf{E}_Q\, \mathrm{I}_{l,m}(\vartheta) + \mathrm{I}_l} + \frac{o_T(1)}{T^{\frac{2k}{2k-1}}}$$

$$= -\frac{16(1+\varepsilon)}{T} \sum_{m=-M}^{M} \sum_{l=1}^{L} \frac{\mathrm{I}_l \, A^2 f_*(a_m)(T^{2\beta}\pi^2 l^2)^{-1}}{2T^{1-\beta}A(1-\varepsilon) + \mathrm{I}_l} + \frac{o_T(1)}{T^{\frac{2k}{2k-1}}}$$

$$\geq -\frac{8(1+\varepsilon)^2 A}{(1-\varepsilon)T^{1+\beta}\pi^2} \sum_{l=1}^{L} \frac{1}{l^2(2T^{1-\beta}A\,\sigma_l(\varepsilon) + 1)} + \frac{o_T(1)}{T^{\frac{2k}{2k-1}}}. \quad (4.156)$$

Further, direct calculations allow us to write the estimate

$$\int_{-A}^{A} [f^{(k)}(\vartheta, x) - f_*^{(k)}(x)]^2 \mathrm{d}x$$

$$= 4(1+o(1)) \int_{-A}^{A} [S^{(k-1)}(\vartheta, x) - S_*^{(k-1)}(x)]^2 f_*^2(x) \mathrm{d}x$$

$$= 4(1+o(1)) \sum_{|m|\leq M} \frac{2A}{T^\beta f_*(a_m)} \sum_{1\leq |l| \leq L} \vartheta_{l,m}^2 f_*^2(a_m) \left(\frac{\pi l A}{T^\beta}\right)^{2(k-1)}.$$

If we evaluate now the mathematical expectation with respect to Q of this expression, using the fact that the partial sums of the function $f_*(\cdot)$ tend to its integral (which equals 1), we obtain

$$\mathbf{E}_Q \int_{\mathscr{R}} [f^{(k)}(\vartheta, x) - f_*^{(k)}(x)]^2 \mathrm{d}x$$

$$= 4(1+o(1)) \sum_{|m|\leq M} \frac{2A f_*(a_m)}{T^\beta} \sum_{1\leq |l|\leq L} \sigma_l(\varepsilon) \left(\frac{\pi l T^\beta}{A}\right)^{2(k-1)}$$

$$= 4(1+o(1)) \sum_{1\leq |l|\leq L} \sigma_l(\varepsilon) \left(\frac{\pi l T^\beta}{A}\right)^{2(k-1)}. \quad (4.157)$$

We can now explain the choice (4.148) of the prior distribution. It is clear from (4.156) that the least favorable prior is the one for which the Fisher informations I_l are minimal under the constraint (4.157) involving only the variances $\{\sigma_l(\varepsilon), l = \pm 1, \ldots, \pm L\}$. It is well known that the minimum of the Fisher information of a random variable with zero mean and fixed variance σ is attained by the Gaussian distribution and this minimum is equal to σ^{-1}. We cannot use here directly the product of normal distributions as a prior distribution, since for Höffding's inequality we need to have bounded random variables. That is why we use an approximation of Gaussian distribution and we pay a factor $1+\varepsilon$ for the error of this approximation.

In order to determine the values of σ_l, we solved the maximization problem related with the functional

$$\Phi(\boldsymbol{y}) = \sum_{l=1}^{L} \frac{1}{l^2(2T^{1-\beta}A\,y_l + 1)},$$

over the set

$$\mathcal{E}(k,R) = \left\{ y \in \mathscr{R}_+^{\mathrm{N}} \,:\, 8 \sum_{l=1}^{L} y_l \left(\frac{\pi l T^\beta}{A} \right)^{2(k-1)} \leq R \right\}.$$

This can be done directly using the method of Lagrange multipliers and the result is given by

$$y_l^* = \frac{1}{2A\,T^{1-\beta}} \left(\left| \frac{\Lambda}{l} \right|^k - 1 \right)_+ .$$

which leads to the definition of σ_l presented in the beginning of the proof.

Now we shall evaluate the sums on the right hand sides of (4.156) and (4.157). For the second one, setting $\Lambda_\varepsilon = \Lambda(1-\varepsilon)$ we have

$$\begin{aligned}
&\sum_{1 \leq |l| \leq L} \sigma_l(\varepsilon) \left(\frac{\pi l T^\beta}{A} \right)^{2(k-1)} \\
&\quad = \frac{1}{2AT^{1-\beta}} \sum_{1 \leq |l| \leq L} \left(\left[\frac{\Lambda_\varepsilon}{l} \right]^k - 1 \right)_+ \left(\frac{\pi l T^\beta}{A} \right)^{2(k-1)} \\
&\quad = \frac{\pi^{2(k-1)} \Lambda_\varepsilon^{2k-1}}{2A^{2k-1}} \sum_{1 \leq |l| \leq L} \left(\left[\frac{l}{\Lambda_\varepsilon} \right]^{k-2} - \left[\frac{l}{\Lambda_\varepsilon} \right]^{2k-2} \right) \frac{1}{\Lambda_\varepsilon}.
\end{aligned}$$

Since Λ_ε tends to infinity, when T tends to infinity, the last sum is close to the integral over $[0,1]$ of the function $x^{k-2} - x^{2k-2}$. This yields

$$\begin{aligned}
&\sum_{1 \leq |l| \leq L} \sigma_l(\varepsilon) \left(\frac{\pi l T^\beta}{A} \right)^{2(k-1)} \\
&\quad = \frac{(1+o(1))\pi^{2(k-1)} \Lambda_\varepsilon^{2k-1}}{2A^{2k-1}} \int_0^1 (x^{k-2} - x^{2k-2})\, dx \\
&\quad = \frac{(1+o(1))\pi^{2(k-1)} \Lambda_\varepsilon^{2k-1}}{2A^{2k-1}} \frac{k}{(k-1)(2k-1)} \leq \frac{R(1-\varepsilon)}{8}. \qquad (4.158)
\end{aligned}$$

This upper estimate permits us to use Höffding's inequality to obtain an exponential bound for Q-probability of the set $\Theta_T \setminus \Sigma_*$ (see Lemma 4.28). Thus we get

$$Q(\Theta_T \setminus \Sigma_*) \leq \frac{C}{T^2}. \qquad (4.159)$$

Evaluating the sum of (4.156) in the same way, we get

$$\begin{aligned}\mathcal{A}_{1T} &\geq -\frac{8(1+\varepsilon)^2 A}{(1-\varepsilon)T^{1+\beta}\pi^2}\sum_{l=1}^{L}\frac{1}{l^2}\left(\frac{l}{\Lambda_\varepsilon}\right)^k\\ &= -\frac{8(1+\varepsilon)^2 A(1+o(1))}{(1-\varepsilon)^2 T^{1+\beta}\pi^2 \Lambda}\int_0^1 x^{k-2}\mathrm{d}x\\ &\geq -\frac{8(1+\varepsilon)^3 A}{(1-\varepsilon)^2 T^{1+\beta}\pi^2\Lambda(k-1)}.\end{aligned}$$

Similarly, due to (4.152)

$$\begin{aligned}\mathcal{A}_{2T} &= -\frac{8A(1+o(1))}{T^{1+\beta}\pi^2}\sum_{l=L+1}^{\infty} l^{-2} = -\frac{8A(1+o(1))}{\Lambda_\varepsilon T^{1+\beta}\pi^2}\int_1^\infty x^{-2}\mathrm{d}x\\ &\geq -\frac{8A(1+\varepsilon)}{\Lambda_\varepsilon T^{1+\beta}\pi^2} \geq -\frac{8(1+\varepsilon)^3 A}{(1-\varepsilon)^2 T^{1+\beta}\pi^2\Lambda}\,.\end{aligned}$$

Putting together the last two estimates we get the following lower bound for the Bayesian risk with prior distribution Q:

$$\mathbf{E}_Q\,\mathscr{R}_T(\bar{f}_T, f_\vartheta) \geq -\frac{8(1+\varepsilon)^3 Ak}{(1-\varepsilon)^2 T^{1+\beta}\pi^2\Lambda(k-1)} = -T^{-\frac{2k}{2k-1}}\hat{\Pi}(k,R)\,\frac{(1+\varepsilon)^3}{(1-\varepsilon)^2}\,.$$

This inequality, (4.148) and (4.158) imply the desired result, since $\bar{f}_T \in \mathcal{W}_T$ and the real number ε can be chosen as small as we want.

4.6.3 Efficient Estimator

Our goal now is to construct an estimator attaining the lower bound obtained in the previous section. Note that in the problem of signal detection with Gaussian white noise, there is a linear estimator which is second order minimax. The linear filter defining this estimator utilizes the so-called Pinsker's weights.

The same holds for the problem of distribution function estimation from i.i.d. observations. In this case the second order minimax estimator is linear with respect to the (global) trigonometric basis on the interval $[-B, B]$, where B tends to infinity when the number of observations tends to infinity.

It turns out that an analogous result holds true in the model of ergodic diffusion, but one has to be careful in the definition of a linear estimator. More precisely, the estimation of the function f in $\mathscr{L}^2$-norm is equivalent to estimating in $\mathscr{L}^2$ its Fourier transform. Therefore, it would be natural to expect that the best linear estimator of the Fourier transform of f (which leads to a kernel estimator) is asymptotically optimal over the set of all possible estimators. But this assertion turned out to be wrong.

A very heuristic explanation of this fact is that when we consider the linear estimators of the Fourier transform, we disperse the information concerning each point over the whole real line and therefore the estimator does not take into account the local structure of the model with respect to the space parameter. That is why the localized bases of functions fit better to our model. In particular it appears that a slightly modified linear estimator with respect to a localized trigonometric basis is second order asymptotically efficient.

To construct our estimator, we introduce the localised orthonormal basis $\{e_{l,m} : l, m \in \mathbb{Z}\}$ of $\mathscr{L}^2(\mathscr{R})$ defined in the following way:

$$e_{l,m}(x) = \frac{1}{\sqrt{2\delta_T}} \exp\left\{\frac{\mathrm{i}\pi l(x - a_m + \delta_T)}{\delta_T}\right\}\chi_{\{x\in \mathbb{I}_m\}} .$$

Above $\mathrm{i} = \sqrt{-1}$ and $\mathbb{I}_m$ is the interval of length $2\delta_T$ with center $a_m = 2m\delta_T$. We call this basis localized since δ_T tends to zero when $T \to \infty$. The Fourier coefficients of the local time and the invariant density with respect to this basis are

$$\varphi^{\circ}_{T,l,m} = \int_{\mathscr{R}} \bar{e}_{l,m}(x) f^{\circ}_T(x)\,\mathrm{d}x = \frac{1}{T}\int_0^T \bar{e}_{l,m}(X_t)\,\mathrm{d}t,$$

$$\varphi_{S,l,m} = \int_{\mathscr{R}} \bar{e}_{l,m}(x) f_S(x)\,\mathrm{d}x = \mathbf{E}_S[\bar{e}_{l,m}(X_t)].$$

Of course, the estimation of the function $f_S(\cdot)$ in $\mathscr{L}^2$ is equivalent to the estimation of the infinite matrix $\{\varphi_{S,l,m} : l, m \in \mathbb{Z}\}$. Since $\varphi^{\circ}_{T,l,m}$ is the empirical estimator of $\varphi_{S,l,m}$, the linear estimator of $\{\varphi_{S,l,m} : l, m \subset \mathbb{Z}\}$ is defined as

$$\bar{\varphi}_{T,l,m} = \varphi_{T,l,m}\varphi^{\circ}_{T,l,m}, \qquad l, m \in \mathbb{Z}, \tag{4.160}$$

where $\varphi_{T,l,m}$ are some numbers between 0 and 1. This leads us to the following definition of the linear estimator of the invariant density:

$$\bar{f}_T(x) = \sum_{l,m\in\mathbb{Z}} \varphi_{T,l,m}\varphi^{\circ}_{T,l,m} e_{l,m}(x).$$

In order to attain the lower bound of the previous section, this linear estimator should be modified. In some sense, we have to correct it at boundaries. Let us discuss it in more detail. By condition the estimated function f_S belongs to the Sobolev class $\Sigma(k-1, R, S_*)$. We would like to derive from this condition an ellipsoid type inequality for the coefficients $\varphi_{S,l,m}$. Using the integration by parts formula, for $l \neq 0$, we get

$$\varphi_{S,l,m} = \varphi_{f^{(k)},l,m}\left(-\frac{\delta_T}{i\pi l}\right)^k - \sum_{j=0}^{k-1}\left(f^{(j)}(a_m+\delta_T) - f^{(j)}(a_m-\delta_T)\right)\left(-\frac{\delta_T}{i\pi l}\right)^{j+1}.$$

This last sum will be denoted $q_{l,m}(f)$. Now the condition

$$\int_{\mathscr{R}} f_S^{(k)}(x)^2 \,\mathrm{d}x \le R$$

can be rewritten as

$$\sum_{m\in\mathbb{Z}}\sum_{l\in\mathbb{Z}} |\varphi_{S,l,m} + q_{l,m}(f_S)|^2 \Big(\frac{\pi l}{\delta_T}\Big)^{2k} \le R. \tag{4.161}$$

Recall that the regularity condition imposed on the estimated function determines the behaviour of the bias and does not influence the variance. That is why we want to modify the linear estimator defined by (4.160) so that the bias of the modified estimator be linear with respect to $\varphi_{S,l,m} + q_{l,m}(f_S)$ and its variance remains unchanged. It is easy to check that the sequence

$$\varphi^*_{T,l,m} = \varphi_{T,l,m}(\varphi^\circ_{T,l,m} + q_{l,m}(f_S)) - q_{l,m}(f_S), \qquad l,m \in \mathbb{Z},$$

satisfies these conditions, but it is not an estimator since it involves the unknown function f_S via the term $q_{l,m}(f_S)$. To overcome this problem, we substitute this term by a rate optimal estimator $\hat q_{T,l,m}$. Thus,

$$\tilde\varphi_{T,l,m} = \varphi_{T,l,m}(\varphi^\circ_{T,l,m} + \hat q_{T,l,m}) - \hat q_{T,l,m}, \qquad l,m \in \mathbb{Z},$$

where

$$\begin{aligned}\hat q_{T,l,m} &= \sum_{j=0}^{k-1} \big(\tilde f_T^{(j)}(a_m+\delta_T) - \tilde f_T^{(j)}(a_m-\delta_T)\big)\Big(-\frac{\delta_T}{i\pi l}\Big)^{j+1} \\ &= \frac{1}{T}\sum_{j=0}^{k-1}\Big(-\frac{\delta_T}{ih_T\pi l}\Big)^{j+1} \int_{\mathbb{I}_m}\int_0^T Q^{(j+1)}\Big(\frac{X_t - x}{h_T}\Big)\,\mathrm{d}t\,\mathrm{d}x \end{aligned} \tag{4.162}$$

and Q is a kernel of order k with support $[-1,1]$. From now on, we set $h_T = T^{-1/(2k-1)}$ and $\delta_T = h_T^{3/4}$. Note that this trick with $q_{l,m}$ has already been exploited by Delattre and Hoffmann [60], it is particularly useful in the case of nonhomogeneous Fisher information.

We show in this section that there exists a modified linear estimator which is second order efficient. The weights $\varphi_{T,l,m}$ of this estimator are defined as follows:

$$\hat\varphi_{T,l,m} = \left(1 - \left|\frac{l}{\hat\nu}\right|^{k+\mu_T}\right)_+,$$

where $\mu_T = 1/\sqrt{\log T}$ tends to zero as $T \to \infty$ and

$$\hat\nu = \hat\nu_T = \delta_T\left(\frac{8k\pi^{2(k-1)}}{RT(k-1)(2k-1)}\right)^{-\frac{1}{2k-1}}.$$

We denote by $\hat f_T(\cdot)$ the estimator of the invariant density defined via these weights. Note that $\hat\nu_T$ tends to infinity as T tends to infinity.

Theorem 4.48. (Dalalyan and Kutoyants [56]) *Let $k > 1$. Then*

$$\lim_{T\to\infty} \sup_{S(\cdot)\in\Sigma_*} T^{\frac{2k}{2k-1}} \mathscr{R}(\hat f_T, f_S) = -\hat\Pi(k,R). \tag{4.163}$$

Proof. Let $S(\cdot) \in \Sigma_*$. Recall that the normalized first order minimax risk asymptotically equals

$$\mathcal{R}(S) = 4\int_{\mathscr{R}} f_S^2(x)\, \mathbf{E}_S\left[\frac{\chi_{\{\xi>x\}} - F_S(\xi)}{f_S(\xi)}\right]^2 dx = \sum_{l,m\in\mathbb{Z}} \mathbf{E}_S\left[\Psi^2_{S,l,m}(\xi)\right]$$

where we applied the Parseval identity to the square integrable function $g_S(\cdot,x)$ having the following Fourier coefficients:

$$\Psi_{S,l,m}(x) = 2\int_{\mathscr{R}} \bar e_{l,m}(y) f_S(y)\left[\frac{\chi_{\{x>y\}} - F_S(x)}{f_S(x)}\right] dy.$$

Using this decomposition, one can rewrite the second order risk function as

$$\begin{aligned}
\mathscr{R}_T(\hat f_T, f_S) &= \mathbf{E}_S\left[\sum_{l,m\in\mathbb{Z}}\left(|\tilde\varphi_{T,l,m} - \varphi_{S,l,m}|^2 - \frac{\mathbf{E}_S\left[\Psi^2_{S,l,m}(\xi)\right]}{T}\right)\right]\\
&= \mathbf{E}_S\left[\sum_{l,m\in\mathbb{Z}}\left(|\hat\varphi_{T,l,m}\varphi^\circ_{T,l,m} - \varphi_{S,l,m}\right.\right.\\
&\qquad \left.\left. + (\hat\varphi_{T,l,m} - 1)\hat q_{T,l,m}|^2 - \frac{\mathbf{E}_S\left[\Psi^2_{S,l,m}(\xi)\right]}{T}\right)\right]\\
&= \mathbf{E}_S\left[\sum_{l,m\in\mathbb{Z}}\left(|\hat\varphi_{T,l,m}(\varphi^\circ_{T,l,m} - q_{l,m}(f_S)) - (\varphi_{S,l,m} - q_{l,m}(f_S))\right.\right.\\
&\qquad \left.\left. + (\hat\varphi_{T,l,m} - 1)(\hat q_{T,l,m} - q_{l,m}(f_S))|^2 - \frac{\mathbf{E}_S\left[\Psi^2_{S,l,m}(\xi)\right]}{T}\right)\right].
\end{aligned}$$

We need the following estimates.

Lemma 4.49. *We have*

$$\mathbf{E}_S\left(\tilde f_T^{(j)}(a) - f_S^{(j)}(a)\right)^2 \le C h_T^{2k-2j}(1+|a|)^{\nu_*} f_S(a), \tag{4.164}$$

$$\mathbf{E}_S\left(\hat q_{T,l,m} - q_{l,m}(f_S)\right)^2$$

$$\le C h_T^{2k+1/2}\int_{\mathbb{I}_m}(1+|a|)^{\nu_*} f_S(a)\,da \sum_{j=0}^{k-1}\left(\frac{\hat\nu_T}{l}\right)^{2j+2}\hat\nu_T^{-1}. \tag{4.165}$$

Proof. The proof of this lemma is quite standard. It is based on a bias-variance decomposition. The bias is estimated in the usual way (like in the i. i. d. case) and the variance is bounded via the Itô formula.

The bias-variance decomposition yields

$$\mathbf{E}_S\left(\tilde{f}_T^{(j)}(a) - f_S^{(j)}(a)\right)^2 \\ \leq 2\left(\mathbf{E}_S[\tilde{f}_T^{(j)}(a)] - f_S^{(j)}(a)\right)^2 + 2\,\mathbf{E}_S\left(\tilde{f}_T^{(j)}(a) - \mathbf{E}_S[\tilde{f}_T^{(j)}(a)]\right)^2.$$

In view of the stationarity of X and the equality $\int Q(u)\mathrm{d}u = 1$, we get

$$\mathbf{E}_S[\tilde{f}_T^{(j)}(a)] - f_S^{(j)}(a) = \int_{\mathscr{R}} Q(u)\left[f_S{}^{(j)}(a + uh_T) - f_S{}^{(j)}(a)\right]\mathrm{d}u.$$

We apply now the Taylor formula with the rest term in the integral form and the fact that Q is a kernel of order k (i.e., $\int u^l Q(u)\mathrm{d}u = 0$ for any integer $l \in [1, k]$):

$$\mathbf{E}_S[\tilde{f}_T^{(j)}(a)] - f_S^{(j)}(a) \\ = \frac{1}{(k-j-1)\,!}\int_{\mathscr{R}} Q(u)\int_0^{uh_T} y^{k-j-1} f_S^{(k)}(a + uh_T - y)\,\mathrm{d}y\,\mathrm{d}u.$$

Since $Q(u) = 0$ if $|u| > 1$, we obtain

$$\left|\mathbf{E}_S[\tilde{f}_T^{(j)}(a)] - f_S^{(j)}(a)\right| \leq \frac{h_T^{k-j-1}}{(k-j-1)\,!}\int_{-h_T}^{h_T}\left|f_S^{(k)}(a+y)\right|\mathrm{d}y.$$

Cauchy–Schwarz inequality yields

$$\left|\mathbf{E}_S[\tilde{f}_T^{(j)}(a)] - f_S^{(j)}(a)\right|^2 \leq \frac{h_T^{2k-2j-1}}{[(k-j-1)\,!]^2}\int_{-h_T}^{h_T}\left[f_S^{(k)}(a+y)\right]^2\mathrm{d}y. \qquad (4.166)$$

We turn to the evaluation of the variance. We have

$$\tilde{f}_T^{(j)}(a) - \mathbf{E}_S[\tilde{f}_T^{(j)}(a)] = \frac{1}{Th_T^{j+1}}\int_0^T\left(Q^{(j)}\left(\frac{X_t - a}{h_T}\right) - \mathbf{E}_S\left[Q^{(j)}\left(\frac{X_t - a}{h_T}\right)\right]\right)\mathrm{d}t.$$

Let us introduce some functions $g(u)$ and $H(u)$ as follows:

$$g(u) = Q^{(j)}\left(\frac{u-a}{h_T}\right) - \mathbf{E}_S\left[Q^{(j)}\left(\frac{X_0 - a}{h_T}\right)\right],$$
$$H(u) = \int_0^u \frac{2}{f_S(y)}\int_{-\infty}^y g(v) f_S(v)\,\mathrm{d}v\,\mathrm{d}y.$$

It can be easily checked that the Itô formula implies

$$\int_0^T g(X_t)\,\mathrm{d}t = H(X_T) - H(X_0) - \int_0^T H'(X_t)\,\mathrm{d}W_t,$$

and therefore, due to stationarity,

$$\mathbf{E}_S\left(\int_0^T g(X_t)\,\mathrm{d}t\right)^2 \le 6\,\mathbf{E}_S\left[H^2(X_0)\right] + 3T\,\mathbf{E}_S\left[\left(H'(X_0)\right)^2\right]$$

By integrating by parts, we get

$$\begin{aligned}
H'(u) &= 2\int_{-\infty}^{u} g(y)\,\mathrm{d}y - \frac{2}{f_S(u)}\int_{-\infty}^{u}\int_{-\infty}^{y} g(v)\,\mathrm{d}v f_S'(y)\,\mathrm{d}y \\
&= 2\int_{-\infty}^{u} Q^{(j)}\left(\frac{y-a}{h_T}\right)\mathrm{d}y \\
&\quad - \frac{2}{f_S(u)}\int_{\mathscr{R}}\int_{-\infty}^{y} Q^{(j)}\left(\frac{v-a}{h_T}\right)\mathrm{d}v f_S'(y)\left(\chi_{\{y<u\}} - F_S(u)\right)\mathrm{d}y \\
&= 2h_T Q^{(j-1)}\left(\frac{u-a}{h_T}\right) \\
&\quad - \frac{2h_T}{f_S(u)}\int_{\mathscr{R}} Q^{(j-1)}\left(\frac{y-a}{h_T}\right) f_S'(y)\left(\chi_{\{y<u\}} - F_S(u)\right)\mathrm{d}y. \qquad (4.167)
\end{aligned}$$

Using the inequality

$$\frac{\chi_{\{y<u\}} - F_S(u)}{f_S(u)} \le C + \frac{C\chi_{\{u\in[0,y]\}}}{f_S(u)} \le C + \frac{C\chi_{\{u\in[0,y]\}}}{\sqrt{f_S(u) f_S(y)}},$$

the equality $f_S'(x) = 2S(x) f_S'(x)$ and that fact that $S(\cdot)$ increases at most like a polynomial, one can prove that the second term in (4.167) is asymptotically negligible with respect to the first one. Therefore,

$$\begin{aligned}
\mathbf{E}_S\left[\left(H'(X_0)\right)^2\right] &\le 4(1+o_T(1))\,\mathbf{E}_S\left(\left[h_T Q^{(j-1)}\left(\frac{X_0-a}{h_T}\right)\right]^2\right) \\
&\le 4(1+o_T(1))\,h_T^3\int_{-1}^{1}\left[Q^{(j-1)}(u)\right]^2 f_S(a+uh_T)\,\mathrm{d}u \\
&\le C h_T^3\int_{a-h_T}^{a+h_T} f_S(x)\,\mathrm{d}x.
\end{aligned}$$

It is evident that the term involving the expectation of $H^2(X_0)$ is asymptotically much smaller than the term $T\,\mathbf{E}_S[H'(X_0)^2]$. Consequently,

$$\begin{aligned}
&\mathbf{E}_S\left(\tilde f_T^{(j)}(a) - f_S^{(j)}(a)\right)^2 \\
&\quad \le C h_T^{2k-2j-1}\int_{a-h_T}^{a+h_T}\left[f_S^{(k)}(y)\right]^2\mathrm{d}y + \frac{C}{T h_T^{2j-1}}\int_{a-h_T}^{a+h_T} f_S(x)\,\mathrm{d}x \\
&\quad \le C h_T^{2k-2j-1}\int_{a-h_T}^{a+h_T} f_S(x)\,\mathrm{d}x \le C h_T^{2k-2j}(1+|a|)^{\nu_*} f_S(a).
\end{aligned}$$

The Cauchy–Schwarz inequality yields

$$\mathbf{E}_S\left(\int_{\mathbb{I}_m}(\tilde f_T^{(j)}(a)-f_S^{(j)}(a))\,\mathrm{d}a\right)^2\le C\delta_T h_T^{2k-2j}\int_{\mathbb{I}_m}(1+|a|)^{\nu_*}f_S(a)\,\mathrm{d}a,$$

and therefore, using the equality

$$\hat q_{T,l,m}-q_{l,m}(f_S)=\sum_{j=0}^{k-1}\left(-\frac{\delta_T}{i\pi l}\right)^{j+1}\int_{\mathbb{I}_m}(\tilde f_T^{(j+1)}(a)-f_S^{(j+1)}(a))\,\mathrm{d}x$$

and the Cauchy–Schwarz inequality, we get

$$\begin{aligned}
&\mathbf{E}_S\left|\hat q_{T,l,m}-q_{l,m}(f_S)\right|^2\\
&\quad\le k\sum_{j=0}^{k-1}\left(\frac{\delta_T}{\pi l}\right)^{2j+2}\mathbf{E}_S\left(\int_{\mathbb{I}_m}(\tilde f_T^{(j+1)}(a)-f_S^{(j+1)}(a))\,\mathrm{d}x\right)^2\\
&\quad\le k\sum_{j=0}^{k-1}\left(\frac{\delta_T}{\pi l}\right)^{2j+2}C\delta_T h_T^{2k-2j-2}\int_{\mathbb{I}_m}(1+|a|)^{\nu_*}f_S(a)\,\mathrm{d}a\\
&\quad\le C h_T^{2k+1/2}\int_{\mathbb{I}_m}(1+|a|)^{\nu_*}f_S(a)\,\mathrm{d}a\sum_{j=0}^{k-1}\left(\frac{\hat\nu_T}{l}\right)^{2j+2}\hat\nu_T^{-1}.
\end{aligned}$$

In the last inequality we used the fact that the step of localization δ_T^2 is equal to the square root of h_T^3 and that $\hat\nu_T$ is of the order of δ_T/h_T.

We return to the proof of the theorem. On the one hand, we show below that the term

$$\sum_{l,m\in\mathbb{Z}}\mathbf{E}_S\left|\hat\varphi_{T,l,m}(\varphi^\circ_{T,l,m}-q_{l,m}(f_S))-(\varphi_{S,l,m}-q_{l,m}(f_S))\right|^2$$

is of the order of T^{-1}. On the other hand, the second inequality of Lemma 4.49 implies that the term

$$\sum_{l,m\in\mathbb{Z}}\mathbf{E}_S\left|(\hat\varphi_{T,l,m}-1)(\hat q_{T,l,m}-q_{l,m}(f_S))\right|^2$$

is of the order of $T^{-(2k+1/2)/(2k-1)}$. So, a simple application of Hölder inequality yields

$$\begin{aligned}
&\mathscr{R}_T(\hat f_T,f_S)\\
&\quad\le\sum_{l,m\in\mathbb{Z}}\Big(\mathbf{E}_S\left|\hat\varphi_{T,l,m}(\varphi^\circ_{T,l,m}-q_{l,m}(f_S))-(\varphi_{S,l,m}-q_{l,m}(f_S))\right|^2\\
&\qquad-T^{-1}\mathbf{E}_S\left[\Psi^2_{S,l,m}(\xi)\right]\Big)+CT^{-(2k+1/4)/(2k-1)}.
\end{aligned}$$

The stationarity of the process X implies that the mathematical expectation of $\varphi^{\circ}_{T,l,m}$ is equal to $\varphi_{S,l,m}$. Therefore

$$\mathscr{R}_T(\hat{f}_T, f_S) = \sum_{l,m\in\mathbb{Z}} \Big(|\hat{\varphi}_{T,l,m} - 1|^2 \, |\varphi_{S,l,m} - q_{l,m}(f_S)|^2 + |\hat{\varphi}_{T,l,m}|^2 \mathbf{Var}_S[\varphi^{\circ}_{T,l,m}] - T^{-1} \mathbf{E}_S \left[\Psi^2_{S,l,m}(\xi)\right] \Big) + CT^{-(2k+1/4)/(2k-1)}.$$

Actually, the first term of this sum is in the appropriate form and we show below that it is of the order of $T^{-2k/(2k+1)}$. To determine how the second and third terms behave, we need several technical results.

Lemma 4.50. *The following relations hold:*

$$\sum_{l,m\in\mathbb{Z}} |\hat{\varphi}_{T,l,m}|^2 \left(\mathbf{Var}_S[\varphi^{\circ}_{T,l,m}] - \frac{\mathbf{E}_S \left[\Psi^2_{S,l,m}(\xi)\right]}{T} \right) \le \frac{o_T(1)}{T^{\frac{2k}{2k-1}}}, \tag{4.168}$$

$$\sum_{l,m\in\mathbb{Z}} \left(1 - |\hat{\varphi}_{T,l,m}|^2\right) \mathbf{E}_S \left[\Psi^2_{S,l,m}(\xi)\right] \ge (1 + o_T(1)) \sum_{l,m\in\mathbb{Z}_*} \frac{8\delta_T^2 f_S(a_m)(1 - |\hat{\varphi}_{T,l,m}|^2)}{\pi^2 l^2}. \tag{4.169}$$

The proof of this lemma is quite technical and relies essentially on the arguments used in the proofs of Lemmas 4.46–4.47. That is why it will be omitted here.

Using this lemma, for any linear filter $\hat{\varphi} = \{\hat{\varphi}_{T,l,m}\}$ such that $|\hat{\varphi}_{T,l,m}| \le 1$, the second order risk can be evaluated as follows:

$$\mathscr{R}_T(\hat{f}_T, f_S) \le \sum_{l,m\in\mathbb{Z}, l\neq 0} \left(|\hat{\varphi}_{T,l,m} - 1|^2 |\varphi_{S,l,m}|^2 - \frac{8\delta_T^2 f_S(a_m)}{T\pi^2 l^2} (1 - |\hat{\varphi}_{T,l,m}|^2) \right) + CT^{-\frac{2k+1/4}{2k-1}}.$$

Recall that the function $f_S(\cdot)$ is such that $S(\cdot)$ belongs to $\Sigma_*(k-1, R)$. Proceeding like in (4.161), we get

$$\int_{\mathscr{R}} \left(f_S^{(k)}(x) - f_*^{(k)}(x)\right)^2 \mathrm{d}x = (\pi\delta_T^{-1})^{2k} \sum_{l,m\in\mathbb{Z}} l^{2k} |\varphi_{S,l,m} - q_{l,m}(f_S) - \varphi_{*,l,m} + q_{l,m}(f_*)|^2 \le R. \tag{4.170}$$

Since the central function $f_*(\cdot)$ is smoother than $f_S(\cdot)$, we have

$$\sum_{l,m\in\mathbb{Z}} l^{2k+2\mu_T}|\varphi_{*,l,m} - q_{l,m}(f_*)|^2 < C < +\infty,$$

where the constant C does not depend on T. If we use this inequality and the convergence $\hat{\nu}_T^{\mu_T} \to \infty$, we obtain

$$\begin{aligned}\sum_{l,m\in\mathbb{Z}} &\left|1-\hat{\varphi}_{T,l,m}\right|^2|\varphi_{S,l,m} - q_{l,m}(f_S)|^2 \\ &\leq \hat{\nu}_T^{-2k}\sum_{l,m\in\mathbb{Z}} l^{2k}|\varphi_{S,l,m} - q_{l,m}(f_S) - \varphi_{*,l,m} + q_{l,m}(f_*)|^2(1+o_T(1)) \\ &\leq (\hat{\nu}_T\delta_T^{-1}\pi)^{-2k}R\,(1+o_T(1)).\end{aligned}$$

Since the coefficients $\hat{\varphi}_{T,l,m}$ do not depend on m, we will omit the index m. We denote by $\mathbb{Z}_*$ the set of all integers different from zero. Using the fact that the integral of $f_S(\cdot)$ is equal to one and the convergence of partial sums, we get

$$\mathscr{R}_T(\hat{f}_T, f_S) \leq \frac{R\,\delta_T^{2k}}{(\pi\,\hat{\nu})^{2k}}\,(1+o_T(1)) - \sum_{l\in\mathbb{Z}_*}\frac{4\delta_T\big(1-|\hat{\varphi}_{T,l}|^2\big)}{\pi^2 T l^2}\,.$$

It is easy to check that

$$\begin{aligned}\sum_{l\in\mathbb{Z}_*}\frac{\big(1-|\hat{\varphi}_{T,l}|^2\big)}{l^2} &\sim \sum_{|l|\leq\hat{\nu}}\frac{1}{l^2}\left(2\left|\frac{l}{\hat{\nu}}\right|^k - \left|\frac{l}{\hat{\nu}}\right|^{2k}\right) + \sum_{|l|>\hat{\nu}}\frac{1}{l^2} \\ &\sim \frac{2}{\hat{\nu}}\int_0^1 (2x^{k-2} - x^{2k-2})\,\mathrm{d}x + \frac{2}{\hat{\nu}}\int_1^\infty x^{-2}\mathrm{d}x \\ &= \frac{2}{\hat{\nu}}\left(\frac{2}{k-1} - \frac{1}{2k-1} + 1\right) = \frac{4k^2}{\hat{\nu}(k-1)(2k-1)}\,.\end{aligned}$$

Above $a_T \sim b_T$ means $a_T = b_T\big(1+o_T(1)\big)$. We are able now to explain the choice of $\hat{\nu}$; it is simply the value minimising the function

$$U(\nu) = \frac{R\,\delta_T^{2k}\,\nu^{-2k}}{\pi^{2k}} - \frac{16\,k^2\delta_T\,\nu^{-1}}{\pi^2\,T\,(k-1)(2k-1)}\,.$$

The substitution of $\hat{\nu}_T$ by its value leads to the desired bound for the second order risk of our estimator:

$$\mathscr{R}_T(\hat{f}_T, f_S) \leq \big(1+o_T(1)\big)\,T^{-\frac{2k}{2k-1}}\,\hat{\Pi}(k,R).$$

This completes the proof of the theorem.

4.6.4 Discussion

Note that the second order asymptotically efficient estimator $\hat{f}_T(x)$ can be written (for $x \in \mathbb{I}_m$) as

$$\hat{f}_T(x) = \frac{1}{2T\delta_T} \int_0^T \sum_{l=-\hat{\nu}_T}^{\hat{\nu}_T} \left(1 - \left|\frac{l}{\hat{\nu}_T}\right|^{k_T}\right) \cos\left(\frac{\pi l\,(x - X_t)}{\delta_T}\right) \chi_{\{X_t \in \mathbb{I}_m\}} \mathrm{d}t$$

where $k_T = k + \mu_T$ or

$$\hat{f}_T(x) = \sum_{l=-\hat{\nu}_T}^{\hat{\nu}_T} \left(1 - \left|\frac{l}{\hat{\nu}_T}\right|^{k_T}\right) \frac{1}{2\delta_T} \int_{a_m-\delta_T}^{a_m+\delta_T} \cos\left(\frac{\pi l\,(x - y)}{\delta_T}\right) f_T^{\circ}(y)\,\mathrm{d}y.$$

We supposed in all statements and proofs that the diffusion coefficient $\sigma(\cdot)$ is identically 1, but this condition can be relaxed. There are two possible issues in this case:

- Either one considers the weighted $\mathscr{L}^2$ risk

$$\mathscr{R}_T(\bar{f}_T, f_S) = \mathbf{E}_S \int_{\mathscr{R}} \left[\bar{f}_T(x) - f_S(x)\right]^2 \sigma(x)^2\,\mathrm{d}x$$
$$-\frac{1}{T} \int_{\mathscr{R}} f_S(x)^2\,\sigma(x)^2\,\mathbf{E}_S \left(\frac{\chi_{\{\xi > x\}} - F_S(\xi)}{\sigma(\xi) f_S(\xi)}\right)^2 \mathrm{d}x$$

  and defines the weighted Sobolev ball

$$\Sigma\,(k-1, R) = \left\{S(\cdot)\ :\ \int_{\mathscr{R}} \left[f_S^{(k)}(x) - f_*^{(k)}(x)\right]^2 \sigma(x)^2\ \mathrm{d}x \le R\right\}$$

  as the parameter space. Then the optimal constant remains unchanged.
- Or one refuses to change the risk definition but accepts to work in local minimax settings, i.e., when the estimated function belongs to a narrowing neighborhood of an unknown (but fixed) function $f_*(\cdot)$. In that case the optimal constant depends on the central function $f_*(\cdot)$ and has the following form:

$$\hat{\Pi}(f_*, k, R) = 2(2k-1)\left(\frac{4k}{\pi(k-1)(2k-1)}\right)^{\frac{2k}{2k-1}} \left(\int_{\mathscr{R}} \frac{f_*(x)}{\sigma(x)^2}\,\mathrm{d}x\right)^{\frac{2k}{2k-1}} R^{\frac{1}{2k-1}}.$$

The optimal constant $\hat{\Pi}(k, R)$ we obtained in this work contains a factor 2 (therefore is twice smaller) which is absent in the second order optimal constant in the problem of distribution function estimation from i.i.d. observations (cf. Golubev and Levit [91]). In some sense, it reveals that the estimation problem in our settings is easier than in the settings of [91]. This phenomenon has a simple explanation.

In the definition of the parameter set, we required that the function $S(\cdot)$ satisfies the inequality $\operatorname{sgn}(x)S(x) \leq \gamma_*$ for $x \notin [-A, A]$. This condition excludes automatically the density functions $f_S(\cdot)$ possessing heavy tails. Moreover, it implies that these densities decrease exponentially fast at infinity. In particular, the densities close to those of uniform distributions over a large interval do not belong to our nonparametric class. But exactly these densities are the most difficult to estimate, because the observations distributed following these laws are quite dispersed on the whole real line and contain very little information about the local variation of the underlying density function.

In the paper [91] this condition is not supposed to be satisfied and the least favorable parametric family is constructed precisely around a uniform distribution over an increasing interval. That is why the optimal constant is worse than in our case. On the other hand, if we consider the problem of distribution function estimation from i.i.d. data in a local vicinity of a fixed function, the factor 2 will appear (as in our problem) and the optimal constant will be

$$\hat{\Pi}(k, R) = 2(2k-1)\left(\frac{k}{\pi(k-1)(2k-1)}\right)^{\frac{2k}{2k-1}} R^{\frac{1}{2k-1}}.$$

Remember that the local time estimator $f_T^\circ(x)$ is asymptotically normal with mean $f_S(x)$ and variance $[T\mathrm{I}_f(S,x)]^{-1}$. Since the local time is a sufficient statistic, this latter is weakly asymptotically equivalent to the Gaussian experiment

$$Y_t = f_S(t) + \frac{2f_S(t)}{\sqrt{T}} \int_{\mathscr{R}} \frac{\chi_{\{u>t\}} - F_S(u)}{\sqrt{f_S(u)}} \, \mathrm{d}B_u,$$

where $B = (B_u, u \in \mathscr{R})$ is a Brownian motion. Let now $v(\cdot)$ be a test function (infinitely differentiable with compact support) and $V(\cdot)$ be its primitive. A formal use of the integration by parts formula yields

$$\begin{aligned}
\int_{\mathscr{R}} v(t) Y_t \, \mathrm{d}t &= \int_{\mathscr{R}} v(t) f_S(t) \, \mathrm{d}t + \frac{2}{\sqrt{T}} \int_{\mathscr{R}} v(t) f_S(t) \int_{\mathscr{R}} \frac{\chi_{\{u>t\}} - F_S(u)}{\sqrt{f_S(u)}} \, \mathrm{d}B_u \, \mathrm{d}t \\
&= -\int_{\mathscr{R}} V(t) f_S'(t) \, \mathrm{d}t - \frac{2}{\sqrt{T}} \int_{\mathscr{R}} V(t) \sqrt{f_S(t)} \, \mathrm{d}B_t \\
&\quad - \frac{2}{\sqrt{T}} \int_{\mathscr{R}} V(t) f_S'(t) \int_{\mathscr{R}} \frac{\chi_{\{u>t\}} - F_S(u)}{\sqrt{f_S(u)}} \, \mathrm{d}B_u \, \mathrm{d}t.
\end{aligned}$$

The third term of this sum can be dropped in the asymptotical results, since in contrast with the second one, it is uniformly bounded in $\mathscr{L}^2$ in the following sense:

$$\sup_{S \in \Sigma_*} \mathbf{E}_S \int_{\mathscr{R}} \left[f_S'(t) \int_{\mathscr{R}} \frac{\chi_{\{u>t\}} - F_S(u)}{\sqrt{f_S(u)}} \, \mathrm{d}B_u \right]^2 \mathrm{d}t < \infty.$$

Therefore, if we drop the third term we come to the Gaussian experiment

$$dY_t = f'_S(t)\,dt + 2\sqrt{f_S(t)T^{-1}}\,dB_t, \qquad t \in \mathscr{R}.$$

As is noted by Golubev and Levit [91], second order minimax estimation of a function, which is a regular functional (invariant density in our case) of the unknown parameter, is closely related to the problem of first order minimax estimation of the derivative of this function. On the other hand, it is known (see, for instance, Golubev [89]) that in the Gaussian shift experiment

$$dY_t = \theta(t)\,dt + \epsilon\sqrt{\mathrm{I}^{-1}(t)}\,dB_t,$$

the optimal constant depends on the Fisher information $\mathrm{I}(t)$ only via the integral $\int \mathrm{I}^{-1}(t)\,dt$. Due to the fact that $f_S(\cdot)$ is a probability density, the integral of the inverse of the Fisher information in our case is equal to one and does not depend on $S(\cdot)$.

This explains why we succeed in our program to obtain a second order minimax (up to constant) bound over a global nonparametric class of the unknown parameter. These arguments show also that the restriction to the $\mathscr{L}^2$-norm in the risk definition is essential for obtaining the optimal constants.

To end up with heuristic reasoning, let us remark that the reason, which made the obtaining of the second order minimax constant possible in the problem of the distribution function $F(\cdot)$ estimation from i.i.d. observations, is the same. Since according to Nussbaum [192], the statistical problem of $F'(\cdot)$ estimation is locally asymptotically equivalent to recovering the function $F'(\cdot)$ from the Gaussian shift experiment

$$dY_t = F'(t)\,dt + \sqrt{n^{-1}F'_0(t)}\,dB_t,$$

where n is the size of the sample and $F_0(\cdot)$ is the center of localization.

5

Hypotheses Testing

We consider several problems of hypotheses testing for ergodic diffusion processes. We start from the simplest one: simple hypothesis against simple alternative and then will check several composite alternatives of increasing complexity: one-sided parametric alternatives (smooth and non smooth), one-sided nonparametric alternatives and some smooth nonparametric alternatives. In every problem we construct an asymptotically optimal test. In all problems we suppose that the hypotheses concern the trend coefficient only and the diffusion coefficient is a known positive function.

5.1 Simple Hypothesis and Alternative

Let the observed diffusion process be

$$\mathrm{d}X_t = S(\vartheta, X_t)\,\mathrm{d}t + \sigma(X_t)\,\mathrm{d}W_t, \quad X_0, \quad 0 \le t \le T, \tag{5.1}$$

and we have to check a simple hypothesis

$$\mathscr{H}_0: \quad \vartheta = \vartheta_0,$$

against a simple alternative

$$\mathscr{H}_1: \quad \vartheta = \vartheta_1.$$

We suppose that $S(\vartheta_i, \cdot) \in \mathscr{S}_\sigma$, $i = 0, 1$. Therefore the process (5.1) has ergodic properties under hypothesis and alternatives and the measures $\mathbf{P}_{\vartheta_0}^{(T)}$ and $\mathbf{P}_{\vartheta_1}^{(T)}$ are equivalent.

Fix a number $\varepsilon \in (0,1)$ and recall that the test

$$\phi_T^*(X^T) = \chi_{\{L(\vartheta_1, \vartheta_0, X^T) \ge c_\varepsilon\}} \tag{5.2}$$

is the most powerful in the class $\mathscr{K}_\varepsilon$ of tests of level $1 - \varepsilon$ (see Proposition 1.69). Here $L(\vartheta_1, \vartheta_0, X^T)$ is the likelihood ratio

$$L\left(\vartheta_1,\vartheta_0,X^T\right)=\frac{G\left(\vartheta_1\right)}{G\left(\vartheta_0\right)}\,\exp\left\{\int_0^T\frac{S\left(\vartheta_1,X_t\right)-S\left(\vartheta_0,X_t\right)}{\sigma\left(X_t\right)^2}\,\mathrm{d}X_t\right.$$
$$\left.-\frac{1}{2}\int_0^T\frac{S\left(\vartheta_1,X_t\right)^2-S\left(\vartheta_0,X_t\right)^2}{\sigma\left(X_t\right)^2}\,\mathrm{d}t+2\int_0^{X_0}\frac{S\left(\vartheta_1,y\right)-S\left(\vartheta_0,y\right)}{\sigma\left(y\right)^2}\,\mathrm{d}y\right\}$$

and the constant $c_\varepsilon=c_\varepsilon\left(T\right)$ is defined by the equation

$$\mathbf{E}_{\vartheta_0}\,\phi_T^*\left(X^T\right)=\mathbf{P}_{\vartheta_0}^{(T)}\left\{L\left(\vartheta_1,\vartheta_0,X^T\right)\geq c_\varepsilon\right\}=1-\varepsilon. \tag{5.3}$$

We are interested in the calculation or approximation of the constant c_ε and of the power $\beta_T\left(\phi_T^*\right)=\mathbf{E}_{\vartheta_1}\,\phi_T^*\left(X^T\right)$. There are at least two ways to simplify the problem. The first one is to introduce the randomization: the test

$$\hat\phi_T\left(X^T\right)=\chi_{\left\{\Delta_T\left(\vartheta_1,\vartheta_0,X^T\right)\geq z_\varepsilon\sqrt{2\,J\left(\vartheta_1,\vartheta_0\right)}\right\}}$$

belongs to $\mathscr{K}_\varepsilon$ and its power

$$\begin{aligned}\beta_T\left(\hat\phi_T\right)&=\mathbf{P}\left\{\zeta\geq z_\varepsilon-\sqrt{2T}\,J\left(\vartheta_1,\vartheta_0\right)^{1/2}\right\}\\&=1-\exp\left\{-T\,J\left(\vartheta_1,\vartheta_0\right)+z_\varepsilon\sqrt{2T}\,J\left(\vartheta_1,\vartheta_0\right)^{1/2}\left(1+o(1)\right)\right\}\end{aligned}$$

(see Proposition 1.70 and (1.179)).

Another way is to study the wider class of tests $\mathscr{K}_\varepsilon'$ of asymptotic level $1-\varepsilon$. Then the constant c_ε can be written as (1.98)

$$c_\varepsilon=\exp\left\{-T\,J\left(\vartheta_0,\vartheta_1\right)+z_\varepsilon\,d\sqrt{T}\right\}.$$

In this case to describe the asymptotics of the power function $\beta_T\left(\phi_T^*\right)$ we need one result by Veretennikov [236] concerning the large deviations principle.

5.1.1 Large Deviations Principle

Let the observed process be

$$\mathrm{d}X_t=S\left(X_t\right)\,\mathrm{d}t+\sigma\left(X_t\right)\mathrm{d}W_t,\quad x_0,\ 0\leq t\leq T, \tag{5.4}$$

where x_0 is a nonrandom initial value and the functions $S\left(\cdot\right)$ and $\sigma\left(\cdot\right)$ are such that the conditions $\mathcal{ES}$ and $\mathcal{RP}$ are fulfilled. Then for any function $H\left(\cdot\right)$ measurable and integrable w.r.t. invariant density $f\left(\cdot\right)$ we have the law of large numbers: with probability 1

$$\lim_{T\to\infty}\frac{1}{T}\int_0^T H\left(X_t\right)\,\mathrm{d}t=\int_{\mathscr{R}}H\left(x\right)\,f\left(x\right)\,\mathrm{d}x\equiv\mathbf{E}\,H(\xi).$$

The large deviations principle (LDP) for ordinary integrals allows us to describe the exponential asymptotics of the probability

$$\mathbf{P}\left\{\frac{1}{T}\int_0^T H\left(X_t\right)\,\mathrm{d}t - \mathbf{E}\,H(\xi) > y\right\}$$

for $y > 0$.

Suppose that $\mathbf{E}\,H(\xi) = 0$ and introduce the following condition.

$\mathcal{U}$. *The functions* $S\left(\cdot\right)$, $\sigma\left(\cdot\right)$ *and* $H\left(\cdot\right)$ *are such that* $\inf_x \sigma(x)^2 > 0$ *and*

$$\overline{\lim_{|x|\to\infty}}\ \operatorname{sgn}\left(x\right)\,\frac{S\left(x\right)}{\sigma\left(x\right)^2} = -\infty, \tag{5.5}$$

$$\overline{\lim_{|x|\to\infty}}\ \operatorname{sgn}\left(x\right)\,\frac{S\left(x\right)}{\left|H\left(x\right)\right|} = -\infty. \tag{5.6}$$

Define the function

$$L\left(\lambda\right) = \lim_{T\to\infty} T^{-1}\ln\mathbf{E}\,\exp\left\{\lambda\int_0^T H\left(X_t\right)\,\mathrm{d}t\right\},\qquad \lambda\in\mathscr{R}. \tag{5.7}$$

This function under conditions (5.5) and (5.6) exists and is convex and its *Fenchel–Legendre transform*

$$\mathrm{R}\left(\alpha\right) = \sup_\lambda\Big(\alpha\lambda - L\left(\lambda\right)\Big),\qquad \alpha\in\mathscr{R}$$

is also a convex function (see Veretennikov [236]).

Proposition 5.1. (Veretennikov [236]) *Let the condition* $\mathcal{U}$ *be fulfilled. Then*

$$\mathbf{P}\left\{\frac{1}{T}\int_0^T H\left(X_t\right)\,\mathrm{d}t > y\right\} = \exp\left\{-\mathrm{R}\left(y\right)\,T\,\left(1+o(1)\right)\right\}. \tag{5.8}$$

Using (5.8) and the Itô formula we can formulate a similar large deviations principle for the stochastic integral

$$\frac{1}{T}\int_0^T h\left(X_t\right)\,\mathrm{d}W_t$$

too.

Proposition 5.2. *Suppose that the functions* $h\left(\cdot\right)$ *and* $\sigma\left(\cdot\right)$ *are continuously differentiable, the function* $h\left(x\right)\sigma\left(x\right)^{-1}$ *is bounded and is such that the condition* $\mathcal{U}$ *is fulfilled with*

$$H\left(x\right) = \frac{h'\left(x\right)\sigma\left(x\right) - h\left(x\right)\sigma'\left(x\right)}{2} + \frac{h\left(x\right)\,S\left(x\right)}{\sigma\left(x\right)},$$

then

$$\mathbf{P}\left\{\frac{1}{T}\int_0^T h\left(X_t\right)\,\mathrm{d}W_t > y\right\} = \exp\left\{-\mathrm{R}\left(y\right)\,T\,\left(1+o(1)\right)\right\}. \tag{5.9}$$

Proof. This follows from the representation

$$\int_0^T h(X_t)\,\mathrm{d}W_t = \int_{x_0}^{X_T} \frac{h(x)}{\sigma(x)}\,\mathrm{d}x - \int_0^T \left[\frac{h(X_t)\,S(X_t)}{\sigma(X_t)} + \frac{h'(X_t)\,\sigma(X_t) - h(X_t)\,\sigma'(X_t)}{2}\right]\,\mathrm{d}t$$

and the following elementary estimates. For any $\delta > 0$

$$\mathbf{P}\left\{\frac{1}{T}\int_0^T h(X_t)\,\mathrm{d}W_t > y\right\} \leq \mathbf{P}\left\{\frac{1}{T}\int_0^T H(X_t)\,\mathrm{d}t > y - \delta\right\} + \mathbf{P}\left\{\frac{1}{T}\int_{x_0}^{X_T} \frac{h(x)}{\sigma(x)}\,\mathrm{d}x > \delta\right\}$$

and

$$\mathbf{P}\left\{\frac{1}{T}\int_0^T h(X_t)\,\mathrm{d}W_t > y\right\} \geq \mathbf{P}\left\{\frac{1}{T}\int_0^T H(X_t)\,\mathrm{d}t > y + \delta\right\} - \mathbf{P}\left\{\frac{1}{T}\int_{x_0}^{X_T} \frac{h(x)}{\sigma(x)}\,\mathrm{d}x > \delta\right\}.$$

For the first probability in these inequalities we have (5.8) and for the second probability, using the boundedness of $h(x)\,\sigma(x)^{-1}$ we obtain the estimate

$$\mathbf{P}\left\{\frac{1}{T}\int_{x_0}^{X_T} \frac{h(x)}{\sigma(x)}\,\mathrm{d}x > \delta\right\} \leq \mathbf{P}\left\{|X_T - x_0| > \frac{\delta T}{C}\right\} \leq C_1\,\exp\{-c\,\delta\,T\} \tag{5.10}$$

where the positive constant c can be chosen large. To obtain the last inequality we used the estimate

$$\int_N^\infty f(x)\,\mathrm{d}x \leq C\int_N^\infty \exp\left\{2\int_0^x \frac{S(y)}{\sigma(y)^2}\mathrm{d}y\right\}\,\mathrm{d}x \leq C\,\exp\{-c\,N\}, \tag{5.11}$$

where the constant $c > 0$ can be chosen as large as we want due to (5.5). In particular we can take $c > \mathrm{R}(y)/\delta$ and this makes the probability (5.10) essentially smaller than (5.9).

Note as well that $\mathbf{E}\,H(\xi) = 0$.

5.1.2 Asymptotic Behavior of Errors

Let us remember the notation of Section 1.3.3 :

$$d^2 = \int_{-\infty}^{\infty} \left[\frac{S(\vartheta_1, x) - S(\vartheta_0, x)}{\sigma(x)} + \frac{g(x)}{2} \right]^2 f(\vartheta_0, x)\, dx,$$

where

$$g(x) = \int_{-\infty}^{x} \left(\left[\frac{S(\vartheta_1, v) - S(\vartheta_0, v)}{\sigma(v)} \right]^2 - 2\, J(\vartheta_0, \vartheta_1) \right) \frac{2 f(\vartheta_0, v)}{\sigma(x)\, f(\vartheta_0, x)}\, dv,$$

and

$$J(\vartheta_0, \vartheta_1) = \frac{1}{2} \int_{-\infty}^{\infty} \left(\frac{S(\vartheta_1, y) - S(\vartheta_0, y)}{\sigma(y)} \right)^2 f(\vartheta_0, y)\, dy.$$

Remember that the Kullback–Leibler distance between two measures $\mathbf{P}_{\vartheta_1}^{(T)}$ and $\mathbf{P}_{\vartheta_0}^{(T)}$ is

$$\mathbf{E}_{\vartheta_0} \ln L\left(\vartheta_0, \vartheta_1, X^T\right) = T\, J(\vartheta_0, \vartheta_1).$$

Introduce the condition $\boldsymbol{\mathcal{V}}$.

$\mathcal{V}_1$. *The functions $S(\vartheta_0, x)$, $S(\vartheta_1, x)$ and $\sigma(x)$ are continuously differentiable (w.r.t. x) and the difference $[S(\vartheta_0, x) - S(\vartheta_1, x)]\, \sigma(x)^{-2}$ is a bounded function.*

$\mathcal{V}_2$. *The condition $\boldsymbol{\mathcal{U}}$ is fulfilled with $S(x) = S(\vartheta_1, x)$ and*

$$H(x) = \frac{1}{2} \left[\left(\frac{S(\vartheta_1, x)}{\sigma(x)} \right)^2 + \left(\frac{S(\vartheta_1, x)}{\sigma(x)^2} \right)' \sigma(x)^2 \right.$$
$$\left. - \left(\frac{S(\vartheta_0, x)}{\sigma(x)} \right)^2 - \left(\frac{S(\vartheta_0, x)}{\sigma(x)^2} \right)' \sigma(x)^2 \right] + J(\vartheta_1, \vartheta_0). \quad (5.12)$$

By this condition the function $L(\lambda)$ defined by

$$L(\lambda) = \lim_{T \to \infty} T^{-1} \ln \mathbf{E}_{\vartheta_1} \exp\left\{ \lambda \int_0^T H(X_t)\, dt \right\}, \quad \lambda \in \mathscr{R}$$

exists and does not depend on the initial value x_0. The function $\mathrm{R}(\cdot)$ is the Fenchel–Legendre transform of $L(\cdot)$.

The asymptotics of the function $\beta_T(\phi_T^*)$ is described in the following proposition.

Proposition 5.3. *Let the condition $\boldsymbol{\mathcal{V}}$ be fulfilled and the function $\mathrm{R}(\cdot)$ be continuous at the point $J_* = J(\vartheta_1, \vartheta_0) + J(\vartheta_0, \vartheta_1)$. Then the test*

$$\phi_T^*\left(X^T\right) = \chi_{\{L(\vartheta_1, \vartheta_0, X^T) \geq c_\varepsilon\}}$$

with

$$c_\varepsilon = \exp\left\{-T\ J\left(\vartheta_0,\vartheta_1\right) + z_\varepsilon\ d\ \sqrt{T}\right\}$$

belongs to the class $\mathscr{K}_\varepsilon'$ and

$$\beta_T\left(\phi_T^*\right) = 1 - \exp\left\{-T\ \mathrm{R}\left(J_*\right)\left(1+o(1)\right)\right\}. \tag{5.13}$$

Proof. Under hypothesis $\mathscr{H}_1$ we have

$$\beta_T\left(\phi_T^*\right) = \mathbf{P}_{\vartheta_1}^{(T)}\left\{\int_0^T \frac{\left[S\left(\vartheta_1,X_t\right)-S\left(\vartheta_0,X_t\right)\right]}{\sigma\left(X_t\right)}\,\mathrm{d}W_t \right.$$
$$\left. +\frac{1}{2}\int_0^T\left(\frac{S\left(\vartheta_1,X_t\right)-S\left(\vartheta_0,X_t\right)}{\sigma\left(X_t\right)}\right)^2\mathrm{d}t > \ln c_\varepsilon\right\}.$$

By the Itô formula

$$\int_0^T \frac{\left[S\left(\vartheta_1,X_t\right)-S\left(\vartheta_0,X_t\right)\right]}{\sigma\left(X_t\right)}\,\mathrm{d}W_t = \int_{x_0}^{X_T}\frac{\left[S\left(\vartheta_1,x\right)-S\left(\vartheta_0,x\right)\right]}{\sigma\left(x\right)^2}\,\mathrm{d}x$$
$$-\int_0^T\left[\frac{\left[S\left(\vartheta_1,X_t\right)-S\left(\vartheta_0,X_t\right)\right]S\left(\vartheta_1,X_t\right)}{\sigma\left(X_t\right)^2}\right.$$
$$\left.+\frac{1}{2}\left(\frac{S\left(\vartheta_1,X_t\right)-S\left(\vartheta_0,X_t\right)}{\sigma\left(X_t\right)^2}\right)'\sigma\left(X_t\right)^2\right]\mathrm{d}t.$$

Hence, elementary calculation gives us

$$\beta_T\left(\phi_T^*\right) = \mathbf{P}_{\vartheta_1}^{(T)}\left\{-\int_0^T H\left(X_t\right)\,\mathrm{d}t + Q\left(X_T\right) + T\,J(\vartheta_1,\vartheta_0) > \ln c_\varepsilon\right\}$$
$$= 1 - \mathbf{P}_{\vartheta_1}^{(T)}\left\{\frac{1}{T}\int_0^T H\left(X_t\right)\,\mathrm{d}t - \frac{Q\left(X_T\right)}{T} > J(\vartheta_1,\vartheta_0) + J\left(\vartheta_0,\vartheta_1\right) - \frac{d\ z_\varepsilon}{\sqrt{T}}\right\},$$

where we denote

$$Q\left(x\right) = \int_{x_0}^{x}\frac{\left[S\left(\vartheta_1,v\right)-S\left(\vartheta_0,v\right)\right]}{\sigma\left(v\right)^2}\,\mathrm{d}v.$$

The probability $\mathbf{P}_{\vartheta_1}^{(T)}\left\{\left|Q\left(X_T\right)\right| > \delta T\right\}$ can be estimated like (5.10) and (5.11). Remember that with $\mathbf{P}_{\vartheta_1}^{(T)}$ probability 1 we have the convergence

$$\frac{1}{T}\int_0^T H\left(X_t\right)\,\mathrm{d}t \rightarrow 0.$$

Example 5.4. Let $q\left(x\right),\ r\left(x\right)$ be two continuously differentiable functions with compact support and we have two hypotheses

$$\mathscr{H}_0: \quad (\vartheta = 0) \quad \mathrm{d}X_t = -\left(a\, X_t + q\left(X_t\right)\right)\, \mathrm{d}t + \mathrm{d}W_t, \quad x_0, \quad 0 \le t \le T$$

and

$$\mathscr{H}_1: \quad (\vartheta = 1) \quad \mathrm{d}X_t = -\left(a\, X_t + r\left(X_t\right)\right)\, \mathrm{d}t + \mathrm{d}W_t, \quad x_0, \quad 0 \le t \le T,$$

where $a > 0$. Then it can be shown that all the conditions of Proposition 5.3 are fulfilled. Hence the test:

to accept the hypothesis $\mathscr{H}_0$ if

$$\int_0^T [q(X_t) - r(X_t)]\mathrm{d}X_t - \frac{1}{2}\int_0^T \left\{r(X_t)^2 - q(X_t)^2 + 2X_t\left[r(X_t) - q(X_t)\right]\right\}\mathrm{d}t$$
$$< z_\varepsilon\,\sqrt{T} - T\,J(\vartheta_0, \vartheta_1)$$

belongs to the class $\mathscr{K}_\varepsilon'$ and its power function has the asymptotics (5.13).

5.2 One-Sided Parametric Alternatives

Suppose that the observed diffusion process is

$$\mathrm{d}X_t = S\left(\vartheta, X_t\right)\, \mathrm{d}t + \sigma\left(X_t\right)\, \mathrm{d}W_t, \quad X_0, \quad 0 \le t \le T, \tag{5.14}$$

and we have to test the following two hypotheses :

$$\begin{aligned} \mathscr{H}_0 &: \quad \vartheta = \vartheta_0, \\ \mathscr{H}_1 &: \quad \vartheta > \vartheta_0. \end{aligned} \tag{5.15}$$

Fix some value $\varepsilon \in (0,1)$ and introduce the class $\mathscr{K}_\varepsilon'$ of tests of asymptotic level $1 - \varepsilon$. Our goals are to construct some usual tests (maximum likelihood and Bayesian) belonging to $\mathscr{K}_\varepsilon'$ and to describe their power functions. The power functions we study for local alternatives only.

Remember what kind of difficulties we had before in this composite hypotheses testing problem. The Neyman–Pearson lemma (Proposition 1.69) says that if we have two simple hypotheses ϑ_0 and ϑ the test based on the likelihood ratio $L(\vartheta, \vartheta_0, X^T)$ is optimal. In composite hypotheses testing we cannot use it directly because we do not know the true value ϑ under alternatives but we can replace this unknown value by one of the estimators $\bar{\vartheta}_T$ studied in Chapter 2. Then, if the estimator $\bar{\vartheta}_T$ is consistent, we can hope that the test based on the statistic $\bar{L}_T\left(X^T\right) = L(\bar{\vartheta}_T, \vartheta_0, X^T)$ will have good properties.

The statistic $\bar{L}_T\left(X^T\right)$ contains the following stochastic integral:

$$\int_0^T \frac{S\left(\bar{\vartheta}_T, X_t\right)}{\sigma\left(X_t\right)^2}\, \mathrm{d}X_t,$$

where the estimator $\bar{\vartheta}_T$ usually depends on the whole trajectory X^T. Hence the function $S\left(\bar{\vartheta}_T, X_t\right)$ is no longer $\mathfrak{F}_t$-measurable and this (extended) integral needs a special definition. Note that to work with such integrals is quite difficult and it is preferable to seek another solution. One way is to replace the stochastic integral in the likelihood ratio formula by an ordinary one, as was done in Section 2.5 (see (1.33)), and then to substitute the estimator. Using the consistency and asymptotic normality of the estimator it is possible to describe the asymptotic behavior of the test (expanding the likelihood ratio statistic in the vicinity of the true value). This way is appropriate for non-MLE in the regular case. In change-point type problems the trend coefficient is not differentiable w.r.t. x therefore if we apply Itô formula to replace the stochastic integral by an ordinary one then its expression will contain the local time estimator of the density. Thus, in the simple switching model we obtain the term $f_T^{\circ}\left(\bar{\vartheta}_T\right)$, but this random function is defined as well with the help of a stochastic integral. Of course, it can be approximated by a kernel-type estimator of the density. Another way (for MLE) is to study directly the test statistic $\hat{L}_T\left(X^T\right) = L(\hat{\vartheta}_T, \vartheta_0, X^T)$ without explicit use of the MLE, because the asymptotic behavior of the supremum of the likelihood ratio function can be described through the weak convergence of the likelihood ratio process. Below we follow the last way (in the case of the likelihood ratio and Bayesian tests).

Note that it is possible as well to construct an *estimator process*

$$\left\{\bar{\vartheta}_t,\ 0 \leq t \leq T\right\}$$

such that the estimator $\bar{\vartheta}_t$ depends on the observations $\{X_s\ 0 \leq s \leq t\}$ only. Then the "likelihood ratio" statistics

$$\ln L_T\left(X^T\right) = \int_0^T \frac{S\left(\bar{\vartheta}_t, X_t\right) - S\left(\vartheta_0, X_t\right)}{\sigma\left(X_t\right)^2}\,\mathrm{d}X_t - \frac{1}{2}\int_0^T \frac{S\left(\bar{\vartheta}_t, X_t\right)^2 - S\left(\vartheta_0, X_t\right)^2}{\sigma\left(X_t\right)^2}\,\mathrm{d}t$$

is well defined. For example, if the observed process is

$$\mathrm{d}X_t = -\vartheta\, X_t\,\mathrm{d}t + \mathrm{d}W_t, \qquad 0 \leq t \leq T,$$

then we can put

$$\hat{\vartheta}_t = -\frac{\int_0^t X_s\,\mathrm{d}X_s}{\int_0^t X_s^2\,\mathrm{d}s}, \qquad 0 \leq t \leq T,$$

and the integral

$$\int_0^T \hat{\vartheta}_t\, X_t\,\mathrm{d}X_t$$

is well defined. Of course, in this linear case it is possible to use directly the product

$$\hat{\vartheta}_T \int_0^T X_t \, dX_t$$

too, but note as well that for the extended stochastic integral

$$\int_0^T \hat{\vartheta}_T \, X_t \, dX_t \neq \hat{\vartheta}_T \int_0^T X_t \, dX_t$$

(see Nualart [191]). The use of the Stratonovich integral, probably, is preferable in such problems.

Another example is the simple switching process

$$dX_t = - \, \operatorname{sgn}(X_t - \vartheta) \, dt + dW_t, \qquad 0 \le t \le T.$$

We can use the estimator of the method of moments

$$\bar{\vartheta}_t = \frac{1}{t} \int_0^t X_s \, ds$$

which is $\sqrt{T}$- consistent and asymptotically normal and to construct the statistic

$$\ln \bar{L}_T\left(X^T\right) = \int_0^T \left[\operatorname{sgn}(X_t - \vartheta_0) - \operatorname{sgn}\left(X_t - \bar{\vartheta}_t\right)\right] \, dX_t.$$

In the present work we do not study this type of tests and we consider another three well-known tests. One is based on the score function and is locally optimal in regular problems, the second test is based on the maximum of the likelihood ratio (in regular and nonregular problems) and the third is a Bayesian test.

Remember that for any reasonable test $\bar{\phi}_T \in \mathscr{K}_\varepsilon$ (or $\bar{\phi}_T \in \mathscr{K}'_\varepsilon$) and any $\vartheta > \vartheta_0$ the power function $\beta_T\left(\bar{\phi}_T, \vartheta\right) = \mathbf{E}_\vartheta \, \bar{\phi}_T\left(X^T\right) \to 1$. Hence to compare different tests we have to study the rates of this convergence to one and this is a more complicated problem (related to the large deviation principle). At the same time there exist neighborhoods Θ_T of ϑ_0, such that for $\vartheta_T \in \Theta_T$ the function $\beta_T\left(\bar{\phi}_T, \vartheta_T\right)$ has a nondegenerate limit ($\neq 0, \neq 1$). Therefore it is interesting to compare the powers of the different tests for these values of ϑ. This approach is due to Pitman [202]. Following it we consider the problem of testing hypotheses

$$\begin{aligned} \mathscr{H}_0 &: \quad \vartheta = \vartheta_0, \\ \mathscr{H}_1 &: \quad \vartheta = \vartheta_0 + \varphi_T(\vartheta_0)\, u, \qquad u > 0, \end{aligned} \tag{5.16}$$

where the function $\varphi_T(\vartheta_0) \to 0$ but with such a rate that the measures $\mathbf{P}^{(T)}_{\vartheta_0}$ and $\mathbf{P}^{(T)}_{\vartheta_0 + \varphi_T(\vartheta_0)\, u}$ for any fixed $u > 0$ corresponding to the hypothesis and any simple alternatives are asymptotically nonsingular (or contiguous). In particular, we seek such $\varphi_T(\vartheta_0)$ that the likelihood ratio $Z_T(u) =$

$L\left(\vartheta_0+\varphi_T\left(\vartheta_0\right)u,\vartheta_0,X^T\right)$ has a nondegenerate limit. We saw in Chapters 2 and 3 that the choice of $\varphi_T\left(\vartheta_0\right)$ depends on the regularity of the problem and on the type of the process (ergodic, null recurrent, transient).

5.2.1 Score Function Test

For an ergodic diffusion process (5.14) with smooth (w.r.t. ϑ) function $S\left(\vartheta,\cdot\right)$ we obtain a nondegenerated limit of the likelihood ratio $Z_T\left(u\right)$ if $\varphi_T\left(\vartheta_0\right)=T^{-1/2}$ (see Proposition 2.2). Therefore, we consider the following problem of testing a simple hypothesis against a one-sided parametric alternative:

$$\begin{aligned}\mathscr{H}_0 :&\quad \vartheta=\vartheta_0,\\ \mathscr{H}_1 :&\quad \vartheta=\vartheta_0+\frac{u}{T^{1/2}},\qquad u>0.\end{aligned}$$

To simplify the notation we denote as $\beta_T\left(u,\bar\phi_T\right)$ the power function of the test $\bar\phi_T\left(X^T\right)$ under "alternative" $\vartheta_0+T^{-1/2}u$, i.e., $\beta_T\left(u,\bar\phi_T\right)=\mathbf{E}_u\,\bar\phi_T\left(X^T\right)$, where $\mathbf{E}_u$ is the mathematical expectation w.r.t. the measure $\mathbf{P}^{(T)}_{\vartheta_0+T^{-1/2}u}$.

Introduce the condition $\mathcal{W}$.

$\mathcal{W}_1$. *The functions $S\left(\vartheta,\cdot\right),\vartheta\geq\vartheta_0$ and $\sigma\left(\cdot\right)$ satisfy the conditions $\mathcal{ES}$ and $\mathcal{EM}$ and the functions $S\left(\vartheta_0,\cdot\right)$ satisfies the condition $\mathcal{RP}$.*

$\mathcal{W}_2$. *The function $S(\vartheta,\cdot)$ is differentiable from the right on ϑ at the point ϑ_0 in the following sense: there exists a function $\dot S\left(\vartheta_0,\cdot\right)\sigma\left(\cdot\right)^{-1}\in\mathscr{L}_2\left(f_{\vartheta_0}\right)$ such that for $h\geq 0$ and $h\to 0$*

$$\mathbf{E}_{\vartheta_0}\left(\frac{S(\vartheta_0+h,\xi)-S(\vartheta_0,\xi)-h\,\dot S\left(\vartheta_0,\xi\right)}{\sigma\left(\xi\right)}\right)^2=o\left(h^2\right)$$

and

$$\mathrm{I}\left(\vartheta_0\right)=\mathbf{E}_{\vartheta_0}\left(\frac{\dot S\left(\vartheta_0,\xi\right)}{\sigma\left(\xi\right)}\right)^2=\int_{-\infty}^{\infty}\left(\frac{\dot S\left(\vartheta_0,x\right)}{\sigma\left(x\right)}\right)^2 f\left(\vartheta_0,x\right)\,\mathrm{d}x>0.$$

Therefore the process $\{X_t,\ t\geq 0\}$ is ergodic under hypothesis $\mathscr{H}_0$ only.

As is usual in such situations (see, e.g., [211]) we compare the tests with the help of the following definition.

Definition 5.5. *Let the condition $\mathcal{W}$ be fulfilled. Then we call a test $\phi_T^*\in\mathscr{K}_\varepsilon'$ locally asymptotically uniformly most powerful in the class $\mathscr{K}_\varepsilon'$ if for any other test $\phi_T\in\mathscr{K}_\varepsilon'$ we have*

$$\lim_{T\to\infty}\ \inf_{0<u\leq K}\left[\beta_T\left(u,\phi_T^*\right)-\beta_T\left(u,\phi_T\right)\right]\geq 0$$

for any $K>0$.

To construct a test which belongs to $\mathscr{K}_\varepsilon$ for all $T>0$ we need a stopping time.

Let us put $\sigma(X_t)=1$ and $S(\vartheta,X_t)=\vartheta\,\sqrt{T\,\mathrm{I}(\vartheta_0)}$ for $t\in(T,T+1]$, (hence $\dot S(\vartheta_0,X_t)=\sqrt{T\,\mathrm{I}(\vartheta_0)}$) and

$$X_t=X_T+\int_T^t S(\vartheta_0,X_s)\,\mathrm{d}s+\tilde W_t,\quad T\le t\le T+1,$$

where $\left\{\tilde W_t, T\le t\le T+1\right\}$ is a Wiener process (*randomization*) independent of $\{W_t, 0\le t\le T\}$. Then we introduce the stopping time

$$\tau_T=\inf\left\{\tau\le T+1:\ \frac{1}{T}\int_0^\tau\left(\frac{\dot S(\vartheta_0,X_t)}{\sigma(X_t)}\right)^2\,\mathrm{d}t\ge \mathrm{I}(\vartheta_0)\right\}$$

and the statistic

$$\Delta_\tau\left(\vartheta_0,X^T\right)=\frac{1}{\sqrt{T}}\int_0^{\tau_T}\frac{\dot S(\vartheta_0,X_t)}{\sigma(X_t)^2}\,\left[\mathrm{d}X_t-S(\vartheta_0,X_t)\,\mathrm{d}t\right].$$

Below z_ε is the $1-\varepsilon$ quantile of standard Gaussian law.

Theorem 5.6. *Let the condition $\mathcal{W}$ be fulfilled. Then the test*

$$\phi_T^*\left(X^T\right)=\chi_{\left\{\Delta_\tau(\vartheta_0,X^T)\ge z_\varepsilon\,\mathrm{I}(\vartheta_0)^{1/2}\right\}}\in\ \mathscr{K}_\varepsilon,$$

is locally asymptotically uniformly most powerful in the class $\mathscr{K}_\varepsilon'$ and its power function

$$\beta_T\left(u,\phi_T^*\right)=\mathbf{P}\left\{\zeta>z_\varepsilon-u\,\mathrm{I}(\vartheta_0)^{1/2}\right\}+o(1),\tag{5.17}$$

where $\zeta\sim\mathcal{N}(0,1)$.

Proof. To prove this theorem we follow Roussas [211], Section 4.3. Fix some value $u>0$. Then for any test $\phi_T\in\mathscr{K}_\varepsilon$ we have the estimate

$$\beta_T\left(u,\phi_T\right)\le\beta_T\left(u,\hat\phi_T(u)\right),$$

where $\hat\phi_T\left(u,X^T\right)\in\mathscr{K}_\varepsilon$ is the likelihood ratio test (the most powerful in the problem of two simple hypotheses testing), i.e.,

$$\hat\phi_T\left(u,X^T\right)=\chi_{\left\{L\left(\vartheta_0+T^{-1/2}u,\vartheta_0,X^T\right)\ge c_\varepsilon\right\}}$$

with corresponding constant c_ε. The likelihood ratio $L\left(\vartheta_0+T^{-1/2}u,\vartheta_0,X^T\right)$ admits the representation (*LAN from the right* $u\ge0$)

$$
\begin{aligned}
L&\left(\vartheta_0+T^{-1/2}u,\vartheta_0,X^T\right)\\
&=\exp\left\{\frac{u}{\sqrt{T}}\int_0^T\frac{\dot S\left(\vartheta_0,X_t\right)}{\sigma\left(X_t\right)^2}\left[{\rm d}X_t-S_0\left(\vartheta_0,X_t\right){\rm d}t\right]\right.\\
&\qquad\left.-\frac{u^2}{2T}\int_0^T\left(\frac{\dot S\left(\vartheta_0,X_t\right)}{\sigma\left(X_t\right)}\right)^2{\rm d}t+r\left(\vartheta_0,u,X^T\right)\right\}\\
&=\exp\left\{u\Delta_T\left(\vartheta_0,X^T\right)-\frac{u^2}{2}{\rm I}\left(\vartheta_0\right)+\tilde r\left(\vartheta_0,u,X^T\right)\right\},
\end{aligned}
\tag{5.18}
$$

where

$$
\mathcal{L}_{\vartheta_0}\left\{\Delta_T\left(\vartheta_0,X^T\right)\right\}\Longrightarrow\mathcal{N}\left(0,{\rm I}\left(\vartheta_0\right)\right)
\tag{5.19}
$$

and for any bounded sequence $\left\{u_T\right\}>0$

$$
\mathbf{P}_{\vartheta_0}-\lim_{T\to\infty}\tilde r\left(\vartheta_0,u_T,X^T\right)=0.
\tag{5.20}
$$

It follows from this representation that we can take the constant

$$
c_\varepsilon=\exp\left\{-\frac{u^2}{2}{\rm I}\left(\vartheta_0\right)+u\,z_\varepsilon\,{\rm I}\left(\vartheta_0\right)^{1/2}\right\},
$$

then $\hat\phi_T\in\mathscr{K}_\varepsilon'$.

Note that (5.20) is a bit more than LAN because we have here the uniform w.r.t. $u\in\mathbb{U}_K$ convergence and remind that for LAN families the sequences of measures $\mathbf{P}_{\vartheta_0}^{(T)}$ and $\mathbf{P}_{\vartheta_0+T^{-1/2}u_T}^{(T)}$ for any $u_T\in\mathbb{U}_K$ with $\mathbb{U}_K=\{u:\ 0\le u\le K\}$ are contiguous.

It can be shown that if $u_T\to u$, then

$$
\mathscr{L}_{\vartheta_0+T^{-1/2}u_T}\left\{\Delta_T\left(\vartheta_0,X^T\right)\right\}\Longrightarrow\mathcal{N}\left(u{\rm I}\left(\vartheta_0\right),{\rm I}\left(\vartheta_0\right)\right).
$$

This convergence allows us to write

$$
\begin{aligned}
\beta_T&\left(u_T,\hat\phi_T\right)\\
&=\mathbf{P}_{\vartheta_0+T^{-1/2}u_T}^{(T)}\left\{u_T\Delta_T\left(\vartheta_0,X^T\right)-\frac{u_T^2}{2}{\rm I}\left(\vartheta_0\right)\ge\ln c_\varepsilon-\tilde r\left(\vartheta_0,u_T,X^T\right)\right\}\\
&=\mathbf{P}\left\{\zeta+u_T{\rm I}\left(\vartheta_0\right)^{1/2}\ge z_\varepsilon\right\}+o\left(1\right)=\hat\beta\left(\vartheta_0,u_T\right)+o\left(1\right),
\end{aligned}
$$

where

$$
\hat\beta\left(\vartheta_0,u\right)=\mathbf{P}\left\{\zeta+u\,{\rm I}\left(\vartheta_0\right)^{1/2}\ge z_\varepsilon\right\}.
$$

Hence, for any test $\phi_T\in\mathscr{K}_\varepsilon'$

$$\lim_{T\to\infty}\inf_{0\le u\le K}\left[\hat\beta\left(\vartheta_0,u\right)-\beta_T\left(u,\phi_T\right)\right]\geq 0.$$

This inequality gives us an uniform on $u\in\mathbb{U}_K$ upper bound on the power function of any test from the class $\mathscr{K}_\varepsilon'$.

We have to show now that the power function of the test ϕ_T^* converges to $\hat\beta\left(\vartheta_0,u\right)$ uniformly in $u\in\mathbb{U}_K$.

This test belongs to the class $\mathscr{K}_\varepsilon$ because under the hypothesis $\mathscr{H}_0$ the stochastic integral

$$\Delta_\tau\left(\vartheta_0,X^T\right)=\frac{1}{\sqrt{T}}\int_0^{\tau_T}\frac{\dot S\left(\vartheta_0,X_t\right)}{\sigma\left(X_t\right)}\mathrm{d}W_t\ \sim\ \mathcal{N}\left(0,\mathrm{I}\left(\vartheta_0\right)\right).$$

Let $\vartheta=\vartheta_0+T^{-1/2}u_T$. Then (under alternative $\mathscr{H}_1$)

$$\begin{aligned}\Delta_\tau\left(\vartheta_0,X^T\right)&=\frac{1}{\sqrt{T}}\int_0^{\tau_T}\frac{\dot S\left(\vartheta_0,X_t\right)}{\sigma\left(X_t\right)}\,\mathrm{d}W_t\\&\quad+\frac{1}{\sqrt{T}}\int_0^{\tau_T}\frac{\dot S\left(\vartheta_0,X_t\right)}{\sigma\left(X_t\right)^2}\left[S\left(\vartheta_0+u_T/\sqrt{T},X_t\right)-S\left(\vartheta_0,X_t\right)\right]\,\mathrm{d}t\\&=\frac{1}{\sqrt{T}}\int_0^{\tau_T}\frac{\dot S\left(\vartheta_0,X_t\right)}{\sigma\left(X_t\right)}\,\mathrm{d}W_t+\frac{u_T}{T}\int_0^{\tau_T}\frac{\dot S\left(\vartheta_0,X_t\right)^2}{\sigma\left(X_t\right)^2}\,\mathrm{d}t+o\left(1\right)\\&=\zeta_T\,\mathrm{I}\left(\vartheta_0\right)^{1/2}+u_T\,\mathrm{I}\left(\vartheta_0\right)+o\left(1\right),\end{aligned}$$

where $\mathcal{L}\left\{\zeta_T\right\}=\mathcal{N}\left(0,1\right)$. Hence

$$\beta_T\left(u_T,\phi_T^*\right)-\hat\beta\left(\vartheta_0,u_T\right)\to 0 \tag{5.21}$$

uniformly in $u_T\in\mathbb{U}_K$.

Remark 5.7. Note that for *local non-contiguous alternatives*, i.e., for the values
$h_T=u_TT^{-1/2}\to 0$ with $u_T\to\infty$ this test is consistent in the following sense. We have under $\mathscr{H}_1$

$$\begin{aligned}\Delta_\tau\left(\vartheta_0,X^T\right)&=\zeta_T\,\mathrm{I}\left(\vartheta_0\right)^{1/2}+u_T\,\mathrm{I}\left(\vartheta_0\right)\\&\quad+\frac{1}{\sqrt{T}}\int_0^{\tau_T}\frac{\dot S\left(\vartheta_0,X_t\right)}{\sigma\left(X_t\right)^2}\left[S\left(\vartheta_0+h_T,X_t\right)-S\left(\vartheta_0,X_t\right)-h_T\dot S\left(\vartheta_0,X_t\right)\right]\mathrm{d}t.\end{aligned}$$

Therefore, if we apply the Cauchy–Schwarz inequality

$$\begin{aligned}&\left(\frac{1}{T}\int_0^{\tau_T}\frac{\dot S\left(\vartheta_0,X_t\right)}{\sigma\left(X_t\right)^2}\left[S\left(\vartheta_0+h_T,X_t\right)-S\left(\vartheta_0,X_t\right)-h_T\dot S\left(\vartheta_0,X_t\right)\right]\mathrm{d}t\right)^2\\&\qquad\leq\mathrm{I}\left(\vartheta_0\right)\left(\frac{1}{T}\int_0^{\tau_T}\frac{S(\vartheta_0+h_T,X_t)-S(\vartheta_0,X_t)-h_T\,\dot S\left(\vartheta_0,X_t\right)}{\sigma\left(X_t\right)}\mathrm{d}t\right)^2,\end{aligned}$$

and verify that (under $\mathscr{H}_1$) the following estimate holds

$$\frac{1}{T}\int_0^{\tau_T} \frac{S(\vartheta_0+h_{_T},X_t)-S(\vartheta_0,X_t)-h_{_T}\,\dot S\,(\vartheta_0,X_t)}{\sigma\,(\xi)}{\rm d}t = h_{_T}\,o\,(1)\,.$$

Then

$$\Delta_\tau\left(\vartheta_0,X^T\right)=\zeta_T\,\mathrm{I}\left(\vartheta_0\right)^{1/2}+u_{_T}\,\mathrm{I}\left(\vartheta_0\right)+u_{_T}\,o(1)$$

and

$$\Delta_\tau\left(\vartheta_0,X^T\right)\longrightarrow\infty$$

in probability as $T\to\infty$. Therefore,

$$\beta_{_T}\left(u_{_T},\phi^*_{_T}\right)\to 1,$$

which together with

$$\hat\beta\left(\vartheta_0,u_{_T}\right)\to 1$$

provides (5.21).

Remark 5.8. The key point of the proof, of course, is the local asymptotic normality of the family of measures at the point ϑ_0. It is possible to have other rates even in an ergodic smooth case. Below we give several examples with contiguous alternatives corresponding to different rates.

Example 5.9. (*Null Fisher information*, Section 2.6.3.) Suppose that the function $S\left(\vartheta,\cdot\right)$ is k-times continuously differentiable w.r.t. ϑ at the point ϑ_0,

$$S^{(l)}\left(\vartheta_0,x\right)=0,\;\;x\in\mathscr{R},l=1,\ldots,k-1,$$

and

$$\mathrm{I}_k\left(\vartheta_0\right)=\int_{-\infty}^{\infty}\left(\frac{S^{(k)}\left(\vartheta_0,x\right)}{k!\,\sigma\left(x\right)}\right)^2\,f\left(\vartheta_0,x\right)\,{\rm d}x>0,$$

Consider the hypotheses testing problem

$$\mathscr{H}_0\,:\,\vartheta=\vartheta_0,$$
$$\mathscr{H}_1\,:\,\vartheta=\vartheta_0+\frac{u}{T^{1/2k}},\qquad u>0.$$

It was shown that the normalized likelihood ratio has the representation (see Section 2.6.3 for details)

$$L\left(\vartheta_0+\frac{u}{T^{1/2k}},\vartheta_0,X^T\right)$$
$$=\exp\left\{u^k\Delta_\tau\left(\vartheta_0,X^T\right)-\frac{u^{2k}}{2}\mathrm{I}_k\left(\vartheta_0\right)+r\left(\vartheta_0,u,X^T\right)\right\},$$

where (under hypothesis $\vartheta = \vartheta_0$)

$$\Delta_\tau\left(\vartheta_0, X^T\right) = \frac{1}{\sqrt{T}} \int_0^{\tau_T} \frac{S^{(k)}\left(\vartheta_0, X_t\right)}{k!\,\sigma\left(X_t\right)^2}\,\left[\mathrm{d}X_t - S\left(\vartheta_0, X_t\right)\mathrm{d}t\right] \sim \mathcal{N}\left(0, \mathrm{I}_k\left(\vartheta_0\right)\right),$$

with the corresponding stopping time τ_T and for any bounded set $\left\{u_T\right\}$

$$\mathbf{P}_{\vartheta_0} - \lim_{T\to\infty} r\left(\vartheta_0, u_T, X^T\right) = 0.$$

It is easy to see that with the help of this representation we can construct a locally asymptotically uniformly most powerful test of the type

$$\tilde{\phi}_T\left(X^T\right) = \chi_{\left\{\Delta_\tau\left(\vartheta_0, X^T\right) > z_\varepsilon\,\mathrm{I}_k(\vartheta_0)^{1/2}\right\}}.$$

The limit power function of this test is obviously

$$\hat{\beta}\left(u, \vartheta_0\right) = \mathbf{P}\left\{\zeta > z_\varepsilon - u^k \mathrm{I}_k\left(\vartheta_0\right)^{1/2}\right\}.$$

Here z_ε is, as before, the $1-\varepsilon$ quantile of the standard Gaussian law.

Example 5.10. (*Polynomial growth process*, Section 3.5.2.) Another example of a nonsquare-root rate in a smooth hypothesis testing problem is the case of the nonergodic diffusion process

$$\mathrm{d}X_t = \vartheta\,\left|X_t\right|^\kappa \mathrm{d}t + \mathrm{d}W_t, \quad X_0, \quad 0 \le t \le T,$$

with $\vartheta > 0$. The contiguous alternative is given by the following hypotheses testing problem:

$$\mathscr{H}_0 : \vartheta = \vartheta_0 > 0,$$
$$\mathscr{H}_1 : \vartheta = \vartheta_0 + \frac{u}{T^{\frac{1+\kappa}{2(1-\kappa)}}}, \qquad u > 0.$$

The score function test

$$\phi_T^*\left(X^T\right) = \chi_{\left\{\Delta_\tau\left(\vartheta_0, X^T\right) > z_\varepsilon\,\mathrm{I}(\vartheta_0)^{1/2}\right\}}$$

with

$$\Delta_\tau\left(\vartheta_0, X^T\right) = T^{-\frac{1+\kappa}{2(1-\kappa)}} \int_0^{\tau_T} \left|X_t\right|^\kappa\ \left[\mathrm{d}X_t - \vartheta_0\,\left|X_t\right|^\kappa \mathrm{d}t\right]$$

and

$$\tau_T = \inf\left\{\tau \le T+1 : T^{-\frac{1+\kappa}{1-\kappa}} \int_0^\tau \left|X_t\right|^{2\kappa} \mathrm{d}t = \mathrm{I}\left(\vartheta_0\right)\right\}$$

belongs to $\mathscr{K}_\varepsilon$ and is asymptotically uniformly most powerful. Of course, we have to define the process X_t on the interval $[T, T+1]$ and introduce an independent Wiener process as was done above. Here

$$\mathrm{I}(\vartheta_0) = \frac{(1-\kappa)^{\frac{1+\kappa}{2(1-\kappa)}}}{1+\kappa}\, \vartheta_0^{\frac{2\kappa}{1-\kappa}}$$

is the Fisher information and the power function

$$\hat\beta(u, \vartheta_0) = \mathbf{P}\left\{\zeta > z_\varepsilon - u\,\mathrm{I}(\vartheta_0)^{1/2}\right\}.$$

Example 5.11. (*Null recurrent* and *Ornstein–Uhlenbeck* processes) In the case of the processes (Section 3.5.1)

$$\mathrm{d}X_t = -\vartheta\, \frac{X_t}{1+X_t^2}\, \mathrm{d}t + \mathrm{d}W_t, \quad X_0, \qquad 0 \le t \le T,$$

with $\vartheta \in (-1/2, 1/2)$ and (Section 3.5.2)

$$\mathrm{d}X_t = \vartheta\, X_t\, \mathrm{d}t + \mathrm{d}W_t, \quad X_0, \qquad 0 \le t \le T$$

with $\vartheta > 0$ the normalizing functions are

$$\varphi_T(\vartheta_0) = T^{-\vartheta_0 - 1/2}, \qquad \text{and} \qquad \varphi_T(\vartheta_0) = \mathrm{e}^{-\vartheta_0 T}$$

respectively. The contiguous alternatives are given by the hypotheses testing problem

$$\begin{aligned} &\mathscr{H}_0 : \vartheta = \vartheta_0, \\ &\mathscr{H}_1 : \vartheta = \vartheta_0 + \varphi_T(\vartheta_0)\, u, \quad u > 0. \end{aligned}$$

The normalized likelihood ratio $Z_T(u) = L\left(\vartheta_0 + \varphi_T(\vartheta_0)\, u, \vartheta_0, X^T\right)$ converges in distribution to the random function

$$Z(u) = \exp\left\{u\,\zeta\,\mathcal{I}^{1/2} - \frac{u^2}{2}\,\mathcal{I}\right\}, \quad u \in \mathscr{R}_+$$

where $\zeta \sim \mathcal{N}(0,1)$ and $\mathcal{I} \ge 0$ is independent of ζ random variable (see details in Sections 3.5.1 and 3.5.2).

Therefore, to find the constant c_ε such that $\hat\phi_T \in \mathscr{K}_\varepsilon'$ we have to solve the following equation:

$$\begin{aligned} \varepsilon &= \mathbf{P}\left\{u\,\zeta\,\mathcal{I}^{1/2} - \frac{u^2}{2}\,\mathcal{I} \ge \ln c_\varepsilon\right\} \\ &= \int_0^\infty \mathbf{P}\left\{\zeta \ge \frac{u\sqrt{y}}{2} + \frac{\ln c_\varepsilon}{u\sqrt{y}}\right\} p(y)\; \mathrm{d}y \\ &= 1 - \int_0^\infty \Phi\left(\frac{u\sqrt{y}}{2} + \frac{\ln c_\varepsilon}{u\sqrt{y}}\right) p(y)\; \mathrm{d}y, \end{aligned}$$

where $p(\cdot)$ is the density function of the variable $\mathcal{I}$ and $\Phi(x) = \mathbf{P}\{\zeta < x\}$. The power function

$$\begin{aligned}
&\beta_T\left(u, \hat{\phi}_T\right) \\
&= \mathbf{P}^{(T)}_{\vartheta_0 + T^{-1/2}u}\left\{u\Delta_T\left(\vartheta_0, X^T\right) - \frac{u^2}{2}\,\mathrm{I}_T\left(\vartheta_0\right) \geq \ln c_\varepsilon\right\} \\
&\longrightarrow \mathbf{P}\left\{u\,\zeta\,\mathcal{I}^{1/2} + \frac{u^2}{2}\,\mathcal{I} \geq \ln c_\varepsilon\right\} + o\left(1\right) \\
&= \int_0^\infty \mathbf{P}\left\{\zeta \geq -\frac{u\sqrt{y}}{2} + \frac{\ln c_\varepsilon}{u\sqrt{y}}\right\} p\left(y\right)\,\mathrm{d}y \\
&= 1 - \int_0^\infty \Phi\left(-\frac{u\sqrt{y}}{2} + \frac{\ln c_\varepsilon}{u\sqrt{y}}\right) p\left(y\right)\,\mathrm{d}y = \hat{\beta}\left(\vartheta_0, u\right).
\end{aligned}$$

The score function test has another limit for its power function. Let us denote

$$h\left(x\right) = -\frac{x}{1+x^2}, \qquad \text{or} \qquad h\left(x\right) = x$$

and introduce the statistics

$$\Delta_T\left(X^T\right) = \varphi_T\left(\vartheta_0\right)\int_0^T h\left(X_t\right)\,\mathrm{d}X_t, \qquad \mathrm{I}_T\left(X^T\right) = \varphi_T\left(\vartheta_0\right)^2\int_0^T h\left(X_t\right)^2\,\mathrm{d}t.$$

We have for the score function test

$$\phi^*_T\left(X^T\right) = \chi_{\left\{\delta_T\left(X^T\right) \geq z_\varepsilon\right\}}, \qquad \delta_T\left(X^T\right) = \frac{\Delta_T\left(X^T\right)}{\sqrt{\mathrm{I}_T\left(X^T\right)}}$$

the convergence

$$\mathbf{E}_{\vartheta_0}\,\phi^*_T\left(X^T\right) = \mathbf{P}_{\vartheta_0}\left\{\delta_T\left(X^T\right) \geq z_\varepsilon\right\} \longrightarrow \mathbf{P}_{\vartheta_0}\left\{\zeta \geq z_\varepsilon\right\} = \varepsilon.$$

Hence $\phi^*_T \in \mathscr{K}'_\varepsilon$. Further,

$$\begin{aligned}
\beta_T\left(u, \phi^*_T\right) &= \mathbf{P}_{\vartheta_0 + \varphi_T(\vartheta_0)u}\left\{\delta_T\left(X^T\right) \geq z_\varepsilon\right\} \longrightarrow \mathbf{P}\left\{\zeta + u\,\mathcal{I}^{1/2} \geq z_\varepsilon\right\} \\
&= 1 - \int_0^\infty \Phi\left(-u\sqrt{y} + z_\varepsilon\right) p\left(y\right)\,\mathrm{d}y < \hat{\beta}\left(\vartheta_0, u\right)
\end{aligned}$$

for all $u > 0$. As the "Fisher information" is random there is no locally asymptotically uniformly most powerful test (see Swensen [223]).

5.2.2 Likelihood Ratio and Bayesian Tests

Suppose that the parameter ϑ in the observations

$$\mathrm{d}X_t = S(\vartheta, X_t)\ \mathrm{d}t + \sigma(X_t)\ \mathrm{d}W_t, \quad X_0, 0 \le t \le T,$$

belongs to the set $\Theta = [\vartheta_0, \vartheta_f]$ with finite ϑ_f and we have to test the hypotheses

$$\mathscr{H}_0 : \vartheta = \vartheta_0,$$
$$\mathscr{H}_1 : \vartheta > \vartheta_0.$$

If we replace the unknown parameter ϑ in the likelihood ratio $L(\vartheta, \vartheta_0, X^T)$ by the MLE $\hat{\vartheta}_T$, then we obtain the *likelihood ratio statistic*

$$\hat{L}_T(X^T) \equiv L\left(\hat{\vartheta}_T, \vartheta_0, X^T\right) = \sup_{\vartheta \in \Theta} L(\vartheta, \vartheta_0, X^T),$$

and the test based on this statistic

$$\hat{\phi}_T(X^T) = \chi_{\left\{\hat{L}_T(X^T) > c_\varepsilon\right\}} \tag{5.22}$$

is called the *likelihood ratio test* (LRT). The constant c_ε have to be chosen such that

$$\lim_{T\to\infty} \mathbf{P}_{\vartheta_0}^{(T)} \left\{\hat{L}_T(X^T) > c_\varepsilon\right\} = \varepsilon.$$

Therefore the test $\hat{\phi}_T \in \mathcal{K}'_\varepsilon$.

Another approach is, following Wald [238], to reduce the composite alternative $\mathscr{H}_1$ to a simple one $\mathscr{H}_Q$ with the help of some prior distribution $\mathbf{Q}$ on $\Theta = [\vartheta_0, \vartheta_f]$. Let $q(\vartheta), \vartheta \in \Theta$ be the density function of $\mathbf{Q}$. We suppose below that $q(\cdot)$ is a continuous positive function on $\mathbf{Q}$. We replace the family of distributions $\left\{\mathbf{P}_\vartheta^{(T)},\ \vartheta \in (\vartheta_0, \vartheta_f]\right\}$ by the following one:

$$\mathbf{P}_Q^{(T)}(\mathbb{B}) = \int_\Theta \mathbf{P}_\vartheta^{(T)}(\mathbb{B})\ q(\vartheta)\ \mathrm{d}\vartheta, \qquad \mathbb{B} \in \mathfrak{B}_T.$$

Then we consider two simple hypotheses

$$\mathscr{H}_0 : \mathbf{P}^{(T)} = \mathbf{P}_{\vartheta_0}^{(T)},$$
$$\mathscr{H}_Q : \mathbf{P}^{(T)} = \mathbf{P}_Q^{(T)},$$

where $\mathbf{P}^{(T)}$ is the distribution of the observed process X^T. Now the most powerful test in this problem is given by the equality

$$\tilde{\phi}_T(X^T) = \chi_{\left\{\tilde{L}_T > b_\varepsilon\right\}},$$

where the statistic

$$\tilde{L}_T\left(X^T\right) = \int_\Theta L\left(\vartheta, \vartheta_0, X^T\right) \, q\left(\vartheta\right) \, \mathrm{d}\vartheta \tag{5.23}$$

and the constant b_ε satisfies the equality

$$\lim_{T\to\infty} \mathbf{P}_{\vartheta_0}^{(T)} \left\{L\left(\vartheta, \vartheta_0, X^T\right) > b_\varepsilon\right\} = \varepsilon. \tag{5.24}$$

Therefore, $\tilde{\phi}_T \in \mathscr{K}_\varepsilon'$ too.

We study these tests in two different situations. The first one corresponds to the smooth (w.r.t. ϑ) trend coefficient and the second situation is similar to that considered in Sections 3.2–3.4. In both cases we suppose that the condition $\boldsymbol{\mathcal{A}}_0$ is fulfilled under hypothesis and alternatives, i.e., the function $\sigma\left(\cdot\right)^{-1} \in \mathscr{P}$ and

$$\lim_{|x|\to\infty} \sup_{\vartheta\in\Theta} \operatorname{sgn}\left(x\right) \frac{S\left(\vartheta, x\right)}{\sigma\left(x\right)^2} < 0. \tag{5.25}$$

Regular Case

In this problem we need more regularity conditions.

Theorem 5.12. *Suppose that the function $S(\theta,\cdot)$ is continuously differentiable in $\mathscr{L}_2(f_{\vartheta_0})$ (see T_2), the derivative $\dot{S}(\vartheta,\cdot)$ satisfies the conditions*

$$0 < \mathrm{I}\left(\vartheta\right) = \mathbf{E}_\vartheta \left(\frac{\dot{S}(\vartheta,\xi)}{\sigma(\xi)}\right)^2 \leq \sup_{\vartheta,\theta\in\Theta} \mathbf{E}_\theta \left(\frac{\dot{S}(\vartheta,\xi)}{\sigma(\xi)}\right)^2 < \infty, \tag{5.26}$$

and for any $\nu > 0$ and $\vartheta \in \Theta$

$$\inf_{|\theta-\vartheta|>\nu} \mathbf{E}_\vartheta \left(\frac{S(\theta,\xi) - S(\vartheta,\xi)}{\sigma(\xi)}\right)^2 > 0.$$

Then the LRT $\hat{\phi}_T$ with $c_\varepsilon = \mathrm{e}^{z_\varepsilon^2/2}$ is locally asymptotically uniformly most powerful in the class $\mathcal{K}_\varepsilon'$ and

$$\beta_T\left(u, \hat{\phi}_T\right) = \mathbf{P}\left\{\zeta > z_\varepsilon - u\,\mathrm{I}\left(\vartheta_0\right)^{1/2}\right\} + o\left(1\right),$$

where $\zeta \sim \mathcal{N}\left(0,1\right)$.

Proof. Let us introduce, as in Section 2.1, the process

$$Z_T\left(u\right) = L\left(\vartheta_0 + u/\sqrt{T}, \vartheta_0, X^T\right), \qquad u \in \mathbb{U}_T,$$

where

$$\mathbb{U}_T = \left\{u:\ 0 \leq u \leq \left(\vartheta_f - \vartheta_0\right)\sqrt{T}\right\}$$

$$\mathbb{U}_T = \left\{u:\ 0 \le u \le (\vartheta_f - \vartheta_0)\sqrt{T}\right\}$$

and define $Z_T(u)$ linear decreasing to zero on the interval

$$\left[(\vartheta_f - \vartheta_0)\sqrt{T}, (\vartheta_f - \vartheta_0)\sqrt{T} + 1\right].$$

Then we put $Z_T(u) = 0$ for $u \ge (\vartheta_f - \vartheta_0)\sqrt{T} + 1$. Now the process $Z_T(\cdot)$ is defined on $\mathscr{R}_+$.

As was noted above the corresponding family of measures is LAN at the point ϑ_0 (see (5.18)-(5.20)). Therefore the marginal distributions of the process $Z_T(\cdot)$ (under hypothesis $\mathscr{H}_0$) converge to the marginal distributions of the process

$$Z(u) = \exp\left\{u\, \mathrm{I}(\vartheta_0)^{1/2}\,\zeta - \frac{u^2}{2}\,\mathrm{I}(\vartheta_0)\right\}, \qquad u \in \mathscr{R}_+, \quad \zeta \sim \mathcal{N}(0,1).$$

Moreover we have the following two estimates. The first one is (see (1.42))

$$\mathbf{E}_{\vartheta_0}\left|Z_T^{1/2}(u) - Z_T^{1/2}(v)\right|^2 \le \frac{T}{2}\,\mathbf{E}_{\vartheta_u}\left(\frac{S(\vartheta_u,\xi) - S(\vartheta_v,\xi)}{\sigma(\xi)}\right)^2$$
$$+ \frac{T}{2}\,\mathbf{E}_{\vartheta_v}\left(\frac{S(\vartheta_u,\xi) - S(\vartheta_v,\xi)}{\sigma(\xi)}\right)^2 \le C\,|u-v|^2. \tag{5.27}$$

Here we have denoted $\vartheta_u = \vartheta_0 + \frac{u}{\sqrt{T}}$ and $\vartheta_v = \vartheta_0 + \frac{v}{\sqrt{T}}$. The second estimate is (see (2.39))

$$\mathbf{E}_{\vartheta_0} Z_T^{1/2}(u) \le \frac{C_N}{|u|^N} \tag{5.28}$$

and can be obtained in the same way as (2.39).

Therefore the distribution of the stochastic process $\{Z_T(u),\ u \in \mathscr{R}_+\}$ converges weakly to the distribution of the process $\{Z(u),\ u \in \mathscr{R}_+\}$ in the measurable space $(\mathscr{C}_0(\mathscr{R}_+), \mathfrak{B}(\mathscr{R}_+))$ of continuous on $\mathscr{R}_+$ functions vanishing in infinity (see Theorem A21, [109]). This weak convergence allows us to write (under hypothesis $\mathscr{H}_0$)

$$\mathbf{P}_{\vartheta_0}^{(T)}\left\{\hat{L}_T(X^T) > c_\varepsilon\right\} = \mathbf{P}_{\vartheta_0}^{(T)}\left\{\sup_{u\ge 0} Z_T(u) > c_\varepsilon\right\}$$
$$\longrightarrow \mathbf{P}\left\{\sup_{u\ge 0} Z(u) > c_\varepsilon\right\} = \mathbf{P}\left\{\sup_{u\ge 0}\left(u\mathrm{I}(\vartheta_0)^{1/2}\,\zeta - \frac{u^2}{2}\mathrm{I}(\vartheta_0)\right) > \ln c_\varepsilon\right\}$$
$$= \mathbf{P}\left\{[\max(0,\zeta)]^2 > 2\ln c_\varepsilon\right\} = \mathbf{P}\left\{\zeta > \sqrt{2\ln c_\varepsilon}\right\} = \mathbf{P}\left\{\zeta > z_\varepsilon\right\} = \varepsilon.$$

Hence the test $\hat{\phi}_T(X^T) \in \mathcal{K}'_\varepsilon$.

To study the power function of the LRT we fix an alternative $\vartheta_1 = \vartheta_0 + u_1\,T^{-1/2}$ and consider the asymptotic of the power function $\beta_T(u_1, \hat{\phi}_T) = \beta_T(\vartheta_0 + u_1\,T^{-1/2}, \hat{\phi}_T)$ as $T \to \infty$. Introduce the normalized likelihood ratios

$$Z_T(u_1) = L(\vartheta_0 + T^{-1/2}u_1, \vartheta_0, X^T)$$

and

$$Z^*_T(u) = L(\vartheta_0 + T^{-1/2}u_1 + T^{-1/2}u, \vartheta_0 + T^{-1/2}u_1, X^T), \qquad u \in \mathbb{U}^*_T,$$

where the set

$$\mathbb{U}^*_T = \left[-u_1, (\vartheta_f - \vartheta_0)\sqrt{T} - u_1\right].$$

Note that the family of measures $\left\{\mathbf{P}^{(T)}_\vartheta, \vartheta \geq \vartheta_0\right\}$ has asymptotically the following local structure. The likelihood ratio $L(\vartheta_0 + T^{-1/2}u, \vartheta_0, X^T)$ admits the representation

$$L(\vartheta_0 + T^{-1/2}u, \vartheta_0, X^T) = Z_T(u_1)\; Z^*_T(u)$$

where

$$Z_T(u_1) = \exp\left\{u_1\, \Delta_T(\vartheta_0, X^T) + \frac{u_1^2}{2}\, \mathrm{I}(\vartheta_0) + r_T\left(\vartheta_0, u_1, X^T\right)\right\},$$

$$Z^*_T(u) = \exp\left\{u\, \Delta_T(\vartheta_0, X^T) - \frac{u^2}{2}\, \mathrm{I}(\vartheta_0) + r^*_T\left(\vartheta_0, u, X^T\right)\right\}, \quad u \in \mathbb{U}^*_T.$$

We define the process $Z^*_T(\cdot)$ linear increasing from 0 to $Z^*_T(-u_1)$ on the interval $[-1 - u_1, -u_1]$ and linear decreasing to zero on the interval

$$\left[(\vartheta_f - \vartheta_0)\sqrt{T} - u_1, (\vartheta_f - \vartheta_0)\sqrt{T} - u_1 + 1\right].$$

Then we put $Z^*_T(u) = 0$ for $u \leq -1 - u_1$ and $u \geq (\vartheta_f - \vartheta_0)\sqrt{T} - u_1 + 1$. The random variable $\Delta_T(\vartheta_0, X^T)$ is asymptotically normal and the random variables $r_T\left(\vartheta_0, u_1, X^T\right), r^*_T\left(\vartheta_0, u, X^T\right)$ tend to zero as $T \to \infty$ (see the proof of (5.18)–(5.20) for details). Therefore the marginal distributions of the process $\{Z_T(u_1),\ Z^*_T(u), u \in \mathscr{R}\}$ converge (under hypothesis $\mathscr{H}_1$) to the marginal distributions of the process $\{Z(u_1),\ Z^*(u), u \in \mathscr{R}\}$ defined as

$$Z(u_1) = \exp\left\{u_1\, \zeta\, \mathrm{I}(\vartheta_0)^{1/2} + \frac{u_1^2}{2}\, \mathrm{I}(\vartheta_0)\right\},$$

$$Z^*(u) = \exp\left\{u\, \zeta\, \mathrm{I}(\vartheta_0)^{1/2} - \frac{u^2}{2}\, \mathrm{I}(\vartheta_0)\right\} \chi_{\{u \geq -u_1\}} + (u + u_1 + 1)\, Z^*(-u_1)\, \chi_{\{-1-u_1 < u < -u_1\}}$$

Moreover, the increments of this process satisfy the estimates similar to (5.27) and (5.28) for all $u, v \in \mathscr{R}$

$$\mathbf{E}_{\vartheta_1} \left|Z^*_T(u)^{1/2} - Z^*_T(v)^{1/2}\right|^2 \leq C\, |u - v|^2 \tag{5.29}$$

and for any $N > 0$

$$\mathbf{E}_{\vartheta_1} Z_T^*(u)^{1/2} \leq \frac{C_N}{|u|^N}, \tag{5.30}$$

with some $C_N > 0$.

Therefore we have (under hypothesis $\mathscr{H}_1$) the weak convergence of the process $\{Z_T(u_1),\ Z_T^*(u), u \in \mathscr{R}\}$ to the process $\{Z(u_1),\ Z^*(u), u \in \mathscr{R}\}$ in the measurable space $(\mathscr{R}_+ \times \mathscr{C}_0, \mathfrak{B}_0)$. Here $\mathscr{C}_0$ is the space of continuous on $\mathscr{R}$ functions decreasing to zero at $\pm\infty$. This convergence allows us to write

$$\begin{aligned}
\beta_T(u_1, \hat\phi_T) &= \mathbf{P}^{(T)}_{\vartheta_0 + u_1 T^{-1/2}} \left\{ \hat L_T\left(X^T\right) > c_\varepsilon \right\} = \\
&= \mathbf{P}^{(T)}_{\vartheta_0 + u_1 T^{-1/2}} \left\{ Z_T(u_1) \sup_{u > -u_1} Z_T^*(u) > c_\varepsilon \right\} \\
&\longrightarrow \mathbf{P}\left\{ Z(u_1) \sup_{u > -u_1} Z^*(u) > c_\varepsilon \right\} \\
&= \mathbf{P}\left\{ u_1\, \zeta\, \mathrm{I}(\vartheta_0)^{1/2} + \frac{u_1^2}{2}\, \mathrm{I}(\vartheta_0) + \sup_{v > -u_1 \mathrm{I}(\vartheta_0)^{1/2}} \left(v\zeta - \frac{v^2}{2} \right) > \frac{z_\varepsilon^2}{2} \right\} \\
&= \mathbf{P}\left\{ 2u_1\, \zeta\, \mathrm{I}(\vartheta_0)^{1/2} + u_1^2\, \mathrm{I}(\vartheta_0) + \zeta^* > z_\varepsilon^2 \right\}
\end{aligned}$$

where the random variable

$$\zeta^* = \begin{cases} \zeta^2 & \text{if} \quad \zeta > -u_1\, \mathrm{I}(\vartheta_0)^{1/2} \\ -2u_1\, \zeta\, \mathrm{I}(\vartheta_0)^{1/2} - u_1^2\, \mathrm{I}(\vartheta_0) & \text{if} \quad \zeta \leq -u_1\, \mathrm{I}(\vartheta_0)^{1/2} \end{cases}.$$

Hence

$$\begin{aligned}
\beta_T(u_1, \hat\phi_T) &= \mathbf{P}\left\{ (\zeta + u_1\, \mathrm{I}(\vartheta_0)^{1/2})^2 > z_\varepsilon^2 \right\} + o(1) \\
&= \mathbf{P}\left\{ \zeta > z_\varepsilon - u_1\, \mathrm{I}(\vartheta_0)^{1/2} \right\} + o(1).
\end{aligned} \tag{5.31}$$

Moreover it can be shown that this convergence is uniform w.r.t. u_1 and comparison with (5.17) shows the desired type of optimality of the LRT.

To describe the asymptotic behavior of a Bayesian test $\tilde\phi_T$ we introduce the $1-\varepsilon$ quantile y_ε of the random variable

$$\tilde\eta = \sqrt{2\pi}\, \mathrm{e}^{\zeta^2/2}\, \Phi\left(\zeta\right),$$

i.e., $\mathbf{P}\left\{\tilde\eta > y_\varepsilon\right\} = 1 - \varepsilon$. Here $\zeta \sim \mathcal{N}\left(0,1\right)$, and $\Phi\left(\cdot\right)$ is its distribution function. Remember that $\Phi\left(\zeta\right)$ is uniform on $[0,1]$ random variable.

Theorem 5.13. *Suppose that the conditions of Theorem 5.12 are fulfilled. Then the Bayesian test*

$$\tilde\phi_T\left(X^T\right) = \chi_{\left\{\tilde L(X^T) > b_\varepsilon\right\}}, \qquad b_\varepsilon = \frac{q\left(\vartheta_0\right)\, y_\varepsilon}{\sqrt{T\, \mathrm{I}\left(\vartheta_0\right)}}$$

belongs to the class $\mathcal{K}'_\varepsilon$ and for any local alternative $\vartheta_1 = \vartheta_0 + T^{-1/2}u_1$ the power function

$$\beta_T\left(u_1, \tilde{\phi}_T\right) = \mathbf{P}\left\{\tilde{\eta}_1 > y_\varepsilon\right\} + o\left(1\right),$$

where

$$\tilde{\eta}_1 = \sqrt{2\pi}\, \mathrm{e}^{(\zeta+v_1)^2/2}\, \Phi\left(\zeta + v_1\right), \qquad v_1 = \mathrm{I}\left(\vartheta_0\right)^{1/2}\, u_1.$$

Proof. The proof is based on the weak convergence of the integrals as follows. Note that under hypotheses $\mathscr{H}_0$

$$\int_{\vartheta_0}^{\vartheta_f} L\left(\vartheta, \vartheta_0, X^T\right)\, q\left(\vartheta\right)\, \mathrm{d}\vartheta$$
$$= T^{-1/2} \int_0^{\sqrt{T}(\vartheta_f - \vartheta_0)} L\left(\vartheta_0 + T^{-1/2}u, \vartheta_0, X^T\right)\, q\left(\vartheta_0 + T^{-1/2}u\right)\, \mathrm{d}u.$$

Using the same arguments as in the proof of the theorem it can be shown that the last integral admits the representation

$$\int_0^{\sqrt{T}(\vartheta_f - \vartheta_0)} Z_T\left(u\right)\, q\left(\vartheta_0 + T^{-1/2}u\right)\, \mathrm{d}u$$
$$= q\left(\vartheta_0\right) \int_0^{\sqrt{T}(\vartheta_f - \vartheta_0)} \mathrm{e}^{u\Delta_n - \frac{u^2}{2}\mathrm{I}(\vartheta_0) + r_n}\, \mathrm{d}u\, \left(1 + o\left(1\right)\right),$$

where due to LAN $\Delta_n \Rightarrow \Delta \sim \mathcal{N}\left(0, \mathrm{I}\left(\vartheta_0\right)\right)$ and $r_n \to 0$. Moreover, we have the weak convergence

$$\int_0^{\sqrt{T}(\vartheta_f - \vartheta_0)} \mathrm{e}^{u\Delta_n - \frac{u^2}{2}\mathrm{I}(\vartheta_0) + r_n}\, \mathrm{d}u \Longrightarrow \int_0^\infty \mathrm{e}^{u\Delta - \frac{u^2}{2}\mathrm{I}(\vartheta_0)}\, \mathrm{d}u$$
$$= \mathrm{I}\left(\vartheta_0\right)^{-1/2} \int_0^\infty \mathrm{e}^{u\zeta - \frac{u^2}{2}}\, \mathrm{d}u = \sqrt{2\pi}\, \mathrm{I}\left(\vartheta_0\right)^{-1/2}\, \mathrm{e}^{\zeta^2/2}\, \Phi\left(\zeta\right).$$

Hence

$$\lim_{T\to\infty} \mathbf{P}_{\vartheta_0}^{(T)}\left\{\tilde{L}\left(X^T\right) > b_\varepsilon\right\} = \lim_{T\to\infty} \mathbf{P}_{\vartheta_0}^{(T)}\left\{\frac{\sqrt{T\mathrm{I}\left(\vartheta_0\right)}}{q\left(\vartheta_0\right)}\, \tilde{L}\left(X^T\right) > y_\varepsilon\right\}$$
$$= \mathbf{P}\left(\tilde{\eta} > y_\varepsilon\right) = \varepsilon$$

and $\tilde{\phi}_T \in \mathscr{K}'_\varepsilon$.

Let us fix an alternative $\vartheta_1 = \vartheta_0 + T^{-1/2}u_1$ with $u_1 > 0$. Then using the same arguments as above we obtain under this alternative the weak convergence

$$\frac{\sqrt{T\mathrm{I}\left(\vartheta_0\right)}}{q\left(\vartheta_0\right)}\, \tilde{L}\left(X^T\right) \Longrightarrow \mathrm{I}\left(\vartheta_0\right)^{1/2} \int_0^\infty \exp\left\{u\Delta + u\, u_1\, \mathrm{I}\left(\vartheta_0\right) - \frac{u^2}{2}\mathrm{I}\left(\vartheta_0\right)\right\}\, \mathrm{d}u$$
$$= \int_0^\infty \exp\left\{v\zeta + vv_1 - \frac{v^2}{2}\right\}\, \mathrm{d}v = \sqrt{2\pi}\, \mathrm{e}^{(\zeta+v_1)^2/2}\, \Phi\left(\zeta + v_1\right) = \tilde{\eta}_1.$$

Of course, the Bayesian test in this problem is not locally asymptotically uniformly most powerful.

Nonregular Cases

Let us consider the same hypotheses testing problem

$$\mathscr{H}_0 : \vartheta = \vartheta_0,$$
$$\mathscr{H}_1 : \vartheta > \vartheta_0,$$

by the observations

$$\mathrm{d}X_t = S\left(\vartheta, X_t\right)\, \mathrm{d}t + \sigma\left(X_t\right)\, \mathrm{d}W_t, \quad X_0 = x_0, \quad 0 \leq t \leq T,$$

where the trend coefficient has one of the singularities studied in Sections 3.2–3.4. The underlying family of measures is no longer LAN and there is no locally uniformly most powerful tests in this problem. We study here the asymptotics of the LRT only.

Cusp Testing

Let the observed process be

$$\mathrm{d}X_t = -\operatorname{sgn}\left(X_t - \vartheta\right) \left|X_t - \vartheta\right|^{\kappa}\, \mathrm{d}t + \mathrm{d}W_t, \quad X_0, \qquad 0 \leq t \leq T$$

where $\kappa \in (0, 1/2)$. According to Section 3.2 the contiguous alternatives are given by the normalizing function $\varphi_T\left(\vartheta_0\right) = T^{-\frac{1}{2\kappa+1}}$, i.e., we have the hypotheses testing problem

$$\mathscr{H}_0 : \vartheta = \vartheta_0,$$
$$\mathscr{H}_1 : \vartheta = \vartheta_0 + \frac{u}{T^{\frac{1}{2\kappa+1}}}, \qquad u > 0.$$

Let us denote by y_ε the $(1-\varepsilon)$-quantile ($\mathbf{P}\left\{\eta < y_\varepsilon\right\} = 1-\varepsilon$) of the random variable

$$\eta = \sup_{u \geq 0} \left(W^H\left(u\right) - \frac{\left|u\right|^{2\kappa+1}}{2} \right),$$

where $W^H\left(\cdot\right)$ is a fractional Brownian motion with Hurst parameter $H = \kappa + 1/2$.

Proposition 5.14. *The LRT*

$$\hat{\phi}_T\left(X^T\right) = \chi_{\left\{\ln \hat{L}_T\left(X^T\right) \geq y_\varepsilon\right\}} \in \mathscr{K}_\varepsilon',$$

is consistent and its power function for the local alternative $\vartheta = \vartheta_0 + T^{-1/(2\kappa+1)}u$ *is*

$$\beta_T\left(u, \hat{\phi}_T\right) = \mathbf{P}\left\{ \sup_{v \geq 0} \left(W^H\left(v\right) - \frac{\left|v - \tilde{u}\right|^{2\kappa+1} - \left|\tilde{u}\right|^{2\kappa+1}}{2} \right) \geq y_\varepsilon \right\} + o\left(1\right),$$

where $\tilde{u} = \Gamma_{\vartheta_0}^{\kappa+1/2} u$.

Proof. We know (see the proof of Theorem 3.8) that the likelihood ratio $Z_T(u) = L(\vartheta_0 + \varphi_T(\vartheta_0)\, u)$ for a fixed $u > 0$ under hypothesis $\mathscr{H}_0$ converges to the random variable

$$Z(u) = \exp\left\{\Gamma_{\vartheta_0}\, W^H(u) - \frac{|u|^{2\kappa+1}}{2}\, \Gamma_{\vartheta_0}^2\right\}$$

where the constant Γ_ϑ is given in (3.19). Moreover, we have the estimates (see Lemmas 3.11 and 3.12)

$$\mathbf{E}_{\vartheta_0}\left|Z_T^{1/2}(u) - Z_T^{1/2}(v)\right|^2 \leq C\, |u-v|^{2\kappa+1}, \tag{5.32}$$

$$\mathbf{P}_{\vartheta_0}^{(T)}\left\{Z_T(u) > \mathrm{e}^{-k|u|^{2\kappa+1}}\right\} \leq \frac{C_N}{|u|^N}. \tag{5.33}$$

Hence, as in the regular case above, we have the weak convergence of the likelihood ratio process $\{Z_T(u),\ u \geq 0\}$ to the process $\{Z(u),\ u \geq 0\}$ and we can write

$$\mathbf{P}_{\vartheta_0}^{(T)}\left\{\hat{L}_T(X^T) > \mathrm{e}^{y_\varepsilon}\right\} = \mathbf{P}_{\vartheta_0}^{(T)}\left\{\sup_{u\geq 0} Z_T(u) > \mathrm{e}^{y_\varepsilon}\right\}$$
$$\longrightarrow \mathbf{P}\left\{\sup_{u\geq 0} Z(u) > \mathrm{e}^{y_\varepsilon}\right\} = \mathbf{P}\left\{\sup_{u\geq 0}\left[\Gamma_{\vartheta_0}\, W^H(u) - \frac{|u|^{2\kappa+1}}{2}\, \Gamma_{\vartheta_0}^2\right] \geq y_\varepsilon\right\}$$
$$= \mathbf{P}\left\{\sup_{u\geq 0}\left[W^H(u) - \frac{|u|^{2\kappa+1}}{2}\right] > y_\varepsilon\right\} = \varepsilon.$$

To study the power function under local alternative $\vartheta_u = \vartheta_0 + T^{-1/(2\kappa+1)}u$ with fixed $u > 0$ we represent the likelihood ratio

$$Z_T(v) = L\left(\vartheta_0 + T^{-1/(2\kappa+1)}v, \vartheta_0, X^T\right)$$

as the product

$$Z_T(v) = L\left(\vartheta_v, \vartheta_u, X^T\right)\, L\left(\vartheta_u, \vartheta_0, X^T\right).$$

Then (with the help of the estimates like (5.32)–(5.33) under alternative) we verify the joint weak convergence of the process $L(\vartheta_v, \vartheta_u, X^T), v \geq 0$ and the random variables $L(\vartheta_u, \vartheta_0, X^T)$ to the random function

$$Z(v) = \exp\left\{\Gamma_{\vartheta_0}\left[W^H(v) - W^H(u)\right] - \frac{|v-u|^{2\kappa+1}}{2}\, \Gamma_{\vartheta_0}^2\right\}, \quad v \geq 0$$

and the random variable $Z_*(u) = \exp\left\{\Gamma_{\vartheta_0}\, W^H(u) + \frac{1}{2}|u|^{2\kappa+1}\, \Gamma_{\vartheta_0}^2\right\}$ respectively. Therefore we can write

$$\beta_T\left(u,\hat\phi_T\right)=\mathbf{P}_{\vartheta_u}^{(T)}\left\{\sup_{v\geq 0} Z_T\left(v\right)>e^{y_\varepsilon}\right\}\longrightarrow\mathbf{P}\left\{Z_*\left(u\right)\sup_{v\geq 0} Z\left(v\right)\geq e^{y_\varepsilon}\right\}$$

$$=\mathbf{P}\left\{\sup_{v\geq 0}\left(W^H\left(v\right)-\frac{\left|v-\tilde u\right|^{2\kappa+1}-\left|\tilde u\right|^{2\kappa+1}}{2}\right)\geq y_\varepsilon\right\},$$

where we changed the variables $v\longrightarrow\Gamma_{\vartheta_0}^{1/(2\kappa+1)}v$.

The Bayesian test in this problem can be described directly through the weak convergence of the corresponding integrals.

Delay and Change-point Testing

Suppose now that the observed process is either linear Gaussian with delay

$$\mathrm{d}X_t=-\gamma\, X_{t-\vartheta}\,\mathrm{d}t+\mathrm{d}W_t,\quad \hat X_0,\qquad 0\leq t\leq T \tag{5.34}$$

where $\gamma>0$ is known, the parameter $\vartheta\in[\vartheta_0,\beta)$, $0\leq\vartheta_0<\beta<\pi/2\gamma$ and $\hat X_0=\{x_s,\ -\beta\leq s\leq 0\}$ is the initial trajectory, or an ergodic diffusion process (3.71) with discontinuous trend coefficient satisfying conditions $\boldsymbol{\mathcal{M}}$, say, a simple switching process

$$\mathrm{d}X_t=-\operatorname{sgn}\left(X_t-\vartheta\right)\,\mathrm{d}t+\mathrm{d}W_t,\quad X_0,\qquad 0\leq t\leq T$$

with $\vartheta\in[\vartheta_0,\beta)$. In both cases the trend coefficients are not differentiable w.r.t. the parameter ϑ.

We consider the problem of the two hypotheses testing

$$\begin{aligned}\mathscr{H}_0:&\quad \vartheta=\vartheta_0,\\ \mathscr{H}_1:&\quad \vartheta=\vartheta_0+\frac{u}{T},\qquad u>0.\end{aligned}$$

In these cases the limit likelihood ratio for

$$Z_T\left(u\right)=L\left(\vartheta_0+T^{-1}u,\vartheta_0,X^T\right),u\geq 0$$

has the same form

$$Z\left(u\right)=\exp\left\{\Gamma_{\vartheta_0}\,W\left(u\right)-\frac{u}{2}\,\Gamma_{\vartheta_0}^2\right\},\qquad u\geq 0$$

but with different constants (see Sections 3.3 and 3.4). In particular, $\Gamma_{\vartheta_0}=\gamma$ for the process (5.34) and if the function $S\left(\vartheta,x\right)$ is discontinuous along the curve $x_*\left(\vartheta\right),\alpha\leq\vartheta\leq\beta$ then

$$\Gamma_\vartheta^2=\left|\dot x_*\left(\vartheta\right)\right|\left(\frac{S\left(\vartheta,x_*\left(\vartheta\right)+\right)-S\left(\vartheta,x_*\left(\vartheta\right)-\right)}{\sigma\left(x_*\left(\vartheta\right)\right)}\right)^2 f\left(\vartheta,x_*\left(\vartheta\right)\right).$$

Here $W\left(u\right),u\geq 0$ is a standard Wiener process.

Let us introduce the independent random variables

$$\zeta_1 = \Gamma_{\vartheta_0} W(u_1) + \frac{u_1}{2}\Gamma_{\vartheta_0}^2 \sim \mathcal{N}\left(\frac{u_1}{2}\Gamma_{\vartheta_0}^2, u_1\Gamma_{\vartheta_0}^2\right),$$

$$\eta_1 = \sup_{0\leq v\leq v_1}\left(W(v) - W(v_1) - \frac{v_1 - v}{2}\right), \quad v_1 = u_1\Gamma_{\vartheta_0}^2,$$

$$\eta = \sup_{v\geq 0}\left(W(v) - \frac{v}{2}\right).$$

Note that (see [16])

$$F_\eta(x) = \mathbf{P}\{\eta < x\} = 1 - e^{-x}, \tag{5.35}$$

$$F_{\eta_1}(x) = \Phi\left(\frac{x}{\sqrt{v_1}} + \frac{\sqrt{v_1}}{2}\right) + e^{-x}\left[1 - \Phi\left(\frac{x}{\sqrt{v_1}} - \frac{\sqrt{v_1}}{2}\right)\right], \tag{5.36}$$

where $x \geq 0$ and

$$\Phi(x) = \frac{1}{\sqrt{2\pi}}\int_{-\infty}^{x} e^{-y^2/2}dy.$$

Proposition 5.15. *Let the condition $\mathcal{L}$ for the process (5.34) or the condition $\mathcal{M}$ for the process (3.71) be fulfilled. Then the LRT*

$$\hat\phi_T(X^T) = \chi_{\left\{\hat L_T(X^T) > \frac{1}{\varepsilon}\right\}}$$

belongs to $\mathscr{K}_\varepsilon'$, is consistent and for any local alternative $\vartheta_1 = \vartheta_0 + T^{-1}u_1$, $u_1 > 0$, the power function

$$\beta_T\left(u_1, \hat\phi_T\right) = \mathbf{P}\left\{\zeta_1 + \max\left[\eta_1, \eta\right] > \ln\frac{1}{\varepsilon}\right\} + o(1).$$

Proof. Let us define the likelihood ratio process (under hypothesis $\mathscr{H}_0$) $Z_T(u)$, $u \in \mathbb{U}_T = [0 \leq u \leq (\beta - \vartheta_0)T]$, linear decreasing to zero on the interval

$$[(\beta - \vartheta_0)T \leq u \leq (\beta - \vartheta_0)T + 1]$$

and put $Z_T(u) = 0$ for all $u > (\beta - \vartheta_0)T + 1$. Now the process $Z_T(\cdot)$ is defined on $\mathscr{R}_+$.

According to Lemmas 3.13 and 3.27 the random process $Z_T(\cdot)$ admits (under hypothesis $\mathscr{H}_0$) the representation

$$Z_T(u) = \exp\left\{\Gamma_{\vartheta_0}\,\Delta_T(\vartheta_0, u, X^T) - \frac{u}{2}\,\Gamma_{\vartheta_0}^2 + r_T\left(\vartheta_0, u, X^T\right)\right\}, \quad u \in \mathscr{R}_+$$

where the marginal distributions of

$$\Delta_T(\vartheta_0, u, X^T) = \int_0^T \frac{S\left(\vartheta_0 + uT^{-1}, X_t\right) - S\left(\vartheta_0, X_t\right)}{\sigma\left(X_t\right)}\, dW_t$$

converge to the marginal distributions of the Wiener process

$$\{\Gamma_{\vartheta_0} W(u),\ u \geq 0\}$$

and $r_T(\vartheta_0, u, X^T) \to 0$ in probability. We have as well the estimates (see Lemmas 3.11, 3.12) for all $|u|, |v| < R$ and all $R > 0$,

$$\mathbf{E}_{\vartheta_0} \left| Z_T^{1/4}(u) - Z_T^{1/4}(v) \right|^4 \leq C \left(1 + R^6\right) |u - v|^2,$$

$$\mathbf{P}_{\vartheta_0}^{(T)} \left\{ Z_T(u) \geq \mathrm{e}^{-\kappa |u|} \right\} \leq \frac{C_N}{u^N},$$

and (see Lemmas 3.28) and 3.29)

$$\mathbf{E}_{\vartheta_0} \left| Z_T^{1/8}(u) - Z_T^{1/8}(v) \right|^4 \leq C \left(1 + R^2\right) |u - v|^2,$$

$$\mathbf{P}_{\vartheta_0}^{(T)} \left\{ Z_T(u) \geq \mathrm{e}^{-\frac{\kappa}{4}|u|} \right\} \leq \frac{C_N}{u^N},$$

respectively. Therefore the stochastic process $\left\{ Z_T(u),\ u \in \mathscr{R}_+ \right\}$ converges weakly in the measurable space $\{\mathscr{C}_+, \mathfrak{B}_+\}$ of continuous on $\mathscr{R}_+$ functions

$$\mathscr{C}_+ = \left\{ z(u) :\ z(0) = 1,\ \lim_{u \to \infty} z(u) = 0, \|z_1 - z_2\| = \sup_{u \geq 0} |z_1(u) - z_2(u)| \right\}$$

to the stochastic process $\{Z(u),\ u \geq 0\}$ defined above.

Hence

$$\mathbf{P}_{\vartheta_0}^{(T)} \left\{ \hat{L}_T\left(X^T\right) > \frac{1}{\varepsilon} \right\} = \mathbf{P}_{\vartheta_0}^{(T)} \left\{ \sup_{u \geq 0} Z_T(u) > \frac{1}{\varepsilon} \right\}$$

$$\longrightarrow \mathbf{P} \left\{ \sup_{u \geq 0} Z(u) > \frac{1}{\varepsilon} \right\} = \mathbf{P} \left\{ \sup_{u \geq 0} \left(\Gamma_{\vartheta_0} W(u) - \frac{u}{2}\, \Gamma_{\vartheta_0}^2 \right) > -\ln \varepsilon \right\}$$

$$= \mathbf{P} \left\{ \sup_{u \geq 0} \left(W(u) - \frac{u}{2} \right) > -\ln \varepsilon \right\} = \varepsilon$$

because for the Wiener process $W(u),\ u \geq 0$ we have the equality (5.35). Therefore $\hat{\phi}_T \in \mathscr{K}'_\varepsilon$.

We study the power function under the local (contiguous) alternative $\vartheta_1 = \vartheta_0 + T^{-1}u_1$, i.e., we consider the asymptotic behavior of the function

$$\beta_T\left(u_1, \hat{\phi}_T\right) = \beta_T\left(\vartheta_0 + u_1 T^{-1}, \hat{\phi}_T\right) = \mathbf{P}_{\vartheta_1}^{(T)} \left\{ \hat{L}_T\left(X^T\right) > \frac{1}{\varepsilon} \right\}.$$

As in the regular case above, we introduce the likelihood ratio

$$L(\vartheta_0 + uT^{-1}, \vartheta_0, X^T) = Z_T(u_1)\ Z_T^*(u, u_1),$$

where the random variable $Z_T(u_1) = L(\vartheta_0 + T^{-1}u_1, \vartheta_0, X^T)$ and the stochastic processes $Z_T^*(u, u_1) = L(\vartheta_0 + uT^{-1}, \vartheta_0 + u_1 T^{-1}, X^T)$ admit the representations

$$Z_T(u_1) = \exp\left\{\Gamma_{\vartheta_0}\ \Delta_T(\vartheta_0, u_1, X^T)\ + \frac{u_1}{2}\ \Gamma_{\vartheta_0}^2 + r_T\left(\vartheta_0, u_1, X^T\right)\right\},$$

$$Z_T^*(u, u_1) = \exp\Big\{\Gamma_{\vartheta_0}\ \Delta_T(\vartheta_0, u, u_1, X^T) - \frac{|u - u_1|}{2}\ \Gamma_{\vartheta_0} + r_T^*\left(\vartheta_0, u, u_1, X^T\right)\Big\}$$

with

$$u \in \mathbb{U}_T^* = [0, (\beta - \vartheta_0)\, T]\,.$$

Here

$$\Delta_T(\vartheta_0, u, u_1, X^T) = \int_0^T \frac{S\left(\vartheta_0 + uT^{-1}, X_t\right) - S\left(\vartheta_0 + u_1T^{-1}, X_t\right)}{\sigma\left(X_t\right)}\ \mathrm{d}W_t.$$

We define the process $Z_T^*(\cdot)$ outside of $\mathbb{U}_T^*$ linear decreasing to zero (as in the similar situation in the regular case). We have the estimates

$$\mathbf{E}_{\vartheta_1}\left|Z_T^*(u, u_1)^{1/8} - Z_T^*(u', u_1)^{1/8}\right|^4 \leq C_1\ (1 + R)\ |u - u'|^2$$

for all $|u|\,, |u'| < R$ and all $R > 0$, and

$$\mathbf{E}_{\vartheta_1}\, Z_T^*(u, u_1)^{1/2} \leq \frac{C_N}{|u - u_1|^N}$$

which provide, together with the convergence of the corresponding marginal distributions, the weak convergence of the stochastic process

$$\left\{Z_T(u_1), Z_T^*(u, u_1),\ u \in \mathscr{R}_+\right\}$$

to the stochastic process

$$\left\{Z(u_1), Z^*(u, u_1),\ u \in \mathscr{R}\right\}$$

in the measurable space $(\mathscr{R}_+ \times \mathscr{C}_0, \mathfrak{B}_0)$. Here

$$Z(u_1) = \exp\left\{\Gamma_{\vartheta_0} W\left(u_1\right) + \frac{u_1}{2}\Gamma_{\vartheta_0}^2\right\},$$

$$Z^*(u, u_1) = \exp\left\{\Gamma_{\vartheta_0}\left[W\left(u\right) - W\left(u_1\right)\right] - \frac{|u - u_1|}{2}\Gamma_{\vartheta_0}^2\right\}.$$

Therefore, if we put $v = u\Gamma_{\vartheta_0}^2,\ v_1 = u_1\Gamma_{\vartheta_0}^2$, then we can write

$$\begin{aligned}
\beta_T\left(u_1, \hat\phi_T\right) &= \mathbf{P}_{\vartheta_1}^{(T)}\left\{Z_T(u_1) \sup_{u\geq 0} Z_T^*(u, u_1) > \frac{1}{\varepsilon}\right\} \\
&\longrightarrow \mathbf{P}\left\{Z(u_1) \sup_{u\geq 0} Z^*(u, u_1) > \frac{1}{\varepsilon}\right\} \\
&= \mathbf{P}\left\{W\left(v_1\right) + \frac{v_1}{2} + \sup_{v\geq 0}\left(W\left(v\right) - W\left(v_1\right) - \frac{|v - v_1|}{2}\right) > -\ln\varepsilon\right\}.
\end{aligned}$$

We denote the last probability as $\beta\left(u_1,\hat{\phi}\right)$ and it can be written as

$$\mathbf{P}\left\{\zeta(u_1)+\max\left[\eta(u_1),\eta\right]>-\ln\varepsilon\right\}$$

where $\zeta(u_1)=\zeta_1,\eta(u_1)=\eta_1$ and η are independent random variables defined above.

Note that the distribution function

$$\mathbf{P}\left\{\max\left[\eta_1,\eta\right]<x\right\}=F_\eta\left(x\right)\ F_{\eta_1}\left(x\right). \tag{5.37}$$

Therefore $\beta\left(u_1,\hat{\phi}\right)$ is quite a cumbersome expression of the convolution of the Gaussian law $\mathcal{N}\left(\frac{u_1}{2}\Gamma_{\vartheta_0}^2,u_1\Gamma_{\vartheta_0}^2\right)$ and the distribution function (5.37).

Remember, that the Bayesian test

$$\tilde{\phi}_T\left(X^T\right)=\chi_{\left\{\tilde{L}_T>b_\varepsilon\right\}}$$

is the most powerful in the problem of testing two simple hypotheses

$$\mathscr{H}_0:\ \mathbf{P}^{(T)}=\mathbf{P}_{\vartheta_0}^{(T)},$$
$$\mathscr{H}_Q:\ \mathbf{P}^{(T)}=\mathbf{P}_Q^{(T)},$$

where $\mathbf{P}^{(T)}$ is the distribution of the observed process X^T (see (5.23) and (5.24)).

Let us denote by y_ε the $1-\varepsilon$ quantile of the random variable

$$\eta^*=\int_0^\infty e^{W(v)-v/2}\,dv$$

and for any $u_1>0$ put

$$\eta_1^*=\int_0^\infty e^{W(v)-v/2+(v\wedge v_1)}\,dv$$

where $v_1=\Gamma_{\vartheta_0}^2u_1$.

Proposition 5.16. *Let the conditions of Proposition 5.15 be fulfilled. Then the test*

$$\tilde{\phi}_T\left(X^T\right)=\chi_{\left\{\tilde{L}_T(X^T)>b_\varepsilon\right\}},\qquad b_\varepsilon=\frac{q\left(\vartheta_0\right)\,y_\varepsilon}{T\,\Gamma_{\vartheta_0}^2}$$

belongs to $\mathscr{K}_\varepsilon'$, is consistent and for any local alternative $\vartheta_1=\vartheta_0+T^{-1}u_1$, $u_1>0$, the power function

$$\beta_T\left(u_1,\tilde{\phi}_T\right)=\mathbf{P}\left\{\eta_1^*>y_\varepsilon\right\}+o\left(1\right).$$

Proof. The proof is similar to that of Theorem 5.13.

Windows

In the problems of cusp testing and switching testing we can restrict our observation window up to $[\alpha,\beta]$ and to use the *pseudo-likelihood ratio test*

$$\hat{\phi}_T\left(X^T\right) = \chi_{\left\{\overline{L}_T\left(X^T\right) > \frac{1}{\varepsilon}\right\}}$$

where the function $\overline{L}_T(\cdot)$ is defined by Equation (3.95) (for a change-point model). The power function of this test has the same asymptotics for the local alternatives as the power function of the LRT in Proposition 5.15.

5.3 One-Sided Nonparametric Alternative

5.3.1 Local Alternatives

Let us consider the problem of hypotheses testing by the observations of an ergodic diffusion process

$$\mathrm{d}X_t = S\left(X_t\right)\,\mathrm{d}t + \sigma\left(X_t\right)\,\mathrm{d}W_t, \quad X_0, \quad 0 \le t \le T \tag{5.38}$$

in the situation when the hypothesis

$$\mathscr{H}_0: \quad S\left(\cdot\right) = S_0\left(\cdot\right) \tag{5.39}$$

is simple ($S_0(\cdot)$ is a known function) and the alternative

$$\mathscr{H}_1: \quad S\left(\cdot\right) = S_0\left(\cdot\right) + \frac{u\left(\cdot\right)}{\sqrt{T}}, \quad u\left(\cdot\right) \in \mathbb{U} \tag{5.40}$$

is composite. The function $u(\cdot) \in \mathbb{U}$ is in certain sense *one-sided nonparametric*. The set $\mathbb{U}$ we define with the help of a specially introduced parameterization and this alternative corresponds to some *small perturbations of the hypothesis*, because for the fixed alternative $S(\cdot) \neq S_0(\cdot)$ the power function of any reasonable test, as in the parametric case, tends to one with exponential rate.

Let us fix some number $\varepsilon \in (0,1)$ and consider the class $\mathscr{K}_\varepsilon$ of tests $\bar{\phi}_T$ of the size ε. As before, $\mathscr{K}_\varepsilon'$ is the class of tests of asymptotic size ε. Our goal is to find a class of local alternatives, such as those like the parametric local alternatives (Section 5.2, regular case) for which we will be able to construct locally asymptotically most powerful tests (in the class $\mathscr{K}_\varepsilon'$).

We suppose that the functions $S_0(\cdot)$, $u(\cdot)$ and $\sigma(\cdot)$ are such that the conditions $\mathcal{ES}$, $\mathcal{EM}$ and $\mathcal{RP}$ are fulfilled. Therefore the observed process is always recurrent positive.

The approach taken here is quite close to one presented in Section 4.3 in semiparametric estimation, i.e., we consider the parameterization

$$\vartheta(S) = \mathbf{E}_S\Big(R(\xi)\,S(\xi) + N(\xi)\Big), \tag{5.41}$$

where ξ is a stationary random variable with the invariant density $f_S(\cdot)$ and $R(\cdot)$ and $N(\cdot)$ are some known functions. We consider one-sided local alternatives expressed in the terms of the values of the parameter $\vartheta(S)$, i.e., $\vartheta(S) > \vartheta(S_0)$. Remember, that this parameterization allows us to treat several reasonable problems simultaneously. For example, if $R(x) = \sigma(x_*)^{-2}\,\mathrm{sgn}(x_* - x)$ and $N(x) = 0$, then the hypothesis $\mathscr{H}_0$ provides $f(x_*) = f_{S_0}(x_*)$ and the alternative $\mathscr{H}_1$ corresponds to $f(x_*) > f_{S_0}(x_*)$.

Note that $f(x_*) = f_{S_0}(x_*)$ is not a simple hypothesis and we just say that under hypothesis $\mathscr{H}_0$ we have such an equality. Similarly if $R(x) = 0$ and $N(x) = |x|^p$, where $p > 0$ is given then $\mathscr{H}_0$ gives us $\mathbf{E}\,|\xi|^p = \mathbf{E}_{S_0}\,|\xi|^p$ and the alternative $\mathbf{E}\,|\xi|^p > \mathbf{E}_{S_0}\,|\xi|^p$. The class of functions $S(\cdot)$ with the same value of ϑ is described below in Remark 5.19.

Remember that the Fisher information in this problem is

$$\mathrm{I}(S) = \left\{\mathbf{E}_S\left(\frac{R(\xi)\,\sigma(\xi)^2 f_S(\xi) + 2M_S(\xi)}{\sigma(\xi)\,f_S(\xi)}\right)^2\right\}^{-1},$$

where the function

$$M_S(x) = \mathbf{E}_S\left(\left[F_S(x) - \chi_{\{\xi<x\}}\right][R(\xi)\,S(\xi) + N(\xi)]\right).$$

Further, let us introduce the sets

$$\mathbb{U} = \left\{u(\cdot):\ 0 < \int\left[R(x)\,f_{S_0}(x) + \frac{2M_{S_0}(x)}{\sigma(x)^2}\right]u(x)\,\mathrm{d}x < \infty\right\},$$

$$\mathbb{U}_K = \left\{u(\cdot):\ 0 < \int\left[R(x)\,f_{S_0}(x) + \frac{2M_{S_0}(x)}{\sigma(x)^2}\right]u(x)\,\mathrm{d}x < K\right\},$$

$$\mathbb{U}_h = \left\{u(\cdot):\ 0 < \int\left[R(x)\,f_{S_0}(x) + \frac{2M_{S_0}(x)}{\sigma(x)^2}\right]u(x)\,\mathrm{d}x = h\right\},$$

where $h \geq 0$. Below $S(\cdot) = S_0(\cdot) + \frac{u(\cdot)}{\sqrt{T}}$.

Condition $\boldsymbol{\mathcal{Y}}$.

$\mathcal{Y}_1$. *For any* $K > 0$

$$\sup_{u(\cdot)\in\mathbb{U}_K} G(S) < \infty, \qquad \sup_{u(\cdot)\in\mathbb{U}_K}\left(\mathbf{E}_S\,|R(\xi)\,S(\xi)| + \mathbf{E}_S\,|N(\xi)|\right) < \infty.$$

$\mathcal{Y}_2$.

$$\mathbf{E}_{S_0}\left(\frac{R(\xi)\,\sigma(\xi)^2 f_{S_0}(\xi)+2M_{S_0}(\xi)}{\sigma(\xi)\,f_{S_0}(\xi)}\right)^2<\infty.$$

We consider the problem of hypotheses testing (5.40) and (5.41) with the functions $u(\cdot)\in\mathbb{U}$ defined above. As before we denote by $\beta_T\left(u,\phi_T\right)$ the power function of the test ϕ_T calculated under a fixed simple alternative $S(x)=S_0(x)+T^{-1/2}u(x)$.

Definition 5.17. *Let the condition $\mathcal{Y}$ be fulfilled. Then we say that a test $\phi_T^*\in\mathcal{K}_\varepsilon'$ is locally asymptotically uniformly most powerful in the class $\mathcal{K}_\varepsilon'$ if for any other test $\bar\phi_T\in\mathcal{K}_\varepsilon'$ and any $K>0$ we have*

$$\varliminf_{T\to\infty}\ \inf_{0\le h\le K}\ \inf_{u(\cdot)\in\mathbb{U}_h}\left[\beta_T\left(u,\phi_T^*\right)-\beta_T\left(u,\bar\phi_T\right)\right]\ge 0.$$

5.3.2 Asymptotically Optimal Test

Let us introduce the stopping time ($\tau_T\le T+1$)

$$\tau_T=\left\{\tau:\inf\int_0^\tau\left(R(X_t)\,\sigma(X_t)+\frac{2M_{S_0}(X_t)}{\sigma(X_t)f_{S_0}(X_t)}\right)^2\mathrm{d}t\ge T\mathrm{I}(S_0)^{-1}\right\}$$

and the statistic

$$\Delta_{\tau_T}\left(S_0,X^T\right)=\frac{1}{\sqrt{T}}\int_0^{\tau_T}\left[R(X_t)+\frac{2M_{S_0}(X_t)}{\sigma(X_t)^2f_0(X_t)}\right]\left[\mathrm{d}X_t-S_0(X_t)\,\mathrm{d}t\right].$$

For the values $\tau_T\in(T,T+1]$, as usual, we denote

$$\sigma(X_t)\equiv 1,\qquad \mathrm{d}X_t-S_0(X_t)\,\mathrm{d}t=\mathrm{d}\tilde W_t,$$

where $\tilde W_t, T\le t\le T+1$ is an independent of $W_t, 0\le t\le T$ Wiener process.

Theorem 5.18. *Let the condition $\mathcal{Y}$ be fulfilled. Then the test*

$$\phi_T^*\left(X^T\right)=\chi_{\left\{\Delta_{\tau_T}\left(S_0,X^T\right)\ge z_\varepsilon\ \mathrm{I}(S_0)^{-1/2}\right\}}$$

belongs to $\mathcal{K}_\varepsilon$, is consistent and is locally asymptotically uniformly most powerful in the class $\mathcal{K}_\varepsilon'$. Moreover, for $u(\cdot)\in\mathbb{U}_h$

$$\beta_T\left(u,\phi_T^*\right)=\mathbf{P}\left\{\zeta>z_\varepsilon-h\,\mathrm{I}(S_0)^{1/2}\right\}+o(1),$$

where $\zeta\sim\mathcal{N}(0,1)$.

Proof. We have to find an upper bound on the power function of all tests from the class $\mathcal{K}_\varepsilon$ and then to show that the test ϕ_T^* asymptotically attains this bound.

Let $u_1(\cdot)$ be some function from the set $\mathbb{U}_h$ having a support compact and $h > 0$. Then for any test $\phi_T \in \mathcal{K}_\varepsilon'$ we have

$$\inf_{u(\cdot)\in\mathbb{U}_h} \beta_T\left(u, \phi_T\right) \leq \beta_T\left(u_1, \phi_T\right) \leq \beta_T\left(u_1, \hat{\phi}_T\right) \tag{5.42}$$

where $\hat{\phi}_T\left(X^T\right)$ is the likelihood ratio test in the problem of two simple hypotheses testing, i.e.,

$$\hat{\phi}_T\left(X^T\right) = \chi_{\left\{L\left(S_0+u_1/\sqrt{T}, S_0, X^T\right)\geq c_\varepsilon\right\}}$$

because this test is optimal by the Neyman–Pearson lemma. This likelihood ratio admits the representation

$$\ln L\left(S_0 + u_1/\sqrt{T}, S_0, X^T\right) = \frac{1}{\sqrt{T}}\int_0^T \frac{u_1(X_t)}{\sigma(X_t)^2}\left[\mathrm{d}X_t - S_0(X_t)\,\mathrm{d}t\right]$$
$$-\frac{1}{2T}\int_0^T \left(\frac{u_1(X_t)}{\sigma(X_t)}\right)^2 \mathrm{d}t + o(1),$$

and the likelihood ratio test is asymptotically equivalent to the test

$$\tilde{\phi}_T\left(X^T\right) = \chi_{\left\{\zeta_T\left(u_1, X^T\right)\geq z_\varepsilon \mathrm{I}_*(u_1)^{1/2}\right\}},$$

where

$$\zeta_T\left(u_1, X^T\right) = \frac{1}{\sqrt{T}}\int_0^T \frac{u_1(X_t)}{\sigma(X_t)^2}\left[\mathrm{d}X_t - S_0(X_t)\,\mathrm{d}t\right].$$

Therefore asymptotically as $T \to \infty$ we have

$$\beta_T\left(u_1, \hat{\phi}_T\right) \longrightarrow \mathbf{P}\left\{\zeta(u_1) \geq z_\varepsilon \mathrm{I}_*(u_1)^{1/2} - \mathrm{I}_*(u_1)\right\}$$
$$= \mathbf{P}\left\{\zeta \geq z_\varepsilon - \mathrm{I}_*(u_1)^{1/2}\right\},$$

where $\zeta(u_1) \sim \mathcal{N}(0, \mathrm{I}_*(u_1))$, $\zeta \sim \mathcal{N}(0,1)$ and the quantity (which we call the Fisher information)

$$\mathrm{I}_*(u_1) = \mathbf{E}_{S_0}\left(\frac{u_1(\xi)}{\sigma(\xi)}\right)^2.$$

The inequality (5.42) is valid for all $u_1(\cdot) \in \mathbb{U}_h$. Hence we have the same inequality for the *least favorable function* $u_*(\cdot)$ too, where the least favorable corresponds to the minimal power function or to the minimal Fisher information. We have to find the function $u_*(\cdot)$ and the corresponding quantity I_* such that

$$\inf_{u_1(\cdot)\in\mathbb{U}_h} \mathrm{I}_*\left(u_1\right)=\mathrm{I}_*\left(u_*\right).$$

But this problem was solved in Section 4.3.1 as follows. We have by the Cauchy–Schwarz inequality

$$\begin{aligned}
h^2 &= \left(\int\left(R(x)\sigma\left(x\right)+\frac{2M_{S_0}(x)}{\sigma\left(x\right)\,f_{S_0}\left(x\right)}\right)\frac{u\left(x\right)}{\sigma\left(x\right)}\,f_{S_0}\left(x\right)\,\mathrm{d}x\right)^2\\
&\leq \mathrm{I}_*\left(u\right)\int\left(R(x)\sigma\left(x\right)+\frac{2M_{S_0}(x)}{\sigma\left(x\right)f_{S_0}\left(x\right)}\right)^2 f_{S_0}\left(x\right)\,\mathrm{d}x
\end{aligned}$$

and equality is possible iff

$$u_*\left(x\right)=h\,\mathrm{I}\left(S_0\right)\left(R(x)\sigma\left(x\right)^2+\frac{2M_{S_0}(x)}{f_{S_0}\left(x\right)}\right).$$

The support of this function is not necessarily compact but it can be approximated in a $\mathscr{L}_2(f_{S_0})$ sense by a sequence of functions with support compact and this will provide us with the same value in the limit. Therefore the worst Fisher information is

$$\mathrm{I}_*\left(u_*\right)=h^2\,\mathrm{I}\left(S_0\right)$$

and the power function of the corresponding test is

$$\beta_T\left(u_*,\dot\phi_T\right)=\mathbf{P}\left\{\zeta\geq z_\varepsilon-h\,\mathrm{I}\left(S_0\right)^{1/2}\right\}+o\left(1\right).$$

Now we show that the test $\phi_T^*\left(X^T\right)$ has the same power function. First, under the hypothesis $\mathscr{H}_0$ the statistic

$$\Delta_{\tau_T}\left(S_0,X^T\right)=\frac{1}{\sqrt{T}}\int_0^{\tau_T}\left[R\left(X_t\right)\sigma\left(X_t\right)+\frac{2M_{S_0}(X_t)}{\sigma(X_t)f_{S_0}(X_t)}\right]\,\mathrm{d}W_t$$

is Gaussian random variable $\mathcal{N}\left(0,\mathrm{I}\left(S_0\right)^{-1}\right)$. Hence

$$\mathbf{P}_{S_0}^{(T)}\left\{\Delta_{\tau_T}\left(S_0,X^T\right)\geq z_\varepsilon\,\mathrm{I}\left(S_0\right)^{-1/2}\right\}=\varepsilon$$

and the test $\phi_T^*\left(X^T\right)\in\mathscr{K}_\varepsilon$. Further, under alternative $S\left(x\right)=S_0\left(x\right)+u(x)/\sqrt{T}$ with $u\left(\cdot\right)\in\mathbb{U}_h$ we have

$$\begin{aligned}
\Delta_{\tau_T}\left(S_0,X^T\right) &= \frac{1}{\sqrt{T}}\int_0^{\tau_T}\left[R\left(X_t\right)\sigma\left(X_t\right)+\frac{2M_{S_0}(X_t)}{\sigma(X_t)f_{S_0}(X_t)}\right]\,\mathrm{d}W_t\\
&\quad+\sqrt{T}\left(\vartheta\left(S_0+u/\sqrt{T}\right)-\vartheta\left(S_0\right)\right)+o\left(1\right)\\
&=\zeta\,\mathrm{I}\left(S_0\right)^{-1/2}+h\,\mathrm{I}\left(S_0\right)^{-1}+o\left(1\right)
\end{aligned}$$

and

$$\Delta_{\tau_T}\left(S_0, X^T\right) \sim \mathcal{N}\left(h\, \mathrm{I}\left(S_0\right)^{-1}, \mathrm{I}\left(S_0\right)^{-1}\right).$$

Therefore

$$\beta_T\left(u, \phi_T^*\right) = \mathbf{P}\left\{\zeta \geq z_\varepsilon - h\, \mathrm{I}\left(S_0\right)^{1/2}\right\} + o\left(1\right).$$

Remark 5.19. The statistic $\Delta_{\tau_T}\left(S_0, X^T\right)$ can be obtained as follows. Let us take the estimator of the functional $\vartheta\left(S\right)$

$$\hat{\vartheta}_T = \frac{1}{T}\int_0^T R\left(X_t\right)\, \mathrm{d}X_t + \frac{1}{T}\int_0^T N\left(X_t\right)\, \mathrm{d}t$$

studied in Section 4.3.2 and put

$$\delta_T\left(S_0, X^T\right) = \sqrt{T}\left(\hat{\vartheta}_T - \vartheta\left(S_0\right)\right).$$

Then under hypothesis $\mathscr{H}_0$ (see Section 4.3.2)

$$\begin{aligned}
\delta_T\left(S_0, X^T\right) &= \frac{1}{\sqrt{T}}\int_0^T \left[R\left(X_t\right) S_0\left(X_t\right) + N\left(X_t\right) - \vartheta\left(S_0\right)\right]\, \mathrm{d}t \\
&\quad + \frac{1}{\sqrt{T}}\int_0^T R\left(X_t\right)\, \sigma\left(X_t\right)\, \mathrm{d}W_t \\
&= \frac{1}{\sqrt{T}}\int_{X_0}^{X_T} \frac{2M_{S_0}\left(v\right)}{\sigma\left(v\right)^2 f_{S_0}\left(v\right)}\mathrm{d}v \\
&\quad + \frac{1}{\sqrt{T}}\int_0^T \left[R\left(X_t\right)\sigma\left(X_t\right) + \frac{2M_{S_0}\left(X_t\right)}{\sigma\left(X_t\right) f_{S_0}\left(X_t\right)}\right]\mathrm{d}W_t \\
&= \frac{1}{\sqrt{T}}\int_{X_0}^{X_T} \frac{2M_{S_0}\left(v\right)}{\sigma\left(v\right)^2 f_{S_0}\left(v\right)}\mathrm{d}v + \Delta_T\left(S_0, X^T\right).
\end{aligned}$$

The first integral tends to zero in probability and the second stopped at the instant τ_T gives us $\Delta_{\tau_T}\left(S_0, X^T\right)$.

Note as well that under hypothesis $\mathscr{H}_1$ we have

$$\begin{aligned}
\delta_T\left(S_0, X^T\right) = \sqrt{T}\left(\hat{\vartheta}_T - \vartheta\left(S_0\right)\right) &= \sqrt{T}\left(\hat{\vartheta}_T - \vartheta\left(S_0 + \frac{u}{\sqrt{T}}\right)\right) \\
&\quad + \sqrt{T}\left(\vartheta\left(S_0 + \frac{u}{\sqrt{T}}\right) - \vartheta\left(S_0\right)\right)
\end{aligned}$$

and the last term is asymptotically equal to h for any $u\left(\cdot\right) \in \mathbb{U}_h$.

Remark 5.20. Note that to one fixed value $\vartheta = \vartheta(S)$ corresponds a class $\mathscr{S}_\vartheta$ of functions $S(\cdot)$, which can be described as follows. Suppose that the function $R(x)$ is continuously differentiable and

$$\lim_{|x|\to\infty} R(x)\,\sigma(x)^2 f_S(x) = 0,$$

then using integration by parts we obtain the equality

$$\mathbf{E}_S\Big(R(\xi)\,S(\xi)\Big) = -\frac{1}{2}\,\mathbf{E}_S\left(R'(\xi)\,\sigma(\xi)^2\right).$$

Fix a value ϑ and a function $S(\cdot)$ such that $\vartheta(S) = \vartheta$. Denote this function as $S(\vartheta,\cdot)$ and describe the other functions of the set $\mathscr{S}_\vartheta$ with the help of this function.

Introduce the set of continuously differentiable functions

$$\mathscr{Q}_\vartheta = \left\{q(\cdot):\ \mathbf{E}_{S(\vartheta)}\,q(\xi) = 0,\ \mathbf{E}_{S(\vartheta)}\,q(\xi)\left[N(\xi) - \frac{\sigma(\xi)^2}{2}R'(\xi)\right] = 0\right\},$$

then for any two functions $S_1(\cdot),\ S_2(\cdot) \in \mathscr{S}_\vartheta$, i.e., $\vartheta(S_1) = \vartheta(S_2)$, we have the equality

$$\int_{-\infty}^{\infty}\left[f_{S_1}(x) - f_{S_2}(x)\right]\left[N(x) - \frac{\sigma(x)^2}{2}R'(x)\right]\,\mathrm{d}x = 0.$$

Therefore we can write

$$f_S(x) = f_{S(\vartheta)}(x)\,(1 + q(x)),\tag{5.43}$$

where we consider only such $q(\cdot) \in \mathscr{Q}_\vartheta$ that

$$1 + q(x) > 0, \quad \text{for all} \quad x \in \mathscr{R}.\tag{5.44}$$

Equation (5.43) can be solved w.r.t. $S(\cdot)$ as follows. We have

$$\frac{1}{G(S)}\exp\left\{2\int_0^x \frac{S(v)}{\sigma(v)^2}\mathrm{d}v\right\}$$
$$= \frac{1}{G(S(\vartheta))}\exp\left\{2\int_0^x \frac{S(\vartheta,v)}{\sigma(v)^2}\mathrm{d}v\right\}(1 + q(x)).$$

Hence after taking logarithms and differentiating we obtain

$$S(x) = S(\vartheta,x) + \frac{q'(x)\,\sigma(x)^2}{2\,(1 + q(x))}.\tag{5.45}$$

The functions $S(\cdot)$ obtained by this formula have to belong to $\mathscr{S}$ and have to satisfy the conditions (5.40) as well, and this imposes another condition on $q(\cdot)$. Say, any function $q(\cdot) \in \mathscr{Q}$ with compact support satisfying (5.44) can be used in (5.45).

Note as well that for any $S(\cdot)$ satisfying (5.45) we obtain the equality

$$G(S) = \frac{G(S(\vartheta))}{1 + q(0)}.$$

5.4 Goodness-of-fit Test

Let the observed process be

$$\mathrm{d}X_t = S(X_t)\ \mathrm{d}t + \mathrm{d}W_t, \qquad 0 \le t \le T,$$

where $S(\cdot)$ satisfies the "usual conditions" and suppose (for simplicity) that it is continuous. Fix some function $S_0(\cdot)$. Then it is easy to see that if

$$\sup_x |S(x) - S_0(x)| > 0,$$

then

$$\sup_x \left| f_S(x) - f_{S_0}(x) \right| > 0 \qquad \text{and} \qquad \sup_x \left| F_S(x) - F_{S_0}(x) \right| > 0.$$

Therefore, it is possible to construct the goodness-of-fit tests based on the statistics

$$\sup_x \sqrt{T}\ \left| f_T^\circ(x) - f_{S_0}(x) \right| \qquad \text{or} \qquad \sup_x \sqrt{T}\ \left| \hat{F}_T(x) - F_{S_0}(x) \right|,$$

where $f_T^\circ(\cdot)$ is the LTE and $\hat{F}_T(\cdot)$ is the EDF.

Theorem 4.13 allows us to construct such a test on the basis of the statistic $\eta_T(x) = \sqrt{T}\ \left(f_T^\circ(x) - f_S(x)\right)$ as follows. Let us introduce a Gaussian process $\eta_f(x)$ with $\mathbf{E}_{S_0}\eta_f(x) = 0$ and the covariance function

$$R_f(x,y) = 4 f_{S_0}(x)\, f_{S_0}(y)\, \mathbf{E}_{S_0}\left(\frac{\left[\chi_{\{\xi > x\}} - F_{S_0}(\xi)\right]\left[\chi_{\{\xi > y\}} - F_{S_0}(\xi)\right]}{f_{S_0}(\xi)^2} \right).$$

Denote by y_ε the value defined by the equation

$$\mathbf{P}\left\{ \sup_x \left|\eta_f(x)\right| > y_\varepsilon \right\} = \varepsilon$$

and recall the condition

$\mathcal{A}_0$. *The function $S(\cdot)$ satisfies the condition*

$$\overline{\lim_{|x|\to\infty}}\ \operatorname{sgn}(x)\ S(x) < 0.$$

Let us consider the following hypotheses testing problem:

$$\begin{aligned} \mathscr{H}_0 &: \qquad S(\cdot) = S_0(\cdot), \\ \mathscr{H}_1 &: \qquad S(\cdot) \neq S_0(\cdot) \end{aligned}$$

This means that under $\mathscr{H}_1$ we have $\sup_x |S(x) - S_0(x)| > 0$.

We would like to have a test ϕ_T^* which rejects $\mathscr{H}_0$ when it is true with a probability asymptotically equal to ε (belongs to the class $\mathscr{K}_\varepsilon'$) and is consistent, i.e., $\beta_T(S, \phi_T^*) \to 1$ under $\mathcal{H}_1$.

Introduce the statistic

$$\delta_T(X^T) = \sup_x \sqrt{T} \left| f_T^\circ(x) - f_{S_0}(x) \right|.$$

We have the following

Proposition 5.21. *Let the condition $\boldsymbol{A}_0$ be fulfilled. Then the test*

$$\phi_T^*\left(X^T\right) = \chi_{\left\{\delta_T(X^T) > y_\varepsilon\right\}}$$

belongs to $\mathscr{K}_\varepsilon'$ and is consistent.

Proof. The proof of $\phi_T^* \in \mathscr{K}_\varepsilon'$ follows immediately from Theorem 4.13 because $\phi_T^*(\cdot)$ is a continuous functional on the space $\mathscr{C}_0$ and from the weak convergence (4.31) follows the convergence of the distribution functions. The consistency is evident as well because under $\mathscr{H}_1$ we can write

$$\delta_T(X^T) \geq \sqrt{T} \sup_x \left| f_S(x) - f_{S_0}(x) \right| - \sup_x \sqrt{T} \left| f_T^\circ(x) - f_S(x) \right| \to \infty,$$

because the quantity $\sqrt{T} \left(f_T^\circ(x) - f_S(x) \right)$ converges to the corresponding Gaussian process.

Similar arguments can be used for construction of another goodness-of-fit test based on the Kolmogorov statistic

$$\Delta_T\left(X^T\right) = \sup_x \sqrt{T} \left| \hat{F}_T(x) - F_{S_0}(x) \right|.$$

Let $\eta_F(x)$ be a Gaussian process with $\mathbf{E}_{S_0} \eta_F(x) = 0$ and the covariance function

$$R_F(x, y) = 4\,\mathbf{E}_{S_0} \left(\frac{\left[F_{S_0}(\xi \wedge x) - F_{S_0}(\xi) F_{S_0}(x)\right] \left[F_{S_0}(\xi \wedge y) - F_{S_0}(\xi) F_{S_0}(y)\right]}{f_{S_0}(\xi)^2} \right).$$

Therefore, if we denote by y_ε the solution of the equation

$$\mathbf{P}\left\{ \sup_x \left| \eta_F(x) \right| > y_\varepsilon \right\} = \varepsilon,$$

then we have the following proposition.

Proposition 5.22. *Let the condition* $\boldsymbol{\mathcal{A}}_0$ *be fulfilled. Then the test*

$$\hat{\phi}_T\left(X^T\right) = \chi_{\left\{\Delta_T(X^T)>y_\varepsilon\right\}}$$

belongs to $\mathcal{K}_\varepsilon'$ *and is consistent.*

Proof. The weak convergence to the stochastic process $\eta_F(\cdot)$ was proved by Negri [186].

Historical Remarks

The bibliography presented in this work is in no sense exhaustive. We mention the publications which are closely related with the exposed results or were used to prove these results.

Chapter 1

Section 1.1. The properties of stochastic integral, diffusion processes, local time and the likelihood ratio formula can be found in many books devoted to Itô calculus, see, e.g., Øksendal [193], Liptser and Shiryayev [175], Gikhman and Skorohod [84],[85], Durret [69], Karatzas and Shreve [119], Revuz and Yor [208], Rogers and Williams [209]. Proposition 1.11 is new. The condition $\mathcal{A}_0$ sometimes is called the *condition of Khasminskii.*

Section 1.2. The law of large numbers for one-dimensional diffusion process was obtained first by Maruyama and Tanaka [180] and can be found as well in [85], [69], [219]. The central limit theorem for stochastic integrals was first established by Taraskin [225], [226], who proved the asymptotic normality (1.64) under (1.63) and an additional condition of the convergence of moments. Theorem 1.19 was published in [132] and I am deeply grateful to R. Liptser who essentially helped me to prove it when I was a PhD student and he was my adviser (together with B.R. Levin). See as well Dufkova [67] for the same CLT. Of course, this theorem is just a particular case of the CLT for continuous martingales [112]. CLT for ordinary integrals (Proposition 1.23) was proved by Mandl [179] (see as well Florence-Zmirou [73]). Proposition 1.25 is from [146].

Section 1.3. This section as well contains mainly known or elementary results. The Cramér–Rao inequality can be found in Liptser and Shiryayev [175] and Kutoyants [136] and the van Trees inequality [230] is taken from Gill and Levit [86] (see as well Borovkov [33] and Iacus [105]) for diffusion processes. For comments concerning parameter estimators and nonparametric estimators see below.

Chapter 2

Section 2.1. The properties of the MLE for linear w.r.t. parameter systems of stochastic differential equations were studied by many authors: Akritas and Johnson [3], Arato [6], Brown and Hewitt [39], Feigin [71], Khasminskii *et al.* [124], [113], [114] and Lee and Kozin [166]. As pioneering works we can mention here Arato, Kolmogorov and Sinai [7], Taraskin [225] and Le Breton [159]–[161]. For nonlinear models of diffusion processes the consistency and asymptotic normality of MLE in the regular case were established by Kutoyants [134], Lanska [154] (note that Dufkova and Lanska are the same person), Prakasa Rao and Rubin [206], Prakasa Rao [204], Borkar and Bagchi [32], Lin'kov [173] and Yoshida [242] (for large deviations of MLE see Florens-Landais and Pham [75] and Bishwal [25]–[27] and Bose [34]). The consistency of the MLE is discussed by van Zanten [232] and Lokianova [177]. Florence-Zmirou [74] used the local time in the problem of parameter estimation for ergodic diffusion processes.

Theorems 2.8 and 2.13 (with identifiability condition (2.18)) are new.

Section 2.2. The first MDE ϑ_T^* was introduced in [138] and studied in [77] and [140]. Theorems 2.19 and 2.23 are new. The proofs follow the usual way (see Wolfowitz [240] and Millar [183]). Robustness of the MDE (for i.i.d. models) is discussed by Beran [19], Millar [183], [184], Donoho and Liu [65] and others.

Section 2.3. The TFE (first called MDE) was introduced in [138] and was studied by Dietz and Kutoyants [63] and Dietz [62]. In the present form Theorem 2.25 is new (see as well Bertrand and Kutoyants [22] and Bertrand [21]).

Section 2.4. Theorem 2.28 is new.

Section 2.5. The one-step MLE is a well known device to improve a consistent estimator up to asymptotically efficient. Theorem 2.31 and Propositions 2.33–2.35 are new. Further development can be found in Kutoyants and Yoshida [152] (the use of one-step MLE in moment estimation problems).

Section 2.6. The properties of the MLE under misspecification were described by McKeague [181], who proved the "consistency" and asymptotic normality (see as well Yoshida [242] and Kutoyants [139]). For the more general class of processes this problem was studied by Chitashvili *et al.* [47] and [48]. For i.i.d. observations the asymptotics of the misspecified MLE was studied by Huber [102] (for the i.i.d. case), Dahlhaus and Wefelmeyer [51] (for time series models) and Patilea [195] (for the i.i.d. case). For incorrect Bayes estimators see Bunke and Milhaud [40] (convergence to the set of the points of the minimum Kullback–Leibler distance, etc., in the i.i.d. case). Proposition 2.38 follows from our joint work with Höpfner [99]. In the nonidentifiable situation it is well known that the MLE converges to the set of all "true values", see, e.g., Bagchi and Borkar [11] and for BE see Bunke and Milhaud [40] and the references therein. Here we have the probabilities of the convergence to each particular value. For small noise diffusion and inhomogeneous Poisson processes see [139] and [145] respectively and the general case was studied by Kutoyants, Vostrikova [151].

The rate of convergence of the MLE in the case of the degenerate information matrix was studied by Rotnitzsky *et al.* [210].

The problem of optimal choice of the observation window for inhomogeneous Poisson processes was considered in our joint work with Spokoiny [150] and the case of ergodic diffusion is treated exactly in the same way.

The works related to asymptotic expansions and recursive estimation are mentioned in Sections 2.6.5 and 2.6.6 respectively.

Chapter 3

Section 3.1. The problem of parameter estimation for partially observed linear systems was considered by Bagchi [10], Bagchi and Borkar [11] (consistency and asymptotic normality of the MLE), Kallianpur and Selukar [116] (two-dimensional linear parameter) and Kutoyants [136] (Examples 3.2). Theorem 3.1 is new. The asymptotic behavior of the TFE for partially observed linear systems was studied by Bertrand and Kutoyants [22]. Gerenczer and Vago [83] showed the consistency of the *minimum cost* (fixed-gain) estimator for the same homogeneous system.

Section 3.2. The problem of cusp parameter estimation for the i.i.d. model of observations was first studied by Prakasa Rao [203]. Ibragimov and Khasminskii [109], Chapter VI, described the properties of the MLE and BE in much more general situations of i.i.d. observations (densities with different types of singularities) and in particular they showed the asymptotical efficiency of the BE (see as well Akahira and Takeuchi [2]). The similar problem for an inhomogeneous Poisson process of observations was studied by Dachian [49]. Proposition 3.6 and Theorems 3.8 are taken from our joint work with Dachian [50].

Section 3.3. Gushchin and Küchler [93], [94] studied the properties of the MLE of the linear system with known delay. The authors showed that the MLE has 11 (!) different limit ($T \to \infty$) distributions depending on the values of the unknown parameter.

The one-dimensional case with τ the only unknown parameter for the model (3.28) was studied in our joint work with Küchler [126].

Section 3.4. Change-point problems for different models are widely discussed in the literature (see, e.g., Shiryayev [215], Basseville and Niliforov [17] and Carlstein *et al.* [44]). For diffusion processes the statistical problems concerning the change-point parameter were discussed by Shiryayev [215] and Ibragimov and Khasminskii [108], Kutoyants [139], Chapter 5 (parameter estimation) and Campillo *et al.* [43] (hypotheses testing). The results presented here are new, and were announced in [147] (see as well van Zanten [234]).

Section 3.5. Propositions 3.39 and 3.43 are taken from Höpfner and Kutoyants [99] and the Proposition 3.45 was published in [133]. Two interesting examples of parameter estimation for null recurrent diffusion processes are given by Kulinich [128]. For the Ornstein–Uhlenbeck process (3.118) the properties of the MLE (Proposition 3.46) are well known (see Basawa and Scott [15]). We just remember here that for the value $\vartheta = 0$ the convergence (3.120) was established by Feigin [72] and for $\vartheta > 0$ the convergence (3.121) was presented by Basawa and Scott [15] (see as well [235]). The properties of the MLE for the model (3.123) are studied by Dietz and Kutoyants [64]. TFE was studied by Dietz [62].

Chapter 4

Section 4.1. Theorems 4.2 and 4.6 are published in [141]. The problem of asymptotically efficient estimation of the distribution function in the i.i.d. case was considered by Dvoretsky *et al.* [66] (see also Levit [170], Millar [183] and references therein). Then Penev [196] established the asymptotic efficiency of the empirical distribution function for an exponentially ergodic Markov chain with the state space $[0,1]$ (the more general case was treated by van der Vaart and Wellner [229] and the further generalizations are given by Greenwood and Wefelmeyer [87]). The difference between these results lies in the types of models, regularity conditions and the choice of the definitions of asymptotic optimality and common to all them is the possibility of $\sqrt{n}$-consistent estimation of the underlying distribution. The idea of the construction of a nonparametric lower bound with the help of a *least favorable parametric family* was expressed by Stein [221] and then realized by Levit [170] (see also Ibragimov and Khasminskii [109], Chapter 4 or Bickel *et al.* [24], Chapter 3 where the detailed exposition can be found). We apply here this approach in the proofs of Theorems 4.2, 4.8 and 4.16. Fournie [76] proved the Glivenko–Cantelli theorem and the asymptotic normality of the empirical distribution function (1.144) for a transformed process to $[0,1]$. Negri [186] showed the asymptotical efficiency of the EDF for the same ergodic diffusion process and the loss function $\ell_T\left(\bar F_T, F_S\right) = \ell\left(\sqrt{T}\sup_x\left|\bar F_T(x) - F_S(x)\right|\right)$. The integral $\mathscr{L}_2$-type lower bound and asymptotical efficiency of EDF were presented in our joint work [149].

Section 4.2. The results of this section are due to the author (see [144] and [146]). The problem of invariant density estimation by observations of ergodic diffusion processes was studied by Banon [13] and Nguyen [189] (consistency of kernel-type density estimator). Castellana and Leadbetter [45] showed that for some stationary continuous-time processes the kernel-type estimators can have the *parametric rate* $\sqrt{T}$. For the wide class of continuous-time stationary processes this problem was studied by Arnaud [8], Blanke and Bosq [29], Bosq

[36], Bosq *et al.* [38], Delecroix [61], Leblanc [158], Sköld [217] and van Zanten [233].

Section 4.3. This section is due to the author [143]. Note that the problem of asymptotically efficient estimation of the moments in a parametric and semiparametric framework was studied in our joint paper with Yoshida [152] and $\mathscr{L}_2$ efficiency of the EDF was proved in our joint work with Negri [149]. Here (as in [149]) in the construction of the lower bound we follow the work by Gill and Levit [86].

Section 4.4. The results presented are from our joint work with Dalalyan [54] (see as well Dalalyan [52]). The schema of the proof is similar to that of the paper by Golubev and Levit [91] devoted to the second order optimality of the EDF for the i.i.d. model. This type of problem with a *sharp lower bound up to the constant* was initiated by Pinsker [201] and then applied to different models and statistical estimation problems (see the review by Nussbaum [192] and references therein). Note specially that the local approach (Theorem 4.34) was inspired by the work of Golubev [89]. The next two sections follow exactly the same schema. The consistency and asymptotic normality of the kernel-type estimators of the derivatives of the invariant density and of the trend coefficient were studied as well by van Zanten [233].

Section 4.5. The results presented are from our joint work with Dalalyan [55]. The problem of trend coefficient estimation was studied by Geman [80] and Pham [200] (consistency and asymptotic normality of the kernel-type estimators). We mention as well the work by McKeague and Tofoni [182] devoted to periodical trend coefficients estimation for partially observed linear systems. The problem of sequential estimation of the trend coefficient was considered by Galtchouk and Pergamenshchikov [78], [79]. The problem of adaptive estimation of the trend (of unknown smoothness) was studied by Spokoiny [220] and Dalalyan [53].

Section 4.5. The results presented are from our joint work with Dalalyan [56]. As was mentioned above, we follow the paper by Golubev and Levit [91], who showed the second order efficiency of the special kernel-type estimator of the distribution function (in the i.i.d. case) and we establish the second-order efficiency of the similar kernel-type estimator of the invariant density.

Chapter 5

This chapter is due to the author.

Section 5.1. Proposition 5.3 is new, but relies heavily on the work by Veretennikov [236]. The asymptotics of the errors in the hypotheses testing problems for diffusion were studied by Lin'kov [172].

Section 5.2. Theorem 5.6 is based on the work [132]. The other results are new, but to obtain them we follow the usual ways in such situations. Misspecified testing is discussed by Bera [18].

Section 5.3. The results presented here are new, but as well follow traditional proofs. The one-sided testing problem with alternatives *stochastically larger* than the hypothesis (in the i.i.d. case) can be found in Pfanzagl [199], see as well van der Vaart [228], Section 25.6.

Section 5.4. In this section we just discuss some possibilities of the construction of such tests. Of course, this problem due to its importance needs more detailed study.

References

1. Afanas'ev, A.A. On estimation of a parameter for systems with physical white noise. *Proc. Steklov Inst. Math.* 1994; Moscow, **4**, 1–6.
2. Akahira, M., Takeuchi, K. (1995) *Non-Regular Statistical Estimation.* New York: Springer.
3. Akritas, M.G. and Johnson, R.A. (1982) The contiguity of probability measures and asymptotic inference in continuous time stationary diffusion and Gaussian processes with known covariance. *J. Multivariate Anal.*, **12**, 123–135.
4. Andersen, P.K, Borgan, O., Gill, R.D., Keiding, N. (1993) *Statistical Models Based on Counting Processes.* New York: Springer-Verlag.
5. Anderson, R. (1955) The integral of a symmetric unimodal function. *Proc. Amer. Math. Soc.*, **6**, 1, 170–176.
6. Arato, M. (1983) *Linear Stochastic Systems with Constant Coefficients A Statistical Approach.* Lecture Notes in Control and Inform. Sci., **45**, New York: Springer-Verlag.
7. Arato, M., Kolmogorov, A.N., Sinai, Ya. G (1962) On estimation of parameters of complex stationary Gaussian processes. *Dokl. Akad. Nauk*, **146(4)**, 747–750.
8. Arnaud, F. (2001) Sur l'estimation de la densité marginale d'un processus à temps continu par projection orthogonale. *Annales de l'ISUP*, XXXXV, fasc **1**, 55–92.
9. Arnold, L. (1973) *Stochastic Differential Equations: Theory and Practice.* New York: Wiley.
10. Bagchi, A. (1981) A note on asymptotically efficient estimates of parameters in continuous-time dynamical systems. *SIAM. J. Control Optim.*, **19**, 2, 169–176.
11. Bagchi, A., Borkar,V. (1984) Parameter identification in infinite dimensional linear systems. *Stochastics*, **12**, 201–213.
12. Banks, H.T. (1975) *Modeling and Control in the Biological Sciences.* Lect. Notes Biomath., **6**, Berlin: Springer-Verlag.
13. Banon, G (1978) Non parametric identification for diffusion processes. *SIAM. J. Control Optim.*, **16**, 3, 380–395.
14. Basawa, I.V., Prakasa Rao, B.L.S. (1980) *Statistical Inference for Stochastic Processes.* London: Academic Press.
15. Basawa, I.V., Scott, D.J. (1983) *Asymptotic Optimal Inference for Non-ergodic Models.* Lecture Notes in Statist., **17**, New York: Springer-Verlag.

16. Baskakova, P.E., Borodin, A.N. (1995) Tables of distributions of functionals of Brownian motion *J. Math. Sci.*, **75**, 5, 1873–1883.
17. Basseville, M., Niliforov, I.V. (1993) *Detection of Abrupt Changes: Theory and Application*. Inform. System Sci. Ser., Prentice-Hall, Englewood Cliff, NJ.
18. Bera, A.K. (2000) Hypothesis testing in the 20-th century with special reference to testing with misspecified models. In *Statistics for the 21st Century*, C.R. Rao and G.J. Székely (Eds), New York: Marcel Dekker
19. Beran, R. (1981) Efficient robust estimates in parametric models. *Z. Wahr. Verw. Geb.*, **55**, 91–108.
20. Bergstrom, A.R. (1990) *Continuous Time Econometric Modelling*. Oxford: Oxford University Press.
21. Bertrand, P. (1994) Une estimation asymptotique des paramètres d'un processus de diffusion ergodique. *CRAS Sér. I*, **318**, 957–960.
22. Bertrand, P., Kutoyants, Yu. A. (1996) A minimum distance estimator for partially observed linear stochastic systems. *Statist. Decisions*, **14**, 323–342.
23. Billingsley, P. (1968) *Convergence of Probability Measures*. New York: J.Wiley.
24. Bickel, P.J., Klaassen, C.A.J., Ritov, Y., Wellner, J.A. (1998) *Efficient and Adaptive Estimation for Semiparametric Models*. New York: Springer-Verlag.
25. Bishwal, J.P.N. (1999) Large deviations inequalities for the maximum likelihood estimator and the Bayes estimators in nonlinear stochastic differential equations. *Statist. Probab. Lett.*, **43**, 207–215.
26. Bishwal, J.P.N. (2001) Accuracy of normal approximation for the maximum likelihood estimator and Bayes estimators in the Ornstein–Uhlenbeck process using random normings. *Statist. Probab Lett.*, **52**, 4, 427-439.
27. Bishwal, J.P.N. (2000) *Asymptotic Theory of Estimation of the Drift Parameter in Diffusion Processes*. Thesis Dr Philosophy, Sambalpur University, Jyoti Vinhar, Sambalpur.
28. Blanke, D., Merlevède, F. (2000) Estimation of the asymptotic variance of kernel density estimators for continuous time processes. *Math. Methods Statist.*, **9**, 3, 270–296.
29. Blanke, D., Bosq, D. (2000) A family of minimax rates for density estimators in continuous time. *Stochastic Anal. Appl.*, 18(6), 871–900.
30. Bobrovsky, B.Z., Mayer-Wolf, E., Zakai, M. (1987) Some classes of global Cramér–Rao bounds. *Ann. Statist*, **15**, 1421–1438.
31. Bochner, S. (1960) *Harmonic Analysis and the Theory of Probability*. Univ. California Press, Berkeley and Los Angeles.
32. Borkar, V., Bagchi, A. (1982) Parameter estimation in continuous time stochastic processes. *Stochastics*, **8**, 193–212.
33. Borovkov, A.A. (1998) *Mathematical Statistics*. Amsterdam: Gordon and Breach Science Publishers.
34. Bose, A. (1986) Berry-Essen bound for the maximum likelihood estimator in the Ornstein–Uhlenbeck process. *Sankhya Ser A*,, **48**, 2, 181–187.
35. Bosq, D., Nguen, H.T. (1996) *A Course in Stochastic Processes*. Dordrecht: Kluwer.
36. Bosq, D. (1998) *Nonparametric Statistics for Stochastic Processes*. (Second edition), Lecture Notes Statist., **110**, New York: Springer-Verlag.
37. Bosq, D., Davydov, Y. (1998) Local time and density estimation in continuous time. *Math. Methods Statist.*, **8**, 1, 22–45.

38. Bosq, D., Merlevèvede, F., Peligrad, M. (1999) Asymptotic normality for density kernel estimation in discrete and continuous time. *J. Multivariate Anal.*, **68**, 78–95.
39. Brown, B.M., Hewitt, J.I. (1975) Asymptotic likelihood theory for diffusion processes. *J. Appl. Prob.*, **12**, 228–238.
40. Bunke, O., Milhaud, X. (1998) Asymptotic behavior of Bayes estimates under possibly incorrect models. *Ann. Statist.*, **26**, 2, 617–644.
41. Burnashev, M.V. (1977) Asymptotic expansion for estimates of a signal in white Gaussian noise. *Math. USSR Sbornik*, **33**, 2, 159–184.
42. Burnashev, M.V. (1981) Study of the second order properties of parameter estimators of a signal in white Gaussian noise. *Probl. Inf. Transm.*, **17**, 1, 17–28.
43. Campillo, F, Kutoyants, Yu. A., Le Gland, F. (2000) Small noise asymptotics of the GLR Test for off-line change detection in misspecified diffusion processes. *Stochastics Stochastics Rep.*, **70**, 109–129.
44. Carlstein, E, Müller, H.-G., Siegmund, D. (Eds) (1994) *Change-Point Problems*. Lecture Notes, **23**, Institute Math. Statist., Hayward, California.
45. Castellana, J.V., Leadbetter, M.R. (1986) On smoothed density estimation for stationary processes. *Stochastic Process. Appl.*, **21**, 179–193.
46. Cavalier, L., Golubev, G. K., Picard, D. and Tsybakov, A. B. (2002). Oracle inequalities for inverse problems. *Ann. Statist.* **30**, 843–874.
47. Chitashvili, R.J., Lazrieva, N.L., Torondjadze, T.A. (1990) Asymptotic theory of M-estimators in general statistical models Part 1: On asymptotic behaviour of estimators in the presence of nuisance parameters. CWI, Amsterdam, Report BS-R9020.
48. Chitashvili, R.J., Lazrieva, N.L., Torondjadze, T.A. (1990) Asymptotic theory of M-estimators in general statistical models Part 2: On asymptotic behaviour of estimators under model disturbance. CWI, Amsterdam, Report BS-R9019.
49. Dachian, S. (2003) Estimation of cusp location by Poisson observations. *Stat. Inference Stoch. Process.*, **6**, 1, 1–14.
50. Dachian, S., Kutoyants, Yu. A. (2003) On cusp estimation of ergodic diffusion process. *J. Statist. Plann. Inference*, **117**, 153–166.
51. Dahlhaus, R., Wefelmeyer, W. (1996) Asymptotically optimal estimation in misspecified time series models. *Ann. Statist.*, **24**, 952–974.
52. Dalalyan, A. (2001) *Estimation non-paramétrique pour les processus de diffusion ergodiques*. PhD Thesis, Université du Maine, Le Mans.
53. Dalalyan, A. (2002) Sharp adaptive estimation of the trend coefficient for ergodic diffusion, *Prépublication no. 1* , Université du Maine, Le Mans (submitted).
54. Dalalyan, A., Kutoyants, Yu. A. (2003) Asymptotically efficient estimation of the derivative of invariant density. *Stat. Inference Stoch. Process.*, **6**, 1, 89–107.
55. Dalalyan, A., Kutoyants, Yu. A. (2002) Asymptotically efficient trend coefficient estimation for ergodic diffusion. *Math. Methods Statist.*, **11**, 4, 402–427.
56. Dalalyan, A., Kutoyants, Yu. A. (2003) On second order asymptotically efficient estimation of invariant density for ergodic diffusion. *Prépublication no. 797*, Universités de Paris 6 & Paris 7, Paris (submitted).
57. Daley, D.J., Vere-Jones, D. (1988) *An Introduction to the Theory of Point Processes*. Springer, New York.

58. Davies, R. B., (1985) Asymptotic inference when the amount of information is random. In L. Le Cam, R. Olshen (Eds), *Proc. Berkley Conf. in honor of J. Neyman and J. Kiefer,* Wadsworth, Monterey, vol 2, 841–864.
59. Dehay, D., Kutoyants, Yu. A. (2003) On confidence intervals for distribution function and density of ergodic diffusion process. *J. Statist. Plann. Inference* (to appear).
60. Delattre, S., Hoffmann, M. (2001) The Pinsker bound in mixed Gaussian white noise. *Math. Methods Statist.*, **10**, 3, 283–316.
61. Delecroix, M. (1980) Sur l'estimation des densités d'un processus stationnaire à temps continu. *Publications de l'ISUP*, XXV, **1-2**, 17–39.
62. Dietz, H.M. (2001) Asymptotic behaviour of trajectory fitting estimators for certain non-ergodic SDE. *Stat. Inference Stoch. Process.*, **4**, 249–258.
63. Dietz, H.M., Kutoyants, Yu. A. (1997) A class of minimum distance estimators for diffusion processes with ergodic properties. *Statist. Decisions*, **15**, 211–227.
64. Dietz, H.M., Kutoyants, Yu. A. (2003) Parameter estimation for some non-recurrent solutions of SDE. *Statist. Decisions*, to appear.
65. Donoho, D.L., Liu, R.C. (1988) The "automatic" robustness of minimum distance functionals. *Ann. Statist.*, **16**, 2, 552–586.
66. Dvoretsky, A., Kiefer, J., Wolfowitz, J. (1956) Asymptotic minimax character of the sample distribution function and the classical multinomial estimator. *Ann. Statist.*, **27**, 642–669.
67. Dufkova, V. (1977) On controlled one-dimensional diffusion processes with unknown parameter. *Adv. Appl. Prob.*, **9**, 105–124.
68. Duflo, M. (1997) *Random Iterative Models.* Springer-Verlag, Berlin
69. Durett, R. (1996) *Stochastic Calculus: A Practical Introduction.* Boca Raton: CRC Press.
70. Farrel, R. (1972) On the best obtainable asymptotic rates of convergence in estimation of a density function at a point. *Ann. Math. Statist.*, **43**, 1, 170–180.
71. Feigin, P. (1976) Maximum likelihood estimation for continuous-time stochastic processes. *Adv. Appl. Prob.*, **8**, 712–736.
72. Feigin, P. (1979) Some comments concerning a curios singularity. *J. Appl. Prob.*, **16**, 440–444.
73. Florens-Zmirou, D. (1984) Approximate discrete-time schemes for statistics of diffusion processes. *Statistics,* **20**, 547–557.
74. Florens-Zmirou, D. (1988) Statistiques de diffusions et temps local. *Ann. Inst. H. Poincaré Probab. Statist.*, **24**, 1, 99–130.
75. Florens-Landais, D., Pham, H. (1999) Large deviations in estimation an Ornstein–Uhlenbeck model. *J. Appl. Probab.*, **36**, 60–67.
76. Fournie, E. (1992) Un test de type Kolmogorov-Smirnov pour processus de diffusions ergodic. *Rapport de Recherche,* **1696**, INRIA, Sophia-Antipolis.
77. Fournie, E., Kutoyants, Yu. A. (1993) Estimateur de la distance minimale pour des processus de diffusion ergodiques. *Rapport de Recherche,* **1952**, INRIA, Sophia-Antipolis
78. Galtchouk, L. and Pergamenshchikov, S. (2001) Sequential nonparametric adaptive estimation of the drift coefficient in diffusion processes. *Math. Methods Statist.*, **10**, 3, 316–330.
79. Galtchouk, L., Pergamenshchikov, S. (2002) Nonparametric sequential minimax estimation of the drift coefficient in diffusion processes. (Submitted)

80. Geman, S. (1980) An application of the method of sieves: functional estimation for the drift of a diffusion. *Colloq. Math. Soc. Janos Bolyai*, **6**, Nonparametric Statistical Inference, Budapest.
81. Genon-Catalot, V., Jacod, J. (1993) On the estimation of the coefficient for multi-dimensional diffusion process. *Ann. Inst. H. Poincaré, Probab. Statist.*, **1**, 119–151.
82. Gerencsér, L, Göngy, I., Michaletzky, Gy. (1984) Continuous-time recursive maximum likelihood method A new approach to Ljung'scheme. In L. Ljung and K.J. Aström (Eds), *A Bridge Between Control Science and Technology*, Proc 9th Trienal World Congress of IFAC, *Identifiation, Adaptive and Stochastic Control*, Oxford: Pergamon Press, 75–77.
83. Gerencsér, L., Vágó, Z (1994) Fixed-gain estimation in continuous time. *Acta Appl. Math.*, **35**, 153-164.
84. Gikhman, I.I., Skorokhod, A.V.. (1969) *Introduction to Theory of Random Processes*. Philadelphia: W.B. Saunders.
85. Gikhman, I.I., Skorohod, A.V. (1972) *Stochastic Differential Equations*. New York: Springer-Verlag.
86. Gill, R.D., Levit, B. Ya. (1995) Application of the van Trees inequality: a Bayesian Cramer–Rao bound. *Bernoulli*, **1**, 59–79.
87. Greenwood, P.E., Wefelmeyer W (1990) Efficiency of estimators for partially specified filtered models. *Stoch. Proc. Appl.*, **36**, 353–370.
88. Golubev, G.K. (1979) On efficiency calculation of MLE by observations of discontinuous signal in white noise. *Problems Inform. Transmission*, **XV**, 3, 61–69.
89. Golubev, G.K. (1991) LAN in problems of non-parametric estimation of functions and lower bounds for quadratic risks. *Theory Probab. Appl.*, **36**, 1, 152–157.
90. Golubev, G. K. (1992). Non-parametric estimation of smooth densities in L^2. *Problems Inform. Transmission*, **28**, 44–54.
91. Golubev, G.K., Levit, B. Ya. (1996) On the second order minimax estimation of distribution functions. *Math. Methods Statist.*, **5**, 1, 1–31.
92. Guillou, A., Merlevède, F. (2001) Estimation of the asymptotic variance of kernel density estimators for continuous time processes. *J. Multivariate Anal.*, **79**, 114–137.
93. Gushchin, A., Küchler, U. (1999) Asymptotic inference for a linear stochastic differential equation with time delay. *Bernoulli*, **5(6)**, 1059-1098.
94. Gushchin, A., Küchler, U. (2001) Addendum to "Asymptotic inference for a linear stochastic differential equation with time delay". *Bernoulli*, **7(4)**, 629–632.
95. Hajek, J. (1972) Local asymptotic minimax and admissibility in estimation. In *Proc6-th Berkeley Symp. Math. Statist. and Probab.*, **1**, 175–194.
96. Hale, J.K., Verduyn Lunel, S.M. (1993) *Introduction to Functional-differential Equations*. New York: Springer-Verlag.
97. Hasminskii, R.Z. (1980) *Stochastic Stability of Differential Equations*. Alpen: Sijthoff and Noordhoff.
98. Holden, A.V. (1976) *Models for Stochastic Activity of Neurones*. New York: Springer-Verlag.
99. Höpfner, R., Kutoyants, Yu. A. (2003) On a problem of statitical inference in null recurrent diffusions. *Stat. Inference Stoch. Process.*, **6**, 1, 25–42.

100. Höpfner, R., Löcherbach, E (1998) Birth and death on a flow: local time and estimation of a position-dependent death rate. *Statist. Inference Stoch. Process.*, **1**, 225–243.
101. Höpfner, R., Löcherbach, E. (2003) *Limit Theorems for Null recurrent Markov Processes.* Memoirs of the AMS, American Mathematical Society, **768**, Providence, Rhode Island.
102. Huber, P.J. (1967) The behavior of maximum likelihood estimates under nonstandard conditions. In *Proc. Fifth Berkeley Symposium Math. Statist. Probab.*, L. Le Cam and J. Neyman (Eds), University of California, Berkeley, 221–233.
103. Hübner, M., Khasminskii, R.Z., Rozovskii, BL. (1992) Two examples of parameter estimation for stochastic partial differential equations. In S. Cambanis, J.K. Ghosh, R.L. Karandrikar, P.K. Sen (Eds) *Stochastic Processes, Festschrift in honor of G. Kallianpur.* New York: Springer-Verlag, 149–160.
104. Hübner, M., Rozovskii, B.L. (1995) On the asymptotic properties of maximum likelihood estimators for parabolic stochastic PDEs *Probab. Theory Related Fields*, **103**, 143–163.
105. Iacus, S.M. (2001) Efficient estimation of dynamical systems. *Nonlinear Dynamics and Econometrics*, MIT. Press, **4**, 4, 213–226.
106. Iacus, S.M. (2002) Statistical analysis of stochastic resonance with ergodic diffusion noise. *Stochastics Stochastics Rep.*, **73**, (3-4), 271–285.
107. Iacus S., Kutoyants Yu. A. (2001) Semiparametric hypotheses testing for dynamical systems with small noise. *Math. Methods Statist.*, **10**, 1, 105–120.
108. Ibragimov, I.A., Khasminskii R.Z. (1975) Estimation of a parameter of a discontinuous signal in a white Gaussian noise. *Problems Inform. Transmission*, **11**,3, 31–43.
109. Ibragimov, I.A., Khasminskii R.Z. (1981) *Statistical Estimation. Asymptotic Theory.* New York: Springer-Verlag.
110. Ibragimov, I.A., Khasminskii, R.Z. (1982) Estimation of distribution density belonging to a class of entire functions. *Theory Probab. Appl.*, **27**, 551–562.
111. Iyengar, S. (2000) Diffusion models for neural activity. In *Statistics for the 21st Century: Methodologies for Applications of the Future.* C.R. Rao and G. Szekely (Eds), Marcel-Dekker 233–250.
112. Jacod, J., Shiryayev, A.N. (1987) *Limit Theorems for Stochastic Processes.* New York: Springer-Verlag.
113. Jankunas, A., Khasminskii, R.Z. (1997) Estimation of parameters of linear homogeneous stochastic differential equations. *Stochastic Process. Appl.*, **72**, 205–219.
114. Jankunas, A., Khasminskii R.Z. (1997) Estimation of parameters of linear homogeneous stochastic differential equations. *Stochastic Process. Appl.*, **72**, 205–219.
115. Jeganathan, P. (1983) Some properties of risk functions in estimation when the limit of experiment is mixed normal. *Sankhya Ser A*, **45**, 66–86.
116. Kallianpur, G., Selukar, R.S. (1991) Parameter estimation in linear filtering. *J. Multivar Anal.*, **39**, 284–304.
117. Kalman, R.E., Bucy, R.S. (1961) New results in linear filtering and prediction theory. *Trans. ASME*, **83D**, 95–100.
118. Karandrikar, R.I. (1995) On pathwise stochastic integration. *Stochastic Process. Appl.*, **57**, 11–18.

119. Karatzas, I., Shreve, S.E. (1991) *Brownian Motion and Stochastic Calculus.* New York: Springer-Verlag.
120. Keller, G., Kersting, G., Rösler, U. (1984) On the asymptotic behaviour of solutions of stochastic differential equations. *Z. Wahr. verw. Geb.*, **68**, 163–189.
121. Kleptsyna, M., Le Breton, A. (2002) Statistical analysis of the fractional Ornstein-Uhlenbeck type processes. *Stat. Inference Stoch. Process.*, **5**, 3, 229–248.
122. Kleptsyna, M., Le Breton, A. (2002) Extension of the Kalman-Bucy filter to elementary linear systems with fractional Brownien motion. *Stat. Inference Stoch. Process.*, **5**, 3, 249–271.
123. Khasminskii, R.Z. (1966) On stochastic processes defined by differential equations with small parameter. *Theory Probab. Appl.*, **11**, 240–259.
124. Khasminskii, R.Z., Krylov, N.V., Moshchuk, N. (1999) On the estimation of parameters for linear stochastic differential equations. *Probab. Theory Related Fields*, **113**, 443–472.
125. Küchler, U., Mensch, B. (1992) Langevins stochastic differential equation extended by a time-delayed term. *Stochastics Stochastics Rep.*, **40**, 23–42.
126. Küchler, U., Kutoyants, Yu. A. (2000) Delay estimation for some stationary diffusion-type processes. *Scand. J. Statist.*, **27**, 3, 405–414.
127. Küchler, U., Sørenson, M. (1997) *Exponential Families of Stochastic Processes.* New York: Springer-Verlag.
128. Kulinich, G.L. (1975) On the estimation of the drift parameter of a stochastic diffusion equation. *Theory Probab. Appl.*, **20**, 384–387.
129. Kunita, H. (1988) Stochastic differential equations and stochastic flows of diffemorfisms. In *Ecole d'Eté de Probab. Saint-Flour XII*, Lecture Notes in Math., **1097**, Berlin: Springer.
130. Kushner, H.J. (1967) *Stochastic Stability and Control.* New York: Academic Press.
131. Kusuoka, S., Yoshida, N. (2000) Malliavin calculus, geometric mixing, and expansion of diffusion functionals. *Probab. Theory Related Fields*, **116**, 457–484.
132. Kutoyants, Yu. A. (1975) On a problem of testing hypotheses and asymptotic normality of stochastic integrals. *Theory Probab. Appl.*, **20**, 376–384.
133. Kutoyants, Yu. A. (1977) On a property of estimator of parameter of trend coefficient. *Izv. Nats. Akad. Nauk Armenii Math.*, **12**, 245–251 (In Russian).
134. Kutoyants, Yu. A. (1977) Estimation of the trend parameter of a diffusion process in smooth case. *Theory Probab. Appl.*, **22**, 399–405.
135. Kutoyants Yu. A. (1984) Expansion of a maximum likelihood estimate by diffusion powers. *Theory Probab. Appl.*, **29**, 465-477.
136. Kutoyants, Yu. A. (1984) *Parameter Estimation for Stochastic Processes.* Berlin: Heldermann.
137. Kutoyants, Yu. A. (1988) An example of estimating a parameter of a nondifferentiable drift coefficient. *Theory Probab. Appl.*, **33**, 1, 175–179.
138. Kutoyants, Yu. A. (1991) Minimum distance parameter estimation for diffusion type observations. *CRAS Sér. I*, **312**, 637–642.
139. Kutoyants, Yu. A. (1994) *Identification of Dynamical Systems with Small Noise.* Dordrecht: Kluwer.

140. Kutoyants, Yu. A. (1997) Minimum distance approach in parametric estimation for diffusion processes. In *Applications of Time Series Analysis in Astronomy and Meteorology*, T. Subba Rao, M.B. Priestly, O. Lessi (Eds), London:Chapman & Hall, 37–46.
141. Kutoyants, Yu. A. (1997) Efficiency of empiric distribution function for ergodic diffusion processes. *Bernoulli*, **3(4)**, 445–456.
142. Kutoyants, Yu. A. (1997) Some problems of nonparametric estimation by the observations of ergodic diffusion processes. *Statist. Probab. Lett.*, **32**, 311–320.
143. Kutoyants Yu. A. (1997) On semiparametric estimation for ergodic diffusion. *Proc. A. Razmadze Math. Inst. (in honor of R. Chitashvili)*, Tbilisi, **115**, 45–58.
144. Kutoyants, Yu. A. (1997) On density estimation by the observations of ergodic diffusion processes. In *Statistics and Control of Stochastic Processes*, The Liptser Festschrift, Yu.M. Kabanov, B.L. Rozovskii, A.N. Shiryaev (Eds), Singapore: World Scientific, 253–274.
145. Kutoyants, Yu. A. (1998) *Statistical Inference for Spatial Poisson Processes.* Lect Notes Statist. **134**, New York: Springer-Verlag.
146. Kutoyants, Yu. A. (1998) Efficient density estimation for ergodic diffusion process. *Stat. Inference Stoch. Process.*, **1**, 2, 131–155.
147. Kutoyants, Yu. A. (2000) On parameter estimation for switching ergodic diffusion processes. *CRAS Sér. 1*, **330**, 925–930.
148. Kutoyants, Yu. A. , Mourid, T., Bosq, D (1992) Estimation paramétrique d'un processus de diffusion avec retards. *Ann. Inst. H. Poincaré Probab. Statist.*, **28**, 1, 95–106.
149. Kutoyants, Yu. A., Negri, I. (2001) On L_2 efficiency of empiric distribution of ergodic diffusion processs. *Theory Probab. Appl.*, **46**, 1, 164–169.
150. Kutoyants, Yu. A., Spokoiny, V. (1999) Optimal choice of observation window for Poisson observations. *Statist. Probab. Lett.*, **44**, 291–298.
151. Kutoyants, Yu. A., Vostrikova, L. Yu. (1995) On non-consistency of estimators. *Stochastics Stochastics Rep.*, **53**, 53–80.
152. Kutoyants, Yu. A., Yoshida, N. (2001) *On moment estimation for ergodic diffusion. Bernoulli*, to appear.
153. Lange, K. (2002) *Mathematical and Statistical Methods for Genetic Analysis.* New York: Springer-Verlag.
154. Lanska, V. (1979) Minimum contrast estimation in diffusion processes. *J. Appl. Probab.*, **16**, 65–75.
155. Lazrieva, N.L., Toronjadze, T.A (1986) The Ito-Ventzel formula for semimartingales and applications to recursive estimation. *Bull. Georgian Acad. Sci.*, **123**, 2, 253–256 (In Russian).
156. Lazrieva, N.L., Toronjadze, T.A. (1987) Itô-Ventzel's formula for semimartingales, asymptotic properties of MLE and recursive estimation. *Lecture Notes In Control and Inform. Sci.*, Springer, **96**, 346–355.
157. Lazrieva, N.L., Toronjadze, T.A. (1997) The Robbins–Monro(RM) type SDE I Convergence of solutions. *Stochastics Stochastics Rep.*, **61**,1+2, 67–87.
158. Leblanc, F. (1997) Density estimation for a class of continuous time process. *Math. Methods Statist.*, **6**, 2, 171–199.
159. Le Breton, A. (1974) Estimation de paramètres d'une équation différentielle stochastique vectorielle linéaire. *CRAS* , **279**(A), 289–292.
160. Le Breton, A. (1975) Estimation de paramètres dans un modéle de système dynamique à état et observation régis pas des équations différentielles stochastiques. *CRAS*, **280**(A), 1377–1380.

161. Le Breton, A. (1976) On continuous and discrete sampling for parameter estimation in diffusion process. *Math. Program. Study*, **18**, 124–144.
162. Le Cam, L. (1960) *Locally asymptotically normal families of distributions.* Univ. Calif. Publ. Statist, **3**, 27–98.
163. Le Cam, L. (1972) Limits of experiments. In *Proc 6-th Berkeley Symp. Math. Statist. Probab.*, Univ. Calif. Press, **1**, 245–261.
164. Le Cam, L. (1986) *Asymptotic Methods in Statistical Decision Theory.* New York: Springer-Verlag
165. Le Cam, L., Yang, G.L. (2000) *Asymptotics in Statistics.* (2nd edition), New York: Springer-Verlag.
166. Lee, T.S., Kozin, F. (1977) Almost sure asymptotic likelihood theory for diffusion processes. *J. Appl. Probab.*, **14**, 527–537.
167. Lehman, E.I. (1986) *Testing Statistical Hypotheses.* (2nd edition), New York: Springer-Verlag.
168. Levanoy, D., Shwartz, A., Zeitouni, O. (1994) Recursive identification in continuous-time stochastic processes. *Stochastic Process. Appl.*, **49**, 245–275.
169. Levanoy, D., Shwartz, A., Zeitouni, O. (1990) Continuous-time recursive identification. In E. Arikan (ed) *Communication, Control and Signal Procesisng.* Amsterdam: Elsevier, 1725–1732.
170. Levit, B.Ya. (1974) On optimality of some statistical estimates. *Proc. Prague Symp. Asymptotic Statist.*, Prague: Charles University Press, **2**, 215–232.
171. Levit, B.Ya. (1978) Infinite-dimensional informational inequalities. *Theory Probab. Appl.*, **23**, 371–377.
172. Lin'kov, Y.N. (1981) On asymptotic power of statistical tests for diffusion type processes. *Theory Stochast. Process.*, **9**, 61–71.
173. Lin'kov, Y.N. (1981) On parameter estimation of diffusion type processes *Theory Stochast. Process.*, **9**, 71–78.
174. Lin'kov, Y.N. (1993) *Asymptotic Methods in The Statistics of Stochastic Processes.* Kiev: Naukova Dumka (translated by AMS, Translation of Math. Monograph, **196**, Providence, Rhode Island, 2001).
175. Liptser, R.S., Shiryayev, A.N. (2001) *Statistics of Random Processes, I, II.* (2nd edition), New York: Springer-Verlag.
176. Ljung, L., Mitter, S.K., Moura, J.M.F. (1988) Optimal recursive maximum likelihood estimation. In *Proc. 10th IFAC Congress*, IX, 241–242.
177. Loukianova, D. (2003) Remark on semigroup techniques and the maximum likelihood estimation. *Statist. Probab. Lett.*, **62**, 111–115.
178. Lucas, A. (1998) Can we estimate the density's derivative with suroptimal rate? *Stat. Inference Stochast. Process.*, **1**, 1, 29–41.
179. Mandl, P. (1968) *Analytical Treatment of One-Dimensional Markov Processes.* Academia, Prague; New York: Springer-Verlag.
180. Maruyama, G., Tanaka, H. (1957) Some properties of one-dimensional diffusion processes. *Mem. Fac. Sci. Kyushu Univ.*, **11**, 2, 117–172.
181. McKeague, I. (1984) Estimation for diffusion processes under misspecified models. *J. Appl. Probab.*, **21**, 511–520.
182. McKeague, I and Tofoni, T. (1991) Nonparametric estimation of trends in linear stochastic systems. In *Statistical Inference in Stochastic Processes*, N.U. Prabhu and I.V. Basawa (Eds), New York: Marcel Dekker, 143–166.
183. Millar, P. W. (1983) The Minimax principle in asymptotic statistical theory. In *Ecole d'Eté de Probabilités de Saint Flour XI 1981* P.L. Hennequin, ed, Lect. Notes Math., **976**, New York: Springer-Verlag, 75–266.

184. Millar, P. W. (1984) A general approach to the optimality of minimum distance estimators. *Trans. Amer. Math. Soc.*, **286,** 377–418.
185. Negri, I. (1998) *Efficacité globale de la fonction de répartition empirique dans le cas d'un processus de diffusion ergodique.* PhD Thesis, Université du Maine, Le Mans.
186. Negri, I. (1998) Stationary distribution function estimation for ergodic diffusion process. *Stat. Inference Stoch. Process.*, **1**, 1, 61–84.
187. Negri, I. (2001) On efficient estimation of invariant density for ergodic diffusion processes. *Statist. Probab. Lett.*, **51**, 1, 79–85.
188. Nevelson, M.V., Khasminskii, R.Z. (1973) *Stochastic Approximation and Recursive Estimation of Stochastic Processes.* Transl. of Math. Monographs, **47**, AMS, Providence, Rhode Island.
189. Nguyen, H.T. (1979) Density estimation in a continuous-time Markov processes. *Ann. Statist.*, **7**, 341–348.
190. Nguyen, H.T., Pham, T.D. (1980) Sur l'utilisation du temps local en statistique des processus. *CRAS,* **290**, ser A, 165–170.
191. Nualart, D. (1995) *The Malliavin Calculus and Related Topics.* New York: Springer-Verlag.
192. Nussbaum, M. (1999) Minimax risk: Pinsker's bound. In *Encyclopedia of Statistical Sciences, Update Volume 3,* S. Kotz (Ed), New York: Wiley, 451–460.
193. Øksendal, O. (1995) *Stochastic Differential Equations.* (4th edition), New York: Springer-Verlag.
194. Papanicolaou, G. (1995) Diffusion in random media. In *Surveys in Applied Mathematics,* J.B. Keller, D. McLaughlin and G. Papanicolaou (Eds), Plenum Press, 205–255.
195. Patilea, V. (2001) Convex models, MLE and misspecification. *Ann. Statist.*, **29**, 1, 94–123.
196. Penev, S. (1991) Efficient estimation of the stationary distribution for exponentially ergodic Markov chains. *J. Statist. Plann. Inference,* **27**, 105–123.
197. Pfanzagl, J. (1985) *Asymptotic Expansions for General Statistical Models.* New York: Springer.
198. Pfanzagl, J (1985) *Asymptotic Expansions for General Statistical Models.* Lect Notes Statist , **31**, New York: Springer, (with the assistance of W. Wefelmeyer).
199. Pfanzagl, J. (1994) *Parametric Statistical Theory.* Berlin: W De Gruyter.
200. Pham, T.D. (1981) Nonparametric estimation of the drift coefficient in the diffusion equation. *Math. Operationsforsch. Statist., Ser Statistics,* **1**, 61–73.
201. Pinsker, M.S. (1980) Optimal filtering of square integrable signals in Gaussian white noise. *Probl. Inf. Transm.*, **16**, 120–133.
202. Pitman, E.J.G. (1948) *Lecture Notes on Nonparametric Statistical Inference.* Lectures given at the University of North Carolina, Institute of Statistics
203. Prakasa Rao, B.L.S. (1968) Estimation of the location of the cusp of a continuous density. *Ann. Math. Statist.*, **39**, 1, 76–87.
204. Prakasa Rao, B.L.S. (1985) Estimation of the drift for diffusion process. *Statistics,* **16**, 263–275.
205. Prakasa Rao, B.L.S. (1999) *Statistical Inference for Diffusion Type Processes.* London: Arnold.
206. Prakasa Rao, B.L.S. and Rubin, H. (1981) Asymptotic theory of estimation in nonlinear stochastic differential equations. *Sankhya Ser A,* **43**, 2, 170–189.
207. Rao, C.R, (1973) *Linear Statistical Methods and their Applications.* (2nd edition), New York: Wiley.

208. Revuz, D., Yor, M. (1991) *Continuous Martingales and Brownian Motion.* New York: Springer-Verlag.
209. Rogers, L.C.G., Williams, D. (2000) *Diffusions, Markov Processes and Martingales.* vol. 2, (2nd Edition), Cambridge: Cambridge University Press.
210. Rotnitzsky, A, Cox, D.R., Bottai, M., Robins, J. (2000) Likelihood based inference with singular information matrix. *Bernoulli*, **6(2)**, 243–284.
211. Roussas, G.G. (1972) *Contiguity of Probability Measures.* Cambridge: Cambridge University Press.
212. Sakamoto, Y., Yoshida, N. (1998) Asymptotic expansion of M-estimator over Wiener space. *Stat. Inference Stoch. Process*, **1**, 85–103.
213. Sakamoto, Y., Yoshida, N. (1999) Higher order asymptotic expansion for a functional of a mixing process with applications to diffusion processes (unpublished manuscript).
214. Schipper, M. (1996) Optimal rates and constants in L_2-minimax estimation of probability density. *Math. Methods Statist.*, **5**, 3, 253–274.
215. Shiryayev, A.N. (1978) *Optimal Stopping Rules.* New York: Springer-Verlag
216. Szimayer, A., Maller, R. (2003) Testing for mean reversion in processes of Ornstein-Uhlenbeck type. *Stat. Inference Stoch. Process*, to appear.
217. Sköld, M. (2001) The asymptotic variance of the continuous-time kernel estimator with applications to bandwidth selection. *Stat. Inference Stochast. Process.*, **4**, 1, 99–117.
218. Sköld, M., Hössjer, O. (1999) On the asymptotic variance of the continuous-time kernel density estimator. *Statist. Probab. Lett.*, **44**, 1, 97–106.
219. Skorohod, A.V. (1987) *Asymptotic Methods in the Theory of Stochastic Differential Equations.* AMS Translation of Mathematical Monographs, **78**, Providence-Rhode Island
220. Spokoiny, V.G. (2000) Adaptive drift estimation for nonparametric diffusion model. *Ann. Statist.*, **28**, 3, 815–836..
221. Stein, C. (1956) Efficient nonparametric testing and estimation. In *Proc. 3-th Berkeley Sympos. Math. Statist. Probab.*, Univ Calif Press, **1**, 187–195.
222. Strasser, H. (1985) *Mathematical Theory of Statistics.* Berlin: W. De Gruyter.
223. Swensen, A. (1980) *Asymptotic Inference for a Class of Stochastic Processes.* PhD Thesis, Univ Calif., Berkeley.
224. Taniguchi, M., Kakizawa, Y. (2000) *Asymptotic Theory of Statistical Inference for Time Series.* New York: Springer-Verlag.
225. Taraskin, A.F. (1970) On asymptotic normality of vector valued stochastic integrals and estimates of a multidimensional diffusion processes. *Theory Probab. Math. Statist.*, **2**, 205–220 (in Russian).
226. Taraskin, A.F. (1973) Some limit theorems for stochastic integrals. *Theory Stochast. Process.*, **1**, 119–133 (in Russian).
227. Terent'yev A.S. (1968) Probability distribution of a time location of an absolute maximum at the output of a sinchronized filter. *Radioengineering and Electronics*, **13**, 4, 652–657.
228. van der Vaart, A. W. (1998) *Asymptotic Statistics.* Cambridge: Cambridge University Press.
229. van der Vaart, A. W., Wellner, J.A. (1996) *Weak Convergence and Empirical Processes.* New York: Springer-Verlag.
230. van Trees, H.L. (1968) *Detection, Estimation and Modulation Theory.* Part 1, New York: Wiley.

231. van Zanten, H. (2000) On the uniform convergence of the empirical density of an ergodic diffusion. *Stat. Inference Stoch. Process.*, **3(3)**, 251–262.
232. van Zanten, H. (2001) A note on consistent estimation of multivariate parameters in ergodic diffusion models. *Scand. J. Statist.* **28(4)**, 617–623.
233. van Zanten, H. (2001) Rates of convergence and asymptotic normality of kernel estimators for ergodic diffusion processes. *Nonparametric Statist.*, 13(6), 833–850.
234. van Zanten, H. (2003) On empirical processes for ergodic diffusions and rates of convergence of M-estimators. *Scand. J. Statist.*, **30(3)**, 443–458.
235. Veres, S (1987) Conditional local asymptotic normality for nonergodic diffusion processes. *Found. Control Engineering*, **2**, 2, 93–101.
236. Veretennikov, A. Yu. (1995) On large deviations for diffusion processes with measurable coefficients. *Russian Math. Surveys*, **50**, 5, 977–987.
237. Veretennikov, A. Yu. (1999) On Castellana- Leadbetter's condition for diffusion density estimation. *Stat. Inference Stoch. Process.*, **2**, 1, 1–9.
238. Wald, A. (1939) Contribution to the theory of statistical estimation and testing hypotheses. *Ann. Math. Statist.*, **10**, 105–118.
239. Watson, G.S. and Leadbetter, M.R. (1963) On the estimation of the probability density. *Ann. Math. Statist.*, **34**, 480–491.
240. Wolfowitz, J. (1957) The minimum distance method. *Ann. Math. Statist.*, **28**, 75–88.
241. Yoshida, N. (1988) Robust M-estimators in diffusion processes. *Ann. Inst. Statist. Math.*, **40**, 4, 799–820.
242. Yoshida, N. (1990) Asymptotic behavior of M-estimator and related random field for diffusion process. *Ann. Inst. Statist. Math.*, **42**, 2, 221–251.
243. Yoshida N. (1992) Asymptotic expansions of maximum likelihood estimators for small diffusions via the theory of Malliavin–Watanabe. *Probab. Theory Related Fields*, **92**, 275-311.
244. Yoshida N. (1993) Asymptotic expansion of Bayes estimators for small diffusion. *Probab Theory Relat Fields*, **95**, 429-450.
245. Yoshida N. (1997) Malliavin calculus and asymptotic expansion for martingales. *Probab. Theory Related Fields*, **109**, 301-342.
246. Yoshida N. (1998) *Malliavin calculus and statistics.* In S. Kotz *et al.* (Eds), *Encyclopedia in Statistics*, vol 3., Wiley Canada.

Index

Springer Series in Statistics

(continued from p. ii)

Jolliffe: Principal Component Analysis.
Knottnerus: Sample Survey Theory: Some Pythagorean Perspectives.
Kolen/Brennan: Test Equating: Methods and Practices.
Kotz/Johnson (Eds.): Breakthroughs in Statistics Volume I.
Kotz/Johnson (Eds.): Breakthroughs in Statistics Volume II.
Kotz/Johnson (Eds.): Breakthroughs in Statistics Volume III.
Küchler/Sørensen: Exponential Families of Stochastic Processes.
Lahiri: Resampling Methods for Dependent Data.
Le Cam: Asymptotic Methods in Statistical Decision Theory.
Le Cam/Yang: Asymptotics in Statistics: Some Basic Concepts, 2nd edition.
Liu: Monte Carlo Strategies in Scientific Computing.
Longford: Models for Uncertainty in Educational Testing.
Manski: Partial Identification of Probability Distributions.
Mielke/Berry: Permutation Methods: A Distance Function Approach.
Pan/Fang: Growth Curve Models and Statistical Diagnostics.
Parzen/Tanabe/Kitagawa: Selected Papers of Hirotugu Akaike.
Politis/Romano/Wolf: Subsampling.
Ramsay/Silverman: Applied Functional Data Analysis: Methods and Case Studies.
Ramsay/Silverman: Functional Data Analysis.
Rao/Toutenburg: Linear Models: Least Squares and Alternatives.
Reinsel: Elements of Multivariate Time Series Analysis, 2nd edition.
Rosenbaum: Observational Studies, 2nd edition.
Rosenblatt: Gaussian and Non-Gaussian Linear Time Series and Random Fields.
Särndal/Swensson/Wretman: Model Assisted Survey Sampling.
Santner/Williams/Notz: The Design and Analysis of Computer Experiments.
Schervish: Theory of Statistics.
Shao/Tu: The Jackknife and Bootstrap.
Simonoff: Smoothing Methods in Statistics.
Singpurwalla and Wilson: Statistical Methods in Software Engineering: Reliability and Risk.
Small: The Statistical Theory of Shape.
Sprott: Statistical Inference in Science.
Stein: Interpolation of Spatial Data: Some Theory for Kriging.
Taniguchi/Kakizawa: Asymptotic Theory of Statistical Inference for Time Series.
Tanner: Tools for Statistical Inference: Methods for the Exploration of Posterior Distributions and Likelihood Functions, 3rd edition.
van der Laan: Unified Methods for Censored Longitudinal Data and Causality.
van der Vaart/Wellner: Weak Convergence and Empirical Processes: With Applications to Statistics.
Verbeke/Molenberghs: Linear Mixed Models for Longitudinal Data.
Weerahandi: Exact Statistical Methods for Data Analysis.
West/Harrison: Bayesian Forecasting and Dynamic Models, 2nd edition.

MIX
Papier aus verantwortungsvollen Quellen
Paper from responsible sources
FSC® C105338

If you have any concerns about our products,
you can contact us on
ProductSafety@springernature.com

In case Publisher is established outside the EU,
the EU authorized representative is:
Springer Nature Customer Service Center GmbH
Europaplatz 3, 69115 Heidelberg, Germany

Printed by Libri Plureos GmbH
in Hamburg, Germany